EUROPE, MIDDLE EAST AND AFRICA EDITION

MANAGERIAL STATISTICS

GERALD KELLER NICOLETA GACIU

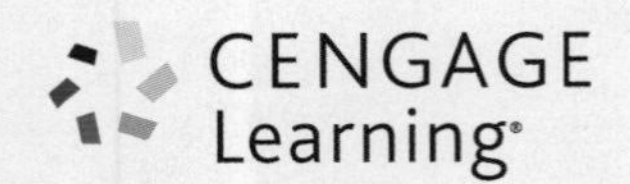

Managerial Statistics, 1st Edition
Gerald Keller and Nicoleta Gaciu

Publisher: Andrew Ashwin
Development Editor: Felix Rowe
Content Project Manager: Sue Povey
Manufacturing Buyer: Elaine Field
Marketing Manager: Vicky Fielding
Typesetter: Integra Software Services Pvt. Ltd.
Cover design: Adam Renvoize Creative
Text design: Design Deluxe Ltd

For product information and technology assistance, contact **emea.info@cengage.com.**

For permission to use material from this text or product, and for permission queries, email **emea.permissions@cengage.com.**

British Library Cataloguing-in-Publication Data
A catalogue record for this book is available from the British Library.

ISBN: 978-1-4737-0480-0

Cengage Learning EMEA
Cheriton House, North Way, Andover, Hampshire, SP10 5BE
United Kingdom

Cengage Learning products are represented in Canada by Nelson Education, Ltd.

For your lifelong learning solutions, visit
www.cengage.co.uk

Purchase your next print book, e-book or e-chapter at
www.cengagebrain.com

Printed in Singapore by Seng Lee Press
Print Number: 01 Print Year: 2015

BRIEF CONTENTS

About the Authors
Preface
Key Features
Acknowledgements
Digital Resources

PART ONE: Introduction to Statistics 1

1 What is Statistics? 002

PART TWO: Descriptive Statistical Methods 009

2 Graphical Descriptive Techniques I 010
3 Graphical Descriptive Techniques II 037
4 Numerical Descriptive Techniques 078
5 Data Collection and Sampling 126

PART THREE: Probability 139

6 Probability 140
7 Random Variables and Discrete Probability Distributions 172
8 Continuous Probability Distributions 209

PART FOUR: Sampling Theory and Inferential Statistics 247

9 Sampling Distributions 248
10 Introduction to Estimation 269
11 Introduction to Hypothesis Testing 289
12 Inference about Populations 322
13 Analysis of Variance 398
14 Chi-Squared Tests 448
15 Simple Linear Regression and Correlation 473
16 Multiple Regression 514

PART FIVE: Nonparametric Statistics for Analysis, Control and Forecasting 553

17 Nonparametric Statistics 554
18 Time-Series Analysis and Forecasting 600
19 Statistical Process Control 628
20 Decision Analysis 653
21 Conclusion 670

Appendix A 672
Credits 702
Index 705

About the Authors
Preface
Key Features
Acknowledgements
Digital Resources

PART ONE: INTRODUCTION TO STATISTICS 1

1 What is statistics? 2
Introduction 2
1-1 Key statistical concepts 5
1-2 Statistical applications in business 5
1-3 Large real data sets 6
1-4 Statistics and the computer 6

PART TWO: DESCRIPTIVE STATISTICAL METHODS 9

2 Graphical descriptive techniques I 10
Introduction 11
2-1 Types of data and information 11
2-2 Describing a set of nominal data 16
2-3 Describing the relationship between two nominal variables and comparing two or more nominal data sets 28

3 Graphical descriptive techniques II 37
Introduction 38
3-1 Graphical techniques to describe a set of interval data 38
3-2 Describing time-series data 55
3-3 Describing the relationship between two interval variables 61
3-4 Art and science of graphical presentations 68

4 Numerical descriptive techniques 78
Introduction 79
Sample statistic or population parameter 79
4-1 Measures of central location 79
4-2 Measures of variability 86
4-3 Measures of relative standing and box plots 94
4-4 Measures of linear relationship 101
4-5 Applications in finance: market model 115
4-6 Comparing graphical and numerical techniques 119
4-7 General guidelines for exploring data 122

5 Data collection and sampling 126
Introduction 127
5-1 Methods of collecting data 127
5-2 Sampling 130
5-3 Sampling plans 131
5-4 Sampling and non-sampling errors 136

PART THREE: PROBABILITY 139

6 Probability 140
Introduction 141
6-1 Assigning probability to events 141
6-2 Joint, marginal and conditional probability 145
6-3 Probability rules and trees 153
6.4 Bayes's law 159
6-5 Identifying the correct method 168

7 Random variables and discrete probability distributions 172
Introduction 173
7-1 Random variables and probability distributions 173
7-2 Bivariate distributions 182

7-3 Applications in finance: portfolio diversification and asset allocation 189
7-4 Binomial distribution 195
7-5 Poisson distribution 201

8 Continuous probability distributions 209
Introduction 210
8-1 Probability density functions 210
8-2 Normal distribution 216
8-3 Exponential distribution 231
8-4 Other continuous distributions 235

PART FOUR: SAMPLING THEORY AND INFERENTIAL STATISTICS 247

9 Sampling distributions 248
Introduction 249
9-1 Sampling distribution of the mean 249
9-2 Sampling distribution of a proportion 259
9-3 Sampling distribution of the difference between two means 264
9-4 From here to inference 266

10 Introduction to estimation 269
Introduction 270
10-1 Concepts of estimation 270
10-2 Estimating the population mean when the population standard deviation is known 273
10-3 Selecting the sample size 285

11 Introduction to hypothesis testing 289
Introduction 290
11-1 Concepts of hypothesis testing 290
11-2 Testing the population mean when the population standard deviation is known 293
11-3 Calculating the probability of a type II error 311
11-4 The road ahead 319

12 Inference about a population 322
Introduction 323
12-1 Inference about a population mean when the standard deviation is unknown 324
12-2 Inference about a population variance 334
12-3 Inference about a population proportion 340
12-4 Inference about the difference between two means: Independent samples 350
12-5 Inference about the difference between two means: matched pairs experiment 367
12-6 Inference about the ratio of two variances 378
12-7 Inference about the difference between two population proportions 383

13 Analysis of variance 398
Introduction 399
13-1 One-way analysis of variance (one-way anova) 399
13-2 Multiple comparisons 413
13-3 Analysis of variance experimental designs 422
13-4 Randomised block (two-way) analysis of variance 423
13-5 Two-factor analysis of variance 431

14 Chi-squared tests 448
Introduction 449
14-1 Chi-squared goodness-of-fit test 449
14-2 Chi-squared test of a contingency table 455
14-3 Summary of tests on nominal data 464
14-4 Chi-squared test for normality 466

15 Simple linear regression and correlation 473
Introduction 474
15-1 Model 475
15-2 Estimating the coefficients 476
15-3 Error variable: required conditions 485
15-4 Assessing the model 487
15-5 Using the regression equation 499
15-6 Regression diagnostics-I 504

16 Multiple regression 514
Introduction 515
16-1 Model and required conditions 515
16-2 Estimating the coefficients and assessing the model 516
16-3 Regression diagnostics-II 529
16-4 Regression diagnostics-III (time series) 531
16-5 Polynomial models 539
16-6 Nominal independent variables 547

PART FIVE: NONPARAMETRIC STATISTICS FOR ANALYSIS, CONTROL AND FORECASTING 553

17 Nonparametric statistics 554

Introduction 555
17-1 Wilcoxon rank sum test 556
17-2 Sign test and wilcoxon signed rank sum test 567
17-3 Kruskal–Wallis test and Friedman test 580
17-4 Spearman rank correlation coefficient 589

18 Time-series analysis and forecasting 600

Introduction 601
18-1 Time-series components 601
18-2 Smoothing techniques 603
18-3 Trend and seasonal effects 613
18-4 Introduction to forecasting 618
18-5 Forecasting models 620

19 Statistical process control 628

Introduction 629
19-1 Process variation 629
19-2 Control charts 630
19-3 Control charts for variables: $\overline{x}$ and *S* charts 636
19-4 Control charts for attributes: *p* chart 648

20 Decision analysis 653

Introduction 654
20-1 Decision problem 654
20-2 Acquiring, using and evaluating additional Information 659

21 Conclusion 670

21-1 Twelve key statistical concepts 670

Appendix A 672
Credits 702
Index 705

ABOUT THE AUTHORS

GERALD KELLER is Emeritus Professor of Business at Wilfrid Laurier University, where he taught statistics, management science and operations management from 1974 to 2011.

NICOLETA GACIU is Senior Lecturer in Education: Secondary Science / SKE Course Leader at Oxford Brookes University. Nicoleta is a member of the Institute of Physics (MInstP), Fellow of Higher Education Academy (FHEA) and BELMAS member.

PREFACE

Businesses are increasingly using statistical techniques to convert data into information. For students preparing for the business world, it is not enough merely to focus on mastering a diverse set of statistical techniques and calculations. A course and its associated textbook must provide a complete picture of statistical concepts and their applications to the real world. *Managerial Statistics*, Europe, Middle East and Africa Edition, is designed to demonstrate that statistics methods are vital tools for today's managers and economists.

Fulfilling this objective requires the several features built into this book, including data-driven examples, exercises and cases that demonstrate statistical applications that are and can be used by marketing managers, financial analysts, accountants, economists, operations managers and others. Many are accompanied by large and either genuine or realistic data sets. The applied nature of the discipline is reinforced by teaching students how to choose the correct statistical technique. Finally, students are taught the concepts that are essential to interpreting the statistical results.

THE APPROACH OF THIS BOOK

Business is complex and requires effective management to succeed. Managing complexity requires many skills. There are more competitors, more places to sell products and more places to locate workers. As a consequence, effective decision-making is more crucial than ever before. On the other hand, managers have more access to larger and more detailed data that are potential sources of information. However, to achieve this potential requires that managers know how to convert data into information. This knowledge extends well beyond the arithmetic of calculating statistics. What is required is a complete approach to applying statistical techniques.

Keller's approach has been refined over the course of ten editions, to emphasise interpretation and decision-making equally. The solution of statistical problems is typically divided into three stages: (1) identify the technique, (2) compute the statistics and (3) interpret the results. The compute stage can be completed in any or all of three ways: manually (with the aid of a calculator), using Excel, and using Minitab. For those courses that use the computer extensively, manual calculations can be played down or omitted completely. Conversely, those that wish to emphasise manual calculations may easily do so, and the computer solutions can be selectively introduced or skipped entirely. This approach is designed to provide maximum flexibility.

This approach offers several advantages:

- An emphasis on identification and interpretation provides students with practical skills they can apply to real problems they will face regardless of whether a course uses manual or computer calculations.
- Students learn that statistics is a method of converting data into information. With hundreds of data files and corresponding problems that ask students to interpret statistical results, students are given ample opportunities to practice data analysis and decision-making.
- The optional use of the computer allows for larger and more realistic exercises and examples.

NEW TO THIS EDITION

- Fully revised and updated from an international perspective, this edition is brimming with exercises and examples from the UK, Europe, the Middle East and Africa. This edition includes a host of completely new data sets to use with these exercises, including a continuous running exercise based on the UK census.

- The text has been streamlined to present a notably slimmer and more accessible text, with content tailored specifically to meet course requirements.
- Parts have been introduced to the text, grouping together topics methodically and systematically for easy reference.
- The Europe, Middle East and Africa edition features a new, user-friendly text design.

KEY FEATURES

DATA DRIVEN: THE BIG PICTURE

Solving statistical problems begins with a problem and data. The ability to select the right method by problem objective and data type is a valuable tool for business. Because business decisions are driven by data, students will leave this course equipped with the tools they need to make effective, informed decisions in all areas of the business world.

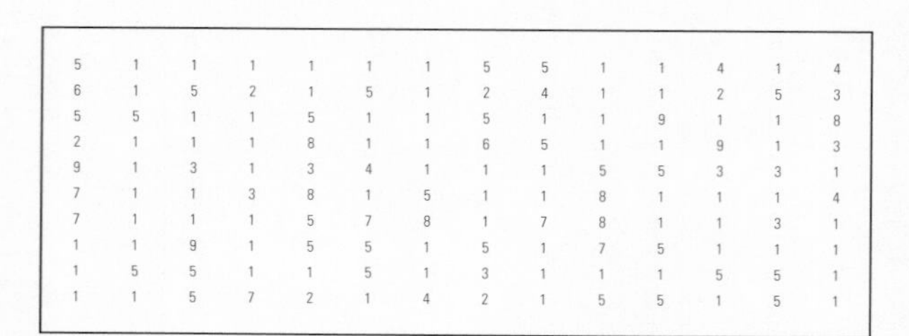

5	1	1	1	1	1	1	5	5	1	1	4	1	4
6	1	5	2	1	5	1	2	4	1	1	2	5	3
5	5	1	1	5	1	1	5	1	1	9	1	1	8
2	1	1	1	8	1	1	6	5	1	1	9	1	3
9	1	3	1	3	4	1	1	1	5	5	3	3	1
7	1	1	3	8	1	5	1	1	8	1	1	1	4
7	1	1	1	5	7	8	1	7	8	1	1	3	1
1	1	9	1	5	5	1	5	1	7	5	1	1	1
1	5	5	1	1	5	1	3	1	1	1	5	5	1
1	1	5	7	2	1	4	2	1	5	5	1	5	1

IDENTIFY THE CORRECT TECHNIQUE

Examples introduce the first crucial step in this approach. Every example's solution begins by examining the data type and problem objective and then identifying the right technique to solve the problem.

COMPUTE THE STATISTICS

Once the correct technique has been identified, examples take students to the next level within the solution by asking them to compute the statistics.

Manual calculation of the problem is presented first in each 'Compute' section of the examples.

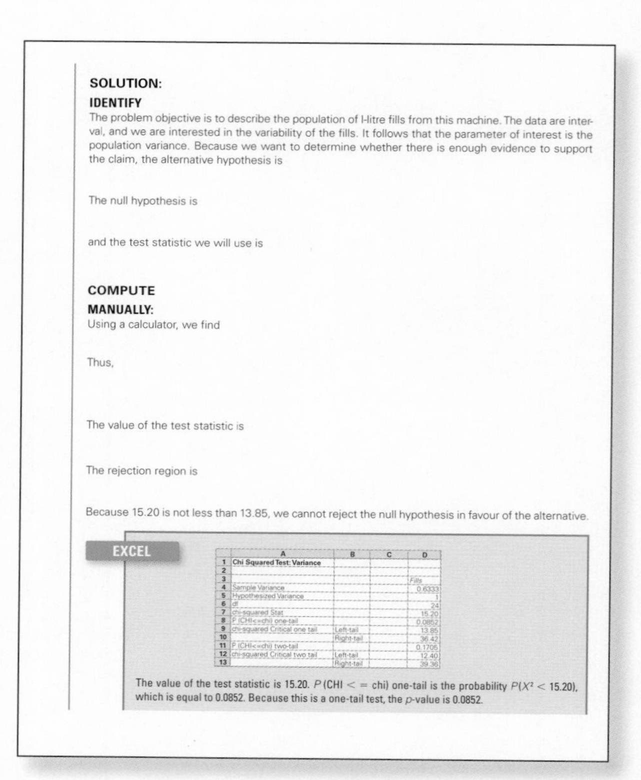

SOLUTION:

IDENTIFY

The problem objective is to describe the population of I-litre fills from this machine. The data are interval, and we are interested in the variability of the fills. It follows that the parameter of interest is the population variance. Because we want to determine whether there is enough evidence to support the claim, the alternative hypothesis is

The null hypothesis is

and the test statistic we will use is

COMPUTE

MANUALLY:

Using a calculator, we find

Thus,

The value of the test statistic is

The rejection region is

Because 15.20 is not less than 13.85, we cannot reject the null hypothesis in favour of the alternative.

EXCEL

	A	B	C	D
1	**Chi Squared Test: Variance**			
2				
3				*Fills*
4	Sample Variance			0.6333
5	Hypothesized Variance			1
6	df			24
7	chi-squared Stat			15.20
8	P (CHI<=chi) one-tail			0.0852
9	chi-squared Critical one tail	Left-tail		13.85
10		Right-tail		36.42
11	P (CHI<=chi) two-tail			0.1705
12	chi-squared Critical two tail	Left-tail		12.40
13		Right-tail		39.36

The value of the test statistic is 15.20. *P* (CHI < = chi) one-tail is the probability $P(X^2 < 15.20)$, which is equal to 0.0852. Because this is a one-tail test, the *p*-value is 0.0852.

INTERPRET THE RESULTS

INTERPRET

In the real world, it is not enough to know how to generate the statistics. To be truly effective, a business person must also know how to **interpret and articulate** the results. Furthermore, students need a framework to understand and apply statistics **within a realistic setting** by using realistic data in exercises, examples and case studies.

Examples round out the final component of the identify–compute–interpret approach by asking students to interpret the results in the context of a business-related decision. This final step motivates and shows how statistics is used in everyday business situations.

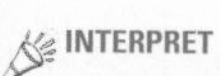

INTERPRET

There is strong evidence to infer that the undergraduate degree and MBA specialism are related. This suggests that the dean can predict the number of optional courses by counting the number of MBA students with each type of undergraduate degree. We can see that BAs favour accounting courses, BEng's prefer finance, BBAs are partial to marketing, and others show no particular preference.

If the null hypothesis is true, undergraduate degree and MBA major are independent of one another. This means that whether an MBA student earned a BA, BEng, BBA, or other degree does not affect his or her choice of programme specialism in the MBA. Consequently, there is no difference in specialism choice among the graduates of the undergraduate programmes. If the alternative hypothesis is true, undergraduate degree does affect the choice of MBA specialism. Thus, there are differences between the four undergraduate degree categories.

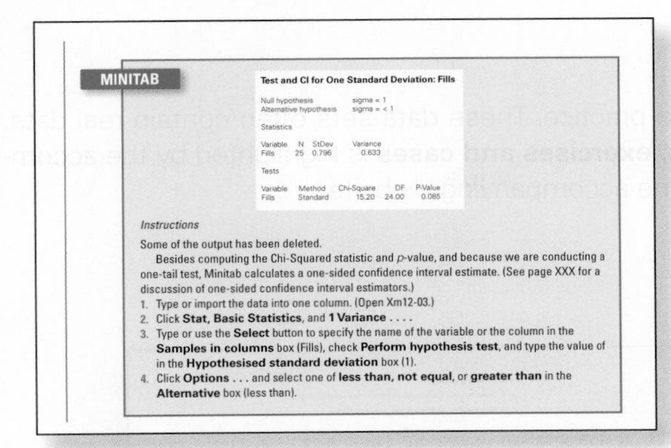
MINITAB

Test and CI for One Standard Deviation: Fills

Null hypothesis	sigma = 1
Alternative hypothesis	sigma = < 1

Statistics

Variable	N	StDev	Variance
Fills	25	0.796	0.633

Tests

Variable	Method	Chi-Square	DF	P-Value
Fills	Standard	15.20	24.00	0.085

Instructions

Some of the output has been deleted.

Besides computing the Chi-Squared statistic and *p*-value, and because we are conducting a one-tail test, Minitab calculates a one-sided confidence interval estimate. (See page XXX for a discussion of one-sided confidence interval estimators.)

1. Type or import the data into one column. (Open Xm12-03.)
2. Click **Stat, Basic Statistics**, and **1 Variance . . . .**
3. Type or use the **Select** button to specify the name of the variable or the column in the **Samples in columns** box (Fills), check **Perform hypothesis test**, and type the value of in the **Hypothesised standard deviation** box (1).
4. Click **Options . . .** and select one of **less than, not equal**, or **greater than** in the **Alternative** box (less than).

Step-by-step instructions in the use of **Excel** and **Minitab** immediately follow the manual presentation. Instruction appears in the book with the printouts – there's no need to incur the extra expense of separate software manuals.

AN APPLIED APPROACH

Applications in ... sections and boxes (in finance, marketing, operations management, human resources, economics and accounting) highlight how statistics is used in those professions. In addition to sections and boxes, **Applications in ... exercises** can be found within the exercise sections to further reinforce the big picture.

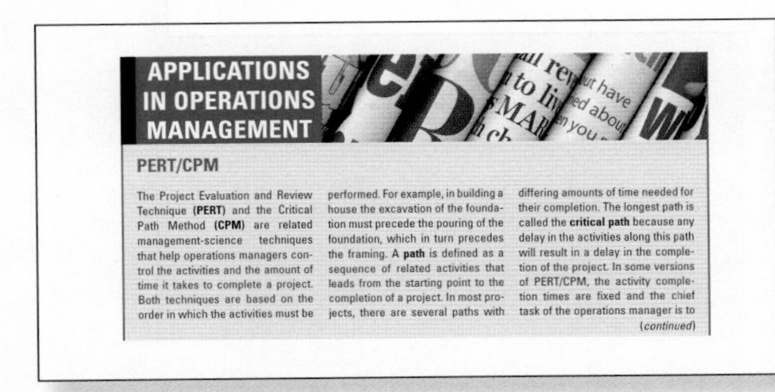
APPLICATIONS IN OPERATIONS MANAGEMENT

PERT/CPM

The Project Evaluation and Review Technique **(PERT)** and the Critical Path Method **(CPM)** are related management-science techniques that help operations managers control the activities and the amount of time it takes to complete a project. Both techniques are based on the order in which the activities must be performed. For example, in building a house the excavation of the foundation must precede the pouring of the foundation, which in turn precedes the framing. A **path** is defined as a sequence of related activities that leads from the starting point to the completion of a project. In most projects, there are several paths with differing amounts of time needed for their completion. The longest path is called the **critical path** because any delay in the activities along this path will result in a delay in the completion of the project. In some versions of PERT/CPM, the activity completion times are fixed and the chief task of the operations manager is to

(continued)

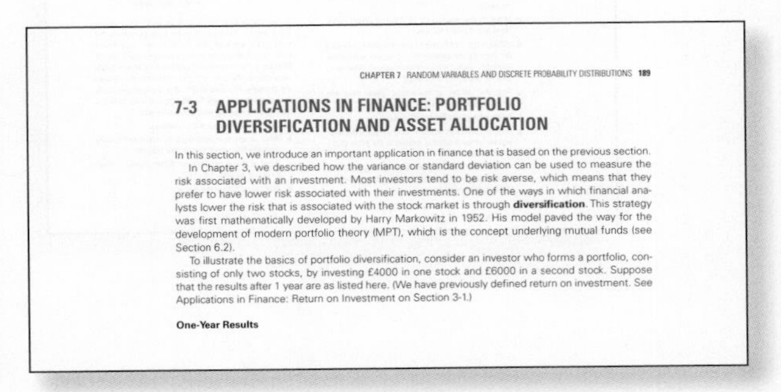
CHAPTER 7 RANDOM VARIABLES AND DISCRETE PROBABILITY DISTRIBUTIONS 189

7-3 APPLICATIONS IN FINANCE: PORTFOLIO DIVERSIFICATION AND ASSET ALLOCATION

In this section, we introduce an important application in finance that is based on the previous section.

In Chapter 3, we described how the variance or standard deviation can be used to measure the risk associated with an investment. Most investors tend to be risk averse, which means that they prefer to have lower risk associated with their investments. One of the ways in which financial analysts lower the risk that is associated with the stock market is through **diversification**. This strategy was first mathematically developed by Harry Markowitz in 1952. His model paved the way for the development of modern portfolio theory (MPT), which is the concept underlying mutual funds (see Section 6.2).

To illustrate the basics of portfolio diversification, consider an investor who forms a portfolio, consisting of only two stocks, by investing £4000 in one stock and £6000 in a second stock. Suppose that the results after 1 year are as listed here. (We have previously defined return on investment. See Applications in Finance: Return on Investment on Section 3-1.)

One-Year Results

EDUCATION AND INCOME: HOW ARE THEY RELATED?

DATA XM15-00

You are probably a student in an undergraduate or graduate business or economics programme. Your plan is to graduate, get a good job, and draw a high salary. You have probably assumed that more education equals better job equals higher income. Is this true? A Social Survey recorded two variables that will help determine whether education and income are related and, if so, what the value of an additional year of education might be. The next statistical procedures will help you to answer these questions.

Chapter-opening examples and solutions present compelling discussions of how the techniques and concepts introduced in that chapter are applied to real-world problems. These examples are then revisited with a solution as each chapter unfolds, applying the methodologies introduced in the chapter.

EXERCISES

16.41 How many indicator variables must be created to represent a nominal independent variable that has five categories?

16.42 Create and identify indicator variables to represent the following nominal variables.

- **a.** Religious affiliation (Catholic, Protestant, and others)
- **b.** Working shift (8 am to 4 pm, 4 pm to 12 midnight, and 12 midnight to 8 am)
- **c.** Supervisor (Jack Jones, Mary Brown, George Fosse, and Elaine Smith)

16.43 In a study of computer applications, a survey asked which microcomputer a number of companies used. The following indicator variables were created.

Which computer is being referred to by each of the following pairs of values?

- **a.** $I_1 = 0; I_2 = 1$
- **b.** $I_1 = 1; I_2 = 0$
- **c.** $I_1 = 0; I_2 = 0$

The following exercises require the use of a computer and software.

undergraduate degree is a factor in determining how well a student performs in the MBA programme?

16.45 Xr16-12* Refer to Exercise 16.12.

- **a.** Predict with 95% confidence the MBA programme GPA of a BEng whose undergraduate GPA was 9.0, whose GMAT score as 700, and who has had 10 years of work experience.
- **b.** Repeat part (a) for a BA student.

16.46 Xr16-09* Refer to Exercise 16.9, where a multiple regression analysis was performed to predict men's longevity based on the parents' and grandparents' longevity. In addition to these data, suppose that the actuary also recorded whether the man was a smoker (1 = yes and 0 = no).

- **a.** Use regression analysis to produce the statistics for this model.
- **b.** Compare the equation you just produced to that produced in Exercise 16.40. Describe the differences.
- **c.** Are smoking and length of life related? Explain.

16.47 Xr16-47 The manager of an amusement park would like to be able to predict daily attendance in order to develop more accurate plans about how much food to order and how many ride operators to hire. After

Hundreds of exercises, many of them new or updated, offer ample practice for students to use statistics in an applied context.

DATA SETS

Many data sets available to be downloaded provide ample practice. These data sets often contain real data, and are typically large. **Prevalent use of data in examples, exercises and cases** is highlighted by the accompanying data set file name, which alerts students to go to the accompanying website.

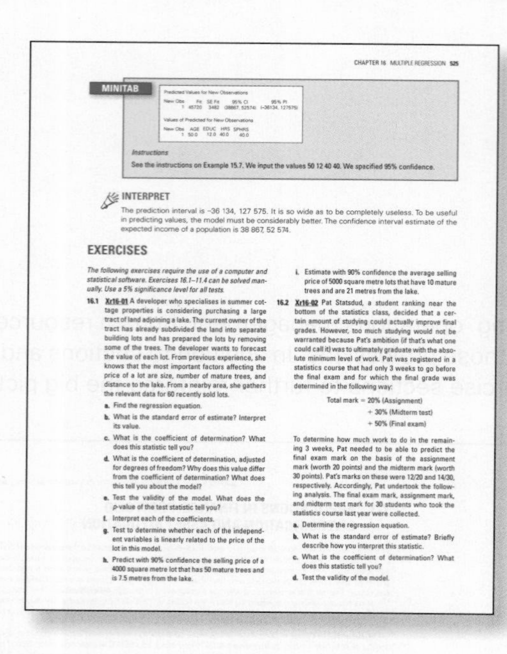

CHAPTER 16 MULTIPLE REGRESSION 525

MINITAB

Predicted Values for New Observations

New Obs	Fit	SE Fit	95% CI	95% PI
1	45720	3482	(38867, 52574)	(−36134, 127575)

Values of Predicted for New Observations

New Obs	AGE	EDUC	HRS	SPHRS
1	50.0	12.0	40.0	40.0

Instructions

See the instructions on Example 15.7. We input the values 50 12 40 40. We specified 95% confidence.

INTERPRET

The prediction interval is −36 134, 127 575. It is so wide as to be completely useless. To be useful in predicting values, the model must be considerably better. The confidence interval estimate of the expected income of a population is 38 867, 52 574.

EXERCISES

The following exercises require the use of a computer and statistical software. Exercises 16.1–11.4 can be solved manually. Use a 5% significance level for all tests.

16.1 Xr16-01 A developer who specialises in summer cottage properties is considering purchasing a large tract of land adjoining a lake. The current owner of the tract has already subdivided the land into separate building lots and has prepared the lots by removing some of the trees. The developer wants to forecast the value of each lot. From previous experience, she knows that the most important factors affecting the price of a lot are size, number of mature trees, and distance to the lake. From a nearby area, she gathers the relevant data for 60 recently sold lots.

a. Find the regression equation.
b. What is the standard error of estimate? Interpret its value.
c. What is the coefficient of determination? What does this statistic tell you?
d. What is the coefficient of determination, adjusted for degrees of freedom? Why does this value differ from the coefficient of determination? What does this tell you about the model?
e. Test the validity of the model. What does the *p*-value of the test statistic tell you?
f. Interpret each of the coefficients.
g. Test to determine whether each of the independent variables is linearly related to the price of the lot in this model.
h. Predict with 90% confidence the selling price of a 4000 square metre lot that has 50 mature trees and is 7.5 metres from the lake.
i. Estimate with 90% confidence the average selling price of 5000 square metre lots that have 10 mature trees and are 21 metres from the lake.

16.2 Xr16-02 Pat Statsdud, a student ranking near the bottom of the statistics class, decided that a certain amount of studying could actually improve final grades. However, too much studying would not be warranted because Pat's ambition (if that's what one could call it) was to ultimately graduate with the absolute minimum level of work. Pat was registered in a statistics course that had only 3 weeks to go before the final exam and for which the final grade was determined in the following way:

Total mark = 20% (Assignment)
+ 30% (Midterm test)
+ 50% (Final exam)

To determine how much work to do in the remaining 3 weeks, Pat needed to be able to predict the final exam mark on the basis of the assignment mark (worth 20 points) and the midterm mark (worth 30 points). Pat's marks on these were 12/20 and 14/30, respectively. Accordingly, Pat undertook the following analysis. The final exam mark, assignment mark, and midterm test mark for 30 students who took the statistics course last year were collected.

a. Determine the regression equation.
b. What is the standard error of estimate? Briefly describe how you interpret this statistic.
c. What is the coefficient of determination? What does this statistic tell you?
d. Test the validity of the model.

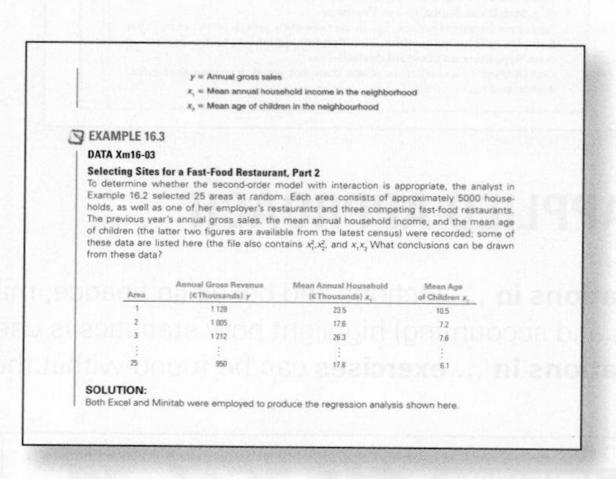

y = Annual gross sales
x_1 = Mean annual household income in the neighborhood
x_2 = Mean age of children in the neighbourhood

EXAMPLE 16.3

DATA Xm16-03

Selecting Sites for a Fast-Food Restaurant, Part 2

To determine whether the second-order model with interaction is appropriate, the analyst in Example 16.2 selected 25 areas at random. Each area consists of approximately 5000 households, as well as one of her employer's restaurants and three competing fast-food restaurants. The previous year's annual gross sales, the mean annual household income, and the mean age of children (the latter two figures are available from the latest census) were recorded; some of these data are listed here (the file also contains x_1^2, x_2^2, and x_1x_2 What conclusions can be drawn from these data?

Area	Annual Gross Revenue (€Thousands) y	Mean Annual Household (€Thousands) x_1	Mean Age of Children x_2
1	1 128	23.5	10.5
2	1 005	17.6	7.2
3	1 212	26.3	7.6
⋮	⋮	⋮	⋮
25	950	17.8	6.1

SOLUTION:

Both Excel and Minitab were employed to produce the regression analysis shown here.

ACKNOWLEDGEMENTS

The number of people who made contributions to various editions of this book is large and unfortunately too long to reproduce comprehensively here. However, we would like to acknowledge the work of all of them, with particular emphasis on the following: Paul Baum (California State University, Northridge, USA), and John Lawrence (California State University, Fullerton, USA) for their contributions to proofing and accuracy. The following individuals played important roles in the production of the US edition of this book: Product Manager Aaron Arnsparger, Content Developer Kendra Brown, Senior Content Project Managers Holly Henjum and Scott Dillon and Media Developer Chris Valentine.

We thank Fernando Rodriguez, Trent Tucker (Wilfrid Laurier University, USA) and Zvi Goldstein (California State University, USA), for their respective contributions to the online materials. We would also like to thank Dr Len Gill (University of Manchester, UK) and Sami Akabawi (The American University in Cairo, Egypt) for their invaluable advice and insight in reviewing the manuscript of this Europe, Middle East and Africa edition.

Digital Resources

Managerial Statistics, Europe, Middle East and Africa Edition, is accompanied by a range of digital resources carefully tailored to the specific needs of its readers. Examples of the kind of resources provided include:

- A password protected area for instructors with, for example, PowerPoint slides and an instructor's manual
- An open-access area for students including a range of datasets, additional exercises and web appendices

Lecturers: to discover the dedicated lecturer digital support resources accompanying this textbook please register here for access: **http://login.cengage.com**

Students: to discover the dedicated student digital support resources accompanying this textbook, please search for MANAGERIAL STATISTICS on: **www.cengagebrain.co.uk**

PART ONE

INTRODUCTION TO STATISTICS

1 What Is Statistics?

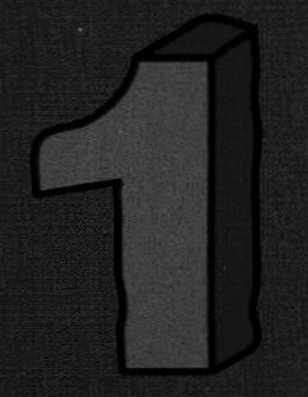

WHAT IS STATISTICS?

1-1 KEY STATISTICAL CONCEPTS

1-2 STATISTICAL APPLICATIONS IN BUSINESS

1-3 LARGE REAL DATA SETS

1-4 STATISTICS AND THE COMPUTER

INTRODUCTION

Statistics is a way to get information from data. That's it! Most of this textbook is devoted to describing how, when, and why managers and statistics practitioners* conduct statistical procedures. The book covers also a range of examples, information and applications of statistics relevant to students and managers. We demonstrate some of these with a case and two examples that are featured in this chapter.

EXAMPLE

Business Statistics Marks (See Chapter 3)

A student enrolled in a business degree programme is attending his first lecture of the required statistics course. The student is somewhat apprehensive because he believes the myth that the course is difficult. To alleviate his anxiety, the student asks the lecturer about last year's marks. Because this lecturer is friendly and helpful, he obliges the student and provides a list of the final marks, which are composed of term work plus the final exam. What information can the student obtain from the list?

This is a typical statistics problem. The student has the data (marks) and needs to apply statistical techniques to get the information he requires. This is a function of **descriptive statistics**.

---

*The term *statistician* is used to describe so many different kinds of occupations that it has ceased to have any meaning. It is used, for example, to describe a person who calculates baseball statistics as well as an individual educated in statistical principles. We will describe the former as a *statistics practitioner* and the latter as a *statistician.* A statistics practitioner is a person who uses statistical techniques properly. Examples of statistics practitioners include the following:

1. A financial analyst who develops stock portfolios based on historical rates of return.
2. An economist who uses statistical models to help explain and predict variables such as inflation rate, unemployment rate and changes in the gross domestic product.
3. A market researcher who surveys consumers and converts the responses into useful information.

Our goal in this book is to convert you into one such capable individual.

The term *statistician* refers to an individual who works with the mathematics of statistics. His or her work involves research that develops techniques and concepts that in the future may help the statistics practitioner. Statisticians are also statistics practitioners, frequently conducting empirical research and consulting. If you are taking a statistics course, your instructor is probably a statistician.

Descriptive Statistics

Descriptive statistics deals with methods of organising, summarising and presenting data in a convenient and informative way. One form of descriptive statistics uses graphical techniques that allow statistics practitioners to present data in ways that make it easy for the reader to extract useful information. In Chapters 2 and 3 we will present a variety of graphical methods.

Another form of descriptive statistics uses numerical techniques to summarise data. One such method that you have already used frequently calculates the average or mean. In the same way that you calculate the average age of the employees of a company, we can compute the mean mark of last year's statistics course. Chapter 4 introduces several numerical statistical measures that describe different features of the data.

The actual technique we use depends on what specific information we would like to extract. In this example, we can see at least three important pieces of information. The first is the 'typical' mark. We call this a *measure of central location*. The *average* is one such measure. In Chapter 4, we will introduce another useful measure of central location, the *median*. Suppose the student was told that the average mark last year was 67. Is this enough information to reduce his anxiety? The student would likely respond 'No' because he would like to know whether most of the marks were close to 67 or were scattered far below and above the average. He needs a *measure of variability*. The simplest such measure is the *range*, which is calculated by subtracting the smallest number from the largest. Suppose the largest mark is 96 and the smallest is 24. Unfortunately, this provides little information since it is based on only two marks. We need other measures—these will be introduced in Chapter 4. Moreover, the student must determine more about the marks. In particular, he needs to know how the marks are distributed between 24 and 96. The best way to do this is to use a graphical technique, the histogram, which will be introduced in Chapter 3.

CASE 1.1 Coca-Cola's Exclusivity Agreement with a South African University

In the last few years, the wellness trend continues to have a major influence on a new healthier product development in the South African soft drink market. Manufacturers across the market, such as Coca-Cola South Africa, continue to develop new low calorie and sugar free variants, as well as drinks made from natural ingredients.

A public university with a total enrollment of about 50 000 students has offered Coca-Cola an exclusivity agreement that would give Coca-Cola exclusive rights to sell its products at all university facilities for the next year with an option for future years. In return, the university would receive 35% of the on-campus revenues and an additional lump sum of 2 000 000 South African rand per year. Coca-Cola has been given 2 weeks to respond.

The management at Coca-Cola quickly reviews what it knows. The market for soft drinks is measured in terms of 330 ml cans. Coca-Cola currently sells an average of 22 000 cans per week over the 40 weeks of the year that the university operates. The cans sell for an average of 5 South African rand each. The costs, including labour, total 150 cents per can. Coca-Cola is unsure of its market share but suspects it is considerably less than 50%. A quick analysis reveals that if its current market share were 25%, then, with an exclusivity agreement, Coca-Cola would sell 88 000 (22 000 is 25% of 88 000) cans per week or 3 520 000 cans per year. The gross revenue would be computed as follows[†]:

Gross revenue = 3 520 000 × R5/can = R17 600 000

This figure must be multiplied by 65% because the university would rake in 35% of the gross. Thus,

Gross revenue after deducting 35% university take
= 65% × R17 600 000 = R11 440 000

[†]We have created an Excel spreadsheet that does the calculations for this case available to download from the accompanying website. (See the 'Digtal Resources' page for instructions on how to access the website to download this spreadsheet in addition to many more data sets and resources.)

The total cost of R1.50 per can (or R5 280 000) and the annual payment to the university of R2 000 000 are subtracted to obtain the net profit:

Net profit = R11 440 000 − R5 280 000 − R2 000 000 = R4 160 000

Coca-Cola's current annual profit is

40 weeks × 22 000 cans/week × R3.50 = R3 080 000

If the current market share is 25%, the potential gain from the agreement is

R4 160 000 − R3 080 000 = R1 080 000

The only problem with this analysis is that Coca-Cola does not know how many soft drinks are sold weekly at the university. Other soft drink companies are not likely to supply Coca-Cola with information about their sales.

Coca-Cola assigned a recent university graduate to survey the university's students to supply the missing information. Accordingly, she organises a survey that asks 500 students to keep track of the number of soft drinks they purchase in the next 7 days. The responses are stored in a file named C1-01 and available to be downloaded (see the Digital Resources page).

Inferential Statistics

The information we would like to acquire in Case 1.1 is an estimate of annual profits from the exclusivity agreement. The data are the numbers of cans of soft drinks consumed in 7 days by the 500 students in the sample. We can use descriptive techniques to learn more about the data. In this case, however, we are not so much interested in what the 500 students are reporting as in knowing the mean number of soft drinks consumed by all 50 000 students on campus. To accomplish this goal we need another branch of statistics: **inferential statistics**.

Inferential statistics is a body of methods used to draw conclusions or inferences about characteristics of populations based on sample data. The population in question in this case is the university's 50 000 students. The characteristic of interest is the soft drink consumption of this population. The cost of interviewing each student in the population would be prohibitive and extremely time-consuming. Statistical techniques make such endeavours unnecessary. Instead, we can sample a much smaller number of students (the sample size is 500) and infer from the data the number of soft drinks consumed by all 50 000 students. We can then estimate annual profits for Coca-Cola.

EXAMPLE 1.1

Global Barometer

The Global Barometer signifies the largest and most systematic comparative survey of values and attitudes towards different aspects of public affairs, governance and social policy in Africa, Asia, Latin America and the Arabic countries. This comprehensive multi-country social survey is designed to seek public opinion on diverse topics including political interest and participation, support for democratic institutions, attitudes toward political violence and terrorist activity, conceptions and interpretations of Islam, the relationship between religion and politics, and attitudes toward Middle East international relations in the Arab world.

Public opinion polls have proven to be helpful in predicting election outcomes for several decades in many countries. For example, when an election for political office takes place, the television networks cancel regular programming to provide election coverage. After the ballots are counted, the results are reported. However, the networks actively compete to see which one will be the first to predict a winner. This is done through **exit polls** in which a **random sample** of voters who exit the polling booth are asked for whom they voted. From the data, the sample proportion of voters supporting the candidates is computed. A statistical technique is applied to determine whether there is enough evidence to infer that the leading candidate will garner enough votes to win.

The Centre for Palestine Research and Studies has conducted several public polls or special polls in the West Bank and the Gaza strip. The results of the latest poll (19–22 December 2013) cover, for example, public attitudes, evaluation of the general West Bank and Gaza conditions, and elections.

The total size of the sample used for this public pool was 1270 adults who were interviewed face to face in 127 randomly selected locations and the margin of error was 3%.

1-1 KEY STATISTICAL CONCEPTS

Statistical inference problems involve three key concepts: the population, the sample and the statistical inference. We now discuss each of these concepts in more detail.

1-1a Population

A **population** is the group of all items of interest to a statistics practitioner. It is frequently very large and may, in fact, be infinitely large. In the language of statistics, *population* does not necessarily refer to a group of people. It may, for example, refer to the population of ball bearings produced at a large plant. In Case 1.1, the population of interest consists of the 50 000 students on campus.

A descriptive measure of a population is called a **parameter.** The parameter of interest in Case 1.1 is the mean number of soft drinks consumed by all the students at the university. In most applications of inferential statistics the parameter represents the information we need.

1-1b Sample

A **sample** is a set of data drawn from the studied population. A descriptive measure of a sample is called a **statistic**. We use statistics to make inferences about parameters. In Case 1.1, the statistic we would compute is the mean number of soft drinks consumed in the last week by the 500 students in the sample. We would then use the sample mean to infer the value of the population mean, which is the parameter of interest in this problem.

1-1c Statistical Inference

Statistical inference is the process of making an estimate, prediction or decision about a population based on sample data. Because populations are almost always very large, investigating each member of the population would be impractical and expensive. It is far easier and cheaper to take a sample from the population of interest and draw conclusions or make estimates about the population on the basis of information provided by the sample. However, such conclusions and estimates are not always going to be correct. For this reason, we build into the statistical inference a measure of reliability. There are two such measures: the **confidence level** and the **significance level**. The *confidence level* is the proportion of times that an estimating procedure will be correct. For example, in Case 1.1, we will produce an estimate of the average number of soft drinks to be consumed by all 50 000 students that has a confidence level of 95%. In other words, estimates based on this form of statistical inference will be correct 95% of the time. When the purpose of the statistical inference is to draw a conclusion about a population, the *significance level* measures how frequently the conclusion will be wrong.

1-2 STATISTICAL APPLICATIONS IN BUSINESS

An important function of statistics courses in business and economics programmes is to demonstrate that statistical analysis plays an important role in virtually all aspects of business and economics. We intend to do so through examples, exercises and cases. However, we assume that most students taking their first statistics course have not taken courses in most of the other subjects in management programmes. To understand fully how statistics is used in these and other subjects, it is necessary to know something about them. To provide sufficient background to understand the statistical application we introduce applications in accounting, economics, finance, human resources management, marketing, environment, population and operations management. We will provide readers with some background to these applications by describing their functions in two ways which are presented below.

1-2a Application Sections and Subsections

We feature five sections that describe statistical applications in the functional areas of business. For example, in Section 7-3, we show an application in finance that describes a financial analyst's use of probability and statistics to construct portfolios that decrease risk. One section and one subsection demonstrate the uses of probability and statistics in specific industries. The opening example in Chapter 4 introduces an interesting application of statistics in professional football. A subsection in Section 6-4, presents an application in medical testing (useful in the medical insurance industry).

1-2b Application Boxes

For other topics that require less detailed description, we provide application boxes with a relatively brief description of the background followed by examples or exercises. These boxes are scattered throughout the book. For example, in Chapter 3 we discuss a job a marketing manager may need to undertake to determine the appropriate price for a product. To understand the context, we need to provide a description of marketing management. The statistical application will follow.

1-3 LARGE REAL DATA SETS

The authors believe that you learn statistics by doing statistics. For their lives after college and university, we expect our students to have access to large amounts of real data that must be summarised to acquire the information needed to make decisions. To provide practise in this vital skill we have created a large real data set (amongst many other smaller ones for individual exercises), available to be downloaded from the accompanying website (see the 'Digital Resources' page for information on access).

1-3a Census

The Census is a complete population count for a given area or place and is an essential tool in the administration of a country. In the UK a census is usually taken every 10 years and the census taken in 1801 is regarded as the first national census. The surveys measure a large numbers of variables and observations. We have included a random sample from the 2011 survey. The records are stored as Census 2011. The survey size is 2155. We have deleted the responses that are known as missing data (don't know, refused, etc.). The source is the 2011 Census from England and Wales. We have included some demographic variables, such as age, gender, race, income and education. The full lists of variables are stored on our website in a document file named *Census2011*.

1-4 STATISTICS AND THE COMPUTER

In virtually all applications of statistics, the statistics practitioner must deal with large amounts of data. For example, Case 1.1 (Coca-Cola) involves 500 observations. To estimate annual profits, the statistics practitioner would have to perform computations on the data. Although the calculations do not require any great mathematical skill, the sheer amount of arithmetic makes this aspect of the statistical method time-consuming and tedious.

Fortunately, numerous commercially prepared computer programs are available to perform the arithmetic. We have chosen to use Microsoft Excel, which is a spreadsheet program, and Minitab, which is a statistical software package. (We use the latest versions of both software: Office 2013 and Minitab 16.)

We chose Excel because we believe that it is and will continue to be the most popular spreadsheet package. One of its drawbacks is that it does not offer a complete set of the statistical techniques we introduce. Consequently, we created add-ins that can be loaded onto your computer to enable you to use Excel for all statistical procedures introduced in this book. The add-ins can be downloaded and, when installed, will appear as *Data Analysis Plus©* on Excel's Add-ins menu. Also available are introductions to Excel and Minitab, and detailed instructions for both software packages. A large proportion of the examples, exercises and cases feature large data sets. These are denoted with the file name next to the exercise number. We demonstrate the solution to the statistical examples in three ways: manually; by employing Excel; and by using Minitab. Moreover, we will provide detailed instructions for all techniques.

The files contain the data needed to produce the solution. However, in many real applications of statistics, additional data are collected. Many other data sets are similarly constructed. In later chapters, we will return to these files and require other statistical techniques to extract the needed information.

The approach we prefer to take is to minimise the time spent on manual computations and to focus instead on selecting the appropriate method for dealing with a problem and on interpreting the output after the computer has performed the necessary computations. In this way, we hope to demonstrate that statistics can be as interesting and as practical as any other subject in your curriculum.

1-4a Excel Spreadsheets

Books written for statistics courses taken by students on mathematics or statistics degrees are considerably different from this one. It is not surprising that such courses feature mathematical proofs of theorems and derivations of most procedures. When the material is covered in this way, the underlying concepts that support statistical inference are exposed and relatively easy to see. This book was created for an applied course in business and economics statistics. Consequently, we do not address directly the mathematical principles of statistics. However, one of the most important functions of statistics practitioners is to properly interpret statistical results, whether produced manually or by computer. And, to correctly interpret statistics, students require an understanding of the principles of statistics.

To help students understand the basic foundation, we will teach readers how to create Excel spreadsheets that allow for *what-if analyses*. By changing some of the input value, students can see for themselves how statistics works. (The term is derived from *what* happens to the statistics *if* I change this value?) These spreadsheets can also be used to calculate many of the same statistics that we introduce later in this book.

CHAPTER EXERCISES

1.1 In your own words, define and give an example of each of the following statistical terms:

a. population
b. sample
c. parameter
d. statistic
e. statistical inference

1.2 Briefly describe the difference between descriptive statistics and inferential statistics.

1.3 A politician who is running for the office of mayor of a city with 25 000 registered voters commissions a survey. In the survey, 48% of the 200 registered voters interviewed say they plan to vote for her.

a. What is the population of interest?
b. What is the sample?
c. Is the value 48% a parameter or a statistic? Explain.

1.4 A manufacturer of computer chips in Cairo claims that less than 10% of its products are defective. When 1000 chips were drawn from a large production, 7.5% were found to be defective.

a. What is the population of interest?
b. What is the sample?

c. What is the parameter?
d. What is the statistic?
e. Does the value 10% refer to the parameter or to the statistic?
f. Is the value 7.5% a parameter or a statistic?
g. Explain briefly how the statistic can be used to make inferences about the parameter to test the claim.

1.5 Suppose you believe that, in general, graduates who have completed a degree in your subject are offered higher salaries upon graduating than are graduates of other programmes. Describe a statistical experiment that could help test your belief.

1.6 The owner of a large fleet of taxis in Johannesburg is trying to estimate his costs for next year's operations. One major cost is petrol purchases. To estimate petrol purchases, the owner needs to know the total distance his taxis will travel next year, the cost of a litre of petrol and the petrol consumption of his taxis. The owner has been provided with the first two figures (distance estimate and cost of a litre of petrol). However, because of the high cost of petrol, the owner has recently converted his taxis to operate on propane. He has measured and recorded the propane consumption (in kilometres per litre) for 50 taxis.

a. What is the population of interest?
b. What is the parameter the owner needs?
c. What is the sample?
d. What is the statistic?
e. Describe briefly how the statistic will produce the kind of information the owner wants.

PART TWO

DESCRIPTIVE STATISTICAL METHODS

2 Graphical Descriptive Techniques I
3 Graphical Descriptive Techniques II
4 Numerical Descriptive Techniques
5 Data Collection and Sampling

2 GRAPHICAL DESCRIPTIVE TECHNIQUES I

2-1 TYPES OF DATA AND INFORMATION

2-2 DESCRIBING A SET OF NOMINAL DATA

2-3 DESCRIBING THE RELATIONSHIP BETWEEN TWO NOMINAL VARIABLES AND COMPARING TWO OR MORE NOMINAL DATA SETS

DO MALE AND FEMALE BRITISH VOTERS DIFFER IN THEIR PARTY AFFILIATION?

The European Social Survey (ESS) is an academically driven cross-national survey that is conducted every 2 years across Europe and measures the attitudes, beliefs and behaviour patterns of diverse population. One question in 2012 survey was 'Which party did you vote for in the election? (United Kingdom)'. Responses were:

1. Conservative
2. Labour
3. Liberal Democrat
4. Scottish National Party
5. Plaid Cymru
6. Green Party
7. Other

Respondents were also identified by gender: 1 = male and 2 = female. Some of the data are listed here.

ID	GENDER	PARTY
1	1	3
2	2	1
3	2	2
.	.	.
.	.	.
1436	1	5
1437	1	2
1438	1	1

Determine whether British female and male voters differ in their political affiliations. **See Section 2-3b for the answer**.

INTRODUCTION

In Chapter 1 we pointed out that statistics is divided into two basic areas: descriptive statistics and inferential statistics. The purpose of this chapter, together with the next, is to present the principal methods that fall under the heading of descriptive statistics. In this chapter, we introduce graphical and tabular statistical methods that allow managers to summarise data visually to produce useful information that is often used in decision-making. Another class of descriptive techniques, numerical methods, is introduced in Chapter 4.

Managers frequently have access to large masses of potentially useful data. But before the data can be used to support a decision, it must be organised and summarised. Consider, for example, the problems faced by managers who have access to the databases created by the use of debit cards. The database consists of the personal information supplied by the customer when he or she applied for the debit card. This information includes age, gender, residence and the cardholder's income. In addition, each time the card is used, the database grows to include a history of the timing, price and brand of each product purchased. Using the appropriate statistical technique, managers can determine which segments of the market are buying their company's brands. Specialised marketing campaigns, including telemarketing, can be developed. Both descriptive and inferential statistics would likely be employed in the analysis.

Descriptive statistics involves arranging, summarising and presenting a set of data in such a way that useful information is produced. Its methods make use of graphical techniques and numerical descriptive measures (such as averages) to summarise and present the data, allowing managers to make decisions based on the information generated. Although descriptive statistical methods are quite straightforward, their importance should not be underestimated. Most management, business and economics students will encounter numerous opportunities to make valuable use of graphical and numerical descriptive techniques when preparing reports and presentations in the workplace.

In Chapter 1, we introduced the distinction between a population and a sample. Recall that a population is the entire set of observations under study, whereas a sample is a subset of a population. The descriptive methods presented in this chapter and in Chapters 3 and 4 apply to both a set of data constituting a population and a set of data constituting a sample.

In both the Preface and Chapter 1, we pointed out that a critical part of your education as statistics practitioners includes an understanding of not only *how* to draw graphs and calculate statistics (manually or by computer) but also *when* to use each technique that we cover. The two most important factors that determine the appropriate method to use are: (1) the type of data and (2) the information that is needed. Both are discussed next.

2-1 TYPES OF DATA AND INFORMATION

The objective of statistics is to extract information from data. There are different types of data and information. To help explain this important principle, we need to define some terms.

A **variable** is some characteristic of a population or sample. For example, the mark on a statistics exam is a characteristic of statistics exams that is certainly of interest to readers of this book. Not all students achieve the same mark. The marks will vary from student to student, thus the name *variable*. The price of a stock is another variable. The prices of most stocks vary daily. We usually represent the name of the variable using uppercase letters such as *X*, *Y* and *Z*.

The **values** of the variable are the possible observations of the variable. The values of statistics exam marks are the integers between 0 and 100 (assuming the exam is marked out of 100). The values of a stock price are real numbers that are usually measured in different currencies, such as the dollar, pound, euro, Saudi riyal or South African rand (sometimes in fractions of the corresponding currency). The values range from 0 to hundreds of the corresponding currency.

Data* are the observed values of a variable. For example, suppose that we observe the following midterm test marks of 10 students:

67 74 71 83 93 55 48 82 68 62

These are the data from which we will extract the information we seek. Incidentally, *data* is plural for **datum.** The mark of one student is a datum.

When most people think of data, they think of sets of numbers. However, there are two types of data: *qualitative* and *quantitative*. Each type has been classified into **two scales of measurement** so that it can be easily interpreted universally. These scales of measurement are particular levels at which outcomes are measured. Each level has a particular set of characteristics and the scales of measurement are **nominal**, **ordinal, interval** and **ratio** data.

The **interval** data measurement is used for real numbers, such as heights, weights, incomes and distances. A distinguishing characteristic of this scale is that the differences between the consecutive numbers are of equal intervals and we can interpret differences in the distance along the scale. We also refer to this type of data as **quantitative** or **numerical.**

Ratio data is a special kind of interval data. The **ratio data** measurement scale is used to express the ratio of some of the values of interval data, so the numbers can be compared as multiples of one another.

The values of **nominal** data are categories. For example, responses to questions about marital status produce nominal data. The values of this variable are: single, married, divorced and widowed. Notice that the values are not numbers but instead are words that describe the categories. We often record nominal data by arbitrarily assigning a number to each category. For example, we could record marital status using the following codes:

Single = 1, Married = 2, Divorced = 3, Widowed = 4

However, any other numbering system is valid provided that each category has a different number assigned to it. Here is another coding system that is just as valid as the previous one.

Single = 7, Married = 4, Divorced = 13, Widowed = 1

Nominal data are also called **qualitative** or **categorical.**

The **ordinal** data appear to be nominal, but the difference is that the order of their values has meaning. For example, at the completion of most college and university courses, students are asked to evaluate the course. The variables are the ratings of various aspects of the course, including the professor. Suppose that in a particular college the values are

poor, fair, good, very good and excellent

The difference between nominal and ordinal types of data is that the order of the values of the latter indicates a higher rating. Consequently, when assigning codes to the values, we should maintain the order of the values. For example, we can record the students' evaluations as

Poor = 1, Fair = 2, Good = 3, Very good = 4, Excellent = 5

Because the only constraint that we impose on our choice of codes is that the order must be maintained, we can use any set of codes that are in order. For example, we can also assign the following codes:

Poor = 6, Fair = 18, Good = 23, Very good =45, Excellent = 88

As we will discuss in Chapter 17, which introduces statistical inference techniques for ordinal data, the use of any code that preserves the order of the data will produce exactly the same result. Thus, it is not the magnitude of the values that is important, it is their order.

Students often have difficulty distinguishing between ordinal and interval data. The critical difference between them is that the intervals or differences between values of interval data are consistent

---

*Unfortunately, the term *data,* like the term *statistician,* has taken on several different meanings. For example, dictionaries define data as facts, information or statistics. In the language of computers, data may refer to any piece of information such as this textbook or an essay you have written. Such definitions make it difficult for us to present *statistics* as a method of converting data into *information.* In this book, we carefully distinguish among the three terms.

and meaningful (which is why this type of data is called *interval*). For example, the difference between marks of 85 and 80 is the same five-mark difference that exists between 75 and 70 – that is, we can calculate the difference and interpret the results.

Because the codes representing ordinal data are arbitrarily assigned except for the order, we cannot calculate and interpret differences. For example, using a 1-2-3-4-5 coding system to represent poor, fair, good, very good and excellent, we note that the difference between excellent and very good is identical to the difference between good and fair. With a 6-18-23-45-88 coding, the difference between excellent and very good is 43, and the difference between good and fair is 5. Because both coding systems are valid, we cannot use either system to compute and interpret differences.

Here is another example. Suppose that you are given the following list of the most active stocks traded on the NASDAQ in descending order of magnitude:

Order	Most Active Stocks
1	Microsoft
2	Cisco Systems
3	Dell Computer
4	Sun Microsystems
5	JDS Uniphase

Does this information allow you to conclude that the difference between the numbers of stocks traded in Microsoft and Cisco Systems is the same as the difference in the number of stocks traded between Dell Computer and Sun Microsystems? The answer is 'no' because we have information only about the order of the numbers of trades, which are ordinal, and not the numbers of trades themselves, which are interval. In other words, the difference between 1 and 2 is not necessarily the same as the difference between 3 and 4.

2-1a Calculations for Types of Data

Interval Data All calculations are permitted on interval data. We often describe a set of interval data by calculating the average. For example, the average of the 10 marks listed on Section 2-1 is 70.3. As you will discover, there are several other important statistics that we will introduce.

Nominal Data Because the codes of nominal data are completely arbitrary, we cannot perform any parametric calculations, such as mean, on these codes. To understand why, consider a survey that asks people to report their marital status. Suppose that the first 10 people surveyed gave the following responses:

single, married, married, married, widowed, single, married, married, single, divorced

Using the codes

Single = 1, Married = 2, Divorced = 3, Widowed = 4,

we would record these responses as

1 2 2 2 4 1 2 2 1 3

The average of these numerical codes is 2.0. Does this mean that the average person is married? Now suppose four more persons were interviewed, of whom three are widowed and one is divorced. The data are given here:

1 2 2 2 4 1 2 2 1 3 4 4 4 3

The average of these 14 codes is 2.5. Does this mean that the average person is married – but halfway to getting divorced? The answer to both questions is an emphatic 'no'. This example illustrates a fundamental truth about nominal data: calculations based on the codes used to store this type of data are meaningless. All that we are permitted to do with nominal data is count or compute the percentages of the occurrences of each category. Thus, we would describe the 14 observations by counting the number of each marital status category and reporting the frequency as shown in the following table.

Category	Code	Frequency
Single	1	3
Married	2	5
Divorced	3	2
Widowed	4	4

The remainder of this chapter deals with nominal data only. In Chapter 3 we introduce graphical techniques that are used to describe interval data.

Ordinal Data The most important aspect of ordinal data is the order of the values. As a result, the only permissible calculations are those involving a ranking process. For example, we can place all the data in order and select the code that lies in the middle. As we discuss in Chapter 4, this descriptive measurement is called the *median*.

2-1b Hierarchy of Data

The data types can be placed in order of the permissible calculations. At the top of the list, we place the interval data type because virtually *all* computations are allowed. The nominal data type is at the 'bottom' because *no* calculations other than determining frequencies are permitted. (We are permitted to perform calculations using the frequencies of codes, but this differs from performing calculations on the codes themselves.) In between interval and nominal data lies the ordinal data type. Permissible calculations are ones that rank the data.

Higher-level data types may be treated as lower-level ones. For example, in universities and colleges, we convert the marks in a course, which are interval, to letter grades, which are ordinal. Some graduate courses feature only a pass or fail designation. In this case, the interval data are converted to nominal. It is important to point out that when we convert higher-level data as lower-level we lose information. For example, a mark of 89 on an accounting course exam gives far more information about the performance of that student than does a letter grade of B, which might be the letter grade for marks between 80 and 90. As a result, we do not convert data unless it is necessary to do so. We will discuss this later.

It is also important to note that we cannot treat lower-level data types as higher-level types. The definitions and hierarchy are summarised in the following box.

Types of Data

Interval

Values are real numbers.

All calculations are valid (e.g., mean, median, percentiles, frequency distributions, etc.).

Data may be treated as ordinal or nominal.

Ordinal

Values must represent the ranked order of the data.

Calculations based on an ordering process are valid (e.g., median and percentiles).

Data may be treated as nominal but not as interval.

Nominal

Values are the arbitrary numbers that represent categories.

Only calculations based on the frequencies or percentages of occurrence are valid.

Data may not be treated as ordinal or interval.

2-1c Interval, Ordinal and Nominal Variables

The variables whose observations constitute our data will be given the same name as the type of data. Thus, for example, interval data are the observations of an interval variable.

2-1d Problem Objectives and Information

In presenting the different types of data, we introduced a critical factor in deciding which statistical procedure to use. A second factor is the type of information we need to produce from our data. We discuss the different types of information in greater detail in Section 11-4a, when we introduce *problem objectives*. However, in this part of the book (Chapters 2–5), we will use statistical techniques to describe a set of data, compare two or more sets of data and describe the relationship between two variables. In Section 2-2, we introduce graphical and tabular techniques employed to describe a set of nominal data. Section 2-3 shows how to describe the relationship between two nominal variables and compare two or more sets of nominal data.

EXERCISES

2.1 Provide two examples each of nominal, ordinal, ratio and interval data.

2.2 For each of the following examples of data, determine the type.

a. The number of kilometres joggers run per week

b. The starting salaries of graduates of MBA programmes

c. The months in which a firm's employees choose to take their holidays

d. The final letter grades received by students in a statistics course

2.3 For each of the following examples of data, determine the type.

a. The weekly closing price of the stock of Amazon.com

b. The month of highest vacancy rate at a Crowne Plaza hotel

c. The size of soft drink (small, medium or large) ordered by a sample of McDonald's customers

d. The number of Toyotas imported monthly by the Middle East over the last 5 years

e. The marks achieved by the students in a statistics course final exam marked out of 100

2.4 The placement office at a university regularly surveys the graduates 1 year after graduation and asks for the following information. For each, determine the type of data.

a. What is your occupation?

b. What is your income?

c. What degree did you obtain?

d. What is the amount of your student loan?

e. How would you rate the quality of instruction? (excellent, very good, good, fair, poor)

2.5 Residents of apartments in Madrid were recently surveyed and asked a series of questions. Identify the type of data for each question.

a. What is your age?

b. On what floor is your apartment?

c. Do you own or rent?

d. How large is your apartment (in square metres)?

e. Does your apartment have a pool?

2.6 A sample of shoppers at a shopping plaza near Milan was asked the following questions. Identify the type of data each question would produce.

a. What is your age?

b. How much did you spend?

c. What is your marital status?

d. Rate the availability of parking: excellent, good, fair or poor

e. How many stores did you enter?

2.7 Information about a magazine's readers is of interest to both the publisher and the magazine's advertisers. A survey of readers asked respondents to complete the following:

a. Age

b. Gender

c. Marital status

d. Number of magazine subscriptions

e. Annual income

f. Rate the quality of our magazine: excellent, good, fair or poor

For each item identify the resulting data type.

2.8 Football fans are regularly asked to offer their opinions about various aspects of the sport. A survey asked the following questions. Identify the type of data.

a. How many games do you attend annually?

b. How would you rate the quality of entertainment? (excellent, very good, good, fair, poor)

c. Do you have season tickets?

d. How would you rate the quality of the food? (edible, barely edible, horrible)

2.9 A survey of golfers were asked the following questions. Identify the type of data each question produces.

a. How many rounds of golf do you play annually?

b. Are you a member of a private club?

c. What brand of clubs do you own?

2.10 At the end of the term, university and college students often complete questionnaires about their courses. Suppose that in one university, students were asked the following.

a. Rate the course (highly relevant, relevant, irrelevant)

b. Rate the professor (very effective, effective, not too effective, not at all effective)

c. What was your midterm grade (A, B, C, D, F)?

Determine the type of data each question produces.

2-2 DESCRIBING A SET OF NOMINAL DATA

As we discussed in Section 2-1, the only allowable calculation on nominal data is to count the frequency or compute the percentage that each value of the variable represents. We can summarise the data in a table, which presents the categories and their counts, called a **frequency distribution.** A **relative frequency distribution** lists the categories and the proportion with which each occurs. We can use graphical techniques to present a picture of the data. There are two graphical methods we can use: the bar chart and the pie chart.

EXAMPLE 2.1

DATA Census2011

Work Status in the Census2011 Survey

Census is an official count of members of a population of people. Since 1841, in the UK, the Census has been every 10 years, providing a snapshot of the population and its characteristics. The Census 2011 data set consists of a random sample of 1% of people in the 2011 Census output database for England and Wales. In the 2011 survey respondents were asked the following.

'Last week, were you: working as an employee, on a government sponsored training scheme, self-employed or freelance, working paid or unpaid, away from work, doing any kind of paid work or none of the above?' The responses were:

1. Employed
2. Self-employed
3. Unemployed
4. Full-time student
5. Retired
6. School
7. Looking after home and family
8. Long-term sick or disabled
9. Other

The responses were recorded using the codes 1, 2, 3, 4, 5, 6, 7, 8 and 9, respectively. The first 150 observations are listed here. The name of the variable is Economic Activity, and the data are stored in the 11th column (column K in Excel, column 11 in Minitab).

Construct a frequency and relative frequency distribution for these data and graphically summarise the data by producing a bar chart and a pie chart.

5	1	1	1	1	1	1	5	5	1	1	4	1	4
6	1	5	2	1	5	1	2	4	1	1	2	5	3
5	5	1	1	5	1	1	5	1	1	9	1	1	8
2	1	1	1	8	1	1	6	5	1	1	9	1	3
9	1	3	1	3	4	1	1	1	5	5	3	3	1
7	1	1	3	8	1	5	1	1	8	1	1	1	4
7	1	1	1	5	7	8	1	7	8	1	1	3	1
1	1	9	1	5	5	1	5	1	7	5	1	1	1
1	5	5	1	1	5	1	3	1	1	1	5	5	1
1	1	5	7	2	1	4	2	1	5	5	1	5	1

SOLUTION:

Scan the data. Have you learned anything about the responses of these 150 respondents? Unless you have special skills you have probably learned little about the numbers. If we had listed all 2155 observations you would be even less likely to discover anything useful about the data. To extract useful information requires the application of a statistical or graphical technique. To choose the appropriate technique we must first identify the type of data. In this example the data are nominal because the numbers represent categories. The only calculation permitted on nominal data is to count the number of occurrences of each category. Hence, we count the number of 1s, 2s, 3s, 4s, 5s, 6s, 7s, 8s and 9s. The list of the categories and their counts constitute the frequency distribution. The relative frequency distribution is produced by converting the frequencies into proportions. The frequency and relative frequency distributions are combined in Table 2.1.

TABLE 2.1 Frequency and Relative Frequency Distributions for Example 2.1

WORK STATUS	CODE	FREQUENCY	RELATIVE FREQUENCY (%)
Employed	1	978	45.38
Self-employed	2	108	5.01
Unemployed	3	155	7.19
Full-time student	4	84	3.9
Retired	5	472	21.9
School	6	97	4.5
Looking after home and family	7	97	4.5
Long-term sick or disabled	8	112	5.2
Other	9	52	2.41
Total		21 555	100

As we promised in Chapter 1 (and the Preface), we demonstrate the solution of all examples in this book using three approaches (where feasible): manually, using Excel and using Minitab. For Excel and Minitab, we provide not only the printout but also instructions to produce them.

EXCEL

Instructions

(Specific commands for this example are highlighted.)

1. Type or import the data into one or more columns. **(Open Census2011.)**
2. Activate any empty cell and type

= **COUNTIF** ([Range], [Criteria])

Input range are the cells containing the data. In this example, the range is K1:K2156. The criteria are the codes you want to count: (1) (2) (3) (4) (5) (6) (7) (8) (9). To count the number of 1s type

= **COUNTIF** (K1:K2156,1)

and the frequency will appear in the dialog box. Change the criteria to produce the frequency of the other categories. Relative Frequency is calculated by dividing the individual frequencies by the total number of data values (e.g. 978/2155 = 45.38%).

MINITAB

Economic Activity	Count	Percent
1	978	45.38
2	108	5.01
3	155	7.19
4	84	3.90
5	472	21.90
6	97	4.50
7	97	4.50
8	112	5.20
9	52	2.41
	N = 2155	

Instructions

(Specific commands for this example are highlighted.)

1. Type or import the data into one column. **(Open Census2011.)**
2. Click **Stat, Tables** and **Tally Individual Variables.**
3. Type or use the **Select** button to specify the name of the variable or the column where the data are stored in the **Variables** box **(Economic Activity)**. Under **Display,** click **Counts** and **Percents.**

INTERPRET

Only 45.38% of respondents are employed, 21.9% are retired, 7.19% are unemployed, 5.01% are self-employed and the remaining 20.52% are divided almost equally among the other five categories.

Bar and Pie Charts

The information contained in the data is summarised well in the table. However, graphical techniques generally catch a reader's eye more quickly than does a table of numbers. Two graphical techniques can be used to display the results shown in the table. A **bar chart** is often used to display frequencies; a **pie chart** graphically shows relative frequencies.

The bar chart is created by drawing a rectangle representing each category. The height of the rectangle represents the frequency. The base is arbitrary. Figure 2.1 depicts the manually drawn bar chart for Example 2.1.

FIGURE 2.1

Bar Chart for Example 2.1

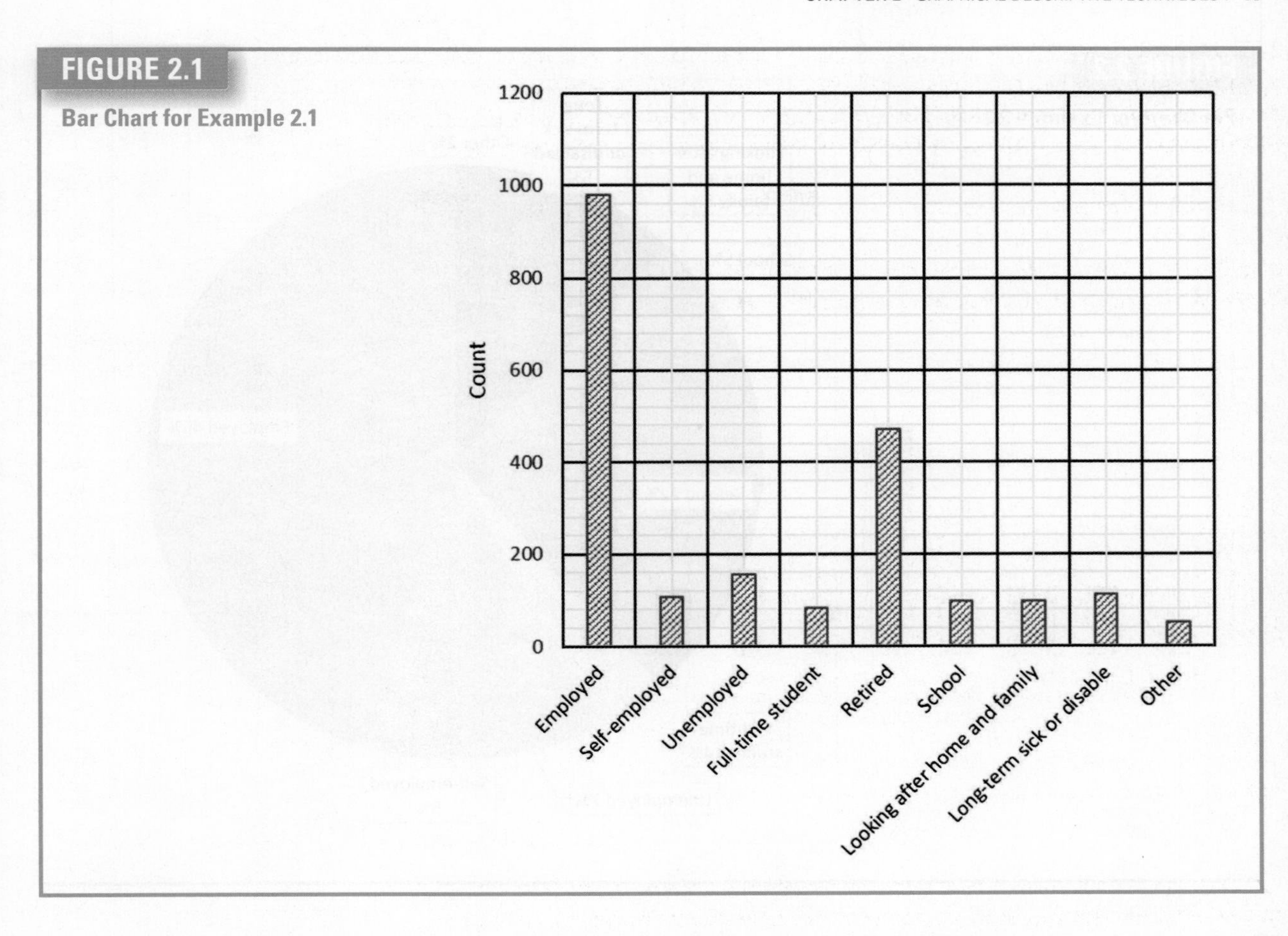

If we wish to emphasise the relative frequencies instead of drawing the bar chart, we draw a pie chart. A pie chart is simply a circle subdivided into slices that represent the categories. It is drawn so that the size of each slice is proportional to the percentage corresponding to that category. For example, because the entire circle is composed of 360 degrees, a category that contains 25% of the observations is represented by a slice of the pie that contains 25% of 360 degrees, which is equal to 90 degrees. The number of degrees for each category in Example 2.1 is shown in Table 2.2.

TABLE 2.2 Proportion in Each Category in Example 2.1

WORK STATUS	RELATIVE FREQUENCY (%)	SLICE OF THE PIE (°)
Employed	45.38	163.4
Self-employed	5.01	18.0
Unemployed	7.19	25.9
Full-time student	3.9	14.0
Retired	21.9	78.8
School	4.5	16.2
Looking after home and family	4.5	16.2
Long-term sick or disabled	5.2	18.7
Other	2.41	8.7
Total	100.0	360

Figure 2.2 was drawn from these results.

FIGURE 2.2

Pie Chart for Example 2.1

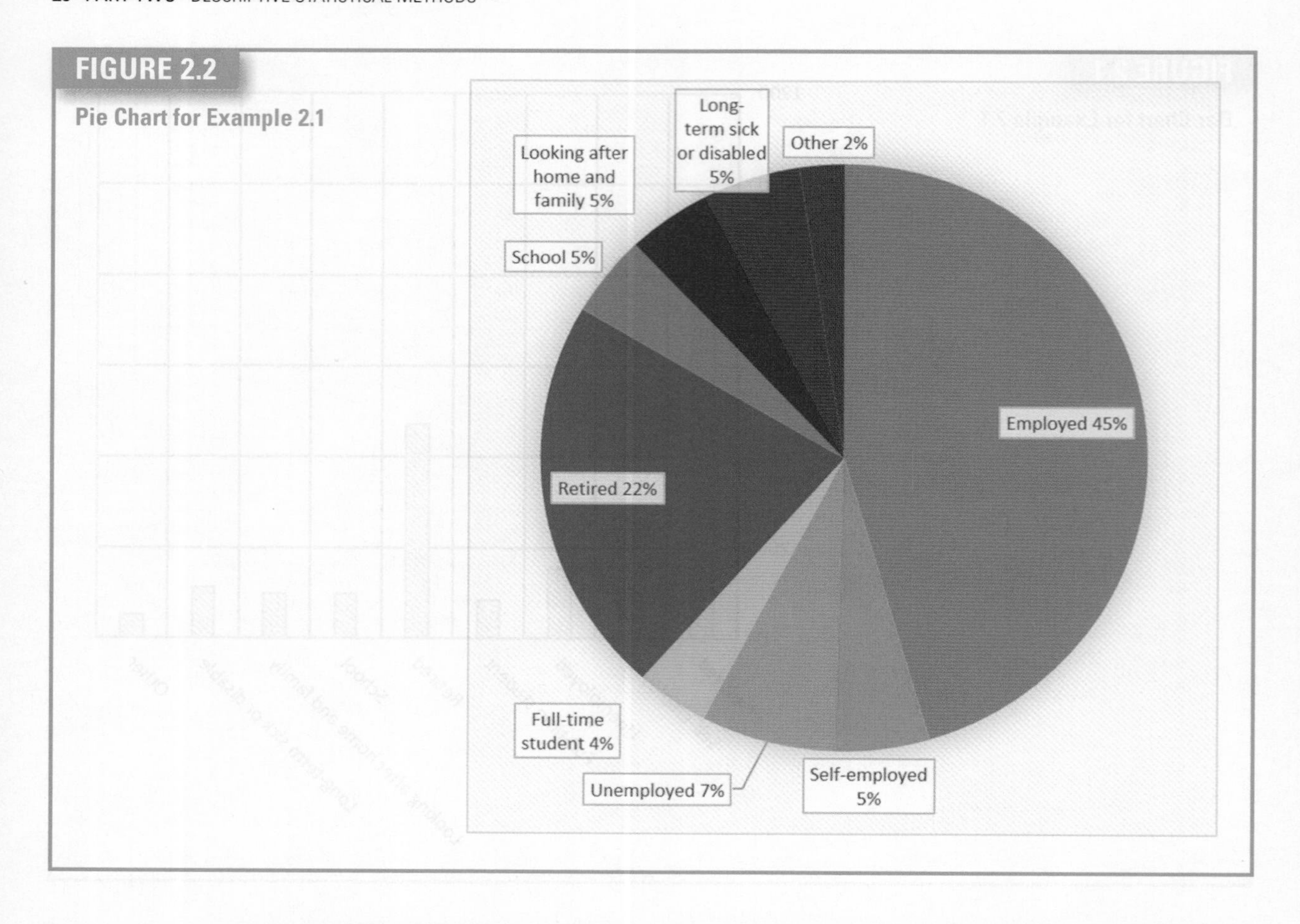

EXCEL **Here are Excel's Bar and Pie Charts.**

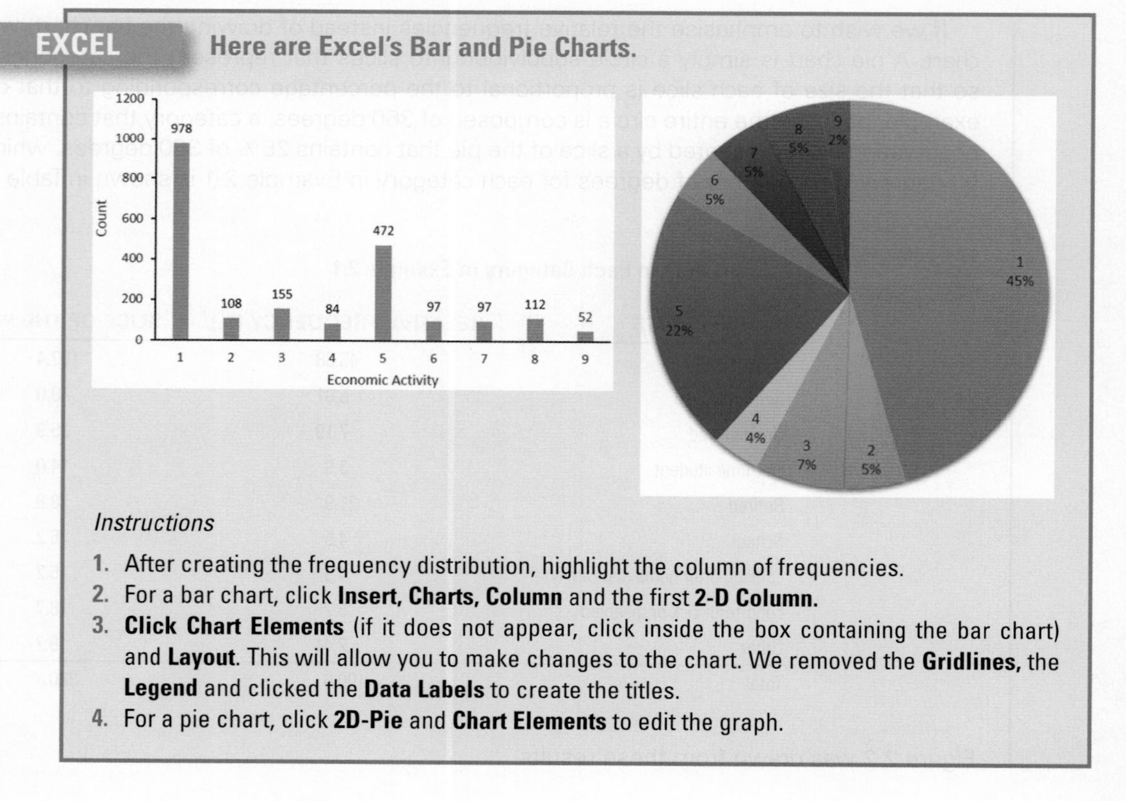

Instructions

1. After creating the frequency distribution, highlight the column of frequencies.
2. For a bar chart, click **Insert, Charts, Column** and the first **2-D Column.**
3. **Click Chart Elements** (if it does not appear, click inside the box containing the bar chart) and **Layout**. This will allow you to make changes to the chart. We removed the **Gridlines,** the **Legend** and clicked the **Data Labels** to create the titles.
4. For a pie chart, click **2D-Pie** and **Chart Elements** to edit the graph.

MINITAB

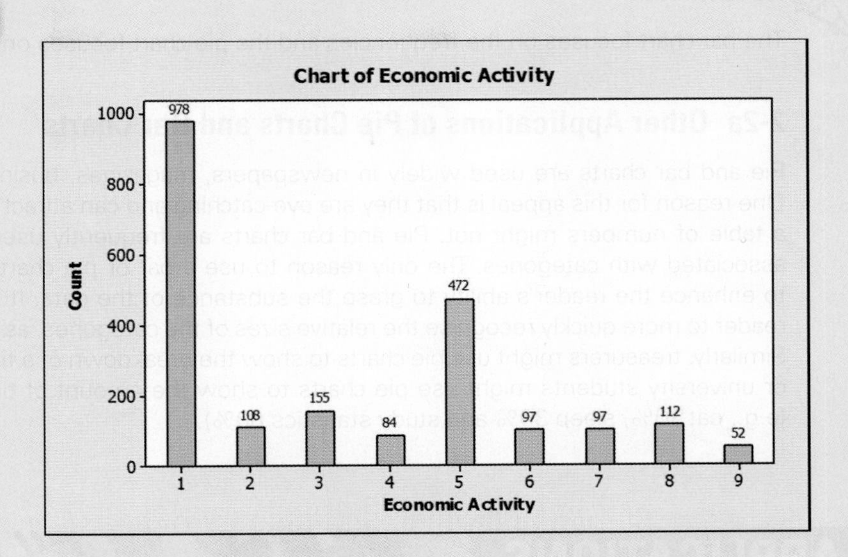

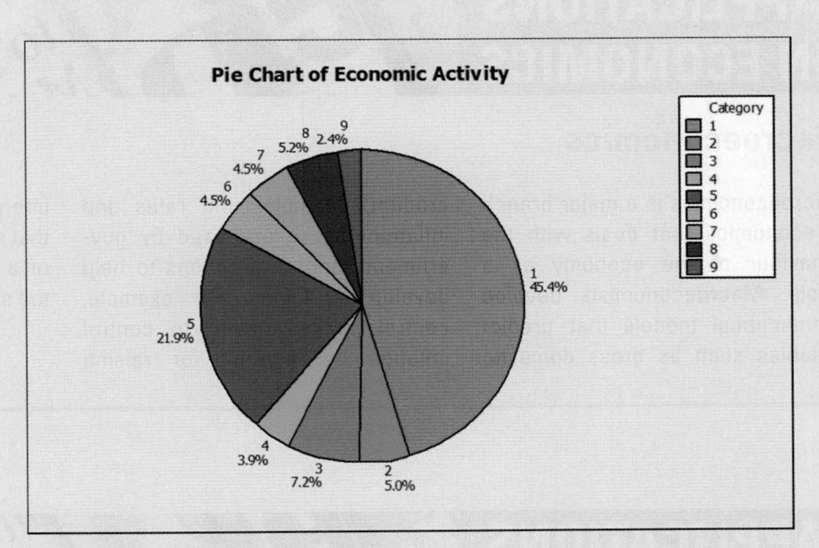

Instructions

1. Type or import the data into one column. (**Open** Census2011.) For a bar chart:
2. Click **Graph** and **Bar Chart.**
3. In the **Bars represent** box, click **Counts of unique values** and select **Simple.**
4. Type or use the **Select** button to specify the variable in the **Categorical variables** box (Economic Activity).

We clicked **Labels** and added the title and clicked **Data Labels** and **Use y-value labels** to display the frequencies at the top of the columns.

For a pie chart:

1. Click **Graph** and **Pie Chart.**
2. Click **Chart, Counts of unique values** and in the **Categorical variables** box type or use the **Select** button to specify the variable (Economic Activity).
3. We clicked **Labels** and added the title. We clicked **Slice Labels** and clicked **Category name** and **Percent.**

INTERPRET

The bar chart focuses on the frequencies and the pie chart focuses on the proportions.

2-2a Other Applications of Pie Charts and Bar Charts

Pie and bar charts are used widely in newspapers, magazines, business and government reports. One reason for this appeal is that they are eye-catching and can attract the reader's interest whereas a table of numbers might not. Pie and bar charts are frequently used to simply present numbers associated with categories. The only reason to use a bar or pie chart in such a situation would be to enhance the reader's ability to grasp the substance of the data. It might, for example, allow the reader to more quickly recognise the relative sizes of the categories, as in the breakdown of a budget. Similarly, treasurers might use pie charts to show the breakdown of a firm's revenues by department, or university students might use pie charts to show the amount of time devoted to daily activities (e.g., eat 10%, sleep 30% and study statistics 60%).

APPLICATIONS IN ECONOMICS

Macroeconomics

Macroeconomics is a major branch of economics that deals with the behaviour of the economy as a whole. Macroeconomists develop mathematical models that predict variables such as gross domestic product, unemployment rates and inflation. These are used by governments and corporations to help develop strategies. For example, central banks attempt to control inflation by lowering or raising interest rates. To do this requires that economists determine the effect of a variety of variables, including the supply and demand for energy.

APPLICATIONS IN ECONOMICS

Energy Economics

One variable that has had a large influence on the economies of virtually every country is energy. The 1973 oil crisis in which the price of oil quadrupled over a short period of time is generally considered to be one of the largest financial shocks to our economy. In fact, economists often refer to two different economies: before the 1973 oil crisis and after.

Unfortunately, the world will be facing more shocks to our economy because of energy, for two primary reasons. The first is the depletion of nonrenewable sources of energy and the resulting price increases. The second is the possibility that burning fossil fuels and the creation of carbon dioxide may be the cause of global warming. One economist predicted that the cost of global warming will be calculated in trillions of dollars. Statistics can play an important role by determining whether the Earth's temperature has been increasing and, if so, whether carbon dioxide is the cause. (See Case 3.1.)

In this chapter, you will encounter other examples and exercises that involve the issue of energy.

EXAMPLE 2.2

DATA Xm02-02

Energy Consumption in Eurasia in 2012

Table 2.3 lists the total energy consumption in Eurasia from all sources in 2012 (latest data available at publication). To make it easier to see the details, the table measures the energy in million tonnes. Use an appropriate graphical technique to depict these figures.

TABLE 2.3 **Energy Consumption in Eurasia by Source, 2012**

ENERGY SOURCES	MILLION TONNES
Non-renewable	
Oil*	900.3
Natural gas	975.0
Coal and coal products	516.9
Nuclear	266.9
Renewable Energy Sources	
Hydroelectric	190.8
Biofuels, Geothermal and other	35.8
Wind	47.2
Solar/photovoltaic	16.1
Total	2949

*Consumption of fuel ethanol and biodiesel is also included.

SOLUTION:

We are interested in describing the proportion of total energy consumption for each source. Thus, the appropriate technique is the pie chart. The next step is to determine the proportions and sizes of the pie slices from which the pie chart is drawn. The following pie chart was created by Excel. Minitab's pie chart would be similar.

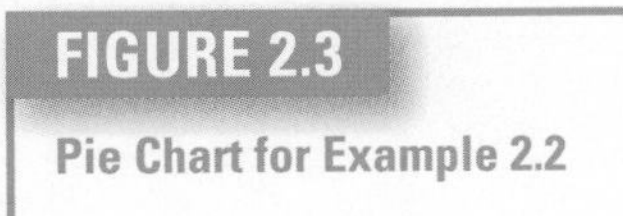

FIGURE 2.3 Pie Chart for Example 2.2

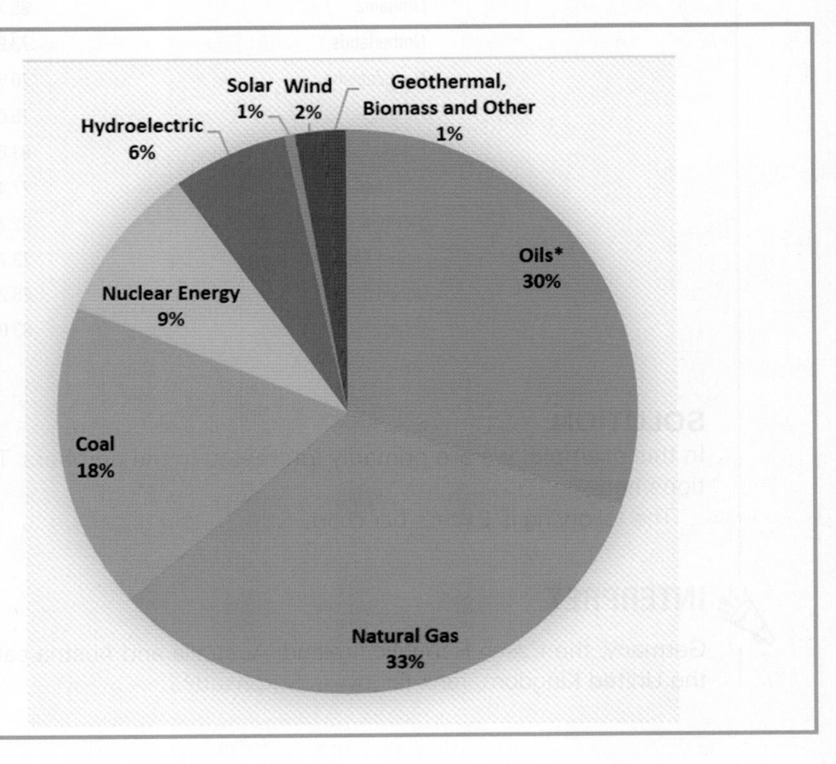

INTERPRET

Eurasia depends heavily on oil, coal and natural gas. More than 82% of energy use is based on these sources. The renewable energy sources amount to less than 10%, of which about two-thirds is hydro-electric and probably cannot be expanded much further. Wind and solar barely appear in the chart. See Exercises 2.11 to 2.14 for more information on energy economics.

EXAMPLE 2.3

DATA Xm02-03

Per Capita Beer Consumption (10 Selected Countries)

Table 2.4 lists the per capita beer consumption for each of 20 countries around the world. Graphically present these numbers.

TABLE 2.4 Per Capita Beer Consumption, 2012

COUNTRY	BEER CONSUMPTION (L/YR)
Australia	83.4
Austria	105.8
Belgium	78.0
Bulgaria	72.8
Croatia	77.8
Czech Republic	131.7
Estonia	90.6
Finland	82.7
Germany	106.8
Ireland	103.7
Lithuania	85.7
Netherlands	73.9
New Zealand	70.5
Panama	75.0
Poland	83.6
Romania	77.4
Slovenia	82.7
United Kingdom	73.7
United States	78.2
Venezuela	83.0

SOLUTION:

In this example, we are primarily interested in the numbers. There is no use in presenting proportions here.

The following is Excel's bar chart.

INTERPRET

Germany, the Czech Republic, Ireland, Australia and Austria came top of the list. Both Belgium and the United Kingdom, rank far lower. Surprised?

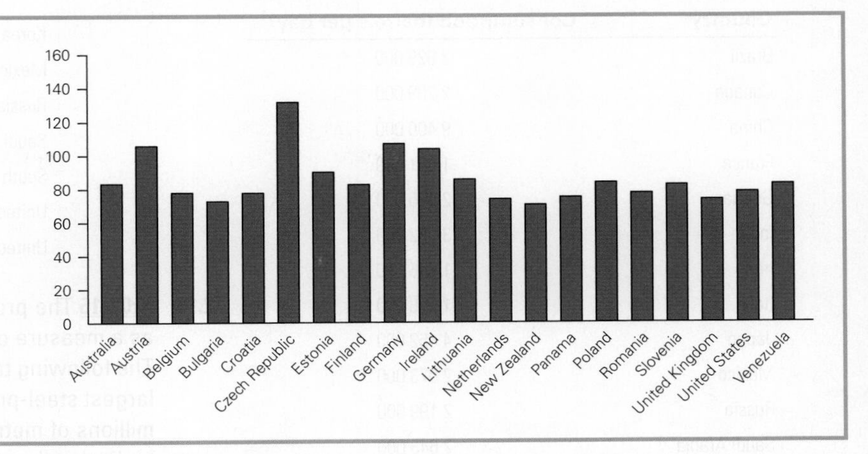

FIGURE 2.4
Excel Bar Chart for Example 2.3

2-2b Describing Ordinal Data

There are no specific graphical techniques for ordinal data. Consequently, when we wish to describe a set of ordinal data, we will treat the data as if they were nominal and use the techniques described in this section. The only criterion is that the bars in bar charts should be arranged in ascending (or descending) ordinal values; in pie charts, the wedges are typically arranged clockwise in ascending or descending order.

We complete this section by describing when bar and pie charts are used to summarise and present data.

Factors That Identify When to Use Frequency and Relative Frequency Tables, Bar and Pie Charts

1. **Objective:** Describe a single set of data.
2. **Data type:** Nominal or ordinal.

EXERCISES

2.11 **Xr02-11** When will the world run out of oil? One way to judge is to determine the oil reserves of the countries around the world. The next table displays the known oil reserves of the top 15 countries. Graphically describe the figures.

Country	Oil Reserves (barrels* of oil)
Brazil	12 860 000 000
Canada	175 200 000 000
China	14 800 000 000
Iran	137 000 000 000
Iraq	115 000 000 000
Kazakhstan	30 000 000 000
Kuwait	104 000 000 000
Libya	46 420 000 000
Nigeria	37 200 000 000
Qatar	25 380 000 000
Russia	60 000 000 000
Saudi Arabia	262 600 000 000
United Arab Emirates	97 800 000 000
United States	20 680 000 000
Venezuela	211 200 000 000

*An oil barrel (abbreviated as *bbl*) is about 159 litres

2.12 Refer to Exercise 2.11. The total oil reserves in the world are 1 481 526 000 000 barrels. The total oil reserves of the top 15 countries are 1 350 140 000 000 barrels. Use a graphical technique that emphasises the percentage breakdown of the top 15 countries plus others.

2.13 **Xr02-13** The following table lists the average oil consumption per day for the top 15 oil-consuming countries. Use a graphical technique to present these figures.

Country	Consumption (barrels per day)
Brazil	2 029 000
Canada	2 209 000
China	9 400 000
France	1 861 000
Germany	2 495 000
India	3 182 000
Iran	1 845 000
Italy	1 528 000
Japan	4 452 000
Mexico	2 073 000
Russia	2 199 000
Saudi Arabia	2 643 000
South Korea	2 195 000
United Kingdom	1 622 000
United States	19 150 000

2.14 Xr02-14 There are 159 litres in a barrel of oil. The number of products produced and the proportion of the total are listed in the following table. Draw a graph to depict these numbers.

Product	Percent of Total
Petrol	51.4
Distillate fuel oil	15.3
Jet fuel	12.6
Still gas	5.4
Marketable coke	5.0
Residual fuel oil	3.3
Liquefied refinery gas	2.8
Asphalt and road oil	1.9
Lubricants	0.9
Other	1.5

2.15 Xr02-15 The planet may be threatened by global warming, possibly caused by burning fossil fuels (petroleum, natural gas and coal) that produce carbon dioxide (CO_2). The following table lists the top 15 producers of CO_2 and the annual amounts (millions of metric tonnes) from fossil fuels. Graphically depict these figures.

Country	CO_2
Australia	417.7
Canada	541.0
China	7706.8
Germany	765.6
India	1591.1
Iran	528.6
Italy	407.9
Japan	1098.0
Korea, South	528.1
Mexico	443.6
Russia	1556.7
Saudi Arabia	438.2
South Africa	451.2
United Kingdom	519.9
United States	5424.5

2.16 Xr02-16 The production of steel has often been used as a measure of the economic strength of a country. The following table lists the steel production in the 20 largest steel-producing nations in 2011. The units are millions of metric tonnes. Use a graphical technique to display these figures.

Country	Steel Production	Country	Steel Production
Australia	6.4	Mexico	18.1
Austria	7.5	Netherlands	6.9
Belgium	8.1	Poland	8.8
Brazil	35.2	Russia	68.7
Canada	13.1	South Africa	6.7
China	683.3	South Korea	68.5
Egypt	6.5	Spain	15.6
France	15.8	Taiwan	22.7
Germany	44.3	Turkey	34.1
India	72.2	Ukraine	35.3
Iran	13	United Kingdom	9.5
Italy	28.7		
Japan	107.6	United States	86.2

2.17 Xr02-17 In 2010 (latest figures available), the total waste generated by 28 EU countries was 220 million tonnes. The following table lists the amount of waste by country. Use one or more graphical techniques to present these figures.

Country	Amount of waste (tonnes)
Belgium	4 678 683
Bulgaria	2 396 337
Czech Republic	3 334 240
Denmark	2 435 921
Germany	36 311 611
Estonia	430 499
Ireland	1 730 028
Greece	5 197 519
Spain	23 198 185
France	29 306 586

Italy	32 478 921
Cyprus	461 227
Latvia	694 013
Lithuania	1 261 402
Luxembourg	385 467
Hungary	2 864 896
Malta	138 099
Netherlands	9 071 995
Austria	4 622 626
Poland	8 889 685
Portugal	5 463 650
Romania	6 127 153
Slovenia	727 708
Slovakia	1 719 012
Finland	1 680 763
Sweden	4 038 272
United Kingdom	28 948 507
Norway	2 228 608

2.18 Xr02-18 The following table lists the top 10 countries and amounts of oil (millions of barrels annually, US oil barrel (abbreviation: bbl) = 158.987295 liters (l)) they exported to the United States in 2010.

Country	Oil Imports (millions of barrels annually)
Algeria	119
Angola	139
Canada	720
Colombia	124
Iraq	151
Kuwait	71
Mexico	416
Nigeria	360
Saudi Arabia	394
Venezuela	333

a. Draw a bar chart.

b. Draw a pie chart.

c. What information is conveyed by each chart?

The following exercises require a computer and software.

2.19 Xr02-19 What are the most important characteristics of colleges and universities? This question was asked of a sample of college-bound high school seniors. The responses included:

1. Location
2. Majors
3. Academic reputation
4. Career focus
5. Community
6. Number of students

The results are stored using the codes. Use a graphical technique to summarise and present the data.

2.20 Xr02-20 A survey asked 392 homeowners which area of the home they would most like to renovate. The responses and frequencies are shown here. Use a graphical technique to present these results. Briefly summarise your findings.

Area	Code
Basement	1
Bathroom	2
Bedroom	3
Kitchen	4
Living/dining room	5

2.21 Xr02-21 Who applies to MBA programmes? To help determine the background of the applicants, a sample of 230 applicants to a university's business school were asked to report their undergraduate degree. The degrees were recorded using the following codes:

1. BA
2. BBA
3. BEng
4. BSc
5. Other

a. Determine the frequency distribution.

b. Draw a bar chart.

c. Draw a pie chart.

d. What do the charts tell you about the sample of MBA applicants?

2.22 Xr02-22 An increasing number of statistics courses use a computer and software rather than manual calculations. A survey of statistics instructors asked each to report the software his or her course uses. The responses included:

1. Excel
2. Minitab
3. SAS
4. SPSS
5. Other

a. Produce a frequency distribution.

b. Graphically summarise the data so that the proportions are depicted.

c. What do the charts tell you about the software choices?

CENSUS EXERCISES

See Chapter 1 for a description of the Census. Exercises 2.23 to 2.25 are based on the Census conducted in 2011.

2.23 Census2011 What is your ethnic group (column I in Excel and column 9 in Minitab – Ethnic Group)?

1. White
2. Mixed
3. Asian and Asian British
4. Black and Black British
5. Chinese or other ethnic group

Summarise the results using an appropriate graphical technique and interpret your findings.

2.24 Census2011 On 27 March 2011, what is your legal marital status (column F in Excel and column 6 in Minitab – Marital Status)?

1. Single
2. Married
3. Separated but still legally married
4. Divorced
5. Widowed

a. Create a frequency distribution.
b. Use a method to present these data and briefly explain what the graph reveals.

2.25 Census2011 To which class do you belong (column N in Excel and column 14 in Minitab – Approximated social grade)?

1. AB
2. C1
3. C2
4. DE

Summarise the data using a graphical method and describe your findings.

2-3 DESCRIBING THE RELATIONSHIP BETWEEN TWO NOMINAL VARIABLES AND COMPARING TWO OR MORE NOMINAL DATA SETS

In Section 2-2, we presented graphical and tabular techniques used to summarise a set of nominal data. Techniques applied to single sets of data are called **univariate analysis.** There are many situations where we wish to depict the relationship between variables; in such cases, **bivariate** methods are required. A **cross-classification table** (also called a **cross-tabulation table** or **contingency table**) is used to describe the relationship between two or more nominal variables. A variation of the bar chart introduced in Section 2-2 is employed to graphically describe the relationship. The same technique is used to compare two or more sets of nominal data.

2-3a Tabular Method of Describing the Relationship between Two Nominal Variables

To describe the relationship between two nominal variables, we must remember that we are permitted only to determine the frequency of the values. As a first step, we need to produce a cross-classification table that lists the frequency of each combination of the values of the two variables.

EXAMPLE 2.4

DATA Xm02-04

Newspaper Readership Survey

A major UK city has four competing newspapers: *The Times, The Independent, The Star* and *The Sun.* To help design advertising campaigns, the advertising managers of the newspapers need to know which segments of the newspaper market are reading their papers. A survey was conducted to analyse the relationship between newspapers readers and their occupation. A sample of newspaper readers was asked to report which newspaper they read – *The Times* (1), *The Independent* (2), *The Star* (3), *The Sun* (4) – and indicate whether they were working class (1), middle class (2) or upper class (3). Some of the data are listed here.

Reader	Occupation	Newspaper
1	2	2
2	1	4
3	2	1
.	.	.
.	.	.
.	.	.
352	3	2
353	1	3
354	2	3

Determine whether the two nominal variables are related.

SOLUTION:

By counting the number of times each of the 12 combinations occurs, we produced the Table 2.5.

TABLE 2.5 Cross-Classification Table of Frequencies for Example 2.4

	NEWSPAPER				
OCCUPATION	TIMES	INDEPENDENT	STAR	SUN	TOTAL
Working class	27	18	38	37	120
Middle class	29	43	21	15	108
Upper class	33	51	22	20	126
Total	89	112	81	72	354

If occupation and newspaper are related, there will be differences in the newspapers readers among the occupations. An easy way to see this is to convert the frequencies in each row (or column) to relative frequencies in each row (or column). That is, compute the row (or column) totals and divide each frequency by its row (or column) total, as shown in Table 2.6. Totals may not equal 1 because of rounding.

TABLE 2.6 Row Relative Frequencies for Example 2.4

	NEWSPAPER				
OCCUPATION	TIMES	INDEPENDENT	STAR	SUN	TOTAL
Working class	0.23	0.15	0.32	0.31	1.00
Middle class	0.27	0.40	0.19	0.14	1.00
Upper class	0.26	0.40	0.17	0.16	1.00
Total	0.25	0.32	0.23	0.20	1.00

EXCEL

Excel can produce the cross-classification table using several methods. We will use and describe the PivotTable in two ways: (1) to create the cross-classification table featuring the counts and (2) to produce a table showing the row relative frequencies.

Count of Reader	Newspaper				
Occupation	Times	Independent	Star	Sun	Grand Total
Working class	27	18	38	37	120
Middle class	29	43	21	15	108
Upper class	33	51	22	20	126
Grand total	89	112	81	72	354

Count of Reader	Newspaper				
Occupation	Times	Independent	Star	Sun	Grand Total
Working class	0.23	0.15	0.32	0.31	1.00
Middle class	0.27	0.40	0.19	0.14	1.00
Upper class	0.26	0.40	0.17	0.16	1.00
Grand total	0.25	0.32	0.23	0.20	1.00

Instructions

The data must be stored in (at least) three columns as we have done in the Excel file named Xm02-04. Put the cursor somewhere in the data range.

1. Click **Insert** and **PivotTable.**
2. Make sure that the Table/Range is correct.
3. Drag the Occupation button to the **ROW** section of the box. Drag the Newspaper button to the **COLUMN** section. Drag the Reader button to the **VALUES** field. Right-click any number in the table, click **Summarise Values By** and check **Count.** To convert to row percentages, right-click any number, click **Summarise Values By, More options ...,** and **Show values as.** Scroll down and click % of **Row Total.** (We then formatted the data into decimals.) To improve both tables, we substituted the names of the occupations and newspapers.

MINITAB

Tabulated Statistics: Occupation, Newspaper

Rows: Occupation				Columns: Newspaper	
	1	2	3	4	All
1	27	18	38	37	120
	22.50	15.00	31.67	30.83	100.00
2	29	43	21	15	108
	26.85	39.81	19.44	13.89	100.00
3	33	51	22	20	126
	26.19	40.48	17.46	15.87	100.00
All	89	112	81	72	354
	25.14	31.64	22.88	20.34	100.00

Cell Contents: Count % of Row

(continued)

Instructions

1. Type or import the data into two columns. (**Open** Xm02-04.)
2. Click **Stat, Tables,** and **Cross Tabulation and Chi-square.**
3. Type or use the **Select** button to specify the **Categorical variables: For rows** (Occupation) and **For columns** (Newspaper).
4. Under **Display,** click **Counts** and **Row percents** (or any you wish).

INTERPRET

Notice that the relative frequencies in the second and third rows are similar and that there are large differences between row 1 and rows 2 and 3. This tells us that working class individuals tend to read different newspapers from both middle class and upper class and that middle class and upper class individuals are quite similar in their newspaper choices.

Graphing the Relationship between Two Nominal Variables

We have chosen to draw three bar charts, one for each occupation depicting the four newspapers. We will use Excel and Minitab for this purpose. The manually drawn charts are identical.

EXCEL

There are several ways to graphically display the relationship between two nominal variables. We have chosen two-dimensional bar charts for each of the three occupations. The charts can be created from the output of the PivotTable (either counts as we have done) or row proportions.

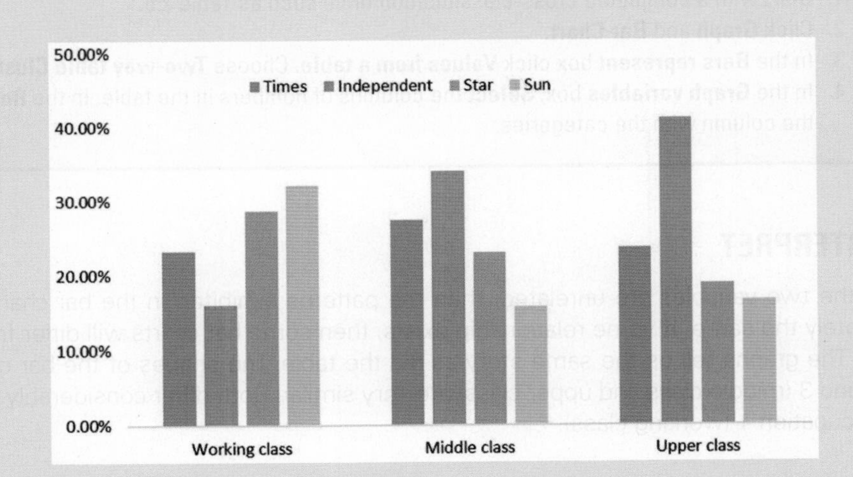

Instructions

From the cross-classification table, click **Insert** and **Column.** You can do the same from any completed cross-classification table.

MINITAB Minitab can draw bar charts from the raw data.

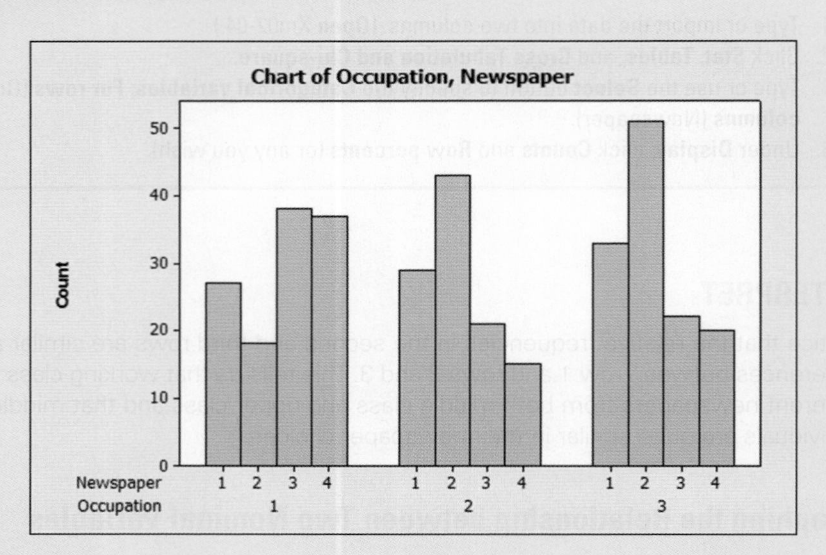

Instructions

1. Click **Graph** and **Bar Chart.**
2. In the **Bars represent** box, specify **Counts of unique values.** Select **Cluster.**
3. In the **Categorical variables** box, type or select the two variables (Newspaper, Occupation).

If you or someone else has created the cross-classification table, Minitab can draw bar charts directly from the table.

Instructions

1. Start with a completed cross-classification table such as Table 2.5.
2. Click **Graph** and **Bar Chart.**
3. In the **Bars represent** box click **Values from a table.** Choose **Two-way table Cluster.**
4. In the **Graph variables** box, **Select** the columns of numbers in the table. In the **Row labels** box, **Select** the column with the categories.

INTERPRET

If the two variables are unrelated, then the patterns exhibited in the bar charts should be approximately the same. If some relationship exists, then some bar charts will differ from others.

The graphs tell us the same story as did the table. The shapes of the bar charts for occupations 2 and 3 (middle class and upper class) are very similar. Both differ considerably from the bar chart for occupation 1 (working class).

2-3b Comparing Two or More Sets of Nominal Data

We can interpret the results of the cross-classification table of the bar charts in a different way. In Example 2.4, we can consider the three occupations as defining three different populations. If differences exist between the columns of the frequency distributions (or between the bar charts), then we can conclude that differences exist among the three populations. Alternatively, we can consider the readership of the four newspapers as four different populations. If differences exist among the frequencies or the bar charts, then we conclude that there are differences between the four populations.

DO MALE AND FEMALE BRITISH VOTERS DIFFER IN THEIR PARTY AFFILIATION?

DATA ESS6-2012, ed.1.2

Using the technique introduced above, we produced the following bar charts.

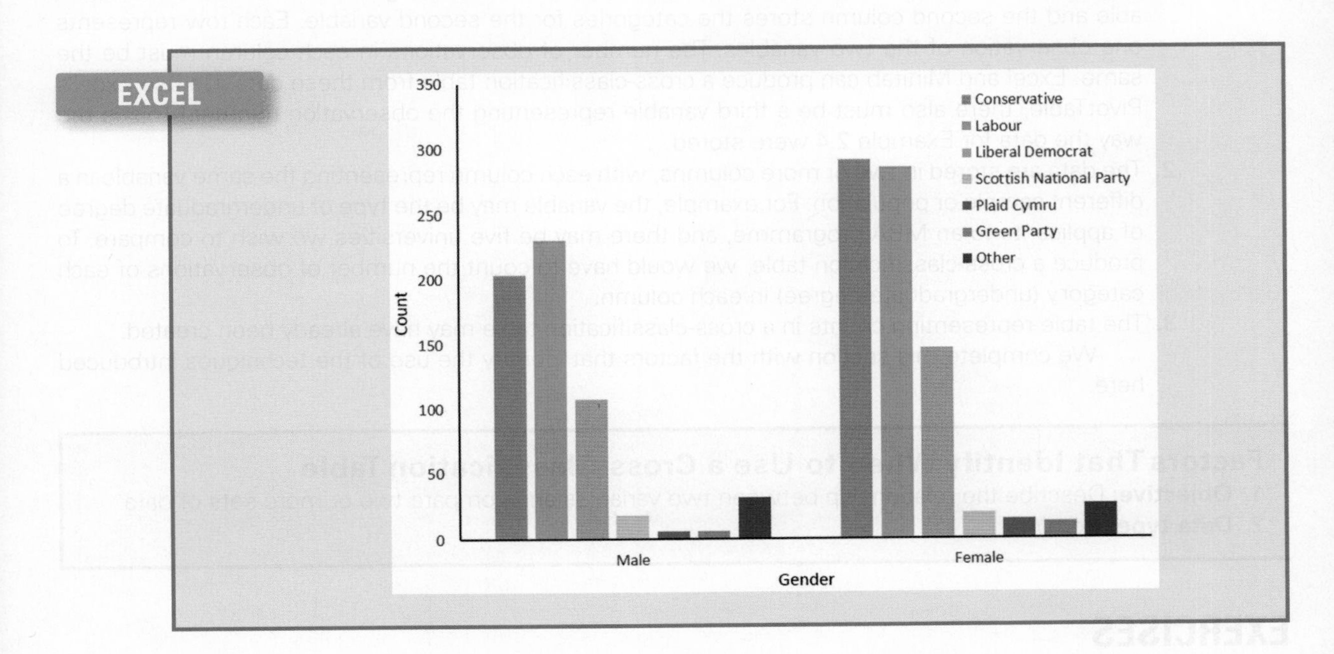

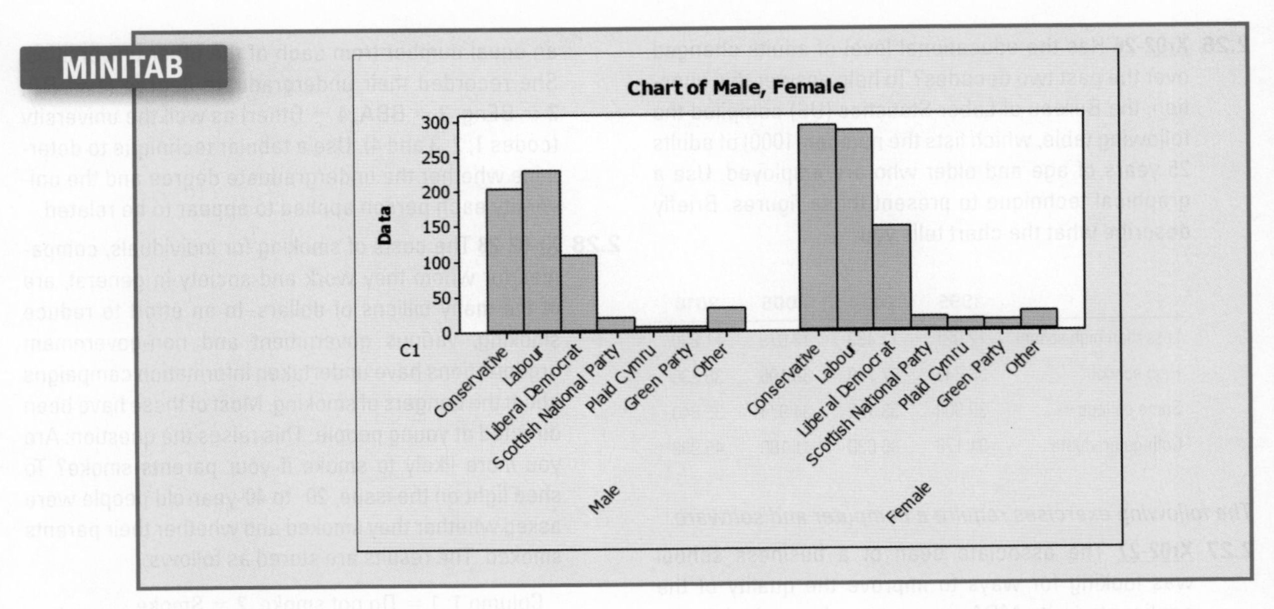

INTERPRET

As you can see, there are substantial differences between the bar charts for men and women. We can conclude that gender and party affiliation are related. However, we can also conclude that differences in party affiliation exist between British male and female voters, specifically, women tend to support the Liberal Democrat party.

2-3c Data Formats

There are several ways to store the data to be used in this section to produce a table or a bar or pie chart.

1. The data are in two columns. The first column represents the categories of the first nominal variable and the second column stores the categories for the second variable. Each row represents one observation of the two variables. The number of observations in each column must be the same. Excel and Minitab can produce a cross-classification table from these data. (To use Excel's PivotTable, there also must be a third variable representing the observation number.) This is the way the data for Example 2.4 were stored.
2. The data are stored in two or more columns, with each column representing the same variable in a different sample or population. For example, the variable may be the type of undergraduate degree of applicants to an MBA programme, and there may be five universities we wish to compare. To produce a cross-classification table, we would have to count the number of observations of each category (undergraduate degree) in each column.
3. The table representing counts in a cross-classification table may have already been created.

We complete this section with the factors that identify the use of the techniques introduced here.

Factors That Identify When to Use a Cross-Classification Table

1. **Objective:** Describe the relationship between two variables and compare two or more sets of data.
2. **Data type:** Nominal.

EXERCISES

2.26 Xr02-26 Has the educational level of adults changed over the past two decades? To help answer this question, the Bureau of Labor Statistics (US) compiled the following table, which lists the number (1000) of adults 25 years of age and older who are employed. Use a graphical technique to present these figures. Briefly describe what the chart tells you.

	1995	2000	2005	2010
Less than high school	12 021	12 483	12 679	11 880
High school	36 746	37 750	38 196	38 236
Some college	30 908	33 198	34 974	36 840
College graduate	31 176	36 630	41 180	45 998

The following exercises require a computer and software.

2.27 Xr02-27 The associate dean of a business school was looking for ways to improve the quality of the applicants to its MBA programme. In particular, she wanted to know whether the undergraduate degree of applicants differed among her school and the three nearby universities with MBA programmes. She sampled 100 applicants to her programme and an equal number from each of the other universities. She recorded their undergraduate degree (1 = BA, 2 = BEng, 3 = BBA, 4 = Other) as well the university (codes 1, 2, 3 and 4). Use a tabular technique to determine whether the undergraduate degree and the university each person applied to appear to be related.

2.28 Xr-02-28 The costs of smoking for individuals, companies for whom they work and society in general, are in the many billions of dollars. In an effort to reduce smoking, various government and non-government organizations have undertaken information campaigns about the dangers of smoking. Most of these have been directed at young people. This raises the question: Are you more likely to smoke if your parents smoke? To shed light on the issue, 20- to 40-year-old people were asked whether they smoked and whether their parents smoked. The results are stored as follows:

Column 1: 1 = Do not smoke, 2 = Smoke
Column 2: 1 = Neither parent smoked,
2 = Father smoked, 3 = Mother smoked,
4 = Both parents smoked

Use a tabular technique to produce the information you need.

2.29 **Xr02-29** In 2013, 1.35 million men and 1.04 million women were unemployed at some time during the year. A statistics practitioner wanted to investigate the reason for that unemployment status and whether the reasons differed by gender. A random sample of people 16 years of age and older was drawn. The reasons given for their status included:

1. Lost job
2. Left job
3. Re-entrants
4. New entrants

Determine whether there are differences between unemployed men and women in terms of the reasons for unemployment.

CHAPTER SUMMARY

Descriptive statistical methods are used to summarise data sets so that we can extract the relevant information. In this chapter, we presented graphical techniques for nominal data.

Bar charts, pie charts and frequency distributions are employed to summarise single sets of nominal data. Because of the restrictions applied to this type of data, all that we can show is the frequency and proportion of each category.

To describe the relationship between two nominal variables, we produce cross-classification tables and bar charts.

CHAPTER EXERCISES

The following exercises require a computer and software.

2.30 **Xr02-30** At Wembley Stadium, 200 people who had purchased food were asked to rate the quality of the food. The responses included:

1. Poor
2. Fair
3. Good
4. Very good
5. Excellent

Draw a graph that describes the data. What does the graph tell you?

2.31 **Xr02-31** There are several ways to teach applied statistics. The most popular approaches include:

1. Emphasise manual calculations.
2. Use a computer combined with manual calculations.
3. Use a computer exclusively with no manual calculations.

A survey of 100 statistics instructors asked each to report his or her approach. Use a graphical method to extract the most useful information about the teaching approaches.

2.32 **Xr02-32** Which Internet search engines are the most popular? A survey undertaken by the Europe Internet Usage Stats asked random samples of Europeans that question. The responses included:

1. Google
2. Bing
3. Yahoo
4. Other

Use a graphical technique that compares the proportions of French and Germans' use of search engines.

2.33 **Xr02-33** Many countries are lowering taxes on corporations in an effort to be more attractive for investment. In the next table, we list the marginal effective corporate tax rates among Organization for Economic Co-operation and Development (OECD) countries. Develop a graph that depicts these figures. Briefly describe your results.

Country	Manufacturers	Services	Aggregate
Australia	27.7	26.6	26.7
Austria	21.6	19.5	19.9
Belgium	−6.0	−4.1	−0.5
Canada	20.0	29.2	25.2
Czech Republic	1.0	7.8	8.4
Denmark	16.5	12.7	13.4
Finland	22.4	22.9	22.8
France	33.0	31.7	31.9
Germany	30.8	29.4	29.7
Greece	18.0	13.2	13.8
Hungary	12.9	12.0	12.2
Iceland	19.5	17.6	17.9
Ireland	12.7	11.7	12.0
Italy	24.6	28.6	27.8
Japan	35.2	30.4	31.3
Korea	32.8	31.0	31.5
Luxembourg	24.1	20.3	20.6
Mexico	17.1	12.1	13.1
Netherlands	18.3	15.0	15.5
New Zealand	27.1	25.4	25.7
Norway	25.8	23.2	23.5
Poland	14.4	15.0	14.9
Portugal	14.8	16.1	15.9
Slovak	13.3	11.7	12.0
Spain	27.2	25.2	25.5
Sweden	19.3	17.5	17.8
Switzerland	14.8	15.0	14.9
Turkey	22.7	20.2	20.8
United Kingdom	22.7	27.8	26.9
United States	32.7	39.9	36.9

2.34 Xr02-34 A survey of the business school graduates undertaken by a university placement office asked, among other questions, in which area each person was employed. The areas of employment included:

1. Accounting
2. Finance
3. General management
4. Marketing/Sales
5. Other

Additional questions were asked, and the responses were recorded in the following way:

Column	Variable
1	Identification number
2	Area
3	Gender (1 = Female, 2 = Male)
4	Job satisfaction (4 = Very, 3 = Quite, 2 = Little, 1 = None)

The placement office wants to know the following:

a. Do female and male graduates differ in their areas of employment? If so, how?

b. Are area of employment and job satisfaction related?

3 GRAPHICAL DESCRIPTIVE TECHNIQUES II

3-1 GRAPHICAL TECHNIQUES TO DESCRIBE A SET OF INTERVAL DATA

3-2 DESCRIBING TIME-SERIES DATA

3-3 DESCRIBING THE RELATIONSHIP BETWEEN TWO INTERVAL VARIABLES

3-4 ART AND SCIENCE OF GRAPHICAL PRESENTATIONS

WERE OIL COMPANIES GOUGING CUSTOMERS 2000–2013?

DATA Xm03-00

The price of oil has been increasing for several reasons. First, oil is a finite resource; the world's reserves will eventually run out. In 2012, the world was consuming more than 100 million barrels of oil and liquid fuels (1 barrel of oil equals 159 litres) per day – more than 36 billion barrels per year. The total proven world reserves of oil in 2012 were 1525 billion barrels. At today's consumption levels, the proven reserves will be exhausted in 37 years. (It should be noted, however, that in 2009 the proven reserves of oil amounted to 1408 billion barrels, indicating that new oil discoveries are offsetting increasing usage.) Second, China's and India's industries are rapidly increasing and require ever-increasing amounts of oil. Third, over the last 10 years, hurricanes have threatened the oil rigs in the Gulf of Mexico.

An increase in oil prices is reflected in the price of petrol and diesel. In January 2000, the average retail price of petrol in the UK was the most expensive in Europe, 77p a litre for unleaded petrol and 79.8p for diesel. Over the next 14 years, the price of both crude oil and total fuel substantially increased. In January 2014, drivers in the UK were paying on average 130.5p a litre for unleaded petrol and 138.2p a litre for diesel. Many drivers complained that the oil companies were guilty of price gouging; that is, they believed that when the price of oil increased, the price of petrol also increased, but when the price of oil decreased, the decrease in the price of petrol seemed to lag behind. To determine whether this perception is accurate, we determined the monthly figures for both commodities. Were oil and petrol prices related? **See Section 3-3 for the answer.**

INTRODUCTION

Chapter 2 introduced graphical techniques used to summarise and present nominal data. In this chapter, we do the same for interval data. Section 3-1 presents techniques to describe a set of interval data, Section 3-2 introduces time series and the method used to present time-series data, and Section 3-3 introduces the technique we use to describe the relationship between two interval variables. We complete this chapter with a discussion of how to properly use graphical methods in Section 3-4.

3-1 GRAPHICAL TECHNIQUES TO DESCRIBE A SET OF INTERVAL DATA

In this section, we introduce several graphical methods that are used when the data are interval. The most important of these graphical methods is the histogram. As you will see, the histogram not only is a powerful graphical technique used to summarise interval data, but also is used to help explain an important aspect of probability (see Chapter 8).

APPLICATIONS IN MARKETING

Pricing

Traditionally, marketing has been defined in terms of the four Ps: product, price, promotion and place. *Marketing management* is the functional area of business that focuses on the development of a product, together with its pricing, promotion and distribution. Decisions are made in these four areas with a view to satisfying the wants and needs of consumers while also satisfying the firm's objective. The pricing decision must be addressed both for a new product and, from time to time, for an existing product. Anyone buying a product such as a personal computer has been confronted with a wide variety of prices, accompanied by a correspondingly wide variety of features. From a vendor's standpoint, establishing the appropriate price and corresponding set of attributes for a product is complicated and must be done in the context of the overall marketing plan for the product.

EXAMPLE 3.1

DATA Xm03-01

Analysis of Long-Distance Telephone Bills

Following deregulation of telephone service, several new companies were created to compete in the business of providing long-distance telephone service. In almost all cases, these companies competed on price because the service each offered is similar. Pricing a service or product in the face of stiff competition is very difficult. Factors to be considered include supply, demand, price elasticity and the actions of competitors. Long-distance packages may employ per minute charges, a flat monthly rate or some combination of the two. Determining the appropriate rate structure is facilitated by acquiring information about the behaviours of customers, especially the size of monthly long-distance bills. As part of a larger study, a long-distance company wanted to acquire information about the monthly bills of new subscribers in the first month after signing with the company. The company's marketing manager conducted a survey of 200 new residential subscribers and recorded the first month's bills. These data are listed here. The general manager planned to present his findings to senior executives. What information can be extracted from these data?

Long-Distance Telephone Bills

42.19	39.21	75.71	8.37	1.62	28.77	35.32	13.9	114.67	15.3
38.45	48.54	88.62	7.18	91.1	9.12	117.69	9.22	27.57	75.49
29.23	93.31	99.5	11.07	10.88	118.75	106.84	109.94	64.78	68.69
89.35	104.88	85.0	1.47	30.62	0.0	8.4	10.7	45.81	35.0
118.04	30.61	0.0	26.4	100.05	13.95	90.04	0.0	56.04	9.12
110.46	22.57	8.41	13.26	26.97	14.34	3.85	11.27	20.39	18.49
0.0	63.7	70.48	21.13	15.43	79.52	91.56	72.02	31.77	84.12
72.88	104.84	92.88	95.03	29.25	2.72	10.13	7.74	94.67	13.68
83.05	6.45	3.2	29.04	1.88	9.63	5.72	5.04	44.32	20.84
95.73	16.47	115.5	5.42	16.44	21.34	33.69	33.4	3.69	100.04
103.15	89.5	2.42	77.21	109.08	104.4	115.78	6.95	19.34	112.94
94.52	13.36	1.08	72.47	2.45	2.88	0.98	6.48	13.54	20.12
26.84	44.16	76.69	0.0	21.97	65.9	19.45	11.64	18.89	53.21
93.93	92.97	13.62	5.64	17.12	20.55	0.0	83.26	1.57	15.3
90.26	99.56	88.51	6.48	19.7	3.43	27.21	15.42	0.0	49.24
72.78	92.62	55.99	6.95	6.93	10.44	89.27	24.49	5.2	9.44
101.36	78.89	12.24	19.6	10.05	21.36	14.49	89.13	2.8	2.67
104.8	87.71	119.63	8.11	99.03	24.42	92.17	111.14	5.1	4.69
74.01	93.57	23.31	9.01	29.24	95.52	21.0	92.64	3.03	41.38
56.01	0.0	11.05	84.77	15.21	6.72	106.59	53.9	9.16	45.77

SOLUTION:

Little information can be developed just by casually reading through the 200 observations. The manager can probably see that most of the bills are under €100, but that is likely to be the extent of the information garnered from browsing through the data. If he examines the data more carefully, he may discover that the smallest bill is €0 and the largest is €119.63. He has now developed some information. However, his presentation to senior executives will be most unimpressive if no other information is produced. For example, someone is likely to ask how the numbers are distributed between 0 and 119.63. Are there many small bills and few large bills? What is the 'typical' bill? Are the bills somewhat similar or do they vary considerably?

To help answer these questions and others like them, the marketing manager can construct a frequency distribution from which a histogram can be drawn. In the previous chapter a frequency distribution was created by counting the number of times each category of the nominal variable occurred. We create a frequency distribution for interval data by counting the number of observations that fall into each of a series of intervals, called **classes** that cover the complete range of observations. We discuss how to decide the number of classes and the upper and lower limits of the intervals later. We have chosen eight classes defined in such a way that each observation falls into one and only one class. These classes are defined as follows:

Classes

Amounts that are less than or equal to 15

Amounts that are more than 15 but less than or equal to 30

Amounts that are more than 30 but less than or equal to 45

Amounts that are more than 45 but less than or equal to 60

Amounts that are more than 60 but less than or equal to 75

Amounts that are more than 75 but less than or equal to 90

Amounts that are more than 90 but less than or equal to 105

Amounts that are more than 105 but less than or equal to 120

TABLE 3.1 Frequency Distribution of the Long-Distance Bills in Example 3.1

CLASS LIMITS	FREQUENCY
0 to 15*	71
15 to 30	37
30 to 45	13
45 to 60	9
60 to 75	10
75 to 90	18
90 to 105	28
105 to 120	14
Total	200

*Classes contain observations greater than their lower limits (except for the first class) and less than or equal to their upper limits.

Notice that the intervals do not overlap, so there is no uncertainty about which interval to assign to any observation. Moreover, because the smallest number is 0 and the largest is 119.63, every observation will be assigned to a class. Finally, the intervals are equally wide. Although this is not essential, it makes the task of reading and interpreting the graph easier.

To create the frequency distribution manually, we count the number of observations that fall into each interval. Table 3.1 presents the frequency distribution.

Although the frequency distribution provides information about how the numbers are distributed, the information is more easily understood and imparted by drawing a picture or graph. The graph is called a **histogram.** A histogram is created by drawing rectangles whose bases are the intervals and whose heights are the frequencies. Figure 3.1 exhibits the histogram that was drawn by hand.

FIGURE 3.1 Histogram for Example 3.1

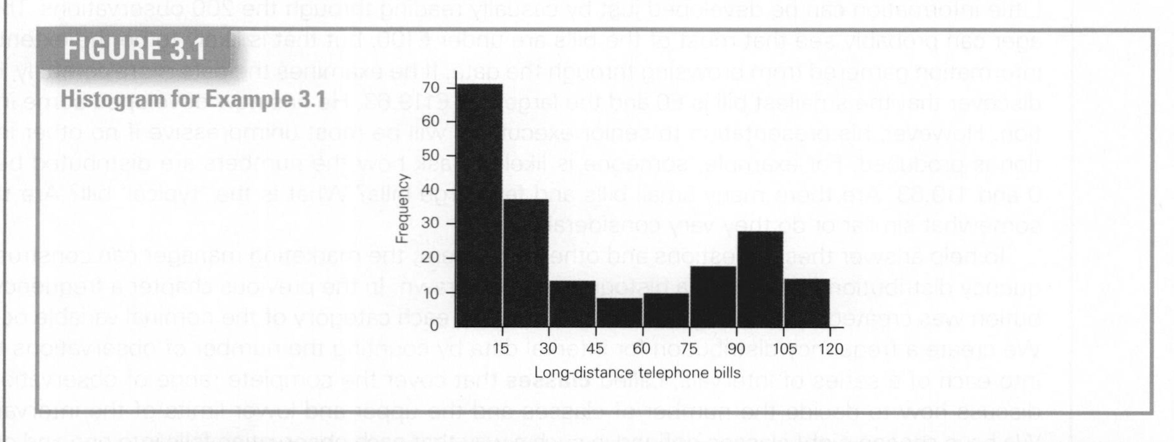

EXCEL

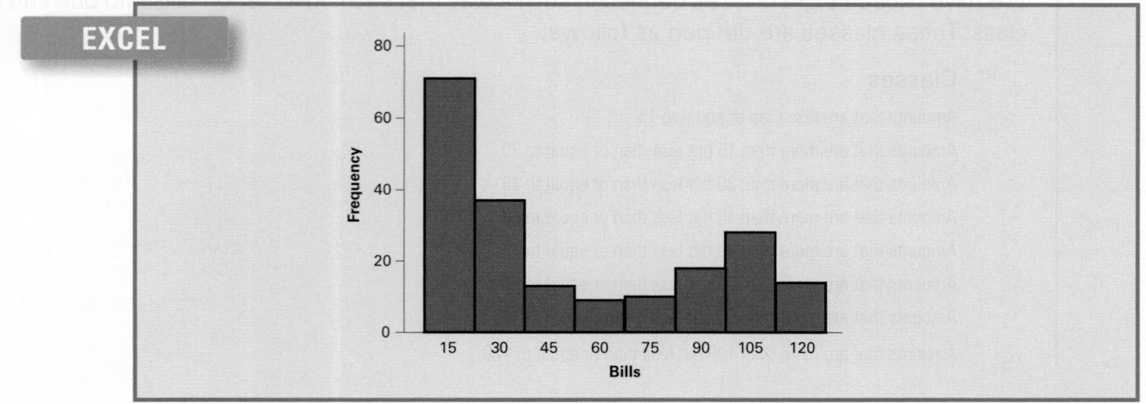

Instructions

1. Type or import the data into one column. (**Open** Xm03-01.) In column B, or any other column, type the upper limits of the class intervals, i.e. 15, 30, 45, 60, 75, 90, 105 and 120. Excel calls them *bins.* (You can put any name in the first row; we typed 'Bills'.)
2. Click **Data, Data Analysis** and **Histogram.** If Data Analysis does not appear in the menu box, see the accompanying website (see the 'Digital Resources' page for information on access).
3. Specify the **Input Range** (Al:A201) and the **Bin Range** (B1:B9). Click **Chart Output**. Click **Labels** if the first row contains names.
4. To remove the gaps, place the cursor over one of the rectangles and click the right button of the mouse. Click (with the left button) **Format Data Series** ... move the pointer to **Gap Width** and use the slider to change the number from 150 to 0.

Except for the first class, Excel counts the number of observations in each class that are greater than the lower limit and less than or equal to the upper limit.

Note that the numbers along the horizontal axis represent the upper limits of each class, although they appear to be placed in the centres. If you wish, you can replace these numbers with the actual midpoints by making changes to the frequency distribution in cells A1:B14 (change 15 to 7.5, 30 to 22.5,..., and 120 to 112.5).

You can also convert the histogram to list relative frequencies instead of frequencies. To do so, change the frequencies to relative frequencies by dividing each frequency by 200; that is, replace 71 by 0.355, 37 by 0.185,..., and 14 by 0.07.

If you have difficulty with this technique, turn to the website Appendix, which provides step-by-step instructions for Excel and provides troubleshooting tips.

MINITAB

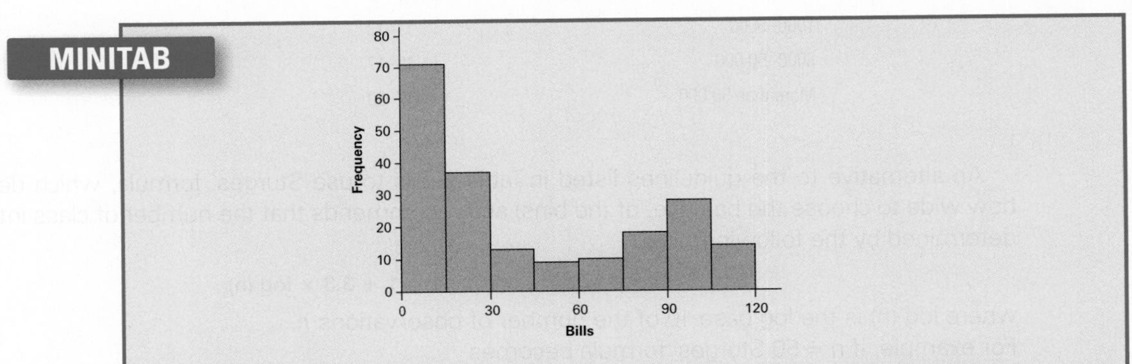

Note that Minitab counts the number of observations in each class that are strictly less than their upper limits.

Instructions

1. Type or import the data into one column. (**Open** Xm03-01.)
2. Click **Graph, Histogram ...** and **Simple.**
3. Type or use the **Select** button to specify the name of the variable in the **Graph Variables** box (Bills). Click **Data View.**
4. Click **Data Display** and **Bars.** Minitab will create a histogram using its own choices of class intervals.
5. To choose your own classes, double-click the horizontal axis. Click **Binning.**
6. Under **Interval Type**, choose **Cutpoint.** Under **Interval Definition,** choose **Midpoint/ Cutpoint positions** and type in your choices (0 15 30 45 60 75 90 105 120) to produce the histogram shown here.

INTERPRET

The histogram gives us a clear view of the way the bills are distributed. About half the monthly bills are small (€0 to €30), a few bills are in the middle range (€30 to €75) and a relatively large number of long-distance bills are at the high end of the range. It would appear from this sample of first-month long-distance bills that the company's customers are split unevenly between light and heavy users of long-distance telephone service. If the company assumes that this pattern will continue, it must address a number of pricing issues. For example, customers who incurred large monthly bills may be targets of competitors who offer flat rates for 15-minute or 30-minute calls. The company needs to know more about these customers. With the additional information, the marketing manager may suggest altering the company's pricing.

3-1a Determining the Number of Class Intervals

The number of class intervals we select depends entirely on the number of observations in the data set. The more observations we have, the larger the number of class intervals we need to use to draw a useful histogram. Table 3.2 provides guidelines on choosing the number of classes. In Example 3.1, we had 200 observations. The table tells us to use 7, 8, 9 or 10 classes.

TABLE 3.2 **Approximate Number of Classes in Histograms**

NUMBER OF OBSERVATIONS	NUMBER OF CLASSES
Less than 50	5–7
50–200	7–9
200–500	9–10
500–1000	10–11
1000–5000	11–13
5000–50 000	13–17
More than 50 000	17–20

An alternative to the guidelines listed in Table 3.2 is to use Sturges' formula, which determines how wide to choose the bars (i.e. of the bins) and recommends that the number of class intervals be determined by the following:

$$\text{Number of class intervals} = 1 + 3.3 \times \log(n),$$

where log (n) is the log base 10 of the number of observations *n*.
For example, if n = 50 Sturges' formula becomes

$$\text{Number of class intervals} = 1 + 3.3 \times \log(50) = 1 + 3.3 \times 1.7 = 6.6$$

which we round to 7, because we take the nearest integer above the calculated value.

Class Interval Widths We determine the approximate width of the classes by subtracting the smallest observation from the largest and dividing the difference by the number of classes. Thus,

$$\text{Class width} = \frac{\textit{Largest observation} - \textit{Smallest observation}}{\textit{Number of classes}}$$

In Example 3.1, we calculated

$$\text{Class width} = \frac{119.63 - 0}{8} = 14.95$$

We often round the result to some convenient value. We then define our class limits by selecting a lower limit for the first class from which all other limits are determined. The only condition we apply is that the first class interval must contain the smallest observation. In Example 3.1, we rounded the class width to 15 and set the lower limit of the first class to 0. Thus, the first class is defined as 'Amounts that are greater than or equal to 0 but less than or equal to 15'. (Minitab users should remember that the classes are defined as the number of observations that are *strictly less* than their upper limits.)

Table 3.2 and Sturges' formula are guidelines only. It is more important to choose classes that are easy to interpret. For example, suppose that we have recorded the marks on an exam of the 100 students registered in the course where the highest mark is 94 and the lowest is 48. Table 3.2 suggests that we use 7, 8 or 9 classes, and Sturges' formula computes the approximate number of classes as:

$$\text{Number of class intervals} = 1 + 3.3 \times \log(100) = 1 + 3.3 \times 2 = 7.6$$

which we round to 8. Thus,

$$\text{Class width} = \frac{94 - 48}{8} = 5.75$$

which we would round to 6. We could then produce a histogram whose upper limits of the class intervals are 50, 56, 62,..., 98. Because of the rounding and the way in which we defined the class limits, the number of classes is 9. However, a histogram that is easier to interpret would be produced using classes whose widths are 5; that is, the upper limits would be 50, 55, 60,..., 95. The number of classes in this case would be 10.

3-1b Shapes of Histograms

The purpose of drawing histograms, like that of all other statistical techniques, is to acquire information. Once we have the information, we frequently need to describe what we have learned to others. We describe the shape of histograms on the basis of the following characteristics.

Symmetry A histogram is said to be symmetric if, when we draw a vertical line down the centre of the histogram, the two sides are identical in shape and size. Figure 3.2 depicts three symmetric histograms.

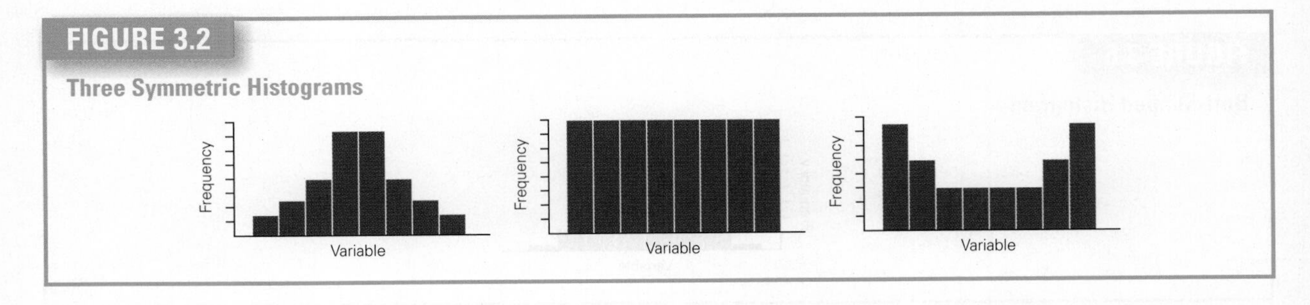

FIGURE 3.2

Three Symmetric Histograms

Skewness A skewed histogram is one with a long tail extending to either the right or the left. The former is called positively skewed, and the latter is called negatively skewed. Figure 3.3 shows examples of both. Incomes of employees in large firms tend to be positively skewed because there is a large number of relatively low-paid workers and a small number of well-paid executives. The time taken by students to write exams is frequently negatively skewed because few students hand in their exams early; most prefer to reread their papers and hand them in near the end of the scheduled test period.

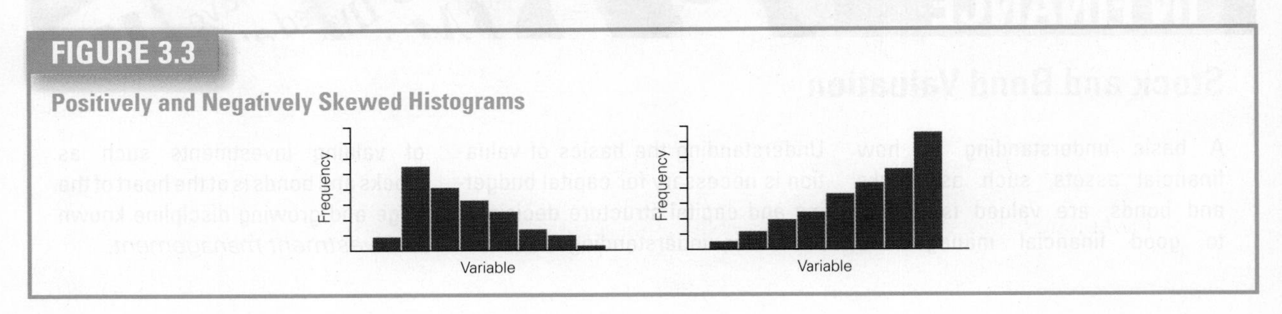

FIGURE 3.3

Positively and Negatively Skewed Histograms

Number of Modal Classes As we discuss in Chapter 4, a *mode* is the observation that occurs with the greatest frequency. A modal class is the class with the largest number of observations. A unimodal histogram is one with a single peak. The histogram in Figure 3.4 is unimodal. A bimodal histogram is one with two peaks, not necessarily equal in height. Bimodal histograms often indicate that two different distributions are present. (See Example 3.4.) Figure 3.5 depicts bimodal histograms.

FIGURE 3.4

A Unimodal Histogram

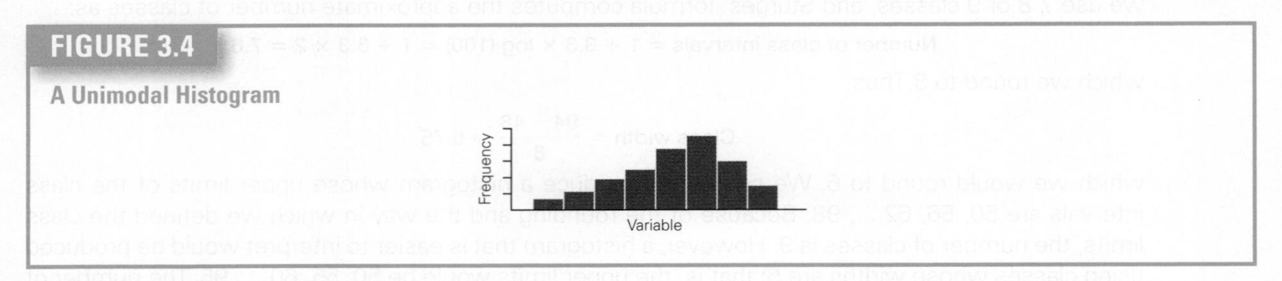

FIGURE 3.5

Bimodal Histograms

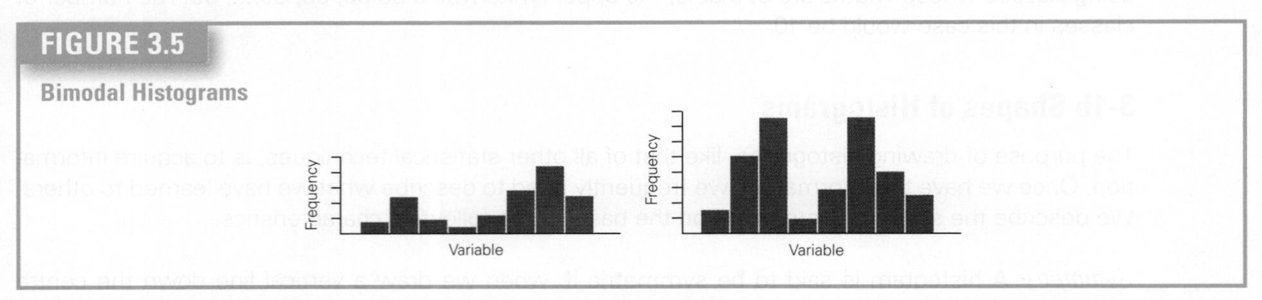

Bell Shape A special type of symmetric unimodal histogram is one that is bell shaped. In Chapter 8, we will explain why this type of histogram is important. Figure 3.6 exhibits a bell-shaped histogram.

FIGURE 3.6

Bell-Shaped Histogram

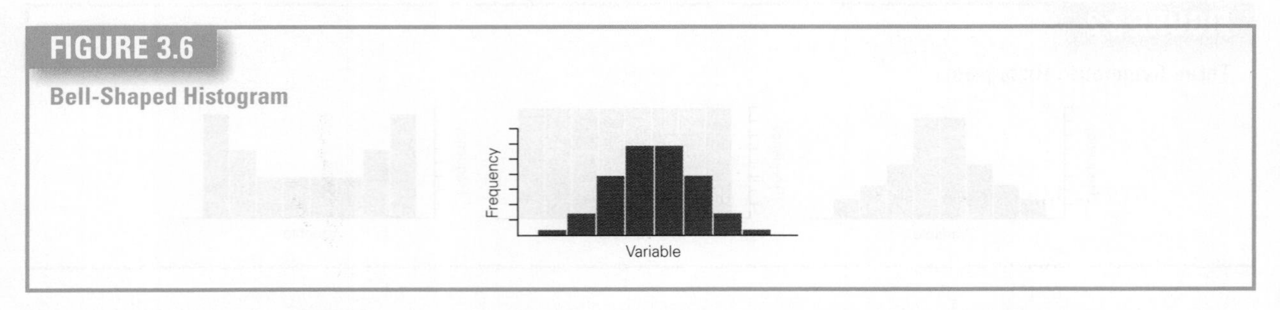

Now that we know what to look for, let's examine some examples of histograms and see what we can discover.

APPLICATIONS IN FINANCE

Stock and Bond Valuation

A basic understanding of how financial assets, such as stocks and bonds, are valued is critical to good financial management. Understanding the basics of valuation is necessary for capital budgeting and capital structure decisions. Moreover, understanding the basics of valuing investments such as stocks and bonds is at the heart of the huge and growing discipline known as *investment management.*

A financial manager must be familiar with the main characteristics of the capital markets where long-term financial assets such as stocks and bonds trade. A well-functioning capital market provides managers with useful information concerning the appropriate prices and rates of return that are required for a variety of financial securities with differing levels of risk. Statistical methods can be used to analyse capital markets and summarise their characteristics, such as the shape of the distribution of stock or bond returns.

APPLICATIONS IN FINANCE

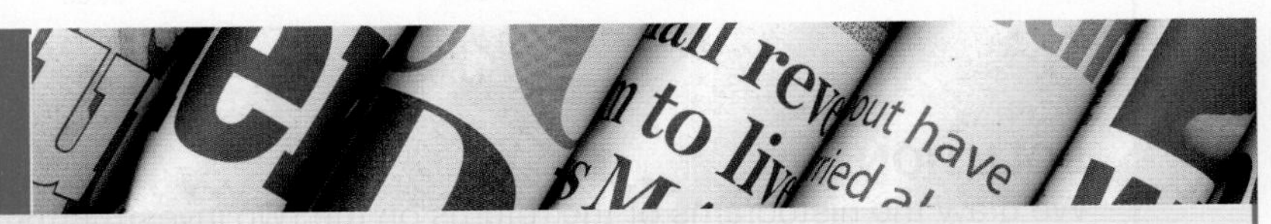

Return on Investment

The return on an investment is calculated by dividing the gain (or loss) by the value of the investment. For example, a €100 investment that is worth €106 after 1 year has a 6% rate of return. A €100 investment that loses €20 has a –20% rate of return. For many investments, including individual stocks and stock portfolios (combinations of various stocks), the rate of return is a variable. In other words, the investor does not know in advance what the rate of return will be. It could be a positive number, in which case the investor makes money – or negative, and the investor loses money.

Investors are torn between two goals. The first is to maximise the rate of return on investment. The second goal is to reduce risk. If we draw a histogram of the returns for a certain investment, the location of the centre of the histogram gives us some information about the return one might expect from that investment. The spread or variation of the histogram provides us with guidance about the risk. If there is little variation, an investor can be quite confident in predicting what his or her rate of return will be. If there is a great deal of variation, the return becomes much less predictable and thus riskier. Minimising the risk becomes an important goal for investors and financial analysts.

EXAMPLE 3.2

DATA Xm03-02

Comparing Returns on Two Investments

Suppose that you are facing a decision about where to invest that small fortune that remains after you have deducted the anticipated expenses for the next year from the earnings from your summer job. A friend has suggested two types of investment, and to help make the decision you acquire some rates of return from each type. You would like to know what you can expect by way of the return on your investment, as well as other types of information, such as whether the rates are spread out over a wide range (making the investment risky) or are grouped tightly together (indicating relatively low risk). Do the data indicate that it is possible that you can do extremely well with little likelihood of a large loss? Is it likely that you could lose money (negative rate of return)?

The returns for the two types of investments are listed here. Draw histograms for each set of returns and report on your findings. Which investment would you choose and why?

Returns on Investment A				Returns on Investment B			
30.00	6.93	13.77	–8.55	30.33	–34.75	30.31	24.3
–2.13	–13.24	22.42	–5.29	–30.37	54.19	6.06	–10.01
4.30	–18.95	34.40	–7.04	–5.61	44.00	14.73	35.24
25.00	9.43	49.87	–12.11	29.00	–20.23	36.13	40.7

12.89	1.21	22.92	12.89	−26.01	4.16	1.53	22.18
−20.24	31.76	20.95	63.00	0.46	10.03	17.61	3.24
1.20	11.07	43.71	−19.27	2.07	10.51	1.2	25.1
−2.59	8.47	−12.83	−9.22	29.44	39.04	9.94	−24.24
33.00	36.08	0.52	−17.00	11.0	24.76	−33.39	−38.47
14.26	−21.95	61.00	17.30	−25.93	15.28	58.67	13.44
−15.83	10.33	−11.96	52.00	8.29	34.21	0.25	68.00
0.63	12.68	1.94		61.00	52.00	5.23	
38.00	13.09	28.45		−20.44	−32.17	66.0	

SOLUTION:

We draw the histograms of the returns on the two investments. We will use Excel and Minitab to do the work.

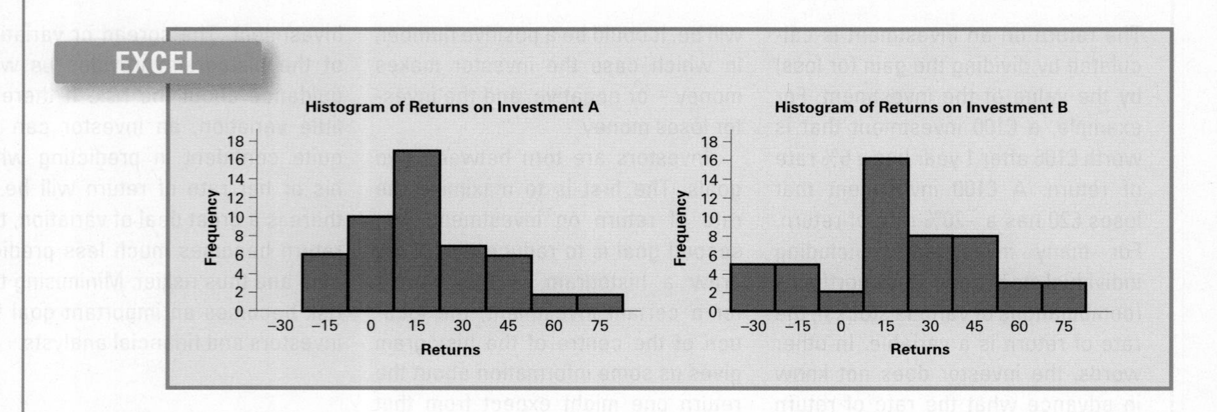

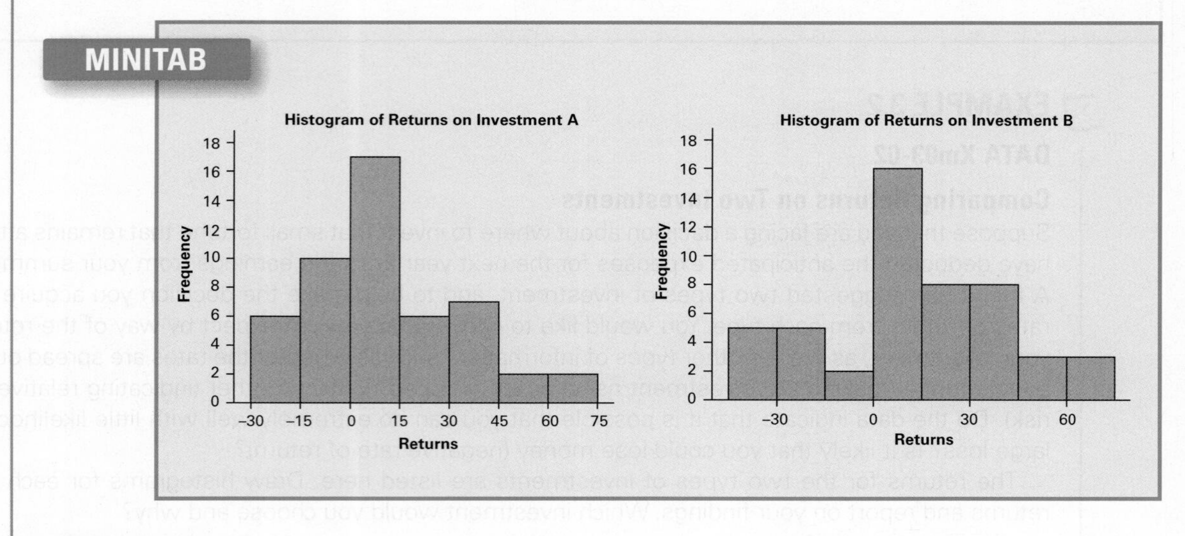

INTERPRET

Comparing the two histograms, we can extract the following information:

1. The centre of the histogram of the returns of investment A is slightly lower than that for investment B.
2. The spread of returns for investment A is considerably less than that for investment B.
3. Both histograms are slightly positively skewed.

These findings suggest that investment A is superior. Although the returns for A are slightly less than those for B, the wider spread for B makes it unappealing to most investors. Both investments allow for the possibility of a relatively large return.

The interpretation of the histograms is somewhat subjective. Other viewers may not concur with our conclusion. In such cases, numerical techniques provide the detail and precision lacking in most graphs. We will redo this example in Chapter 4 to illustrate how numerical techniques compare to graphical ones.

EXAMPLE 3.3

DATA Xm03-03

Business Statistics Marks

A student enrolled in a business programme is attending the first class of the required statistics course. The student is somewhat apprehensive believing that the course is difficult, so to alleviate his anxiety he asks the lecturer about last year's marks. The lecturer obliges and provides a list of the final marks, which is composed of term work plus the final exam. Draw a histogram and describe the result, based on the following marks:

65	81	72	59
71	53	85	66
66	70	72	71
79	76	77	68
65	73	64	72
82	73	77	75
80	85	89	74
86	83	87	77
67	80	78	69
64	67	79	60
62	78	59	92
74	68	63	69
67	67	84	69
72	62	74	73
68	83	74	65

SOLUTION:

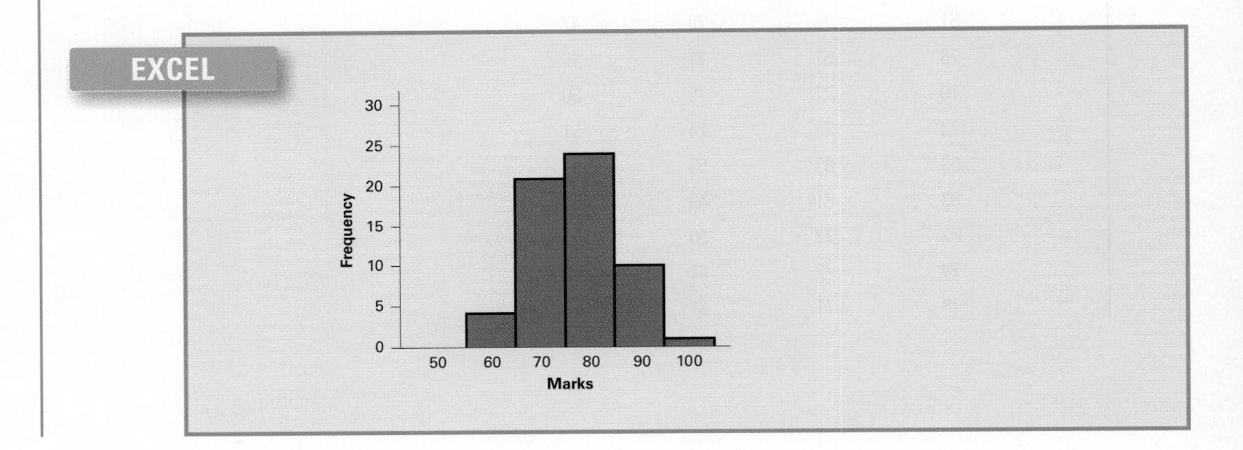

MINITAB

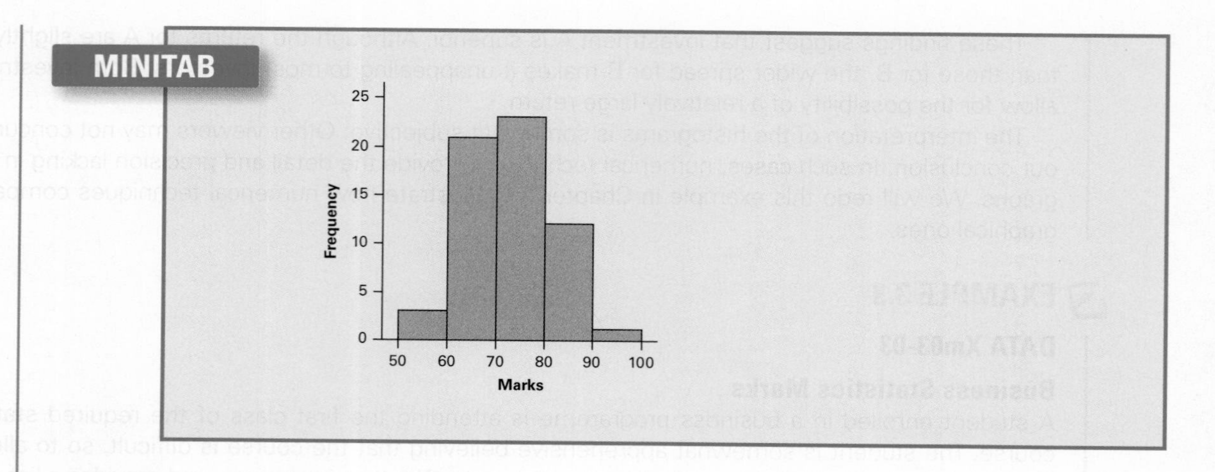

INTERPRET

The histogram is unimodal and approximately symmetric. There are no marks below 50, with the great majority of marks between 60 and 90. The modal class is 70 to 80 and the centre of the distribution is approximately 75.

EXAMPLE 3.4

DATA Xm03-04

Mathematical Statistics Marks

Suppose the student in Example 3.3 obtained a list of last year's marks in a mathematical statistics course. This course emphasises derivations and proofs of theorems. Use the accompanying data to draw a histogram and compare it to the one produced in Example 3.3. What does this histogram tell you?

77	67	53	54
74	82	75	44
75	55	76	54
75	73	59	60
67	92	82	50
72	75	82	52
81	75	70	47
76	52	71	46
79	72	75	50
73	78	74	51
59	83	53	44
83	81	49	52
77	73	56	53
74	72	61	56
78	71	61	53

SOLUTION:

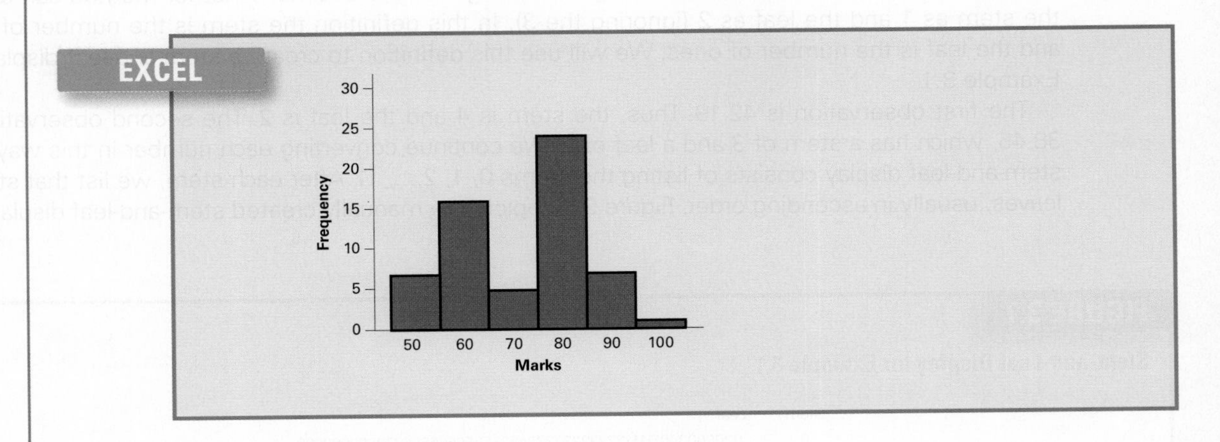

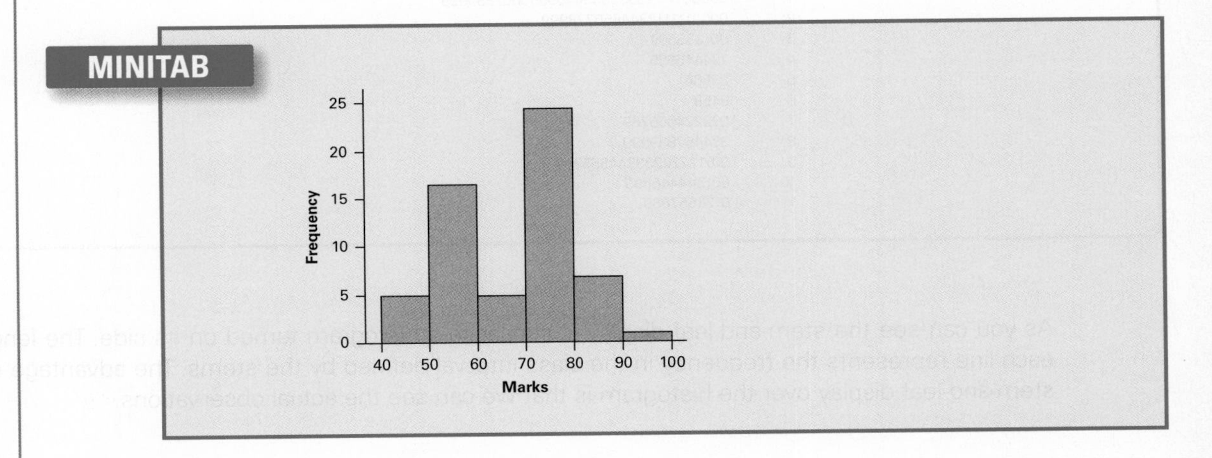

INTERPRET

The histogram is bimodal. The larger modal class is composed of the marks in the 70s. The smaller modal class includes the marks that are in the 50s. There appear to be few marks in the 60s. This histogram suggests that there are two groups of students. Because of the emphasis on mathematics in the course, one may conclude that those who performed poorly in the course are weaker mathematically than those who performed well. The histograms in this example and in Example 3.3 suggest that the courses are quite different from one another and have a completely different distribution of marks.

3-1c Stem-and-Leaf Display

One of the drawbacks of the histogram is that we lose potentially useful information by classifying the observations. In Example 3.1, we learned that there are 71 observations that fall between 0 and 15. By classifying the observations we did acquire useful information. However, the histogram focuses our attention on the frequency of each class and by doing so sacrifices whatever information was contained in the actual observations. A statistician named John Tukey introduced the **stem-and-leaf display**, which is a method that to some extent overcomes this loss.

The first step in developing a stem-and-leaf display is to split each observation into two parts, a stem and a leaf. There are several different ways of doing this. For example, the number 12.3 can be split so that the stem is 12 and the leaf is 3. In this definition, the stem consists of the digits to the left of the decimal and the leaf is the digit to the right of the decimal. Another method can define the stem as 1 and the leaf as 2 (ignoring the 3). In this definition the stem is the number of tens and the leaf is the number of ones. We will use this definition to create a stem-and-leaf display for Example 3.1.

The first observation is 42.19. Thus, the stem is 4 and the leaf is 2. The second observation is 38.45, which has a stem of 3 and a leaf of 8. We continue converting each number in this way. The stem-and-leaf display consists of listing the stems 0, 1, 2,..., 11. After each stem, we list that stem's leaves, usually in ascending order. Figure 3.7 depicts the manually created stem-and-leaf display.

FIGURE 3.7

Stem-and-Leaf Display for Example 3.1

Stem	Leaf
0	0000000001111122222233333455555566666667788889999999
1	00000111123333333344555555667889999
2	00001111123446667789999
3	001335589
4	124445589
5	33566
6	3458
7	022224556789
8	334457889999
9	00112222233344555999
10	001344446699
11	0124557889

As you can see the stem-and-leaf display is similar to a histogram turned on its side. The length of each line represents the frequency in the class interval defined by the stems. The advantage of the stem-and-leaf display over the histogram is that we can see the actual observations.

EXCEL

	A	B	C	D	E	F	G
1	**Stem & Leaf Display**						
2							
3	**Stems**	**Leaves**					
4	0	->0000000001111122222233333455555566666667788889999999					
5	1	->00000111123333333344555555667889999					
6	2	->00001111123446667789999					
7	3	->001335589					
8	4	->12445589					
9	5	->33566					
10	6	->3458					
11	7	->022224556789					
12	8	->334457889999					
13	9	->00112222233344555999					
14	10	->001344446699					
15	11	->0124557889					
16							
17							

Instructions

1. Type or import the data into one column. (**Open** Xm03-01.)
2. Click **Add-ins, Data Analysis Plus** and **Stem-and-Leaf Display.**
3. Specify the **Input Range** (A1:A201). Click one of the values of **Increment** (the increment is the difference between stems) (10).

MINITAB

```
Stem-and-Leaf Display: Bills

Stem-and-leaf of Bills  N  = 200
Leaf Unit = 1.0

 52  0  0000000001111222222333334555555666666677888899999999
 85  1  00000111123333333445555566788 9999
(23) 2  00001111123446667789999
 92  3  001335589
 83  4  12445589
 75  5  33566
 70  6  3458
 66  7  022224556789
 54  8  334457889999
 42  9  00112222233344555999
 22  10 001344446699
 10  11 0124557889
```

The numbers in the left column are called **depths**. Each depth counts the number of observations that are on its line or beyond. For example, the second depth is 85, which means that there are 85 observations that are less than 20. The third depth is displayed in parentheses, which indicates that the third interval contains the observation that falls in the middle of all the observations, a statistic we call the *median* (to be presented in Chapter 4). For this interval, the depth tells us the frequency of the interval; that is, 23 observations are greater than or equal to 20 but less than 30. The fourth depth is 92, which tells us that 92 observations are greater than or equal to 30. Notice that for classes below the median, the depth reports the number of observations that are less than the upper limit of that class. For classes that are above the median, the depth reports the number of observations that are greater than or equal to the lower limit of that class.

Instructions

1. Type or import the data into one column. (**Open** Xm03-01.)
2. Click **Graph** and **Stem-and-Leaf....**
3. Type or use the **Select** button to specify the variable in the **Variables** box (Bills). Type the increment in the **Increment** box (10).

3-1d Ogive

The frequency distribution lists the number of observations that fall into each class interval. We can also create a **relative frequency distribution** by dividing the frequencies by the number of observations. Table 3.3 displays the relative frequency distribution for Example 3.1.

TABLE 3.3 Relative Frequency Distribution for Example 3.1

CLASS LIMITS	RELATIVE FREQUENCY
0 to 15	71/200 = 0.355
15 to 30	37/200 = 0.185
30 to 45	13/200 = 0.065
45 to 60	9/200 = 0.045
60 to 75	10/200 = 0.050
75 to 90	18/200 = 0.090
90 to 105	28/200 = 0.140
105 to 120	14/200 = 0.070
Total	200/200 = 1.0

As you can see, the relative frequency distribution highlights the proportion of the observations that fall into each class. In some situations, we may wish to highlight the proportion of observations that lie below each of the class limits. In such cases, we create the **cumulative relative frequency distribution**. Table 3.4 displays this type of distribution for Example 3.1.

From Table 3.4, you can see that, for example, 54% of the bills were less than or equal to €30 and that 79% of the bills were less than or equal to €90.

Another way of presenting this information is the **ogive**, sometimes called a **cumulative line graph**, which is a graphical representation of the cumulative relative frequencies. Figure 3.8 is the manually drawn ogive for Example 3.1.

TABLE 3.4 Cumulative Relative Frequency Distribution for Example 3.1

CLASS LIMITS	RELATIVE FREQUENCY	CUMULATIVE RELATIVE FREQUENCY
0 to15	71/200 = 0.355	71/200 = 0.355
15 to 30	37/200 = 0.185	108/200 = 0.340
30 to 45	13/200 = 0.065	121/200 = 0.605
45 to 60	9/200 = 0.045	130/200 = 0.650
60 to 75	10/200 = 0.05	140/200 = 0.700
75 to 90	18/200 = 0.09	158/200 = 0.790
90 to 105	28/200 = 0.14	186/200 = 0.930
105 to 120	14/200 = 0.07	200/200 = 1.00

FIGURE 3.8

Ogive for Example 3.1

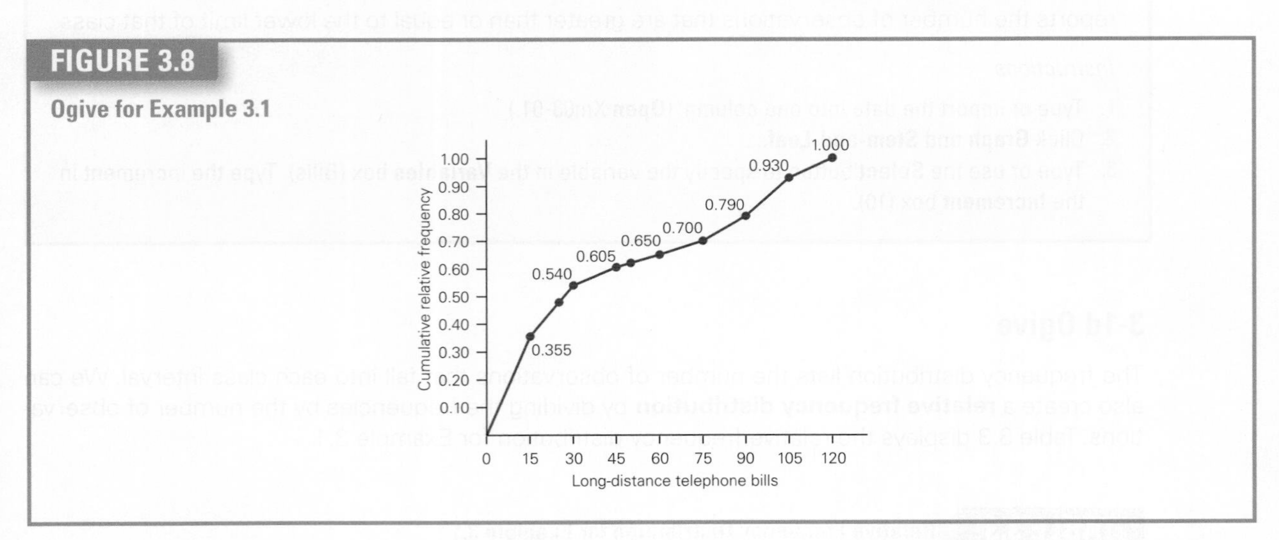

EXCEL

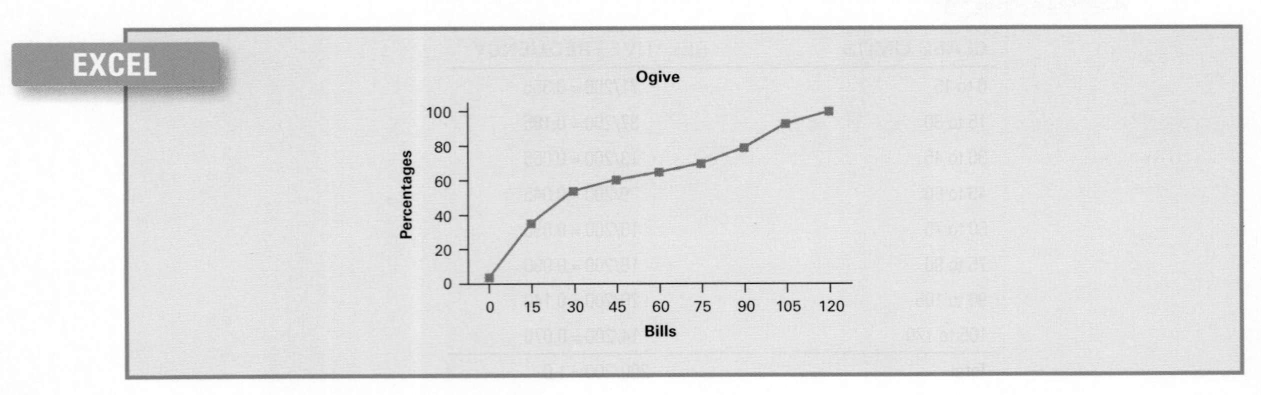

Instructions

Follow instructions to create a histogram. Make the first bin's upper limit a number that is slightly smaller than the smallest number in the data set. Move the cursor to **Chart Output** and click. Do the same for **Cumulative Percentage.** Remove the 'More' category. Click on any of the rectangles and click Delete. Change the **Scale,** if necessary. (Right-click the vertical or horizontal axis, click **Format Axis...** and change the **Maximum** value of Y equal to 1.0.)

MINITAB

Minitab does not draw ogives.

We can use the ogive to estimate the cumulative relative frequencies of other values. For example, we estimate that about 62% of the bills lie below €50 and that about 48% lie below €25. (See Figure 3.9.)

FIGURE 3.9

Ogive with Estimated Relative Frequencies for Example 3.1

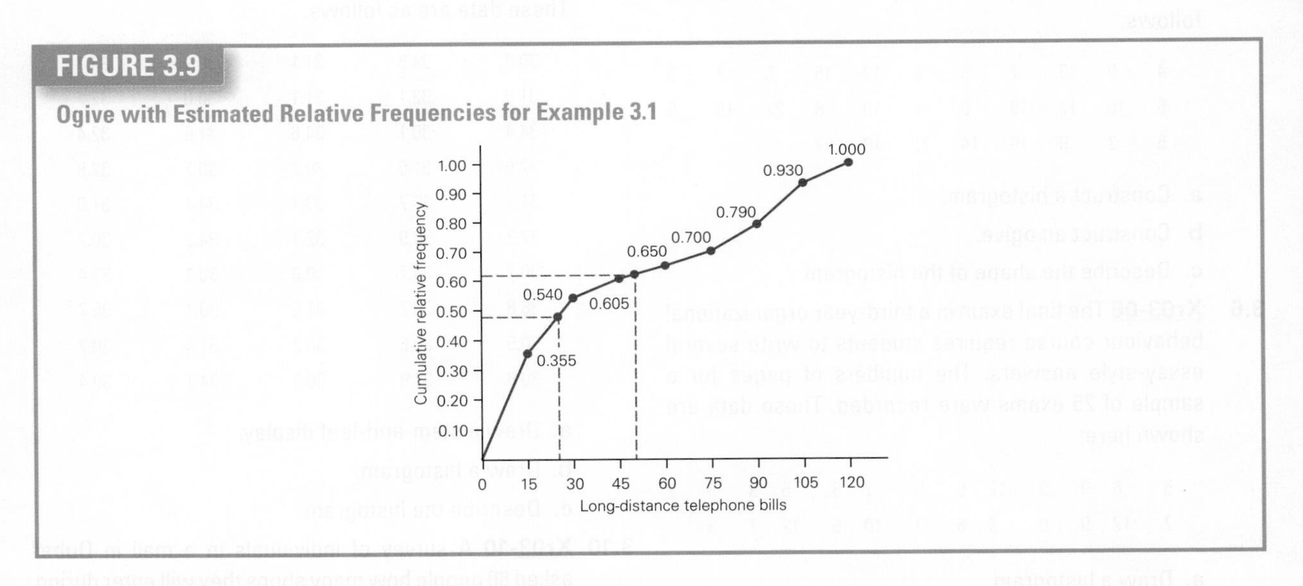

Here is a summary of this section's techniques.

Factors That Identify When to Use a Histogram, Ogive or Stem-and-Leaf Display

1. **Objective:** Describe a single set of data.
2. **Data type:** Interval.

EXERCISES

3.1 How many classes should a histogram contain if the number of observations is 250?

3.2 Determine the number of classes of a histogram for 700 observations.

3.3 A data set consists of 125 observations that range between 37 and 188.

a. What is an appropriate number of classes to have in the histogram?

b. What class intervals would you suggest?

3.4 A statistics practitioner would like to draw a histogram of 62 observations that range from 5.2 to 6.1.

a. What is an appropriate number of class intervals?

b. Define the upper limits of the classes you would use.

3.5 Xr03-05 The number of items rejected daily by a manufacturer in China because of defects was recorded for the past 30 days. The results are as follows:

4	9	13	7	5	6	12	15	5	7	3
6	15	17	19	6	4	10	8	22	16	9
5	3	9	19	14	13	18	7			

a. Construct a histogram.

b. Construct an ogive.

c. Describe the shape of the histogram.

3.6 Xr03-06 The final exam in a third-year organizational behaviour course requires students to write several essay-style answers. The numbers of pages for a sample of 25 exams were recorded. These data are shown here:

5	6	9	3	12	6	5	7	3	6	9	5	2
7	12	9	6	3	6	7	10	9	12	7	3	

a. Draw a histogram.

b. Draw an ogive.

c. Describe what you have learned from the answers to parts (a) and (b).

3.7 Xr03-07 A large investment firm wants to review the distribution of ages of its stockbrokers. The firm believes that this information can be useful in developing plans to recruit new brokers. The ages of a sample of 40 brokers are shown here:

46	28	51	34	29	40	38	33	41	52
53	40	50	33	36	41	25	38	37	41
36	50	46	33	61	48	32	28	30	49
46	28	51	34	29	40	38	33	41	52
41	37	26	39	35	39	46	26	31	35

a. Draw a stem-and-leaf display.

b. Draw a histogram.

c. Draw an ogive.

d. Describe what you have learned.

3.8 Xr03-08 The numbers of weekly sales calls by a sample of 30 telemarketers are listed here. Draw a histogram of these data and describe it.

4	8	6	12	21	4	9	3	25	17
9	5	8	18	16	3	17	19	10	15
5	20	17	14	19	7	10	15	10	8

3.9 Xr03-09 The amount of time (in seconds) needed to complete a critical task on an assembly line in Mumbai was measured for a sample of 50 assemblies. These data are as follows:

30.3	34.5	31.1	30.9	33.7
31.9	33.1	31.1	30.0	32.7
34.4	30.1	34.6	31.6	32.4
32.8	31.0	30.2	30.2	32.8
31.1	30.7	33.1	34.4	31.0
32.2	30.9	32.1	34.2	30.7
30.7	30.7	30.6	30.2	33.4
36.8	30.2	31.5	30.1	35.7
30.5	30.6	30.2	31.4	30.7
30.6	37.9	30.3	34.1	30.4

a. Draw a stem-and-leaf display.

b. Draw a histogram.

c. Describe the histogram.

3.10 Xr03-10 A survey of individuals in a mall in Dubai asked 60 people how many shops they will enter during this visit to the mall. The responses are listed here:

3	2	4	3	3	9
2	4	3	6	2	2
8	7	6	4	5	1
5	2	3	1	1	7
3	4	1	1	4	8
0	2	5	4	4	4
6	2	2	5	3	8
4	3	1	6	9	1
4	4	1	0	4	6
5	5	5	1	4	3

a. Draw a histogram.

b. Draw an ogive.

c. Describe your findings.

3.11 **Xr03-11** A survey asked 50 cricket fans to report the number of matches they attended last year. The results are listed here. Use an appropriate graphical technique to present these data and describe what you have learned.

5	15	14	7	8
16	26	6	15	23
11	15	6	4	7
8	19	16	9	9
8	7	10	5	8
8	6	6	21	10
5	24	5	28	9
11	20	24	5	13
14	9	25	10	24
10	18	22	12	17

3.12 **Xr03-12** To help determine the need for more golf courses, a survey was undertaken. A sample of 75 self-declared golfers was asked how many rounds of golf they played last year. These data are as follows:

18	26	16	35	30
15	18	15	18	29
25	30	35	14	20
18	24	21	25	18
29	23	15	19	27
28	9	17	28	25
23	20	24	28	36
20	30	26	12	31
13	26	22	30	29
26	17	32	36	24
29	18	38	31	36
24	30	20	13	23
3	28	5	14	24
13	18	10	14	16
28	19	10	42	22

a. Draw a histogram.

b. Draw a stem-and-leaf display.

c. Draw an ogive.

d. Describe what you have learned.

CENSUS2011 EXERCISE

3.13 **Census2011** Employ a graphical technique to present the ages (column E in Excel and column 5 in Minitab) of the respondents in the 2011 survey. Describe your results.

3-2 DESCRIBING TIME-SERIES DATA

Besides classifying data by type, we can also classify them according to whether the observations are measured at the same time or whether they represent measurements at successive points in time. The former are called **cross-sectional data** and the latter **time-series data**.

The techniques described in Section 3-1 are applied to cross-sectional data. All the data for Example 3.1 were probably determined within the same day. We can probably say the same thing for Examples 3.2 to 3.4. To give another example, consider a real estate consultant who feels that the selling price of a house is a function of its size, age and plot size. To estimate the specific form of the function she samples, say, 100 homes recently sold and records the price, size, age and plot size for each home. These data are cross-sectional in that they all are observations at the same point in time. The real estate consultant is also working on a separate project to forecast the monthly housing starts in the northeastern region over the next year. To do so, she collects the monthly housing starts in this region for each of the past 5 years. These 60 values (housing starts) represent time-series data, because they are observations taken over time.

Note that the original data may be interval or nominal. All of these illustrations deal with interval data. A time series can also list the frequencies and relative frequencies of a nominal variable over a number of time periods. For example, a brand-preference survey asks consumers to identify their favourite brand. These data are nominal. If we repeat the survey once a month for several years, the proportion of consumers who prefer a certain company's product each month would constitute a time series.

3-2a Line Chart

Time-series data are often graphically depicted on a **line chart**, which is a plot of the variable over time. It is created by plotting the value of the variable on the vertical axis and the time periods on the horizontal axis.

The chapter-opening example addresses the issue of the relationship between the price of petrol and the price of oil. We will introduce the technique we need in order to answer the question in Section 3-3. Another question that arises, is the recent price of petrol high compared to the past prices?

EXAMPLE 3.5

DATA Xm03-05

Price of Petrol

We recorded the monthly average retail price of petrol (in pence per litre) since January 1989. The first 12 months and the last 12 months are displayed below. Draw a line chart to describe these data and briefly describe the results.

Year	Month	Price of Petrol (Pence per Litre)
1989	1	36.02
1989	2	36.88
1989	3	37.30
1989	4	39.09
1989	5	40.81
1989	6	40.74
1989	7	39.26
1989	8	37.40
1989	9	38.19
1989	10	38.30
1989	11	38.08
1989	12	37.38
2013	1	131.71
2013	2	136.37
2013	3	137.25
2013	4	136.81
2013	5	132.75
2013	6	134.06
2013	7	134.74
2013	8	136.87
2013	9	137.19
2013	10	131.48
2013	11	129.73
2013	12	130.79

SOLUTION:

Here are the line charts produced by Excel and Minitab.

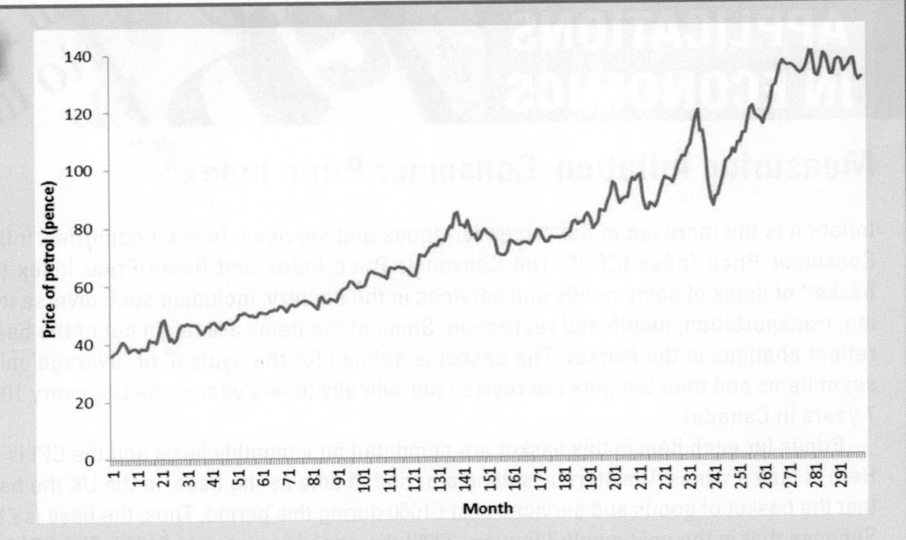

Instructions

1. Type or import the data into one column. (**Open** Xm03-05.)
2. Highlight the column of data. Click **Insert, Line** and the first **2-D Line.** Click **Chart Tools** and **Layout** to make whatever changes you wish.

You can draw two or more line charts (for two or more variables) by highlighting all columns of data you wish to graph.

MINITAB

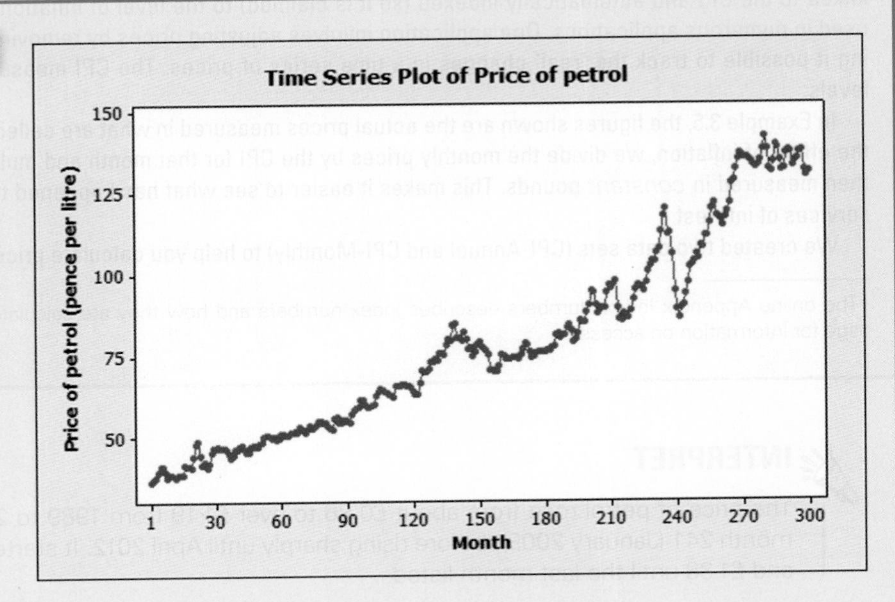

Instructions

1. Type or import the data into one column. (**Open** Xm03-05.)
2. Click **Graph** and **Time-Series Plot** ... Click **Simple.**
3. In the **Series** box type or use the **Select** button to specify the variable (Price). Click **Time/Scale.**
4. Click the **Time** tab and under **Time Scale** click **Index.**

APPLICATIONS IN ECONOMICS

Measuring Inflation: Consumer Price Index*

Inflation is the increase in the prices for goods and services. In most countries, inflation is measured using the Consumer Price Index (CPI)*. The Consumer Price Index and Retail Price Index (RPI) work with a 'shopping basket' of items of some goods and services in the country, including such diverse items as food, housing, clothing, transportation, health and recreation. Some of the items are taken out of the basket, some are brought in, to reflect changes in the market. The basket is defined for the 'typical' or 'average' middle-income family, and the set of items and their weights are revised periodically (every year in the UK, every 10 years in the USA, and every 7 years in Canada).

Prices for each item in this basket are computed on a monthly basis and the CPI is computed from these prices. Here is how it works. We start by setting a period of time as the base. In the UK the base is the year 2005. Suppose that the basket of goods and services cost £1000 during this period. Thus, the base is £1000 and the CPI is set at 100. Suppose that in the next month (January 2006) the price increases to £1010. The CPI for January 2006 is calculated in the following way:

$$CPI(January\ 2006) = \frac{1010}{1000} \times 100 = 101$$

If the price increases to £1050 in the next month, the CPI is

$$CPI(February\ 2006) = \frac{1050}{1000} \times 100 = 105$$

The CPI, despite never really being intended to serve as the official measure of inflation, has come to be interpreted in this way by the general public. Pension-plan payments and some labour contracts are automatically linked to the CPI and automatically indexed (so it is claimed) to the level of inflation. Despite its flaws, the CPI is used in numerous applications. One application involves adjusting prices by removing the effect of inflation, making it possible to track the 'real' changes in a time series of prices. The CPI measures price changes, not price levels.

In Example 3.5, the figures shown are the actual prices measured in what are called *current pounds*. To remove the effect of inflation, we divide the monthly prices by the CPI for that month and multiply by 100. These prices are then measured in *constant* pounds. This makes it easier to see what has happened to the prices of the goods and services of interest.

We created two data sets (CPI-Annual and CPI-Monthly) to help you calculate prices in constant pounds.

*The online Appendix Index Numbers describes index numbers and how they are calculated (see the 'Digital Resources' page for information on access).

INTERPRET

The price of petrol rose from about £0.86 to over £1.19 from 1989 to 2008, then dropped rapidly to month 241 (January 2009), before rising sharply until April 2012. It started to fluctuate between £1.30 and £1.38 until the last month listed.

EXAMPLE 3.6

DATA Xm03-06

Adjusted Price of Petrol

Remove the effect of inflation in Example 3.5 to determine whether petrol prices are higher than they have been in the past.

SOLUTION:

Here are the 1989 to 2013 average monthly prices of petrol, the CPI and the adjusted prices.

The adjusted figures for all months were used in the line chart produced by Excel. Minitab's chart is similar.

Year	Month	Price of Petrol (Pence per Litre)	CPI	Adjusted Price
1989	1	36.02	73.7	48.87
1989	2	36.88	74	49.84
1989	3	37.30	74.3	50.20
1989	4	39.09	75.3	51.91
1989	5	40.81	75.8	53.84
1989	6	40.74	76	53.61
1989	7	39.26	75.9	51.73
1989	8	37.40	76.1	49.15
1989	9	38.19	76.6	49.86
1989	10	38.30	77.1	49.68
1989	11	38.08	77.4	49.20
1989	12	37.38	77.5	48.23
2013	1	131.71	124.4	105.88
2013	2	136.37	125.2	108.92
2013	3	137.25	125.6	109.28
2013	4	136.81	125.9	108.66
2013	5	132.75	126.1	105.27
2013	6	134.06	125.9	106.48
2013	7	134.74	125.8	107.11
2013	8	136.87	126.3	108.28
2013	9	137.19	126.8	108.19
2013	10	131.48	126.9	103.61
2013	11	129.73	127.0	102.15
2013	12	130.79	127.5	102.58

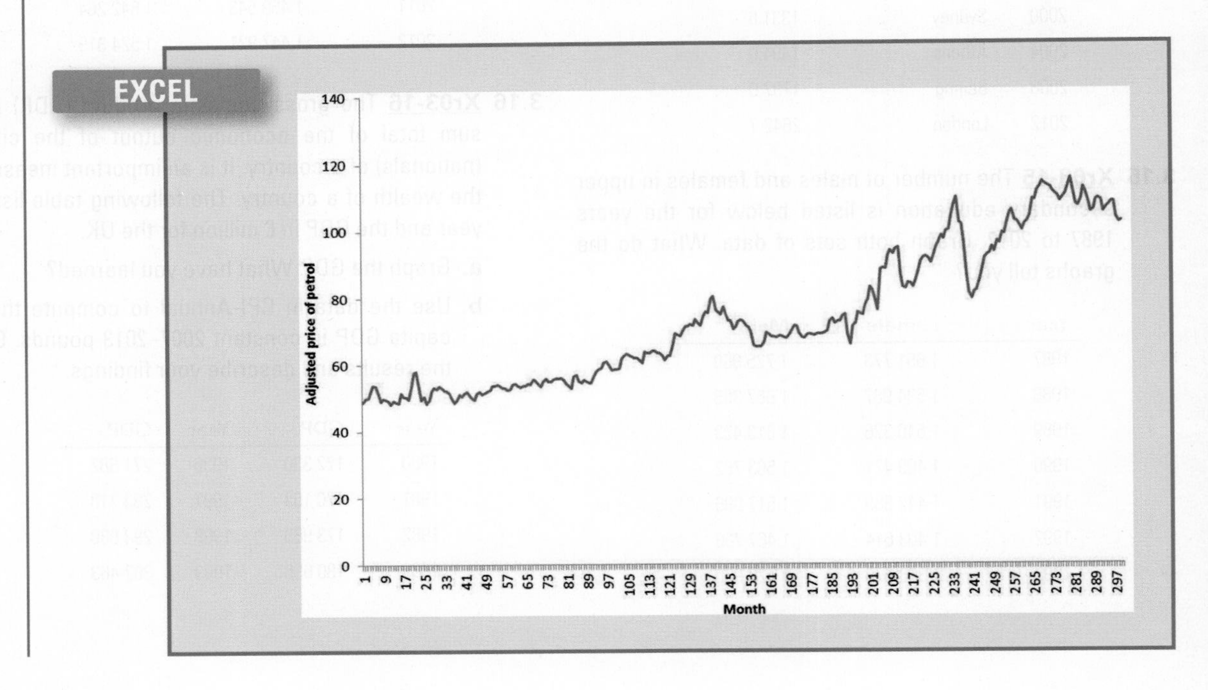

INTERPRET

We can see that the average price of a litre of petrol hits its peak in April 2012 (month 280). At that point, the adjusted price was the highest. The petrol prices increases were a consequence of rising oil prices, the weaker pound and increases in duty and VAT.

There are two more factors to consider in judging whether the price of petrol is high. The first is distance travelled and the second is fuel consumption. This issue is dealt with later in the Exercises section.

EXERCISES

3.14 Xr03-14 The fees television broadcasters pay to cover the summer Olympic Games has become the largest source of revenue for the host country. Below we list the year, city and revenue in millions of US dollars paid by television broadcasters around the world. Draw a chart to describe these prices paid by the networks.

Year	City	Broadcast Revenue
1960	Rome	1.2
1964	Tokyo	1.6
1968	Mexico City	9.8
1972	Munich	17.8
1976	Montreal	34.9
1980	Moscow	88.0
1984	Los Angeles	266.9
1988	Seoul	402.6
1992	Barcelona	636.1
1996	Atlanta	898.3
2000	Sydney	1331.6
2004	Athens	1494.0
2008	Beijing	1737.0
2012	London	2642.7

3.15 Xr03-15 The number of males and females in upper secondary education is listed below for the years 1987 to 2012. Graph both sets of data. What do the graphs tell you?

Year	Female	Male
1987	1 651 773	1 725 960
1988	1 594 987	1 667 305
1989	1 540 326	1 613 423
1990	1 489 421	1 563 762
1991	1 442 889	1 517 096
1992	1 403 614	1 467 766
1993	1 368 670	1 423 980
1994	1 344 119	1 393 410
1995	1 335 963	1 382 494
1996	1345586	1 392 726
1997	1 372 364	1 418 890
1998	1 399 008	1 447 758
1999	1 420 247	1 472 991
2000	1 437 869	1 495 640
2001	1 454 677	1 518 014
2002	1 488 123	1 557 101
2003	1 512 599	1 588 224
2004	1527079	1 609 817
2005	1 531 792	1 620 542
2006	1 528 838	1 621 259
2007	1 531 263	1 620 341
2008	1 522 556	1 609 819
2009	1 504 938	1 591 507
2010	1 482 449	1 567 963
2011	1 459 543	1 542 264
2012	1 447 921	1 524 319

3.16 Xr03-16 The gross domestic product (GDP) is the sum total of the economic output of the citizens (nationals) of a country. It is an important measure of the wealth of a country. The following table lists the year and the GDP in £ million for the UK.

a. Graph the GDP. What have you learned?

b. Use the data in CPI-Annual to compute the per capita GDP in constant 2007–2013 pounds. Graph the results and describe your findings.

Year	GDP	Year	GDP
1980	172 330	1996	271 882
1981	170 163	1997	283 710
1982	173 985	1998	293 830
1983	180 655	1999	302 463

1984	185 972	2000	315 658
1985	193 166	2001	322 554
1986	201 478	2002	329 958
1987	211 864	2003	342 987
1988	223 664	2004	353 871
1989	229 468	2005	365 318
1990	233 648	2006	375 382
1991	230 628	2007	388 248
1992	233 614	2008	385 260
1993	241 768	2009	365 341
1994	253 744	2010	371 404
1995	262 710	2011	375 554

3.17 Xr03-17 The table below lists the value (£ million) of new public housing construction in Great Britain quarterly between 2010 and 2013. Graph these data and describe what you have learned.

Period	Value (£ million)
Q1 2010	1 072
Q2 2010	1 230
Q3 2010	1 311
Q4 2010	1 279
Q1 2011	1 244
Q2 2011	1 308
Q3 2011	1 199
Q4 2011	1 169
Q1 2012	1 010
Q2 2012	1 009
Q3 2012	1 005
Q4 2012	973
Q1 2013	884
Q2 2013	1 092
Q3 2013	1 102

The following exercise requires a computer and software.

3.18 Xr03-18 The value of the US dollar to one British pound was recorded monthly for the period 1971 to 2014. Draw a graph of these figures and interpret your findings.

3-3 DESCRIBING THE RELATIONSHIP BETWEEN TWO INTERVAL VARIABLES

Statistics practitioners frequently need to know how two interval variables are related. For example, financial analysts need to understand how the returns of individual stocks are related to the returns of the entire market. Marketing managers need to understand the relationship between sales and advertising. Economists develop statistical techniques to describe the relationship between such variables as unemployment rates and inflation. The technique is called a **scatter diagram**.

To draw a scatter diagram, we need data for two variables. In applications where one variable depends to some degree on the other variable, we label the *dependent variable* Y and the other, called the *independent variable*, X. For example, an individual's income depends somewhat on the number of years of education. Accordingly, we identify income as the dependent variable and label it Y, and we identify years of education as the independent variable and label it X. In other cases, where no dependency is evident, we label the variables arbitrarily.

EXAMPLE 3.7

DATA Xm03-07

Analyzing the Relationship between Price and Size of House

An estate agent wanted to know to what extent the selling price of a home is related to its size. To acquire this information, he took a sample of 12 homes that had recently sold, recording the price in thousands of euros and the size in square metres. These data are listed in the accompanying table. Use a graphical technique to describe the relationship between size and price.

Size (m^2)	Price (euro 1000)
219	230
168	168
245	260
189	191
209	171
139	158
314	225
263	224
214	211
193	149
253	194
171	143

SOLUTION:

Using the guideline just stated, we label the price of the house *Y* (dependent variable) and the size *X* (independent variable). Figure 3.10 depicts the scatter diagram.

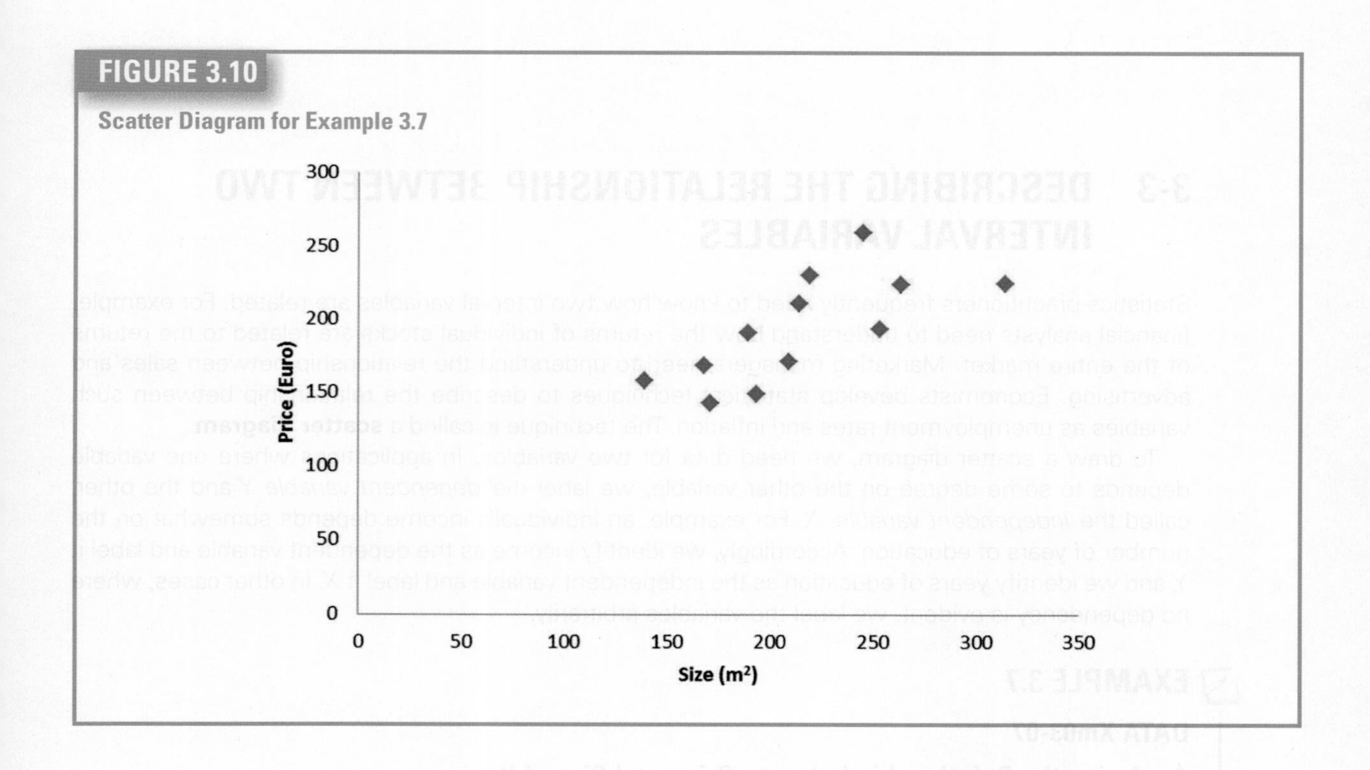

FIGURE 3.10

Scatter Diagram for Example 3.7

EXCEL

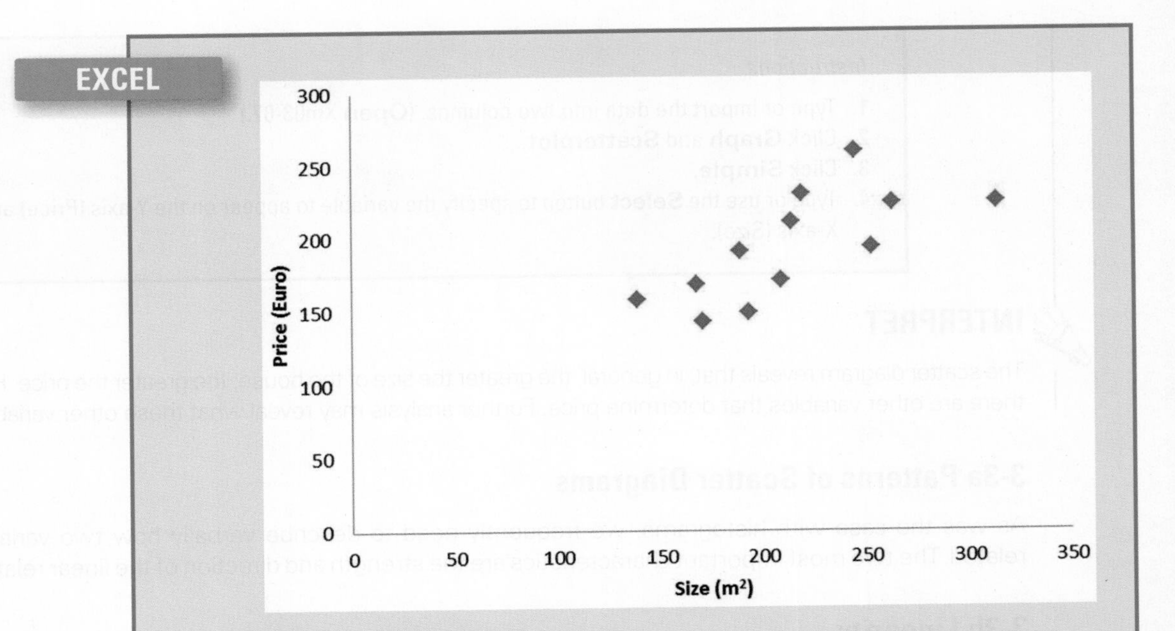

Instructions

1. Type or import the data into two adjacent columns. Store variable X in the first column and variable Y in the next column. (**Open** Xm03-07.)
2. Click **Insert** and **Scatter**.
3. To make cosmetic changes, click **Chart Tools** and **Layout.** (We chose to add titles and remove the gridlines.) If you wish to change the scale, click **Axes, Primary Horizontal Axis** or **Primary Vertical Axis, More Primary Horizontal** or **Vertical Axis Options** ... and make the changes you want.

MINITAB

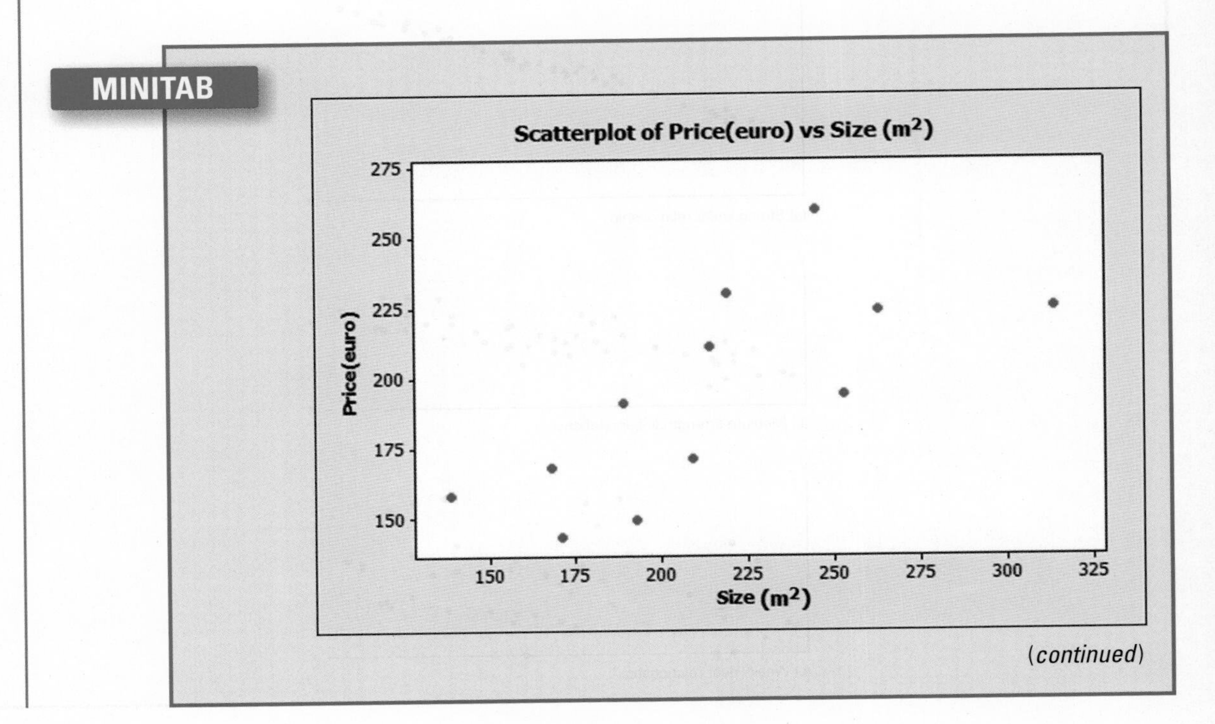

(continued)

Instructions

1. Type or import the data into two columns. (**Open** Xm03-07.)
2. Click **Graph** and **Scatterplot**....
3. Click **Simple**.
4. Type or use the **Select** button to specify the variable to appear on the Y-axis (Price) and the X-axis (Size).

INTERPRET

The scatter diagram reveals that, in general, the greater the size of the house, the greater the price. However, there are other variables that determine price. Further analysis may reveal what these other variables are.

3-3a Patterns of Scatter Diagrams

As was the case with histograms, we frequently need to describe verbally how two variables are related. The two most important characteristics are the strength and direction of the linear relationship.

3-3b Linearity

To determine the strength of the linear relationship, we draw a straight line through the points in such a way that the line represents the relationship. If most of the points fall close to the line, we say that there is a **linear relationship**. If most of the points appear to be scattered randomly with only a semblance of a straight line, there is no, or at best, a weak linear relationship. Figure 3.11 depicts several scatter diagrams that exhibit various levels of linearity.

FIGURE 3.11

Scatter Diagrams Depicting Linearity

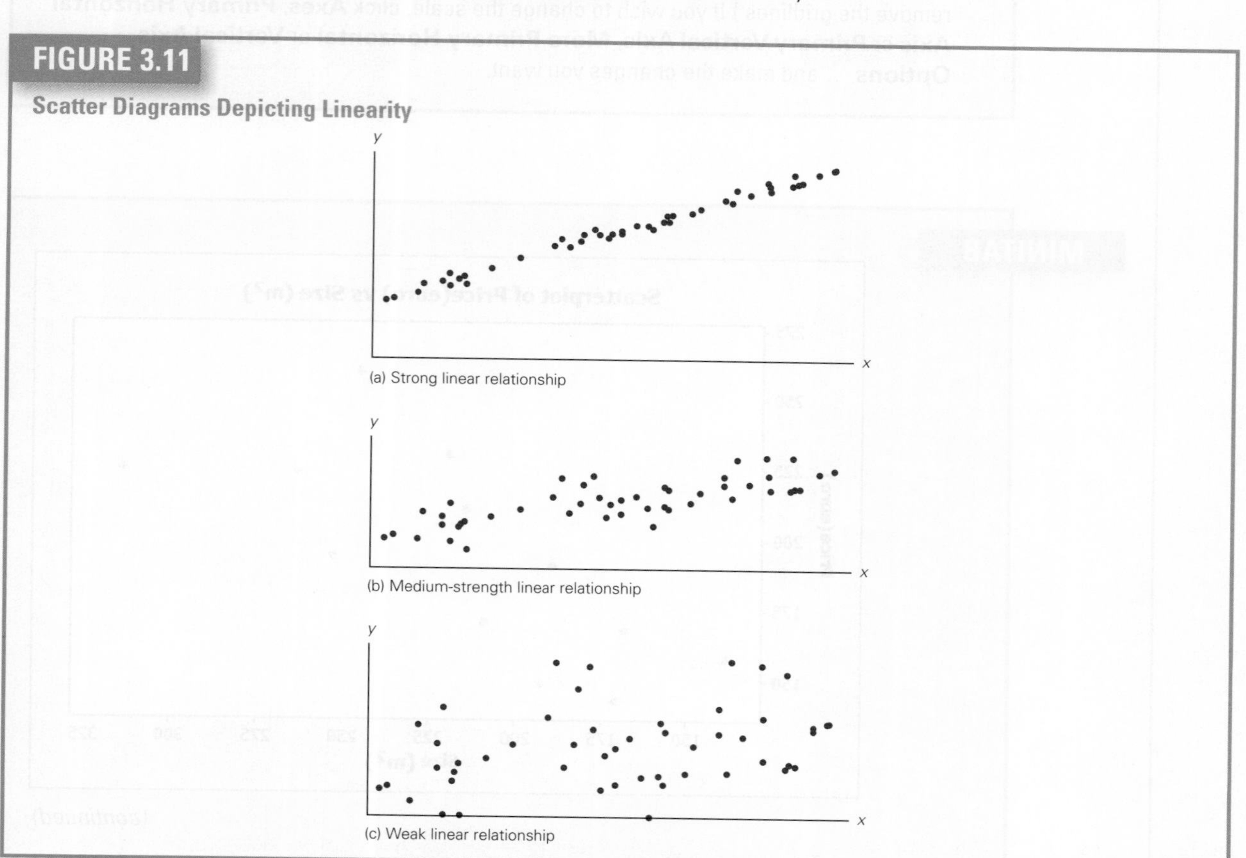

In drawing the line freehand, we would attempt to draw it so that it passes through the middle of the data. Unfortunately, different people drawing a straight line through the same set of data will produce somewhat different lines. Fortunately, statisticians have produced an objective way to draw the straight line. The method is called the *least squares method*, and it will be presented in Chapter 4 and employed in Chapters 15 and 16.

Note that there may well be some other type of relationship, such as a quadratic or exponential one.

3-3c Direction

In general, if one variable increases when the other does, we say that there is a **positive linear relationship**. When the two variables tend to move in opposite directions, we describe the nature of their association as a **negative linear relationship**. (The terms *positive* and *negative* will be explained in Chapter 4.) See Figure 3.12 for examples of scatter diagrams depicting a positive linear relationship, a negative linear relationship, no relationship and a non-linear relationship.

3-3d Interpreting a Strong Linear Relationship

In interpreting the results of a scatter diagram it is important to understand that if two variables are linearly related it does not mean that one is causing the other. In fact, we can never conclude that one variable causes another variable. We can express this more eloquently as:

Correlation is not causation

FIGURE 3.12

Scatter Diagrams Describing Direction

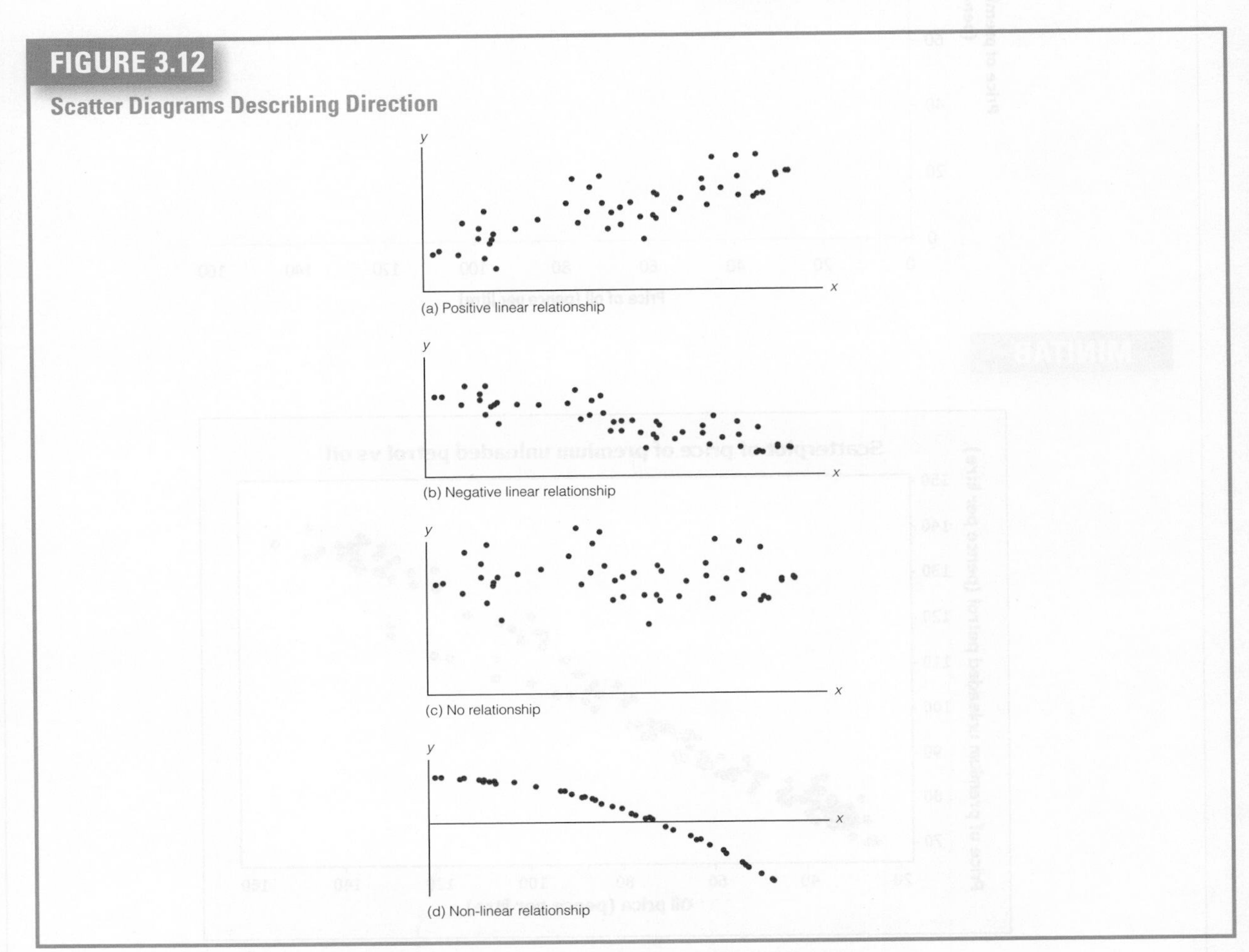

Now that we know what to look for, we can answer the chapter-opening example.

WERE OIL COMPANIES GOUGING CUSTOMERS 2000–2013?: SOLUTION

To determine whether drivers' perceptions that oil companies were gouging consumers we need to determine whether and to what extent the two variables are related. The appropriate statistical technique is the scatter diagram. We label the price of petrol *Y* and the price of oil *X*.

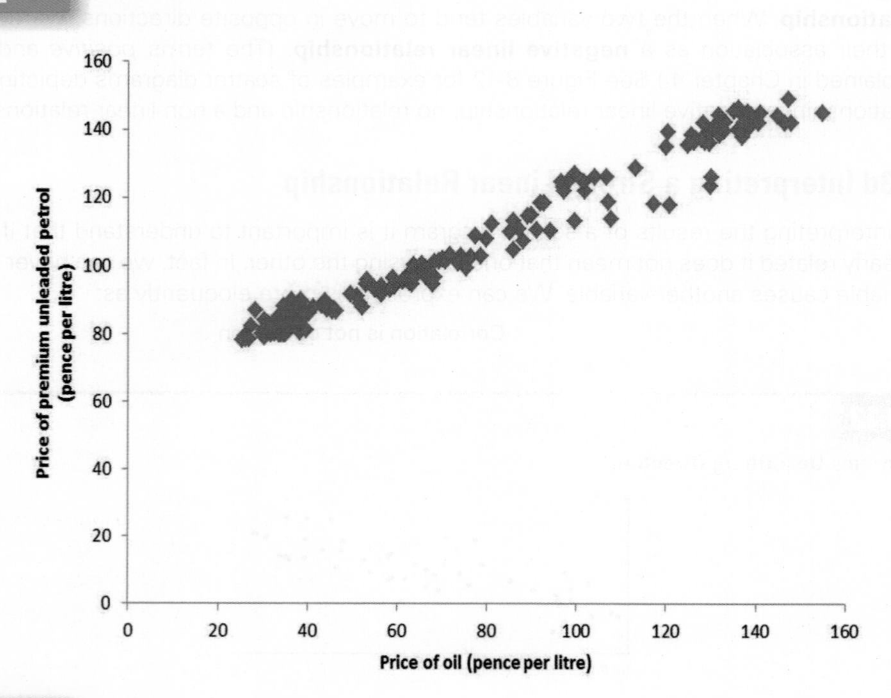

MINITAB

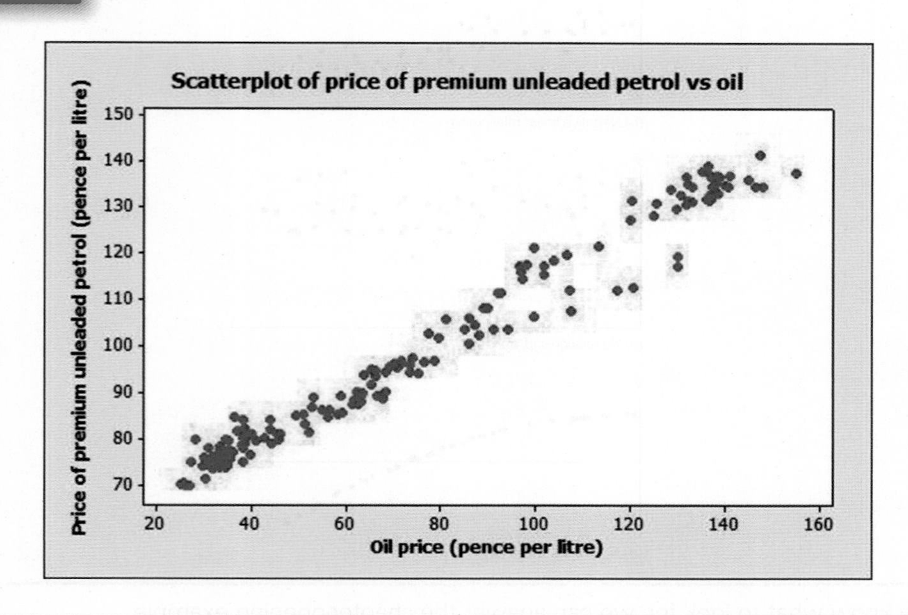

INTERPRET

The scatter diagram reveals that the two prices are strongly related linearly. When the price of oil was below £0.80, the relationship between the two was stronger than when the price of oil exceeded £0.80.

We close this section by reviewing the factors that identify the use of the scatter diagram.

Factors that Identify When to Use a Scatter Diagram

1. **Objective:** Describe the relationship between two variables.
2. **Data type:** Interval.

EXERCISES

3.19 Xr03-19 Because inflation reduces the purchasing power of the euro, investors seek investments in France that will provide higher returns when inflation is higher. It is frequently stated that common stocks provide just such a hedge against inflation. The annual percentage rates of return on common stock and annual inflation rates for a recent 10-year period (from 2004 to 2013) are listed here.

Year	1	2	3	4	5	6	7	8	9	10
Returns	25	6	6	11	21	−15	12	−1	33	0
Inflation	2.1	1.5	1.5	2.6	1.1	0.9	1.8	2.5	1.3	0.7

a. Use a graphical technique to depict the relationship between the two variables.

b. Does it appear that the returns on common stocks and inflation are linearly related?

3.20 Xr03-20 In a university in Saudi Arabia, where calculus is a prerequisite for the statistics course, a sample of 15 students was drawn. The marks for calculus and statistics were recorded for each student. The data are as follows:

Calculus	65	58	93	68	74	81	58	85
Statistics	74	72	84	71	68	85	63	73
Calculus	88	75	63	79	80	54	72	
Statistics	79	65	62	71	74	68	73	

3.21 Xr03-21 The cost of repairing cars involved in accidents is one reason that insurance premiums are so high. In an experiment, ten cars were driven into a wall. The speeds were varied between 2 and 20 km/h. The costs of repair were estimated and are listed here. Draw an appropriate graph to analyze the relationship between the two variables. What does the graph tell you?

Speed	2	4	6	8	10	12
Cost of repair (€)	88	124	358	519	699	816
Speed	14	16	18	20		
Cost of repair (€)	905	1 521	1 888	2 201		

3.22 Xr03-22 The growing interest in and use of the Internet have forced many companies into considering ways to sell their products on the Web. Therefore, it is of interest to these companies to determine who is using the Web. A statistics practitioner undertook a study to determine how education and Internet use are connected. She took a random sample of 15 adults (20 years of age and older) and asked each to report the years of education they had completed and the number of hours of Internet use in the previous week. These data follow.

Education	11	11	8	13	17	11	11	11
Internet use	10	5	0	14	24	0	15	12
Education	19	13	15	9	15	15	11	
Internet use	20	10	5	8	12	15	0	

a. Employ a suitable graph to depict the data.

b. Does it appear that there is a linear relationship between the two variables? If so, describe it.

3.23 Xr03-23 A statistics lecturer formed the opinion that students who handed in assignments and exams early outperformed students who handed in their assignments later. To develop data to decide whether

her opinion is valid, she recorded the amount of time (in minutes) taken by students to submit their midterm tests (time limit 90 minutes) and the subsequent mark for a sample of 12 students.

Time	90	73	86	85	80	87	90	78	84	71	72	88
Mark	68	65	58	94	76	91	62	81	75	83	85	74

a. Draw a scatter diagram of the data.

b. What does the graph tell you about the relationship between the marks in calculus and statistics?

The following exercises require a computer and software.

3.24 Xr03-24 In an attempt to determine the factors that affect the amount of energy used, 200 households were analysed. The number of occupants and the amount of electricity used were measured for each household.

a. Draw a graph of the data.

b. What have you learned from the graph?

3.25 Xr03-25 Many downhill skiers in countries that have colder climates or high mountain ranges eagerly look forwards to the winter months and fresh snowfalls. However, winter also entails cold days. How does the temperature affect skiers' desire? To answer this question, a local ski resort recorded the temperature for 50 randomly selected days and the number of lift tickets they sold. Use a graphical technique to describe the data and interpret your results.

3.26 Xr03-26 Are younger workers less likely to stay with their jobs? To help answer this question, a random sample of workers was selected. All were asked to report their ages and how many months they had been employed with their current employers. Use a graphical technique to summarise these data.

3.27 Xr03-27 A. very large contribution to profits for a cinema is the sales of popcorn, soft drinks and sweets. A cinema manager speculated that the longer the time between showings of a film, the greater the sales of concession items. To acquire more information, the manager conducted an experiment. For a month he varied the amount of time between movie showings and calculated the sales. Use a graphical technique to help the manager determine whether a longer time gap produces higher concession stand sales.

3.28 Xr03-28 It is generally believed that higher interest rates result in less employment because companies are more reluctant to borrow to expand their business. To determine whether there is a relationship between interest rate and unemployment, an economist collected the monthly interest rate and the monthly unemployment rate for the years 1971 to 2013. Use a graphical technique to supply your answer.

CENSUS EXERCISES

3.29 Census2011 Do more educated people tend to be single? Draw a scatter diagram of OCCUPATION (column L in Excel and column 12 in Minitab) and MARITAL STATUS (column F in Excel and column 6 in Minitab) to answer the question.

3.30 Census2011 Is there a positive linear relationship between occupation (OCCUPATION – column L in Excel and column 12 in Minitab) and the social grade (APPOXIMATIVE SOCIAL GRADE – column N in Excel and column 14 in Minitab)? Draw a scatter diagram to answer the question.

3-4 ART AND SCIENCE OF GRAPHICAL PRESENTATIONS

In this chapter and in Chapter 2, we introduced a number of graphical techniques. The emphasis was on how to construct each one manually and how to command the computer to draw them. In this section, we discuss how to use graphical techniques effectively. We introduce the concept of **graphical excellence**, which is a term we apply to techniques that are informative and concise and that impart information clearly to their viewers. Additionally, we discuss an equally important concept: graphical integrity and its enemy **graphical deception**.

3-4a Graphical Excellence

Graphical excellence is achieved when the following characteristics apply.

1. **The graph presents large data sets concisely and coherently.** Graphical techniques were created to summarise and describe large data sets. Small data sets are easily summarised with a table. One or two numbers can best be presented in a sentence.
2. **The ideas and concepts the statistics practitioner wants to deliver are clearly understood by the viewer.** The chart is designed to describe what would otherwise be described in words. An excellent chart is one that can replace a thousand words and still be clearly comprehended by its readers.
3. **The graph encourages the viewer to compare two or more variables.** Graphs displaying only one variable provide very little information. Graphs are often best used to depict relationships between two or more variables or to explain how and why the observed results occurred.
4. **The display induces the viewer to address the substance of the data and not the form of the graph.** The form of the graph is supposed to help present the substance. If the form replaces the substance, the chart is not performing its function.
5. **There is no distortion of what the data reveal.** You cannot make statistical techniques say whatever you like. A knowledgeable reader will easily see through distortions and deception. We will endeavour to make you a knowledgeable reader by describing graphical deception later in this section.

Edward Tufte, professor of statistics at Yale University, summarised graphical excellence this way:

1. Graphical excellence is the well-designed presentation of interesting data – a matter of substance, statistics and design.
2. Graphical excellence is that which gives the viewer the greatest number of ideas in the shortest time with the least ink in the smallest space.
3. Graphical excellence is nearly always multivariate.
4. Graphical excellence requires telling the truth about the data.

FIGURE 3.13

Chart Depicting Napoleon's Invasion and Retreat from Russia in 1812

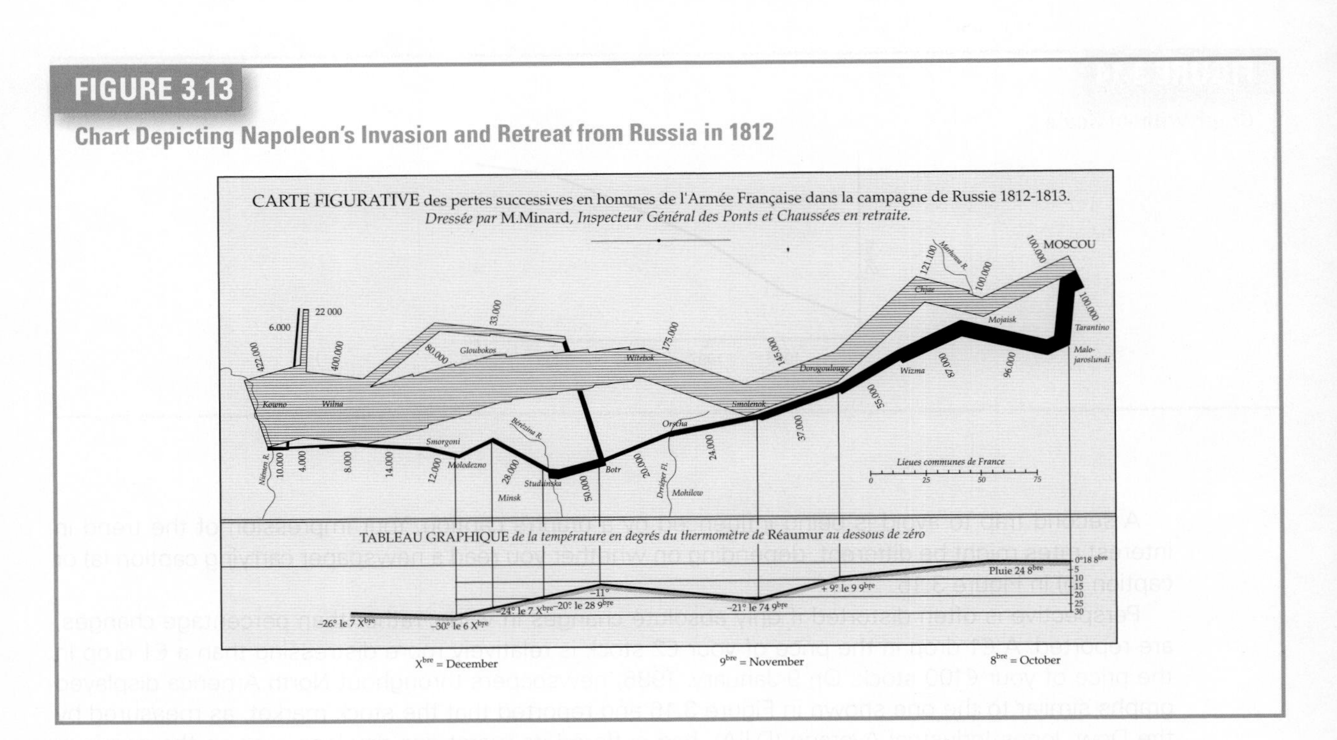

Now let's examine the chart that has been acclaimed the best chart ever drawn. Figure 3.13 depicts Minard's graph. The striped band is a time series depicting the size of the army at various places on the map, which is also part of the chart.

When Napoleon invaded Russia by crossing the Niemen River on 21 June, 1812, there were 422 000 soldiers. By the time the army reached Moscow, the number had dwindled to 100 000. At that point, the army started its retreat. The black band represents the army in retreat. At the bottom of the chart, we see the dates starting with October 1813. Just above the dates, Minard drew another time series, this one showing the temperature. It was bitterly cold during the fall, and many soldiers died of exposure. As you can see, the temperature dipped to –30 degrees Celsius on 6 December. The chart is effective because it depicts five variables clearly and succinctly.

3-4b Graphical Deception

The use of graphs and charts is pervasive in newspapers, magazines, business and economic reports, and seminars, in large part because of the increasing availability of computers and software that allow the storage, retrieval, manipulation and summary of large masses of raw data. It is therefore more important than ever to be able to evaluate critically the information presented by means of graphical techniques. In the final analysis, graphical techniques merely create a visual impression, which is easy to distort. In fact, distortion is so easy and commonplace that in 1992 the Canadian Institute of Chartered Accountants found it necessary to begin setting guidelines for financial graphics, after a study of hundreds of the annual reports of major corporations found that 8% contained at least one misleading graph that covered up bad results. Although the heading for this section mentions deception, it is quite possible for an inexperienced person inadvertently to create distorted impressions with graphs. In any event, you should be aware of possible methods of graphical deception. This section illustrates a few of them.

The first thing to watch for is a graph without a scale on one axis. The line chart of a company's sales in Figure 3.14 might represent a growth rate of 100% or 1% over the 5 years depicted, depending on the vertical scale. It is best simply to ignore such graphs.

FIGURE 3.14
Graph without Scale

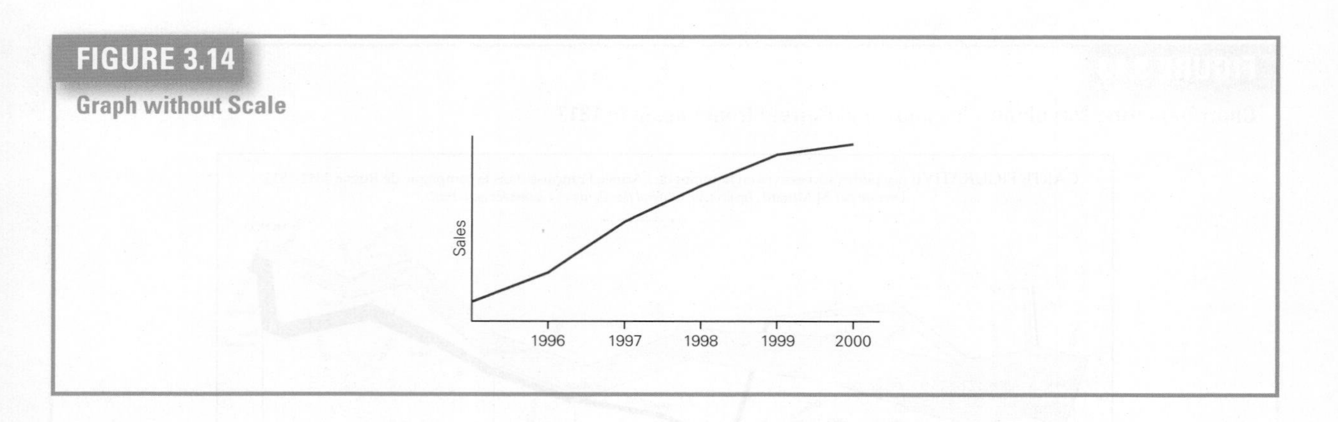

A second trap to avoid is being influenced by a graph's caption. Your impression of the trend in interest rates might be different, depending on whether you read a newspaper carrying caption (a) or caption (b) in Figure 3.15.

Perspective is often distorted if only absolute changes in value, rather than percentage changes, are reported. A €1 drop in the price of your €2 stock is relatively more distressing than a €1 drop in the price of your €100 stock. On 9 January, 1986, newspapers throughout North America displayed graphs similar to the one shown in Figure 3.16 and reported that the stock market, as measured by the Dow Jones Industrial Average (DJIA), had suffered its worst one-day loss ever on the previous day. The loss was 39 points, exceeding even the loss of Black Tuesday: 28 October, 1929. While the loss was indeed a large one, many news reports failed to mention that the 1986 level of the DJIA was much higher than the 1929 level. A better perspective on the situation could be gained by noticing that the loss on 8 January, 1986, represented a 2.5% decline, whereas the decline in 1929 was

FIGURE 3.15

Graphs with Different Captions

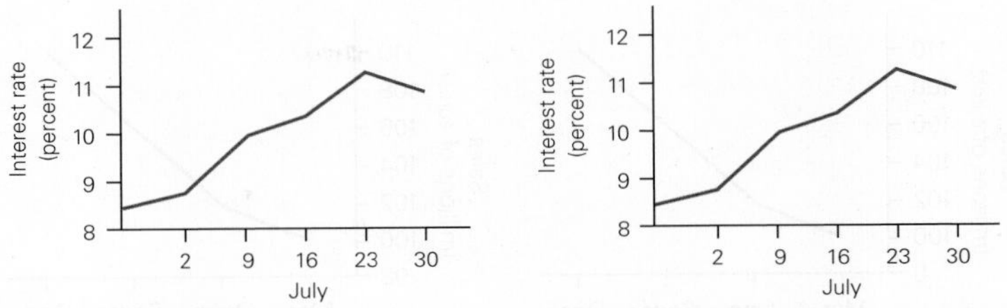

(a) Interest rates have finally begun to turn downwards.

(b) Last week provided temporary relief from the upwards trend in interest rates.

12.8%. As a point of interest, we note that the stock market was 12% higher within 2 months of this historic drop and 40% higher 1 year later. The largest 1-day percentage drop in the DJIA is 24.4% (12 December, 1914).

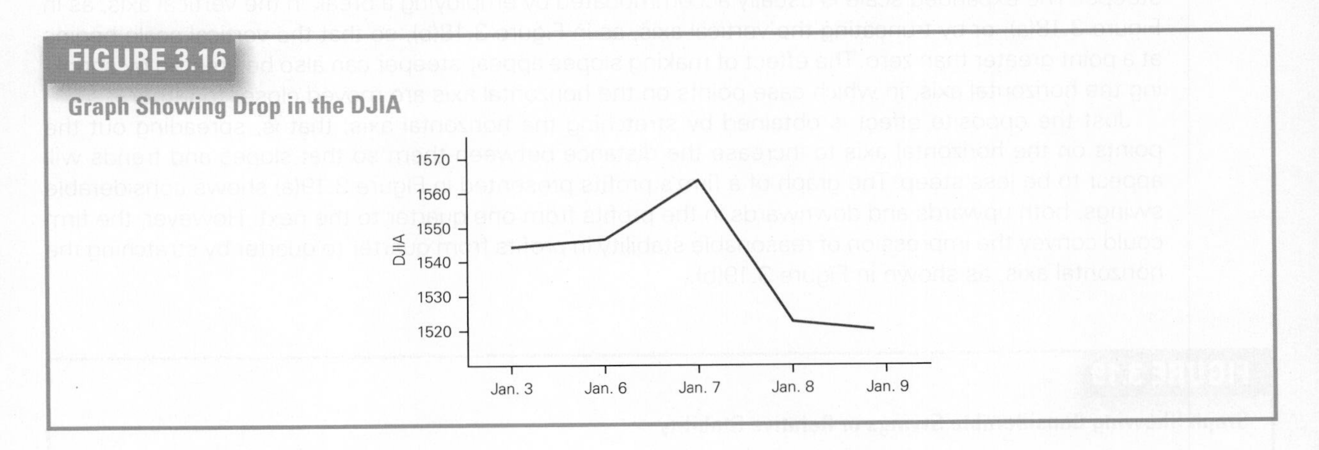

FIGURE 3.16

Graph Showing Drop in the DJIA

We now turn to some rather subtle methods of creating distorted impressions with graphs. Consider the graph in Figure 3.17, which depicts the growth in a company's quarterly sales during the past year, from €100 million to €110 million. This 10% growth in quarterly sales can be made to appear more dramatic by stretching the vertical axis – a technique that involves changing the scale on the

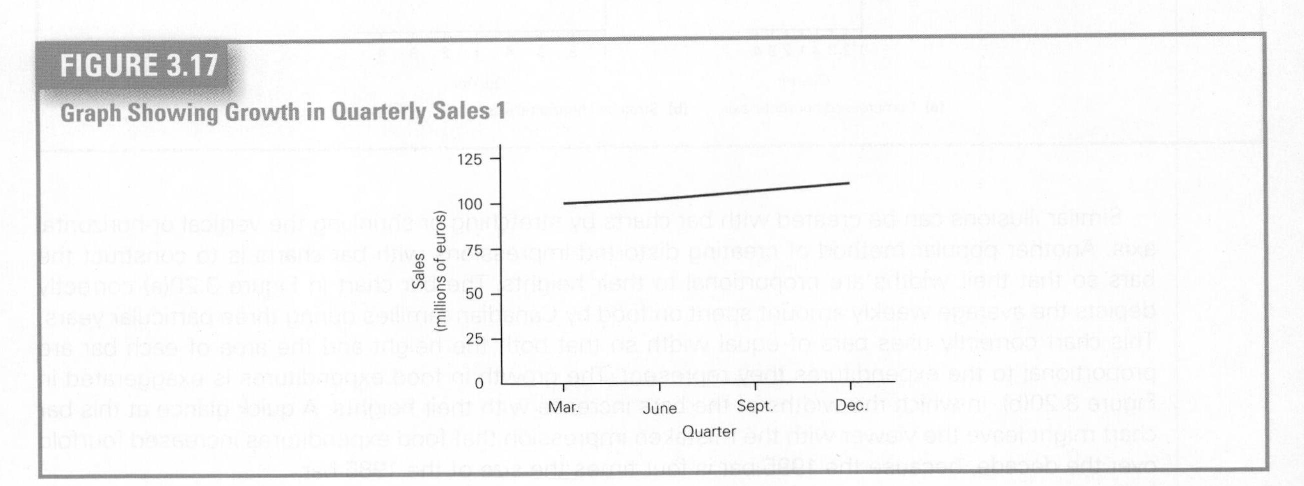

FIGURE 3.17

Graph Showing Growth in Quarterly Sales 1

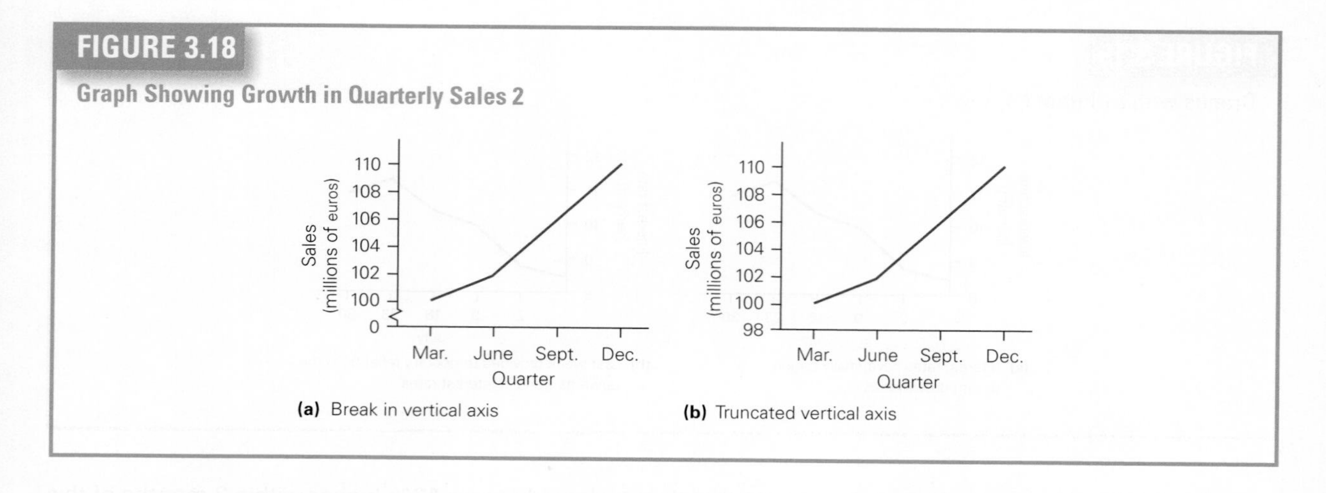

FIGURE 3.18

Graph Showing Growth in Quarterly Sales 2

(a) Break in vertical axis **(b)** Truncated vertical axis

vertical axis so that a given dollar amount is represented by a greater height than before. As a result, the rise in sales appears to be greater because the slope of the graph is visually (but not numerically) steeper. The expanded scale is usually accommodated by employing a break in the vertical axis, as in Figure 3.18(a), or by truncating the vertical axis, as in Figure 3.18(b), so that the vertical scale begins at a point greater than zero. The effect of making slopes appear steeper can also be created by shrinking the horizontal axis, in which case points on the horizontal axis are moved closer together.

Just the opposite effect is obtained by stretching the horizontal axis; that is, spreading out the points on the horizontal axis to increase the distance between them so that slopes and trends will appear to be less steep. The graph of a firm's profits presented in Figure 3.19(a) shows considerable swings, both upwards and downwards in the profits from one quarter to the next. However, the firm could convey the impression of reasonable stability in profits from quarter to quarter by stretching the horizontal axis, as shown in Figure 3.19(b).

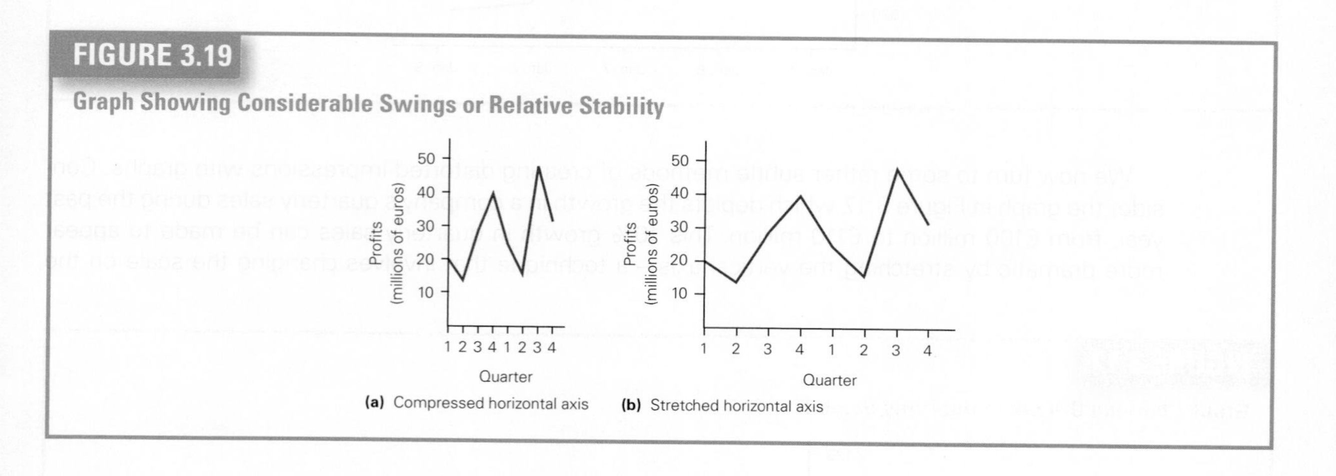

FIGURE 3.19

Graph Showing Considerable Swings or Relative Stability

(a) Compressed horizontal axis **(b)** Stretched horizontal axis

Similar illusions can be created with bar charts by stretching or shrinking the vertical or horizontal axis. Another popular method of creating distorted impressions with bar charts is to construct the bars so that their widths are proportional to their heights. The bar chart in Figure 3.20(a) correctly depicts the average weekly amount spent on food by Canadian families during three particular years. This chart correctly uses bars of equal width so that both the height and the area of each bar are proportional to the expenditures they represent. The growth in food expenditures is exaggerated in Figure 3.20(b), in which the widths of the bars increase with their heights. A quick glance at this bar chart might leave the viewer with the mistaken impression that food expenditures increased fourfold over the decade, because the 1995 bar is four times the size of the 1985 bar.

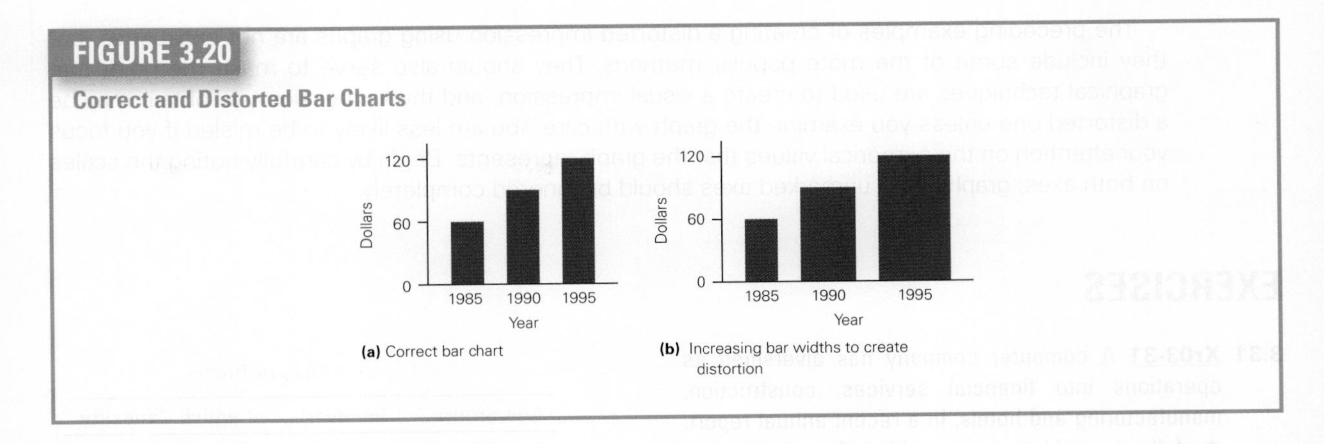

FIGURE 3.20
Correct and Distorted Bar Charts

You should be on the lookout for size distortions, particularly in pictograms, which replace the bars with pictures of objects (such as bags of money, people or animals) to enhance the visual appeal. Figure 3.21 displays the misuse of a pictogram – the snowman grows in width as well as height. The proper use of a pictogram is shown in Figure 3.22, which effectively uses pictures of Coca-Cola bottles.

FIGURE 3.21
Misuse of Pictogram

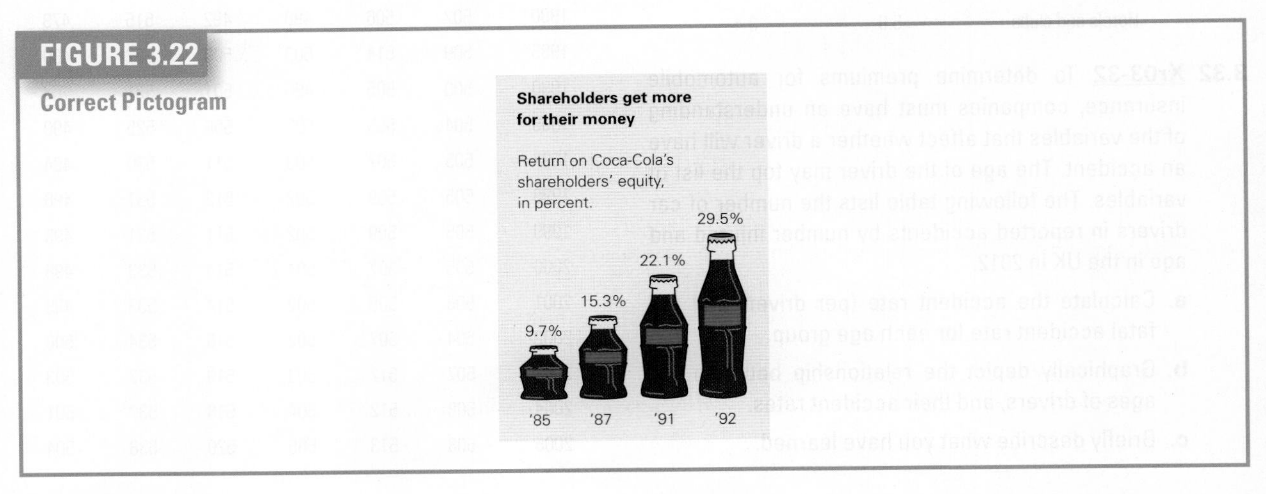

FIGURE 3.22
Correct Pictogram

The preceding examples of creating a distorted impression using graphs are not exhaustive, but they include some of the more popular methods. They should also serve to make the point that graphical techniques are used to create a visual impression, and the impression you obtain may be a distorted one unless you examine the graph with care. You are less likely to be misled if you focus your attention on the numerical values that the graph represents. Begin by carefully noting the scales on both axes; graphs with unmarked axes should be ignored completely.

EXERCISES

3.31 Xr03-31 A computer company has diversified its operations into financial services, construction, manufacturing and hotels. In a recent annual report, the following tables were provided. Create charts to present these data so that the differences between last year and the previous year are clear. (*Note:* It may be necessary to draw the charts manually.)

Sales (millions of dollars) by Region

Region	Last year	Previous year
United States	67.3	40.4
Canada	20.9	18.9
Europe	37.9	35.5
Australasia	26.2	10.3
Total	152.2	105.1

Sales (millions of dollars) by division

Division	Last year	Previous year
Customer service	54.6	43.8
Library systems	49.3	30.5
Construction/property management	17.5	7.7
Manufacturing and distribution	15.4	8.9
Financial systems	9.4	10.9
Hotels and clubs	5.9	3.4

3.32 Xr03-32 To determine premiums for automobile insurance, companies must have an understanding of the variables that affect whether a driver will have an accident. The age of the driver may top the list of variables. The following table lists the number of car drivers in reported accidents by number injured and age in the UK in 2012.

a. Calculate the accident rate (per driver) and the fatal accident rate for each age group.

b. Graphically depict the relationship between the ages of drivers, and their accident rates.

c. Briefly describe what you have learned.

	All Drivers or Riders	
Age group	Involved	of which Causality
Under 17	72	42
17–19	10 235	5 758
20–24	22 805	12 174
25–29	20 772	10 123
30–34	19 680	8 636
35–39	17 089	7 507
40–49	35 315	15 376
50–59	23 799	10 109
60–69	14 011	5 607
70 and over	10 927	4 882
Age not reported	22 683	627
Total	197 388	80 841

3.33 Xr03-33 The accompanying table lists the average test scores in the Scholastic Assessment Test (SAT) for the years 1967, 1970, 1975, 1980, 1985, 1990, 1995 and 1997 to 2007.

Year	Verbal All	Verbal Male	Verbal Female	Math All	Math Male	Math Female
1967	543	540	545	516	535	595
1970	537	536	538	512	531	493
1975	512	515	509	498	518	479
1980	502	506	498	492	515	473
1985	509	514	503	500	522	480
1990	500	505	496	501	521	483
1995	504	505	502	506	525	490
1997	505	507	503	511	530	494
1998	505	509	502	512	531	496
1999	505	509	502	511	531	495
2000	505	507	504	514	533	498
2001	506	509	502	514	533	498
2002	504	507	502	516	534	500
2003	507	512	503	519	537	503
2004	508	512	504	518	537	501
2005	508	513	505	520	538	504

2006	503	505	502	518	536	502
2007	502	504	502	515	533	499
2008	502	504	500	515	533	500
2009	501	503	498	515	534	499

Draw a chart for each of the following.

a. You wish to show that both verbal and mathematics test scores for all students have not changed much over the years.

b. The exact opposite of part (a).

c. You want to claim that there are no differences between genders.

d. You want to 'prove' that differences between genders exist.

3.34 Xr03-34 The monthly unemployment rate for those age 16 or over in the UK for the past 12 months is listed here.

Month	1	2	3	4	5	6	7	8	9	10	11	12
Rate	7.8	7.8	7.9	7.8	7.8	7.8	7.8	7.7	7.7	7.6	7.4	7.1

a. Draw a bar chart of these data with 7.0% as the lowest point on the vertical axis.

b. Draw a bar chart of these data with 0.0% as the lowest point on the vertical axis.

c. Discuss the impression given by the two charts.

d. Which chart would you use? Explain.

CHAPTER SUMMARY

Histograms are used to describe a single set of interval data. Statistics practitioners examine several aspects of the shapes of histograms. These are symmetry, number of modes and its resemblance to a bell shape.

We described the difference between time-series data and cross-sectional data. Time series are graphed by line charts.

To analyse the relationship between two interval variables, we draw a scatter diagram. We look for the direction and strength of the linear relationship.

CHAPTER EXERCISES

The following exercises require a computer and software.

3.35 Xr03-35 The monthly values of one euro measured in American dollars since 2002 were recorded. Draw a graph that shows how the exchange rate has varied over the 12-year period.

3.36 Xr03-36 Studies of twins may reveal more about the 'nature or nurture' debate. The issue being debated is whether nature or the environment has more of an effect on individual traits such as intelligence. Suppose that a sample of identical twins was selected and their IQs measured. Use a suitable graphical technique to depict the data, and describe what it tells you about the relationship between the IQs of identical twins.

3.37 Xr03-37 One-hundred students who had reported that they use their computers for at least 20 hours per week were asked to keep track of the number of crashes their computers incurred during a 12-week period. Using an appropriate statistical method, summarise the data. Describe your findings.

3.38 Xr03-38 In Chapters 15 and 16, we introduce regression analysis, which addresses the relationships among variables. One of the first applications of regression analysis was to analyse the relationship between the heights of fathers and sons. Suppose that a sample of 80 fathers and sons was drawn. The heights of the fathers and of the adult sons were measured.

a. Draw a scatter diagram of the data. Draw a straight line that describes the relationship.

b. What is the direction of the line?

c. Does it appear that there is a linear relationship between the two variables?

3.39 Xr03-39 Do better golfers play faster than poorer ones? To determine whether a relationship exists, a sample of 125 foursomes was selected. Their total scores and the amount of time taken to complete the 18 holes were recorded. Graphically depict the data, and describe what they tell you about the relationship between score and time.

3.40 Xr-03-40 An increasing number of consumers prefer to use debit cards in place of cash or credit cards. To analyse the relationship between the amounts of purchases made with debit and credit cards, 240 people were interviewed and asked to report the amount of money spent on purchases using debit cards and the amount spent using credit cards during the last month. Draw a graph of the data and summarise your findings.

3.41 Xr03-41 Is airline travel becoming safer? To help answer this question, a student recorded the number of fatal accidents and the number of deaths that occurred in the years 1986 to 2008 for scheduled airlines. Use a graphical method to answer the question.

3.42 Xr03-42 Most car-hire companies keep their cars for about a year and then sell them to used car dealerships. Suppose one company decided to sell the used cars themselves. Because most used car buyers make their decision on what to buy and how much to spend based on the car's odometer reading, this would be an important issue for the car-rental company. To develop information about the mileage shown on the company's hire cars, the general manager took a random sample of 658 customers and recorded the average number of miles driven per day. Use a graphical technique to display these data.

3.43 Xr03-43 Several years ago, a large art exhibition toured major cities all over the world, with millions of people flocking to see it. Because of the size and value of the collection, it was predicted (correctly) that in each city a large number of people would come to view the paintings. Because space was limited, most galleries had to sell tickets that were valid at one time (much like a play). In this way, they were able to control the number of visitors at any one time. To judge how many people to let in at any time, it was necessary to know the length of time people would spend at the exhibit; longer times would dictate smaller audiences; shorter times would allow for the sale of more tickets. The manager of a gallery that will host the exhibit realised her facility can comfortably and safely hold about 250 people at any one time. Although the demand will vary throughout the day and from weekday to weekend, she believes that the demand will not drop below 500 at any time. To help make a decision about how many tickets to sell, she acquired the amount of time a sample of 400 people spent at the exhibit from another city. What ticket procedure should the museum management institute?

The following exercises are based on data sets that include additional data referenced in previously presented examples and exercises.

3.44 Xm03-03 Xm03-04 Examples 3.3 and 3.4 listed the final marks in the business statistics course and the mathematical statistics course. The professor also provided the final marks in the first-year required calculus course. Graphically describe the relationship between calculus and statistics marks. What information were you able to extract?

3.45 Xm03-03 Xm03-04 In addition to the previously discussed data in Examples 3.3 and 3.4, the professor listed the midterm mark. Conduct an analysis of the relationship between final exam mark and midterm mark in each course. What does this analysis tell you?

3.46 Xr02-34 Two other questions were asked in Exercise 2.34:

Number of weeks job searching?
Salary (thousands)?

The placement office wants the following:

a. Graphically describe salary.
b. Is salary related to the number of weeks needed to land the job?

CASE 3.1

DATA C03-01a C03-01b

The Question of Global Warming

In the last part of the 20th century, scientists developed the theory that the planet was warming and the primary cause was the increasing amounts of carbon dioxide (CO_2), which is the product of burning oil, natural gas and coal (fossil fuels). Although many climatologists believe in the so-called greenhouse effect, there are many others who do not subscribe to this theory. There are three critical questions that need to be answered in order to resolve the issue.

1. Is the Earth actually warming? To answer this question, we need accurate temperature measurements over a large number of years. But how do we measure the temperature before the invention of accurate thermometers? Moreover, how do we go about measuring the Earth's temperature even with accurate thermometers?
2. If the planet is warming, is there a man-made cause or is it natural fluctuation? The temperature of the Earth has increased and decreased many times in its long history. We have had higher temperatures and we have had lower temperatures, including various ice ages. In fact, a period called the 'little ice age' ended around the middle to the end of the 19th century. Then the temperature rose until about 1940, at which point it decreased until 1975. A 28 April, 1975, *Newsweek* article discussed the possibility of global cooling, which seemed to be the consensus among scientists.
3. If the planet is warming, is CO_2 the cause? There are greenhouse gases in the atmosphere without which the Earth would be considerably colder. These gases include methane, water vapour and carbon dioxide. All occur naturally in nature. Carbon dioxide is vital to our life on Earth because it is necessary for growing plants. The amount of CO_2 produced by fossil fuels is relatively a small proportion of all the CO_2 in the atmosphere.

The generally accepted procedure is to record monthly temperature anomalies. To do so, we calculate the average for each month over many years. We then calculate any deviations between the latest month's temperature reading and its average. A positive anomaly would represent a month's temperature that is above the average. A negative anomaly indicates a month in which the temperature is less than the average. One key question is how we measure the temperature.

Although there are many different sources of data, we have chosen to provide you with one, the National Climatic Data Center (NCDC), which is affiliated with the National Oceanic and Atmospheric Administration (NOAA). (Other sources tend to agree with the NCDC's data.) C03-01a stores the monthly temperature anomalies from 1880 to 2013.

The best measures of CO_2 levels in the atmosphere come from the Mauna Loa Observatory in Hawaii, which started measuring this variable in December 1958 and continues to do so. However, attempts to estimate CO_2 levels prior to 1958 are as controversial as the methods used to estimate temperatures. These techniques include taking ice-core samples from the Arctic and measuring the amount of CO_2 trapped in the ice from which estimates of atmospheric CO_2 are produced. To avoid this controversy, we will use the Mauna Loa Observatory numbers only. These data are stored in file C03-01b. (Note that some of the original data are missing and were replaced by interpolated values.)

a. Use whichever techniques you wish to determine whether there is global warming.

b. Use a graphical technique to determine whether there is a relationship between temperature anomalies and CO_2 levels.

CASE 3.2

DATA C03-02

Economic Freedom and Prosperity

Adam Smith published *The Wealth of Nations* in 1776. In it, he argued that when institutions protect the liberty of individuals, greater prosperity results for all. Since 1995, the *Wall Street Journal* and the Heritage Foundation, a think tank in Washington, DC, have produced the Index of Economic Freedom for all countries in the world. The index is based on a subjective score for ten freedoms. These are business freedom, trade freedom, fiscal freedom, government size, monetary freedom, investment freedom, financial freedom, property rights, freedom from corruption and labour freedom. We downloaded the scores for 2013. From the *CIA Factbook,* we determined the per capita gross domestic product (GDP), measured in terms of purchasing power parity (PPP), which makes it possible to compare the GDP for all countries. The per capita GDP PPP figures for 2012 were used. Both sets of data are stored in C03-02. Use a graphical technique to see how freedom and prosperity are related.

NUMERICAL DESCRIPTIVE TECHNIQUES

4-1 MEASURES OF CENTRAL LOCATION

4-2 MEASURES OF VARIABILITY

4-3 MEASURES OF RELATIVE STANDING AND BOX PLOTS

4-4 MEASURES OF LINEAR RELATIONSHIP

4-5 APPLICATIONS IN FINANCE: MARKET MODEL

4-6 COMPARING GRAPHICAL AND NUMERICAL TECHNIQUES

4-7 GENERAL GUIDELINES FOR EXPLORING DATA

THE COST OF ONE MORE WIN IN THE WORLD'S BIGGEST FOOTBALL LEAGUES

DATA Xm04-00

In the era of free agency, professional sports teams must compete for the services of the best players. It is generally believed that only teams whose salaries place them in the top quarter have a chance of winning the championship. The world's most in-demand footballers, such as Real Madrid's Christiano Ronaldo and Barcelona's Lionel Messi, for example, can command huge salaries to retain them at the club. Efforts have been made to provide balance by establishing salary caps or some form of equalization. To examine the problem, we gathered data from the 2011 football season. For each team in major league football, we recorded the number of points and the average annual salary per player.

To make informed decisions, we need to know how the number of points and the average annual salary are related. After the statistical technique is presented, we return to this problem and solve it. **See Section 4-4e for the answer.**

INTRODUCTION

In Chapters 2 and 3, we presented several graphical techniques that describe data. In this chapter we introduce numerical descriptive techniques that allow the statistics practitioner to be more precise in describing various characteristics of a sample or population. These techniques are critical to the development of statistical inference.

As we pointed out in Chapter 2, arithmetic calculations can be applied to interval data only. Consequently, most of the techniques introduced here may be used only to numerically describe interval data. However, some of the techniques can be used for ordinal data, and one of the techniques can be employed for nominal data.

When we introduced the histogram, we commented that there are several bits of information that we look for. The first is the location of the centre of the data. In Section 4-1 we will present **measures of central location**. Another important characteristic that we seek from a histogram is the spread of the data. The spread will be measured more precisely by measures of variability, which we present in Section 4-2. Section 4-3 introduces measures of relative standing and another graphical technique, the box plot.

In Section 3-3, we introduced the scatter diagram, which is a graphical method that we use to analyse the relationship between two interval variables. The numerical counterparts to the scatter diagram are called *measures of linear relationship,* and they are presented in Section 4-4.

In Section 4-6, we compare the information provided by graphical and numerical techniques. Finally, we complete this chapter by providing guidelines on how to explore data and retrieve information.

SAMPLE STATISTIC OR POPULATION PARAMETER

Recall the terms introduced in Chapter 1: population, sample, parameter and statistic. A parameter is a descriptive measurement about a population, and a statistic is a descriptive measurement about a sample. In this chapter, we introduce a dozen descriptive measurements. For each one, we describe how to calculate both the population parameter and the sample statistic. However, in most realistic applications, populations are very large – in fact, virtually infinite. The formulas describing the calculation of parameters are not practical and are seldom used. They are provided here primarily to teach the concept and the notation. In Chapter 7 we introduce probability distributions, which describe populations. At that time we show how parameters are calculated from probability distributions. In general, small data sets of the type we feature in this book are samples.

4-1 MEASURES OF CENTRAL LOCATION

4-1a Arithmetic Mean

There are three different measures that we use to describe the centre of a set of data. The first is the best known, the *arithmetic mean,* which we will refer to simply as the **mean**. Students may be more familiar with its other name, the *average.* The mean is computed by summing the observations and dividing by the number of observations. We label the observations in a sample $x_1, x_2, \ldots, x_n$, where x_1 is the first observation, x_2 is the second and so on until x_n, where n is the sample size. The Greek letter Σ (sigma) represents **summation**. As a result, the sample mean is denoted $\bar{x}$. In a population, the number of observations is labelled N and the population mean is denoted by μ (Greek letter *mu*).

Mean

Population mean: $\mu = \frac{\sum_{i=1}^{N} x_i}{N}$

Sample mean: $\bar{x} = \frac{\sum_{i=1}^{n} x_i}{n}$

EXAMPLE 4.1

Mean Time Spent on the Internet

A sample of ten adults was asked to report the number of hours they spent on the Internet the previous month. The results are listed here. Manually calculate the sample mean.

0 7 12 5 33 14 8 0 9 22

SOLUTION:

Using our notation, we have $x_1 = 0$, $x_2 = 7, \ldots, x_{10} = 22$ and $n = 10$. The sample mean is

$$\bar{x} = \frac{\sum_{i=1}^{n} x_i}{n} = \frac{0 + 7 + 12 + 5 + 33 + 14 + 8 + 0 + 9 + 22}{10} = \frac{110}{10} = 11.0$$

EXAMPLE 4.2

DATA Xm03-01

Mean Long-Distance Telephone Bill

Refer to Example 3.1. Find the mean long-distance telephone bill.

SOLUTION:

To calculate the mean, we add the observations and divide the sum by the size of the sample. Thus,

$$\bar{x} = \frac{\sum_{i=1}^{n} x_i}{n} = \frac{42.19 + 38.45 + \cdots + 45.77}{200} = \frac{8717.52}{200} = 43.59$$

Using the Computer

There are several ways to command Excel and Minitab to compute the mean. If we simply want to compute the mean and no other statistics, we can proceed as follows.

EXCEL

Instructions

Type or import the data into one or more columns. (**Open** Xm03-01.) Type into any empty cell

= **AVERAGE** ([Input Range])

For Example 4.2, we would type into any cell

= **AVERAGE** (A1:A201)

The active cell would store the mean as 43.5876.

MINITAB

Instructions

1. Type or import the data into one column. (Open Xm03-01.)
2. Click **Calc** and **Column Statistics** .... Specify **Mean** in the **Statistic** box. Type or use the **Select** button to specify the **Input variable** and click OK. The sample mean is outputted in the session window as 43.5876.

4-1b Median

The second most popular measure of central location is the *median.*

Median

The **median** is calculated by placing all the observations in order (ascending or descending). The observation that falls in the middle is the median. The sample and population medians are computed in the same way.

When there is an even number of observations, the median is determined by averaging the two observations in the middle.

EXAMPLE 4.3

Median Time Spent on Internet

Find the median for the data in Example 4.1.

SOLUTION:

When placed in ascending order, the data appear as follows:

0 0 5 7 8 9 12 14 22 33

The median is the average of the fifth and sixth observations (the middle two), which are 8 and 9, respectively. Thus, the median is 8.5.

EXAMPLE 4.4

DATA Xm03-01

Median Long-Distance Telephone Bill

Find the median of the 200 observations in Example 3.1.

SOLUTION:

All the observations were placed in order. We observed that the 100th and 101st observations are 26.84 and 26.97, respectively. Thus, the median is the average of these two numbers:

$$\text{Median} = \frac{26.84 + 26.97}{2} = 26.905$$

EXCEL

Instructions

To calculate the median, substitute **MEDIAN** in place of **AVERAGE** in the instructions for the mean. The median is reported as 26.905.

MINITAB

Instructions

Follow the instructions to compute the mean except click **Median** instead of **Mean.** The median is outputted as 26.905 in the session window.

INTERPRET

Half the observations are below 26.905, and half the observations are above 26.905.

4-1c Mode

The third and last measure of central location that we present here is the *mode.*

Mode

The **mode** is defined as the observation (or observations) that occurs with the greatest frequency. Both the statistic and parameter are computed in the same way.

For populations and large samples, it is preferable to report the modal class, which we defined in Chapter 2.

There are several problems with using the mode as a measure of central location. First, in a small sample it may not be a very good measure. Second, it may not be unique.

EXAMPLE 4.5

Mode Time Spent on Internet

Find the mode for the data in Example 4.1.

SOLUTION:

All observations except 0 occur once. There are two 0s. Thus, the mode is 0. As you can see, this is a poor measure of central location. It is nowhere near the centre of the data. Compare this with the mean 11.0 and median 8.5 and you can appreciate that in this example the mean and median are superior measures.

EXAMPLE 4.6

DATA Xm03-01

Mode of Long-Distance Bill

Determine the mode for Example 3.1.

SOLUTION:

An examination of the 200 observations reveals that, except for 0, it appears that each number is unique. However, there are eight zeroes, which indicate that the mode is 0.

EXCEL

Instructions

To compute the mode, substitute **MODE** in place of **AVERAGE** in the previous instructions. Note that, if there is more than one mode, Excel prints only the smallest one, without indicating whether there are other modes. In this example, Excel reports that the mode is 0.

MINITAB

Follow the instructions to compute the mean except click **Mode** instead of **Mean**. The mode is outputted as 0 in the session window. (See Example 2.1.)

Excel and Minitab: Printing All the Measures of Central Location plus Other Statistics Both Excel and Minitab can produce the measures of central location plus a variety of others that we will introduce in later sections.

EXCEL

Excel Output for Examples 4.2, 4.4 and 4.6

	A	B
1	*Bills*	
2		
3	Mean	43.59
4	Standard Error	2.76
5	Median	26.91
6	Mode	0
7	Standard Deviation	38.97
8	Sample Variance	1518.64
9	Kurtosis	-1.29
10	Skewness	0.54
11	Range	119.63
12	Minimum	0
13	Maximum	119.63
14	Sum	8717.5
15	Count	200

Excel reports the mean, median and mode as the same values we obtained previously. Most of the other statistics will be discussed later.

Instructions

1. Type or import the data into one column. (**Open** Xm03-01.)
2. Click **Data, Data Analysis** and **Descriptive Statistics.**
3. Specify the **Input Range** (A1:A201) and click **Summary Statistics.**

MINITAB

Minitab Output for Examples 4.2, 4.4, and 4.6

Descriptive Statistics: Bills

Variable	Mean	Median	Mode	N for Mode
Bills	43.59	26.91	0	8

Instructions

1. Type or import the data into one column. (**Open** Xm03-01.)
2. Click **Stat, Basic Statistics** and **Display Descriptive Statistics**. . .
3. Type or use **Select** to identify the name of the variable or column (Bills). Click **Statistics** . . . to add or delete particular statistics.

4-1d Mean, Median, Mode: Which Is Best?

With three measures from which to choose, which one should we use? There are several factors to consider when making our choice of measure of central location. The mean is generally our first selection. However, there are several circumstances when the median is better. The mode is seldom the best measure of central location. One advantage the median holds is that it is not as sensitive to extreme values as is the mean. To illustrate, consider the data in Example 4.1. The mean was 11.0 and the median was 8.5. Now suppose that the respondent who reported 33 hours actually reported 133 hours (obviously an Internet addict). The mean becomes

$$\bar{x} = \frac{\sum_{i=1}^{n} x_i}{n} = \frac{0 + 7 + 12 + 5 + 133 + 14 + 8 + 0 + 22}{10} = \frac{210}{10} = 21.0$$

This value is exceeded by only two of the ten observations in the sample, making this statistic a poor measure of *central* location. The median stays the same. When there is a relatively small number of extreme observations (either very small or very large, but not both), the median usually produces a better measure of the centre of the data.

To see another advantage of the median over the mean, suppose you and your classmates have written a statistics test and the instructor is returning the graded tests. What piece of information is most important to you? The answer, of course, is *your* mark. What is the next important bit of information? The answer is how well you performed relative to the class. Most students ask their instructor for the class mean. This is the wrong statistic to request. You want the *median* because it divides the class into two halves. This information allows you to identify which half of the class your mark falls into. The median provides this information; the mean does not. Nevertheless, the mean can also be useful in this scenario. If there are several sections of the course, the section means can be compared to determine whose class performed best (or worst).

4-1e Measures of Central Location for Ordinal and Nominal Data

When the data are interval, we can use any of the three measures of central location. However, for ordinal and nominal data, the calculation of the mean is not valid. Because the calculation of the median begins by placing the data in order, this statistic is appropriate for ordinal data. The mode, which is determined by counting the frequency of each observation, is appropriate for nominal data. However, nominal data do not have a 'centre', so we cannot interpret the mode of nominal data in that way. It is generally pointless to compute the mode of nominal data.

APPLICATIONS IN FINANCE

Geometric Mean

The arithmetic mean is the single most popular and useful measure of central location. We noted certain situations where the median is a better measure of central location. However, there is another circumstance where neither the mean nor the median is the best measure. When the variable is a growth rate or rate of change, such as the value of an investment over periods of time, we need another measure. This will become apparent from the following illustration. Suppose you make a 2-year investment of £1000, and it grows by 100% to £2000 during the first year. During the second year, however, the investment suffers a 50% loss, from £2000 back to £1000. The rates of return for years 1 and 2 are $R_1 = 100\%$ and $R_2 = -50\%$, respectively. The arithmetic mean (and the median) is computed as

$$\bar{R} = \frac{R_1 + R_2}{2} = \frac{100 + (-50)}{2} = 25\%$$

But this figure is misleading. Because there was no change in the value of the investment from the beginning to the end of the 2-year period, the 'average' compounded rate of return is 0%. As you will see, this is the value of the *geometric mean.*

Let R_i denote the rate of return (in decimal form) in period $i (i = 1, 2, \ldots, n)$. The **geometric mean** R_g of the returns $R_1, R_2, \ldots, R_n$ is defined such that

$$(1 + R_g)^n = (1 + R_1)(1 + R_2) \ldots (1 + R_n)$$

Solving for R, we produce the following formula:

$$R_g = \sqrt[n]{(1 + R_1)(1 + R_2) \cdots (1 + R_n)} - 1$$

The geometric mean of our investment illustration is

$$R_g = \sqrt[n]{0(1 + R_1)(1 + R_2) \cdots (1 + R_n)} - 1$$
$$= \sqrt[2]{(1 + 1)(1 + [-.50])} - 1 = 1 - 1 = 0$$

The geometric mean is therefore 0%. This is the single 'average' return that allows us to compute the value of the investment at the end of the investment period from the beginning value. Thus, using the formula for compound interest with the rate = 0%, we find:

Value at the end of the investment period $= 1000(1 + R_g)^2$
$= 1000(1 + 0)^2 = 1000$

The geometric mean is used whenever we wish to find the 'average' growth rate, or rate of change, in a variable *over time.* However, the arithmetic mean of n returns (or growth rates) is the appropriate mean to calculate if you wish to estimate the mean rate of return (or growth rate) for any *single* period in the future; that is, in the illustration above if we wanted to estimate the rate of return in year 3, we would use the arithmetic mean of the two annual rates of return, which we found to be 25%.

EXCEL

Instructions

1. Type or import the values of $1 + R_i$ into a column.
2. Follow the instructions to produce the mean (Section 4-1a) except substitute **GEOMEAN** in place of **AVERAGE**.
3. To determine the geometric mean, subtract 1 from the number produced.

MINITAB

Minitab does not compute the geometric mean.

Here is a summary of the numerical techniques introduced in this section and when to use them.

Factors That Identify When to Compute the Mean

1. **Objective:** Describe a single set of data.
2. **Type of data:** Interval.
3. **Descriptive measurement:** Central location.

Factors That Identify When to Compute the Median

1. **Objective:** Describe a single set of data.
2. **Type of data:** Ordinal or interval (with extreme observations).
3. **Descriptive measurement:** Central location.

Factors That Identify When to Compute the Mode

1. **Objective:** Describe a single set of data.
2. **Type of data:** Nominal, ordinal, interval.

Factors That Identify When to Compute the Geometric Mean

1. **Objective:** Describe a single set of data.
2. **Type of data:** Interval; growth rates.

EXERCISES

4.1 Xr04-01 The number of sick days due to colds and flu last year in a company in Manchester was recorded by a sample of 15 adults. The data are:

5 7 0 3 15 6 5 9
3 8 10 5 2 0 12

Compute the mean, median and mode.

4.2 Xr04-02 A random sample of 12 joggers was asked to keep track and report the number of kilometres they ran last week. The responses are:

8.9 11.6 2.6 35.4 14.0 4.5 8.5 5.5
20.1 29.9 13.4 10.6

a. Compute the three statistics that measure central location.

b. Briefly describe what each statistic tells you.

4.3 Xr04-03 The midterm test for a statistics course has a time limit of 1 hour. However, like most statistics exams this one was quite easy. To assess how easy, the professor recorded the amount of time taken by a sample of nine students to hand in their test papers. The times (rounded to the nearest minute) are:

33 29 45 60 42 19 52 38 36

a. Compute the mean, median and mode.

b. What have you learned from the three statistics calculated in part (a)?

4.4 What is the geometric mean of the following rates of return?

0.50 0.30 −0.50 −0.25

4.5 The following returns were realised on an investment over a 5-year period.

Year	1	2	3	4	5
Rate of return	10	0.22	0.06	−0.05	0.20

a. Compute the mean and median of the returns.

b. Compute the geometric mean.

c. Which one of the three statistics computed in parts (a) and (b) best describes the return over the 5-year period? Explain.

4.6 An investment of £1000 you made 4 years ago was worth £1200 after the first year, £1200 after the second year, £1500 after the third year and £2000 today.

a. Compute the annual rates of return.

b. Compute the mean and median of the rates of return.

c. Compute the geometric mean.

d. Discuss whether the mean, median or geometric mean is the best measure of the performance of the investment.

4.7 Suppose that you bought a stock 6 years ago at €12. The stock's price at the end of each year is shown here.

Year	1	2	3	4	5	6
Price	10	14	15	22	30	25

a. Compute the rate of return for each year.

b. Compute the mean and median of the rates of return.

c. Compute the geometric mean of the rates of return.

d. Explain why the best statistic to use to describe what happened to the price of the stock over the 6-year period is the geometric mean.

The following exercises require the use of a computer and software.

4.8 Xr04-08 The starting salaries (in euros) of a sample of 125 recent MBA graduates are recorded.

a. Determine the mean and median of these data.

b. What do these two statistics tell you about the starting salaries of MBA graduates?

4.9 Xr04-09 To determine whether changing the colour of its invoices would improve the speed of payment, a company selected 200 customers at random and sent their invoices on blue paper. The number of days until the bills were paid was recorded. Calculate the mean and median of these data. Report what you have discovered.

4-2 MEASURES OF VARIABILITY

The statistics introduced in Section 4-1 serve to provide information about the central location of the data. However, as we have already discussed in Chapter 2, there are other characteristics of data that are of interest to practitioners of statistics. One such characteristic is the spread or variability of the data. In this section, we introduce four **measures of variability.** We begin with the simplest.

4-2a Range

Range

Range = Largest observation − Smallest observation

The advantage of the range is its simplicity. The disadvantage is also its simplicity. Because the range is calculated from only two observations, it tells us nothing about the other observations. Consider the following two sets of data.

Set1:	4	4	4	4	4	50
Set 2:	4	8	15	24	39	50

The range of both sets is 46. The two sets of data are completely different, yet their ranges are the same. To measure variability, we need other statistics that incorporate all the data and not just two observations.

4-2b Variance

The variance and its related measure, the **standard deviation**, are arguably the most important statistics. They are used to measure variability, but, as you will discover, they play a vital role in almost all statistical inference procedures.

Variance

Population variance: $\sigma^2 = \dfrac{\sum_{i=1}^{N}(x_i - \mu)^2}{N}$

Sample variance: $s^2 = \dfrac{\sum_{i=1}^{n}(x_i - \bar{x})^2}{n-1}$

The population variance is represented by σ^2 (Greek letter *sigma* squared).

Examine the formula for the sample variance s^2. It may appear to be illogical that in calculating s^2 we divide by $n - 1$ rather than by n.* However, we do so for the following reason. Population parameters in practical settings are seldom known. One objective of statistical inference is to estimate the parameter from the statistic. For example, we estimate the population mean μ from the sample mean $\bar{x}$. Although it is not obviously logical, the statistic created by dividing $\sum(x_i - \bar{x})^2$ by $n - 1$ is a better estimator than the one created by dividing by n. We will discuss this issue in greater detail in Section 10-1.

To compute the sample variance s^2, we begin by calculating the sample mean $\bar{x}$. Next we compute the difference (also called the **deviation**) between each observation and the mean. We square the deviations and sum. Finally, we divide the sum of squared deviations by $n - 1$.

We will illustrate with a simple example. Suppose that we have the following observations of the numbers of hours five students spent studying statistics last week:

8 4 9 11 3

* Technically, the variance of the sample is calculated by dividing the sum of squared deviations by n. The statistic computed by dividing the sum of squared deviations by $n - 1$ is called the *sample variance corrected for the mean*. Because this statistic is used extensively, we will shorten its name to *sample variance*.

The mean is

$$\bar{x} = \frac{8 + 4 + 9 + 11 + 3}{5} = \frac{35}{5} = 7$$

For each observation, we determine its deviation from the mean. The deviation is squared, and the sum of squares is determined as shown in Table 4.1.

TABLE 4.1 Calculation of Sample Variance

x_i	$(x_i - \bar{x})$	$(x_i - \bar{x})^2$
8	$(8 - 7) = 1$	$(1)^2 = 1$
4	$(4 - 7) = -3$	$(-3)^2 = 9$
9	$(9 - 7) = 2$	$(2)^2 = 4$
11	$(11 - 7) = 4$	$(4)^2 = 16$
3	$(3 - 7) = -4$	$(-4)^2 = 16$
	$\sum_{i=1}^{5}(x_i - \bar{x}) = 0$	$\sum_{i=1}^{5}(x_i - \bar{x})^2 = 46$

The sample variance is

$$s^2 = \frac{\sum_{i=1}^{n}(x_i - \bar{x})^2}{n - 1} = \frac{46}{5 - 1} = 11.5$$

The calculation of this statistic raises several questions. Why do we square the deviations before averaging? If you examine the deviations, you will see that some of the deviations are positive and some are negative. When you add them together, the sum is 0. This will always be the case because the sum of the positive deviations will always equal the sum of the negative deviations. Consequently, we square the deviations to avoid the 'cancelling effect'.

Is it possible to avoid the cancelling effect without squaring? We could average the *absolute* value of the deviations. In fact, such a statistic has already been invented. It is called the **mean absolute deviation** or MAD. However, this statistic has limited utility and is seldom calculated.

What is the unit of measurement of the variance? Because we squared the deviations, we also squared the units. In this illustration the units were hours (of study). Thus, the sample variance is 11.5 hours².

EXAMPLE 4.7

Summer Jobs

The following are the number of summer jobs a sample of six students applied for. Find the mean and variance of these data.

17 15 23 7 9 13

SOLUTION:

The mean of the six observations is

$$\bar{x} = \frac{17 + 15 + 23 + 7 + 9 + 13}{6} = \frac{84}{6} = 14 \text{ jobs}$$

The sample variance is

$$s^2 = \frac{\sum_{i=1}^{n}(x_i - \bar{x})^2}{n - 1}$$

$$= \frac{(17 - 14)^2 + (15 - 14)^2 + (23 - 14)^2 + (7 - 14)^2 + (9 - 14)^2 + (13 - 14)^2}{6 - 1}$$

$$= \frac{9 + 1 + 81 + 49 + 25 + 1}{5} = \frac{166}{5} = 33.2 \text{ jobs}^2$$

(Optional) Shortcut Method for Variance The calculations for larger data sets are quite time-consuming. The following shortcut for the sample variance may help lighten the load.

Shortcut for Sample Variance

$$s^2 = \frac{1}{n-1}\left[\sum_{i=1}^{n} x_i^2 - \frac{\left(\sum_{i=1}^{n} x_i\right)^2}{n}\right]$$

To illustrate, we will do Example 4.7 again.

$$\sum_{i=1}^{n} x_i^2 = 17^2 + 15^2 + 23^2 + 7^2 + 9^2 + 13^2 = 1342$$

$$\sum_{i=1}^{n} x_i = 17 + 15 + 23 + 7 + 9 + 13 = 84$$

$$\left(\sum_{i=1}^{n} x_i\right)^2 = 84^2 = 7056$$

$$s^2 = \frac{1}{n-1}\left[\sum_{i=1}^{n} x_i^2 - \frac{\left(\sum_{i=1}^{n} x_i\right)^2}{n}\right] = \frac{1}{6-1}\left[1342 - \frac{7056}{6}\right] = 33.2 \text{ jobs}^2$$

EXCEL

Instructions

Follow the instructions to compute the mean (Example 4.2) except type VAR instead of AVERAGE.

MINITAB

Instructions

1. Type or import data into one column.
2. Click **Stat, Basic Statistics, Display Descriptive Statistics** ... and select the variable.
3. Click **Statistics** and **Variance.**

Notice that we produced the same exact answer.

4-2c Interpreting the Variance

We calculated the variance in Example 4.7 to be 33.2 jobs^2. What does this statistic tell us? Unfortunately, the variance provides us with only a rough idea about the amount of variation in the data. However, this statistic is useful when comparing two or more sets of data of the same type of variable. If the variance of one data set is larger than that of a second data set, we interpret that to mean that the observations in the first set display more variation than the observations in the second set.

The problem of interpretation is caused by the way the variance is calculated. Because we squared the deviations from the mean, the unit attached to the variance is the square of the unit attached to the original observations. In other words, in Example 4.7 the unit of the data is jobs; the unit of the variance is jobs squared. This contributes to the problem of interpretation. We resolve this difficulty by calculating another related measure of variability.

4-2d Standard Deviation

Standard Deviation

Population standard deviation: $\sigma = \sqrt{\sigma^2}$

Sample standard deviation: $s = \sqrt{s^2}$

The standard deviation is simply the positive square root of the variance. Thus, in Example 4.7 the sample standard deviation is

$$s = \sqrt{s^2} = \sqrt{33.2} = 5.76 \text{ jobs}$$

Notice that the unit associated with the standard deviation is the unit of the original data set.

EXAMPLE 4.8

DATA Xm04-08

Comparing the Consistency of Two Types of Golf Clubs

Consistency is the hallmark of a good golfer. Golf equipment manufacturers are constantly seeking ways to improve their products. Suppose that a recent innovation is designed to improve the consistency of its users. As a test, a golfer was asked to hit 150 shots using a 7 iron, 75 of which were hit with his current club and 75 with the new innovative 7 iron. The distances were measured and recorded. Which 7 iron is more consistent?

SOLUTION:

To gauge the consistency, we must determine the standard deviations. (We could also compute the variances, but as we just pointed out, the standard deviation is easier to interpret.) We can get Excel and Minitab to print the sample standard deviations. Alternatively, we can calculate all the descriptive statistics, a course of action we recommend because we often need several statistics. The printouts for both 7 irons are shown here.

EXCEL

	A	B	C	D	E
1	*Current*			*Innovation*	
2					
3	Mean	150.55		Mean	150.15
4	Standard Error	0.67		Standard Error	0.36
5	Median	151		Median	150
6	Mode	150		Mode	149
7	Standard Deviation	5.79		Standard Deviation	3.09
8	Sample Variance	33.55		Sample Variance	9.56
9	Kurtosis	0.13		Kurtosis	-0.89
10	Skewness	-0.43		Skewness	0.18
11	Range	28		Range	12
12	Minimum	134		Minimum	144
13	Maximum	162		Maximum	156
14	Sum	11291		Sum	11261
15	Count	75		Count	75

MINITAB

Descriptive Statistics: Current, Innovation

Variable	N	N*	Mean	StDev	Variance	Minimum	Q1	Median	Q3
Current	75	0	150.55	5.79	33.55	134.00	148.00	151.00	155.00
Innovation	75	0	150.15	3.09	9.56	144.00	148.00	150.00	152.00

Variable	Maximum
Current	162.00
Innovation	156.00

INTERPRET

The standard deviation of the distances of the current 7 iron is 5.79 yards (1 yard = 0.9144 metres) whereas that of the innovative 7 iron is 3.09 yards. Based on this sample, the innovative club is more consistent. Because the mean distances are similar it would appear that the new club is indeed superior.

Interpreting the Standard Deviation Knowing the mean and standard deviation allows the statistics practitioner to extract useful bits of information. The information depends on the shape of the histogram. If the histogram is bell shaped, we can use the **Empirical Rule**.

Empirical Rule

1. Approximately 68% of all observations fall within one standard deviation of the mean.
2. Approximately 95% of all observations fall within two standard deviations of the mean.
3. Approximately 99.7% of all observations fall within three standard deviations of the mean.

EXAMPLE 4.9

Using the Empirical Rule to Interpret Standard Deviation

After an analysis of the returns on an investment, a statistics practitioner discovered that the histogram is bell shaped and that the mean and standard deviation are 10% and 8%, respectively. What can you say about the way the returns are distributed?

SOLUTION:

Because the histogram is bell shaped, we can apply the Empirical Rule:

1. Approximately 68% of the returns lie between 2% (the mean minus one standard deviation = 10 − 8) and 18% (the mean plus one standard deviation = 10 + 8).
2. Approximately 95% of the returns lie between −6% [the mean minus two standard deviations = 10 − 2(8)] and 26% [the mean plus two standard deviations = 10 + 2(8)].
3. Approximately 99.7% of the returns lie between −14% [the mean minus three standard deviations = 10 − 3(8)] and 34% [the mean plus three standard deviations = 10 + 3(8)].

A more general interpretation of the standard deviation is derived from ***Chebyshev's*** (*transliterated as Chebysheff's Theorem*), which applies to all shapes of histograms.

Chebysheff's Theorem

The proportion of observations in any sample or population that lie within k standard deviations of the mean is at least

$$1 - \frac{1}{k^2} \text{ for } k > 1$$

When $k = 2$, **Chebysheff's Theorem** states that at least three-quarters (75%) of all observations lie within two standard deviations of the mean. With $k = 3$, Chebysheff's Theorem states that at least eight-ninths (88.9%) of all observations lie within three standard deviations of the mean.

Note that the Empirical Rule provides approximate proportions, whereas Chebysheff's Theorem provides lower bounds on the proportions contained in the intervals.

EXAMPLE 4.10

Using Chebysheff's Theorem to Interpret Standard Deviation

The annual salaries of the employees of a chain of computer stores produced a positively skewed histogram. The mean and standard deviation are £28 000 and £3000, respectively. What can you say about the salaries at this chain?

SOLUTION:

Because the histogram is not bell shaped, we cannot use the Empirical Rule. We must employ Chebysheff's Theorem instead.

The intervals created by adding and subtracting two and three standard deviations to and from the mean are as follows:

1. At least 75% of the salaries lie between £22 000 [the mean minus two standard deviations = 28 000 – 2(3000)] and £34 000 [the mean plus two standard deviations = 28 000 + 2(3000)].
2. At least 88.9% of the salaries lie between £19 000 [the mean minus three standard deviations = 28 000 – 3(3000)] and £37 000 [the mean plus three standard deviations =28 000 + 3(3000)].

4-2e Coefficient of Variation

Is a standard deviation of 10 a large number indicating great variability or a small number indicating little variability? The answer depends somewhat on the magnitude of the observations in the data set. If the observations are in the millions, then a standard deviation of 10 will probably be considered a small number. On the other hand, if the observations are less than 50, then the standard deviation of 10 would be seen as a large number. This logic lies behind yet another measure of variability, the *coefficient of variation*.

Coefficient of Variation

The **coefficient of variation** of a set of observations is the standard deviation of the observations divided by their mean:

$$\text{Population coefficient of variation: } CV = \frac{\sigma}{\mu}$$

$$\text{Sample coefficient of variation: } cv = \frac{s}{\bar{x}}$$

4-2f Measures of Variability for Ordinal and Nominal Data

The measures of variability introduced in this section can be used only for interval data. The next section will feature a measure that can be used to describe the variability of ordinal data. There are no measures of variability for nominal data.

4-2g Approximating the Mean and Variance from Grouped Data

The statistical methods presented in this chapter are used to compute descriptive statistics from data. However, in some circumstances, the statistics practitioner does not have the raw data but instead has a frequency distribution. This is often the case when data are supplied by government organizations. In Appendix Approximating Means and Variances for Grouped Data on the website we provide the formulas used to approximate the sample mean and variance (see the 'Digital Resources' page for information on access).

We complete this section by reviewing the factors that identify the use of measures of variability.

Factors That Identify When to Compute the Range, Variance, Standard Deviation and Coefficient of Variation

1. **Objective:** Describe a single set of data.
2. **Type of data:** Interval.
3. **Descriptive measurement:** Variability.

EXERCISES

4.10 Xr04-10 Calculate the variance of the following data.

9 3 7 4 1 7 5

4.11 Xr04-11 Determine the variance and standard deviation of the following sample.

12 6 22 31 23 13 15 17 21

4.12 Xr04-12 Find the variance and standard deviation of the following sample.

0 −5 −3 6 4 −4 1 −5 0 3

4.13 Xr04-13 Examine the three samples listed here. Without performing any calculations, indicate which sample has the largest amount of variation and which sample has the smallest amount of variation. Explain how you produced your answer.

a. 17 29 12 16 11

b. 22 18 23 20 17

c. 24 37 6 39 29

4.14 Refer to Exercise 4.13. Calculate the variance for each part. Was your answer in Exercise 4.13 correct?

4.15 A friend calculates a variance and reports that it is −25.0. How do you know that he has made a serious calculation error?

4.16 Create a sample of five numbers whose mean is 6 and whose standard deviation is 0.

4.17 A set of data whose histogram is bell shaped yields a mean and standard deviation of 50 and 4, respectively. Approximately what proportion of observations

a. are between 46 and 54?

b. are between 42 and 58?

c. are between 38 and 62?

4.18 Refer to Exercise 4.17. Approximately what proportion of observations

a. are less than 46?

b. are less than 58?

c. are greater than 54?

4.19 A set of data whose histogram is extremely skewed yields a mean and standard deviation of 70 and 12, respectively. What is the minimum proportion of observations that

a. are between 46 and 94?

b. are between 34 and 106?

The following exercises require a computer and software.

4.20 Xr04-20 There has been much media coverage of the high cost of medicinal drugs in France. One concern is the large variation from pharmacy to pharmacy. To investigate, a consumer advocacy group took a random sample of 100 pharmacies around the country and recorded the price (in euro per 120 pills) of Prozac. Compute the range, variance and standard deviation of the prices. Discuss what these statistics tell you.

4.21 Xr04-21 Many traffic experts argue that the most important factor in accidents is not the average speed of cars but the amount of variation. Suppose that the speeds of a sample of 200 cars were taken over a stretch of highway that has seen numerous accidents. Compute the variance and standard deviation of the speeds (in km/h), and interpret the results.

4.22 Xr04-22 Variance is often used to measure quality in production-line products. Suppose that a sample of steel rods that are supposed to be exactly 100 cm long is taken. The length of each is determined and the results are recorded. Calculate the variance and the standard deviation. Briefly describe what these statistics tell you.

4.23 Xr04-23 Everyone is familiar with waiting lines or queues. For example, people wait in line at a supermarket to go through the checkout counter. There are two factors that determine how long the queue becomes. One is the speed of service. The other is the number of arrivals at the checkout counter.

The mean number of arrivals is an important number, but so is the standard deviation. Suppose that a consultant for the supermarket counts the number of arrivals per hour during a sample of 150 hours.

a. Compute the standard deviation of the number of arrivals.

b. Assuming that the histogram is bell shaped, interpret the standard deviation.

CENSUS EXERCISE

4.24 **Census2011** The Census in 2011 asked respondents to state their ages stored as AGE (column E in Excel and column 5 in Minitab).

a. Calculate the mean, variance and standard deviation.

b. Draw a histogram.

c. Use the Empirical Rule, if applicable, or Chebysheff's Theorem to interpret the mean and standard deviation.

4-3 MEASURES OF RELATIVE STANDING AND BOX PLOTS

Measures of relative standing are designed to provide information about the position of particular values relative to the entire data set. We have already presented one measure of relative standing, the median, which is also a measure of central location. Recall that the median divides the data set into halves, allowing the statistics practitioner to determine which half of the data set each observation lies in. The statistics we are about to introduce will give you much more detailed information.

Percentile

The *P*th **percentile** is the value for which *P* percent are less than that value and (100 – *P*)% are greater than that value.

The scores and the percentiles of the International Baccalaureate Diploma and the European Baccalaureate, as well as various other admissions tests, such as Hong Kong Certificate of Education Examination or Abitur, are reported to students taking them. Suppose for example, that your European Baccalaureate score is reported to be at the 60th percentile. This means that 60% of all the other marks are below yours and 40% are above it. You now know exactly where you stand relative to the population of SAT scores.

We have special names for the 25th, 50th and 75th percentiles. Because these three statistics divide the set of data into quarters, these measures of relative standing are also called **quartiles**. The *first* or *lower quartile* is labelled Q_1. It is equal to the 25th percentile. The *second quartile,* Q_2, is equal to the 50th percentile, which is also the median. The *third* or *upper quartile,* Q_3, is equal to the 75th percentile. Incidentally, many people confuse the terms *quartile* and *quarter.* A common error is to state that someone is in the lower *quartile* of a group when they actually mean that someone is in the lower *quarter* of a group.

Besides quartiles, we can also convert percentiles into quintiles and deciles. *Quintiles* divide the data into fifths, and *deciles* divide the data into tenths.

4-3a Locating Percentiles

The following formula allows us to approximate the location of any percentile.

Location of a Percentile

$$L_P = (n + 1)\frac{P}{100}$$

where L_P is the location of the *P*th percentile.

Example 4.11

Percentiles of Time Spent on Internet

Calculate the 25th, 50th and 75th percentiles (first, second and third quartiles) of the data in Example 4.1.

SOLUTION:

Placing the ten observations in ascending order we get

0 0 5 7 8 9 12 14 22 33

The location of the 25th percentile is

$$L_{25} = (10 + 1)\frac{25}{100} = (11)(0.25) = 2.75$$

The 25th percentile is three-quarters of the distance between the second (which is 0) and the third (which is 5) observations. Three-quarters of the distance is

$$(0.75)(5 - 0) = 3.75$$

Because the second observation is 0, the 25th percentile is 0 + 3.75 = 3.75.

To locate the 50th percentile, we substitute $P = 50$ into the formula and produce

$$L_{50} = (10 + 1)\frac{50}{100} = (11)(0.5) = 5.5$$

which means that the 50th percentile is halfway between the fifth and sixth observations. The fifth and sixth observations are 8 and 9, respectively. The 50th percentile is 8.5. This is the median calculated in Example 4.3. The 75th percentile's location is

$$L_{75} = (10 + 1)\frac{75}{100} = (11)(0.75) = 8.25$$

Thus, it is located one-quarter of the distance between the eighth and ninth observations, which are 14 and 22, respectively. One-quarter of the distance is

$$(0.25)(22 - 14) = 2$$

which means that the 75th percentile is

$$14 + 2 = 16$$

EXAMPLE 4.12

DATA Xm03-01

Quartiles of Long-Distance Telephone Bills

Determine the quartiles for Example 3.1.

SOLUTION:

EXCEL

	A	B
1	Bills	
2		
3	Mean	43.59
4	Standard Error	2.76
5	Median	26.91
6	Mode	0
7	Standard Deviation	38.97
8	Sample Variance	1518.64
9	Kurtosis	-1.29
10	Skewness	0.54
11	Range	119.63
12	Minimum	0
13	Maximum	119.63
14	Sum	8717.52
15	Count	200
16	Largest(50)	85
17	Smallest(50)	9.22

(continued)

Instructions

Follow the instructions for **Descriptive Statistics** (earlier in this chapter). In the dialog box, click **Kth Largest** and type in the integer closest to $n/4$. Repeat for **Kth Smallest,** typing in the integer closest to $n/4$.

Excel approximates the third and first quartiles in the following way. The **Largest(50)** is 85, which is the number such that 150 numbers are below it and 49 numbers are above it. The **Smallest(50)** is 9.22, which is the number such that 49 numbers are below it and 150 numbers are above it. The median is 26.91, a statistic we discussed in Example 4.4.

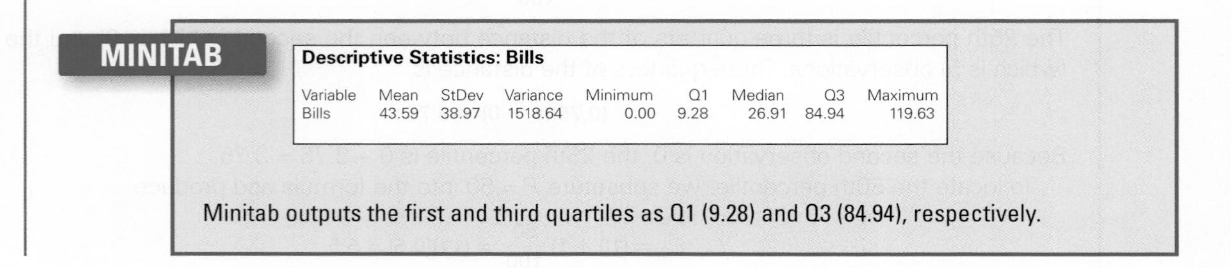
MINITAB

Descriptive Statistics: Bills

Variable	Mean	StDev	Variance	Minimum	Q1	Median	Q3	Maximum
Bills	43.59	38.97	1518.64	0.00	9.28	26.91	84.94	119.63

Minitab outputs the first and third quartiles as Q1 (9.28) and Q3 (84.94), respectively.

We can often get an idea of the shape of the histogram from the quartiles. For example, if the first and second quartiles are closer to each other than are the second and third quartiles, then the histogram is positively skewed. If the first and second quartiles are further apart than the second and third quartiles, then the histogram is negatively skewed. If the difference between the first and second quartiles is approximately equal to the difference between the second and third quartiles, then the histogram is approximately symmetric. The box plot described subsequently is particularly useful in this regard.

4-3b Interquartile Range

The quartiles can be used to create another measure of variability, the **interquartile range**, which is defined as follows.

Interquartile Range

$$\text{Interquartile range} = Q_3 - Q_1$$

The interquartile range measures the spread of the middle 50% of the observations. Large values of this statistic mean that the first and third quartiles are far apart, indicating a high level of variability.

EXAMPLE 4.13

DATA Xm03-01

Interquartile Range of Long-Distance Telephone Bills

Determine the interquartile range for Example 3.1.

SOLUTION:

Using Excel's approximations of the first and third quartiles, we find

$$\text{Interquartile range} = Q_3 - Q_1 = 85 - 9.22 = 75.78$$

4-3c Box Plots

Now that we have introduced quartiles we can present one more graphical technique, the **box plot**. This technique graphs five statistics: the minimum and maximum observations, and the first, second and third quartiles. It also depicts other features of a set of data. Figure 4.1 exhibits the box plot of the data in Example 4.1.

FIGURE 4.1

Box Plot for Example 4.1

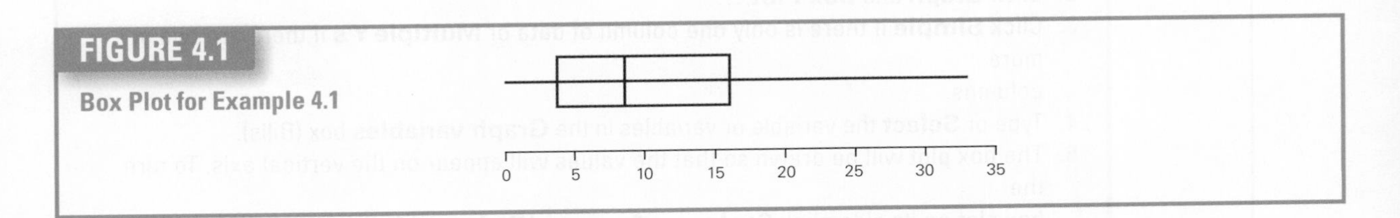

The three vertical lines of the box are the first, second and third quartiles. The lines extending to the left and right are called *whiskers*. Any points that lie outside the whiskers are called *outliers*. The whiskers extend outward to the smaller of 1.5 times the interquartile range or to the most extreme point that is not an outlier.

Outliers are unusually large or small observations. Because an outlier is considerably removed from the main body of the data set, its validity is suspect. Consequently, outliers should be checked to determine that they are not the result of an error in recording their values. Outliers can also represent unusual observations that should be investigated. For example, if a salesperson's performance is an outlier on the high end of the distribution, the company could profit by determining what sets that salesperson apart from the others.

Example 4.14

DATA Xm03-01

Box Plot of Long-Distance Telephone Bills

Draw the box plot for Example 3.1.

SOLUTION:

EXCEL

Instructions

1. Type or import the data into one column or two or more adjacent columns. (**Open** Xm03-01.)
2. Click **Add-Ins, Data Analysis Plus** and **Box Plot.**
3. Specify the **Input Range** (A1:A201).

A box plot will be created for each column of data that you have specified or highlighted.

Notice that the quartiles produced in the **Box Plot** are not exactly the same as those produced by **Descriptive Statistics.** The **Box Plot** command uses a slightly different method than the **Descriptive Methods** command.

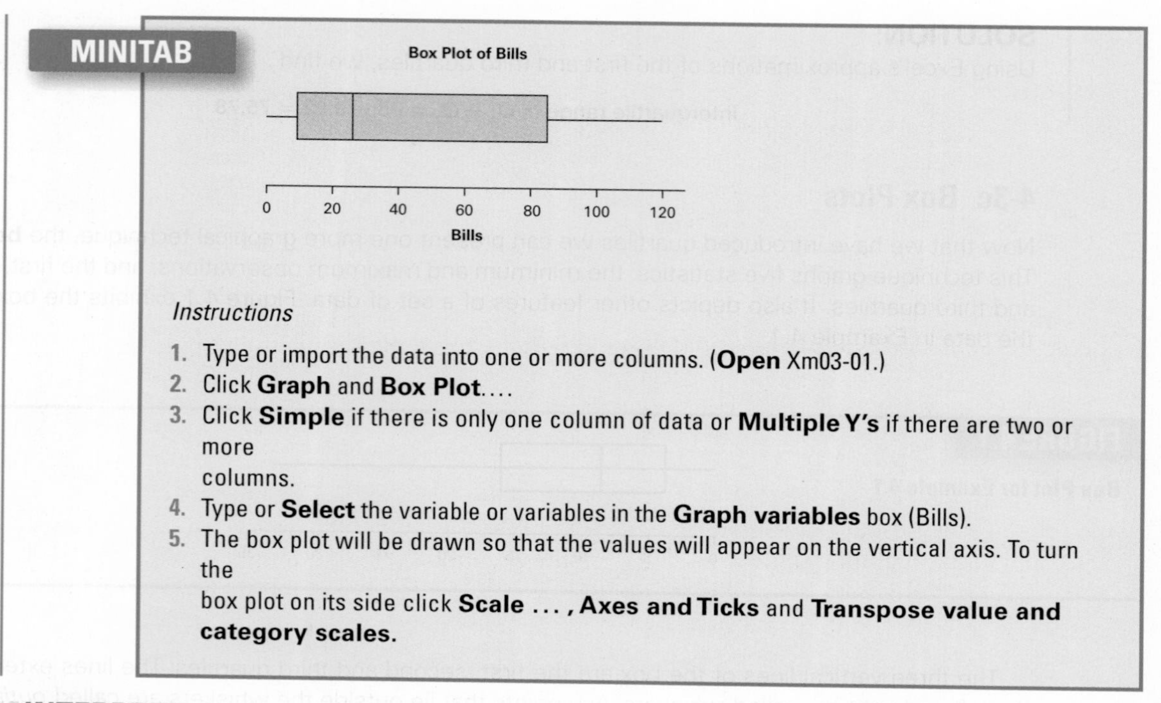

Instructions

1. Type or import the data into one or more columns. (**Open** Xm03-01.)
2. Click **Graph** and **Box Plot**....
3. Click **Simple** if there is only one column of data or **Multiple Y's** if there are two or more
columns.
4. Type or **Select** the variable or variables in the **Graph variables** box (Bills).
5. The box plot will be drawn so that the values will appear on the vertical axis. To turn the
box plot on its side click **Scale ... , Axes and Ticks** and **Transpose value and category scales.**

INTERPRET

The smallest value is 0 and the largest is 119.63. The first, second and third quartiles are 9.275, 26.905 and 84.9425, respectively. The interquartile range is 75.6675. One and one-half times the interquartile range is 1.5 × 75.6675 = 113.5013. Outliers are defined as any observations that are less than 9.275 − 113.5013 = −104.226 and any observations that are larger than 84.9425 + 113.5013 = 198.4438. The whisker to the left extends only to 0, which is the smallest observation that is not an outlier. The whisker to the right extends to 119.63, which is the largest observation that is not an outlier. There are no outliers.

The box plot is particularly useful when comparing two or more data sets.

EXAMPLE 4.15

DATA Xm04-15

Comparing Service Times of Fast-Food Restaurants on Driving Route

A large number of fast-food restaurants on a driving route offer drivers and their passengers the advantages of quick service. To measure how good the service is, an organization called QSR planned a study in which the amount of time taken by a sample of customers at each of five restaurants was recorded. Compare the five sets of data using a box plot and interpret the results.

SOLUTION:

We use the computer and our software to produce the box plots.

EXCEL

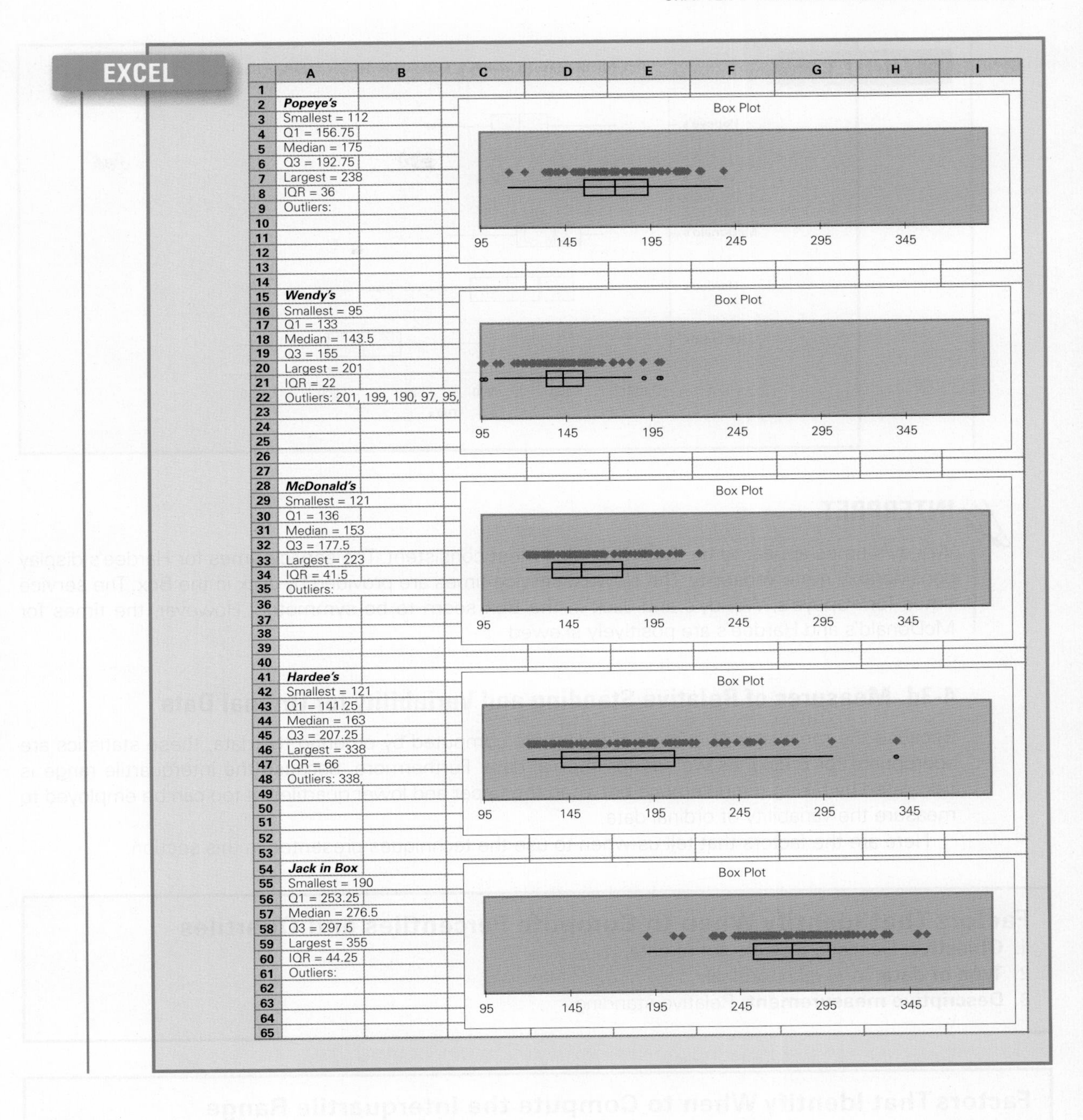
Popeye's
Smallest = 112
Q1 = 156.75
Median = 175
Q3 = 192.75
Largest = 238
IQR = 36
Outliers:
Wendy's
Smallest = 95
Q1 = 133
Median = 143.5
Q3 = 155
Largest = 201
IQR = 22
Outliers: 201, 199, 190, 97, 95,
McDonald's
Smallest = 121
Q1 = 136
Median = 153
Q3 = 177.5
Largest = 223
IQR = 41.5
Outliers:
Hardee's
Smallest = 121
Q1 = 141.25
Median = 163
Q3 = 207.25
Largest = 338
IQR = 66
Outliers: 338,
Jack in Box
Smallest = 190
Q1 = 253.25
Median = 276.5
Q3 = 297.5
Largest = 355
IQR = 44.25
Outliers:
Box Plot
95
145
195
245
295
345

MINITAB

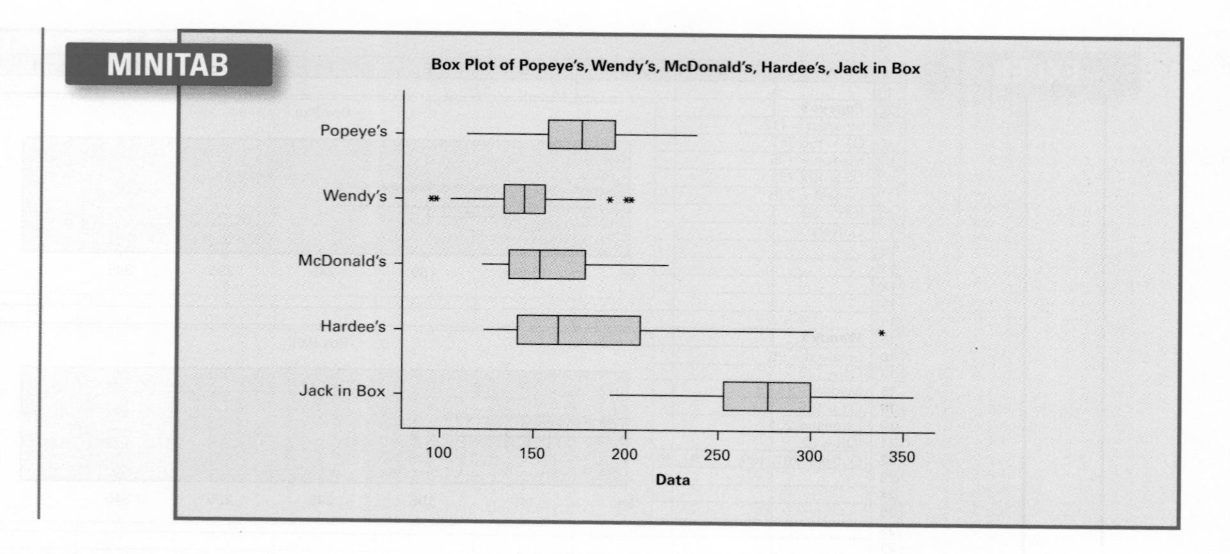

INTERPRET

Wendy's times appear to be the lowest and most consistent. The service times for Hardee's display considerably more variability. The slowest service times are provided by Jack in the Box. The service times for Popeye's, Wendy's and Jack in the Box seem to be symmetric. However, the times for McDonald's and Hardee's are positively skewed.

4-3d Measures of Relative Standing and Variability for Ordinal Data

Because the measures of relative standing are computed by ordering the data, these statistics are appropriate for ordinal as well as for interval data. Furthermore, because the interquartile range is calculated by taking the difference between the upper and lower quartiles, it too can be employed to measure the variability of ordinal data.

Here are the factors that tell us when to use the techniques presented in this section.

Factors That Identify When to Compute Percentiles and Quartiles

1. **Objective:** Describe a single set of data.
2. **Type of data:** Interval or ordinal.
3. **Descriptive measurement:** Relative standing.

Factors That Identify When to Compute the Interquartile Range

1. **Objective:** Describe a single set of data.
2. **Type of data:** Interval or ordinal.
3. **Descriptive measurement:** Variability.

EXERCISES

4.25 Calculate the first, second and third quartiles of the following sample.

8 2 9 5 3 7 4 2 7 4 1 4 3 5 5 0

4.26 Find the third and eighth deciles (30th and 80th percentiles) of the following data set.

26	23	29	31	24
22	15	31	30	20

4.27 Find the first and second quintiles (20th and 40th percentiles) of the data shown here.

52	61	88	43	64
71	39	73	51	60

4.28 Determine the first, second and third quartiles of the following data.

10.5	14.7	15.3	17.7	15.9	12.2	10.0
14.1	13.9	18.5	13.9	15.1	14.7	

4.29 Calculate the 3rd and 6th deciles of the accompanying data.

7	18	12	17	29	18	4	27	30	2
4	10	21	5	8					

4.30 Refer to Exercise 4.28. Determine the interquartile range.

4.31 Refer to Exercise 4.25. Determine the interquartile range.

4.32 Xr04-32 Compute the interquartile range from the following data.

5 8 14 6 21 11 9 10 18 2

4.33 Xr04-33 Draw the box plot of the following set of data.

9	28	15	21	12	22	29	
20	23	31	11	19	24	16	13

The following exercises require a computer and software.

4.34 Xr04-34 Many automotive experts believe that speed limits on roads are too low. One particular expert has stated that he thinks that most drivers drive at speeds that they consider safe. He suggested that the 'correct' speed limit should be set at the 85th percentile. Suppose that a random sample of 400 speeds on any UK road where the limit is 70 mph (113 km/h) was recorded. Find the 'correct' speed limit.

4.35 Xr04-35 Accountemps, a company that supplies temporary workers, sponsored a survey of 100 executives. Each was asked to report the number of minutes they spend screening each job résumé they receive.

a. Compute the quartiles.

b. What information did you derive from the quartiles? What does this suggest about writing your resumé?

4.36 Xr04-36 How much do pets cost? A random sample of dog and cat owners was asked to compute the amounts of money spent on their pets (exclusive of pet food). Draw a box plot for each data set and describe your findings.

4.37 Xr04-37 The career-counselling centre at a university wanted to learn more about the starting salaries of the university's graduates. They asked each graduate to report the highest salary offer received. The survey also asked each graduate to report the degree and starting salary (column 1 = BA (Bachelor of Arts), column 2 = BSc (Bachelor of Science), column 3 = BBA (Bachelor of Business Administration), column 4 = other). Draw box plots to compare the four groups of starting salaries. Report your findings.

4.38 Xr04-38 A random sample of London Marathon runners was drawn and the times to complete the race were recorded.

a. Draw the box plot.

b. What are the quartiles?

c. Identify outliers.

d. What information does the box plot deliver?

4.39 Xr04-39 For many restaurants, the amount of time customers linger over coffee and dessert negatively affects profits. To learn more about this variable, a sample of 200 restaurant groups was observed, and the amount of time customers spent in the restaurant was recorded.

a. Calculate the quartiles of these data.

b. What do these statistics tell you about the amount of time spent in this restaurant?

CENSUS EXERCISE

4.40 Census2011 Draw a box plot of the ages (column E for Excel and column 4 for Minitab) of respondents from the 2011 survey. Briefly describe the graph.

4-4 MEASURES OF LINEAR RELATIONSHIP

In Chapter 3 we introduced the scatter diagram, a graphical technique that describes the relationship between two interval variables. At that time we pointed out that we were particularly interested in the direction and strength of the linear relationship. We now present three numerical measures of linear relationship that provide this information: *covariance, coefficient of correlation* and *coefficient of determination*. Later in this section, we discuss another related numerical technique, the *least squares line*.

4-4a Covariance

As we did in Chapter 3, we label one variable X and the other Y.

Covariance

Population covariance: $\sigma_{xy} = \frac{\sum_{i=1}^{N}(x_i - \mu_x)(y_i - \mu_y)}{N}$

Sample covariance: $s_{xy} = \frac{\sum_{i=1}^{n}(x_i - \bar{x})(y_i - \bar{y})}{n-1}$

The denominator in the calculation of the sample **covariance** is $n - 1$, not the more logical n for the same reason we divide by $n - 1$ to calculate the sample variance (see Section 4-2). If you plan to compute the sample covariance manually, here is a shortcut calculation.

Shortcut for Sample Covariance

$$s_{xy} = \frac{1}{n-1}\left[\sum_{i=1}^{n} x_i y_i - \frac{\sum_{i=1}^{n} x_i \sum_{i=1}^{n} y_i}{n}\right]$$

To illustrate how covariance measures the linear relationship, examine the following three sets of data.

Set 1

x_i	y_i	$(x_i - \bar{x})$	$(y_i - \bar{y})$	$(x_i - \bar{x})(y_i - \bar{y})$
2	13	−3	−7	21
6	20	1	0	0
7	27	2	7	14
$\bar{x} = 5$	$\bar{y} = 20$			$s_{xy} = 35/2 = 17.5$

Set 2

x_i	y_i	$(x_i - \bar{x})$	$(y_i - \bar{y})$	$(x_i - \bar{x})(y_i - \bar{y})$
2	27	−3	−7	−21
6	20	1	0	0
7	13	2	7	−14
$\bar{x} = 5$	$\bar{y} = 20$			$s_{xy} = -35/2 = -17.5$

Set 3

x_i	y_i	$(x_i - \bar{x})$	$(y_i - \bar{y})$	$(x_i - \bar{x})(y_i - \bar{y})$
2	20	−3	0	0
6	27	1	7	7
7	13	2	−7	−14
$\bar{x} = 5$	$\bar{y} = 20$			$s_{xy} = -7/2 = -3.5$

Notice that the values of x are the same in all three sets and that the values of y are also the same. The only difference is the *order* of the values of y.

In set 1, as x increases so does y. When x is larger than its mean, y is at least as large as its mean. Thus $(x_i - \bar{x})$ and $(y_i - \bar{y})$ have the same sign or 0. Their product is also positive or 0. Consequently, the covariance is a positive number. Generally, when two variables move in the same direction (both increase or both decrease), the covariance will be a large positive number.

If you examine set 2, you will discover that as x increases, y decreases. When x is larger than its mean, y is less than or equal to its mean. As a result when $(x_i - \bar{x})$ is positive, $(y_i - \bar{y})$ is negative or 0. Their products are either negative or 0. It follows that the covariance is a negative number. In general, when two variables move in opposite directions, the covariance is a large negative number.

In set 3, as x increases, y does not exhibit any particular direction. One of the products $(x_i - \bar{x})(y_i - \bar{y})$ is 0, one is positive and one is negative. The resulting covariance is a small number. In general, when there is no particular pattern, the covariance is a small number.

We would like to extract two pieces of information. The first is the sign of the covariance, which tells us the nature of the relationship. The second is the magnitude, which describes the strength of the association. Unfortunately, the magnitude may be difficult to judge. For example, if you are told that the covariance between two variables is 500, does this mean that there is a strong linear relationship? The answer is that it is impossible to judge without additional statistics. Fortunately, we can improve on the information provided by this statistic by creating another one.

4-4b Coefficient of Correlation

The **coefficient of correlation** is defined as the covariance divided by the standard deviations of the variables.

Coefficient of Correlation

Population coefficient of correlation: $\rho = \dfrac{\sigma_{xy}}{\sigma_x \sigma_y}$

Sample coefficient of correlation: $r = \dfrac{s_{xy}}{s_x s_y}$

The population parameter is denoted by the Greek letter *rho*.

The advantage that the coefficient of correlation has over the covariance is that the former has a set lower and upper limit. The limits are −1 and +1, respectively – that is,

$$-1 \leq r \leq +1 \quad \text{and} \quad -1 \leq \rho \leq +1$$

When the coefficient of correlation equals −1, there is a negative linear relationship and the scatter diagram exhibits a straight line. When the coefficient of correlation equals +1, there is a perfect positive relationship. When the coefficient of correlation equals 0, there is no linear relationship. All other values of correlation are judged in relation to these three values. The drawback to the coefficient of correlation is that – except for the three values −1, 0, and +1 – we cannot interpret the correlation. For example, suppose that we calculated the coefficient of correlation to be −0.4. What does this tell us? It tells us two things. The minus sign tells us the relationship is negative and because 0.4 is closer to 0 than to 1, we judge that the linear relationship is weak. In many applications, we need a better interpretation than the 'linear relationship is weak'. Fortunately, there is yet another measure of the strength of a linear relationship, which gives us more information. It is the *coefficient of determination*, which we introduce later in this section.

Example 4.16

Calculating the Coefficient of Correlation

Calculate the coefficient of correlation for the three sets of data on Section 4-4.

SOLUTION:

Because we have already calculated the covariance we need to compute only the standard deviations of *X* and *Y*.

$$\bar{x} = \frac{2+6+7}{3} = 5.0$$

$$\bar{y} = \frac{13+20+27}{3} = 20.0$$

$$s_x^2 = \frac{(2-5)^2+(6-5)^2+(7-5)^2}{3-1} = \frac{9+1+4}{2} = 7.0$$

$$s_y^2 = \frac{(13-20)^2+(20-20)^2+(27-20)^2}{3-1} = \frac{49+0+49}{2} = 49.0$$

The standard deviations are

$$s_x = \sqrt{7.0} = 2.65$$

$$s_y = \sqrt{49.0} = 7.00$$

The coefficients of correlation are

$$\textbf{Set 1:} \quad r = \frac{s_{xy}}{s_x s_y} = \frac{17.5}{(2.65)(7.0)} = 0.943$$

$$\textbf{Set 2:} \quad r = \frac{s_{xy}}{s_x s_y} = \frac{-17.5}{(2.65)(7.0)} = -0.943$$

$$\textbf{Set 3:} \quad r = \frac{s_{xy}}{s_x s_y} = \frac{-3.5}{(2.65)(7.0)} = -0.189$$

It is now easier to see the strength of the linear relationship between *X* and *Y*.

4-4c Comparing the Scatter Diagram, Covariance and Coefficient of Correlation

The scatter diagram depicts relationships graphically; the covariance and the coefficient of correlation describe the linear relationship numerically. Figures 4.2, 4.3 and 4.4 depict three scatter diagrams. To show how the graphical and numerical techniques compare, we calculated the covariance and the coefficient of correlation for each. As you can see, Figure 4.2 depicts a strong positive relationship between the two variables. The covariance is 36.87 and the coefficient of correlation is 0.9641. The variables in Figure 4.3 produced a relatively strong negative linear relationship; the covariance and coefficient of correlation are –34.18 and –0.8791, respectively. The covariance and coefficient of correlation for the data in Figure 4.4 are 2.07 and 0.1206, respectively. There is no apparent linear relationship in this figure.

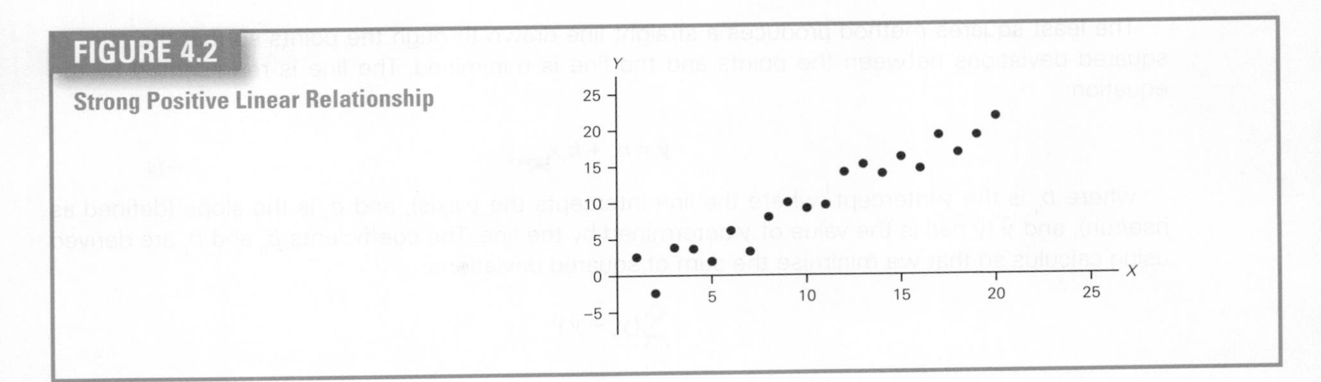

FIGURE 4.2
Strong Positive Linear Relationship

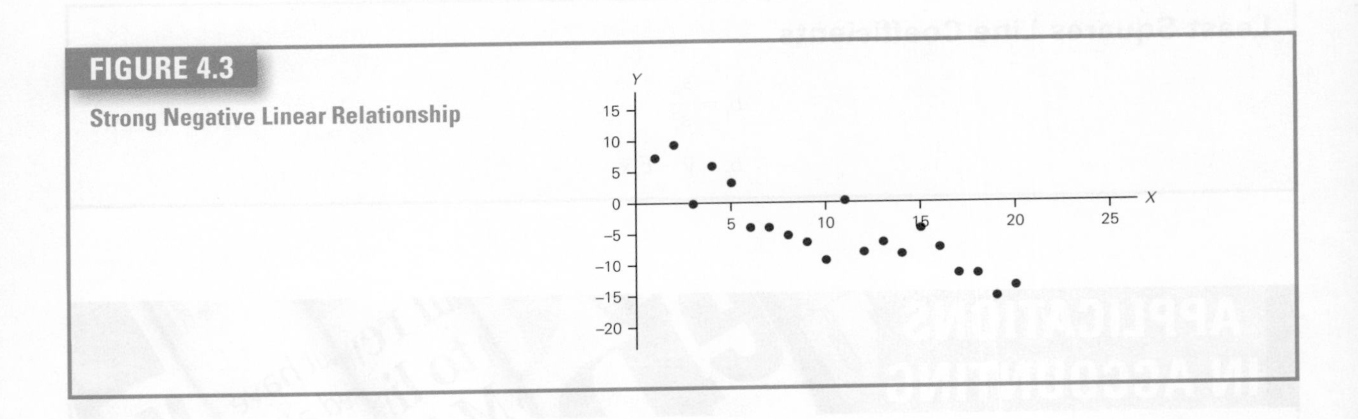

FIGURE 4.3
Strong Negative Linear Relationship

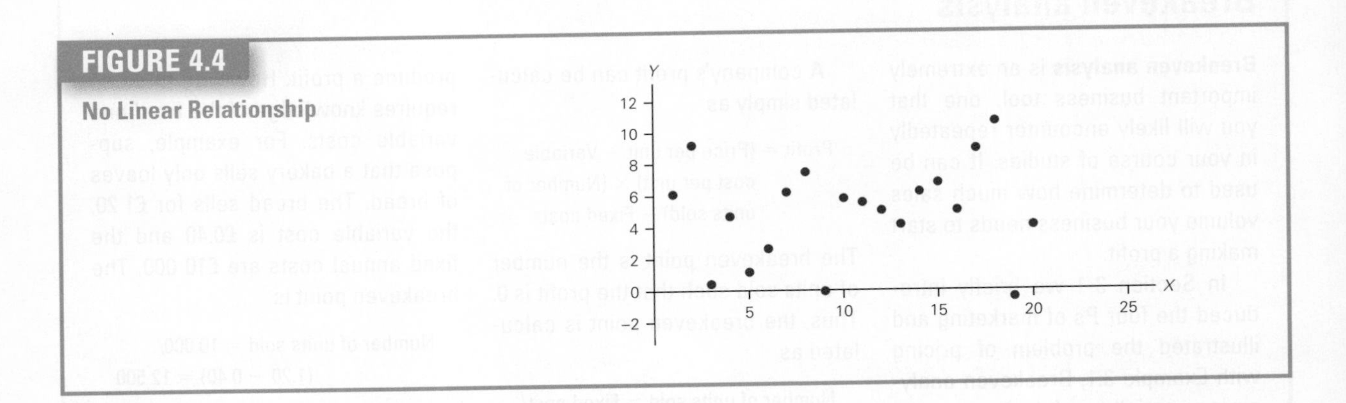

FIGURE 4.4
No Linear Relationship

4-4d Least Squares Method

When we presented the scatter diagram in Section 3-3, we pointed out that we were interested in measuring the strength and direction of the linear relationship. Both can be more easily judged by drawing a straight line through the data. However, if different people draw a line through the same data set, it is likely that each person's line will differ from all the others. Moreover, we often need to know the equation of the line. Consequently, we need an objective method of producing a straight line. Such a method has been developed; it is called the **least squares method**.

The least squares method produces a straight line drawn through the points so that the sum of squared deviations between the points and the line is minimised. The line is represented by the equation:

$$\hat{y} = b_0 + b_1x$$

where b_0 is the *y*-intercept (where the line intercepts the *y*-axis), and b_1 is the slope (defined as rise/run), and $\hat{y}$ (*y hat*) is the value of *y* determined by the line. The coefficients b_0 and b_1 are derived using calculus so that we minimise the sum of squared deviations:

$$\sum_{i=1}^{n}(y_1 - \hat{y}_i)^2$$

Least Squares Line Coefficients

$$b_1 = \frac{s_{xy}}{s_x^2}$$

$$b_0 = \bar{y} - b_1\bar{x}$$

APPLICATIONS IN ACCOUNTING

Breakeven analysis

Breakeven analysis is an extremely important business tool, one that you will likely encounter repeatedly in your course of studies. It can be used to determine how much sales volume your business needs to start making a profit.

In Section 3-1 we briefly introduced the four Ps of marketing and illustrated the problem of pricing with Example 3.1. Breakeven analysis is especially useful when managers are attempting to determine the appropriate price for the company's products and services.

A company's profit can be calculated simply as

Profit = (Price per unit − Variable cost per unit) × (Number of units sold) − Fixed costs

The breakeven point is the number of units sold such that the profit is 0. Thus, the breakeven point is calculated as

Number of units sold = Fixed cost/(Price − Variable cost)

Managers can use the formula to help determine the price that will produce a profit. However, to do so requires knowledge of the fixed and variable costs. For example, suppose that a bakery sells only loaves of bread. The bread sells for £1.20, the variable cost is £0.40 and the fixed annual costs are £10 000. The breakeven point is

Number of units sold = 10 000/(1.20 − 0.40) = 12 500

The bakery must sell more than 12 500 loaves per year to make a profit.

In the next application box, we discuss fixed and variable costs.

APPLICATIONS IN ACCOUNTING

Fixed and Variable costs

Fixed costs are costs that must be paid whether or not any units are produced. These costs are 'fixed' over a specified period of time or range of production. Variable costs are costs that vary directly with the number of products produced. For the previous bakery example, the fixed costs would include rent and maintenance of the shop, wages paid to employees, advertising costs, telephone and any other costs that are not related to the number of loaves baked. The variable cost is primarily the cost of ingredients, which rises in relation to the number of loaves baked.

Some expenses are mixed. For the bakery example, one such cost is the cost of electricity. Electricity is needed for lights, which is considered a fixed cost, but also for the ovens and other equipment, which are variable costs.

There are several ways to break the mixed costs into fixed and variable components. One such method is the least squares line; that is, we express the total costs of some component as

$$y = b_0 + b_1 x$$

where y = total mixed cost, b_0 = fixed cost, b_1 = variable cost, and x is the number of units.

EXAMPLE 4.17

DATA Xm04-17

Estimating Fixed and Variable Costs

A tool and die maker operates out of a small shop making specialised tools. He is considering increasing the size of his business and needs to know more about his costs. One such cost is electricity, which he needs to operate his machines and lights. (Some jobs require that he turn on extra bright lights to illuminate his work.) He keeps track of his daily electricity costs and the number of tools that he made that day. These data are listed next. Determine the fixed and variable electricity costs.

Day	1	2	3	4	5	6	7	8	9	10
Number of tools	7	3	2	5	8	11	5	15	3	6
Electricity cost	23.80	11.89	15.98	26.11	31.79	39.93	12.27	40.06	21.38	18.65

SOLUTION:

The dependent variable is the daily cost of electricity, and the independent variable is the number of tools. To calculate the coefficients of the least squares line and other statistics (calculated below), we need the sum of X, Y, XY, X^2 and Y^2.

Day	X	Y	XY	X^2	Y^2
1	7	23.80	166.60	49	566.44
2	3	11.89	35.67	9	141.37
3	2	15.98	31.96	4	255.36
4	5	26.11	130.55	25	681.73
5	8	31.79	254.32	64	1010.60
6	11	39.93	439.23	121	1594.40
7	5	12.27	61.35	25	150.55
8	15	40.06	600.90	225	1604.80
9	3	21.38	64.14	9	457.10
10	6	18.65	111.90	36	347.82
Total	65	241.86	1896.62	567	6810.20

Covariance

$$s_{xy} = \frac{1}{n-1}\left[\sum_{i=1}^{n} x_i y_i - \frac{\sum_{i=1}^{n} x_i \sum_{i=1}^{n} y_i}{n}\right] = \frac{1}{10-1}\left[1896.62 - \frac{(65)(241.86)}{10}\right] = 36.06$$

Variance of X

$$s_x^2 = \frac{1}{n-1}\left[\sum_{i=1}^{n} x_i^2 - \frac{\left(\sum_{i=1}^{n} x_i\right)^2}{n}\right] = \frac{1}{10-1}\left[567 - \frac{(65)^2}{10}\right] = 16.06$$

Sample means

$$\bar{x} = \frac{\sum x_i}{n} = \frac{65}{10} = 6.5$$

$$\bar{y} = \frac{\sum y_i}{n} = \frac{241.86}{10} = 24.186$$

The coefficients of the least squares line are:
Slope

$$b_1 = \frac{s_{xy}}{s_x^2} = \frac{36.06}{16.06} = 2.246$$

y-Intercept

$$b_0 = \bar{y} - b_1\bar{x} = 24.19 - (2.25)(6.5) = 9.587$$

The least squares line is

$$\hat{y} = 9.587 + 2.245x$$

EXCEL

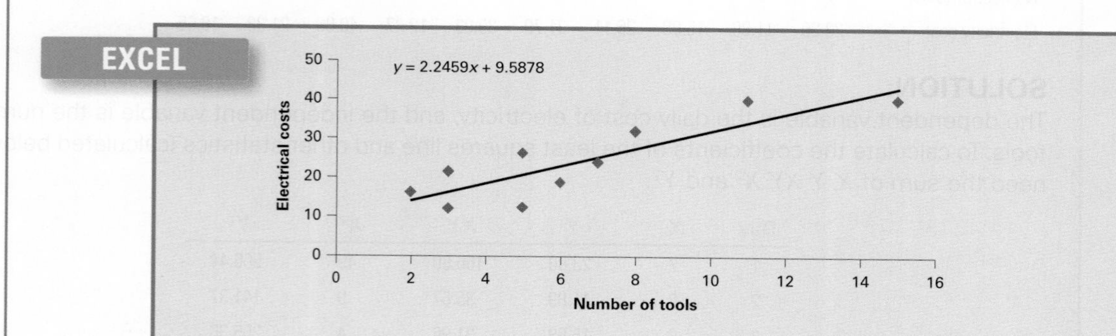

Instructions

1. Type or import the data into two columns where the first column stores the values of X and the second stores Y. (**Open** Xm04-17.) Highlight the columns containing the variables. Follow the instructions to draw a scatter diagram (Section 3-3).
2. In the **Chart Elements**, click **Trendline** and **Linear Trendline.**
3. Click **Trendline** and **More Trendline Options** .... Click **Display Equation on Chart.**

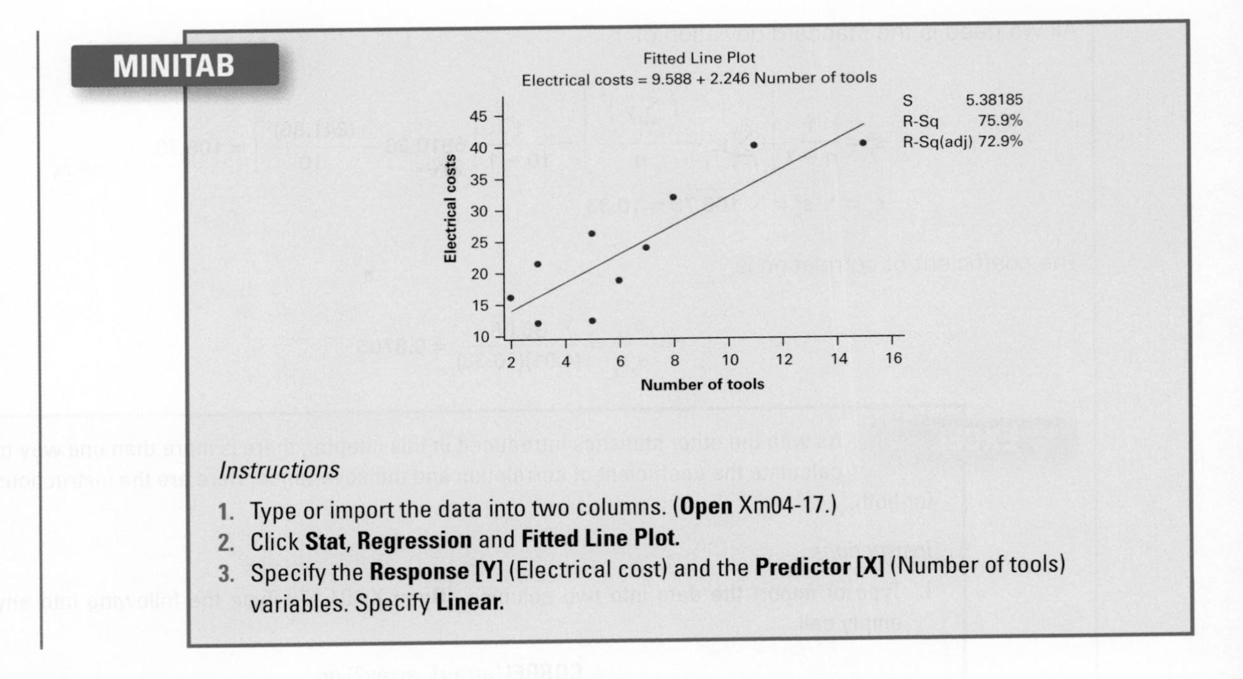

Instructions

1. Type or import the data into two columns. (**Open** Xm04-17.)
2. Click **Stat, Regression** and **Fitted Line Plot.**
3. Specify the **Response [Y]** (Electrical cost) and the **Predictor [X]** (Number of tools) variables. Specify **Linear.**

INTERPRET

The slope is defined as rise/run, which means that it is the change in y (rise) for a one-unit increase in x (run). Put less mathematically, the slope measures the *marginal* rate of change in the dependent variable. The marginal rate of change refers to the effect of increasing the independent variable by one additional unit. In this example, the slope is 2.25, which means that in this sample, for each one-unit increase in the number of tools, the marginal increase in the electricity cost is £2.25.Thus, the estimated variable cost is £2.25 per tool.

The y-Intercept is 9.57; that is, the line strikes the y-axis at 9.57. This is simply the value of y when $x = 0$. However, when $x = 0$, we are producing no tools and hence the estimated fixed cost of electricity is £9.57 per day.

Because the costs are estimates based on a straight line, we often need to know how well the line fits the data.

EXAMPLE 4.18

DATA Xm04-17

Measuring the Strength of the Linear Relationship

Calculate the coefficient of correlation for Example 4.17.

SOLUTION:

To calculate the coefficient of correlation, we need the covariance and the standard deviations of both variables. The covariance and the variance of X were calculated in Example 4.17. The covariance is

$$s_{xy} = 36.06$$

and the variance of X is $= 16.06$

$$s_x^2 = 16.06$$

Standard deviation of X is

$$s_x = \sqrt{s_x^2} = \sqrt{16.06} = 4.01$$

All we need is the standard deviation of Y

$$s_y^2 = \frac{1}{n-1}\left[\sum_{i=1}^{n} y_i^2 - \frac{\left(\sum_{i=1}^{n} y_i\right)^2}{n}\right] = \frac{1}{10-1}\left[6810.20 - \frac{(241.86)^2}{10}\right] = 106.73$$

$$s_y = \sqrt{s_y^2} = \sqrt{106.73} = 10.33$$

The coefficient of correlation is

$$r = \frac{s_{XY}}{s_x s_y} = \frac{36.06}{(4.01)(10.33)} = 0.8705$$

EXCEL

As with the other statistics introduced in this chapter, there is more than one way to calculate the coefficient of correlation and the covariance. Here are the instructions for both.

Instructions

1. Type or import the data into two columns. (**Open** Xm04-17.) Type the following into any empty cell.

= **CORREL**(array1, array2) or

= **CORREL**([Input range of one variable], [Input range of second variable])

In this example, we would enter

= **CORREL**(B1:B11,C1:C11)

To calculate the covariance, replace **CORREL** with **COVAR**.

Another method, which is also useful if you have more than two variables and would like to compute the coefficient of correlation or the covariance for each pair of variables, is to produce the correlation matrix and the variance–covariance matrix. We do the correlation matrix first.

	A	B	C
1		*Number of tools*	*Electrical costs*
2	Number of tools	1	
3	Electrical costs	0.8711	1

Instructions

1. Type or import the data into adjacent columns. (**Open** Xm04-17.)
2. Click **Data, Data Analysis** and **Correlation.**
3. Specify the **Input Range** (B1:C11).

The coefficient of correlation between number of tools and electrical costs is 0.8711 (slightly different from the manually calculated value). (The two 1s on the diagonal of the matrix are the coefficients of number of tools and number of tools, and electrical costs and electrical costs, telling you the obvious.)

Incidentally, the formula for the population parameter ρ (Greek letter *rho*) and for the sample statistic r produce exactly the same value.

The variance–covariance matrix is shown next.

	A	B	C
1		*Number of tools*	*Electrical costs*
2	Number of tools	14.45	
3	Electrical costs	32.45	96.06

Instructions

1. Type or import the data into adjacent columns. (**Open** Xm04-17.)
2. Click **Data**, **Data Analysis** and **Covariance.**
3. Specify the **Input Range** (B1:C11).

Unfortunately, Excel computes the population parameters. In other words, the variance of the number of tools is $\sigma_x^2 = 14.45$, the variance of the electrical costs is $\sigma_y^2 = 96.06$, and the covariance is $\sigma_{xy} = 32.45$. You can convert these parameters to statistics by multiplying each by $n/(n-1)$.

	D	E	F
1		*Number of tools*	*Electrical costs*
2	Number of tools	16.06	
3	Electrical costs	36.06	106.73

MINITAB

Correlations: Number of tools, Electrical costs

Pearson correlation of Number of tools and Electrical costs = 0.871

Instructions

1. Type or import the data into two columns. (**Open** Xm04-17.)
2. Click **Stat, Basic Statistics** and **Correlation . . . .**
3. In the **Variables** box, type or use the **Select** button to specify the variables **(Number of tools, Electrical costs).**

Covariances: Number of tools, Electrical costs

	Number of tools	Electrical costs
Number of tools	16.0556	
Electrical costs	36.0589	106.7301

Instructions

Click **Covariance** . . . instead of **Correlation** . . . in step 2 above.

INTERPRET

The coefficient of correlation is 0.8711, which tells us that there is a positive linear relationship between the number of tools and the electricity cost. The linear relationship is quite strong because the correlation coefficient is close to +1 and thus the estimates of the fixed and variable costs should be good.

4-4e Coefficient of Determination

When we introduced the coefficient of correlation, we pointed out that except for −1, 0 and +1 we cannot precisely interpret its meaning. We can judge the coefficient of correlation in relation to its proximity to only −1, 0 and +1. Fortunately, we have another measure that can be precisely interpreted. It is the coefficient of determination, which is calculated by squaring the coefficient of correlation. For this reason, we denote it R^2.

The coefficient of determination measures the amount of variation in the dependent variable that is explained by the variation in the independent variable. For example, if the coefficient of correlation is −1 or +1, a scatter diagram would display all the points lining up in a straight line. The coefficient of determination is 1, which we interpret to mean that 100% of the variation in the dependent variable Y is explained by the variation in the independent variable X. If the coefficient of correlation is 0, then there is no linear relationship between the two variables, $R^2 = 0$, and none of the variation in Y is explained by the variation in X. In Example 4.18, the coefficient of correlation was calculated to be $r = 0.8711$. Thus, the coefficient of determination is

$$R^2 = (0.8711)^2 = 0.7588$$

This tells us that 75.88% of the variation in electrical costs is explained by the number of tools. The remaining 24.12% is unexplained.

Using the Computer

EXCEL

You can use Excel to calculate the coefficient of correlation and then square the result. Alternatively, use Excel to draw the least squares line. After doing so, click **Trendline, Trendline Options** and **Display R-squared value on chart.**

MINITAB

Minitab automatically prints the coefficient of determination.

The concept of explained variation is an extremely important one in statistics. We return to this idea repeatedly in Chapters 12, 13, 15 and 16. In Chapter 15, we explain why we interpret the coefficient of determination in the way that we do.

COST OF ONE MORE WIN IN FOOTBALL: SOLUTION

To determine the cost of an additional win or additional points, we must describe the relationship between two variables. To do so, we use the least squares method to produce a straight line through the data. Because we believe that the number of points a football team wins depends to some extent on its team average annual salary per player, we label Points as the dependent variable and Average Annual Salary per Player as the independent variable. Because of rounding problems, we expressed the average annual salary in the number of millions of euros.

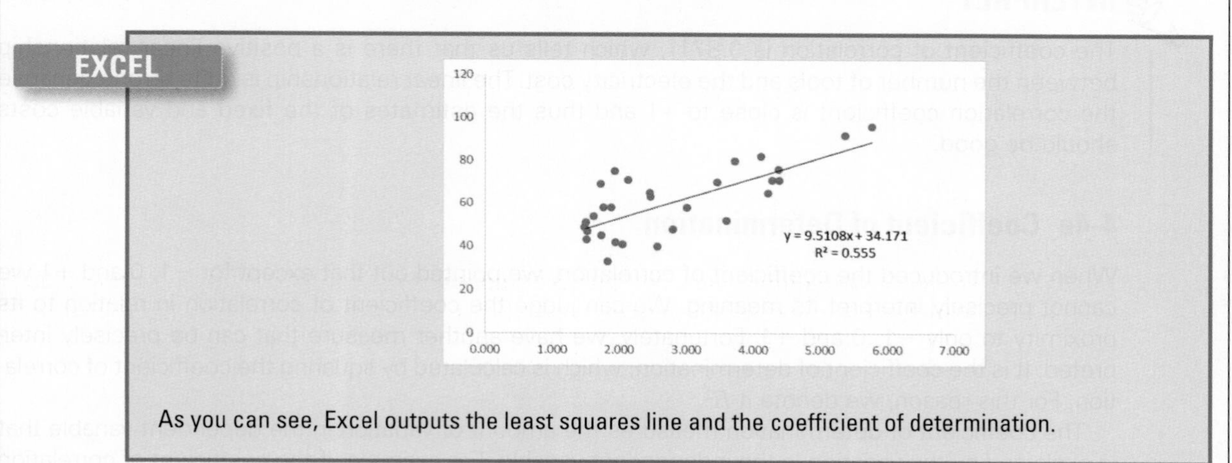

As you can see, Excel outputs the least squares line and the coefficient of determination.

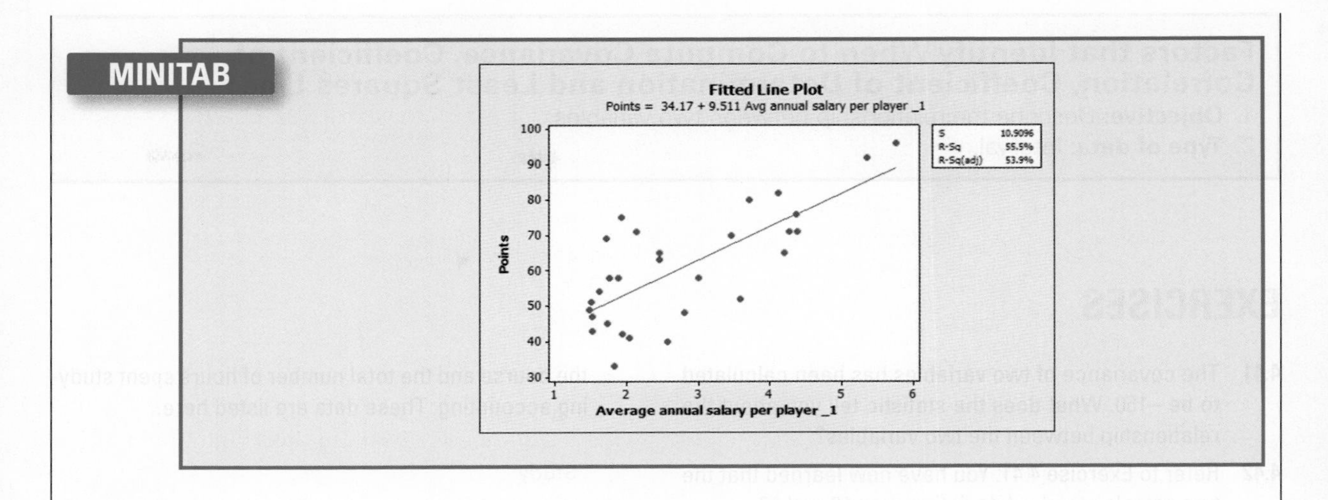

INTERPRET

The least squares line is

$$\hat{y} = 34.17 + 9.5108x$$

The slope is equal to 9.5108, which is the marginal rate of change in games won for each 1-unit increase in salary. Because annual salary is measured in millions of euros, we estimate that for each €1 million increase in the payroll the number of games won increases on average by 9.5108. Thus, to obtain one more point requires on average an additional expenditure of an incredible €105 144 (calculated as 1 million/9.5108).

In addition to analysing the least squares line, we should determine the strength of the linear relationship. The coefficient of determination is 0.744983, which means that the variation in the players' salaries explains 74.505% of the variation in the teams' number of points won. This tells us that there is virtually no linear relationship between players' annual salary and the number of points won in the 2011 season. This suggests that there are some teams that win a small number of points with large payrolls (e.g. Liverpool, €3.6 million annual salary per player, 52 points), whereas others win a large number of points with small annual salary per player (e.g. Borussia Dortmund, €1.93 million payroll and 75 points).

4-4f Interpreting Correlation

Because of its importance, we remind you about the correct interpretation of the analysis of the relationship between two interval variables that we discussed in Chapter 3. That is, if two variables are linearly related, it does not mean that *X* causes *Y*. It may mean that another variable causes both *X* and *Y* or that *Y* causes *X*. Remember:

Correlation is not causation

We complete this section with a review of when to use the techniques introduced in this section.

Factors that Identify When to Compute Covariance, Coefficient of Correlation, Coefficient of Determination and Least Squares Line

1. **Objective:** Describe the relationship between two variables.
2. **Type of data:** Interval.

EXERCISES

4.41 The covariance of two variables has been calculated to be −150. What does the statistic tell you about the relationship between the two variables?

4.42 Refer to Exercise 4.41. You have now learned that the two sample standard deviations are 16 and 12.

a. Calculate the coefficient of correlation. What does this statistic tell you about the relationship between the two variables?

b. Calculate the coefficient of determination and describe what this says about the relationship between the two variables.

4.43 Xr04-43 A retailer wanted to estimate the monthly fixed and variable selling expenses. As a first step, she collected data from the past 8 months. The total selling expenses (and the total sales) were recorded and listed below.

Total Sales	Selling Expenses
20	14
40	16
60	18
50	17
50	18
55	18
60	18
70	20

a. Calculate the covariance, the coefficient of correlation and the coefficient of determination and describe what these statistics tell you.

b. Determine the least squares line and use it to produce the estimates the retailer wants.

4.44 Xr04-44 Are the marks one receives in a course related to the amount of time spent studying the subject? To analyze this possibility, a student took a random sample of ten students who had enrolled in an accounting class last semester. He asked each to report his or her mark (values range from 0 to 100) in the course and the total number of hours spent studying accounting. These data are listed here.

Study Time	40	42	37	47	25	44	41	48	35	28
Marks	71	63	79	86	51	78	83	90	65	47

a. Calculate the covariance.

b. Calculate the coefficient of correlation.

c. Calculate the coefficient of determination.

d. Determine the least squares line.

e. What do the statistics calculated above tell you about the relationship between marks and study time?

4.45 Xr04-45 In some regions, students who apply to MBA degree programmes must take the Graduate Management Admission Test (GMAT). University admissions committees use the GMAT score as one of the critical indicators of how well a student is likely to perform in the MBA programme. However, the GMAT may not be a very strong indicator for all MBA programmes. Suppose that an MBA programme designed for middle managers who wish to upgrade their skills was launched 3 years ago. To judge how well the GMAT score predicts MBA performance, a sample of 12 graduates was taken. Their grade point averages in the MBA programme (values from 0 to 12) and their GMAT score (values range from 200 to 800) are listed here. Compute the covariance, the coefficient of correlation and the coefficient of determination. Interpret your findings.

GMAT and GPA Scores for 12 MBA Students

GMAT	599	689	584	631	594	643
GPA	9.6	8.8	7.4	10.0	7.8	9.2
GMAT	656	594	710	611	593	683
GPA	9.6	8.4	11.2	7.6	8.8	8.0

The following exercises require a computer and software.

4.46 Xr04-46 The unemployment rate is an important measure of a country's economic health. The unemployment rate measures the percentage of people who are looking for work and who are without jobs. Another way of measuring this economic variable is to calculate the employment rate, which is the percentage of adults who are employed. Here are the unemployment rates and employment rates of 19 countries. Calculate the coefficient of determination and describe what you have learned.

Country	Unemployment Rate	Employment Rate
Australia	6.7	70.7
Austria	3.6	74.8
Belgium	6.6	59.9
Canada	7.2	72.0
Denmark	4.3	77.0
Finland	9.1	68.1
France	8.6	63.2
Germany	7.9	69.0
Hungary	5.8	55.4
Ireland	3.8	67.3
Japan	5.0	74.3
Netherlands	2.4	65.4
New Zealand	5.3	62.3
Poland	18.2	53.5
Portugal	4.1	72.2
Spain	13.0	57.5
Sweden	5.1	73.0
United Kingdom	5.0	72.2
United States	4.8	73.1

4.47 Xr04-47 When the price of crude oil increases, do oil companies drill more oil wells? To determine the strength and nature of the relationship, an economist recorded the price of a barrel of domestic crude oil and the number of exploratory oil wells drilled for each month from 1973 to 2010. Analyse the data and explain what you have discovered.

4.48 Xr04-48 A German manufacturing firm produces its products in batches using sophisticated machines and equipment. The general manager wanted to investigate the relationship between direct labour costs and the number of units produced per batch. He recorded the data from the last 30 batches. Determine the fixed and variable labour costs (in euros).

4.49 Xr04-49 A manufacturer has recorded its cost of electricity and the total number of hours of machine time for each of 52 weeks. Estimate the fixed and variable electricity costs.

4.50 Xr04-50 The chapter-opening example showed that there is a very weak linear relationship between a football average annual salary per player and the number of points. This raises the question: Are success on price of the tickets per match and the number of points related? If the answer is no, then profit-driven owners may not be inclined to increase the price of the tickets per match. The statistics practitioner recorded the number of points and the price of matchday tickets per 2013 football season.

Calculate whichever parameters you wish to help guide football owners.

4-5 APPLICATIONS IN FINANCE: MARKET MODEL

In the Applications in Finance boxes, we introduced the terms *return on investment* and *risk*. We described two goals of investing. The first is to maximise the expected or mean return and the second is to minimise the risk. Financial analysts use a variety of statistical techniques to achieve these goals. Most investors are risk-averse, which means that for them minimising risk is of paramount importance. In Section 4-2, we pointed out that variance and standard deviation are used to measure the risk associated with investments.

APPLICATIONS IN FINANCE

Stock Markets Indexes

Stock markets like the New York Stock Exchange (NYSE), NASDAQ, Financial Times, the London Stock Exchange (FTSE) and many others around the world calculate indexes to provide information about the prices of stocks on their exchanges. A stock market index is composed of a number of stocks that more or less represent the entire market. For example, the Dow Jones Industrial Average (DJIA) is the average price of a group of 30 NYSE stocks of large publicly traded companies. The Standard and Poor's 500 (S&P) is the average price of 500 NYSE stocks. These indexes represent their stock exchanges and give readers a quick view of how well the exchange is doing, as well the economy of the country as a whole. The NASDAQ 100 is the average price of the 100 largest non-financial companies on the NASDAQ exchange. The FTSE is similar to the S&P/TSX Composite Index and the FTSE 100 Index is composed of 100 large companies listed on the London Stock Exchange.

In this section, we describe one of the most important applications of the use of a least squares line. It is the well-known and often applied *market model*. This model assumes that the rate of return on a stock is linearly related to the rate of return on the stock market index. The return on the index is calculated in the same way the return on a single stock is computed. For example, if the index at the end of last year was 10 000 and the value at the end of this year is 11 000, the market index annual return is 10%. The return on the stock is the dependent variable Y and the return on the index is the independent variable X.

We use the least squares line to represent the linear relationship between X and Y. The coefficient β_1 is called the stock's *beta coefficient*, which measures how sensitive the stock's rate of return is to changes in the level of the overall market. For example, if β_1 is greater than 1, the stock's rate of return is more sensitive to changes in the level of the overall market than is the average stock. To illustrate, suppose that $\beta_1 = 2$. Then a 1% increase in the index results in an average increase of 2% in the stock's return. A 1% decrease in the index produces an average 2% decrease in the stock's return. Thus, a stock with a beta coefficient greater than 1 will tend to be more volatile than the market.

EXAMPLE 4.19

DATA Xm04-19

Market Model for Daimler

The monthly rates of return for Daimler (Symbol DA) and the FTSE index were recorded for each month between January 2009 and December 2013. Some of these data are shown here. Estimate the market model and analyze the results.

Year	Month	S&P 500 Index	GE
2009	January	−0.07700	−0.18123
	February	0.02506	0.05894
	March	0.08089	0.45887
	April	0.04105	−0.04748
	May	−0.03819	−0.00267
	June	0.08453	0.25926
	July	0.06521	−0.02906

Year	Month	S&P 500 Index	GE
	August	0.04584	0.09124
	September	−0.01741	−0.03846
	October	0.02898	0.01948
	November	0.04281	0.10406
	December	−0.04146	−0.10229
2013	January	0.01337	0.06517
	February	0.00800	−0.07042
	March	0.00287	0.04347
	April	0.02379	0.17615
	May	−0.05584	−0.05910
	June	0.06526	0.12325
	July	−0.02720	−0.00574
	August	0.00329	0.10978
	September	0.04166	0.04894
	October	−0.01200	0.00927
	November	0.01481	0.03115
	December	−0.00273	0.00000

SOLUTION:

Excel's scatter diagram and least squares line are shown below. (Minitab produces a similar result.) We added the equation and the coefficient of determination to the scatter diagram.

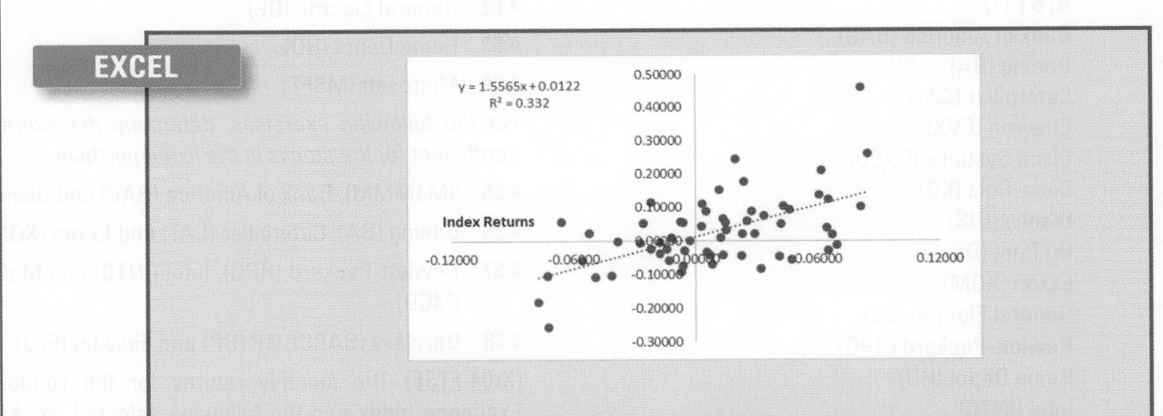

We note that the slope coefficient for Daimler is 1.5565. We interpret this to mean that in this sample for each 1% increase in the FTSE Index return, the average increase in Daimler's return is 1.5565%. Because b_1 is greater than 1, we conclude that the return on investing in Daimler is more volatile and therefore riskier than the FTSE Index.

4-5a Systematic and Firm-Specific Risk

The slope coefficient b_1 is a measure of the stock's *market-related* (or *systematic) risk* because it measures the volatility of the stock price that is related to the overall market volatility. The slope coefficient only informs us about the nature of the relationship between the two sets of returns. It tells us nothing about the *strength* of the linear relationship.

The coefficient of determination measures the proportion of the total risk that is market-related. In this case, we see that 33.2% of DA's total risk is market-related. That is, 33.2% of the variation in DA's returns is explained by the variation in the returns of the FTSE Index. The remaining 66.8% is the proportion of the risk that is associated with events specific to DA, rather than the market. Financial analysts (and nearly everyone else) call this the *firm-specific* (or *nonsystematic*) *risk*. The firm-specific risk is attributable to variables and events not included in the market model, such as the effectiveness of Daimler's sales force and managers. This is the part of the risk that can be 'diversified away' by creating a portfolio of stocks as discussed in Section 7-3. We cannot, however, diversify away the part of the risk that is market-related.

When a portfolio has been created, we can estimate its beta by averaging the betas of the stocks that compose the portfolio. If an investor believes that the market is likely to rise, a portfolio with a beta coefficient greater than 1 is desirable. Risk-averse investors or ones who believe that the market will fall will seek out portfolios with betas less than 1.

EXERCISES

The following exercises require the use of a computer and software.

Xr04-S&P500 We have recorded the monthly returns for the S&P 500 index and the following Dow Jones Industrials 30 stocks listed on the New York Stock Exchange for the period January 2008 to December 2012:

3M (MMM)
Alcoa (AA)
American Express (AXP)
AT&T (T)
Bank of America (BAC)
Boeing (BA)
Caterpillar (CAT)
Chevron (CVX)
Cisco Systems (CSCO)
Coca-Cola (KO)
Disney (DIS)
Du Pont (DD)
Exxon (XOM)
General Electric (GE)
Hewlett-Packard (HPQ)
Home Depot (HD)
Intel (INTC)
International Business Machines (IBM)
Johnson & Johnson (JNJ)
JP Morgan Chase (JPM)
McDonald's (MCD)
Merck (MRK)
Microsoft (MSFT)
Pfizer (PFE)
Proctor & Gamble (PG)
Travelers (TRV)
United Technologies (UTX)
United Health (UNH)
Verizon Communications (VZ)
Wal-Mart Stores (WMT)

For the following exercises, calculate the beta coefficient and the coefficient of determination for the listed stock and interpret their values.

(Excel users: Note that to use the scatter diagram to compute the beta coefficient, the data must be stored in two adjacent columns. The first must contain the returns on the index and the second stores the returns for the coefficient of whichever stock you wish to calculate.)

4.51 American Express (AXP)

4.52 General Electric (GE)

4.53 Home Depot (HD)

4.54 Microsoft (MSFT)

For the following exercises, determine the average beta coefficient for the stocks in the listed portfolio.

4.55 3M (MMM), Bank of America (BAC) and Disney (DIS)

4.56 Boeing (BA), Caterpillar (CAT) and Exxon (XOM)

4.57 Hewlett-Packard (HPQ), Intel (INTC) and McDonald's (MCD)

4.58 Barclays (BARC), BP (BP) and EasyJet (EZJ)

(Xr04-FTSE) The monthly returns for the London Stock Exchange Index and the following selected stocks on the Toronto Stock Exchange were recorded for the years 2009 to 2013.

Aberdeen Asset Management (AND)
BAE Systems (BA)
Barclays (BARC)
BP (BP)
British American Tobacco (BTI)
Centamin (CEY)
Capita (CPI)
City of London Investment Group (CLIG)
CREAT RESOURCES (CRHL)
EasyJet (EZJ)
Experian (EXPN)

- Fresnillo (FRES)
- Marshalls (MSLH)
- Mondi (MNDI)
- Playtech (PTEC)
- Pearson (PSON)
- Reed Elsevier (REL)
- Rio Tinto (RIO)
- The Royal Bank of Scotland Group (RBS)
- Shire (SHP)
- SSE (SSE)
- Trinity Mirror (TNI)
- Tesco (TSCO)

Calculate the beta coefficient and the coefficient of determination for the listed stock and interpret their values.

4.59 Aberdeen Asset Management (AND)

4.60 Trinity Mirror (TNI)

4.61 CREAT RESOURCES (CRHL) Determine the average beta coefficient for the stocks in the listed portfolio.

4.62 Barclays (BARC) and The Royal Bank of Scotland Group (RBS)

4.63 BAE Systems (BA) and EasyJet (EZJ)

4.64 Fresnillo (FRES) and Rio Tinto (RIO)

Xr-04-NASDAQ We calculated the returns on the NASDAQ Index and the following selected stocks on the NASDAQ Exchange for the period January 2008 to December 2012:

- Adobe Systems (ADBE)
- Amazon (AMZN)
- Amgen (AMGN)
- Apple (AAPL)
- Bed Bath & Beyond (BBBY)
- Cisco Systems (CSCO)
- Comcast (CMCSA)
- Costco Wholesale (COST)
- Dell (DELL)
- Dollar Tree (DLTR)
- Expedia (EXPE)
- Garmin (GRMN)
- Google (GOOG)
- Intel (INTC)
- Mattel (MAT)
- Microsoft (MSFT)
- Netflix (NFLX)
- Oracle (ORCL)
- Research in Motion (RIMM)
- ScanDisk (SNDK)
- Sirius XM Radio (SIRI)
- Staples (SPLS)
- Starbucks (SBUX)
- Whole Foods Market (WFM)
- Wynn Resorts (WYNN)
- Yahoo (YHOO)

Calculate the beta coefficient and the coefficient of determination for the listed stock and interpret their values.

4.65 Amazon (AMZN)

4.66 Expedia (EXPE)

4.67 Netflix (NFLX)

Determine the average beta coefficient for the stocks in the listed portfolio.

4.68 Apple (AAPL), Dell (DELL) and Google (GOOG)

4.69 Costco Wholesale (COST) and Dollar Tree (DLTR)

4.70 Cisco Systems (CSCO), Intel (INTC) and Oracle (ORCL)

4-6 COMPARING GRAPHICAL AND NUMERICAL TECHNIQUES

As we mentioned before, graphical techniques are useful in producing a quick picture of the data. For example, you learn something about the location, spread and shape of a set of interval data when you examine its histogram. Numerical techniques provide the same approximate information. We have measures of central location, measures of variability and measures of relative standing that do what the histogram does. The scatter diagram graphically describes the relationship between two interval variables, but so do the numerical measures covariance, coefficient of correlation, coefficient of determination and least squares line. Why then do we need to learn both categories of techniques? The answer is that they differ in the information each provides. We illustrate the difference between graphical and numerical methods by redoing four examples we used to illustrate graphical techniques in Chapter 3.

EXAMPLE 3.2

Comparing Returns on Two Investments

In Example 3.2, we wanted to judge which investment appeared to be better. As we discussed in the Applications in Finance: Return on Investment (Chapter 3), we judge investments in terms of the return we can expect and its risk. We drew histograms and attempted to interpret them. The centres of the histograms provided us with information about the expected return and their spreads gauged the risk. However, the histograms were not clear. Fortunately, we can use numerical measures. The mean and median provide us with information about the return we can expect, and the variance or standard deviation tells us about the risk associated with each investment.

Here are the descriptive statistics produced by Excel. Minitab's descriptive statistics are similar. (We combined the output into one worksheet.)

Microsoft Excel Output for Example 3.2

	A	B	C	D	E
1	*Return A*			*Return B*	
2					
3	Mean	10.95		Mean	12.76
4	Standard Error	3.10		Standard Error	3.97
5	Median	9.88		Median	10.76
6	Mode	12.89		Mode	#N/A
7	Standard Deviation	21.89		Standard Deviation	28.05
8	Sample Variance	479.35		Sample Variance	786.62
9	Kurtosis	-0.32		Kurtosis	-0.62
10	Skewness	0.54		Skewness	0.01
11	Range	84.95		Range	106.47
12	Minimum	-21.95		Minimum	-38.47
13	Maximum	63		Maximum	68
14	Sum	547.27		Sum	638.01
15	Count	50		Count	50

We can now see that investment B has a larger mean and median but that investment A has a smaller variance and standard deviation. If an investor were interested in low-risk investments, then he or she would choose investment A. If you re-examine the histograms from Example 3.2 (Chapter 3), you will see that the precision provided by the numerical techniques (mean median and standard deviation) provides more useful information than did the histograms.

EXAMPLES 3.3 AND 3.4

Business Statistics Marks; Mathematical Statistical Marks

In these examples we wanted to see what differences existed between the marks in the two statistics classes. Here are the descriptive statistics. (We combined the two printouts in one worksheet.)

Microsoft Excel Output for Examples 3.3 and 3.4

	A	B	C	D	E
1	*Marks (Example 3.3)*			*Marks (Example 3.4)*	
2					
3	Mean	72.67		Mean	66.40
4	Standard Error	1.07		Standard Error	1.610
5	Median	72		Median	71.5
6	Mode	67		Mode	75
7	Standard Deviation	8.29		Standard Deviation	12.470
8	Sample Variance	68.77		Sample Variance	155.498
9	Kurtosis	-0.36		Kurtosis	-1.241
10	Skewness	0.16		Skewness	-0.217
11	Range	39		Range	48
12	Minimum	53		Minimum	44
13	Maximum	92		Maximum	92
14	Sum	4360		Sum	3984
15	Count	60		Count	60
16	Largest(15)	79		Largest(15)	76
17	Smallest(15)	67		Smallest(15)	53

The statistics tell us that the mean and median of the marks in the business statistics course (Example 3.3) are higher than in the mathematical statistics course (Example 3.4). We found that the histogram of the mathematical statistics marks was bimodal, which we interpreted to mean that this type of approach created differences between students. The unimodal histogram of the business statistics marks informed us that this approach eliminated those differences.

Chapter 3 Opening Example: Were Oil Companies Gouging Customers 2000–2013?

In this earlier example, recall that we wanted to know whether the prices of petrol and oil were related. The scatter diagram did reveal a strong positive linear relationship. We can improve on the quality of this information by computing the coefficient of correlation and drawing the least squares line.

Excel Output for Chapter 3 Opening Example: Coefficient of Correlation

	A	B	C
1		*Price of oil*	*Price of petrol*
2	Price of oil	1	
3	Price of petrol	0.9687	1

The coefficient of correlation seems to confirm what we learned from the scatter diagram. That is, there is a strong positive linear relationship between the two variables.

Excel Output for Chapter 3 Opening Example: Least Squares Line

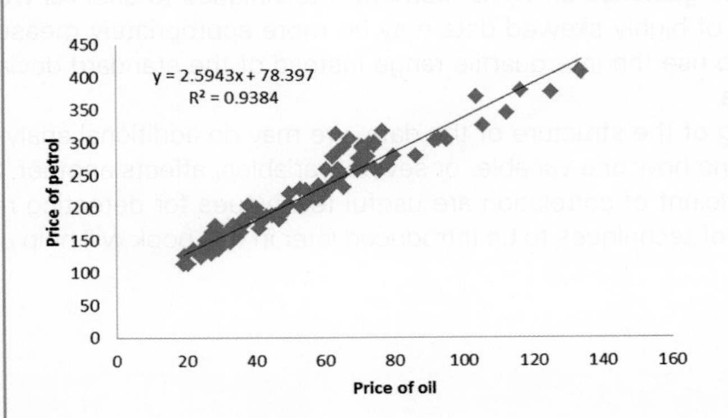

The slope coefficient tells us that for each pound increase in the price of a barrel of oil, the price of a gallon of petrol (one US gallon is equal to 3.78541 litres) increases on average by 2.6 pence. However, because there are 42 gallons per barrel, we would expect a pound increase in a barrel of oil to yield a 2.4† cents per gallon (calculated as £1.00/42) increase. It does appear that the oil companies are taking some small advantage by adding an extra half cent per gallon. The coefficient of determination is 0.9384, which indicates that 93.84% of the variation in petrol prices is explained by the variation in oil prices.

EXERCISES

The following exercises require a computer and statistical software.

4.71 **Xr03-20** Calculate the coefficient of determination for Exercise 3.20. Is this more informative than the scatter diagram?

4.72 **Xr03-21** Refer to Exercise 3.21. Compute the coefficients of the least squares line and compare your results with the scatter diagram.

4.73 **Xr03-27** Refer to Exercise 3.27. Calculate the coefficient of determination and the least squares line. Is this more informative than the scatter diagram?

4.74 **Xm03-07**

a. Calculate the coefficients of the least squares line for the data in Example 3.7.

b. Interpret the coefficients.

c. Is this information more useful than the information extracted from the scatter diagram?

4.75 **Xr04-37** In Exercise 4.37 you drew box plots. Draw histograms instead and compare the results. What have you learned?

† This is a simplification. In fact, a barrel of oil yields a variety of other profitable products. See Exercise 2.14.

4-7 GENERAL GUIDELINES FOR EXPLORING DATA

The purpose of applying graphical and numerical techniques is to describe and summarise data. Statisticians usually apply graphical techniques as a first step because we need to know the shape of the distribution. The shape of the distribution helps answer the following questions:

1. Where is the approximate centre of the distribution?
2. Are the observations close to one another, or are they widely dispersed?
3. Is the distribution unimodal, bimodal or multimodal? If there is more than one mode, where are the peaks and where are the valleys?
4. Is the distribution symmetric? If not, is it skewed? If symmetric, is it bell shaped?

Histograms and box plots provide most of the answers. We can frequently make several inferences about the nature of the data from the shape. For example, we can assess the relative risk of investments by noting their spreads. We can attempt to improve the teaching of a course by examining whether the distribution of final grades is bimodal or skewed.

The shape can also provide some guidance on which numerical techniques to use. As we noted in this chapter, the central location of highly skewed data may be more appropriately measured by the median. We may also choose to use the interquartile range instead of the standard deviation to describe the spread of skewed data.

When we have an understanding of the structure of the data, we may do additional analysis. For example, we often want to determine how one variable, or several variables, affects another. Scatter diagrams, covariance and the coefficient of correlation are useful techniques for detecting relationships between variables. A number of techniques to be introduced later in this book will help uncover the nature of these associations.

CHAPTER SUMMARY

This chapter extended our discussion of descriptive statistics, which deals with methods of summarising and presenting the essential information contained in a set of data. After constructing a frequency distribution to obtain a general idea about the distribution of a data set, we can use numerical measures to describe the central location and variability of interval data. Three popular measures of central location, or averages, are the mean, the median and the mode. Taken by themselves, these measures provide an inadequate description of the data because they say nothing about the extent to which the data vary. Information regarding the variability of interval data is conveyed by such numerical measures as the range, variance and standard deviation.

For the special case in which a sample of measurements has a mound-shaped distribution, the Empirical Rule provides a good approximation of the percentages of measurements that fall within one, two and three standard deviations of the mean. Chebysheff's Theorem applies to all sets of data no matter the shape of the histogram.

Measures of relative standing that were presented in this chapter are percentiles and quartiles. The box plot graphically depicts these measures as well as several others. The linear relationship between two interval variables is measured by the covariance, the coefficient of correlation, the coefficient of determination and the least squares line.

SYMBOLS

Symbol	Pronounced	Represents
μ	mu	Population mean
σ^2	sigma squared	Population variance
σ	sigma	Population standard deviation

Symbol	Pronounced	Represents
ρ	rho	Population coefficient of correlation
Σ	sum of	Summation
$\sum_{i=1}^{n} x_i$	sum of x_i from 1 to n	Summation of n numbers
$\hat{y}$	y hat	Fitted or calculated value of y
b_0	b zero	y-Intercept
b_1	b one	Slope coefficient

FORMULAS

Population mean

$$\mu = \frac{\sum_{i=1}^{N} x_i}{N}$$

Sample mean

$$\bar{x} = \frac{\sum_{i=1}^{n} x_i}{n}$$

Range

Largest observation − Smallest observation

Population variance

$$\sigma^2 = \frac{\sum_{i=1}^{N} (x_i - \mu)^2}{N}$$

Sample variance

$$s^2 = \frac{\sum_{i=1}^{n} (x_i - \bar{x})^2}{n-1}$$

Population standard deviation

$$\sigma = \sqrt{\sigma^2}$$

Sample standard deviation

$$s = \sqrt{s^2}$$

Population covariance

$$\sigma_{xy} = \frac{\sum_{i=1}^{N} (x_i - \mu_x)(y_i - \mu_y)}{N}$$

Sample covariance

$$s_{xy} = \frac{\sum_{i=1}^{n} (x_i - \bar{x})(y_i - \bar{y})}{n-1}$$

Population coefficient of correlation

$$\rho = \frac{\sigma_{xy}}{\sigma_x \sigma_y}$$

Sample coefficient of correlation

$$r = \frac{s_{xy}}{s_x s_y}$$

Coefficient of determination

$$R^2 = r^2$$

Slope coefficient

$$b_1 = \frac{s_{xy}}{s_x^2}$$

y-Intercept

$$b_0 = \bar{y} - b_1 \bar{x}$$

CHAPTER EXERCISES

4.76 Xr04-76 Osteoporosis is a condition in which bone density decreases, often resulting in broken bones. Bone density usually peaks at age 30 and decreases thereafter. To understand more about the condition, a random sample of women aged 50 and over were recruited. Each woman's bone density loss was recorded.

a. Compute the mean and median of these data.

b. Compute the standard deviation of the bone density losses.

c. Describe what you have learned from the statistics.

4.77 Refer to Exercise 4.76. In addition to the bone density losses, the ages of the women were also recorded. Compute the coefficient of determination and describe what this statistic tells you.

4.78 Xr04-78 Chris Golfnut loves the game of golf. Chris also loves statistics. Combining both passions, Chris records a sample of 100 scores.

a. What statistics should Chris compute to describe the scores?

b. Calculate the mean and standard deviation of the scores.

c. Briefly describe what the statistics computed in part (b) divulge.

4.79 Xr04-79 The Internet is growing rapidly with an increasing number of regular users. However, among people older than 50, Internet use is still relatively low. To learn more about this issue, a sample of 250 men and women older than 50 who had used the Internet at least once were selected. The number of hours on the Internet during the past month was recorded.

a. Calculate the mean and median.

b. Calculate the variance and standard deviation.

c. Draw a box plot.

d. Briefly describe what you have learned from the statistics you calculated.

4.80 Refer to Exercise 4.79. In addition to Internet use, the numbers of years of education were recorded.

a. Compute the coefficient of determination.

b. Determine the coefficients of the least squares line.

c. Describe what these statistics tell you about the relationship between Internet use and education.

d. Discuss the information obtained here and in Exercise 4.79.

4.81 Xr04-81 A sample was drawn of 1-acre plots of land planted with corn. The crop yields were recorded. Calculate the descriptive statistics you judge to be useful. Interpret these statistics.

4.82 Refer to Exercise 4.81. For each plot, the amounts of rainfall were also recorded.

a. Compute the coefficient of determination.

b. Determine the coefficients of the least squares line.

c. Describe what these statistics tell you about the relationship between crop yield and rainfall.

4.83 Refer to Exercise 4.81. For each plot, the amounts of fertilizer were recorded.

a. Compute the coefficient of determination.

b. Determine the coefficients of the least squares line.

c. Describe what these statistics tell you about the relationship between crop yield and the amount of fertilizer.

d. Discuss the information obtained here and in Exercise 4.81.

4.84 Xr-04-84 Increasing tuition has resulted in some students being saddled with large debts on graduation. To examine this issue, a random sample of recent graduates was asked to report whether they had student loans, and, if so, how much was the debt at graduation.

a. Compute all three measures of central location.

b. What do these statistics reveal about student loan debt at graduation?

CASE 4.1

DATA C04-01a C04-01b

Return to the Global Warming Question

Now that we have presented techniques that allow us to conduct more precise analyses, we will return to Case 3.1. Recall that there are two issues in this discussion. First, is there global warming and, second, if so, is carbon dioxide the cause? The only tools available at the end of Chapter 3 were graphical techniques including line charts and scatter diagrams. You are now invited to apply the more precise techniques used in this chapter to answer the same questions.

Here are the data sets you can work with:

C04-01 a: Column 1: Months numbered 1 to 1597
Column 2: Temperature anomalies produced by the National Climatic Data Center

C04-01b: Column 1: Year
Column 2: Month
Column 3: Monthly carbon dioxide levels measured by the Mauna Loa Observatory
Column 4: Temperature anomalies produced by the National Climatic Data Center

a. Use the least squares method to estimate average monthly changes in temperature anomalies.

b. Calculate the least squares line and the coefficient of correlation between CO_2 levels and temperature anomalies and describe your findings.

CASE 4.2

DATA C04-02a C04-02b C04-02c C04-02d

Another Return to the Global Warming Question

Did you conclude in Case 4.1 that the Earth has warmed since 1880 and that there is some linear relationship between CO_2 and temperature anomalies? If so, here is another look at the same data. C04-02a lists the temperature anomalies from 1880 to 1940, C04-02b lists the data from 1941 to 1975, C04-02c stores the temperature anomalies from 1976 to 1997 and C04-02d contains the data from 1998 to 2013. For each set of data, draw a line chart and report your findings.

5 DATA COLLECTION AND SAMPLING

5-1 METHODS OF COLLECTING DATA

5-2 SAMPLING

5-3 SAMPLING PLANS

5-4 SAMPLING AND NON-SAMPLING ERRORS

SAMPLING AND THE CENSUS

Censuses, which are conducted every 10 years in the European Union countries, provide valuable statistical input for formulating and evaluating policies. In addition, the censuses serve an important function if their results can be compared between Member States. The European Union census programmes from 1980, 1990, 2001 and 2011 rounds compare and harmonise the national censuses outputs. The information is of great value to policymaking, administration, businesses and the citizens. For example, businesses often use the information derived from the census to help make decisions about products, advertising and plant locations.

One of the problems with the census is the issue of undercounting, which occurs when some people are not included, or Member States might use data sources that are based only on statistically controlled samples of the statistical units of the target population. Regulation (EU) No. 1151/2010 makes sure that 'the size of the complementary set of statistical units is calculated according to the sampling design', and 'when the size of the census population is calculated, it is not affected by the sampling'.

Later in this chapter we will discuss how the sampling is conducted and how the adjustments are made.

INTRODUCTION

In Chapter 1, we briefly introduced the concept of statistical inference – the process of inferring information about a population from a sample. Because information about populations can usually be described by parameters, the statistical technique used generally deals with drawing inferences about population parameters from sample statistics. (Recall that a parameter is a measurement about a population, and a statistic is a measurement about a sample.)

Working within the covers of a statistics textbook, we can assume that population parameters are known. In real life, however, calculating parameters is virtually impossible because populations tend to be very large. As a result, most population parameters are not only unknown but also unknowable. The problem that motivates the subject of statistical inference is that we often need information about the value of parameters in order to make decisions. Only the context of the investigation can determine what the relevant population is, and indicate what might be an appropriate sample to draw from it. For example, to make decisions about whether to expand a line of clothing, we may need to know the mean annual expenditure on clothing by European adults. Because the size of this population is approximately 740 million, determining the mean is prohibitive. However, if we are willing to accept less than 100% accuracy, we can use statistical inference to obtain an estimate. Rather than investigating the entire population, we select a sample of people, determine the annual expenditures on clothing in this group and calculate the sample mean. Although the probability that the sample mean will equal the population mean is very small, we would expect them to be close. For many decisions, we need to know how close. We postpone that discussion until Chapters 10 and 11. In this chapter, we will discuss the basic concepts and techniques of sampling itself. But first we take a look at various sources for collecting data.

5-1 METHODS OF COLLECTING DATA

Most of this book addresses the problem of converting data into information. The question arises, where do data come from? The answer is that a large number of methods produce data. Before we proceed, however, we will remind you of the definition of data introduced in Section 2-1. Data are the observed values of a variable; that is, we define a variable or variables that are of interest to us and then proceed to collect observations of those variables.

5-1a Direct Observation

The simplest method of obtaining data is by direct observation. When data are gathered in this way, they are said to be **observational**. For example, suppose that a researcher for a pharmaceutical company wants to determine whether aspirin actually reduces the incidence of heart attacks. Observational data may be gathered by selecting a sample of men and women and asking each whether he or she has taken aspirin regularly over the past 2 years. Each person would be asked whether he or she had suffered a heart attack over the same period. The proportions reporting heart attacks would be compared and a statistical technique that is introduced in Chapter 12 would be used to determine whether aspirin is effective in reducing the likelihood of heart attacks. There are many drawbacks to this method. One of the most critical is that it is difficult to produce useful information in this way. For example, if the statistics practitioner concludes that people who take aspirin suffer fewer heart attacks, can we conclude that aspirin is effective? It may be that people who take aspirin tend to be more health conscious, and health-conscious people tend to have fewer heart attacks. The one advantage of direct observation is that it is relatively inexpensive.

5-1b Experiments

A more expensive but better way to produce data is through experiments. Data produced in this manner are called **experimental**. In the aspirin illustration, a statistics practitioner can randomly select men and women. The sample would be divided into two groups. One group would take aspirin regularly, and the other would not. After 2 years, the statistics practitioner would determine the proportion of people in each group who had suffered heart attacks, and statistical methods again would be used to determine whether aspirin works. If we find that the aspirin group suffered fewer heart attacks, then we may more confidently conclude that taking aspirin regularly is a healthy decision.

5-1c Surveys

One of the most familiar methods of collecting data is the **survey**, which solicits information from people concerning such things as their income, family size and opinions on various issues. We are all familiar, for example, with opinion polls that accompany each political election. The Gallup Poll and the European Commission are two well-known surveys of public opinion whose results are often reported by the media. But the majority of surveys are conducted for private use. Private surveys are used extensively by market researchers to determine the preferences and attitudes of consumers and voters. The results can be used for a variety of purposes, from helping to determine the target market for an advertising campaign to modifying a candidate's platform in an election campaign. As an illustration, consider a television network that has hired a market research firm to provide the network with a profile of owners of luxury automobiles, including what they watch on television and at what times. The network could then use this information to develop a package of recommended time slots for Audi commercials, including costs, which it would present to the Volkswagen Group. It is quite likely that many students reading this book will one day be marketing executives who will 'live and die' by such market research data.

An important aspect of surveys is the **response rate**. The response rate is the proportion of all people who were selected who complete the survey. As we discuss in the next section, a low response rate can destroy the validity of any conclusion resulting from the statistical analysis. Statistics practitioners need to ensure that data are reliable.

Mode of data collection

Personal (or Face-to-Face) Interview Many researchers feel that the best way to survey people is by means of a personal interview, which involves an interviewer soliciting information from a respondent by asking prepared questions. A personal interview has the advantage of having a higher expected response rate than other methods of data collection. In addition, there will probably be fewer incorrect responses resulting from respondents misunderstanding some questions because the interviewer can clarify misunderstandings when asked to. But the interviewer must also be careful not to say too much for fear of biasing the response. To avoid introducing such biases, as well as to reap the potential benefits of a personal interview, the interviewer must be well-trained in proper interviewing techniques and well-informed on the purpose of the study. The main disadvantage of personal interviews is that they are expensive, especially when travel is involved.

Telephone Survey A telephone survey is usually less expensive, can be conducted fairly quickly, permits a greater amount of follow-up, but it is also less personal and has a lower expected response rate. Unless the issue is of interest, many people will refuse to respond to telephone surveys. This problem is exacerbated by telemarketers trying to sell something.

Self-Administered (or Mail) Survey A third popular method of data collection is the self-administered questionnaire, which is usually mailed to a sample of people. This is an inexpensive method of conducting a survey and is therefore attractive when the number of people to be surveyed is large.

But self-administered questionnaires usually have a low response rate and may have a relatively high number of incorrect responses due to respondents misunderstanding some questions.

Questionnaire Design Whether a questionnaire is self-administered or completed by an interviewer, it must be well designed. Proper questionnaire design takes knowledge, experience, time and money. Some basic points to consider regarding questionnaire design follow.

1. First and foremost, the questionnaire should be kept as short as possible to encourage respondents to complete it. Most people are unwilling to spend much time filling out a questionnaire.
2. The questions themselves should also be short, as well as simply and clearly worded, to enable respondents to answer quickly, correctly and without ambiguity. Even familiar terms such as *'unemployed'* and *'family'* must be defined carefully because several interpretations are possible. Is this a question that will mean the same thing to everyone?
3. Questionnaires often begin with simple demographic questions to help respondents get started and become comfortable quickly.
4. Dichotomous questions (questions with only two possible responses such as 'yes' and 'no' and multiple-choice questions) are useful and popular because of their simplicity, but they also have possible shortcomings. For example, a respondent's choice of 'yes' or 'no' to a question may depend on certain assumptions not stated in the question. In the case of a multiple-choice question, a respondent may feel that none of the choices offered is suitable.
5. Open-ended questions provide an opportunity for respondents to express opinions more fully, but they are time-consuming and more difficult to tabulate and analyse.
6. Avoid using leading questions, such as 'Wouldn't you agree that the statistics exam was too difficult?' These types of questions tend to lead the respondent to a particular answer.
7. Time permitting, it is useful to pretest a questionnaire on a small number of people in order to uncover potential problems such as ambiguous wording.
8. Finally, when preparing the questions, think about how you intend to tabulate and analyse the responses. First, determine whether you are soliciting values (i.e. responses) for an interval variable or a nominal variable. Then consider which type of statistical techniques – descriptive or inferential – you intend to apply to the data to be collected, and note the requirements of the specific techniques to be used. Thinking about these questions will help ensure that the questionnaire is designed to collect the data you need.

Whatever method is used to collect primary data, we need to know something about sampling, the subject of the next section.

EXERCISES

5.1 Briefly describe the difference between observational and experimental data.

5.2 A soft drinks manufacturer has been supplying its cola drink in bottles to grocery stores and in cans to small convenience stores. The company is analysing sales of this cola drink to determine which type of packaging is preferred by consumers.

a. Is this study observational or experimental?

Explain your answer.

b. Outline a better method for determining whether a store will be supplied with cola in bottles or in cans so that future sales data will be more helpful in assessing the preferred type of packaging.

5.3 **a.** Briefly describe how you might design a study to investigate the relationship between smoking and lung cancer.

b. Is your study in part (a) observational or experimental? Explain why.

5.4 **a.** List three methods of conducting a survey of people.

b. Give an important advantage and disadvantage of each of the methods listed in part (a).

5.5 List five important points to consider when designing a questionnaire.

5-2 SAMPLING

The chief motive for examining a sample rather than a population is cost. Statistical inference permits us to draw conclusions about a population parameter based on a sample that is quite small in comparison to the size of the population. For example, television executives want to know the proportion of television viewers who watch a network's programmes. Because 100 million people may be watching television in the country on a given evening, determining the actual proportion of the population that is watching certain programmes is impractical and prohibitively expensive. The Broadcasters' Audience Research Board (BARB) ratings provide approximations of the desired information by observing what is watched by a sample of 5000 television viewers in the UK. The proportion of households watching a particular programme can be calculated for the households in the BARB sample. This sample proportion is then used as an **estimate** of the proportion of all households (the population proportion) that watched the programme.

Another illustration of sampling can be taken from the field of quality management. To ensure that a production process is operating properly, the operations manager needs to know what proportion of items being produced is defective. If the quality technician must destroy the item to determine whether it is defective, then there is no alternative to sampling: a complete inspection of the product population would destroy the entire output of the production process.

We know that the sample proportion of television viewers or of defective items is probably not exactly equal to the population proportion we want to estimate. Nonetheless, the sample statistic can come quite close to the parameter it is designed to estimate if the **target population** (the population about which we want to draw inferences or generalise) and the **sampled population** (the actual population from which the sample has been taken) are the same. In practice, these may not be the same. One of statistics' most famous failures illustrates this phenomenon.

The *Literary Digest* was a popular magazine of the 1920s and 1930s that had correctly predicted the outcomes of several presidential elections. In 1936, the *Digest* predicted that the Republican candidate, Alfred Landon, would defeat the Democratic incumbent, Franklin D. Roosevelt, by a 3 to 2 margin. But, in that election, Roosevelt defeated Landon in a landslide victory, garnering the support of 62% of the electorate. The source of this blunder was the sampling procedure, and there were two distinct mistakes.* First, the *Digest* sent out 10 million sample ballots to prospective voters. However, most of the names of these people were taken from the *Digest*'s subscription list and from telephone directories. Subscribers to the magazine and people who owned telephones tended to be wealthier than average and such people then, as today, tended to vote Republican. In addition, only 2.3 million ballots were returned resulting in a self-selected sample.

Self-selected samples are almost always biased because the individuals who participate in them are more keenly interested in the issue than are the other members of the population. You often find similar surveys conducted today when radio and television stations ask people to call and give their opinion on an issue of interest. Again, only listeners who are concerned about the topic and have enough patience to get through to the station will be included in the sample. Hence, the sampled population is composed entirely of people who are interested in the issue, whereas the target population is made up of all the people within the listening radius of the radio station. As a result, the conclusions drawn from such surveys are frequently wrong.

An excellent example of this phenomenon occurred on ABC's *Nightline* in 1984. Viewers were given a 900 telephone number (cost: 50 cents) and asked to telephone in their responses to the question of whether the United Nations should continue to be located in the USA. More than 186 000 people called, with 67% responding 'no'. At the same time, a (more scientific) market research poll of 500 people revealed that 72% wanted the United Nations to remain in the USA. In general, because the true value of the parameter being estimated is never known, these surveys give the impression of

---

*Many statisticians ascribe the *Literary Digest*'s statistical debacle to the wrong causes. For an understanding of what really happened, read Maurice C. Bryson, 'The Literary Digest Poll: Making of a Statistical Myth', *American Statistician* 30(4) (November 1976): 184–185.

providing useful information. In fact, the results of such surveys are likely to be no more accurate than the results of the 1936 *Literary Digest* poll or *Nightline*'s phone-in show. Statisticians have coined two terms to describe these polls: SLOP (self-selected opinion poll) and *Oy vey* (from the Yiddish lament), both of which convey the contempt that statisticians have for such data-gathering processes.

EXERCISES

5.6 For each of the following sampling plans, indicate why the target population and the sampled population are not the same.

a. To determine the opinions and attitudes of customers who regularly shop at a particular mall, a surveyor stands outside a large department store in the mall and randomly selects people to participate in the survey.

b. A library wants to estimate the proportion of its books that have been damaged. The librarians decide to select one book per shelf as a sample by measuring 12 in (31 cm) from the left edge of each shelf and selecting the book in that location.

c. Political surveyors visit 200 residences during one afternoon to ask eligible voters present in the house at the time whom they intend to vote for.

5.7 **a.** Describe why the *Literary Digest* poll of 1936 has become infamous.

b. What caused this poll to be so wrong?

5.8 **a.** What is meant by *self-selected sample*?

b. Give an example of a recent poll that involved a self-selected sample.

c. Why are self-selected samples not desirable?

5.9 A regular feature in a newspaper asks readers to respond via email to a survey that requires a 'yes' or 'no' response. In the following day's newspaper, the percentage of 'yes' and 'no' responses are reported. Discuss why we should ignore these statistics.

5.10 Suppose your statistics professor distributes a questionnaire about the course. One of the questions asks, 'Would you recommend this course to a friend?' Can the professor use the results to infer something about all statistics courses? Explain.

5-3 SAMPLING PLANS

There are two main types of sampling: random sampling and non-random sampling. Our objective in this section is to introduce the three most common ways of obtaining random sampling: simple random sampling, stratified random sampling and cluster sampling. We begin our presentation with the most basic design.

5-3a Simple Random Sampling

> **Simple Random Sample**
> A **simple random sample** is a sample selected in such a way that every possible sample with the same number of observations is equally likely to be chosen.

One way to conduct a simple random sample is to assign a number to each element in the population, write these numbers on individual slips of paper, toss them into a hat and draw the required number of slips (the sample size, *n*) from the hat. This is the kind of procedure that occurs in raffles, when all the ticket stubs go into a large rotating drum from which the winners are selected.

Sometimes elements of the population are already numbered. For example, virtually all adults have National Insurance numbers (in the UK) or Personal Identification numbers (in Denmark); all employees of large corporations have employee numbers; many people have driver's licence numbers, medical plan numbers, student numbers and so on. In such cases, choosing which sampling procedure to use is simply a matter of deciding how to select from among these numbers.

In other cases, the existing form of numbering has built-in flaws that make it inappropriate as a source of samples. Not everyone has a telephone number, for example, so the telephone book does not list all the people in a given area. Many households have two (or more) adults but only one telephone listing. It seems that everyone in the world has a mobile phone. Many of these people have no landline telephone, so they do not appear on any list. Some people do not have telephones, some have unlisted telephone numbers, and some have more than one telephone; these differences mean that each element of the population does not have an equal probability of being selected.

After each element of the chosen population has been assigned a unique number, sample numbers can be selected at random. A random number table can be used to select these sample numbers. (See, for example, *CRC Standard Management Tables*, W. H. Beyer, ed., Boca Raton, FL: CRC Press.) Alternatively, we can use Excel to perform this function (i.e. RAND() or RANDBETWEEN()).

Example 5.1

Random Sample of Income Tax Returns

A government income tax auditor has been given responsibility for 1000 tax returns. A computer is used to check the arithmetic of each return. However, to determine whether the returns have been completed honestly, the auditor must check each entry and confirm its veracity. Because it takes, on average, 1 hour to completely audit a return and she has only 1 week to complete the task, the auditor has decided to randomly select 40 returns. The returns are numbered from 1 to 1000. Use a computer random-number generator to select the sample for the auditor.

SOLUTION:

We generated 50 numbers between 1 and 1000 even though we needed only 40 numbers. We did so because it is likely that there will be some duplicates. We will use the first 40 unique random numbers to select our sample. The following numbers were generated by Excel. The instructions for both Excel and Minitab are provided here. [Notice that the 24th and 36th (counting down the columns) numbers generated were the same – 467.]

Computer-Generated Random Numbers

383	246	372	952	75
101	46	356	54	199
597	33	911	706	65
900	165	467	817	359
885	220	427	973	488
959	18	304	467	512
15	286	976	301	374
408	344	807	751	986
864	554	992	352	41
139	358	257	776	231

EXCEL

Instructions

1. Click **Data, Data Analysis** and **Random Number Generation.**
2. Specify the **Number of Variables** (1) and the **Number of Random Numbers** (50).
3. Select **Uniform Distribution.**
4. Specify the range of the uniform distribution (**Parameters**) (0 and 1).
5. Click **OK**. Column A will fill with 50 numbers that range between 0 and 1.
6. Multiply column A by 1000 and store the products in column B.
7. Make cell C1 active, and click f_x, **Math & Trig, ROUNDUP(number,num_digits)** and **OK.**
8. Specify the first number to be rounded (B1).
9. Type the **number of digits** (decimal places) (0). Click **OK.**
10. Complete column C.

The first five steps command Excel to generate 50 uniformly distributed random numbers between 0 and 1 to be stored in column A. Steps 6 through 10 convert these random numbers to integers between 1and 1000. Each tax return has the same probability 1/1000 = 0.001 of being selected. Thus, each member of the population is equally likely to be included in the sample.

MINITAB

Instructions

1. Click **Calc, Random Data** and **Integer . . . .**
2. Type the number of random numbers you wish (50).
3. Specify where the numbers are to be stored (**C1**).
4. Specify the **Minimum value** (1).
5. Specify the **Maximum value** (1000). Click **OK.**

INTERPRET

The auditor would examine the tax returns selected by the computer. She would pick returns numbered 383, 101, 597, . . . ,352, 776 and 75 (the first 40 unique numbers). Each of these returns would be audited to determine whether it is fraudulent. If the objective is to audit these 40 returns, no statistical procedure would be employed. However, if the objective is to estimate the proportion of all 1000 returns that are dishonest, then she would use one of the inferential techniques presented later in this book.

5-3b Stratified Random Sampling

In making inferences about a population, we attempt to extract as much information as possible from a sample. The basic sampling plan, simple random sampling, often accomplishes this goal at low cost. Other methods, however, can be used to increase the amount of information about the population and the likelihood of representativeness, especially if one's sample is not very large. One such procedure is *stratified random sampling.*

Stratified Random Sample

A **stratified random sample** is obtained by separating the population into mutually exclusive sets, or strata, and then drawing simple random samples from each stratum.

Examples of criteria for separating a population into strata (and of the strata themselves) follow.

1. Gender
   - male
   - female
2. Age
   - under 20
   - 20–30
   - 31–40
   - 41–50
   - 51–60
   - over 60
3. Occupation
   - professional
   - clerical
   - service
   - other
4. Household income
   - under £25 000
   - £25 000–£39 999
   - £40 000–£60 000
   - over £60 000

To illustrate, suppose a public opinion survey is to be conducted to determine how many people favour a tax increase. A stratified random sample could be obtained by selecting a random sample of people from each of the four income groups we just described. We usually stratify in a way that enables us to obtain particular kinds of information. In this example, we would like to know whether people in the different income categories differ in their opinions about the proposed tax increase, because the tax increase will affect the strata differently. We avoid stratifying when there is no connection between the survey and the strata. For example, little purpose is served in trying to determine whether people within religious strata have divergent opinions about the tax increase.

One advantage of stratification is that, besides acquiring information about the entire population, we can also make inferences within each stratum or compare strata. For instance, we can estimate what proportion of the lowest income group favours the tax increase, or we can compare the highest and lowest income groups to determine whether they differ in their support of the tax increase.

Any stratification must be done in such a way that the strata are mutually exclusive: Each member of the population must be assigned to exactly one stratum. After the population has been stratified in this way, we can use simple random sampling to generate the complete sample. There are several ways to do this. For example, we can draw random samples from each of the four income groups according to their proportions in the population. Thus, if in the population the relative frequencies of the four groups are as listed here, our sample will be stratified in the same proportions. If a total sample of 1000 is to be drawn, then we will randomly select 250 from stratum 1, 400 from stratum 2, 300 from stratum 3 and 50 from stratum 4.

Stratum	Income Categories (£)	Population Proportions (%)
1	Less than 25 000	25
2	25 000–39 999	40
3	40 000–60 000	30
4	More than 60 000	5

The problem with this approach, however, is that if we want to make inferences about the last stratum, a sample of 50 may be too small to produce useful information. In such cases, we usually increase the sample size of the smallest stratum to ensure that the sample data provide enough information for our purposes. An adjustment must then be made before we attempt to draw inferences about the entire population. The required procedure is beyond the level of this book. We recommend that anyone planning such a survey consult an expert statistician or a reference book on the subject. Better still, become an expert statistician yourself by taking additional statistics courses.

5-3c Cluster Sampling

Cluster Sample
A **cluster sample** is a simple random sample of groups or clusters of elements.

Cluster sampling is particularly useful when it is difficult or costly to develop a complete list of the population members (making it difficult and costly to generate a simple random sample). It is also useful whenever the population elements are widely dispersed geographically. For example, suppose we wanted to estimate the average annual household income in a large city. To use simple random sampling, we would need a complete list of households in the city from which to sample. To use stratified random sampling, we would need the list of households, and we would also need to have each household categorised by some other variable (such as age of household head) in order to develop the strata. A less-expensive alternative would be to let each block within the city represent a cluster. A sample of clusters could then be randomly selected, and every household within these clusters could be questioned to determine income. By reducing the distances the surveyor must cover to gather data, cluster sampling reduces the cost.

But cluster sampling also increases sampling error (see Section 5-4) because households belonging to the same cluster are likely to be similar in many respects, including household income. This can be partially offset by using some of the cost savings to choose a larger sample than would be used for a simple random sample.

5-3d Sample Size

Whichever type of sampling plan you select, you still have to decide what size sample to use. Determining the appropriate sample size will be addressed in detail in Chapters 10 and 12. Until then, samples should be as large as the researcher can obtain with a reasonable expenditure of time and energy. The larger the sample size is, the more accurate we can expect the estimates to be. A recommended minimum number of subjects is 100 for a descriptive study, 50 for a correlation study and 30 in each group for experimental or causal–comparative studies.

SAMPLING AND THE CENSUS

Statistical offices such as Eurostat (Statistical Office of the European Communities) conduct cluster sampling to adjust the undercounting. To carry out cluster sampling, a sample of clusters, which is a formed group of population elements such as clusters of people in households, is first drawn from the population of clusters by using one of the basic sampling techniques.

Cluster sampling is often used for populations that have a large regional spread. Suppose that for the 2011 Census, a statistical office from Germany randomly sampled 10 000 blocks, which contained 250 000 housing units. Each unit was intensively revisited to ensure that all residents were counted. From the results of this survey, the statistical office estimated the

(continued)

number of people missed by the first census in various subgroups, defined by several variables including gender, race and age. Because of the importance of determining state populations, adjustments were made to state totals. For example, by comparing the results of the census and of the sampling, the statistical office determined that the undercount in Germany was 1.5079%. The official census produced a state population of 80 586 000. Taking 1.5079% of this total produced an adjustment of 1 215 156. Using this method changed the population of Germany to 81 801 156.

It should be noted that this process is contentious. The controversy concerns the way in which subgroups are defined. Changing the definition alters the undercounts, making this statistical technique subject to politicking.

EXERCISES

5.11 A statistics practitioner would like to conduct a survey to ask people their views on a proposed new shopping mall in their community. According to the latest census, there are 500 households in the community. The statistician has numbered each household (from 1 to 500), and she would like to randomly select 25 of these households to participate in the study. Use Excel or Minitab to generate the sample.

5.12 A German safety expert wants to determine the proportion of cars in his state (i.e. Stuttgart) with worn tyre treads. The state licence plate contains seven digits. Use Excel or Minitab to generate a sample of 20 cars to be examined.

5.13 A large university campus has 60 000 students. The president of the students' association wants to conduct a survey of the students to determine their views on an increase in the student activity fee. She would like to acquire information about all the students but would also like to compare the school of business, the faculty of arts and sciences, and the graduate school. Describe a sampling plan that accomplishes these goals.

5.14 A telemarketing firm has recorded the households that have purchased one or more of the company's products. The firm would like to conduct a survey of purchasers to acquire information about their attitude concerning the timing of the telephone calls. The president of the company would like to know the views of all purchasers but would also like to compare the attitudes of people in the West, South, North and East. Describe a suitable sampling plan.

5.15 The operations manager of a large plant in Singapore with four departments wants to estimate the person-hours lost per month from accidents. Describe a sampling plan that would be suitable for estimating the plant-wide loss and for comparing departments.

5.16 A statistics practitioner wants to estimate the mean age of children in his city. Unfortunately, he does not have a complete list of households. Describe a sampling plan that would be suitable for his purposes.

5-4 SAMPLING AND NON-SAMPLING ERRORS

Two major types of error can arise when a sample of observations is taken from a population: *sampling error* and *non-sampling error*. Anyone reviewing the results of sample surveys and studies, as well as statistics practitioners conducting surveys and applying statistical techniques, should understand the sources of these errors.

5-4a Sampling Error

Sampling error refers to differences between the sample and the population that exist only because of the observations that happened to be selected for the sample. Sampling error is an error that we expect to occur when we make a statement about a population that is based only on the observations contained in a sample taken from the population. Furthermore, no two samples will be the same in their characteristics.

To illustrate, suppose that we wish to determine the mean annual income of middle-class workers. To determine this parameter we would have to ask each middle-class worker what his or her income is and then calculate the mean of all the responses. Because the size of this population is several million, the task is both expensive and impractical. We can use statistical inference to estimate the mean income μ of the population if we are willing to accept less than 100% accuracy. We record the incomes of a sample of the workers and find the mean $\overline{x}$ of this sample of incomes. This sample mean is an estimate of the desired population mean. But the value of the sample mean will deviate from the population mean simply by chance because the value of the sample mean depends on which incomes just happened to be selected for the sample. The difference between the true (unknown) value of the population mean and its estimate, the sample mean, is the sampling error. The size of this deviation may be large simply because of bad luck – bad luck that a particularly unrepresentative sample happened to be selected. The only way we can reduce the expected size of this error is to take a larger sample.

Given a fixed sample size, the best we can do is to state the probability that the sampling error is less than a certain amount (as we will discuss in Chapter 10). It is common today for such a statement to accompany the results of an opinion poll. If an opinion poll states that, based on sample results, the incumbent candidate for mayor has the support of 54% of eligible voters in an upcoming election, the statement may be accompanied by the following explanatory note: 'This percentage is correct to within three percentage points, 19 times out of 20'. This statement means that we estimate that the actual level of support for the candidate is between 51% and 57%, and that in the long run this type of procedure is correct 95% of the time.

5-4b Non-sampling Error

Non-sampling error is more serious than sampling error because taking a larger sample won't diminish the size, or the possibility of occurrence, of this error. Even a census can (and probably will) contain non-sampling errors. **Non-sampling errors** result from mistakes made in the acquisition of data or from the sample observations being selected improperly.

1. *Errors in data acquisition.* This type of error arises from the recording of incorrect responses. Incorrect responses may be the result of incorrect measurements being taken because of faulty equipment, mistakes made during transcription from primary sources, inaccurate recording of data because terms were misinterpreted, or inaccurate responses were given to questions concerning sensitive issues such as sexual activity or possible tax evasion.
2. *Non-response error.* **Non-response error** refers to error (or **bias**) introduced when responses are not obtained from some members of the sample. When this happens, the sample observations that are collected may not be representative of the target population, resulting in biased results (as was discussed in Section 5-2). Non-response can occur for a number of reasons. An interviewer may be unable to contact a person listed in the sample, or the sampled person may refuse to respond for some reason. In either case, responses are not obtained from a sampled person, and bias is introduced. The problem of non-response is even greater when self-administered questionnaires are used rather than an interviewer, who can attempt to reduce the non-response rate by means of callbacks. As noted previously, the *Literary Digest* fiasco was largely the result of a high non-response rate, resulting in a biased, self-selected sample.
3. *Selection bias.* **Selection bias** occurs when the sampling plan is such that some members of the target population cannot possibly be selected for inclusion in the sample. Together with non-response error, selection bias played a role in the *Literary Digest* poll being so wrong, as voters without telephones or without a subscription to *Literary Digest* were excluded from possible inclusion in the sample taken.

EXERCISES

5.17 **a.** Explain the difference between sampling error and non-sampling error.

b. Which type of error in part (a) is more serious? Why?

5.18 Briefly describe three types of non-sampling error.

5.19 Is it possible for a sample to yield better results than a census? Explain.

CHAPTER SUMMARY

Because most populations are very large, it is extremely costly and impractical to investigate each member of the population to determine the values of the parameters. As a practical alternative, we take a sample from the population and use the sample statistics to draw inferences about the parameters. Care must be taken to ensure that the **sampled population** is the same as the **target population.**

We can choose from among several different sampling plans, including **simple random sampling, stratified random sampling** and **cluster sampling**. Whatever sampling plan is used, it is important to realise that both **sampling error** and **non-sampling error** will occur and to understand what the sources of these errors are.

PART THREE

PROBABILITY

6 Probability
7 Random Variables and Discrete Probability Distributions
8 Continuous Probability Distributions

6 PROBABILITY

6-1 ASSIGNING PROBABILITY TO EVENTS

6-2 JOINT, MARGINAL AND CONDITIONAL PROBABILITY

6-3 PROBABILITY RULES AND TREES

6-4 BAYES'S LAW

6-5 IDENTIFYING THE CORRECT METHOD

AUDITING TAX RETURNS

Government auditors routinely check tax returns to determine whether calculation errors were made. They also attempt to detect fraudulent returns. There are several methods that dishonest taxpayers use to evade income tax. One method is not to declare various sources of income. Auditors have several detection methods, including spending patterns. Another form of tax fraud is to invent deductions that are not real. After analysing the returns of thousands of self-employed taxpayers, an auditor has determined that 45% of fraudulent returns contain two suspicious deductions, 28% contain one suspicious deduction and the rest no suspicious deductions. Among honest returns the rates are 11% for two deductions, 18% for one deduction and 71% for no deductions. The auditor believes that 5% of the returns of self-employed individuals contain significant fraud. The auditor has just received a tax return for a self-employed individual that contains one suspicious expense deduction. What is the probability that this tax return contains significant fraud? **See Section 6-4a for the answer.**

INTRODUCTION

In Chapters 2, 3 and 4, we introduced graphical and numerical descriptive methods. Although the methods are useful on their own, we are particularly interested in developing statistical inference. As we pointed out in Chapter 1, statistical inference is the process by which we acquire information about populations from samples. A critical component of inference is *probability* because it provides the link between the population and the sample, and it is indispensable to the learning and understanding of interference concepts, such as sample statistics or significance levels.

Our primary objective in this and the following two chapters is to develop the probability-based tools that are at the basis of statistical inference. However, probability can also play a critical role in decision-making, a subject we explore in Chapter 20.

6-1 ASSIGNING PROBABILITY TO EVENTS

To introduce probability, we must first define a *random experiment*.

Random Experiment

A **random experiment** is an action or process that leads to one of several possible outcomes.

Here are six illustrations of random experiments and their outcomes.

Illustration 1.	Experiment: Outcomes:	Flip a coin. Heads and tails.
Illustration 2.	Experiment: Outcomes:	Record marks on a statistics test (out of 100). Numbers between 0 and 100.
Illustration 3.	Experiment: Outcomes:	Record grade on a statistics test. A, B, C, D and F.
Illustration 4.	Experiment: Outcomes:	Record student evaluations of a course. Poor, fair, good, very good and excellent.
Illustration 5.	Experiment: Outcomes:	Measure the time to assemble a computer. Number whose smallest possible value is 0 seconds with no predefined upper limit.
Illustration 6.	Experiment: Outcomes:	Record the party that a voter will vote for in an upcoming election. Party A, Party B, …

The first step in assigning probabilities is to produce a list of the outcomes. The listed outcomes must be **exhaustive**, which means that all possible outcomes must be included. In addition, the outcomes must be **mutually exclusive**, which means that no two outcomes can occur at the same time.

To illustrate the concept of exhaustive outcomes consider this list of the outcomes of the toss of a die:

1   2   3   4   5

This list is not exhaustive, because we have omitted 6.

The concept of mutual exclusiveness can be seen by listing the following outcomes in illustration 2:

0–50   50–60   60–70   80–100

If these intervals include both the lower and upper limits, then these outcomes are not mutually exclusive because two outcomes can occur for any student. For example, if a student receives a mark of 70, both the third and fourth outcomes occur.

Note that we could produce more than one list of exhaustive and mutually exclusive outcomes. For example, here is another list of outcomes for illustration 3:

Pass and Fail

A list of exhaustive and mutually exclusive outcomes is called a *sample space* and is denoted by S. The outcomes are denoted by $O_1, O_2, \ldots, O_k$

Sample Space

A **sample space** of a random experiment is a list of all possible outcomes of the experiment. The outcomes must be exhaustive and mutually exclusive.

Using set notation, we represent the sample space and its outcomes as

$$S = \{O_1, O_2, \ldots, O_k\}$$

Once a sample space has been prepared we begin the task of assigning probabilities to the outcomes. There are three ways to assign probability to outcomes. However it is done, there are two rules governing probabilities as stated in the next box.

Requirements of Probabilities

Given a sample space $S = \{O_1, O_2, \ldots, O_k\}$, the probabilities assigned to the outcomes must satisfy two requirements.

1. The probability of any outcome must lie between 0 and 1; that is,

$$0 \le P(O_i) \le 1 \quad \text{for each } i$$

[Note: $P(O_i)$ is the notation we use to represent the probability that outcome O_i will occur.]

2. The sum of the probabilities of all the outcomes in a sample space must be 1. That is,

$$\sum_{i=1}^{k} P(O_i) = 1$$

6-1a Three Approaches to Assigning Probabilities

The **classical approach** is used by mathematicians to help determine probability associated with games of chance. For example, the classical approach specifies that the probabilities of heads and tails in the flip of a balanced coin are equal to each other. Because the sum of the probabilities must be 1, the probability of heads and the probability of tails are both 50%. Similarly, the six possible outcomes of the toss of a balanced die have the same probability; each is assigned a probability of 1/6. In some experiments, it is necessary to develop mathematical ways to count the number of outcomes. For example, to determine the probability of winning a lottery, we need to determine the number of possible combinations. For details on how to count events, see the website Appendix Counting Formulas.

The **relative frequency approach** defines probability as the long-run relative frequency with which an outcome occurs. For example, suppose that we know that of the last 1000 students who took the statistics course you are now taking, 200 received a grade of *A*. The relative frequency of *A*s is then 200/1000 or 20%. This figure represents an estimate of the probability of obtaining a grade of *A* in the course. It is only an estimate because the relative frequency approach defines probability as the 'long-run' relative frequency. One thousand students do not constitute the long run. The larger the number of students whose grades we have observed, the better the estimate becomes. In theory, we would have to observe an infinite number of grades to determine the exact probability.

When it is not reasonable to use the classical approach and there is no history of the outcomes, we have no alternative but to employ the **subjective approach**. In the subjective approach, we define probability as the degree of belief that we hold in the occurrence of an event. An excellent example is derived from the field of investment. An investor would like to know the probability that a particular stock will increase in value. Using the subjective approach, the investor would analyse a number of factors associated with the stock and the stock market in general and, using his or her judgement, assign a probability to the outcomes of interest.

6-1b Defining Events

An individual outcome of a sample space is called a *simple event*. All other events are composed of the simple events in a sample space. For example, a *string event* is an event which involves a sequence of simple events and a *compound event* is one comprising two or more string events.

Event

An **event** is a collection or set of one or more simple events in a sample space.

In illustration 2, we can define the event; achieve a grade of *A*, as the set of numbers that lie between 80 and 100, inclusive. Using set notation, we have

$$A = \{80, 81, 82, \ldots, 99, 100\}$$

Similarly,

$$F = \{0, 1, 2, \ldots, 48, 49\}$$

6-1c Probability of Events

We can now define the probability of any event.

Probability of an Event

The probability of an event is the sum of the probabilities of the simple events that constitute the event.

For example, suppose that in illustration 3, we employed the relative frequency approach to assign probabilities to the simple events as follows:

$$P(A) = 0.20$$
$$P(B) = 0.30$$
$$P(C) = 0.25$$
$$P(D) = 0.15$$
$$P(F) = 0.10$$

The probability of the event, pass the course, is

$$P(\text{Pass the course}) = P(A) + P(B) + P(C) + P(D) = 0.20 + 0.30 + 0.25 + 0.15 = 0.90$$

6-1d Interpreting Probability

No matter what method was used to assign probability, we interpret it using the relative frequency approach for an infinite number of experiments. For example, an investor may have used the subjective approach to determine that there is a 65% probability that a particular stock's price will increase over the next month. However, we interpret the 65% figure to mean that if we had an infinite number of stocks with exactly the same economic and market characteristics as the one the investor will buy, 65% of them will increase in price over the next month. Similarly, we can determine that the probability of throwing a 5 with a balanced die is 1/6. We may have used the classical approach to determine this probability. However, we interpret the number as the proportion of times that a 5 is observed on a balanced die thrown an infinite number of times.

This relative frequency approach is useful to interpret probability statements such as those heard from weather forecasters or scientists. You will also discover that this is the way we link the population and the sample in statistical inference.

EXERCISES

6.1 A weather forecaster in Aberdeen reports that the probability of rain tomorrow is 10%.

a. Which approach was used to arrive at this number?

b. How do you interpret the probability?

6.2 A sportscaster states that he believes that the probability that Chelsea Football Club will win the European Cup this year is 25%.

a. Which method was used to assign that probability?

b. How would you interpret the probability?

6.3 A quiz contains a multiple-choice question with five possible answers, only one of which is correct. A student plans to guess the answer because he knows absolutely nothing about the subject.

a. Produce the sample space for each question.

b. Assign probabilities to the simple events in the sample space you produced.

c. Which approach did you use to answer part (b)?

d. Interpret the probabilities you assigned in part (b).

6.4 An investor tells you that in her estimation there is a 60% probability that the FTSE 100 Index will increase tomorrow.

a. Which approach was used to produce this figure?

b. Interpret the 60% probability.

6.5 The sample space of the toss of a fair die is

$$S = \{1, 2, 3, 4, 5, 6\}$$

If the die is balanced each simple event has the same probability. Find the probability of the following events.

a. An even number

b. A number less than or equal to 4

c. A number greater than or equal to 5

6.6 Four candidates are running for mayor. The four candidates are Adams, Brown, Collins and Dalton. Determine the sample space of the results of the election.

6.7 Refer to Exercise 6.6. Employing the subjective approach a political scientist has assigned the following probabilities:

$$P(\text{Adams wins}) = 0.42$$
$$P(\text{Brown wins}) = 0.09$$
$$P(\text{Collins wins}) = 0.27$$
$$P(\text{Dalton wins}) = 0.22$$

Determine the probabilities of the following events.

a. Adams loses

b. Either Brown or Dalton wins

c. Adams, Brown or Collins wins

6.8 The manager of a computer store has kept track of the number of computers sold per day. On the basis of this information, the manager produced the following list of the number of daily sales.

Number of Computers Sold	Probability
0	0.08
1	0.17
2	0.26
3	0.21
4	0.18
5	0.10

a. If we define the experiment as observing the number of computers sold tomorrow, determine the sample space.

b. Use set notation to define the event, sell more than three computers.

c. What is the probability of selling five computers?

d. What is the probability of selling two, three or four computers?

e. What is the probability of selling six computers?

6.9 Three contractors (call them contractors 1, 2 and 3) bid on a project to build a new bridge. What is the sample space?

6.10 Refer to Exercise 6.9. Suppose that you believe that contractor 1 is twice as likely to win as contractor 3 and that contractor 2 is three times as likely to win as contactor 3. What are the probabilities of winning for each contractor?

6.11 Shoppers can pay for their purchases with cash, a credit card or a debit card. Suppose that the proprietor of a shop determines that 60% of her customers use a credit card, 30% pay with cash and the rest use a debit card.

a. Determine the sample space for this experiment.

b. Assign probabilities to the simple events.

c. Which method did you use in part (b)?

6.12 Refer to Exercise 6.11.

a. What is the probability that a customer does not use a credit card?

b. What is the probability that a customer pays in cash or with a credit card?

6.13 A survey asks adults to report their marital status. The sample space is $S = \{\text{single, married, divorced, widowed}\}$. Use set notation to represent the event the adult is not married.

6.14 Refer to Exercise 6.13. Suppose that in the city in which the survey is conducted, 50% of adults are married, 15% are single, 25% are divorced and 10% are widowed.

a. Assign probabilities to each simple event in the sample space.

b. Which approach did you use in part (a)?

6.15 Refer to Exercises 6.13 and 6.14. Find the probability of each of the following events.

a. The adult is single

b. The adult is not divorced

c. The adult is either widowed or divorced

6-2 JOINT, MARGINAL AND CONDITIONAL PROBABILITY

In the previous section, we described how to produce a sample space and assign probabilities to the simple events in the sample space. Although this method of determining probability is useful, we need to develop more sophisticated methods. In this section, we discuss how to calculate the probability of more complicated events from the probability of related events. Here is an illustration of the process.

The sample space for the toss of a die is

$$S = \{1, 2, 3, 4, 5, 6\}$$

If the die is balanced, the probability of each simple event is 1/6. In most parlour games and casinos, players toss two dice. To determine playing and wagering strategies, players need to compute the probabilities of various totals of the two dice. For example, the probability of tossing a total of 3 with two dice is 2/36. This probability was derived by creating combinations of the simple events. There are several different types of combinations. One of the most important types is the *intersection* of two events.

6-2a Intersection

Intersection of Events *A* and *B*

The **intersection** of events *A* and *B* is the event that occurs when both *A* and *B* occur. It is denoted as

$$A \text{ and } B$$

The probability of the intersection is called the **joint probability**.

For example, one way to toss a 3 with two dice is to toss a 1 on the first die *and* a 2 on the second die, which is the intersection of two simple events. Incidentally, to compute the probability of a total of 3, we need to combine this intersection with another intersection, namely, a 2 on the first die and a 1 on the second die. This type of combination is called a *union* of two events, and it will be described later in this section. Here is another illustration.

APPLICATIONS IN FINANCE

Mutual Funds

A mutual fund is a pool of investments made on behalf of people who share similar objectives. In most cases, a professional manager who has been educated in finance and statistics, manages the fund. He or she makes decisions to buy and sell individual stocks and bonds in accordance with a specified investment philosophy. For example, there are funds that concentrate on other publicly traded mutual fund companies. Other mutual funds specialise in Internet stocks (so-called dot. coms), whereas others buy stocks of biotechnology firms. Surprisingly, most mutual funds do not outperform the market; that is, the increase in the net asset value (NAV) of the mutual fund is often less than the increase in the value of stock indexes that represent their stock markets. One reason for this is the management expense ratio (MER) which is a measure of the costs charged to the fund by the manager to cover expenses, including the salary and bonus of the managers. The MERs for most funds range from 5% to more than 4%. The ultimate success of the fund depends on the skill and knowledge of the fund manager. This raises the question: Which managers do best?

EXAMPLE 6.1

Data Xm06-01

Determinants of Success among Mutual Fund Managers – Part 1

Why are some mutual fund managers more successful than others? One possible factor is the university where the manager earned his or her master of business administration (MBA). Suppose that a potential investor examined the relationship between how well the mutual fund performs and where the fund manager earned his or her MBA. After the analysis, Table 6.1, a table of joint probabilities, was developed. Analyse these probabilities and interpret the results.

TABLE 6.1 Determinants of Success among Mutual Fund Managers, Part 1*

	MUTUAL FUND OUTPERFORMS MARKET	MUTUAL FUND DOES NOT OUTPERFORM MARKET
Top-20 MBA programme	0.11	0.29
Not top-MBA programme	0.06	0.54

*This example is adapted from 'Are Some Mutual Fund Managers Better than Others? Cross-Sectional Patterns in Behavior and Performance' by Judith Chevalier and Glenn Ellison, Working paper 5852, National Bureau of Economic Research.

Table 6.1 tells us the joint probability that a mutual fund outperforms the market *and* that its manager graduated from a top-20 MBA programme is 0.11; that is, 11% of all mutual funds outperform the market and their managers graduated from a top-20 MBA programme. The other three joint probabilities are defined similarly:

The probability that a mutual fund outperforms the market and that its manager did not graduate from a top-20 MBA programme is 0.06.

The probability that a mutual fund does not outperform the market and that its manager graduated from a top-20 MBA programme is 0.29.

The probability that a mutual fund does not outperform the market and that its manager did not graduate from a top-20 MBA programme is 0.54.

To help make our task easier, we will use notation to represent the events. Let

A_1 = **Fund manager graduated from a top-20 MBA programme**

A_2 = **Fund manager did not graduate from a top-20 MBA programme**

B_1 = **Fund outperforms the market**

B_2 = **Fund does not outperform the market**

Thus,

$$P(A_1 \text{ and } B_1) = 0.11$$
$$P(A_2 \text{ and } B_1) = 0.06$$
$$P(A_1 \text{ and } B_2) = 0.29$$
$$P(A_2 \text{ and } B_2) = 0.54$$

6-2b Marginal Probability

The joint probabilities in Table 6.1 allow us to compute various probabilities. **Marginal probabilities**, computed by adding across rows or down columns, are so named because they are calculated in the margins of the table.

$$P(A_1 \text{ and } B_1) + P(A_1 \text{ and } B_2) = 0.11 + 0.29 = 0.40$$

Notice that both intersections state that the manager graduated from a top-20 MBA programme (represented by A_1). Thus, when randomly selecting mutual funds, the probability that its manager graduated from a top-20 MBA programme is 0.40. Expressed as relative frequency, 40% of all mutual fund managers graduated from a top-20 MBA programme.

Adding across the second row:

$$P(A_2 \text{ and } B_1) + P(A_2 \text{ and } B_2) = 0.06 + 0.54 = 0.60$$

This probability tells us that 60% of all mutual fund managers did not graduate from a top-20 MBA programme (represented by A_2). Notice that the probability that a mutual fund manager graduated from a top-20 MBA programme and the probability that the manager did not graduate from a top-20 MBA programme add to 1.

Adding down the columns produces the following marginal probabilities.

$$\text{Column 1: } P(A_1 \text{ and } B_1) + P(A_2 \text{ and } B_1) = 0.11 + 0.06 = 0.17$$
$$\text{Column 2: } P(A_1 \text{ and } B_2) + P(A_2 \text{ and } B_2) = 0.29 + 0.54 = 0.83$$

These marginal probabilities tell us that 17% of all mutual funds outperform the market and that 83% of mutual funds do not outperform the market.

Table 6.2 lists all the joint and marginal probabilities.

TABLE 6.2 Joint and Marginal Probabilities

	MUTUAL FUND OUTPERFORMS MARKET	MUTUAL FUND DOES NOT OUTPERFORM MARKET	TOTALS
Top-20 MBA programme	$P(A_1 \text{ and } B_1) = 0.11$	$P(A_1 \text{ and } B_2) = 0.29$	$P(A_1) = 0.40$
Not top-20 MBA programme	$P(A_2 \text{ and } B_1) = 0.06$	$P(A_2 \text{ and } B_2) = 0.54$	$P(A_2) = 0.60$
Totals	$P(B_1) = 0.17$	$P(B_2) = 0.83$	1.00

6-2c Conditional Probability

We frequently need to know how two events are related. In particular, we would like to know the probability of one event given the occurrence of another related event. For example, we would certainly like to know the probability that a fund managed by a graduate of a top-20 MBA programme will outperform the market. Such a probability will allow us to make an informed decision about where to invest our money. This probability is called a **conditional probability** because we want to know the probability that a fund will outperform the market *given* the condition that the manager graduated from a top-20 MBA programme. The conditional probability that we seek is represented by

$$P(B_1|A_1)$$

where the '|' represents the word *given*. Here is how we compute this conditional probability.

The marginal probability that a manager graduated from a top-20 MBA programme is 0.40, which is made up of two joint probabilities. They are (1) the probability that the mutual fund outperforms the market and the manager graduated from a top-20 MBA programme 0.11 and (2) the probability that the fund does not outperform the market and the manager graduated from a top-20 MBA programme [$P(A_1 \text{ and } B_2)$]. Their joint probabilities are 0.11 and 0.29, respectively. We can interpret these numbers in the following way. On average, for every 100 mutual funds, 40 will be managed by a graduate of a top-20 MBA programme. Of these 40 managers, on average 11 of them will manage a mutual fund that will outperform the market. Thus, the conditional probability is 0.11/0.40 = 0.275. Notice that this ratio is the same as the ratio of the joint probability to the marginal probability 0.11/0.40. All conditional probabilities can be computed this way.

Conditional Probability

The probability of event *A* given event *B* is

$$P(A|B) = \frac{P(A \text{ and } B)}{P(B)}$$

The probability of event *B* given event *A* is

$$P(B|A) = \frac{P(A \text{ and } B)}{P(A)}$$

EXAMPLE 6.2

Determinants of Success among Mutual Fund Managers – Part 2

Suppose that in Example 6.1 we select one mutual fund at random and discover that it did not outperform the market. What is the probability that a graduate of a top-20 MBA programme manages it?

SOLUTION:

We wish to find a conditional probability. The condition is that the fund did not outperform the market (event B_2), and the event whose probability we seek is that the fund is managed by a graduate of a top-20 MBA programme (event A_1). Thus, we want to compute the following probability:

$$P(A_1|B_2)$$

Using the conditional probability formula, we find

$$P(A_1|B_2) = \frac{P(A_1 \text{ and } B_2)}{P(B_2)} = \frac{0.29}{0.83} = 0.349$$

Thus, 34.9% of all mutual funds that do not outperform the market are managed by top-20 MBA programme graduates.

The calculation of conditional probabilities raises the question of whether the two events, the fund outperformed the market and the manager graduated from a top-20 MBA programme, are related, a subject we tackle next.

6-2d Independence

One of the objectives of calculating conditional probability is to determine whether two events are related. In particular, we would like to know whether they are **independent events**.

Independent Events

Two events A and B are said to be independent if

$$P(A|B) = P(A)$$

or

$$P(B|A) = P(B)$$

Put another way, two events are independent if the probability of one event is not affected by the occurrence of the other event.

EXAMPLE 6.3

Determinants of Success among Mutual Fund Managers – Part 3

Determine whether the event that the manager graduated from a top-20 MBA programme and the event the fund outperforms the market are independent events.

SOLUTION:

We wish to determine whether A_1 and B_1 are independent. To do so, we must calculate the probability of A_1 given B_1 that is,

$$P(A_1|B_1) = \frac{P(A_1 \text{ and } B_1)}{P(B_1)} = \frac{0.11}{0.17} = 0.647$$

The marginal probability that a manager graduated from a top-20 MBA programme is

$$P(A_1) = 0.40$$

Since the two probabilities are not equal, we conclude that the two events are dependent.

Incidentally, we could have made the decision by calculating $P(B_1|A_1) = 0.275$ and observing that it is not equal to $P(B_1) = 0.17$.

Note that there are three other combinations of events in this problem. They are (A_1 and B_2) and (A_2 and B_1), (A_2 and B_2) ignoring mutually exclusive combinations (A_1 and A_2) and (B_1 and B_2), which are dependent. In each combination, the two events are dependent. In this type of problem, where there are only four combinations, if one combination is dependent, then all four will be dependent. Similarly, if one combination is independent, then all four will be independent. This rule does not apply to any other situation.

6.2e Union

Another event that is the combination of other events is the *union*.

Union of Events *A* and *B*

The **union** of events A and B is the event that occurs when either A or B or both occur. It is denoted as

$$A \text{ or } B$$

EXAMPLE 6.4

Determinants of Success among Mutual Fund Managers – Part 4

Determine the probability that a randomly selected fund outperforms the market or the manager graduated from a top-20 MBA programme.

SOLUTION:

We want to compute the probability of the union of two events

$$P(A_1 \text{ or } B_1)$$

The union A_1 or B_1 consists of three events; That is, the union occurs whenever any of the following joint events occurs:

1. Fund outperforms the market and the manager graduated from a top-20 MBA programme.
2. Fund outperforms the market and the manager did not graduate from a top-20 MBA programme.
3. Fund does not outperform the market and the manager graduated from a top-20 MBA programme. Their probabilities are

$$P(A_1 \text{ and } B_1) = 0.11$$
$$P(A_2 \text{ and } B_1) = 0.06$$
$$P(A_1 \text{ and } B_2) = 0.29$$

Thus, the probability of the union – the fund outperforms the market or the manager graduated from a top-20 MBA programme – is the sum of the three probabilities; that is,

$$P(A_1 \text{ or } B_1) = P(A_1 \text{ and } B_1) + P(A_2 \text{ and } B_1) + P(A_1 \text{ and } B_2) = 0.11 + 0.06 + 0.29 = 0.46$$

Notice that there is another way to produce this probability. Of the four probabilities in Table 6.1, the only one representing an event that is not part of the union is the probability of the event that the fund does not outperform the market and the manager did not graduate from a top-20 MBA programme. That probability is

$$P(A_2 \text{ and } B_2) = 0.54$$

which is the probability that the union *does not* occur. Thus, the probability of the union is

$$P(A_1 \text{ or } B_1) = 1 - P(A_2 \text{ and } B_2) = 1 - 0.54 = 0.46$$

Thus, we determined that 46% of mutual funds either outperform the market or are managed by a top-20 MBA programme graduate or have both characteristics.

EXERCISES

6.16 Xr06-16 Calculate the marginal probabilities from the following table of joint probabilities.

	A_1	A_2
B_1	0.4	0.3
B_2	0.2	0.1

6.17 Refer to Exercise 6.16.

a. Determine $P(A_1|B_1)$
b. Determine $P(A_2|B_1)$
c. Did you expect the answers to parts (a) and (b) to be reciprocals? In other words, did you expect that $P(A_1|B_2) = 1/P(B_2|A_1)$? Why is this impossible (unless both probabilities are 1)?

6.18 Refer to Exercise 6.16. Calculate the following probabilities.

a. $P(A_1|B_2)$
b. $P(B_2|A_1)$
c. Did you expect the answers to parts (a) and (b) to be reciprocals? In other words, did you expect that $P(A_1|B_2) = 1/P(B_2|A_1)$? Why is this impossible (unless both probabilities are 1)?

6.19 Are the events in Exercise 6.16 independent? Explain.

6.20 Refer to Exercise 6.16. Compute the following.

a. $P(A_1 \text{ or } B_1)$
b. $P(A_1 \text{ or } B_2)$
c. $P(A_1 \text{ or } A_2)$

6.21 Xr06-21 Determine whether the events are independent from the following joint probabilities.

	A_1	A_2
B_1	0.20	0.15
B_2	0.60	0.05

6.22 Xr06-22 Suppose we have the following joint probabilities.

	A_1	A_2	A_3
B_1	0.15	0.20	0.10
B_2	0.25	0.25	0.05

Compute the marginal probabilities.

6.23 Refer to Exercise 6.22.

a. Compute $P(A_2|B_2)$

b. Compute $P(B_2|A_2)$

c. Compute. $P(B_1|A_2)$

6.24 Refer to Exercise 6.22.

a. Compute $P(A_1|A_2)$

b. Compute $P(A_2|B_2)$

c. Compute $P(A_3|B_1)$

6.25 Xr06-25 Discrimination in the workplace is illegal, and companies that discriminate are often sued. The female instructors at a large university recently lodged a complaint about the most recent round of promotions from assistant professor to associate professor. An analysis of the relationship between gender and promotion produced the following joint probabilities.

	Promoted	Not Promoted
Female	0.03	0.12
Male	0.17	0.68

a. What is the rate of promotion among female assistant professors?

b. What is the rate of promotion among male assistant professors?

c. Is it reasonable to accuse the university of gender bias?

6.26 A department store analysed its most recent sales and determined the relationship between the way the customer paid for the item and the price category of the item. The joint probabilities in the following table were calculated.

	Cash	Credit Card	Debit Card
Less than €20	0.09	0.03	0.04
€20–€100	0.05	0.21	0.18
More than €100	0.03	0.23	0.14

a. What proportion of purchases was paid by debit card?

b. Find the probability that a credit card purchase was more than €100.

c. Determine the proportion of purchases made by credit card or by debit card.

6.27 Xr06-27 The method of instruction in college and university-applied statistics courses is changing. Historically, most courses were taught with an emphasis on manual calculation. The alternative is to employ a computer and a software package to perform the calculations. An analysis of applied statistics courses investigated whether the instructor's educational background is primarily mathematics (or statistics) or some other field. The result of this analysis is the accompanying table of joint probabilities.

	Statistics Course Emphasises Manual Calculations	Statistics Course Employs Computer and Software
Mathematics or statistics education	0.23	0.36
Other education	0.11	0.30

a. What is the probability that a randomly selected applied statistics course instructor, whose education was in statistics, emphasises manual calculations?

b. What proportion of applied statistics courses employ a computer and software?

c. Are the educational background of the instructor and the way his or her course is taught independent?

6.28 Xr06-28 A restaurant chain routinely surveys its customers. Among other questions, the survey asks each customer whether he or she would return and to rate the quality of food. Summarising hundreds of thousands of questionnaires produced this table of joint probabilities.

Rating	Customer will Return	Customer will not Return
Poor	0.02	0.10
Fair	0.08	0.09
Good	0.35	0.14
Excellent	0.20	0.02

a. What proportion of customers say that they will return and rate the restaurant's food as good?

b. What proportion of customers who say that they will return rate the restaurant's food as good?

c. What proportion of customers who rate the restaurant's food as good say that they will return?

d. Discuss the differences in your answers to parts (a), (b) and (c).

6.29 Xr06-29 To determine whether drinking alcoholic beverages has an effect on the bacteria that cause ulcers, researchers developed the following table of joint probabilities.

Number of Alcoholic Drinks per day	Ulcer	No Ulcer
None	0.01	0.22
One	0.03	0.19
Two	0.03	0.32
More than two	0.04	0.16

a. What proportion of people have ulcers?

b. What is the probability that a teetotaller (no alcoholic beverages) develops an ulcer?

c. What is the probability that someone who has an ulcer does not drink alcohol?

d. What is the probability that someone who has an ulcer drinks alcohol?

6.30 Many critics of television claim that there is too much violence and that it has a negative effect on society. There may also be a negative effect on advertisers. To examine this issue, researchers developed two versions of a cops-and-robbers made-for-television movie. One version depicted several violent crimes, and the other removed these scenes. In the middle of the movie, one 60-second commercial was shown advertising a new product and brand name. At the end of the movie, viewers were asked to name the brand. After observing the results, the researchers produced the following table of joint probabilities.

	Watch Violent Movie	Watch Non-Violent Movie
Remember the brand name	0.15	0.18
Do not remember the brand name	0.35	0.32

a. What proportion of viewers remember the brand name?

b. What proportion of viewers who watch the violent movie remember the brand name?

c. Does watching a violent movie affect whether the viewer will remember the brand name? Explain.

6.31 Xr06-31 A firm has classified its customers in two ways: (1) according to whether the account is overdue and (2) whether the account is new (less than 12 months) or old. An analysis of the firm's records provided the input for the following table of joint probabilities.

	Overdue	Not Overdue
New	0.06	0.13
Old	0.52	0.29

One account is randomly selected.

a. If the account is overdue, what is the probability that it is new?

b. If the account is new, what is the probability that it is overdue?

c. Is the age of the account related to whether it is overdue? Explain.

6.32 Xr06-32 Credit scorecards are used by financial institutions to help decide to whom loans should be granted. An analysis of the records of one bank produced the following probabilities.

Loan Performance	Score: Under 400	400 or more
Fully repaid	0.19	0.64
Defaulted	0.13	0.04

a. What proportion of loans is fully repaid?

b. What proportion of loans given to scorers of less than 400 fully repay?

c. What proportion of loans given to scorers of 400 or more fully repay?

d. Are score and whether the loan is fully repaid independent? Explain.

6.33 Xr06-33 A retail outlet wanted to know whether its weekly advertisement in the daily newspaper works. To acquire this critical information, the store manager surveyed the people who entered the store and determined whether each individual saw the ad and whether a purchase was made. From the information developed, the manager produced the following table of joint probabilities. Are the ads effective? Explain.

	Purchase	No Purchase
See ad	0.18	0.42
Do not see ad	0.12	0.28

6-3 PROBABILITY RULES AND TREES

In Section 6-2, we introduced intersection and union and described how to determine the probability of the intersection and the union of two events. In this section, we present other methods of determining these probabilities. We introduce three rules that enable us to calculate the probability of more complex events from the probability of simpler events.

6-3a Complement Rule

The **complement** of event *A* is the event that occurs when event *A* does not occur. The complement of event *A* is denoted by A^C. The **complement rule** defined here derives from the fact that the probability of an event and the probability of the event's complement must sum to 1.

Complement Rule

$$P(A^C) = 1 - P(A)$$

for any event *A*.

We will demonstrate the use of this rule after we introduce the next rule.

6-3b Multiplication Rule

The **multiplication rule** is used to calculate the joint probability of two events. It is based on the formula for conditional probability supplied in the previous section; that is, from the following formula

$$P(A|B) = \frac{P(A \text{ and } B)}{P(B)}$$

we derive the multiplication rule simply by multiplying both sides by $P(B)$.

Multiplication Rule

The joint probability of any two events *A* and *B* is

$$P(A \text{ and } B) = P(B)P(A|B)$$

or, altering the notation,

$$P(A \text{ and } B) = P(B)P(B|A)$$

If *A* and *B* are independent events, $P(A|B) = P(A)$ and $P(B|A) = P(B)$. It follows that the joint probability of two independent events is simply the product of the probabilities of the two events. We can express this as a special form of the multiplication rule.

Multiplication Rule for Independent Events

The joint probability of any two independent events *A* and *B* is

$$P(A \text{ and } B) = P(A)P(B)$$

EXAMPLE 6.5*

Selecting two Students without Replacement

A graduate statistics course has seven male and three female students. The professor wants to select two students at random to help her conduct a research project. What is the probability that the two students chosen are female?

SOLUTION:

Let A represent the event that the first student chosen is female and B represent the event that the second student chosen is also female. We want the joint probability $P(A \text{ and } B)$. Consequently, we apply the multiplication rule:

$$P(A \text{ and } B) = P(A)P(B|A)$$

Because there are three female students in a class of ten, the probability that the first student chosen is female is

$$P(A) = 3/10$$

After the first student is chosen, there are only nine students left. Given that the first student chosen was female, there are only two female students left. It follows that

$$P(B|A) = 2/9$$

Thus, the joint probability is

$$P(A \text{ and } B) = P(A)P(B|A) = \left(\frac{3}{10}\right)\left(\frac{2}{9}\right) = \frac{6}{90} = 0.67$$

EXAMPLE 6.6

Selecting two Students with Replacement

Refer to Example 6.5. The professor who teaches the course is suffering from the flu and will be unavailable for two classes. The professor's replacement will teach the next two classes. His style is to select one student at random and pick on him or her to answer questions during that class. What is the probability that the two students chosen are female?

SOLUTION:

The form of the question is the same as in Example 6.5. We wish to compute the probability of choosing two female students. However, the experiment is slightly different. It is now possible to choose the *same* student in each of the two classes taught by the replacement. Thus, A and B are independent events, and we apply the multiplication rule for independent events:

$$P(A \text{ and } B) = P(A)P(B)$$

The probability of choosing a female student in each of the two classes is the same; that is,

$$P(A) = 3/10 \text{ and } P(B) = 3/10$$

Hence,

$$P(A \text{ and } B) = P(A)P(B) = \left(\frac{3}{10}\right)\left(\frac{3}{10}\right) = \frac{9}{100} = 0.09$$

6-3c Addition Rule

The **addition rule** enables us to calculate the probability of the union of two events.

*This example can be solved using the Hypergeometric distribution, which is described in the website Appendix Hypergeometric Distribution.

Addition Rule

The probability that event *A*, or event *B*, or both occur is

$$P(A \text{ or } B) = P(A) + P(B) - P(A \text{ and } B)$$

If you are like most students, you are wondering why we subtract the joint probability from the sum of the probabilities of *A* and *B*. To understand why this is necessary, examine Table 6.2 (Section 6-2b), which we have reproduced here as Table 6.3.

TABLE 6.3 Joint and Marginal Probabilities

	B_1	B_2	TOTALS
A_1	$P(A_1 \text{ and } B_1) = 0.11$	$P(A_1 \text{ and } B_2) = 0.29$	$P(A_1) = 0.40$
A_2	$P(A_2 \text{ and } B_1) = 0.06$	$P(A_2 \text{ and } B_2) = 0.54$	$P(A_2) = 0.60$
Totals	$P(B_1) = 0.17$	$P(B_2) = 0.83$	1.00

This table summarises how the marginal probabilities were computed. For example, the marginal probability of A_1 and the marginal probability of B_1 were calculated as

$$P(A_1) = P(A_1 \text{ and } B_1) + P(A_1 \text{ and } B_2) = 0.11 + 0.29 = 0.40$$
$$P(B_1) = P(A_1 \text{ and } B_1) + P(A_2 \text{ and } B_1) = 0.11 + 0.06 = 0.17$$

If we now attempt to calculate the probability of the union of A_1 and B_1 by summing their probabilities, we find

$$P(A_1) + P(B_1) = 0.11 + 0.29 + 0.11 + 0.06$$

Notice that we added the joint probability of A_1 and B_1 (which is 0.11) twice. To correct the double counting, we subtract the joint probability from the sum of the probabilities of A_1 and B_1. Thus,

$$P(A_1 \text{ or } B_1) = P(A_1) + P(B_1) - P(A_1 \text{ and } B_1)$$
$$= [0.11 + 0.29] + [0.11 + 0.06] - 0.11$$
$$= 0.40 + 0.17 - 0.11 = 0.46$$

This is the probability of the union of A_1 and B_1, which we calculated in Example 6.4 (Section 6-2e).

As was the case with the multiplication rule, there is a special form of the addition rule. When two events are mutually exclusive (which means that the two events cannot occur together), their joint probability is 0.

Addition Rule for Mutually Exclusive Events

The probability of the union of two mutually exclusive events *A* and *B* is

$$P(A \text{ or } B) = P(A) + P(B)$$

EXAMPLE 6.7

Applying the Addition Rule

In a large city, two newspapers are published, the *Sun* and the *Mail*. The circulation departments report that 22% of the city's households have a subscription to the *Sun* and 35% subscribe to the *Mail*. A survey reveals that 6% of all households subscribe to both newspapers. What proportion of the city's households subscribe to either newspaper?

SOLUTION:

We can express this question as: What is the probability of selecting a household at random that subscribes to the *Sun*, the *Mail*, or both? Another way of asking the question is, what is the probability that a randomly selected household subscribes to *at least one* of the newspapers? It is now clear that we seek the probability of the union, and we must apply the addition rule. Let A = household subscribes to the *Sun* and B = household subscribes to the *Mail*. We perform the following calculation:

$$P(A \text{ or } B) = P(A) + P(B) - P(A \text{ and } B) = 0.22 + 0.35 - 0.06 = 0.51$$

The probability that a randomly selected household subscribes to either newspaper is 0.51. Expressed as relative frequency, 51% of the city's households subscribe to either newspaper.

6-3d Probability Trees

An effective and simpler method of applying the probability rules is the probability tree, wherein the events in an experiment are represented by lines. The resulting figure resembles a tree, hence the name. We will illustrate the probability tree with several examples, including two that we addressed using the probability rules alone.

In Example 6.5, we wanted to find the probability of choosing two female students, where the two choices had to be different. The tree diagram in Figure 6.1 describes this experiment. Notice that the first two branches represent the two possibilities, female and male students, on the first choice. The second set of branches represents the two possibilities on the second choice. The probabilities of female and male student chosen first are 3/10 and 7/10, respectively. The probabilities for the second set of branches are conditional probabilities based on the choice of the first student selected.

We calculate the joint probabilities by multiplying the probabilities on the linked branches. Thus, the probability of choosing two female students is $P(F \text{ and } F) = (3/10)(2/9) = 6/90$. The remaining joint probabilities are computed similarly.

FIGURE 6.1

Probability Tree for Example 6.5

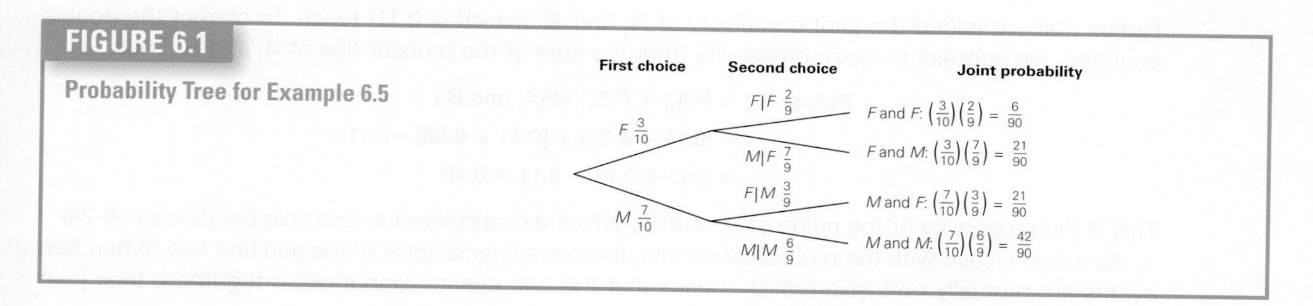

In Example 6.6, the experiment was similar to that of Example 6.5. However, the student selected on the first choice was returned to the pool of students and was eligible to be chosen again. Thus, the probabilities on the second set of branches remain the same as the probabilities on the first set, and the probability tree is drawn with these changes, as shown in Figure 6.2.

FIGURE 6.2

Probability Tree for Example 6.6

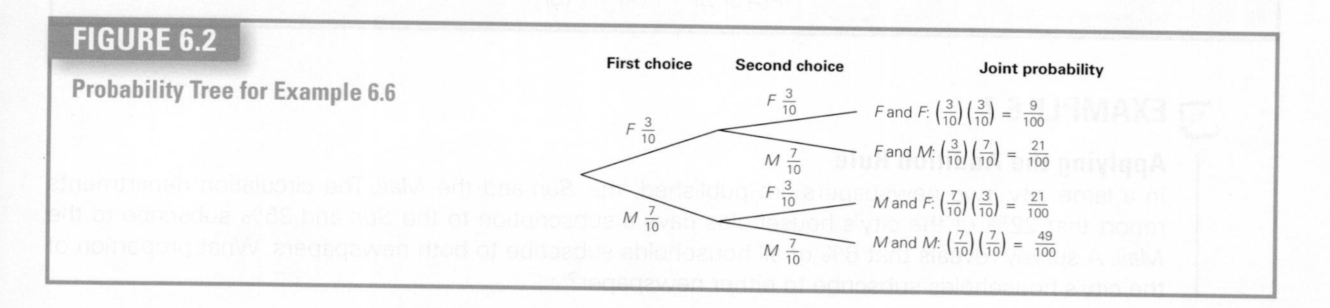

The advantage of a probability tree on this type of problem is that it restrains its users from making the wrong calculation. Once the tree is drawn and the probabilities of the branches inserted, virtually the only allowable calculation is the multiplication of the probabilities of linked branches. An easy check on those calculations is available. The joint probabilities at the ends of the branches must sum to 1 because all possible events are listed. In both figures, notice that the joint probabilities do indeed sum to 1.

The special form of the addition rule for mutually exclusive events can be applied to the joint probabilities. In both probability trees, we can compute the probability that one student chosen is female and one is male simply by adding the joint probabilities. For the tree in Example 6.5, we have

$$P(F \text{ and } M) + P(M \text{ and } F) = 21/90 + 21/90 = 42/90$$

In the probability tree in Example 6.6, we find

$$P(F \text{ and } M) + P(M \text{ and } F) = 21/100 + 21/100 = 42/100$$

EXAMPLE 6.8

Probability of Passing the Bar Exam

Students who graduate from law schools must still pass a bar exam before becoming lawyers. Suppose that in a particular jurisdiction the pass rate for first-time test takers is 72%. Candidates who fail the first exam may take it again several months later. Of those who fail their first test, 88% pass their second attempt. Find the probability that a randomly selected law school graduate becomes a lawyer. Assume that candidates cannot take the exam more than twice.

SOLUTION:

The probability tree in Figure 6.3 is employed to describe the experiment. Note that we use the complement rule to determine the probability of failing each exam.

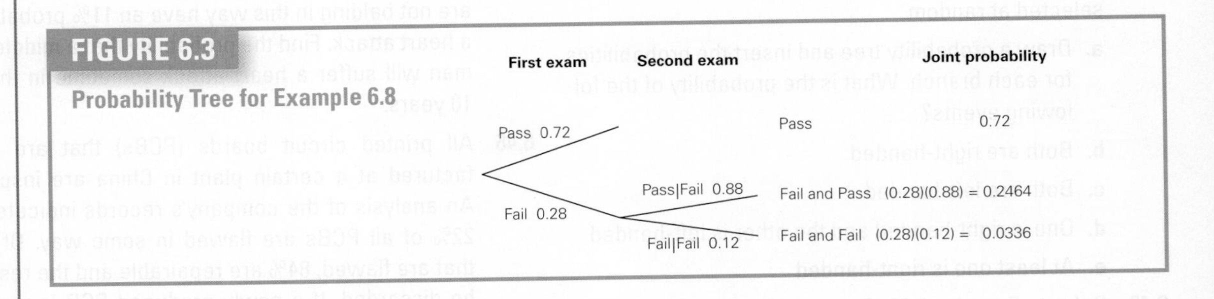

FIGURE 6.3 Probability Tree for Example 6.8

We apply the multiplication rule to calculate P(Fail and Pass), which we find to be 0.2464. We then apply the addition rule for mutually exclusive events to find the probability of passing the first or second exam:

$$P(\text{Pass [on first exam]}) + P(\text{Fail [on first exam] and Pass [on second exam]})$$
$$= 0.72 + 0.2464 = 0.9664$$

Thus, 96.64% of applicants become lawyers by passing the first or second exam.

EXERCISES

6.34 Given the following probabilities, compute all joint probabilities.

$P(A) = 0.9$ $\quad P(A^C) = 0.1$
$P(B|A) = 0.4$ $\quad P(B|A^C) = 0.7$

6.35 Draw a probability tree to compute the joint probabilities from the following probabilities.

$P(A) = 0.5$ $\quad P(A^C) = 0.2$
$P(B|A) = 0.4$ $\quad P(B|A^C) = 0.7$

6.36 Given the following probabilities, find the joint probability.

$P(A) = 0.7$ $P(B|A^C) = 0.3$

6.37 Approximately 10% of people are left-handed. If two people are selected at random, what is the probability of the following events?

a. Both are right-handed

b. Both are left-handed

c. One is right-handed and the other is left-handed

d. At least one is right-handed

6.38 Refer to Exercise 6.37. Suppose that three people are selected at random.

a. Draw a probability tree to depict the experiment.

b. If we use the notation RRR to describe the selection of three right-handed people, what are the descriptions of the remaining seven events? (Use L for left-hander.)

c. How many of the events yield no right-handers, one right-hander, two right-handers and three right-handers?

d. Find the probability of no right-handers, one right-hander, two right-handers and three right-handers.

6.39 Suppose there are 100 students in an accounting class, ten of whom are left-handed. Two students are selected at random.

a. Draw a probability tree and insert the probabilities for each branch. What is the probability of the following events?

b. Both are right-handed

c. Both are left-handed

d. One is right-handed and the other is left-handed

e. At least one is right-handed

6.40 Refer to Exercise 6.39. Suppose that three people are selected at random.

a. Draw a probability tree and insert the probabilities of each branch.

b. What is the probability of no right-handers, one right-hander, two right-handers and three right-handers?

6.41 An aerospace company has submitted bids on two separate government defence contracts. The company president believes that there is a 40% probability of winning the first contract. If they win the first contract, the probability of winning the second is 70%. However, if they lose the first contract, the managing director thinks that the probability of winning the second contract decreases to 50%.

a. What is the probability that they win both contracts?

b. What is the probability that they lose both contracts?

c. What is the probability that they win only one contract?

6.42 A telemarketer calls people and tries to sell them a subscription to a daily newspaper. On 20% of her calls, there is no answer or the line is busy. She sells subscriptions to 5% of the remaining calls. For what proportion of calls does she make a sale?

6.43 A foreman for an injection-moulding firm admits that on 10% of his shifts he forgets to shut off the injection machine on his line. This causes the machine to overheat, increasing the probability from 2% to 20% that a defective moulding will be produced during the early morning run. What proportion of mouldings from the early morning run is defective?

6.44 A study undertaken by the University of Essex between May 2005 and May 2010 revealed the trends in vote intentions in May 2010: 36% people interviewed are Conservative, 28% are Labour, 27% are Liberal Democrat and 9% are others. If two registered voters are selected at random, what is the probability that both of them have the same party affiliation?

6.45 A survey of middle-aged men reveals that 28% of them are balding at the crown of their heads. Moreover, it is known that such men have an 18% probability of suffering a heart attack in the next 10 years. Men who are not balding in this way have an 11% probability of a heart attack. Find the probability that a middle-aged man will suffer a heart attack sometime in the next 10 years.

6.46 All printed circuit boards (PCBs) that are manufactured at a certain plant in China are inspected. An analysis of the company's records indicates that 22% of all PCBs are flawed in some way. Of those that are flawed, 84% are repairable and the rest must be discarded. If a newly produced PCB is randomly selected, what is the probability that it does not have to be discarded?

6.47 A financial analyst has determined that there is a 22% probability that a mutual fund will outperform the market over a 1-year period provided that it outperformed the market the previous year. If only 15% of mutual funds outperform the market during any year, what is the probability that a mutual fund will outperform the market 2 years in a row?

6.48 An investor believes that on a day when the Dow Jones Industrial Average (DJIA) increases, the probability that the NASDAQ also increases is 77%. If the investor believes that there is a 60% probability that the DJIA will increase tomorrow, what is the probability that the NASDAQ will increase as well?

6.49 The controls of an airplane have several backup systems or redundancies so that if one fails the plane will continue to operate. Suppose that the mechanism that controls the flaps has two backups. If the probability that the main control fails is 0.0001 and the probability that each backup will fail is 0.01, what is the probability that all three fail to operate?

6.50 According to TNS Intersearch, 69% of wireless Web users use it primarily for receiving and sending email. Suppose that three wireless Web users are selected at random. What is the probability that all of them use it primarily for email?

6.51 A financial analyst estimates that the probability that the economy will experience a recession in the next 12 months is 25%. She also believes that if the economy encounters a recession, the probability that her mutual fund will increase in value is 20%. If there is no recession, the probability that the mutual fund will increase in value is 75%. Find the probability that the mutual fund's value will increase.

6.4 BAYES'S LAW

Conditional probability is often used to gauge the relationship between two events. In many of the examples and exercises you have already encountered, conditional probability measures the probability that an event occurs given that a possible cause of the event has occurred. In Example 6.2, we calculated the probability that a mutual fund outperforms the market (the effect) given that the fund manager graduated from a top-20 MBA programme (the possible cause). There are situations, however, where we witness a particular event and we need to compute the probability of one of its possible causes. **Bayes's Law** (alternatively Bayes' theorem or Bayes' rule) is the technique we use.

EXAMPLE 6.9

Should an MBA Applicant Take a Preparatory Course?

The Graduate Management Admission Test (GMAT), which originated in the USA, is a requirement for applicants of MBA programmes in certain countries. A variety of preparatory courses are designed to help applicants improve their GMAT scores, which range from 200 to 800. Suppose that a survey of MBA students reveals that among GMAT scorers above 650, 52% took a preparatory course; whereas among GMAT scorers of less than 650 only 23% took a preparatory course. An applicant to an MBA programme has determined that he needs a score of more than 650 to get into a certain MBA programme, but he feels that his probability of getting that high a score is quite low – 10%. He is considering taking a preparatory course that costs £500. He is willing to do so only if his probability of achieving 650 or more doubles. What should he do?

SOLUTION:

The easiest way to address this problem is to draw a tree diagram. The following notation will be used:

A = GMAT score is 650 or more

A^C = GMAT score less than 650

B = Took preparatory course

B^C = Did not take preparatory course

The probability of scoring 650 or more is

$$P(A) = 0.10$$

The complement rule gives us

$$P(A) = 0.10$$

Conditional probabilities are

$$P(B|A) = 0.52$$

and

$$P(B|A^C) = 0.23$$

Again using the complement rule, we find the following conditional probabilities:

$$P(B^C|A) = 1 - 0.52 = 0.48$$

and

$$P(B^C|A^C) = 1 - 0.23 = 0.77$$

We would like to determine the probability that he would achieve a GMAT score of 650 or more given that he took the preparatory course; that is, we need to compute

$$P(A)|B)$$

Using the definition of conditional probability (Section 6-2c), we have

$$P(A|B) = \frac{P(A \text{ and } B)}{P(B)}$$

Neither the numerator nor the denominator is known. The probability tree (Figure 6.4) will provide us with the probabilities.

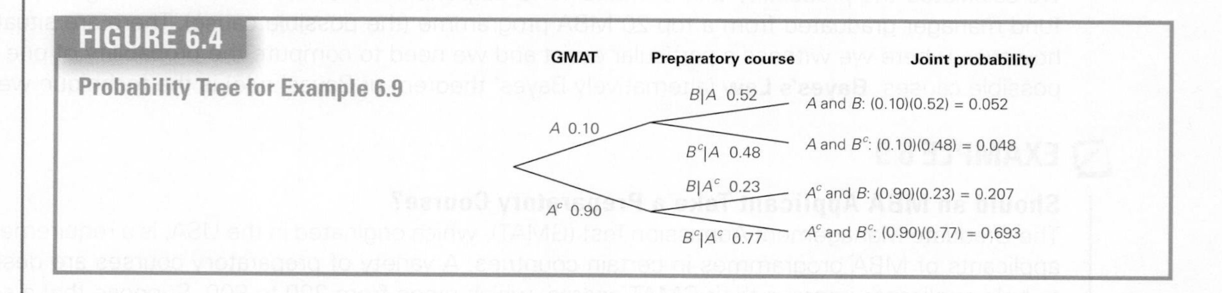

FIGURE 6.4
Probability Tree for Example 6.9

As you can see,

$$P(A \text{ and } B) = (0.10)(0.52) = 0.052$$
$$P(A^C \text{ and } B) = (0.90)(0.23) = 0.207$$

and

$$P(B) = P(A \text{ and } B) + P(A^C \text{ and } B) = 0.052 + 0.207 = 0.259$$

Thus,

$$P(A|B) = \frac{P(A \text{ and } B)}{P(B)} = \frac{0.052}{0.259} = 0.201$$

The probability of scoring 650 or more on the GMAT doubles when the preparatory course is taken.

Thomas Bayes first employed the calculation of conditional probability during the 18th century, as shown in Example 6.9. Accordingly, it is called Bayes's Law.

The probabilities $P(A)$ and $P(A^C)$ are called **prior probabilities** because they are determined *prior* to the decision about taking the preparatory course. The conditional probabilities are called **likelihood probabilities** for reasons that are beyond the mathematics in this book. Finally, the conditional probability $P(A|B)$ and similar conditional probabilities $P(A^C|B)$, $P(A|BC)$, and $P(A^C|B^C)$ are called **posterior probabilities** or **revised probabilities** because the prior probabilities are revised *after* the decision about taking the preparatory course.

You may be wondering why we did not get $P(A|B)$ directly. In other words, why not survey people who took the preparatory course and ask whether they received a score of 650 or more? The answer is that using the likelihood probabilities and using Bayes's Law allows individuals to set their own prior probabilities, which can then be revised. For example, another MBA applicant may assess her probability of scoring 650 or more as 0.40. Inputting the new prior probabilities produces the following probabilities:

$$P(A \text{ and } B) = (.40)(.52) = 0.208$$

$$P(A^C \text{ and } B) = (.60)(.23) = 0.138$$

$$P(B) = P(A \text{ and } B) + P(A^C \text{ and } B) = 0.208 + 0.138 = 0.346$$

$$P(A|B) = \frac{P(A \text{ and } B)}{P(B)} = \frac{0.208}{0.346} = 0.601$$

The probability of achieving a GMAT score of 650 or more increases by a more modest 50% (from 0.40 to 0.601).

6-4a Bayes's Law Formula

Bayes's Law can be expressed as a formula for those who prefer an algebraic approach rather than a probability tree. We use the following notation.

The event B is the given event and the events

$$A_1, A_2, \ldots, A_k$$

are the events for which prior probabilities are known; that is,

$$P(A_1), P(A_2), \ldots, P(A_k)$$

are the prior probabilities.

The likelihood probabilities are

$$P(B|A_1), P(B|A_2), \ldots, P(B|A_k)$$

and

$$P(A_1|B), P(A_2|B), \ldots, P(A_k|B)$$

are the posterior probabilities, which represent the probabilities we seek.

Bayes's Law Formula

$$P(A_i|B) = \frac{P(A_i)P(B|A_i)}{(P(A_1)PB|A_1) + P(A_2)P(B|A_2) + \cdots + P(A_k)P(B|A_k)}$$

To illustrate the use of the formula, we will redo Example 6.9. We begin by defining the events.

A_1 = GMAT score is 650 or more

A_2 = GMAT score less than 650

B = Take preparatory course

The probabilities are

$$P(A_1) = 0.10$$

The complement rule gives us

$$P(A_2) = 1 - 0.10 = 0.90$$

Conditional probabilities are

$$P(B|A_1) = 0.52$$

and

$$P(B|A_1) = 0.23$$

Substituting the prior and likelihood probabilities into the Bayes's Law formula yields the following:

$$P(A_1|B) = \frac{P(A_1)P(B|A_1)}{P(A_1)P(B|A_1) + P(A_2)P(B|A_2)} = \frac{(0.10)(0.52)}{(0.10)(0.52) + (0.90)(0.23)}$$

$$= \frac{0.052}{0.52 + 0.207} = \frac{0.052}{0.259} = 0.201$$

As you can see, the calculation of the Bayes's Law formula produces the same results as the probability tree.

AUDITING TAX RETURNS: SOLUTION

We need to revise the prior probability that this return contains significant fraud. The tree shown in Figure 6.5 details the calculation.

F = Tax return is fraudulent

F^C = Tax return is honest

E_0 = Tax return contains no expense deductions

E_1 = Tax return contains one expense deduction

E_2 = Tax return contains two expense deductions

$$P(E_1) = P(F \text{ and } E_1) + P(F^C \text{ and } E_1) = 0.0140 + 0.1710 = 0.1850$$

$$P(F|E_1) = P(F \text{ and } E_1)/P(E_1) = 0.0140/0.1850 = 0.0757$$

The probability that this return is fraudulent is 0.0757.

FIGURE 6.5

Probability Tree for Auditing Tax Returns

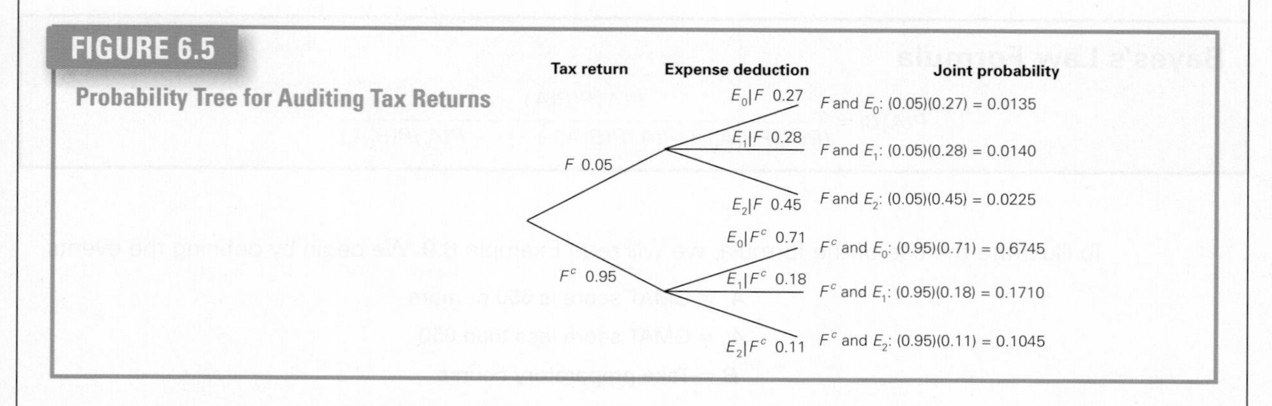

6-4b Applications in Medicine and Medical Insurance

Physicians routinely perform medical tests, called *screenings*, on their patients. Screening tests are conducted for all patients in a particular age and gender group, regardless of their symptoms. For example, men in their 50s are advised to take a prostate-specific antigen (PSA) test to determine whether there is evidence of prostate cancer. Women undergo a Pap test for cervical cancer.

Unfortunately, few of these tests are 100% accurate. Most can produce *false-positive* and *false-negative* results. A **false-positive** result is one in which the patient does not have the disease, but the test shows positive. A **false-negative** result is one in which the patient does have the disease, but the test produces a negative result. The consequences of each test are serious and costly. A false-negative test results in not detecting a disease in a patient, therefore postponing treatment, perhaps indefinitely. A false-positive test leads to apprehension and fear for the patient. In most cases, the patient is required to undergo further testing such as a biopsy. The unnecessary follow-up procedure can pose medical risks.

False-positive test results have financial repercussions. The cost of the follow-up procedure, for example, is usually far more expensive than the screening test. Medical insurance companies as well as government-funded plans are all adversely affected by false-positive test results. Compounding the problem is that physicians and patients are incapable of properly interpreting the results. A correct analysis can save both lives and money.

Bayes's Law is the vehicle we use to determine the true probabilities associated with screening tests. Applying the complement rule to the false-positive and false-negative rates produces the conditional probabilities that represent correct conclusions. Prior probabilities are usually derived by looking at the overall proportion of people with the diseases. In some cases, the prior probabilities may themselves have been revised because of heredity or demographic variables such as age or race. Bayes's Law allows us to revise the prior probability after the test result is positive or negative.

Example 6.10 is based on the actual false-positive and false-negative rates. Note, however, that different sources provide somewhat different probabilities. The differences may be the result of the way positive and negative results are defined or the way technicians conduct the tests. Students who are affected by the diseases described in the example and exercises should seek clarification from their physicians.

EXAMPLE 6.10

Probability of Prostate Cancer

Prostate cancer is the most common form of cancer found in men. The probability of developing prostate cancer over a lifetime is 16%. (This figure may be higher since many prostate cancers go undetected.) Many physicians routinely perform a PSA test, particularly for men over age 50. PSA is a protein produced only by the prostate gland and thus is fairly easy to detect. Normally, men have PSA levels between 0 and 4 mg/ml. Readings above 4 may be considered high and potentially indicative of cancer. However, PSA levels tend to rise with age even among men who are cancer free. Studies have shown that the test is not very accurate. In fact, the probability of having an elevated PSA level, given that the man does not have cancer (false positive), is 0.135. If the man does have cancer, the probability of a normal PSA level (false negative) is almost 0.300. (This figure may vary by age and by the definition of *high* PSA level.) If a physician concludes that the PSA is high, a biopsy is performed. Besides the concerns and health needs of the men, there are also financial costs. The cost of the blood test is low (approximately €50). However, the cost of the biopsy is considerably higher (approximately €1000). A false-positive PSA test will lead to an unnecessary biopsy. Because the PSA test is so inaccurate, some private and public medical plans do not pay for it. Suppose you are a manager in a medical insurance company and must decide on guidelines for whom should be routinely screened for prostate cancer. An analysis of prostate cancer incidence and age produces the following table of probabilities. (The probability of a man under 40 developing prostate cancer is less than 0.0001, or small enough to treat as 0.)

Age	Probability of Developing Prostate Cancer
40–49	0.010
50–59	0.022
60–69	0.046
70 and older	0.079

Assume that a man in each of the age categories undergoes a PSA test with a positive result. Calculate the probability that each man actually has prostate cancer and the probability that he does not. Perform a cost–benefit analysis to determine the cost per cancer detected.

SOLUTION:

As we did in Example 6.9 and the chapter-opening example, we will draw a probability tree (Figure 6.6). The notation is

$$C = \text{Has prostate cancer}$$

$$C^C = \text{Does not have prostate cancer}$$

$$P^T = \text{Positive test result}$$

$$NT = \text{Negative test result}$$

Starting with a man between 40 and 50 years old, we have the following probabilities

Prior

$$P(C) = 0.010$$

$$P(C^C) = 1 - 0.010 = 0.990$$

Likelihood probabilities

False negative: $P(NT|C) = 0.300$

True positive: $P(PT|C) = 1 - 0.300 = 0.700$

False positive: $P(PT|C^C) = 0.135$

True negative: $P(NT|C^C) = 1 - 0.135 = 0.865$

FIGURE 6.6

Probability Tree for Example 6.10

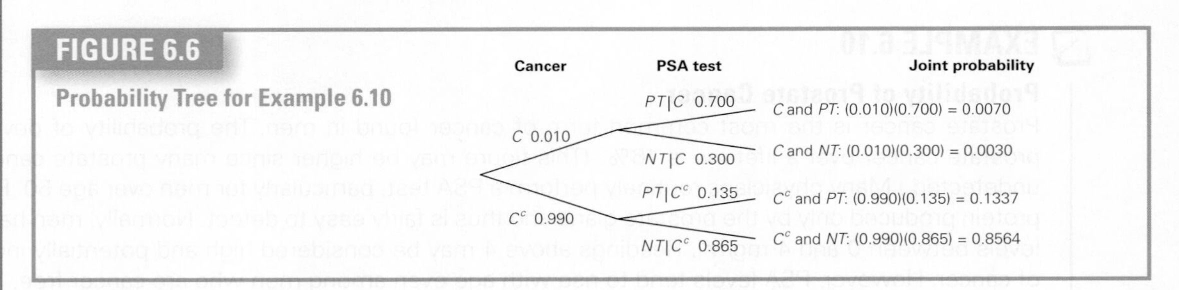

The tree allows you to determine the probability of obtaining a positive test result. It is

$$P(PT) = P(C \text{ and } PT) + P(C^C \text{ and } PT) = 0.0070 + 0.1337 = 0.1407$$

We can now compute the probability that the man has prostate cancer given a positive test result:

$$P(C|PT) = \frac{P(C \text{ and } PT)}{P(PT)} = \frac{0.0070}{0.1407} = 0.0498$$

The probability that he does not have prostate cancer is

$$P(C^C|PT) = 1 - P(C|PT) = 1 - 0.0498 = 0.9502$$

We can repeat the process for the other age categories. Here are the results.

Probabilities Given a Positive PSA Test

Age	Has Prostate Cancer	Does not have Prostate Cancer
40–49	0.0498	0.9502
50–59	0.1045	0.8955
60–69	0.2000	0.8000
70 and older	0.3078	0.6922

The following table lists the proportion of each age category wherein the PSA test is positive [$P(PT)$]

Age	Proportion of Tests that are Positive	Number of Biopsies Performed per million	Number of Cancers Detected	Number of Biopsies per Cancer Detected
40–49	0.1407	140 700	0.0498(140 700) = 7 007	20.10
50–59	0.1474	147 400	0.1045(147 400) = 15 403	9.57
60–79	0.1610	161 000	0.2000(161 000) = 32 200	5.00
70 and older	0.1796	179 600	0.3078(179 600) = 55 281	3.25

If we assume a cost of €1000 per biopsy, the cost per cancer detected is €20 100 for 40 to 50, €9570 for 50 to 60, €5000 for 60 to 70 and €3250 for over 70.

We have created an Excel spreadsheet to help you perform the calculations in Example 6.10. Open the **Excel workbooks** folder and select **Medical screening**. There are three cells that you may alter. In cell B5, enter a new prior probability for prostate cancer. Its complement will be calculated in cell B15. In cells D6 and D15, type new values for the false-negative and false-positive rates, respectively. Excel will do the rest. We will use this spreadsheet to demonstrate some terminology standard in medical testing.

Terminology We will illustrate the terms using the probabilities calculated for the 40 to 50 age category.

The false-negative rate is 0.300. Its complement is the likelihood probability $P(PT|C)$, called the *sensitivity*. It is equal to $1 - 0.300 = 0.700$. Among men with prostate cancer, this is the proportion of men who will get a positive test result.

The complement of the false-positive rate (0.135) is $P(NT|C^C$, which is called the *specificity*. This likelihood probability is $1 - 0.135 = 0.865$

The posterior probability that someone has prostate cancer given a positive test result [$P(C|PT) = 0.0498$] is called the *positive predictive value*. Using Bayes's Law, we can compute the other three posterior probabilities.

The probability that the patient does not have prostate cancer given a positive test result is

$$P(C^C|PT) = 0.9502$$

The probability that the patient has prostate cancer given a negative test result is

$$P(C|NT) = 0.0035$$

The probability that the patient does not have prostate cancer given a negative test result

$$P(C^C|NT) = 0.9965$$

This revised probability is called the *negative predictive value*.

6-4c Developing an Understanding of Probability Concepts

If you review the computations made previously, you will realise that the prior probabilities are as important as the probabilities associated with the test results (the likelihood probabilities) in determining the posterior probabilities. The following table shows the prior probabilities and the revised probabilities.

Age	Prior Probabilities for Prostate Cancer	Posterior Probabilities Given a Positive PSA Test
40–49	0.010	0.0498
50–59	0.022	0.1045
60–69	0.046	0.2000
70 and older	0.079	0.3078

As you can see, if the prior probability is low then, unless the screening test is quite accurate, the revised probability will still be quite low.

To see the effects of different likelihood probabilities, suppose the PSA test is a perfect predictor. In other words, the false-positive and false-negative rates are 0. Figure 6.7 displays the probability tree.

FIGURE 6.7

Probability Tree for Example 6.10 with a Perfect Predictor Test

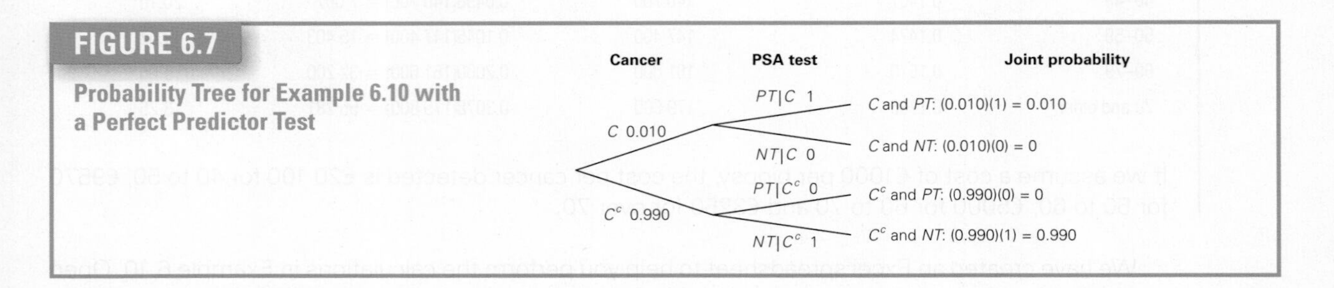

We find

$$P(PT) = P(C \text{ and } PT) + P(C^c \text{ and } PT) = 0.01 + 0 = 0.01$$

$$P(C|PT) = \frac{P(C \text{ and } PT)}{P(PT)} = \frac{0.01}{0.01} = 1.00$$

Now we calculate the probability of prostate cancer when the test is negative.

$$P(NT) = P(C \text{ and } NT) + P(C^c \text{ and } NT) = 0 + 0.99 = 0.99$$

$$P(C|NT) = \frac{P(C \text{ and } NT)}{P(NT)} = \frac{0}{0.99} = 0$$

Thus, if the test is a perfect predictor and a man has a positive test, then as expected the probability that he has prostate cancer is 1.0. The probability that he does not have cancer when the test is negative is 0.

Now suppose that the test is always wrong; that is, the false-positive and false-negative rates are 100%. The probability tree is shown in Figure 6.8.

FIGURE 6.8

Probability Tree for Example 6.10 with a Test that is Always Wrong

Cancer	PSA test	Joint probability
C 0.010	$PT\|C$ 0	C and PT: (0.010)(0) = 0
	$NT\|C$ 1	C and NT: (0.010)(1) = 0.010
C^c 0.990	$PT\|C^c$ 1	C^c and PT: (0.990)(1) = 0.990
	$NT\|C^c$ 0	C^c and NT: (0.990)(0) = 0

$$P(PT) = P(C \text{ and } PT) + P(C^c \text{ and } PT) = 0 + 0.99 = 0.99$$

$$P(C|PT) = \frac{P(C \text{ and } PT)\ 0}{P(PT)\ 0.99} = 0$$

And

$$P(NT) = P(C \text{ and } NT) + P(C^c \text{ and } NT) = 0.01 + 0 = 0.01$$

$$P(C|NT) = \frac{P(C \text{ and } NT)}{P(NT)} = \frac{0.01}{0.01} = 1.00$$

Notice we have another perfect predictor except that it is reversed. The probability of prostate cancer given a positive test result is 0, but the probability becomes 1.00 when the test is negative.

Finally we consider the situation when the set of likelihood probabilities are the same. Figure 6.9 depicts the probability tree for a 40- to 50-year-old male and the probability of a positive test is (say) 0.3 and the probability of a negative test is 0.7.

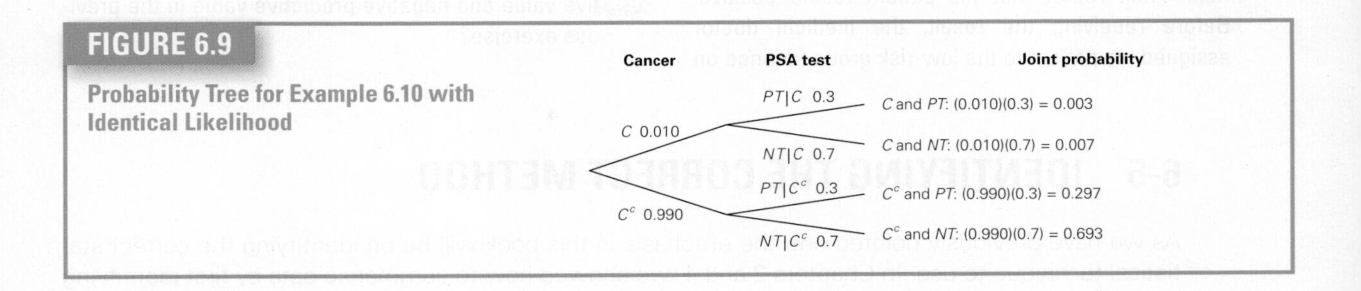

FIGURE 6.9 Probability Tree for Example 6.10 with Identical Likelihood

$$P(PT) = P(C \text{ and } PT) + P(C^c \text{ and } PT) = 0.003 + 0.297 = 0.300$$

$$P(C|PT) = \frac{P(C \text{ and } PT)}{P(PT)} = \frac{0.003}{0.300} = 0.01$$

and

$$P(NT) = P(C \text{ and } NT) + P(C^c \text{ and } NT) = 0.007 + 0.693 = 0.700$$

$$P(C|NT) = \frac{P(C \text{ and } NT)}{P(NT)} = 0.007 + 0.700 = 0.01$$

As you can see, the posterior and prior probabilities are the same. That is, the PSA test does not change the prior probabilities. Obviously, the test is useless.

We could have used any probability for the false-positive and false-negative rates, including 0.5. If we had used 0.5, then one way of performing this PSA test is to flip a fair coin. One side would be interpreted as positive and the other side as negative. It is clear that such a test has no predictive power.

The exercises and Case 6.4 offer the probabilities for several other screening tests.

EXERCISES

6.52 Refer to Exercise 6.34. Determine $P(A|B)$.

6.53 Refer to Example 6.9. An MBA applicant believes that the probability of scoring more than 650 on the GMAT without the preparatory course is 0.95. What is the probability of attaining that level after taking the preparatory course?

6.54 Refer to Exercise 6.43. The plant manager randomly selects a moulding from the early morning run and discovers it is defective. What is the probability that the foreman forgot to shut off the machine the previous night?

6.55 Bad gums may mean a bad heart. Researchers discovered that 85% of people who have suffered a heart attack had periodontal disease, an inflammation of the gums. Only 29% of healthy people have this disease. Suppose that in a certain community, heart attacks are quite rare, occurring with only 10% probability. If someone has periodontal disease, what is the probability that he or she will have a heart attack?

6.56 Refer to Exercise 6.55. If 40% of the people in a community will have a heart attack, what is the probability that a person with periodontal disease will have a heart attack?

6.57 Your favourite football team is in the final playoffs of a major tournament. You have assigned a probability of 60% that they will win the tournament. Past records indicate that when teams win the tournament, they win the first game 70% of the time. When they lose, they win the first game 25% of the time. The first game is over; your team has lost. What is the probability that it will win the tournament?

The following exercises are based on the Applications in Medical Screening and Medical Insurance subsection.

6.58 The Rapid Test is used to determine whether someone has HIV (the virus that causes AIDS). The false-positive and false-negative rates are 0.027 and 0.080, respectively. A medical doctor has just received the Rapid Test report that his patient tested positive. Before receiving the result, the medical doctor assigned his patient to the low-risk group (defined on the basis of several variables) with only a 0.5% probability of having HIV. What is the probability that the patient actually has HIV?

6.59 What are the sensitivity, specificity, positive predictive value and negative predictive value in the previous exercise?

6-5 IDENTIFYING THE CORRECT METHOD

As we have previously pointed out, the emphasis in this book will be on identifying the correct statistical technique to use. In Chapters 2 and 4, we showed how to summarise data by first identifying the appropriate method to use. Although it is difficult to offer strict rules on which probability method to use, we can still provide some general guidelines.

In the examples and exercises in this text (and most other introductory statistics books), the key issue is whether joint probabilities are provided or are required.

6-5a Joint Probabilities Are Given

In Section 6-2 we addressed problems where the joint probabilities were given. In these problems, we can compute marginal probabilities by adding across rows and down columns. We can use the joint and marginal probabilities to compute conditional probabilities, for which a formula is available. This allows us to determine whether the events described by the table are independent or dependent.

We can also apply the addition rule to compute the probability that either of two events occur.

6-5b Joint Probabilities Are Required

The previous section introduced three probability rules and probability trees. We need to apply some or all of these rules in circumstances where one or more joint probabilities are required. We apply the multiplication rule (either by formula or through a probability tree) to calculate the probability of intersections. In some problems, we are interested in adding these joint probabilities. We are actually applying the addition rule for mutually exclusive events here. We also frequently use the complement rule. In addition, we can also calculate new conditional probabilities using Bayes's Law.

CHAPTER SUMMARY

The first step in assigning probability is to create an **exhaustive** and **mutually exclusive** list of outcomes. The second step is to use the **classical, relative frequency** or **subjective approach** and assign probability to the outcomes. A variety of methods are available to compute the probability of other events. These methods include **probability rules** and **trees**.

An important application of these rules is **Bayes's Law**, which allows us to compute conditional probabilities from other forms of probability.

FORMULAS:

Conditional probability

$$P(A|B) = \frac{P(A \text{ and } B)}{P(B)}$$

Complement rule

$$P(A^C) = 1 - P(A)$$

Multiplication rule

$$P(A \text{ and } B) = P(A|B)P(B)$$

Addition rule

$$P(A \text{ or } B) = P(A) + P(B) - P(A \text{ and } B)$$

CHAPTER EXERCISES

6.60 Xr06-60 The following table lists the joint probabilities of achieving grades of A and not achieving grades of A in two MBA courses.

	Achieve a Grade of A in Marketing	Does not Achieve a Grade of A in Marketing
Achieve a grade of A in statistics	0.053	0.130
Does not achieve a grade of A in statistics	0.237	0.580

a. What is the probability that a student achieves a grade of A in marketing?

b. What is the probability that a student achieves a grade of A in marketing, given that he or she does not achieve a grade of A in statistics?

c. Are achieving grades of A in marketing and statistics independent events? Explain.

6.61 A construction company has bid on two contracts. The probability of winning contract A is 0.3. If the company wins contract A, then the probability of winning contract B is 0.4. If the company loses contract A, then the probability of winning contract B decreases to 0.2. Find the probability of the following events.

a. Winning both contracts

b. Winning exactly one contract

c. Winning at least one contract

6.62 Xr06-62 Laser surgery to fix shortsightedness is becoming more popular. However, for some people, a second procedure is necessary. The following table lists the joint probabilities of needing a second procedure and whether the patient has a corrective lens with a factor (diopter) of minus 8 or less.

	Vision Corrective Factor of more than minus 8	Vision Corrective Factor of minus 8 or less
First procedure is successful	0.66	0.15
Second procedure is required	0.05	0.14

a. Find the probability that a second procedure is required.

b. Determine the probability that someone whose corrective lens factor is minus 8 or less does not require a second procedure.

c. Are the events independent? Explain your answer.

6.63 The effect of an antidepressant drug varies from person to person. Suppose that the drug is effective on 80% of women and 65% of men. It is known that 66% of the people who take the drug are women. What is the probability that the drug is effective?

6.64 Refer to Exercise 6.63. Suppose that you are told that the drug is effective. What is the probability that the drug taker is a man?

6.65 A telemarketer sells magazine subscriptions over the telephone. The probability of a busy signal or no answer is 65%. If the telemarketer does make contact, the probability of 0, 1, 2 or 3 magazine subscriptions is 5, 25, 20 and 0.05, respectively. Find the probability that in one call she sells no magazines.

6.66 Xr06-66 A statistics lecturer believes that there is a relationship between the number of missed classes and the grade on his midterm test. After examining his records, he produced the following table of joint probabilities.

	Student Fails the Test	Student Passes the Test
Student misses fewer than five classes	0.02	0.86
Student misses five or more classes	0.09	0.03

a. What is the pass rate on the midterm test?
b. What proportion of students who miss five or more classes passes the midterm test?
c. What proportion of students who miss fewer than five classes passes the midterm test?
d. Are the events independent?

6.67 A customer service supervisor regularly conducts a survey of customer satisfaction. The results of the latest survey indicate that 8% of customers were not satisfied with the service they received at their last visit to the store. Of those who are not satisfied, only 22% return to the store within a year. Of those who are satisfied, 64% return within a year. A customer has just entered the store. In response to your question, he informs you that it is less than 1 year since his last visit to the store. What is the probability that he was satisfied with the service he received?

6.68 A statistics lecturer and his wife are planning to take a 2-week holiday in Hawaii, but they cannot decide whether to spend 1 week on each of the islands of Maui and Oahu, 2 weeks on Maui or 2 weeks on Oahu. Placing their faith in random chance, they insert two Maui brochures in one envelope, two Oahu brochures in a second envelope and one brochure from each island in a third envelope. The lecturer's wife will select one envelope at random, and their vacation schedule will be based on the brochures of the islands so selected. After his wife randomly selects an envelope, the professor removes one brochure from the envelope (without looking at the second brochure) and observes that it is a Maui brochure. What is the probability that the other brochure in the envelope is a Maui brochure? (Proceed with caution: the problem is more difficult than it appears.)

6.69 Researchers have developed statistical models based on financial ratios that predict whether a company will go bankrupt over the next 12 months. In a test of one such model, the model correctly predicted the bankruptcy of 85% of firms that did in fact fail, and it correctly predicted no bankruptcy for 74% of firms that did not fail. Suppose that we expect 8% of the firms in a particular city to fail over the next year. Suppose that the model predicts bankruptcy for a firm that you own. What is the probability that your firm will fail within the next 12 months?

6.70 In a class on probability, a statistics lecturer flips two balanced coins. Both fall to the floor and roll under his desk. A student in the first row informs the lecturer that he can see both coins. He reports that at least one of them shows tails. What is the probability that the other coin is also tails? (Beware the obvious.)

6.71 Refer to Exercise 6.70. Suppose the student informs the lecturer that he can see only one coin and it shows tails. What is the probability that the other coin is also tails?

CASE 6.1

Let's Make a Deal

A number of years ago, there was a popular television game show called *Let's Make a Deal*, which originated in the USA and was subsequently reproduced in many countries throughout the world, for example, *Le Bigdil* (France), *Gehaufs Ganze* (Germany). The host would randomly select contestants from the audience and, as the title suggests, he would make deals for prizes. Contestants would be given relatively modest prizes and would then be offered the opportunity to risk those prizes to win better ones.

Suppose that you are a contestant on this show. The host has just given you a free trip touring toxic waste sites around the country. He now offers you a trade: Give up the trip in exchange for a gamble. On the stage are three curtains, A, B and C. Behind one of them is a brand new car. Behind the other two curtains, the stage is empty. You decide to gamble and select curtain A. In an attempt to make things more interesting, the host then exposes an empty stage by opening curtain C (he knows there is nothing behind curtain C). He then offers you the free trip again if you quit now or, if you like, he will propose another deal (i.e. you can keep your choice of curtain A or perhaps switch to curtain B). What do you do?

To help you answer that question, first try answering these questions.

1. Before the host shows you what's behind curtain C, what is the probability that the car is behind curtain A? What is the probability that the car is behind curtain B?
2. After the host shows you what's behind curtain C, what is the probability that the car is behind curtain A? What is the probability that the car is behind curtain B?

CASE 6.2

Maternal Serum Screening Test for Down Syndrome

Pregnant women are screened for a condition in the baby called Down syndrome. Down syndrome babies are mentally and physically challenged. The most common screening is maternal serum screening, a blood test that looks for markers in the blood to indicate whether the birth defect may occur. The false-positive and false-negative rates vary according to the age of the mother.

Mother's Age	False-Positive Rate	False-Negative Rate
Under 30	0.04	0.376
30–34	0.082	0.290
35–37	0.178	0.269
Over 38	0.343	0.029

The probability that a baby has Down syndrome is primarily a function of the mother's age. The probabilities are listed here.

Age	Probability of Down Syndrome
25	1/1300
30	1/900
35	1/350
40	1/100
45	1/25
49	1/12

a. For each of the ages 25, 30, 35, 40, 45, determine the probability of Down syndrome if the maternity serum screening produces a positive result.

b. Repeat for a negative result.

CASE 6.3

Probability that at Least Two People in the Same Room have the Same Birthday

Suppose that there are two people in a room. The probability that they share the same birthday (date, not necessarily year) is 1/365, and the probability that they have different birthdays is 364/365. To illustrate, suppose that you are in a room with one other person and that your birthday is 1 July. The probability that the other person does not have the same birthday is 364/365 because there are 364 days in the year that are not 1 July. If a third person now enters the room, the probability that he or she has a different birthday from the first two people in the room is 363/365. Thus, the probability that three people in a room having different birthdays is (364/365) (363/365). You can continue this process for any number of people.

Find the number of people in a room so that there is about a 50% probability that at least two have the same birthday.

Hint 1: Calculate the probability that they do not have the same birthday.

Hint 2: Excel users can employ the **product** function to calculate joint probabilities.

7 RANDOM VARIABLES AND DISCRETE PROBABILITY DISTRIBUTIONS

7-1 RANDOM VARIABLES AND PROBABILITY DISTRIBUTIONS

7-2 BIVARIATE DISTRIBUTIONS

7-3 APPLICATIONS IN FINANCE: PORTFOLIO DIVERSIFICATION AND ASSET ALLOCATION

7-4 BINOMIAL DISTRIBUTION

7-5 POISSON DISTRIBUTION

INVESTING TO MAXIMISE RETURNS AND MINIMISE RISK

DATA Xm07-00

An investor has €100 000 to invest in the stock market. She is interested in developing a stock portfolio made up of stocks on the DAX Index, CAC 40 Index, AEX Amsterdam Index and the Thomson Reuters Equity Europe Index (TRXFLDEUPU). The stocks are Allianz SE (ALV), KPN NV (KPN), BNP Paribas (BNP) and Merck (MRK). However, she doesn't know how much to invest in each one. She wants to maximise her return, but she would also like to minimise the risk. She has computed the monthly returns for all four stocks during a 60-month period (January 2009 to December 2013). After some consideration, she narrowed her choices down to the following three. What should she do?

1. €25 000 in each stock
2. Allianz: €10 000, KPN: €20 000, BNP Paribas: €30 000, Merck: €40 000
3. Allianz: €10 000, KPN: €50 000, BNP Paribas: €20 000, Merck: €20 000

See Section 7-3b for the answer.

INTRODUCTION

In this chapter, we extend the concepts and techniques of probability introduced in Chapter 6. We present random variables and probability distributions, which are essential in the development of statistical inference.

Here is a brief glimpse into the wonderful world of statistical inference. Suppose that you flip a coin 100 times and count the number of heads. The objective is to determine whether we can infer from the count that the coin is not balanced. It is reasonable to believe that observing a large number of heads (say, 90) or a small number (say, 15) would be a statistical indication of an unbalanced coin. However, where do we draw the line? At 75 or 65 or 55? Without knowing the probability of the frequency of the number of heads from a balanced coin, we cannot draw any conclusions from the sample of 100 coin flips.

The concepts and techniques of probability introduced in this chapter will allow us to calculate the probability we seek. As a first step, we introduce random variables and probability distributions.

7-1 RANDOM VARIABLES AND PROBABILITY DISTRIBUTIONS

Consider an experiment where we flip two balanced coins and observe the results. We can represent the events as

Heads on the first coin and heads on the second coin
Heads on the first coin and tails on the second coin
Tails on the first coin and heads on the second coin
Tails on the first coin and tails on the second coin

However, we can list the events in a different way. Instead of defining the events by describing the outcome of each coin, we can count the number of heads (or, if we wish, the number of tails). Thus, the events are now

2 heads
1 head
1 head
0 head

The number of heads is called the **random variable**. We often label the random variable X, and we are interested in the probability of each value of X. Thus, in this illustration, the values of X are 0, 1 and 2.

Here is another example. In many parlour games, as well as in the game of craps played in casinos, the player tosses two dice. One way of listing the events is to describe the number on the first die and the number on the second die as follows.

1,1	1,2	1,3	1,4	1,5	1,6
2,1	2,2	2,3	2,4	2,5	2,6
3,1	3,2	3,3	3,4	3,5	3,6
4,1	4,2	4,3	4,4	4,5	4,6
5,1	5,2	5,3	5,4	5,5	5,6
6,1	6,2	6,3	6,4	6,5	6,6

However, in almost all games, the player is primarily interested in the total. Accordingly, we can list the totals of the two dice instead of the individual numbers.

2	3	4	5	6	7
3	4	5	6	7	8
4	5	6	7	8	9
5	6	7	8	9	10
6	7	8	9	10	11
7	8	9	10	11	12

If we define the random variable X as the total of the two dice, then X can equal 2, 3, 4, 5, 6, 7, 8, 9, 10, 11 and 12.

Random Variable

A **random variable** is a variable whose value is a numerical outcome of a random process or random experiment.

In some processes or experiments, the outcomes are numbers. For example, when we observe the return on an investment or measure the amount of time to assemble a computer, the experiment produces events that are numbers. Simply stated, the value of a random variable is a numerical event.

There are two types of random variables: discrete and continuous. A **discrete random variable** is one that can take on a countable number of values. Discrete applies when all possible values are separated from each other by impossible values. For example, if we define X as the number of heads observed in an experiment that flips a coin ten times, then the values of X are 0, 1, 2, . . . , 10. The variable X can assume a total of 11 values. Obviously, we counted the number of values and we cannot have any value between 0 and 1 or 1 and 2, and so on; hence, X is discrete.

A **continuous random variable** is one whose values vary and takes an infinite number of possible values. An excellent example of a continuous random variable is the amount of time to complete a task. For example, let $X =$ time to write a statistics exam in a university where the time limit is 3 hours and students cannot leave before 30 minutes. The smallest value of X is 30 minutes. If we attempt to count the number of values that X can take on, we need to identify the next value. Is it 30.1 minutes? 30.01 minutes? 30.001 minutes? None of these is the second possible value of X because there exist numbers larger than 30 and smaller than 30.001. It becomes clear that we cannot identify the second, or third, or any other values of X (except for the largest value 180 minutes). Thus, we cannot count the number of values, and X is continuous.

A **probability distribution** is a table, formula, or graph that describes the values of a random variable and the probability associated with these values. We will address discrete probability distributions in the rest of this chapter and cover continuous distributions in Chapter 8.

As we noted earlier, an upper case letter will represent the *name* of the random variable, usually X. Its lower case counterpart will represent the value of the random variable. Thus, we represent the probability that the random variable X will equal x as

$$P(X = x)$$

or more simply

$$P(x)$$

7-1a Discrete Probability Distributions

The discrete probability distribution is a list of probabilities associated with each discrete random variable. It is also known as probability function or the probability mass function. The probabilities of the values of a discrete random variable may be derived by means of probability tools such as tree diagrams or by applying one of the definitions of probability. However, two fundamental requirements apply as stated in the box.

Requirements for a Distribution of a Discrete Random Variable

1. $0 \leq P(x) \leq 1$ for all x
2. $\sum_{\text{all } x} P(x) = 1$

where the random variable can assume values x and $P(x)$ is the probability that the random variable is equal to x.

These requirements are equivalent to the rules of probability provided in Chapter 6. To illustrate, consider the following example.

EXAMPLE 7.1

Probability Distribution of Persons per Household

The *2011 Census: Population Estimates by 5-year age bands, and Household Estimates, for Local Authorities in the UK* provides information for administrative areas within the UK at the time of Census day, 27 March 2011. The following table summarises the data regarding the number of persons living in the household in the UK. Develop the probability distribution of the random variable defined as the number of persons per household.

Number of Persons	Number of Households (millions)
1	8.1
2	9.1
3	4.1
4	3.4
5 or more	1.8
Total	26.5

SOLUTION:

The probability of each value of *X*, the number of persons per household, is computed as the relative frequency. We divide the frequency for each value of *X* by the total number of households, producing the following probability distribution (**DATA Xm07-01**).

X	*P*(*x*)
1	8.1/26.5 = 0.306
2	9.1/26.5 = 0.343
3	4.1/26.5 = 0.155
4	3.4/26.5 = 0.128
5 or more	1.8/26.5 = 0.068
Total	1.000

As you can see, the requirements are satisfied. Each probability lies between 0 and 1, and the total is 1.

We interpret the probabilities in the same way we did in Chapter 6. For example, if we select one household at random, the probability that it has three persons is

$$P(3) = 0.155$$

We can also apply the addition rule for mutually exclusive events. (The values of *X* are mutually exclusive; a household can have 1, 2, 3, 4 or 5 or more persons.) The probability that a randomly selected household has four or more persons is

$$P(X \geq 4) = P(4) + P(5 \text{ or more})$$
$$= 0.128 + 0.068 = 0.196$$

In Example 7.1, we calculated the probabilities using census information about the entire population. The next example illustrates the use of the techniques introduced in Chapter 6 to develop a probability distribution.

EXAMPLE 7.2

Probability Distribution of the Number of Sales

A mutual fund salesperson has arranged to call on three people tomorrow. Based on past experience, the salesperson knows there is a 20% chance of closing a sale on each call. Determine the probability distribution of the number of sales the salesperson will make.

SOLUTION:

We can use the probability rules and trees introduced in Section 6-3. Figure 7.1 displays the probability tree for this example. Let X = the number of sales (**DATA Xm07-02**).

FIGURE 7.1

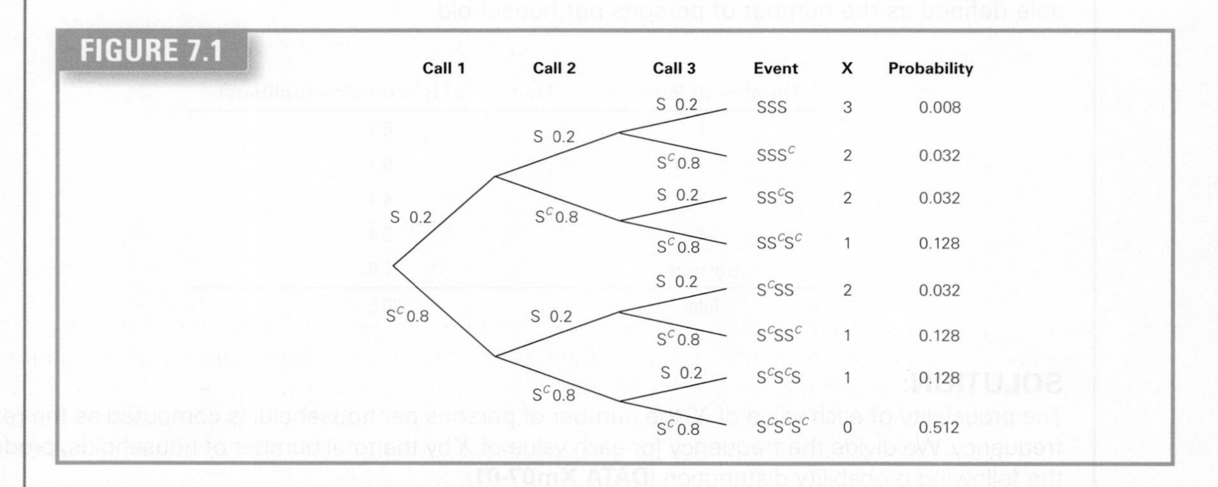

The tree exhibits each of the eight possible outcomes and their probabilities. We see that there is one outcome that represents no sales, and its probability is $P(0) = 0.512$. There are three outcomes representing one sale, each with probability 0.128, so we add these probabilities. Thus,

$$P(1) = 0.128 + 0.128 + 0.128 = 3(0.128) = 0.384$$

The probability of two sales is computed similarly:

$$P(X) = 3\ (0.032) = 0.096$$

There is one outcome where there are three sales:

$$P(3) = 0.008$$

The probability distribution of X is listed in Table 7.1.

TABLE 7.1 Probability Distribution of the Number of Sales in Example 7.2

X	$P(x)$
0	0.512
1	0.384
2	0.096
3	0.008

7-1b Probability Distributions and Populations

The importance of probability distributions derives from their use as representatives of populations. In Example 7.1, the distribution provided us with information about the population of numbers of persons per household. In Example 7.2, the population was the number of sales made in three calls by the salesperson. And, as we noted before, statistical inference deals with inference about populations.

7-1c Describing the Population/Probability Distribution

In Chapter 4 we showed how to calculate the mean, variance and standard deviation of a population. The formulas we provided were based on knowing the value of the random variable for each member of the population. For example, if we want to know the mean and variance of annual income of all European middle-class workers, we would record each of their incomes and use the formulas introduced in Chapter 4:

$$\mu = \frac{\sum_{i=1}^{N} X_i}{N}$$

$$\sigma^2 = \frac{\sum_{i=1}^{N} (X_i - \mu)^2}{N}$$

where X_1 is the income of the first middle-class worker, X_2 is the second worker's income and so on. It is likely that N equals several million. As you can appreciate, these formulas are seldom used in practical applications because populations are so large. It is unlikely that we would be able to record all the incomes in the population of European middle-class workers. However, probability distributions often represent populations. Rather than record each of the many observations in a population, we list the values and their associated probabilities as we did in deriving the probability distribution of the number of persons per household in Example 7.1 and the number of successes in three calls by the mutual fund salesperson. These can be used to compute the mean and variance of the population.

The population mean is the weighted average of all of its values. The weights are the probabilities. This parameter is also called the **expected value** of X and is represented by $E(X)$.

Population Mean

$$E(X) = \mu = \sum_{\text{all } x} xP(x)$$

The population variance is calculated similarly. It is the weighted average of the squared deviations from the mean.

Population Variance

$$V(X) = \sigma^2 = \sum_{\text{all } x} (x - \mu)^2 P(x)$$

There is a shortcut calculation that simplifies the calculations for the population variance. This formula is not an approximation; it will yield the same value as the formula above.

Shortcut Calculation for Population Variance

$$V(X) = \sigma^2 = \sum_{\text{all } x} x^2 P(x) - \mu^2$$

The standard deviation is defined as in Chapter 4.

Population Standard Deviation

$$\sigma = \sqrt{\sigma^2}$$

EXAMPLE 7.3

Describing the Population of the Number of Persons per Household

Find the mean, variance and standard deviation for the population of the number of persons per household as given in Example 7.1.

SOLUTION (DATA Xm07-03):

For this example, we will assume that the last category is exactly five persons. The mean of X is

$$E(x) = \mu = \sum_{\text{all } x} xP(x) = 1P(1) + 2P(2) + 3P(3) + 3P(4) + 5P(5)$$
$$= 1(0.306) + 2(0.343) + 3(0.155) + 4(0.128) + 5(0.068)$$
$$= 2.309$$

Notice that the random variable can assume integer values only, yet the mean is 2.309. The variance of X is

$$V(X) = \sigma^2 = \sum_{\text{all } x} (x - \mu)^2 P(x)$$
$$= (1 - 2.309)^2(0.306) + (2 - 2.309)^2(0.343) + (3 - 2.309)^2(0.155)$$
$$+ (4 - 2.309)^2(0.128) + (5 - 2.309)^2(0.068) = 1.490$$

To demonstrate the shortcut method, we will use it to recompute the variance:

$$\sum_{\text{all } x} x^2 P(x) = 1^2(0.306) + 2^2(0.343) + 3^2(0.155) + 4^2(0.128) + 5^2(0.068)$$
$$= 6.821$$

and

$$\mu = 2.309$$

Thus,

$$\sigma^2 = \sum_{\text{all } x} x^2 P(x) - \mu^2 = 6.821 - (2.309)^2 = 1.490$$

The standard deviation is

$$\sigma = \sqrt{\sigma^2} = \sqrt{1.490} = 1.220$$

These parameters tell us that the mean and standard deviation of the number of persons per household are 2.309 and 1.220, respectively.

7-1d Laws of Expected Value and Variance

As you will discover, we often create new variables that are functions of other random variables. The formulas given in the next two boxes allow us to quickly determine the expected value and variance of these new variables. In the notation used here, X is the random variable and c is a constant.

Laws of Expected Value

1. $E(c) = c$
2. $E(X + c) = E(X) + c$
3. $E(cX) = cE(X)$

Laws of Variance

1. $V(c) = 0$
2. $V(X + c) = V(X)$
3. $V(cX) = c^2 V(X)$

EXAMPLE 7.4

Describing the Population of Monthly Profits

The monthly sales at a computer store have a mean of £25 000 and a standard deviation of £4000. Profits are calculated by multiplying sales by 30% and subtracting fixed costs of £6000. Find the mean and standard deviation of monthly profits.

SOLUTION:

We can describe the relationship between profits and sales by the following equation:

$$\text{Profit} = 0.30(\text{Sales}) - 6000$$

The expected or mean profit is

$$E(\text{Profit}) = E[0.30(\text{Sales}) - 6000]$$

Applying the second law of expected value, we produce

$$E(\text{Profit}) = E[0.30(\text{Sales})] - 6000$$

Applying law 3 yields

$$E(\text{Profit}) = E0.30(\text{Sales}) - 6000 = 0.30(25\,000) - 6000 - 1500$$

Thus, the mean monthly profit is £1500.

The variance is

$$V(\text{Profit}) = V[0.30(\text{Sales}) - 6\,000]$$

The second law of variance states that

$$V(\text{Profit}) = V[0.\ 30(\text{Sales})] \text{ and law 3 yields}$$

$$V(\text{Profit}) = (0.30)^2 V(\text{Sales}) = 0.09(4000)^2 = 1\,440\,000$$

Thus, the standard deviation of monthly profits is

$$\sigma_{Profit} = \sqrt{1\,440\,000} = £1200$$

EXERCISES

7.1 The number of accidents that occur on a busy stretch of highway is a random variable.

a. What are the possible values of this random variable?

b. Are the values countable? Explain.

c. Is there a finite number of values? Explain.

d. Is the random variable discrete or continuous? Explain.

7.2 The distance a car travels on a tank of petrol is a random variable.

a. What are the possible values of this random variable?

b. Are the values countable? Explain.

c. Is there a finite number of values? Explain.

d. Is the random variable discrete or continuous? Explain.

7.3 The amount of money students earn on their summer jobs is a random variable.

a. What are the possible values of this random variable?

b. Are the values countable? Explain.

c. Is there a finite number of values? Explain.

d. Is the random variable discrete or continuous? Explain.

7.4 The mark on a statistics exam that consists of 100 multiple-choice questions is a random variable.

a. What are the possible values of this random variable?

b. Are the values countable? Explain.

c. Is there a finite number of values? Explain.

d. Is the random variable discrete or continuous? Explain.

7.5 Determine whether each of the following is a valid probability distribution.

a.

X	0	1	2	3
$P(x)$	0.1	0.3	0.4	0.1

b.

X	5	−6	10	0
$P(x)$	0.01	0.01	0.01	0.97

c.

X	14	12	−7	13
$P(x)$	0.25	0.46	0.04	0.24

7.6 Xr07-06 In a recent census, the number of colour televisions per household was recorded

Number of colour televisions	0	1	2	3	4	5
Number of households (thousands)	1218	32 379	37 961	19 387	7714	2842

a. Develop the probability distribution of X, the number of colour televisions per household.

b. Determine the following probabilities.

$P(X \leq 2)$
$P(X > 2)$
$P(X \geq 4)$

7.7 Xr07-07 Using historical records, the personnel manager of a plant in Hong Kong has determined the probability distribution of X, the number of employees absent per day. It is

X	0	1	2	3	4	5	6	7
$P(x)$	0.005	0.025	0.310	0.340	0.220	0.080	0.019	0.001

a. Find the following probabilities.

$P(2 \leq X \leq 5)$
$P(X > 5)$
$P(X < 4)$

b. Calculate the mean of the population.

c. Calculate the standard deviation of the population.

7.8 Xr07-08 The random variable X has the following probability distribution.

X	−3	2	6	8
$P(x)$	0.2	0.3	0.4	0.1

Find the following probabilities.

a. $P(X > 0)$

b. $P(X \geq 1)$

c. $P(X \geq 2)$

d. $P(2 \leq X \leq 5)$

7.9 Xr07-09 A pharmacy advertises on the Internet that it will deliver the over-the-counter products that customers purchase in 3 to 6 days. The manager of the company wanted to be more precise in its advertising. Accordingly, she recorded the number of days it took to deliver to customers. From the data, the following probability distribution was developed.

Number of days	0	1	2	3	4	5	6	7	8
Probability	0	0	0.01	0.04	0.28	0.42	0.21	0.02	0.02

a. What is the probability that a delivery will be made within the advertised 3- to 6-day period?

b. What is the probability that a delivery will be late?

c. What is the probability that a delivery will be early?

7.10 The probability that a university graduate will be offered no jobs within a month of graduation is estimated to be 5%. The probability of receiving one, two and three job offers has similarly been estimated to be 43%, 31% and 21%, respectively. Determine the following probabilities.

a. A graduate is offered fewer than two jobs.

b. A graduate is offered more than one job.

7.11 Use a probability tree to compute the probability of the following events when flipping two fair coins.

a. Heads on the first coin and heads on the second coin

b. Heads on the first coin and tails on the second coin

c. Tails on the first coin and heads on the second coin

d. Tails on the first coin and tails on the second coin

7.12 Refer to Exercise 7.11. Find the following probabilities.

a. No heads

b. One head

c. Two heads

d. At least one head

7.13 Draw a probability tree to describe the flipping of three fair coins.

7.14 Refer to Exercise 7.13. Find the following probabilities.

a. Two heads

b. One head

c. At least one head

d. At least two heads

7.15 **Xr07-15** The random variable X has the following distribution.

X	−2	5	7	8
$P(x)$	0.59	0.15	0.25	0.01

a. Find the mean and variance for the probability distribution below.

b. Determine the probability distribution of Y where $Y = 5X$.

c. Use the probability distribution in part (b) to compute the mean and variance of Y.

d. Use the laws of expected value and variance to find the expected value and variance of Y from the parameters of X.

7.16 **Xr07-16** We are given the following probability distribution.

X	0	1	2	3
$P(x)$	0.4	0.3	0.2	0.1

a. Calculate the mean, variance and standard deviation.

b. Suppose that $Y = 3X + 2$. For each value of X, determine the value of Y. What is the probability distribution of Y?

c. Calculate the mean, variance and standard deviation from the probability distribution of Y.

d. Use the laws of expected value and variance to calculate the mean, variance and standard deviation of Y from the mean, variance and standard deviation of X. Compare your answers in parts (c) and (d). Are they the same (except for rounding)?

7.17 **Xr07-17** The number of pizzas delivered to university students each month is a random variable with the following probability distribution.

X	0	1	2	3
$P(x)$	0.1	0.3	0.4	0.2

a. Find the probability that a student has received delivery of two or more pizzas this month.

b. Determine the mean and variance of the number of pizzas delivered to students each month.

7.18 Refer to Exercise 7.17. If the pizzeria makes a profit of £3 per pizza, determine the mean and variance of the profits per student.

7.19 **Xr07-19** A survey of Amazon.com shoppers reveals the following probability distribution of the number of books purchased per hit.

X	0	1	2	3	4	5	6	7
$P(x)$	0.35	0.25	0.20	0.08	0.06	0.03	0.02	0.01

a. What is the probability that an Amazon.com visitor will buy four books?

b. What is the probability that an Amazon.com visitor will buy eight books?

c. What is the probability that an Amazon.com visitor will not buy any books?

d. What is the probability that an Amazon.com visitor will buy at least one book?

7.20 **Xr07-20** A university librarian produced the following probability distribution of the number of times a student walks into the library over the period of a semester.

X	0	5	10	15	20	25	30	40	50	75	100
$P(x)$	0.22	0.29	0.12	0.09	0.08	0.05	0.04	0.04	0.03	0.03	0.01

Find the following probabilities.

a. $P(X \geq 20)$

b. $P(X = 60)$

c. $P(X > 50)$

d. $P(X > 100)$

7.21 **Xr07-21** A shopping mall in Liverpool estimates the probability distribution of the number of shops that customers actually enter within the mall, as shown in the table.

X	0	1	2	3	4	5	6
$P(x)$	0.04	0.19	0.22	0.28	0.12	0.09	0.06

Find the mean and standard deviation of the number of stores entered.

7.22 Refer to Exercise 7.21. Suppose that, on average, customers spend 10 minutes in each shop they enter. Find the mean and standard deviation of the total amount of time customers spend in stores.

7.23 **Xr07-23** When parking in a city centre car park in Southampton, drivers pay according to the number of hours parked. The probability distribution of the number of hours cars are parked has been estimated as follows.

X	1	2	3	4	5	6	7	8
$P(x)$	0.24	0.18	0.13	0.10	0.07	0.04	0.04	0.20

Find the mean and standard deviation of the number of hours cars are parked in the car park.

7.24 Refer to Exercise 7.23. The cost of parking is £2.50 per hour. Calculate the mean and standard deviation of the amount of revenue each car generates.

7.25 **Xr07-25** The manager of a bookshop in Rotterdam recorded the number of customers who arrive at a checkout counter every 5 minutes from which the following distribution was calculated. Calculate the mean and standard deviation of the random variable.

X	0	1	2	3	4
P(x)	0.10	0.20	0.25	0.25	0.20

7.26 The owner of a small business in Hamburg has just purchased a personal computer, which she expects will serve her for the next 2 years. The owner has been told that she 'must' buy a surge suppressor to provide protection for her new hardware against possible surges or variations in the electrical current, which have the capacity to damage the computer. The amount of damage to the computer depends on the strength of the surge. It has been estimated that there is a 1% chance of incurring €400 damage, a 2% chance of incurring €200 damage and 10% chance of €100 damage. An inexpensive suppressor, which would provide protection for only one surge can be purchased. How much should the owner be willing to pay if she makes decisions on the basis of expected value?

7.27 It cost £2 to buy a lottery ticket, which has five prizes. The prizes and the probability that a player wins the prize are listed here. Calculate the expected value of the payoff.

Prize (£)	1 million	200 000	50 000
Probability	1/10 million	1/1 million	1/500 000

Prize (£)	10 000	1 000
Probability	1/50 000	1/10 000

7.28 Xr07-28 After an analysis of incoming faxes, the manager of an accounting firm determined the probability distribution of the number of pages per facsimile as follows:

X	1	2	3	4	5	6	7
P(x)	0.05	0.12	0.20	0.30	0.15	0.10	0.08

Compute the mean and variance of the number of pages per fax.

7.29 Refer to Exercise 7.28. Further analysis by the manager revealed that the cost of processing each page of a fax is £0.25. Determine the mean and variance of the cost per fax.

7.30 Xr07-30 To examine the effectiveness of its four annual advertising promotions, a mail-order company has sent a questionnaire to each of its customers, asking how many of the previous year's promotions prompted orders that would not otherwise have been made. The table lists the probabilities that were derived from the questionnaire, where *X* is the random variable representing the number of promotions that prompted orders. If we assume that overall customer behaviour next year will be the same as last year, what is the expected number of promotions that each customer will take advantage of next year by ordering goods that otherwise would not be purchased?

X	0	1	2	3	4
P(x)	0.10	0.25	0.40	0.20	0.05

7.31 Refer to Exercise 7.30. A previous analysis of historical records found that the mean value of orders for promotional goods is £20, with the company earning a gross profit of 20% on each order. Calculate the expected value of the profit contribution next year.

7-2 BIVARIATE DISTRIBUTIONS

Thus far, we have dealt with the distribution of a *single* variable. However, there are circumstances where we need to know about the relationship between two variables. Recall that we have addressed this problem statistically in Chapter 3 by drawing the scatter diagram and in Chapter 4 by calculating the covariance and the coefficient of correlation. In this section, we present the **bivariate distribution**, which provides probabilities of combinations of two variables. Incidentally, when we need to distinguish between the bivariate distributions and the distributions of one variable, we will refer to the latter as *univariate* distributions.

The joint probability that two variables will assume the values *x* and *y* is denoted $P(x, y)$. A bivariate (or joint) probability distribution of *X* and *Y* can be written in the form of a table or formula that lists the joint probabilities for all pairs of values of *x* and *y*. As was the case with univariate distributions, the joint probability must satisfy two requirements.

Requirements for a Discrete Bivariate Distribution

1. $0 \leq P(x, y) \leq 1$ for all pairs of values (x, y)
2. $\sum_{\text{all } x} \sum_{\text{all } y} P(x, y) = 1$

EXAMPLE 7.5

Bivariate Distribution of the Number of House Sales

Xavier and Yvette are real estate agents. Let *X* denote the number of houses that Xavier will sell in a month and let *Y* denote the number of houses Yvette will sell in a month. An analysis of their past monthly performances has the following joint probabilities.

Bivariate Probability Distribution

		X		
		0	**1**	**2**
	0	0.12	0.42	0.06
Y	1	0.21	0.06	0.03
	2	0.07	0.02	0.01

We interpret these joint probabilities in the same way we did in Chapter 6. For example, the probability that Xavier sells 0 houses and Yvette sells 1 house in the month is $P(0, 1) = 0.21$.

7-2a Marginal Probabilities

As we did in Chapter 6, we can calculate the marginal probabilities by summing across rows or down columns.

Marginal Probability Distribution of X in Example 7.5

$$P(X = 0) = P(0, 0) + P(0,1) + P(0, 2) = 0.12 + 0.21 + 0.07 = 0.4$$

$$P(X = 1) = P(1, 0) + P(1, 1) + P(1, 2) = 0.42 + 0.06 + 0.02 = 0.5$$

$$P(X = 2) = P(2, 0) + P(2, l) + P(2, 2) = 0.06 + 0.03 + 0.01 = 0.1$$

The marginal probability distribution *X is*

X	*P*(*X*)
0	0.4
1	0.5
2	0.1

Marginal Probability Distribution of *Y* in Example 7.5

$$P(Y = 0) = P(0, 0) + P(l, 0) + P(2, 0) = 0.12 + 0.42 + 0.06 = 0.6$$

$$P(Y = 1) = P(0, 1) + P(l, 1) + P(2, 1) = 0.21 + 0.06 + 0.03 = 0.3$$

$$P(Y = 2) = P(0, 2) + P(1, 2) + P(2, 2) = 0.07 + 0.02 + 0.01 = 0.1$$

The marginal probability distribution of *Y* is

Y	*P*(*Y*)
0	0.6
1	0.3
2	0.1

Notice that both marginal probability distributions meet the requirements; the probabilities are between 0 and 1, and they add to 1.

7-2b Describing the Bivariate Distribution

As we did with the univariate distribution, we often describe the bivariate distribution by computing the mean, variance and standard deviation of each variable. We do so by utilizing the marginal probabilities.

Expected Value, Variance and Standard Deviation of *X* in Example 7.5

$$E(X) = \mu_X = \sum xP(x) = 0(0.4) + 1(0.5) + 2(0.1) = 0.7$$
$$V(X) = \sigma_X^2 = \sum (x - \mu_X)^2 P(x) = (0 - 0.7)^2(0.4) + (1 - 0.7)^2(0.5) + (2 - 0.7)^2(0.1) = 0.41$$
$$\sigma_X = \sqrt{\sigma_X^2} = \sqrt{0.41} = 0.64$$

Expected Value, Variance and Standard Deviation of *Y* in Example 7.5

$$E(Y) = \mu_Y = \sum yP(y) = 0(0.6) + 1(0.3) + 2(0.1) = 0.5$$
$$V(Y) = \sigma_Y^2 = \sum (y - \mu_Y)^2 P(y) = (0 - 0.5)^2(0.6) + (1 - 0.5)^2(0.3) + (2 - 0.5)^2(0.1) = 0.45$$
$$\sigma_Y = \sqrt{\sigma_Y^2} = \sqrt{0.45} = 0.67$$

There are two more parameters we can and need to compute. Both deal with the relationship between the two variables. They are the covariance and the coefficient of correlation. Recall that both were introduced in Chapter 4, where the formulas were based on the assumption that we knew each of the *N* observations of the population. In this chapter, we compute parameters like the covariance and the coefficient of correlation from the bivariate distribution.

Covariance

The covariance of two discrete variables is defined as

$$\text{COV}(X, Y) = \sigma_{xy} = \sum_{\text{all } x} \sum_{\text{all } y} (x - \mu_X)(y - \mu_Y)P(x, y)$$

Notice that we multiply the deviations from the mean for both *X* and *Y* and then multiply by the joint probability.

The calculations are simplified by the following shortcut method.

Shortcut Calculation for Covariance

$$\text{COV}(X, Y) = \sigma_{xy} = \sum_{\text{all } x} \sum_{\text{all } y} xyP(x, y) - \mu_X \mu_Y$$

The coefficient of correlation is calculated in the same way as in Chapter 4.

Coefficient of Correlation

$$\rho = \frac{\sigma_{xy}}{\sigma_x \sigma_y}$$

EXAMPLE 7.6

Describing the Bivariate Distribution

Compute the covariance and the coefficient of correlation between the numbers of houses sold by the two agents in Example 7.5.

SOLUTION:

We start by computing the covariance.

$$\sigma_{xy} = \sum_{\text{all } x}\sum_{\text{all } y}(x - \mu_X)(y - \mu_Y)P(x, y)$$
$$= (0 - 0.7)(0 - 0.5)(0.12) + (1 - 0.7)(0 - 0.5)(0.42) + (2 - 0.7)(0 - 0.5)(0.06)$$
$$+ (0 - 0.7)(1 - 0.5)(0.21) + (1 - 0.7)(1 - 0.5)(0.06) + (2 - 0.7)(1 - 0.5)(0.03)$$
$$+ (0 - 0.7)(2 - 0.5)(0.07) + (1 - 0.7)(2 - 0.5)(0.02) + (2 - 0.7)(2 - 0.5)(0.01)$$
$$= -0.15$$

As we did with the shortcut method for the variance, we will recalculate the covariance using its shortcut method.

$$\sum_{\text{all } x}\sum_{\text{all } y} xyP(x, y) = (0)(0)(0.12) + (1)(0)(0.42) + (2)(0)(0.06)$$
$$+ (0)(1)(0.21) + (1)(1)(0.06) + (2)(1)(0.03)$$
$$+ (0)(2)(0.07) + (1)(2)(0.02) + (2)(2)(0.01)$$
$$= 0.2$$

Using the expected values computed above, we find

$$\sigma_{xy} = \sum_{\text{all } x}\sum_{\text{all } y} xyP(x, y) - \mu_X\mu_Y = 0.2 - (0.7)(0.5) = -0.15$$

We also computed the standard deviations above. Thus, the coefficient of correlation is

$$\rho = \frac{\sigma_{xy}}{\sigma_X\sigma_Y} = \frac{-0.15}{(0.64)(0.67)} = -0.35$$

There is a weak negative relationship between the two variables: the number of houses Xavier will sell in a month (X) and the number of houses Yvette will sell in a month (Y).

7-2c Sum of Two Variables

The bivariate distribution allows us to develop the probability distribution of any combination of the two variables. Of particular interest to us is the sum of two variables. The analysis of this type of distribution leads to an important statistical application in finance, which we present in the next section.

To see how to develop the probability distribution of the sum of two variables from their bivariate distribution, return to Example 7.5. The sum of the two variables X and Y is the total number of houses sold per month. The possible values of $X + Y$ are 0, 1, 2, 3 and 4. The probability that $X + Y = 2$, for example, is obtained by summing the joint probabilities of all pairs of values of X and Y that sum to 2:

$$P(X + Y + 2) = P(0, 2) + P(1, 1) + P(2, 0) = 0.07 + 0.06 + 0.06 = 0.19$$

We calculate the probabilities of the other values of $X + Y$ similarly, producing the following table.

Probability Distribution of $X + Y$ in Example 7.5

$X + Y$	0	1	2	3	4
$P(x + y)$	0.12	0.63	0.19	0.05	0.01

We can compute the expected value, variance and standard deviation of $X + Y$ in the usual way.

$$E(X + Y) = 0(0.12) + 1(0.63) + 2(0.19) + 3(0.05) + 4(0.01) = 1.2$$
$$V(X + Y) = \sigma^2_{X+Y} = (0 - 1.2)^2(0.12) + (1 - 1.2)^2(0.63) + (2 - 1.2)^2(0.19)$$
$$+ (3 - 1.2)^2(0.05) + (4 - 1.2)^2(0.01)$$
$$= 0.56$$
$$\sigma_{X+Y} = \sqrt{0.56} = 0.75$$

We can derive a number of laws that enable us to compute the expected value and variance of the sum of two variables.

Laws of Expected Value and Variance of the Sum of Two Variables

1. $E(X + Y) = E(X) + E(Y)$
2. $V(X + Y) = V(X) + V(Y) + 2COV(X, Y)$

If X and Y are independent, $COV(X, Y) = 0$ and thus $V(X + Y) = V(X) + V(Y)$

EXAMPLE 7.7

Describing the Population of the Total Number of House Sales

Use the rules of expected value and variance of the sum of two variables to calculate the mean and variance of the total number of houses sold per month in Example 7.5.

SOLUTION:

Using law 1 we compute the expected value of $X + Y$:

$$E(X + Y) = E(X) + E(Y) = 0.7 + 0.5 = 1.2$$

which is the same value we produced directly from the probability distribution of $X + Y$. We apply law 3 to determine the variance:

$$V(X + Y) = V(X) + V(Y) + 2COV(X, Y) = 0.41 + 0.45 + 2(-0.15) = 0.56$$

This is the same value we obtained from the probability distribution of $X + Y$.

We will encounter several applications where we need the laws of expected value and variance for the sum of two variables. Additionally, we will demonstrate an important application in operations management where we need the formulas for the expected value and variance of the sum of more than two variables.

EXERCISES

7.32 Xr07-32 The following table lists the bivariate distribution of X and Y.

	X	
Y	**1**	**2**
1	0.5	0.1
2	0.1	0.3

a. Find the marginal probability distribution of X.

b. Find the marginal probability distribution of Y.

c. Compute the mean and variance of X.

d. Compute the mean and variance of Y.

7.33 Refer to Exercise 7.32. Compute the covariance and the coefficient of correlation.

7.34 Refer to Exercise 7.32. Use the laws of expected value and variance of the sum of two variables to compute the mean and variance of $X + Y$.

7.35 Refer to Exercise 7.32.

a. Determine the distribution of $X + Y$.

b. Determine the mean and variance of $X + Y$.

c. Does your answer to part (b) equal the answer to Exercise 7.34?

7.36 Xr07-36 The bivariate distribution of X and Y is described here.

	X	
Y	**1**	**2**
1	0.28	0.42
2	0.12	0.18

a. Find the marginal probability distribution of X.

b. Find the marginal probability distribution of Y.

c. Compute the mean and variance of X.

d. Compute the mean and variance of Y.

7.37 Refer to Exercise 7.36. Compute the covariance and the coefficient of correlation.

7.38 Refer to Exercise 7.36. Use the laws of expected value and variance of the sum of two variables to compute the mean and variance of $X + Y$.

7.39 Refer to Exercise 7.36.

a. Determine the distribution of $X + Y$.

b. Determine the mean and variance of $X + Y$.

c. Does your answer to part (b) equal the answer to Exercise 7.49?

7.40 Xr07-40 The joint probability distribution of X and Y is shown in the following table.

	X		
Y	1	2	3
1	0.42	0.12	0.06
2	0.28	0.08	0.04

a. Determine the marginal distributions of X and Y.

b. Compute the covariance and coefficient of correlation between X and Y.

c. Develop the probability distribution of $X + Y$.

7.41 The following distributions of X and of Y have been developed. If X and Y are independent, determine the joint probability distribution of X and Y.

X	0	1	2
P(*x*)	0.6	0.3	0.1

Y	1	2
P(*y*)	0.7	0.3

a. Compute the mean and variance of the number of refrigerators sold daily.

b. Compute the mean and variance of the number of stoves sold daily.

c. Compute the covariance and the coefficient of correlation.

7.42 The distributions of X and of Y are described here. If X and Y are independent, determine the joint probability distribution of X and Y.

X	0	1
P(*x*)	0.2	0.8

Y	1	2	3
P(*y*)	0.2	0.4	0.4

7.43 Xr07-43 After analysing several months of sales data, the owner of an appliance store produced the following joint probability distribution of the number of refrigerators and cookers sold daily.

	Refrigerators		
Cookers	0	1	2
0	0.08	0.14	0.12
1	0.09	0.17	0.13
2	0.05	0.18	0.04

a. Find the marginal probability distribution of the number of refrigerators sold daily.

b. Find the marginal probability distribution of the number of cookers sold daily.

7.44 Refer to Exercise 7.43. Find the following conditional probabilities.

a. P(1 refrigerator|0 cookers)

b. P(0 cookers|1 refrigerator)

c. P(2 refrigerators|2 cookers)

APPLICATIONS IN OPERATIONS MANAGEMENT

PERT/CPM

The Project Evaluation and Review Technique **(PERT)** and the Critical Path Method **(CPM)** are related management-science techniques that help operations managers control the activities and the amount of time it takes to complete a project. Both techniques are based on the order in which the activities must be performed. For example, in building a house the excavation of the foundation must precede the pouring of the foundation, which in turn precedes the framing. A **path** is defined as a sequence of related activities that leads from the starting point to the completion of a project. In most projects, there are several paths with differing amounts of time needed for their completion. The longest path is called the **critical path** because any delay in the activities along this path will result in a delay in the completion of the project. In some versions of PERT/CPM, the activity completion times are fixed and the chief task of the operations manager is to

(*continued*)

determine the critical path. In other versions, each activity's completion time is considered to be a random variable, where the mean and variance can be estimated. By extending the laws of expected value and variance for the sum of two variables to more than two variables, we produce the following, where $X_1, X_2, \ldots X_k$ are the times for the completion of activities 1, 2, ..., k, respectively. These times are independent random variables.

Laws of Expected Value and Variance for the Sum of More than Two Independent Variables

1. $E(X_1 + X_2 + \cdots + X_k) = E(X_1) + E(X_2) + \cdots + E(X_k)$
2. $V(X_1 + X_2 + \cdots + X_k) = V(X_1) + V(X_2) + V(X_k)$

Using these laws, we can then produce the expected value and variance for the complete project. Exercises 7.45–7.48 address this problem.

7.45 There are four activities along the critical path for a project. The expected values and variances of the completion times of the activities are listed here. Determine the expected value and variance of the completion time of the project.

Activity	Expected Completion Time (days)	Variance
1	18	8
2	12	5
3	27	6
4	8	2

7.46 The operations manager of a large plant wishes to overhaul a machine. After conducting a PERT/CPM analysis he has developed the following critical path.

1. Disassemble machine.
2. Determine parts that need replacing.
3. Find needed parts in inventory.
4. Reassemble machine.
5. Test machine.

He has estimated the mean (in minutes) and variances of the completion times as follows.

Activity	Mean	Variance
1	35	8
2	20	5
3	20	4
4	50	12
5	20	2

Determine the mean and variance of the completion time of the project.

7.47 In preparing to launch a new product, a marketing manager has determined the critical path for her department. The activities and the mean and variance of the completion time for each activity along the critical path are shown in the accompanying table. Determine the mean and variance of the completion time of the project.

Activity	Expected Completion Time (days)	Variance
Develop survey questionnaire	8	2
Pretest the questionnaire	14	5
Revise the questionnaire	5	1
Hire survey company	3	1
Conduct survey	30	8
Analyze data	30	10
Prepare report	10	3

7.48 A professor of business statistics is about to begin work on a new research project. Because his time is quite limited, he has developed a PERT/CPM critical path, which consists of the following activities:

1. Conduct a search for relevant research articles.
2. Write a proposal for a research grant.
3. Perform the analysis.
4. Write the article and send to journal.
5. Wait for reviews.
6. Revise on the basis of the reviews and resubmit.

The mean (in days) and variance of the completion times are as follows

Activity	Mean	Variance
1	10	9
2	3	0
3	30	100
4	5	1
5	100	400
6	20	64

Compute the mean and variance of the completion time of the entire project.

7-3 APPLICATIONS IN FINANCE: PORTFOLIO DIVERSIFICATION AND ASSET ALLOCATION

In this section, we introduce an important application in finance that is based on the previous section.

In Chapter 3, we described how the variance or standard deviation can be used to measure the risk associated with an investment. Most investors tend to be risk averse, which means that they prefer to have lower risk associated with their investments. One of the ways in which financial analysts lower the risk that is associated with the stock market is through **diversification**. This strategy was first mathematically developed by Harry Markowitz in 1952. His model paved the way for the development of modern portfolio theory (MPT), which is the concept underlying mutual funds (see Section 6.2).

To illustrate the basics of portfolio diversification, consider an investor who forms a portfolio, consisting of only two stocks, by investing £4000 in one stock and £6000 in a second stock. Suppose that the results after 1 year are as listed here. (We have previously defined return on investment. See Applications in Finance: Return on Investment on Section 3-1.)

One-Year Results

Stock	Initial Investment (£)	Value of Investment After 1 year (£)	Rate of Return on Investment
1	4 000	5 000	$R_1 = 0.25$ (25%)
2	6 000	5 400	$R_2 = -0.10$ (−10%)
Total	10 000	10 400	$R_p = 0.04$ (4%)

Another way of calculating the portfolio return R is to compute the weighted average of the individual stock returns R_1 and R_2, where the weights w_1 and w_2 are the proportions of the initial £10 000 invested in stocks 1 and 2, respectively. In this illustration, $w_1 = 0.4$ and $w_2 = 0.6$. (Note that w_1 and w_2 must always sum to 1 because the two stocks constitute the entire portfolio.) The weighted average of the two returns is

$$R_p = w_1R_1 + w_2R_2$$
$$= (0.4)\,(0.25) + (0.6)\,(-10) = 0.04$$

This is how portfolio returns (R_p) are calculated. However, when the initial investments are made, the investor does not know what the returns will be. In fact, the returns are random variables. We are interested in determining the expected value and variance of the portfolio. The formulas in the box were derived from the laws of expected value and variance introduced in the two previous sections.

Mean and Variance of a Portfolio of Two Stocks

$$E(R_p) = w_1E(R_1) + w_2E(R_2)$$
$$V(R_p) = w_1^2V(R_1) + w_2^2V(R_2) + 2w_1\,w_2\text{COV}(R_1,\,R_2)$$
$$= w_1^2\sigma_1^2 + w_2^2\sigma_2^2 + 2w_1w_2\rho\sigma_1\sigma_2$$

where w_1 and w_2 are the proportions or weights of investments 1 and 2, $E(R_1)$ and $E(R_2)$ are their expected values, σ_1 and σ_2 are their standard deviations, $\text{COV}(R_1,R_2)$ covariance and ρ is the coefficient of correlation, which ranges between −1 (negative correlation) and +1 (positive correlation).

(Recall that $\rho = \dfrac{\text{COV}(R_1,R_2)}{\sigma_1\sigma_2}$, which means that $\text{COV}(R_1,\,R_2) = \rho\sigma_1\sigma_2$.)

EXAMPLE 7.8

Describing the Population of the Returns on a Portfolio

An investor has decided to form a portfolio by putting 25% of his money into McDonald's stock and 75% into Cisco Systems stock. The investor assumes that the expected returns will be 8% and 15%, respectively, and that the standard deviations will be 12% and 22%, respectively.

a. Find the expected return on the portfolio.

b. Compute the standard deviation of the returns on the portfolio assuming that

i. the two stocks' returns are perfectly positively correlated

ii. the coefficient of correlation is 0.5

iii. the two stocks' returns are uncorrelated.

SOLUTION:

a. The expected values of the two stocks are

$$E(R_1) = 0.08 \text{ and } E(R_2) = 0.15$$

The weights are $w_1 = 0.25$ and $w_2 = 0.75$. Thus, the expected value of the portfolio is

$$E(R_p) = w_1R_1 + w_2E(R_2) = 0.25(0.08) + 0.75(0.15) = 0.1325$$

b. The standard deviations are

$$\sigma_1 = 0.12 \quad \text{and} \quad \sigma_2 = 0.22$$

Thus,

$$\begin{aligned} V(R_p) &= w_1^2\sigma_1^2 + w_2^2\sigma_2^2 + 2w_1w_2\rho\sigma_1\sigma_2 \\ &= (0.25^2)(0.12^2) + (0.75^2)(0.22^2) + 2(0.25)(0.75)\rho(0.12)(0.22) \\ &= 0.0281 + 0.0099\rho \end{aligned}$$

When $\rho = 1$

$$V(R_p) = 0.0281 + 0.0099(1) = 0.0380$$
$$\text{Standard deviation} = \sqrt{V(R_p)} = \sqrt{0.0380} = 0.1949$$

When $\rho = 0.5$

$$V(R_p) = 0.0281 + 0.0099(0.5) = 0.0331$$
$$\text{Standard deviation} = \sqrt{V(R_p)} = \sqrt{0.0331} = 0.1819$$

When $\rho = 0$

$$V(R_p) = 0.0281 + 0.0099(0) = 0.0281$$
$$\text{Standard deviation} = \sqrt{V(R_p)} = \sqrt{0.0281} = 0.1676$$

Notice that the variance and standard deviation of the portfolio returns decrease as the coefficient of correlation decreases.

7-3a Portfolio Diversification in Practice

The formulas introduced in this section require that we know the expected values, variances and covariance (or coefficient of correlation) of the investments we are interested in. The question arises, how do we determine these parameters? (Incidentally, this question is rarely addressed in finance textbooks!) The most common procedure is to estimate the parameters from historical data, using sample statistics.

7-3b Portfolios with More than Two Stocks

We can extend the formulas that describe the mean and variance of the returns of a portfolio of two stocks to a portfolio of any number of stocks.

Mean and Variance of a Portfolio of *k* Stocks

$$E(R_p) = \sum_{i=1}^{k} w_i E(R_i)$$

$$V(R_p) = \sum_{i=1}^{k} w_i^2 \sigma_i^2 + 2\sum_{i=1}^{k} \sum_{j=i+1}^{k} w_i w_j COV(R_i, R_j)$$

Where R_i is the return of the ith stock, w_i is the proportion of the portfolio invested in stock i, and k is the number of stocks in the portfolio.

When k is greater than 2, the calculations can be tedious and time-consuming. For example, when $k = 3$, we need to know the values of the three weights, three expected values, three variances and three covariances. When $k = 4$, there are four expected values, four variances and six covariances. [The number of covariances required in general is $k(k - 1)/2$.] To assist you, we have created an Excel worksheet to perform the computations when $k = 2$, 3 or 4. To demonstrate, we will return to the problem described in this chapter's introduction.

INVESTING TO MAXIMISE RETURNS AND MINIMISE RISK: SOLUTION

Because of the large number of calculations, we will solve this problem using only Excel. From the file, we compute the means of each stock's returns.

Excel Means

	A	B	C	D
1	0.01778	0.01881	0.02944	0.01458

Next we compute the variance–covariance matrix. (The commands are the same as those described in Chapter 4 – simply include all the columns of the returns of the investments you wish to include in the portfolio.)

Excel Variance–Covariance Matrix

	A	B	C	D	E
1		ALV	BNP	KNP	MRK
2	ALV	0.00637			
3	BNP	0.00642	0.01075		
4	KNP	−0.00141	−0.00173	0.02081	
5	MRK	0.00239	0.00264	0.00016	0.00397

Notice that the variances of the returns are listed on the diagonal. Thus, for example, the variance of the 60 monthly returns of Allianz is 0.00637. The covariances appear below the diagonal. The covariance between the returns of BNP Paribas and KPN is −0.00173.

The means and the variance-covariance matrix are copied to the spreadsheet using the commands described here. The weights are typed producing the accompanying output.

(continued)

Excel Worksheet: Portfolio Diversification-Plan 1

	A	B	C	D	E	F
1	Portfolio of 4 Stocks					
2			ALV	BNP	KNP	MRK
3	Variance-Covariance Matrix	ALV	0.00637			
4		BNP	0.00642	0.01075		
5		KNP	−0.00141	−0.00173	0.02081	
6		MRK	0.00239	0.00264	0.00016	0.00397
7						
8	Expected Returns		0.01778	0.01881	0.02944	0.01458
9						
10	Weights		0.25000	0.25000	0.25000	0.25000
11						
12	Portfolio Return					
13	Expected Value	0.0202				
14	Variance	0.0037				
15	Standard Deviation	0.0606				

The expected return on the portfolio is 0.0202 and the variance is 0.0037.

Instructions

1. Open the file containing the returns. In this example, open file **Xm07-00.**
2. Compute the means of the columns containing the returns of the stocks in the portfolio.
3. Using the commands described in Chapter 4 compute the variance–covariance matrix.
4. Open the **Portfolio Diversification** workbook. Use the tab to select the 4 **Stocks** worksheet. DO NOT CHANGE ANY CELLS THAT APPEAR IN BOLD PRINT. DO NOT SAVE ANY WORKSHEETS.
5. Copy the means into cells C8 to F8. (Use **Copy, Paste Special** with **Values and Number Formats**.)
6. Copy the variance-covariance matrix (including row and column labels) into columns B, C, D, E and F.
7. Type the weights into cells C10 to F10.

The mean, variance and standard deviation of the portfolio will be printed. Use similar commands for 2-stock and 3-stock portfolios.

The results for Plan 2 are:

	A	B
12	Portfolio Return	
13	Expected Value	0.0202
14	Variance	0.0036
15	Standard Deviation	0.0601

Plan 3:

	A	B
12	Portfolio Return	
13	Expected Value	0.0200
14	Variance	0.0046
15	Standard Deviation	0.0679

Plan 1 has high expected value and the second largest variance. As a result, Plan 1 is not the best choice. Plan 2 has the same expected value as Plan 1 and the smallest variance. Plan 3's expected value is the smallest and the variance is the largest. If the investor is like most investors, she would select Plan 2 because of its lower risk.

In this example, we showed how to compute the expected return, variance and standard deviation from a sample of returns on the investments for any combination of weights. (We illustrated the process with three sets of weights.) It is possible to determine the 'optimal' weights that minimise risk for a given expected value or maximise expected return for a given standard deviation. This is an extremely important function of financial analysts and investment advisors. Solutions can be determined using a management science technique called *linear programming,* a subject taught by most schools of business and faculties of management.

EXERCISES

7.49 Describe what happens to the expected value and standard deviation of the portfolio returns when the coefficient of correlation decreases.

7.50 A portfolio is composed of two stocks. The proportion of each stock, their expected values and standard deviations are listed next.

Stock	1	2
Proportion of portfolio	0.30	0.70
Mean	0.12	0.25
Standard deviation	0.02	0.15

For each of the following coefficients of correlation, calculate the expected value and standard deviation of the portfolio:

a. $\rho = 0.5$

b. $\rho = 0.2$

c. $\rho = 0.2$

7.51 An investor is given the following information about the returns on two stocks:

Stock	1	2
Mean	0.09	0.13
Standard deviation	0.15	0.21

a. If he is most interested in maximising his returns, which stock should he choose?

b. If he is most interested in minimising his risk, which stock should he choose?

7.52 Refer to Exercise 7.51. Compute the expected value and standard deviation of the portfolio composed of 60% stock 1 and 40% stock 2. The coefficient of correlation is 0.4.

7.53 Refer to Exercise 7.51. Compute the expected value and standard deviation of the portfolio composed of 30% stock 1 and 70% stock 2.

The following exercises require a computer and software.

Xr07-NYSE *We have recorded the monthly returns for the following Dow Jones Industrials 30 stocks listed on the New York Stock Exchange for the period January 2008 to December 2012:*

3M (MMM), Alcoa (AA), American Express (AXP), AT&T (T), Bank of America (BAC), Boeing (BA), Caterpillar (CAT), Chevron (CVX), Cisco Systems (CSCO), Coca-Cola (KO), Disney (DIS), Du Pont (DD), Exxon (XOM), General Electric (GE), Hewlett-Packard (HPQ), Home Depot (HD), Intel (INTC), International Business Machines (IBM), Johnson & Johnson (JNJ), JP Morgan Chase (JPM), McDonald's (MCD), Merck (MRK), Microsoft (MSFT), Pfizer (PFE), Proctor & Gamble (PG), Travelers (TRV), United Technologies (UTX), United Health (UNH), Verizon Communications (VZ), Wal-Mart Stores (WMT).

For Exercises 7.54 to 7.60, calculate the mean and standard deviation of the portfolio. The proportions invested in each stock are shown in parentheses.

7.54 **Xr07-54** **a.** American Express (AXP) (20%), Bank of America (BAC) (30%), JP Morgan Chase (JPM) (50%)

b. AXP (20%), BAC (60%), JPM (20%)

c. AXP (50%), BAC (30%), JPM (20%)

d. Which portfolio would an investor who likes to gamble choose? Explain.

e. Which portfolio would a risk-averse investor choose? Explain.

7.55 **a.** Alcoa (AA) (25%), Home Depot (HD) (25%), International Business Machines (IBM) (25%), Travelers (TRV) (25%)

b. AA (10%), HD (50%), IBM (20%), TRV (20%)

c. AA (30%), HD (20%),IBM (10%), TRV (40%)

d. Which portfolio would a gambler choose? Explain.

e. Which portfolio would a risk-averse investor choose? Explain.

7.56 **a.** Cisco Systems (CSCO) (25%), Disney (DIS) (25%), General Electric (GE) (25%), Wal-Mart Stores (WMT) (25%)

b. CSCO (10%), DIS (20%), GE (30%), WMT (40%)

c. CSCO (55%), DIS (15%), GE (15%), WMT (15%)

d. Explain why the choice of which portfolio to invest in is obvious.

7.57 **a.** Boeing (BA) (25%), Caterpillar (CAT) (25%), Chevron (CVX) (25%), Coca-Cola (KO) (25%)

b. BA (10%), CAT (20%), CVX (20%), KO (50%)

c. BA(15%), CAT(55%), CVX(15%), KO (15%)

d. Which portfolio would a gambler choose? Explain.

e. Which portfolio would a risk-averse investor choose? Explain.

7.58 **a.** Du Pont (DD) (25%), Exxon (XOM) (25%), Verizon Communications (VZ), (25%), Intel (INTC) (25%)

b. DD (40%), XOM (30%), VZ (20%), INTC (10%)

c. DD (10%), XOM (20%), VZ (30%), INTC (40%)

d. Which portfolio would a gambler choose? Explain.

e. Which portfolio would a risk-averse investor choose? Explain.

7.59 **a.** Johnson & Johnson (JNJ) (25%), McDonald's (MCD) (25%), Merck (MRK) (25%), Microsoft (MSFT) (25%)

b. JNJ (10%), MCD (30%), MRK (10%), MSFT (50%)

c. JNJ (10%), MCD (70%), MRK (10%), MSFT (0%)

d. Explain why the choice of which portfolio to invest in is obvious.

7.60 **a.** Pfizer (PFE) (25%), Proctor & Gamble (PG) (25%), United Technologies (UTX) (25%), United Health (UNH) (25%)

b. PFE (40%), PG (30%), UTX (20%), UNH (10%)

c. PFE (15%), PG (55%), UTX (15%), UNH (15%)

d. Which portfolio would a gambler choose? Explain.

e. Which portfolio would a risk-averse investor choose? Explain.

Xr07-FTSE *Monthly returns for the following selected stocks on the FTSE 100 Index were recorded for the years 2009 to 2013:*

Aberdeen Asset Management (ADN), BAE Systems (BA), Barclays (BARC), BP (BP), British American Tobacco (BTI), Centamin (CEY), Capita (CPI), City of London Investment Group (CLIG), CREAT RESOURCES (CRHL), EasyJet (EZJ), Experian (EXPN), Fresnillo (FRES), Marshalls (MSLH), Mondi (MNDI), Playtech (PTEC), Pearson (PSON), Reed Elsevier (REL), Rio Tinto (RIO), The Royal Bank of Scotland Group (RBS), Shire (SHP), SSE (SSE), Trinity Mirror (TNI), Tesco (TSCO)

For Exercises 7.61 to 7.64, calculate the mean and standard deviation of the portfolio. The proportions invested in each stock are shown in parentheses.

7.61 **Xr07-61** **a.** Aberdeen Asset Management (ADN) (25%), BAE Systems (BA) (25%), Barclays (BARC) (25%), BP (BP) (25%)

b. ADN (50%), BA (10%), BARC (10%), BP (30%)

c. ADN (10%), BA (40%), BARC (40%), BP (10%)

d. Which portfolio would a gambler choose? Explain.

e. Which portfolio would a risk-averse investor choose? Explain.

7.62 **a.** Rio Tinto (RIO) (25%), The Royal Bank of Scotland Group (RBS) (25%), Shire (SHP) (25%), SSE (SSE) (25%)

b. RIO (10%), RBS (50%), SHP (30%), SSE (10%)

c. RIO (20%), RBS (40%), SHP (30%), SSE (10%)

d. Which portfolio would a gambler choose? Explain.

e. Which portfolio would a risk-averse investor choose? Explain.

7.63 **a.** Centamin (CEY) (25%), Experian (EXPN) (25%), Playtech (PTEC) (25%), Pearson (PSON) (25%)

b. CEY (15 %), EXPN (5 5 %), PTEC (15%), PSON (15%)

c. CEY (10%), EXPN (20%), PTECT (30%), PSON (40%)

d. Explain why the choice of which portfolio to invest in is obvious.

7.64 **a.** British American Tobacco (BTI) (25%), City of London Investment Group (CLIG) (25%), CREAT RESOURCES (CRHL), (25%), Mondi (MNDI) (25%)

b. BTI (20%), CLING (60%), CRHL (10%), MDNI (10%)

c. BTI (40%), CLING (30%), CRHL (20%), MDNI (10%)

d. Explain why the choice of which portfolio to invest in is obvious.

7.65 Refer to Exercise 7.64. Suppose you want the expected value to be at least 1%. Try several sets of proportions (remember they must add to 1.0) to see if you can find the portfolio with the smallest variance.

7.66 Refer to Exercise 7.64.

a. Compute the expected value and variance of the portfolio described next.

TI (4.9%), CLING (65.1%), CRHL (27.5%), MDNI (2.5%)

b. Can you do better? That is, can you find a portfolio whose expected value is greater than or equal to 1% and whose variance is less than the one you calculated in part (a)? (*Hint:* Do not spend too much time at this. You won't be able to do better. If you want to learn how we produced the portfolio above, take a course that teaches linear and non-linear programming.)

Xr07-NASDAQ *We calculated the returns on the following selected stocks on the NASDAQ Exchange for the period January 2008 to December 2012:*

Adobe Systems (ADBE), Amazon (AMZN), Amgen (AMGN), Apple (AAPL), Bed Bath & Beyond (BBBY), Cisco Systems (CSCO), Comcast (CMCSA), Costco Wholesale (COST), Dell (DELL), Dollar Tree (DLTR), Expedia (EXPE), Garmin (GRMN),

Google (*GOOG*), Intel (*INTC*), *Mattel* (*MAT*), *Microsoft* (*MSFT*), *Netflix* (*NFLX*), *Oracle* (*ORCL*), *Research in Motion* (*RIMM*), *ScanDisk* (*SNDK*), *Sirius XM Radio* (*SIRI*), *Staples* (*SPLS*), *Starbucks* (*SBUX*), *Whole Foods Market* (*WFM*), *Wynn Resorts* (*WYNN*), *Yahoo* (*YHOO*)

For Exercises 7.67 to 7.69 calculate the mean and standard deviation of the portfolio. The proportions invested in each stock are shown in parentheses.

7.67 **a.** Bed Bath & Beyond (BBBY) (25%), Expedia (EXPE) (25%), Mattel (MAT) (25%), Starbucks (SBUX) (25%)

b. BBBY (40%), EXPE (30%), MAT (20%), SBUX (10%)

c. BBBY (10%), EXPE (20%), MAT (30%), SBUX (40%)

d. Explain why the choice of which portfolio to invest in is obvious.

7.68 **a.** Netflix (NFLX) (25%), ScanDisk (SNDK) (25%), Sirius XM Radio (SIRI) (25%), Wynn Resorts (WYNN) (25%)

b. NFLX (15%), SNDK (15%), SIRI (15%), WYNN (55%)

c. NFLX (20%), SNDK (20%), SIRI (50%), WYNN (10%)

d. Which portfolio would a gambler choose? Explain.

e. Which portfolio would a risk-averse investor choose? Explain.

7.69 **a.** Amazon (AMZN) (25%), Apple (AAPL) (25%), Dell (DELL) (25%), Dollar Tree (DLTR) (25%)

b. AMZN (10%), AAPL (20%), DELL (30%), DLTR (40%)

c. AMZN (40%), AAPL (40%), DELL (10%), DLTR (10%)

d. Which portfolio would a gambler choose? Explain.

e. Which portfolio would a risk-averse investor choose? Explain.

7.70 Refer to Exercise 7.69. Suppose you want the expected value to be at least 2%. Try several sets of proportions (remember they must add to 1.0) to see if you can find the portfolio with the smallest variance.

7.71 Refer to Exercise 7.69.

a. Compute the expected value and variance of the portfolio described next.

AMZN (20.4%), AAPL (8.3%), DELL (9.4%), DLTR (62%)

b. Can you do better? That is, can you find a portfolio whose expected value is greater than or equal to 2% and whose variance is less than the one you calculated in part (a)? (*Hint:* Don't spend too much time at this. You won't be able to do better. If you want to learn how we produced the portfolio above, take a course that teaches linear and non-linear programming.)

7-4 BINOMIAL DISTRIBUTION

Now that we have introduced probability distributions in general, we need to introduce several specific probability distributions. In this section, we present the *binomial distribution*.

The binomial distribution is the result of a *binomial experiment,* which has the following properties.

Binomial Experiment

1. The **binomial experiment** consists of a fixed number of trials. We represent the number of trials by n.
2. Each trial has two possible outcomes. We label one outcome a *success* and the other *a failure.*
3. The probability of success is p. The probability of failure is $1 - p$.
4. The trials are independent, which means that the outcome of one trial does not affect the outcomes of any other trials.

If properties 2, 3 and 4 are satisfied, we say that each trial is a **Bernoulli process**. Adding property 1 yields the binomial experiment. The random variable of a binomial experiment is defined as the number of successes in the n trials. It is called the **binomial random variable**. Here are several examples of binomial experiments.

1. Flip a coin 10 times. The two outcomes per trial are heads and tails. The terms *success* and *failure* are arbitrary. We can label either outcome success. However, generally, we call success anything we are looking for. For example, if we were betting on heads, we would label heads a success. If the coin is fair, the probability of heads is 50%. Thus, $p = 0.5$. Finally, we can see that the trials are independent because the outcome of one coin flip cannot possibly affect the outcomes of other flips.
2. Draw five cards out of a shuffled deck. We can label as success whatever card we seek. For example, if we wish to know the probability of receiving five clubs, a club is labelled a success. On the first draw, the probability of a club is $13/52 = 0.25$. However, if we draw a second card without replacing the first card and shuffling, the trials are not independent. To see why, suppose that the first draw is a club. If we draw again without replacement the probability of drawing a second club is 12/51, which is not 0.25. In this experiment, the trials are *not* independent.* Hence, this is not a binomial experiment. However, if we replace the card and shuffle before drawing again, the experiment is binomial. Note that in most card games we do not replace the card, and as a result the experiment is not binomial.
3. A political survey asks 1500 voters who they intend to vote for in an approaching election. In most elections in the USA, there are only two candidates, the Republican and Democratic nominees. Thus, we have two outcomes per trial. The trials are independent because the choice of one voter does not affect the choice of other voters. In the UK, and in other countries with parliamentary systems of government, there are usually several candidates in the race. However, we can label a vote for our favoured candidate (or the party that is paying us to do the survey) a success and all the others are failures.

As you will discover, the third example is a very common application of statistical inference. The actual value of p is unknown, and the job of the statistics practitioner is to estimate its value. By understanding the probability distribution that uses p, we will be able to develop the statistical tools to estimate p.

7-4a Binomial Random Variable

The binomial random variable is the number of successes in the experiment's n trials. It can take on values 0, 1, 2, …, n. Thus, the random variable is discrete. To proceed, we must be capable of calculating the probability associated with each value.

Using a probability tree, we draw a series of branches as depicted in Figure 7.2. The stages represent the outcomes for each of the n trials. At each stage, there are two branches representing success and failure. To calculate the probability that there are x successes in n trials, we note that for

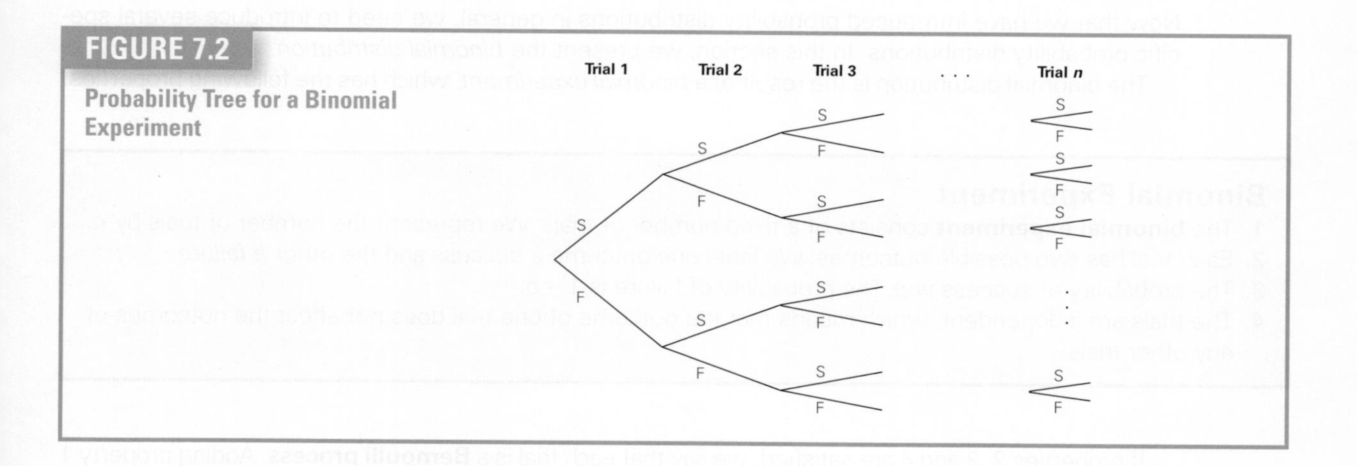

FIGURE 7.2 Probability Tree for a Binomial Experiment

*The hypergeometric distribution described in the website Appendix of the same name is used to calculate probabilities in such cases.

each success in the sequence, we must multiply by p. And if there are x successes, there must be $n - x$ failures. For each failure in the sequence, we multiply by $1 - p$. Thus, the probability for each sequence of branches that represent x successes and $n - x$ failures has probability

$$p^x(1 - p)^{n-x}$$

There are a number of branches that yield x successes and $n - x$ failures. For example, there are two ways to produce exactly one success and one failure in two trials: SF and FS. To count the number of branch sequences that produce x successes and $n - x$ failures, we use the combinatorial formula

$$C_x^n = \frac{n!}{x!(n - x)!}$$

where $n! = n(n - 1)(n - 2) \ldots (2)(1)$ is called *n factorial*. For example, $3! = 3(2)(1) = 6$. Incidentally, although it may not appear to be logical $0! = 1$.

Pulling together the two components of the probability distribution yields the following.

Binomial Probability Distribution

The probability of x successes in a binomial experiment with n trials and probability of success is

$$P(x) = \frac{n!}{x!(n - x)!} p^x(1 - p)^{n-x} \text{ for } x = 0, 1, 2, \ldots, n$$

EXAMPLE 7.9

Pat Statsdud and the Statistics Quiz

Pat Statsdud is a student taking a statistics course. Unfortunately, Pat is not a good student. Pat does not read the textbook before class, does not do homework, and regularly misses class. Pat intends to rely on luck to pass the next quiz. The quiz consists of ten multiple-choice questions. Each question has five possible answers, only one of which is correct. Pat plans to guess the answer to each question.

a. What is the probability that Pat gets no answers correct?

b. What is the probability that Pat gets two answers correct?

SOLUTION:

The experiment consists of ten identical trials, each with two possible outcomes and where success is defined as a correct answer. Because Pat intends to guess, the probability of success is 1/5 or 0.2. Finally, the trials are independent because the outcome of any of the questions does not affect the outcomes of any other questions. These four properties tell us that the experiment is binomial with $n = 10$ and $p = 0.2$.

a. From

$$P(x) = \frac{n!}{x!(n = x)!} p^x(1 - p)^{n-x}$$

we produce the probability of no successes by letting $n = 10$, $p = 0.2$ and $x = 0$. Hence,

$$P(0) = \frac{10!}{0!(10 - 0)!}(0.2)^0(1 - 0.2)^{10-0}$$

The combinatorial part of the formula is $\frac{10!}{0!10!}$, which is 1. This is the number of ways to get 0 correct and ten incorrect. Obviously, there is only one way to produce $X = 0$.

And because $(0.2)^0 = 1$,

$$P(X = 0) = 1(1)(0.8)^{10} = 0.1074$$

b. The probability of two correct answers is computed similarly by substituting $n = 10$, $p = 0.2$ and $x = 2$:

$$P(x) = \frac{n!}{x!(n-x)!} p^x(1-p)^{n-x}$$

$$P(0) = \frac{10!}{2!(10-2)!}(0.2)^2(1-0.2)^{10-2}$$

$$= \frac{(10)(9)(8)(7)(6)(5)(4)(3)(2)(1)}{(2)(1)(8)(7)(6)(5)(4)(3)(2)(1)}(0.04)(0.1678)$$

$$= 45(0.006712)$$

$$= 0.3020$$

In this calculation, we discovered that there are 45 ways to get exactly two correct and eight incorrect answers, and that each such outcome has probability 0.006712. Multiplying the two numbers produces a probability of 0.3020.

7-4b Cumulative Probability

The formula of the binomial distribution allows us to determine the probability that X equals individual values. In Example 7.9, the values of interest were 0 and 2. There are many circumstances where we wish to find the probability that a random variable is less than or equal to a value; that is, we want to determine $P(X \leq x)$, where x is that value. Such a probability is called a **cumulative probability**.

 **EXAMPLE 7.10**

Will Pat Fail the Quiz?

Find the probability that Pat fails the quiz. A mark is considered a failure if it is less than 50%.

SOLUTION:

In this quiz, a mark of less than 5 is a failure. Because the marks must be integers, a mark of 4 or less is a failure. We wish to determine $P(X \leq 4)$. So,

$$P(X \leq 4) = P(0) + P(1) + P(2) + P(3) + P(4)$$

From Example 7.9, we know $P(0) = 0.1074$ and $P(2) = 0.3020$. Using the binomial formula, we find $P(1) = 0.2684$, $P(3) = 0.2013$ and $P(4) = 0.0881$. Thus

$$P(X \leq 4) = 0.1074 + 0.2684 + 0.3020 + 0.2013 + 0.0881 = 0.9672$$

There is a 96.72% probability that Pat will fail the quiz by guessing the answer for each question.

7-4c Binomial Table

There is another way to determine binomial probabilities. Table 1 in Appendix A (See the 'Digital Resources' page.) provides cumulative binomial probabilities for selected values of n and p. We can use this table to answer the question in Example 7.10, where we need $P(X \leq 4)$. Refer to Table 1, find $n = 10$ and in that table find $p = 0.20$. The values in that column are $P(X \leq x)$ for $x = 0, 1, 2, \ldots, 10$, which are shown in Table 7.2.

TABLE 7.2 **Cumulative Binomial Probabilities with $n = 10$ and $p = 0.2$**

X	$P(X \le x)$
0	0.1074
1	0.3758
2	0.6778
3	0.8791
4	0.9672
5	0.9936
6	0.9991
7	0.9999
8	1.000
9	1.000
10	1.000

The first cumulative probability is $P(X \le 0)$, which is $P(0) = 0.1074$. The probability we need for Example 7.10 is $P(X \le 4) = 0.9672$, which is the same value we obtained manually.

We can use the table and the complement rule to determine probabilities of the type $P(X \ge x)$. For example, to find the probability that Pat will pass the quiz, we note that

$$P(X \le 4) + P(X \ge 5) = 1$$

Thus,

$$P(X \ge 5) = 1 - P(X \le 4) = 1 - 0.9672 = 0.0328$$

Using Table 1 to find the Binomial Probability $P(X \ge x)$

$$P(X \le x) = -P(X \le [x - 1])$$

The table is also useful in determining the probability of an individual value of X. For example, to find the probability that Pat will get exactly two right answers we note that

$$P(X \le 2) = P(0) + P(1) + P(2)$$

and

$$P(X \le 1) = P(0) + P(l)$$

The difference between these two cumulative probabilities is $p(2)$. Thus,

$$P(2) = P(X \le 2) - P(X \le 1) = 0.6778 - 0.3758 = 0.3020$$

Using Table 1 to Find the Binomial Probability $P(X = x)$

$$P(x) = P(X \le x) - P(X \le [x - 1])$$

Using the Computer

EXCEL *Instructions*

Type the following into any empty cell:

= **BINOM.DIST**(number_s, trials, probability_s, cumulative)

or with Excel 2007 and earlier:

= **BINOMDIST**([x], [n], [p], [True] or [False])

Typing 'True' calculates a cumulative probability and typing 'False' computes the probability of an individual value of X. For Example 7.9(a), type

= **BINOM.DIST**(0,10, 0.2, False)

For Example 7.10, enter

= **BINOM.DIST**(4,10, 0.2, True)

MINITAB *Instructions*

This is the first of seven probability distributions for which we provide instructions. All work in the same way. Click **Calc, Probability Distributions** and the specific distribution whose probability you wish to compute. In this case, select **Binomial**. . . . Check either **Probability** or **Cumulative probability**. If you wish to make a probability statement about one value of x, specify **Input constant** and type the value of x.

If you wish to make probability statements about several values of x from the same binomial distribution, type the values of x into a column before checking Calc. Choose **Input column** and type the name of the column. Finally, enter the components of the distribution. For the binomial, enter the **Number of trials** n and the **Event Probability** p.

For the other six distributions, we list the distribution (here it is **Binomial**) and the components only (for this distribution it is n and p).

7-4d Mean and Variance of a Binomial Distribution

Statisticians have developed general formulas for the mean, variance and standard deviation of a binomial random variable. They are

$$\mu = np$$
$$\sigma^2 = np(1 - p)$$
$$\sigma = \sqrt{np(1 - p)}$$

 **EXAMPLE 7.11**

Pat Statsdud Has Been Cloned!

Suppose that a professor has a class full of students like Pat (a nightmare!). What is the mean mark? What is the standard deviation?

SOLUTION:

The mean mark for a class of Pat Statsdud is

$$\mu = np = 10(0.2) = 2$$

The standard deviation is

$$\sigma = \sqrt{np(1-p)} = \sqrt{10(0.2)(1-0.2)} = 1.26$$

EXERCISES

7.72 Given a binomial random variable with $n = 10$ and $p = 0.3$, use the formula to find the following probabilities.

a. $P(X = 3)$

b. $P(X = 5)$

c. $P(X = 8)$

7.73 Repeat Exercise 7.72 using Excel or Minitab.

7.74 A sign on the petrol pumps of a chain of petrol stations encourages customers to have their oil checked with the claim that one out of four cars needs to have oil added. If this is true, what is the probability of the following events?

a. One out of the next four cars needs oil

b. Two out of the next eight cars need oil

c. Three out of the next 12 cars need oil

7.75 The leading brand of dishwasher detergent has a 30% market share. A sample of 25 dishwasher detergent customers was taken. What is the probability that 10 or fewer customers chose the leading brand?

7.76 A certain type of tomato seed germinates 90% of the time. An amateur gardener planted 25 seeds.

a. What is the probability that exactly 20 germinate?

b. What is the probability that 20 or more germinate?

c. What is the probability that 24 or fewer germinate?

d. What is the expected number of seeds that germinate?

7.77 A student taking a degree in accounting is trying to decide on the number of firms to which he should apply. Given his work experience and grades, he can expect to receive a job offer from 70% of the firms to which he applies. The student decides to apply to only four firms. What is the probability that he receives no job offers?

7.78 Major software manufacturers offer a helpline that allows customers to call and receive assistance in solving their problems. However, because of the volume of calls, customers frequently are put on hold. One software manufacturer claims that only 20% of callers are put on hold. Suppose that 100 customers call. What is the probability that more than 25 of them are put on hold?

The following exercises are best solved with a computer.

7.79 In a *Bon Appetit* poll, 38% of people said that chocolate was their favourite flavour of ice cream. A sample of 20 people was asked to name their favourite flavour of ice cream. What is the probability that half or more of them prefer chocolate?

7.80 According to the last census, 94% of working women held full-time jobs between June to August 2013 and September to November 2013. If a random sample of 50 working women is drawn, what is the probability that 19 or more hold full-time jobs?

7-5 POISSON DISTRIBUTION

Another useful discrete probability distribution is the **Poisson distribution**, named after the French mathematician Siméon Denis Poisson. Like the binomial random variable, the **Poisson random variable** is the number of occurrences of events, which we will continue to call *successes*. The difference between the two random variables is that a binomial random variable is the number of successes in a set number of trials, whereas a Poisson random variable is the number of successes in an interval of time or specific region of space. Here are several examples of Poisson random variables.

1. The number of cars arriving at a service station in 1 hour. (The interval of time is 1 hour.)
2. The number of flaws in a bolt of cloth. (The specific region is a bolt of cloth.)
3. The number of accidents in 1 day on a particular stretch of highway. (The interval is defined by both time, 1 day, and space, the particular stretch of highway.)

The Poisson experiment is described in the box.

Poisson Experiment

A **Poisson experiment** is characterised by the following properties:

1. The number of successes that occur in any interval is independent of the number of successes that occur in any other interval.
2. The probability of a success in an interval is the same for all equal-size intervals.
3. The probability of a success in an interval is proportional to the size of the interval.
4. The probability of more than one success in an interval approaches 0 as the interval becomes smaller.

Poisson Random Variable

The **Poisson random variable** is the number of successes that occur in a period of time or an interval of space in a Poisson experiment.

There are several ways to derive the probability distribution of a Poisson random variable. However, all are beyond the mathematical level of this book. We simply provide the formula and illustrate how it is used.

Poisson Probability Distribution

The probability that a Poisson random variable assumes a value of x in a specific interval is

$$P(x) = \frac{e^{-\mu}\mu^x}{x!} \text{ for } x = 0, 1, 2, \ldots$$

where μ is the mean number of successes in the interval or region and e is the base of the natural logarithm (approximately 2.71828). Incidentally, the variance of a Poisson random variable is equal to its mean; that is, $\sigma^2 = \mu$.

EXAMPLE 7.12

Probability of the Number of Typographical Errors in Textbooks

A statistics instructor has observed that the number of typographical errors in new editions of textbooks varies considerably from book to book. After some analysis, he concludes that the number of errors is Poisson distributed with a mean of 1.5 per 100 pages. The instructor randomly selects 100 pages of a new book. What is the probability that there are no typographical errors?

SOLUTION:

We want to determine the probability that a Poisson random variable with a mean of 1.5 is equal to 0. Using the formula

$$P(x) = \frac{e^{-\mu}\mu^x}{x}$$

and substituting $x = 0$ and $\mu = 1.5$, we get

$$P(0) = \frac{e^{-1.5}1.5^0}{0} = \frac{(2.71828)^{-1.5}(1)}{1} = 0.2231$$

The probability that in the 100 pages selected there are no errors is 0.2231.

Notice that in Example 7.12, we wanted to find the probability of 0 typographical errors in 100 pages given a mean of 1.5 typos in 100 pages. The next example illustrates how we calculate the probability of events where the intervals or regions do not match.

EXAMPLE 7.13

Probability of the Number of Typographical Errors in 400 Pages

Refer to Example 7.12. Suppose that the instructor has just received a copy of a new statistics book. He notices that there are 400 pages.

a. What is the probability that there are no typos?

b. What is the probability that there are five or fewer typos?

SOLUTION:

The specific region that we are interested in is 400 pages. To calculate Poisson probabilities associated with this region, we must determine the mean number of typos per 400 pages. Because the mean is specified as 1.5 per 100 pages, we multiply this figure by 4 to convert to 400 pages. Thus, μ = six typos per 400 pages.

a. The probability of no typos is

$$P(0) = \frac{e^{-6}6^0}{0!} = \frac{(2.71828)^{-6}(1)}{1} = 0.002479$$

b. We want to determine the probability that a Poisson random variable with a mean of 6 is 5 or less; that is, we want to calculate

$$P(X < 5) = P(0) + P(1) + P(2) + P(3) + P(4) + P(5)$$

To produce this probability, we need to compute the six probabilities in the summation.

$$P(0) = 0.002479$$

$$P(1) = \frac{e^{-\mu}\mu^x}{x!} = \frac{e^{-6}6^1}{1!} = \frac{(2.71828)^{-6}(6)}{1} = 0.01487$$

$$P(2) = \frac{e^{-\mu}\mu^x}{x!} = \frac{e^{-6}6^2}{2!} = \frac{(2.71828)^{-6}(6)}{2} = 0.04462$$

$$P(3) = \frac{e^{-\mu}\mu^x}{x!} = \frac{e^{-6}6^3}{3!} = \frac{(2.71828)^{-6}(216)}{6} = 0.08924$$

$$P(4) = \frac{e^{-\mu}\mu^x}{x!} = \frac{e^{-6}6^4}{4!} = \frac{(2.71828)^{-6}(1296)}{24} = 0.1339$$

$$P(5) = \frac{e^{-\mu}\mu^x}{x!} = \frac{e^{-6}6^5}{5!} = \frac{(2.71828)^{-6}(7776)}{120} = 0.1606$$

Thus,

$$P(X \le 5) = 0.002479 + 0.01487 + 0.04462 + 0.08924 + 0.1339 + 0.1606 = 0.4457$$

The probability of observing five or fewer typos in this book is 0.4457.

7-5a Poisson Table

As was the case with the binomial distribution, a table is available that makes it easier to compute Poisson probabilities of individual values of x as well as cumulative and related probabilities.

Table 2 in Appendix A (See the 'Digital Resources' page.) provides cumulative Poisson probabilities for selected values of μ. This table makes it easy to find cumulative probabilities like those in Example 7.13, part (b), where we found $P(X \le 5)$.

To do so, find $\mu = 6$ in Table 2. The values in that column are $P(X \le x)$ for $x = 0, 1, 2, \ldots, 18$ which are shown in Table 7.3.

TABLE 7.3 **Cumulative Poisson Probabilities for $\mu = 6$**

X	$P(X \le x)$
0	0.0025
1	0.0174
2	0.0620
3	0.1512
4	0.2851
5	0.4457
6	0.6063
7	0.7440
8	0.8472
9	0.9161
10	0.9574
11	0.9799
12	0.9912
13	0.9964
14	0.9986
15	0.9995
16	0.9998
17	0.9999
18	1.0000

Theoretically, a Poisson random variable has no upper limit. The table provides cumulative probabilities until the sum is 1.0000 (using four decimal places).

The first cumulative probability is $P(X \le 0)$, which is $P(0) = 0.0025$. The probability we need for Example 7.13, part (b), is $P(X \le 5) = 0.4457$, which is the same value we obtained manually.

Like Table 1 for binomial probabilities, Table 2 can be used to determine probabilities of the type $P(X \ge x)$. For example, to find the probability that in Example 7.13 there are six or more typos, we note that $P(X \ge 5) + P(X \ge 6) = 1$. Thus,

$$P(X \ge 6) = 1 - P(X \le 5) = 1 - 0.4457 = 0.5543$$

Using Table 2 to Find the Poisson Probability $P(X \ge x)$

$$P(X \ge x) = 1 - P(X \le [x - 1])$$

We can also use the table to determine the probability of one individual value of X. For example, to find the probability that the book contains exactly ten typos, we note that

$$P(X \le 10) = P(0) + P(1) + \cdots + P(9) + P(10)$$

and

$$P(X \le 9) = P(0) + P(1) + \cdots + P(9)$$

The difference between these two cumulative probabilities is $P(10)$. Thus,

$$P(10) = P(X \le 10) - P(X \le 9) = 0.9574 - 0.9161 = 0.0413$$

Using Table 2 to Find the Poisson Probability $P(X = x)$

$$P(x) = P(X \leq x) - P(X \leq [x - 1])$$

Using the Computer

EXCEL *Instructions*

Type the following into any empty cell:

= **POISSON.DIST** (*x*, mean, cumulative)

We calculate the probability in Example 7.12 by typing

= **POISSON.DIST**(0,1.5, False)

For Example 7.13, we type

= **POISSON.DIST**(5, 6, True)

MINITAB *Instructions*

Click **Calc, Probability Distributions** and **Poisson** ... and type the mean.

EXERCISES

7.81 Given that X is a Poisson random variable with $\mu = 0.5$, use the formula to determine the following probabilities.

a. $P(X = 0)$

b. $P(X = 1)$

c. $P(X = 2)$

7.82 The number of accidents that occur at a busy intersection is Poisson distributed with a mean of 3.5 per week. Find the probability of the following events.

a. No accidents in 1 week

b. Five or more accidents in 1 week

c. One accident today

7.83 Snowfalls occur randomly and independently over the course of winter in Piemonte ski resort in Italy. The average is one snowfall every 3 days.

a. What is the probability of five snowfalls in 2 weeks?

b. Find the probability of a snowfall today.

7.84 Hits on a personal website occur quite infrequently. They occur randomly and independently with an average of five per week.

a. Find the probability that the site gets ten or more hits in a week.

b. Determine the probability that the site gets 20 or more hits in 2 weeks.

7.85 Complaints about an Internet brokerage firm occur at a rate of five per day. The number of complaints appears to be Poisson distributed.

a. Find the probability that the firm receives ten or more complaints in a day.

b. Find the probability that the firm receives 25 or more complaints in a 5-day period.

APPLICATIONS IN OPERATIONS MANAGEMENT

Queues

Everyone is familiar with queues (or waiting lines). We queue at banks, supermarkets and fast-food restaurants. There are also queues in companies where lorries wait to load and unload and on assembly lines where stations wait for new parts. Management scientists have developed mathematical models that allow managers to determine the operating characteristics of queues. Some of the operating characteristics are:

- The probability that there are no units in the system.
- The average number of units in the queue.
- The average time a unit spends in the queue.
- The probability that an arriving unit must wait for service.

The Poisson probability distribution is used extensively in queuing (also called *waiting-line*) models. Many models assume that the arrival of units for service is Poisson distributed with a specific value of μ. In the next chapter, we will discuss the operating characteristics of queues. Exercises 7.86–7.88 require the calculation of the probability of a number of arrivals.

7.86 The number of lorries crossing at the Oresund Bridge connecting Sweden and Demark is Poisson distributed with a mean of 1.5 per minute.

a. What is the probability that in any 1-minute time span two or more lorries will cross the bridge?

b. What is the probability that fewer than four lorries will cross the bridge over the next 4 minutes?

7.87 Cars arriving for petrol at a particular petrol station follow a Poisson distribution with a mean of 5 per hour.

a. Determine the probability that over the next hour only one car will arrive.

b. Compute the probability that in the next 3 hours more than 20 cars will arrive.

7.88 The number of users of an ATM, or 'cash machine', is Poisson distributed. The mean number of users per 5-minute interval is 1.5. Find the probability of the following events.

a. No users in the next 5 minutes

b. Five or fewer users in the next 15 minutes

c. Three or more users in the next 10 minutes

CHAPTER SUMMARY

There are two types of random variables. A **discrete random variable** is one whose values are countable. A **continuous random variable** can assume an uncountable number of values. In this chapter we discussed discrete random variables and their **probability distributions.** We defined the **expected value, variance** and **standard deviation** of a population represented by a discrete probability distribution. Also introduced in this chapter were **bivariate discrete distributions** on which an important application in finance was based. Finally, the two most important discrete distributions – the **binomial** and the **Poisson** – were presented.

SYMBOLS:

Symbol	Pronounced	Represents
$\sum_{\text{all } x} x$	Sum of x for all values of x	Summation
C_x^n	n choose x	Number of combinations
$n!$	n factorial	$n(n-1)(n-2)\cdots(3)(2)(1)$
e		2.71828...

FORMULAS:

Expected value (mean)

$$E(X) = \mu = \sum_{\text{all } x} x\,P(x)$$

Variance

$$V(x) = \sigma^2 = \sum_{\text{all } x} (x - \mu)^2 P(x)$$

Standard deviation

$$\sigma = \sqrt{\sigma^2}$$

Covariance

$$COV(X, Y) = \sigma_{xy} = \sum (x - \mu_x)(y - \mu_y)P(x, y)$$

Coefficient of correlation

$$\rho = \frac{COV(X,Y)}{\sigma_x \sigma_y} = \frac{\sigma_{xy}}{\sigma_x \sigma_y}$$

Laws of expected value

1. $E(c) = c$
2. $E(X + c) = E(X) + c$
3. $E(cX) = cE(X)$

Laws of variance

1. $V(c) = 0$
2. $V(X + c) = V(X)$
3. $V(cX) = c^2V(X)$

Laws of expected value and variance of the sum of two variables

1. $E(X + Y) = E(X) + E(Y)$
2. $V(X + Y) = V(X) + V(Y) + 2COV(X, Y)$

Laws of expected value and variance for the sum of k variables, where $k \geq 2$

1. $E(X_1 + X_2 + \ldots + X_k) = E(X_1) + E(X_1) + \ldots + E(X_k)$
2. $V(X_1 + X_2 + \ldots + X_k) = V(X_1) + V(X_2) + \cdots + V(X_k)$

if the variables are independent

Mean and variance of a portfolio of two stocks

$$E(R_p) = w_1E(R_1) + w_2E(R_2)$$

$$V(R_p) = w_1^2V(R_1) + w_2^2V(R_2) + 2w_1 w_2COV(R_1, R_2) = w_1^2\sigma_1^2 + w_2^2\sigma_2^2 + 2w_1 w_2\rho\sigma_1\sigma_2$$

Mean and variance of a portfolio of k stocks

$$E(R_p) = \sum_{i=1}^{k} w_iE(R_i)$$

$$V(R_p) = \sum_{i=1}^{k} w_i^2\sigma_i^2 + 2\sum_{i=1}^{k}\sum_{j=i+1}^{k} w_iw_jCOV(R_i, R_j)$$

Binomial probability

$$P(X = x) = \frac{n!}{x!(n-x)!}p^x(1-p)^{n-x}$$

$$\mu = np$$

$$\sigma^2 = np(1-p)$$

$$\sigma = \sqrt{np(1-p)}$$

Poisson probability

$$P(X = x) = \frac{e^{-\mu}\mu^x}{x!}$$

CHAPTER EXERCISES

7.89 The number of magazine subscriptions per household is represented by the following probability distribution.

Magazine subscriptions per household	0	1	2	3	4
Probability	0.48	0.35	0.08	0.05	0.04

a. Calculate the mean number of magazine subscriptions per household.
b. Find the standard deviation.

7.90 The number of arrivals at a car wash is Poisson distributed with a mean of 8 per hour.
a. What is the probability that ten cars will arrive in the next hour?
b. What is the probability that more than five cars will arrive in the next hour?
c. What is the probability that fewer than twelve cars will arrive in the next hour?

7.91 Lotteries are an important income source for various governments around the world. However, the availability of lotteries and other forms of gambling have created a social problem: gambling addicts. A critic of government-controlled gambling contends that 30% of people who regularly buy lottery tickets are gambling addicts. If we randomly select ten people among those who report that they regularly buy lottery tickets, what is the probability that more than five of them are addicts?

7.92 An auditor is preparing for a physical count of inventory as a means of verifying its value. Items counted are reconciled with a list prepared by the storeroom supervisor. In one particular firm, 20% of the items counted cannot be reconciled without reviewing invoices. The auditor selects ten items. Find the probability that six or more items cannot be reconciled.

7.93 Most London restaurants offer 'early-bird' specials. These are lower-priced meals that are available only from 5 to 7 pm. However, not all customers who arrive between 5 and 7 pm order the special. In fact, only 70% do.
a. Find the probability that of 80 customers between 5 and 7 pm, more than 65 order the special.
b. What is the expected number of customers who order the special?
c. What is the standard deviation?

7.94 According to climatologists, the long-term average for Atlantic storms is 9.6 per season (1 June to 30 November), with six becoming hurricanes and 2.3 becoming intense hurricanes. Find the probability of the following events.
a. Ten or more Atlantic storms
b. Five or fewer hurricanes
c. Three or more intense hurricanes

7.95 Advertising researchers have developed a theory that states that commercials that appear in violent television shows are less likely to be remembered and will thus be less effective. After examining samples of viewers who watch violent and non-violent programmes and asking them a series of five questions about the commercials, the researchers produced the following probability distributions of the number of correct answers.

Viewers of Violent Shows

X	0	1	2	3	4	5
$P(x)$	0.36	0.22	0.20	0.09	0.08	0.05

Viewers of Non-Violent Shows

X	0	1	2	3	4	5
$P(x)$	0.15	0.18	0.23	0.26	0.10	0.08

a. Calculate the mean and standard deviation of the number of correct answers among viewers of violent television programmes.
b. Calculate the mean and standard deviation of the number of correct answers among viewers of non-violent television programmes.

7.96 In a recent election, the mayor received 60% of the vote. Last week, a survey was undertaken that asked 100 people whether they would vote for the mayor. Assuming that her popularity has not changed, what is the probability that more than 50 people in the sample would vote for the mayor?

7.97 When Earth travelled through the storm of meteorites trailing the comet Tempel-Tuttle on 17 November, 1998, the storm was 1000 times as intense as the average meteor storm. Before the comet arrived, telecommunication companies worried about the potential damage that might be inflicted on the approximately 650 satellites in orbit. It was estimated that each satellite had a 1% chance of being hit, causing damage to the satellite's electronic system. One company had five satellites in orbit at the time. Determine the probability distribution of the number of the company's satellites that would be damaged.

8 CONTINUOUS PROBABILITY DISTRIBUTIONS

8-1 PROBABILITY DENSITY FUNCTIONS

8-2 NORMAL DISTRIBUTION

8-3 EXPONENTIAL DISTRIBUTION

8-4 OTHER CONTINUOUS DISTRIBUTIONS

MINIMUM GMAT SCORE TO ENTER EXECUTIVE MBA PROGRAMME

A university has just approved a new Executive MBA Programme. The new director believes that to maintain the prestigious image of the business school, the new programme must be seen as having high standards. Accordingly, the Faculty Council decides that one of the entrance requirements will be that applicants must score in the top 1% of Graduate Management Admission Test (GMAT) scores. The director knows that GMAT scores are normally distributed with a mean of 490 and a standard deviation of 61. The only thing she does not know is what the minimum GMAT score for admission should be.

After introducing the normal distribution, we will return to this question and answer it. **See Section 8-2b for the answer.**

INTRODUCTION

This chapter completes our presentation of probability by introducing continuous random variables and their distributions. In Chapter 7 we introduced discrete probability distributions that are employed to calculate the probability associated with discrete random variables. In Section 7-4 we introduced the binomial distribution, which allows us to determine the probability that the random variable equals a particular value (the number of successes). In this way we connected the population represented by the probability distribution with a sample of nominal data. In this chapter, we introduce continuous probability distributions, which are used to calculate the probability associated with an interval variable. By doing so, we develop the link between a population and a sample of interval data.

Section 8-1 introduces probability density functions and uses the uniform density function to demonstrate how probability is calculated. In Section 8-2, we focus on the normal distribution, one of the most important distributions because of its role in the development of statistical inference. Section 8-3 introduces the exponential distribution, a distribution that has proven to be useful in various management-science applications. Finally, in Section 8-4 we introduce three additional continuous distributions. They will be used in statistical inference throughout the book.

8-1 PROBABILITY DENSITY FUNCTIONS

A continuous random variable is one that can assume an uncountable number of values. Because this type of random variable is so different from a discrete variable, we need to treat it completely differently. Whether data are discrete or continuous depends solely upon the real nature of the data and not upon how it is collected. First, we cannot list the possible values because there is an infinite number of them. Second, because there is an infinite number of values, the probability of each individual value is virtually 0. Consequently, we can determine the probability of only a range of values. To illustrate how this is done, consider the histogram we created for the long-distance telephone bills (Example 3.1), which is depicted in Figure 8.1.

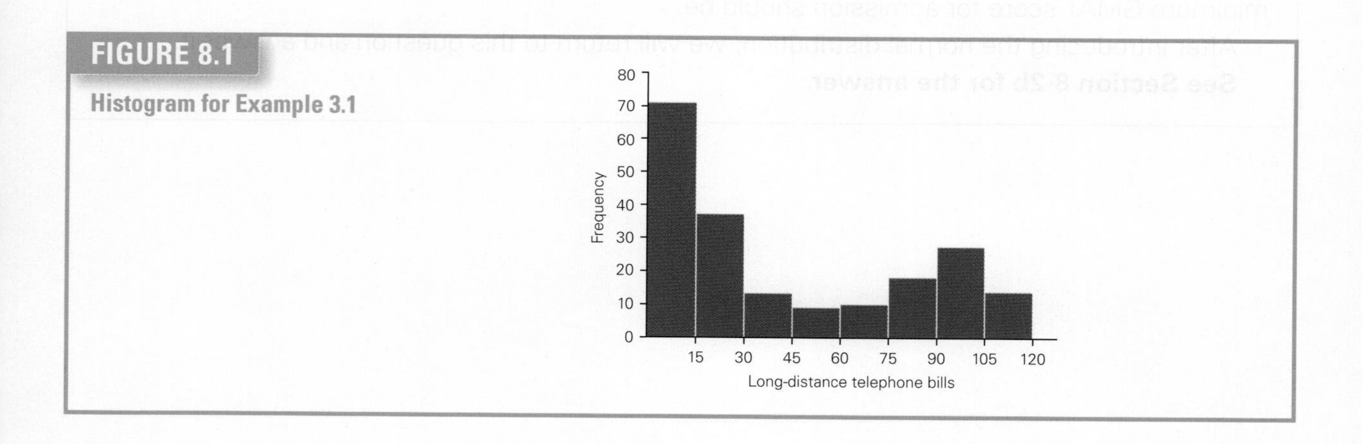

FIGURE 8.1
Histogram for Example 3.1

We found, for example, that the relative frequency of the interval 15 to 30 was 37/200. Using the relative frequency approach, we estimate that the probability that a randomly selected long-distance bill will fall between €15 and €30 is 37/200 = 0.185. We can similarly estimate the probabilities of the other intervals in the histogram.

Interval	Relative Frequency
$0 \le X \le 15$	71/200
$15 < X \le 30$	37/200
$30 < X \le 45$	13/200
$45 < X \le 60$	9/200
$60 < X \le 75$	10/200
$75 < X \le 90$	18/200
$90 < X \le 105$	28/200
$105 < X \le 120$	14/200

Notice that the sum of the probabilities equals 1. To proceed, we set the values along the vertical axis so that the *area* in all the rectangles together adds to 1. We accomplish this by dividing each relative frequency by the width of the interval, which is 15. The result is a rectangle over each interval whose *area* equals the probability that the random variable will fall into that interval.

To determine probabilities of ranges other than the ones created when we drew the histogram, we apply the same approach. For example, the probability that a long-distance bill will fall between €50 and €80 is equal to the area between 50 and 80 as shown in Figure 8.2.

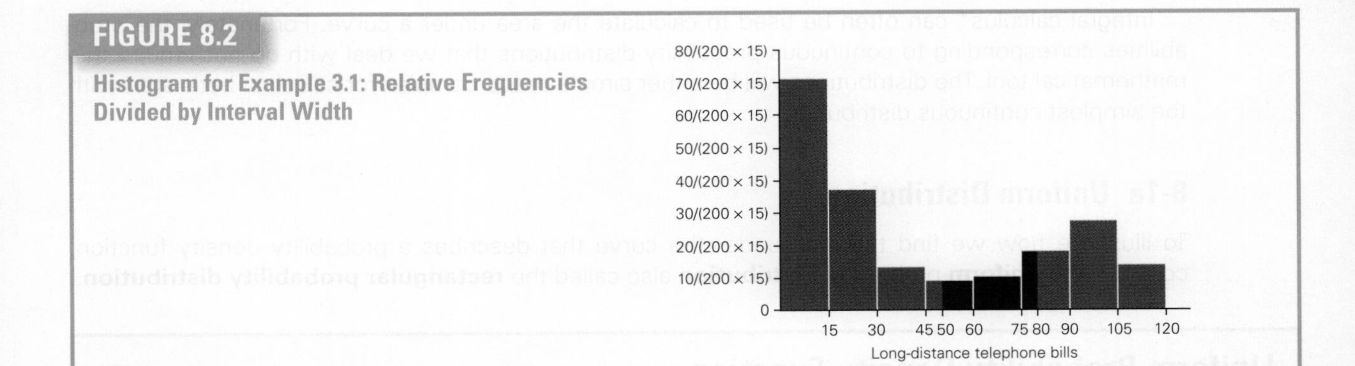

FIGURE 8.2 Histogram for Example 3.1: Relative Frequencies Divided by Interval Width

The areas in each shaded rectangle are calculated and added together as follows:

Interval	Height of Rectangle	Base Multiplied by Height
$50 < X \le 60$	$9/(200 \times 15) = 0.00300$	$(60 - 50) \times 0.00300 = 0.030$
$60 < X \le 75$	$10/(200 \times 15) = 0.00333$	$(75 - 60) \times 0.00333 = 0.050$
$75 < X \le 80$	$18/(200 \times 15) = 0.00600$	$(80 - 75) \times 0.00600 = 0.030$
		Total = 0.110

We estimate that the probability that a randomly selected long-distance bill falls between €50 and €80 is 0.11.

If the histogram is drawn with a large number of small intervals, we can smooth the edges of the rectangles to produce a smooth curve as shown in Figure 8.3. In many cases, it is possible to determine a function $f(x)$ that approximates the curve. The function is called a **probability density function**. Its requirements are stated in the following box.

FIGURE 8.3

Density Function for Example 3.1

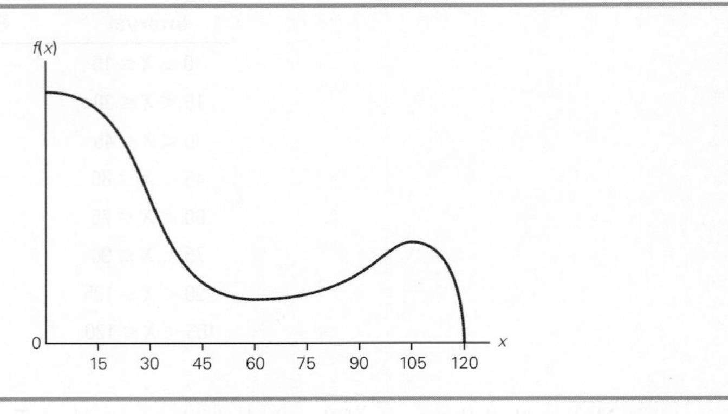

Requirements for a Probability Density Function

The following requirements apply to a probability density function $f(x)$ whose range is $a \le x \le b$.

1. $f(x) \ge 0$ for all x between a and b.
2. The total area under the curve between a and b is 1.0.

Integral calculus* can often be used to calculate the area under a curve. Fortunately, the probabilities corresponding to continuous probability distributions that we deal with do not require this mathematical tool. The distributions will be either simple or too complex for calculus. Let's start with the simplest continuous distribution.

8-1a Uniform Distribution

To illustrate how we find the area under the curve that describes a probability density function, consider the **uniform** probability **distribution** also called the **rectangular probability distribution**.

Uniform Probability Density Function

The uniform distribution is described by the function

$$f(x) = \frac{1}{b-a} \quad \text{where } a \le x \le b$$

The function is graphed in Figure 8.4. You can see why the distribution is called *rectangular*.

FIGURE 8.4

Uniform Distribution

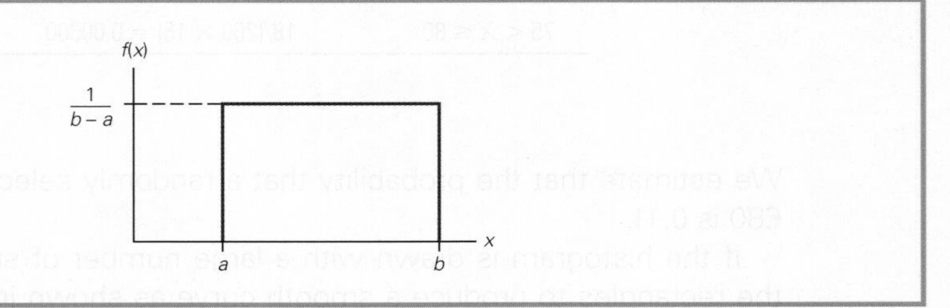

*The website Appendix Continuous Probability Distributions: Calculus Approach demonstrates how to use integral calculus to determine probabilities and parameters for continuous random variables.

To calculate the probability of any interval, simply find the area under the curve. For example, to find the probability that X falls between x_1 and x_2 determine the area in the rectangle whose base is $x_2 - x_1$ and whose height is $1/(b - a)$. Figure 8.5 depicts the area we wish to find. As you can see, it is a rectangle and the area of a rectangle is found by multiplying the base times the height.

FIGURE 8.5
$P(X_1 < X < X_2)$

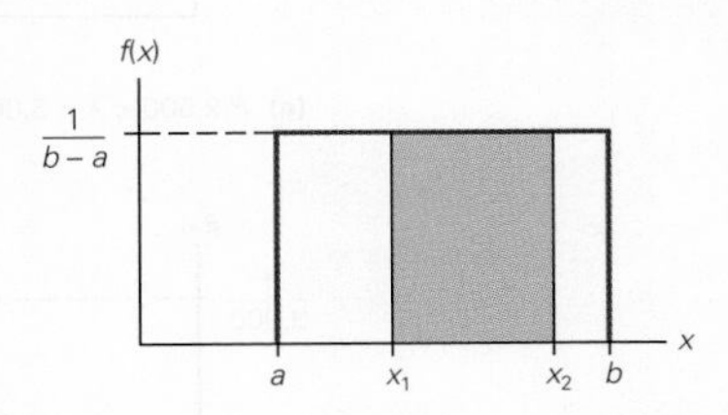

Thus,

$$P(x_1 < X < x_2) = \text{Base} \times \text{Height} = (x_2 - x_1) \times \frac{1}{b-a}$$

EXAMPLE 8.1

Uniformly Distributed Petrol Sales

The amount of petrol sold daily at a service station is uniformly distributed with a minimum of 2000 litres and a maximum of 5000 litres.

a. Find the probability that daily sales will fall between 2500 and 3000 litres.

b. What is the probability that the service station will sell at least 4000 litres?

c. What is the probability that the station will sell exactly 2500 litres?

SOLUTION:

The probability density function is

$$f(x) = \frac{1}{5000 - 2000} = \frac{1}{3000}, \quad 2000 \le x \le 5000$$

a. The probability that X falls between 2500 and 3000 is the area under the curve between 2500 and 3000 as depicted in Figure 8.6(a).The area of a rectangle is the base times the height. Thus,

$$P(2500 \le X \le 3000) = (3000 - 2500) \times \left(\frac{1}{3000}\right) = 0.1667$$

b.

$$P(X \ge 4000) = (5000 - 4000) \times \left(\frac{1}{3000}\right) = 0.3333 \text{ [See Figure 8.6(b).]}$$

c. $P(X = 2500) = 0$

Because there is an uncountable infinite number of values of X, the probability of each individual value is zero. Moreover, as you can see from Figure 8.6(c), the area of a line is 0.

Because the probability that a continuous random variable equals any individual value is 0, there is no difference between $P(2500 \le X \le 3000)$ and $P(2500 < X < 3000)$. Of course, we cannot say the same thing about discrete random variables.

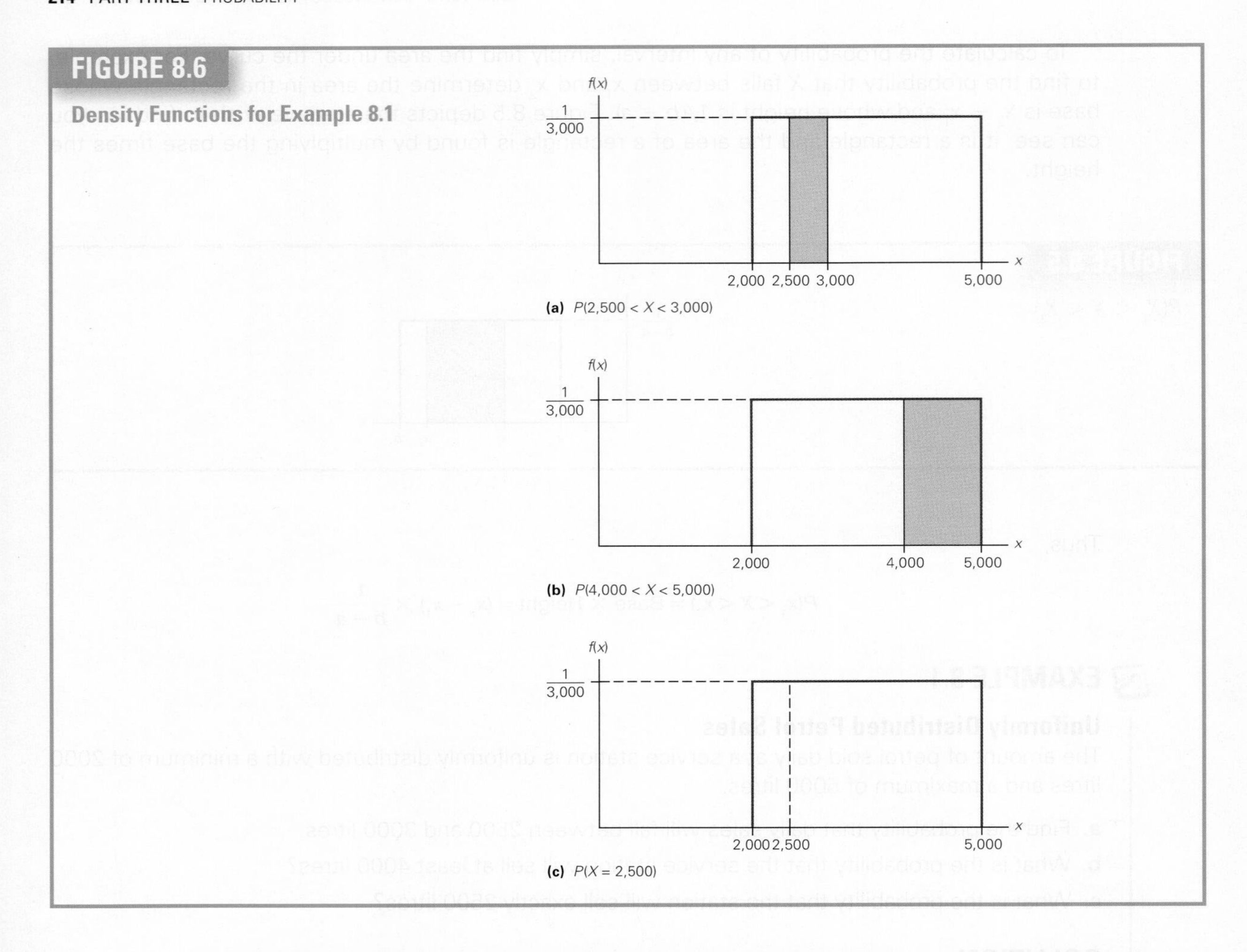

FIGURE 8.6

Density Functions for Example 8.1

8-1b Using a Continuous Distribution to Approximate a Discrete Distribution

In our definition of discrete and continuous random variables, we distinguish between them by noting whether the number of possible values is countable or uncountable. However, in practice, we frequently use a continuous distribution to approximate a discrete one when the number of values the variable can assume is countable but large. For example, the number of possible values of weekly income is countable. The values of weekly income expressed in any specific currency are 0, .01, .02, ... Although there is no set upper limit, we can easily identify (and thus count) all the possible values. Consequently, weekly income is a discrete random variable. However, because it can assume such a large number of values, we prefer to employ a continuous probability distribution to determine the probability associated with such variables. In the next section, we introduce the normal distribution, which is often used to describe discrete random variables that can assume a large number of values.

EXERCISES

8.1 Refer to Example 3.2 (Data Xm03-02). From the histogram for investment A, estimate the following probabilities.

a. $P(X > 45)$

b. $P(10 < X < 40)$

c. $P(X < 25)$

d. $P(35 < X < 65)$

8.2 Refer to Example 3.2. Estimate the following from the histogram of the returns on investment B.

a. $P(X > 45)$

b. $P(10 < X < 40)$

c. $P(X < 25)$

d. $P(35 < X < 65)$

8.3 Refer to Example 3.3 (Data Xm03-03). From the histogram of the marks, estimate the following probabilities.

a. $P(55 < X < 80)$

b. $P(X > 65)$

c. $P(X < 85)$

d. $P(75 < X < 85)$

8.4 A random variable is uniformly distributed between 5 and 25.

a. Draw the density function.

b. Find $P(X > 25)$.

c. Find $P(10 < X < 15)$.

d. Find $P(5.0 < X < 5.1)$.

8.5 A uniformly distributed random variable has minimum and maximum values of 20 and 60, respectively.

a. Draw the density function.

b. Determine $P(35 < X < 45)$.

c. Draw the density function including the calculation of the probability in part (b).

8.6 The amount of time it takes for a student to complete a statistics test is uniformly distributed between 30 and 60 minutes. One student is selected at random. Find the probability of the following events.

a. The student requires more than 55 minutes to complete the test.

b. The student completes the test in a time between 30 and 40 minutes.

c. The student completes the test in exactly 37.23 minutes.

8.7 Refer to Exercise 8.6. The lecturer wants to reward (with bonus marks) students who are in the lowest quarter of completion times. What completion time should she use for the cutoff for awarding bonus marks?

8.8 Refer to Exercise 8.6. The lecturer would like to track (and possibly help) students who are in the top 10% of completion times. What completion time should she use?

8.9 The weekly output of a steel mill is a uniformly distributed random variable that lies between 110 and 175 metric tonnes.

a. Calculate the probability that the steel mill will produce more than 150 metric tonnes next week.

b. Determine the probability that the steel mill will produce between 120 and 160 metric tonnes next week.

8.10 Refer to Exercise 8.9. The operations manager at a business in Tokyo labels any week that is in the bottom 20% of production a 'bad week'. How many metric tonnes should be used to define a bad week?

8.11 A random variable has the following density function.

$$f(x) = 1 - 0.5x, \quad 0 < x < 2$$

a. Graph the density function.

b. Verify that $f(x)$ is a density function.

c. Find $P(X > 1)$.

d. Find $P(X < 0.5)$.

e. Find $P(X = 1.5)$.

8.12 The following function is the density function for the random variable X:

$$f(x) = \frac{x-1}{8} \quad 1 < x < 5.$$

a. Graph the density function.

b. Find the probability that X lies between 2 and 4.

c. What is the probability that X is less than 3?

8.13 The following density function describes the random variable X.

$$f(x) = \begin{cases} \dfrac{x}{25} & 0 < x < 5 \\ \dfrac{10 - x}{25} & 5 < x < 10 \end{cases}$$

a. Graph the density function.

b. Find the probability that X lies between 1 and 3.

c. What is the probability that X lies between 4 and 8?

d. Compute the probability that X is less than 7.

e. Find the probability that X is greater than 3.

8.14 The following is a graph of a density function.

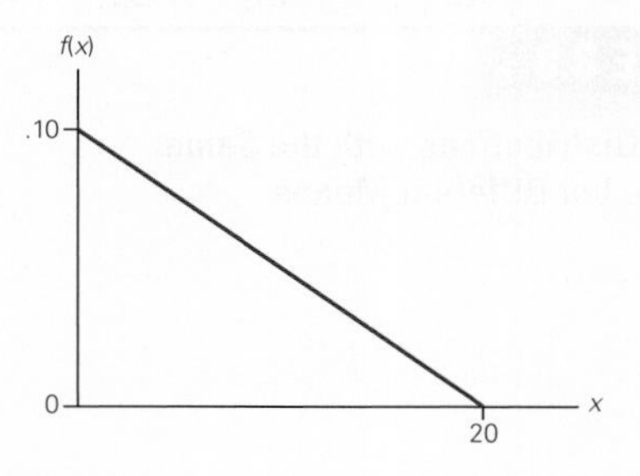

a. Determine the density function.

b. Find the probability that X is greater than 10.

c. Find the probability that X lies between 6 and 12.

8-2 NORMAL DISTRIBUTION

The **normal distribution** is the most important of all probability distributions because of its crucial role in statistical inference.

Normal Density Function

The probability density function of a **normal random variable** is

$$f(x) = \frac{1}{\sigma\sqrt{2\pi}} e^{-\frac{1}{2}\left(\frac{x-\mu}{\sigma}\right)^2}, \quad -\infty < x < \infty$$

where $e = 2.71828 \ldots$ and $\pi = 3.14159 \ldots$ and with parameters μ and σ.

Figure 8.7 depicts a normal distribution. Notice that the curve is symmetric about its mean and the random variable ranges between $-\infty$ and $+\infty$.

FIGURE 8.7

Normal Distribution

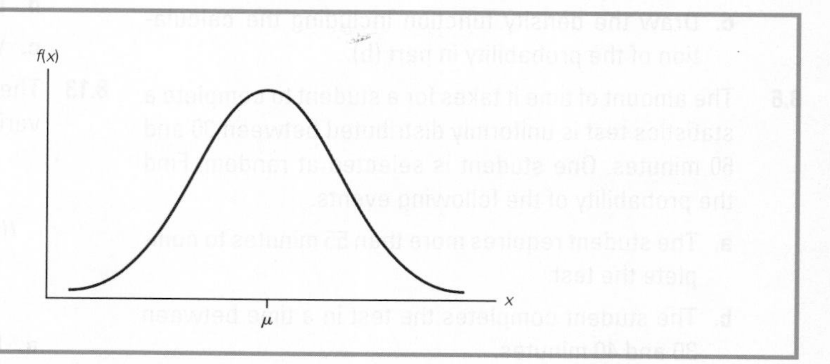

The normal distribution is described by two parameters, the mean μ and the standard deviation σ. In Figure 8.8, we demonstrate the effect of changing the value of μ. Obviously, increasing μ shifts the curve to the right and decreasing μ shifts it to the left.

FIGURE 8.8

Normal Distributions with the Same Variance but Different Means

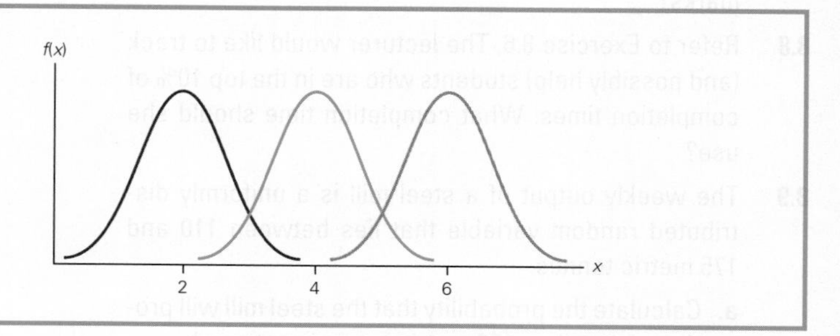

Figure 8.9 describes the effect of σ. Larger values of σ widen the curve and smaller ones make it narrower.

FIGURE 8.9

Normal Distributions with the Same Means but Different Variances

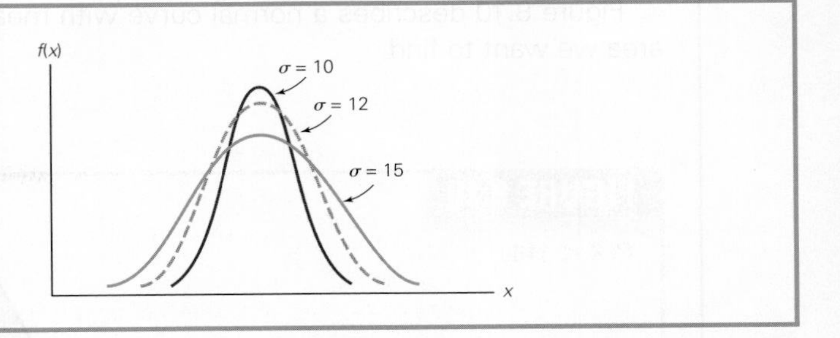

8-2a Calculating Normal Probabilities

To calculate the probability that a normal random variable falls into any interval, we must compute the area in the interval under the curve. Unfortunately, the function is not as simple as the uniform precluding the use of simple mathematics or even integral calculus. Instead we will resort to using a probability table similar to Tables 1 and 2 in Appendix A, which are used to calculate binomial and Poisson probabilities, respectively. Recall that to determine binomial probabilities from Table 1 we needed probabilities for selected values of n and p. Similarly, to find Poisson probabilities we needed probabilities for each value of μ that we chose to include in Table 2. It would appear then that we will need a separate table for normal probabilities for a selected set of values of μ and σ. Fortunately, this won't be necessary. Instead, we reduce the number of tables needed to one by standardising the random variable. We standardise a random variable by subtracting its mean and dividing by its standard deviation. When the variable is normal, the transformed variable is called a **standard normal random variable** and denoted by Z; that is,

$$Z = \frac{X - \mu}{\sigma}$$

The probability statement about X is transformed by this formula into a statement about Z. To illustrate how we proceed, consider the following example.

Example 8.2

Normally Distributed Petrol Sales

Suppose that the daily demand for regular petrol at another petrol station is normally distributed with a mean of 1000 litres and a standard deviation of 100 litres. The station manager has just opened the station for business and notes that there is exactly 1100 litres of regular petrol in storage. The next delivery is scheduled later today at the close of business. The manager would like to know the probability that he will have enough regular petrol to satisfy today's demands.

SOLUTION:

The amount of petrol on hand will be sufficient to satisfy the demand if the demand is less than the supply. We label the demand for regular petrol as X and we want to find the probability:

$$P(X < 1100)$$

Note that because X is a continuous random variable, we can also express the probability as

$$P(X < 1100)$$

because the area for $X = 1100$ is 0.

Figure 8.10 describes a normal curve with mean of 1000 and standard deviation of 100, and the area we want to find.

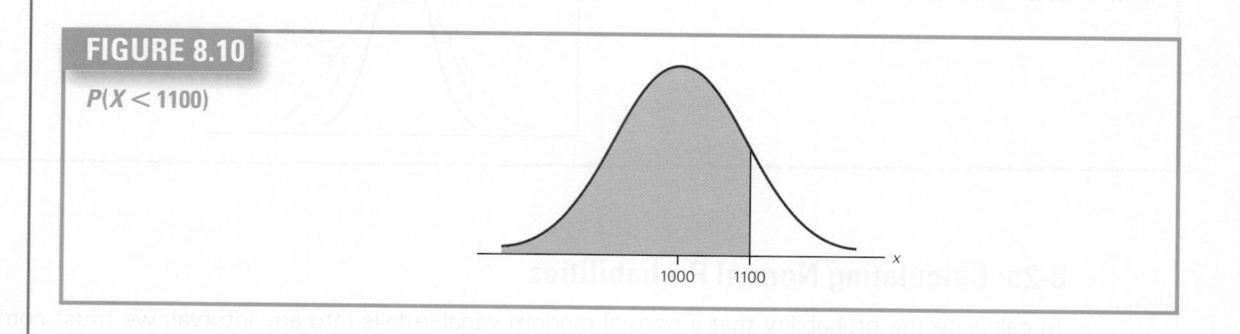

FIGURE 8.10
$P(X < 1100)$

The first step is to standardise X. However, if we perform any operations on X, we must perform the same operations on 1100. Thus,

$$P(X < 1100) = P\left(\frac{X - \mu}{\sigma} < \frac{1100 - 1000}{100}\right) = P(Z < 1.00)$$

Figure 8.11 describes the transformation that has taken place. Notice that the variable X was transformed into Z, and 1100 was transformed into 1. However, the area has not changed. In other words, the probability that we wish to compute $P(X < 1100)$ is identical to $P(Z < 1)$.

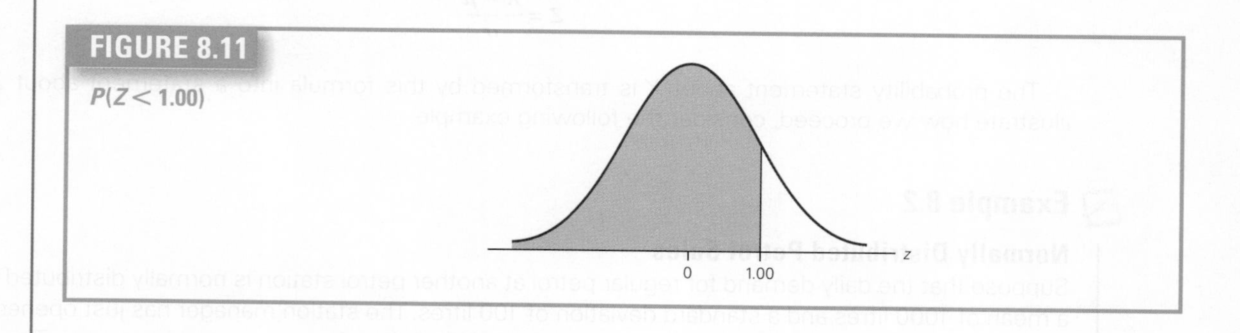

FIGURE 8.11
$P(Z < 1.00)$

The values of Z specify the location of the corresponding value of X. A value of $Z = 1$ corresponds to a value of X that is 1 standard deviation above the mean. Notice as well that the mean of Z, which is 0, corresponds to the mean of X.

If we know the mean and standard deviation of a normally distributed random variable, we can always transform the probability statement about X into a probability statement about Z. Consequently, we need only one table, Table 3 in Appendix A, the standard normal probability table, which is reproduced here as Table 8.1*.

---

*Supplementary Appendix Determining Normal Probabilities using $P(0 < Z < z)$ provides instructions and examples using this table.

TABLE 8.1 Normal Probabilities (Table 3 in Appendix A)

Z	0.00	0.01	0.02	0.03	0.04	0.05	0.06	0.07	0.08	0.09
−3.0	0.0013	0.0013	0.0013	0.0012	0.0012	0.0011	0.0011	0.0011	0.0010	0.0010
−2.9	0.0019	0.0018	0.0018	0.0017	0.0016	0.0016	0.0015	0.0015	0.0014	0.0014
−2.8	0.0026	0.0025	0.0024	0.0023	0.0023	0.0022	0.0021	0.0021	0.0020	0.0019
−2.7	0.0035	0.0034	0.0033	0.0032	0.0031	0.0030	0.0029	0.0028	0.0027	0.0026
−2.6	0.0047	0.0045	0.0044	0.0043	0.0041	0.0040	0.0039	0.0038	0.0037	0.0036
−2.5	0.0062	0.0060	0.0059	0.0057	0.0055	0.0054	0.0052	0.0051	0.0049	0.0048
−2.4	0.0082	0.0080	0.0078	0.0075	0.0073	0.0071	0.0069	0.0068	0.0066	0.0064
−2.3	0.0107	0.0104	0.0102	0.0099	0.0096	0.0094	0.0091	0.0089	0.0087	0.0084
−2.2	0.0139	0.0136	0.0132	0.0129	0.0125	0.0122	0.0119	0.0116	0.0113	0.0110
−2.1	0.0179	0.0174	0.0170	0.0166	0.0162	0.0158	0.0154	0.0150	0.0146	0.0143
−2.0	0.0228	0.0222	0.0217	0.0212	0.0207	0.0202	0.0197	0.0192	0.0188	0.0183
−1.9	0.0287	0.0281	0.0274	0.0268	0.0262	0.0256	0.0250	0.0244	0.0239	0.0233
−1.8	0.0359	0.0351	0.0344	0.0336	0.0329	0.0322	0.0314	0.0307	0.0301	0.0294
−1.7	0.0446	0.0436	0.0427	0.0418	0.0409	0.0401	0.0392	0.0384	0.0375	0.0367
−1.6	0.0548	0.0537	0.0526	0.0516	0.0505	0.0495	0.0485	0.0475	0.0465	0.0455
−1.5	0.0668	0.0655	0.0643	0.0630	0.0618	0.0606	0.0594	0.0582	0.0571	0.0559
−1.4	0.0808	0.0793	0.0778	0.0764	0.0749	0.0735	0.0721	0.0708	0.0694	0.0681
−1.3	0.0968	0.0951	0.0934	0.0918	0.0901	0.0885	0.0869	0.0853	0.0838	0.0823
−1.2	0.1151	0.1131	0.1112	0.1093	0.1075	0.1056	0.1038	0.1020	0.1003	0.0985
−1.1	0.1357	0.1335	0.1314	0.1292	0.1271	0.1251	0.1230	0.1210	0.1190	0.1170
−1.0	0.1587	0.1562	0.1539	0.1515	0.1492	0.1469	0.1446	0.1423	0.1401	0.1379
−0.9	0.1841	0.1814	0.1788	0.1762	0.1736	0.1711	0.1685	0.1660	0.1635	0.1611
−0.8	0.2119	0.2090	0.2061	0.2033	0.2005	0.1977	0.1949	0.1922	0.1894	0.1867
−0.7	0.2420	0.2389	0.2358	0.2327	0.2296	0.2266	0.2236	0.2206	0.2177	0.2148
−0.6	0.2743	0.2709	0.2676	0.2643	0.2611	0.2578	0.2546	0.2514	0.2483	0.2451
−0.5	0.3085	0.3050	0.3015	0.2981	0.2946	0.2912	0.2877	0.2843	0.2810	0.2776
−0.4	0.3446	0.3409	0.3372	0.3336	0.3300	0.3264	0.3228	0.3192	0.3156	0.3121
−0.3	0.3821	0.3783	0.3745	0.3707	0.3669	0.3632	0.3594	0.3557	0.3520	0.3483
−0.2	0.4207	0.4168	0.4129	0.4090	0.4052	0.4013	0.3974	0.3936	0.3897	0.3859
−0.1	0.4602	0.4562	0.4522	0.4483	0.4443	0.4404	0.4364	0.4325	0.4286	0.4247
−0.0	0.5000	0.4960	0.4920	0.4880	0.4840	0.4801	0.4761	0.4721	0.4681	0.4641
0.0	0.5000	0.5040	0.5080	0.5120	0.5160	0.5199	0.5239	0.5279	0.5319	0.5359
0.1	0.5398	0.5438	0.5478	0.5517	0.5557	0.5596	0.5636	0.5675	0.5714	0.5753
0.2	0.5793	0.5832	0.5871	0.5910	0.5948	0.5987	0.6026	0.6064	0.6103	0.6141
0.3	0.6179	0.6217	0.6255	0.6293	0.6331	0.6368	0.6406	0.6443	0.6480	0.6517
0.4	0.6554	0.6591	0.6628	0.6664	0.6700	0.6736	0.6772	0.6808	0.6844	0.6879
0.5	0.6915	0.6950	0.6985	0.7019	0.7054	0.7088	0.7123	0.7157	0.7190	0.7224
0.6	0.7257	0.7291	0.7324	0.7357	0.7389	0.7422	0.7454	0.7486	0.7517	0.7549
0.7	0.7580	0.7611	0.7642	0.7673	0.7704	0.7734	0.7764	0.7794	0.7823	0.7852
0.8	0.7881	0.7910	0.7939	0.7967	0.7995	0.8023	0.8051	0.8078	0.8106	0.8133
0.9	0.8159	0.8186	0.8212	0.8238	0.8264	0.8289	0.8315	0.8340	0.8365	0.8389
1.0	0.8413	0.8438	0.8461	0.8485	0.8508	0.8531	0.8554	0.8577	0.8599	0.8621
1.1	0.8643	0.8665	0.8686	0.8708	0.8729	0.8749	0.8770	0.8790	0.8810	0.8830
1.2	0.8849	0.8869	0.8888	0.8907	0.8925	0.8944	0.8962	0.8980	0.8997	0.9015

(continued)

Z	0.00	0.01	0.02	0.03	0.04	0.05	0.06	0.07	0.08	0.09
1.3	0.9032	0.9049	0.9066	0.9082	0.9099	0.9115	0.9131	0.9147	0.9162	0.9177
1.4	0.9192	0.9207	0.9222	0.9236	0.9251	0.9265	0.9279	0.9292	0.9306	0.9319
1.5	0.9332	0.9345	0.9357	0.9370	0.9382	0.9394	0.9406	0.9418	0.9429	0.9441
1.6	0.9452	0.9463	0.9474	0.9484	0.9495	0.9505	0.9515	0.9525	0.9535	0.9545
1.7	0.9554	0.9564	0.9573	0.9582	0.9591	0.9599	0.9608	0.9616	0.9625	0.9633
1.8	0.9641	0.9649	0.9656	0.9664	0.9671	0.9678	0.9686	0.9693	0.9699	0.9706
1.9	0.9713	0.9719	0.9726	0.9732	0.9738	0.9744	0.9750	0.9756	0.9761	0.9767
2.0	0.9772	0.9778	0.9783	0.9788	0.9793	0.9798	0.9803	0.9808	0.9812	0.9817
2.1	0.9821	0.9826	0.9830	0.9834	0.9838	0.9842	0.9846	0.9850	0.9854	0.9857
2.2	0.9861	0.9864	0.9868	0.9871	0.9875	0.9878	0.9881	0.9884	0.9887	0.9890
2.3	0.9893	0.9896	0.9898	0.9901	0.9904	0.9906	0.9909	0.9911	0.9913	0.9916
2.4	0.9918	0.9920	0.9922	0.9925	0.9927	0.9929	0.9931	0.9932	0.9934	0.9936
2.5	0.9938	0.9940	0.9941	0.9943	0.9945	0.9946	0.9948	0.9949	0.9951	0.9952
2.6	0.9953	0.9955	0.9956	0.9957	0.9959	0.9960	0.9961	0.9962	0.9963	0.9964
2.7	0.9965	0.9966	0.9967	0.9968	0.9969	0.9970	0.9971	0.9972	0.9973	0.9974
2.8	0.9974	0.9975	0.9976	0.9977	0.9977	0.9978	0.9979	0.9979	0.9980	0.9981
2.9	0.9981	0.9982	0.9982	0.9983	0.9984	0.9984	0.9985	0.9985	0.9986	0.9986
3.0	0.9987	0.9987	0.9987	0.9988	0.9988	0.9989	0.9989	0.9989	0.9990	0.9990

This table is similar to the ones we used for the binomial and Poisson distributions; that is, this table lists cumulative probabilities

$$P(Z < z)$$

for values of z ranging from -3.09 to $+3.09$.

To use the table, we simply find the value of z and read the probability. For example, the probability $P(Z < 2.00)$ is found by finding 2.0 in the left margin and under the heading 0.00 finding 0.9772.The probability $P(Z < 2.01)$ is found in the same row but under the heading 0.0 it is 0.9778.

Returning to Example 8.2, the probability we seek is found in Table 8.1 by finding 1.0 in the left margin. The number to its right under the heading 0.00 is 0.8413. See Figure 8.12.

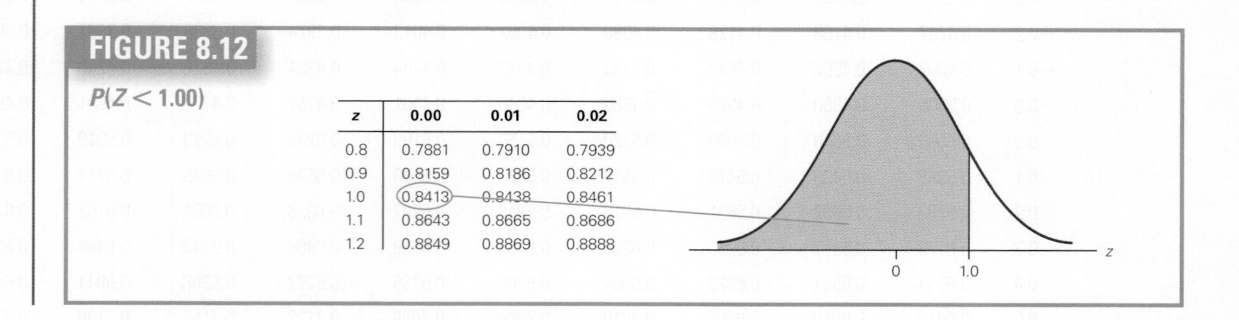

FIGURE 8.12 $P(Z < 1.00)$

As was the case with Tables 1 and 2, we can also determine the probability that the standard normal random variable is greater than some value of Z. For example, we find the probability that Z is greater than 1.80 by determining the probability that Z is less than 1.80 and subtracting that value from 1. By applying the complement rule, we get

$$P(Z > 1.80) = 1 - P(Z < 1.80) = 1 - 0.9641 = 0.0359$$

See Figure 8.13.

FIGURE 8.13
$P(Z < 1.80)$

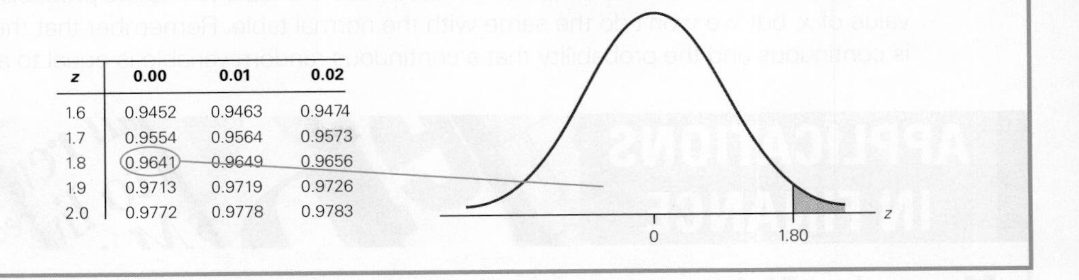

z	0.00	0.01	0.02
1.6	0.9452	0.9463	0.9474
1.7	0.9554	0.9564	0.9573
1.8	0.9641	0.9649	0.9656
1.9	0.9713	0.9719	0.9726
2.0	0.9772	0.9778	0.9783

We can also easily determine the probability that a standard normal random variable lies between two values of z. For example, we find the probability

$$P(-0.71 < Z < 0.92)$$

by finding the two cumulative probabilities and calculating their difference; that is,

$$P(Z < -0.71) = 0.2389$$

and

$$P(Z < 0.92) = 0.8212$$

Hence,

$$P(-0.71 < Z < 0.92) = P(Z < 92) - P(Z < -0.71) = 0.8212 - 0.2389 = 0.5823$$

Figure 8.14 depicts this calculation.

FIGURE 8.14
$P(-0.71 < Z < 0.92)$

z	0.00	0.01	0.02
−0.8	0.2119	0.2090	0.2061
−0.7	0.2420	0.2389	0.2358
−0.6	0.2743	0.2709	0.2676
−0.5	0.3085	0.3050	0.3015
−0.4	0.3446	0.3409	0.3372
−0.3	0.3821	0.3783	0.3745
−0.2	0.4207	0.4168	0.4129
−0.1	0.4602	0.4562	0.4522
−0.0	0.5000	0.4960	0.4920
0.0	0.5000	0.5040	0.5080
0.1	0.5398	0.5438	0.5478
0.2	0.5793	0.5832	0.5871
0.3	0.6179	0.6217	0.6255
0.4	0.6554	0.6591	0.6628
0.5	0.6915	0.6950	0.6985
0.6	0.7257	0.7291	0.7324
0.7	0.7580	0.7611	0.7642
0.8	0.7881	0.7910	0.7939
0.9	0.8159	0.8186	0.8212
1.0	0.8413	0.8438	0.8461

−0.71 0 0.92 z

Notice that the largest value of z in the table is 3.09 and that $P(Z < 3.09) = 0.9990$. This means that

$$P(Z > 3.09) = 1 - 0.9990 = 0.0010$$

However, because the table lists no values beyond 3.09, we approximate any area beyond 3.10 as 0. In other words,

$$P(Z > 3.10) = P(Z < -3.10) \approx 0$$

Recall that in Tables 1 and 2 we were able to use the table to find the probability that X is *equal* to some value of x, but we won't do the same with the normal table. Remember that the normal random variable is continuous and the probability that a continuous random variable is equal to any single value is 0.

APPLICATIONS IN FINANCE

Measuring Risk

In previous chapters we discussed several probability and statistical applications in finance where we wanted to measure and perhaps reduce the risk associated with investments. In Example 3.2 we drew histograms to gauge the spread of the histogram of the returns on two investments. We repeated this example in Chapter 4, where we computed the standard deviation and variance as numerical measures of risk. In Section 7-3, we developed an important application in finance in which we emphasised reducing the variance of the returns on a portfolio. However, we have not demonstrated why risk is measured by the variance and standard deviation. The following example corrects this deficiency.

EXAMPLE 8.3

Probability of a Negative Return on Investment

Consider an investment whose return is normally distributed with a mean of 10% and a standard deviation of 5%.

a. Determine the probability of losing money.

b. Find the probability of losing money when the standard deviation is equal to 10%.

SOLUTION:

a. The investment loses money when the return is negative. Thus, we wish to determine

$$P(X < 0)$$

The first step is to standardise both X and 0 in the probability statement:

$$P(X < 0) = P\left(\frac{X - \mu}{\sigma} < \frac{0 - 10}{5}\right) = P(Z < -2.00) = 0.0228$$

Therefore, the probability of losing money is 0.0228.

b. If we increase the standard deviation to 10%, the probability of suffering a loss becomes

$$P(X < 0) = P\left(\frac{X - \mu}{\sigma} < \frac{0 - 10}{10}\right) = P(Z < -1.00) = 0.1587$$

As you can see, increasing the standard deviation increases the probability of losing money. Note that increasing the standard deviation will also increase the probability that the return will exceed some relatively large amount. However, because investors tend to be risk averse, we emphasise the increased probability of negative returns when discussing the effect of increasing the standard deviation.

8-2b Finding Values of Z

There is a family of problems that require us to determine the value of Z given a probability. We use the notation Z_A to represent the value of z such that the area to its right under the standard normal curve is A; that is, Z_A is a value of a standard normal random variable such that

$$P(Z > Z_A) = A$$

FIGURE 8.15

$P(Z > Z_A) = A$

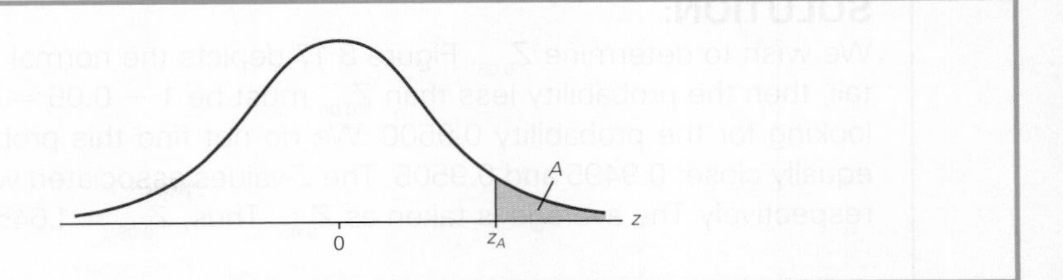

To find Z_A for any value of A requires us to use the standard normal table backwards. As you saw in Example 8.2, to find a probability about Z, we must find the value of 2 in the table and determine the probability associated with it. To use the table backwards, we need to specify a probability and then determine the z-value associated with it. We will demonstrate by finding $Z_{0.025}$. Figure 8.16 depicts the standard normal curve and $Z_{0.025}$. Because of the format of the standard normal table, we begin by determining the area *less than* $Z_{0.025}$, which is $1 - 0.025 = 0.9750$. (Notice that we expressed this probability with four decimal places to make it easier for you to see what you need to do.) We now search through the probability part of the table looking for 0.9750. When we locate it, we see that the *z*-value associated with it is 1.96.

Thus, $Z_{0.025} = 1.96$, which means that $P(Z > 1.96) = 0.025$.

FIGURE 8.16

$Z_{.025}$

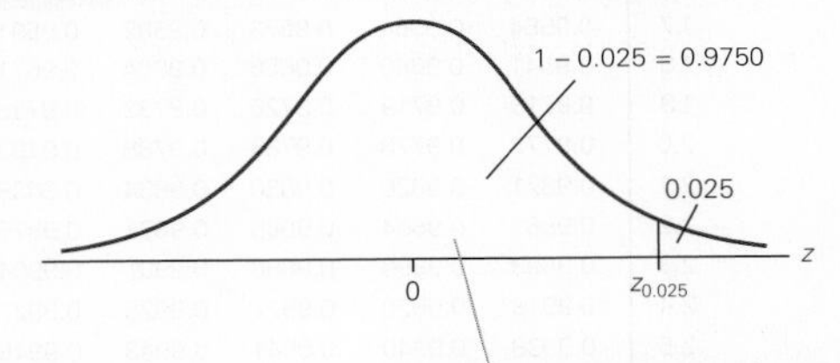

z	0.00	0.01	0.02	0.03	0.04	0.05	0.06	0.07	0.08	0.09
1.0	0.8413	0.8438	0.8461	0.8485	0.8508	0.8531	0.8554	0.8577	0.8599	0.8621
1.1	0.8643	0.8665	0.8686	0.8708	0.8729	0.8749	0.8770	0.8790	0.8810	0.8830
1.2	0.8849	0.8869	0.8888	0.8907	0.8925	0.8944	0.8962	0.8980	0.8997	0.9015
1.3	0.9032	0.9049	0.9066	0.9082	0.9099	0.9115	0.9131	0.9147	0.9162	0.9177
1.4	0.9192	0.9207	0.9222	0.9236	0.9251	0.9265	0.9279	0.9292	0.9306	0.9319
1.5	0.9332	0.9345	0.9357	0.9370	0.9382	0.9394	0.9406	0.9418	0.9429	0.9441
1.6	0.9452	0.9463	0.9474	0.9484	0.9495	0.9505	0.9515	0.9525	0.9535	0.9545
1.7	0.9554	0.9564	0.9573	0.9582	0.9591	0.9599	0.9608	0.9616	0.9625	0.9633
1.8	0.9641	0.9649	0.9656	0.9664	0.9671	0.9678	0.9686	0.9693	0.9699	0.9706
1.9	0.9713	0.9719	0.9726	0.9732	0.9738	0.9744	0.9750	0.9756	0.9761	0.9767
2.0	0.9772	0.9778	0.9783	0.9788	0.9793	0.9798	0.9803	0.9808	0.9812	0.9817
2.1	0.9821	0.9826	0.9830	0.9834	0.9838	0.9842	0.9846	0.9850	0.9854	0.9857
2.2	0.9861	0.9864	0.9868	0.9871	0.9875	0.9878	0.9881	0.9884	0.9887	0.9890
2.3	0.9893	0.9896	0.9898	0.9901	0.9904	0.9906	0.9909	0.9911	0.9913	0.9916
2.4	0.9918	0.9920	0.9922	0.9925	0.9927	0.9929	0.9931	0.9932	0.9934	0.9936
2.5	0.9938	0.9940	0.9941	0.9943	0.9945	0.9946	0.9948	0.9949	0.9951	0.9952
2.6	0.9953	0.9955	0.9956	0.9957	0.9959	0.9960	0.9961	0.9962	0.9963	0.9964
2.7	0.9965	0.9966	0.9967	0.9968	0.9969	0.9970	0.9971	0.9972	0.9973	0.9974
2.8	0.9974	0.9975	0.9976	0.9977	0.9977	0.9978	0.9979	0.9979	0.9980	0.9981
2.9	0.9981	0.9982	0.9982	0.9983	0.9984	0.9984	0.9985	0.9985	0.9986	0.9986
3.0	0.9987	0.9987	0.9987	0.9988	0.9988	0.9989	0.9989	0.9989	0.9990	0.9990

EXAMPLE 8.4

Finding $Z_{0.05}$

Find the value of a standard normal random variable such that the probability that the random variable is greater than it is 5%.

SOLUTION:

We wish to determine $Z_{0.05}$. Figure 8.17 depicts the normal curve and $Z_{0.05}$. If 0.05 is the area in the tail, then the probability less than $Z_{0.05}$ must be $1 - 0.05 = 0.9500$. To find $Z_{0.05}$ we search the table looking for the probability 0.9500. We do not find this probability, but we find two values that are equally close: 0.9495 and 0.9505. The Z-values associated with these probabilities are 1.64 and 1.65, respectively. The average is taken as $Z_{0.05}$. Thus, $Z_{0.05} = 1.645$.

FIGURE 8.17

$Z_{0.05}$

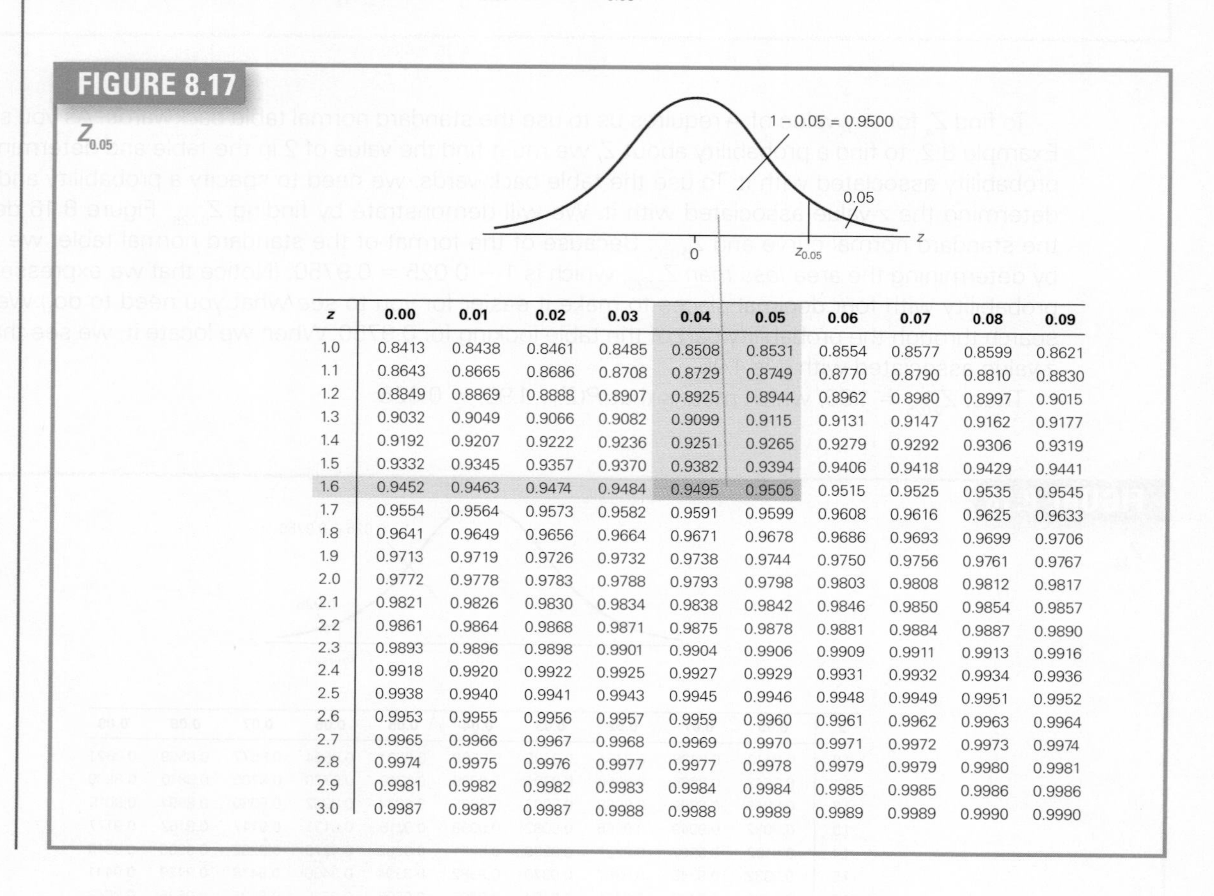

z	0.00	0.01	0.02	0.03	0.04	0.05	0.06	0.07	0.08	0.09
1.0	0.8413	0.8438	0.8461	0.8485	0.8508	0.8531	0.8554	0.8577	0.8599	0.8621
1.1	0.8643	0.8665	0.8686	0.8708	0.8729	0.8749	0.8770	0.8790	0.8810	0.8830
1.2	0.8849	0.8869	0.8888	0.8907	0.8925	0.8944	0.8962	0.8980	0.8997	0.9015
1.3	0.9032	0.9049	0.9066	0.9082	0.9099	0.9115	0.9131	0.9147	0.9162	0.9177
1.4	0.9192	0.9207	0.9222	0.9236	0.9251	0.9265	0.9279	0.9292	0.9306	0.9319
1.5	0.9332	0.9345	0.9357	0.9370	0.9382	0.9394	0.9406	0.9418	0.9429	0.9441
1.6	0.9452	0.9463	0.9474	0.9484	0.9495	0.9505	0.9515	0.9525	0.9535	0.9545
1.7	0.9554	0.9564	0.9573	0.9582	0.9591	0.9599	0.9608	0.9616	0.9625	0.9633
1.8	0.9641	0.9649	0.9656	0.9664	0.9671	0.9678	0.9686	0.9693	0.9699	0.9706
1.9	0.9713	0.9719	0.9726	0.9732	0.9738	0.9744	0.9750	0.9756	0.9761	0.9767
2.0	0.9772	0.9778	0.9783	0.9788	0.9793	0.9798	0.9803	0.9808	0.9812	0.9817
2.1	0.9821	0.9826	0.9830	0.9834	0.9838	0.9842	0.9846	0.9850	0.9854	0.9857
2.2	0.9861	0.9864	0.9868	0.9871	0.9875	0.9878	0.9881	0.9884	0.9887	0.9890
2.3	0.9893	0.9896	0.9898	0.9901	0.9904	0.9906	0.9909	0.9911	0.9913	0.9916
2.4	0.9918	0.9920	0.9922	0.9925	0.9927	0.9929	0.9931	0.9932	0.9934	0.9936
2.5	0.9938	0.9940	0.9941	0.9943	0.9945	0.9946	0.9948	0.9949	0.9951	0.9952
2.6	0.9953	0.9955	0.9956	0.9957	0.9959	0.9960	0.9961	0.9962	0.9963	0.9964
2.7	0.9965	0.9966	0.9967	0.9968	0.9969	0.9970	0.9971	0.9972	0.9973	0.9974
2.8	0.9974	0.9975	0.9976	0.9977	0.9977	0.9978	0.9979	0.9979	0.9980	0.9981
2.9	0.9981	0.9982	0.9982	0.9983	0.9984	0.9984	0.9985	0.9985	0.9986	0.9986
3.0	0.9987	0.9987	0.9987	0.9988	0.9988	0.9989	0.9989	0.9989	0.9990	0.9990

EXAMPLE 8.5

Finding $-Z_{0.05}$

Find the value of a standard normal random variable such that the probability that the random variable is less than it is 5%.

SOLUTION:

Because the standard normal curve is symmetric about 0, we wish to find $-Z_{0.05}$. In Example 8.4 we found $Z_{0.05} = 1.645$. Thus, $-Z_{0.05} = -1.645$. See Figure 8.18.

FIGURE 8.18

$-Z_{0.05}$

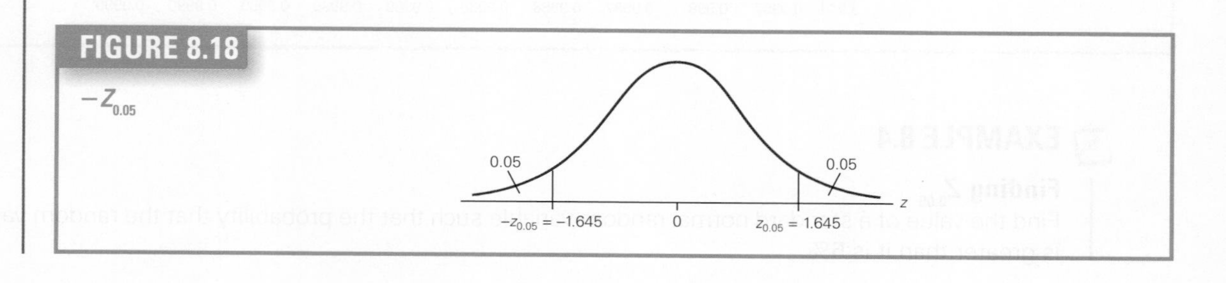

MINIMUM GMAT SCORE TO ENTER EXECUTIVE MBA PROGRAMME: SOLUTION

Figure 8.19 depicts the distribution of GMAT scores. We have labelled the minimum score needed to enter the new MBA programme X_m such that

$$P(X > X_{0.01}) = 0.01$$

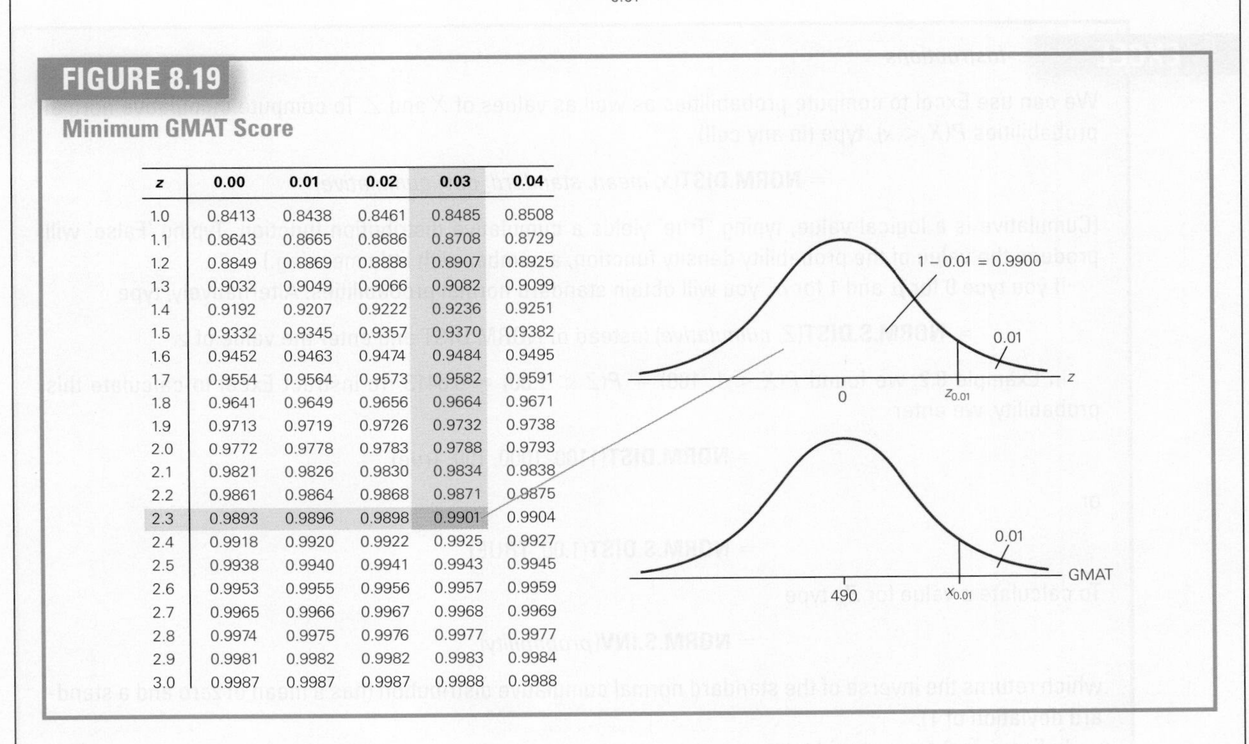

z	0.00	0.01	0.02	0.03	0.04
1.0	0.8413	0.8438	0.8461	0.8485	0.8508
1.1	0.8643	0.8665	0.8686	0.8708	0.8729
1.2	0.8849	0.8869	0.8888	0.8907	0.8925
1.3	0.9032	0.9049	0.9066	0.9082	0.9099
1.4	0.9192	0.9207	0.9222	0.9236	0.9251
1.5	0.9332	0.9345	0.9357	0.9370	0.9382
1.6	0.9452	0.9463	0.9474	0.9484	0.9495
1.7	0.9554	0.9564	0.9573	0.9582	0.9591
1.8	0.9641	0.9649	0.9656	0.9664	0.9671
1.9	0.9713	0.9719	0.9726	0.9732	0.9738
2.0	0.9772	0.9778	0.9783	0.9788	0.9793
2.1	0.9821	0.9826	0.9830	0.9834	0.9838
2.2	0.9861	0.9864	0.9868	0.9871	0.9875
2.3	0.9893	0.9896	0.9898	0.9901	0.9904
2.4	0.9918	0.9920	0.9922	0.9925	0.9927
2.5	0.9938	0.9940	0.9941	0.9943	0.9945
2.6	0.9953	0.9955	0.9956	0.9957	0.9959
2.7	0.9965	0.9966	0.9967	0.9968	0.9969
2.8	0.9974	0.9975	0.9976	0.9977	0.9977
2.9	0.9981	0.9982	0.9982	0.9983	0.9984
3.0	0.9987	0.9987	0.9987	0.9988	0.9988

FIGURE 8.19
Minimum GMAT Score

Above the normal curve, we depict the standard normal curve and $Z_{0.01}$. We can determine the value of $Z_{0.01}$ as we did in Example 8.4. In the standard normal table, we find $1 - 0.01 = 0.9900$ (its closest value in the table is 0.9901) and the Z-value 2.33. Thus, the standardised value of $X_{0.01}$ is $Z_{0.01} = 2.33$. To find $X_{0.01}$ we must unstandardise $Z_{0.01}$. We do so by solving for $X_{0.01}$ in the equation

$$Z_{0.01} = \frac{X_{0.01} - \mu}{\sigma}$$

Substituting $Z_{0.01} = 2.33$, $\mu = 490$ and $\sigma = 61$, we find

$$2.33 = \frac{X_{0.01} - 490}{61}$$

Solving, we get

$$X_{0.01} = 2.33(61) + 490 = 632.13$$

Rounding up (GMAT scores are integers), we find that the minimum GMAT score to enter the Executive MBA Programme is 633.

8-2c Z_A and Percentiles

In Chapter 4 we introduced percentiles, which are measures of relative standing. The values of Z_A are the 100(1 – *A*)*th* percentiles of a standard normal random variable. For example, $Z_{0.05} = 1.645$, which means that 1.645 is the 95th percentile: 95% of all values of *Z* are below it, and 5% are above it. We interpret other values of Z_A similarly.

Using the Computer

EXCEL

Instructions

We can use Excel to compute probabilities as well as values of *X* and *Z*. To compute cumulative normal probabilities $P(X < x)$, type (in any cell)

= **NORM.DIST**(*x, mean, standard_dev, cumulative*)

(Cumulative is a logical value, typing 'True' yields a cumulative distribution function. Typing 'False' will produce the value of the probability density function, a number with little meaning.)

If you type 0 for μ and 1 for σ, you will obtain standard normal probabilities. Alternatively, type

= **NORM.S.DIST**(*Z, cumulative*) instead of NORM.DIST and enter the value of *z*.

In Example 8.2, we found $P(X < 1, 100) = P(Z < 1.00) = 0.8413$. To instruct Excel to calculate this probability, we enter

= **NORM.DIST**(1100, 1000, 100, True)

or

= **NORM.S.DIST**(1.00, TRUE)

To calculate a value for Z_A, type

= **NORM.S.INV**(*probability*)

which returns the inverse of the standard normal cumulative distribution (has a mean of zero and a standard deviation of 1).

In Example 8.4, we would type

= **NORM.S.INV**(0.95)

and produce 1.6449. We calculated $Z_{0.05} = 1.645$.

To calculate a value of *x* given the probability $P(X > x) = A$, enter

= **NORM.INV**(*probability, mean, standard deviation*)

The chapter-opening example would be solved by typing

= **NORM.INV**(0.99, 490, 61)

which yields 632.

MINITAB

Instructions

We can use Minitab to compute probabilities as well as values of *X* and *Z*. Check **Calc, Probability Distributions** and **Normal** . . . and either **Cumulative probability** [to determine $P(X < x)$] or **Inverse cumulative probability** to find the value of *x*. Specify the **Mean** and **Standard deviation.**

APPLICATIONS IN OPERATIONS MANAGEMENT

Inventory Management

Every organization maintains some inventory, which is defined as a stock of items. For example, grocery stores hold inventories of almost all the products they sell. When the total number of products drops to a specified level, the manager arranges for the delivery of more products. An automobile repair shop keeps an inventory of a large number of replacement parts. A school keeps stock of items that it uses regularly, including chalk, pens, envelopes, file folders and paper clips. There are costs associated with inventories. These include the cost of capital, losses (theft and obsolescence) and warehouse space, as well as maintenance and record keeping. Management scientists have developed many models to help determine the optimum inventory level that balances the cost of inventory with the cost of shortages and the cost of making many small orders. Several of these models are deterministic – that is, they assume that the demand for the product is constant. However, in most realistic situations, the demand is a random variable. One commonly applied probabilistic model assumes that the demand during lead time is a normally distributed random variable. *Lead time* is defined as the amount of time between when the order is placed and when it is delivered.

The quantity ordered is usually calculated by attempting to minimise the total costs, including the cost of ordering and the cost of maintaining inventory. (This topic is discussed in most management-science courses.) Another critical decision involves the *reorder point*, which is the level of inventory at which an order is issued to its supplier. If the reorder point is too low, the company will run out of product, suffering the loss of sales and potentially customers who will go to a competitor. If the reorder point is too high, the company will be carrying too much inventory, which costs money to buy and store. In some companies, inventory has a tendency to walk out the back door or become obsolete. As a result, managers create a *safety stock*, which is the extra amount of inventory to reduce the times when the company has a shortage. They do so by setting a service level, which is the probability that the company will not experience a shortage. The method used to determine the reorder point will be demonstrated with Example 8.6.

EXAMPLE 8.6

Determining the Reorder Point

During the onset of hot weather, the demand for electric fans at a large home and garden shopping outlet in Madrid is quite strong. The company tracks inventory using a computer system so that it knows how many fans are in the inventory at any time. The policy is to order a new shipment of 250 fans when the inventory level falls to the reorder point, which is 150. However, this policy has resulted in frequent shortages and thus lost sales because both lead time and demand are highly variable. The manager would like to reduce the incidence of shortages so that only 5% of orders will arrive after inventory drops to 0 (resulting in a shortage). This policy is expressed as a 95% service level. From previous periods, the company has determined that demand during lead time is normally distributed with a mean of 200 and a standard deviation of 50. Find the reorder point.

SOLUTION:

The reorder point is set so that the probability that demand during lead time exceeds this quantity is 5%. Figure 8.20 depicts demand during lead time and the reorder point. As we did in the solution to the chapter-opening example, we find the standard normal value such that the area to its right is 0.05. The standardised value of the reorder point (ROP) is $Z_{0.05} = 1.645$. To find ROP, we must unstandardise $Z_{0.05}$.

$$Z_{0.05} = \frac{\text{ROP} - \mu}{\sigma}$$

$$1.645 = \frac{\text{ROP} - 200}{50}$$

$$\text{ROP} = 50(1.645) + 200 = 282.25$$

which we round up to 283. The policy is to order a new batch of fans when there are 283 fans left in inventory.

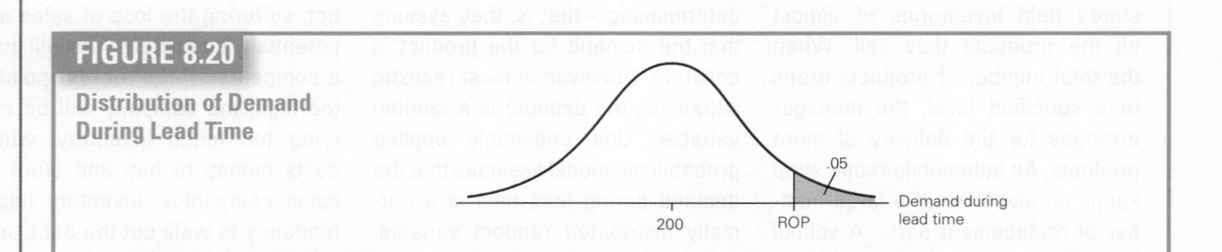

FIGURE 8.20
Distribution of Demand During Lead Time

EXERCISES

In Exercises 8.15 to 8.30, find the probabilities.

8.15 $P(Z < 1.50)$

8.16 $P(Z < 1.51)$

8.17 $P(Z < 1.55)$

8.18 $P(Z < -1.59)$

8.19 $P(Z < -1.60)$

8.20 $P(Z < -2.30)$

8.21 $P(-1.40 < Z < 0.60)$

8.22 $P(Z > -1.44)$

8.23 $P(Z < 2.03)$

8.24 $P(Z > 1.67)$

8.25 $P(Z < 2.84)$

8.26 $P(1.14 < Z < 2.43)$

8.27 $P(-0.91 < Z < -0.33)$

8.28 $P(Z > 3.09)$

8.29 $P(Z > 0)$

8.30 $P(Z > 4.0)$

8.31 Find $z_{0.02}$

8.32 Find $z_{0.45}$

8.33 Find $z_{0.20}$

8.34 X is normally distributed with mean 100 and standard deviation 20. What is the probability that X is greater than 145?

8.35 X is normally distributed with mean 250 and standard deviation 40. What value of X does only the top 15% exceed?

8.36 X is normally distributed with mean 1000 and standard deviation 250. What is the probability that X lies between 800 and 1100?

8.37 X is normally distributed with mean 50 and standard deviation 8. What value of X is such that only 8% of values are below it?

8.38 The long-distance calls made by the employees of a company are normally distributed with a mean of 6.3 minutes and a standard deviation of 2.2 minutes. Find the probability that a call

a. lasts between 5 and 10 minutes.

b. lasts more than 7 minutes.

c. lasts less than 4 minutes.

8.39 Refer to Exercise 8.38. How long do the longest 10% of calls last?

8.40 The lifetimes of light bulbs that are advertised to last for 5000 hours are normally distributed with a mean of 5100 hours and a standard deviation of 200 hours. What is the probability that a bulb lasts longer than the advertised figure?

8.41 Refer to Exercise 8.40. If we wanted to be sure that 98% of all bulbs last longer than the advertised figure, what figure should be advertised?

8.42 Travel-by-us is an Internet-based travel agency wherein customers can see videos of the cities they plan to visit. The number of hits daily is a normally distributed random variable with a mean of 10 000 and a standard deviation of 2400.

a. What is the probability of getting more than 12 000 hits?

b. What is the probability of getting fewer than 9000 hits?

8.43 Refer to Exercise 8.42. Some Internet sites have bandwidths that are not sufficient to handle all their traffic, often causing their systems to crash. Bandwidth can be measured by the number of hits a system can handle. How large a bandwidth should Travel-by-us have in order to handle 99.9% of daily traffic?

8.44 A new gas–electric hybrid car has recently hit the market. The distance travelled on 4 litres of petrol is normally distributed with a mean of 105 km and a standard deviation of 6km. Find the probability of the following events.

a. The car travels more than 113 km per litre.

b. The car travels less than 95 km per litre.

c. The car travels between 88 and 113km per litre.

8.45 The top-selling Michelin tyre is rated 75 000 miles (120 000km), which means nothing. In fact, the distance the tyres can run until they wear out is a normally distributed random variable with a mean of 82 000 miles (131 966 km) and a standard deviation of 6 400 miles (10 300 km).

a. What is the probability that a tyre wears out before 75 000 miles (120 000 km)?

b. What is the probability that a tyre lasts more than 100 000 miles (160 934 km)?

8.46 The heights of 2-year-old children are normally distributed with a mean of 81 cm and a standard deviation of 3.8 cm. Paediatricians regularly measure the heights of toddlers to determine whether there is a problem. There may be a problem when a child is in the top or bottom 5% of heights. Determine the heights of 2-year-old children that could be a problem.

8.47 Refer to Exercise 8.46. Find the probability of these events.

a. A 2-year-old child is taller than 91 cm.

b. A 2-year-old child is shorter than 86 cm.

c. A 2-year-old child is between 76 and 84 cm tall.

8.48 University and college students average 7.2 hours of sleep per night, with a standard deviation of 40 minutes. If the amount of sleep is normally distributed, what proportion of university and college students sleep for more than 8 hours?

8.49 Refer to Exercise 8.48. Find the amount of sleep that is exceeded by only 25% of students.

8.50 The amount of time devoted to studying statistics each week by students who achieve a grade of A in the course is a normally distributed random variable with a mean of 7.5 hours and a standard deviation of 2.1 hours.

a. What proportion of A students study for more than 10 hours per week?

b. Find the probability that an A student spends between 7 and 9 hours studying.

c. What proportion of A students spend fewer than 3 hours studying?

d. What is the amount of time below which only 5% of all A students spend studying?

8.51 The number of pages printed before replacing the cartridge in a laser printer is normally distributed with a mean of 11 500 pages and a standard deviation of 800 pages. A new cartridge has just been installed.

a. What is the probability that the printer produces more than 12 000 pages before this cartridge must be replaced?

b. What is the probability that the printer produces fewer than 10 000 pages?

8.52 Refer to Exercise 8.51. The manufacturer wants to provide guidelines to potential customers advising them of the minimum number of pages they can expect from each cartridge. How many pages should it advertise if the company wants to be correct 99% of the time?

8.53 Battery manufacturers compete on the basis of the amount of time their products last in cameras and toys. A manufacturer of alkaline batteries has observed that its batteries last for an average of 26 hours when used in a toy racing car. The amount of time is normally distributed with a standard deviation of 2.5 hours.

a. What is the probability that the battery lasts between 24 and 28 hours?

b. What is the probability that the battery lasts longer than 28 hours?

c. What is the probability that the battery lasts less than 24 hours?

8.54 Because of the relatively high interest rates, most consumers attempt to pay off their credit card bills promptly. However, this is not always possible. An analysis of the amount of interest paid monthly by a bank's Visa cardholders reveals that the amount is normally distributed with a mean of £27 and a standard deviation of £7.

a. What proportion of the bank's Visa cardholders pay more than £30 in interest?

b. What proportion of the bank's Visa cardholders pay more than £40 in interest?

c. What proportion of the bank's Visa cardholders pay less than £15 in interest?

d. What interest payment is exceeded by only 20% of the bank's Visa cardholders?

8.55 It is said that sufferers of a cold virus experience symptoms for 7 days. However, the amount of time is actually a normally distributed random variable whose mean is 7.5 days and whose standard deviation is 1.2 days.

a. What proportion of cold sufferers experience fewer than 4 days of symptoms?

b. What proportion of cold sufferers experience symptoms for between 7 and 10 days?

8.56 How much money does a typical family of four spend at a Subway restaurant per visit? Suppose the amount is a normally distributed random variable with a mean of £16.40 and a standard deviation of £2.75.

a. Find the probability that a family of four spends less than £10.

b. What is the amount below which only 10% of families of four spend at Subway?

8.57 The final marks in a statistics course are normally distributed with a mean of 70 and a standard deviation of 10. The lecturer must convert all marks to letter grades. She decides that she wants 10% As, 30% Bs, 40% Cs, 15% Ds and 5% Fs. Determine the cutoffs for each letter grade.

8.58 Mensa is an organization whose members possess IQs that are in the top 2% of the population. It is known that IQs are normally distributed with a mean of 100 and a standard deviation of 16. Find the minimum IQ needed to be a Mensa member.

8.59 The lifetimes of televisions produced by the Hishobi Company are normally distributed with a mean of 75 months and a standard deviation of 8 months. If the manufacturer wants to have to replace only 1% of its televisions, what should its warranty be?

8.60 A retailer of computing products in Pretoria, South Africa, sells a variety of computer-related products. One of his most popular products is an HP laser printer. The average weekly demand is 200. Lead time for a new order from the manufacturer to arrive is 1 week. If the demand for printers were constant, the retailer would reorder when there were exactly 200 printers in inventory. However, the demand is a random variable. An analysis of previous weeks reveals that the weekly demand standard deviation is 30. The retailer knows that if a customer wants to buy an HP laser printer but he has none available, he will lose that sale plus possibly additional sales. He wants the probability of running short in any week to be no more than 6%. How many HP laser printers should he have in stock when he reorders from the manufacturer?

8.61 The demand for a daily newspaper at a newsstand at a busy intersection of Delhi is known to be normally distributed with a mean of 150 and a standard deviation of 25. How many newspapers should the newsstand operator order to ensure that he runs short on no more than 20% of days?

8.62 Every day a bakery prepares its famous marble rye. A statistically savvy customer determined that daily demand is normally distributed with a mean of 850 and a standard deviation of 90. How many loaves should the bakery make if it wants the probability of running short on any day to be no more than 30%?

8.63 Refer to Exercise 8.62. Any marble ryes that are unsold at the end of the day are marked down and sold for half-price. How many loaves should the bakery prepare so that the proportion of days that result in unsold loaves is no more than 60%?

8.64 Refer to Exercise 7.46. Find the probability that the project will take more than 60 days to complete.

8.65 The mean and variance of the time to complete the project in Exercise 7.47 was 145 minutes and 31 minutes^2. What is the probability that it will take less than 2.5 hours to overhaul the machine?

8.66 The annual rate of return on a mutual fund is normally distributed with a mean of 14% and a standard deviation of 18%.

a. What is the probability that the fund returns more than 25% next year?

b. What is the probability that the fund loses money next year?

8.67 In Exercise 7.49, we discovered that the expected return is 0.1060 and the standard deviation is 0.1456. Working with the assumption that returns are normally distributed, determine the probability of the following events.

a. The portfolio loses money.

b. The return on the portfolio is greater than 20%.

APPLICATIONS IN OPERATIONS MANAGEMENT

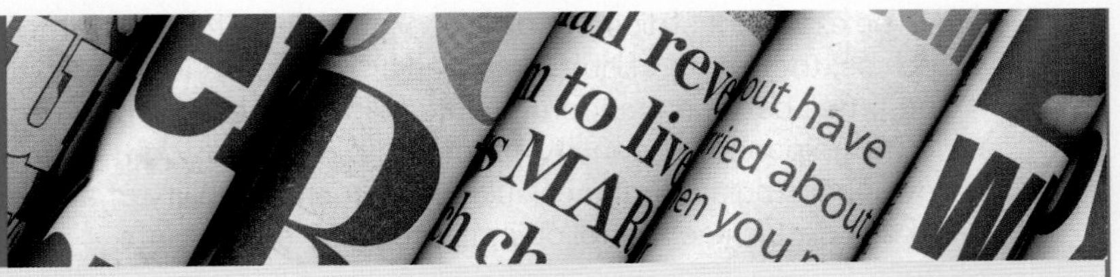

PERT/CPM

In the Applications in Operations Management box after Section 7-2, we introduced PERT/CPM. The purpose of this powerful management-science procedure is to determine the critical path of a project. The expected value and variance of the completion time of the project are based on the expected values and variances of the completion times of the activities on the critical path. Once we have the expected value and variance of the completion time of the project, we can use these figures to determine the probability that the project will be completed by a certain date. Statisticians have established that the completion time of the project is approximately normally distributed, enabling us to compute the needed probabilities.

8-3 EXPONENTIAL DISTRIBUTION

Another important continuous distribution is the **exponential distribution**.

Exponential Probability Density Function

A random variable X is exponentially distributed if its probability density function is given by

$$f(x) = \lambda e^{-\lambda x}, \quad x \geq 0$$

where $e = 2.71828\ldots$ and λ is the parameter of the distribution, which is called *rate*.

Statisticians have shown that the mean and standard deviation of an exponential random variable are equal to each other:

$$\mu = \sigma = 1/\lambda$$

Recall that the normal distribution is a two-parameter distribution. The distribution is completely specified once the values of the two parameters μ and σ are known. In contrast, the exponential distribution is a one-parameter distribution. The distribution is completely specified once the value of the parameter λ is known. Figure 8.21 depicts three exponential distributions, corresponding to three different values of the parameter X. Notice that for any exponential density function $f(x)$, $f(0) = \lambda$ and $f(x)$ approaches 0 as x approaches infinity.

FIGURE 8.21

Exponential Distributions

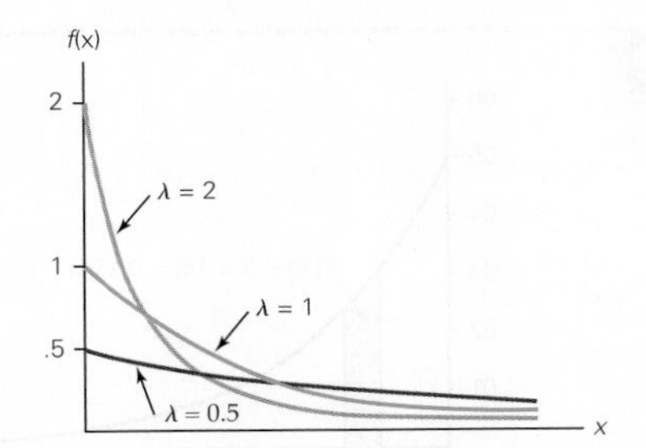

By looking at the shape of the curves, the greater the rate, the faster the curve drops.

The exponential distribution is very often used to model time interval between some random events, such as: the length of time between arrivals at a check point, the lifetime of an electronic device, etc. The times between random events follow the exponential distribution with rate λ.

The exponential density function is easier to work with than the normal. As a result, we can develop formulas for the calculation of the probability of any range of values. Using integral calculus, we can determine the following probability statements.

Probability Associated with an Exponential Random Variable

If X is an exponential random variable,

$$P(X > x) = e^{-\lambda x}$$

$$P(X < x) = 1 - e^{-\lambda x}$$

$$P(x_1 < X < x_2) = P(X < x_2) - P(X < x_1) = e^{-\lambda x_1} - e^{-\lambda x_2}$$

The value of $e^{-\lambda x}$ can be obtained with the aid of a calculator. The above formulae correspond to the areas under the density curve to the left of x, to the right of x, and between x_1 and x_2.

EXAMPLE 8.7

Lifetimes of Alkaline Batteries

The lifetime of an alkaline battery (measured in hours) is exponentially distributed with $\lambda = 0.05$.

a. What is the mean and standard deviation of the battery's lifetime?

b. Find the probability that a battery will last between 10 and 15 hours.

c. What is the probability that a battery will last for more than 20 hours?

SOLUTION:

a. The mean and standard deviation are equal to $1/\lambda$. Thus,

$$\mu = \sigma = 1/\lambda = 1/0.05 = 20 \text{ hours}$$

b. Let X denote the lifetime of a battery. The required probability is

$$\begin{aligned} P(10 < X < 15) &= e^{-0.05(10)} - e^{-0.05(15)} \\ &= e^{-0.5} - e^{-0.75} \\ &= 0.6065 - 0.4724 \\ &= 0.1341 \end{aligned}$$

c. $$\begin{aligned} P(X > 20) &= e^{-0.5(20)} \\ &= e^{-1} \\ &= 0.3679 \end{aligned}$$

Figure 8.22 depicts these probabilities.

FIGURE 8.22 Probabilities for Example 8.7

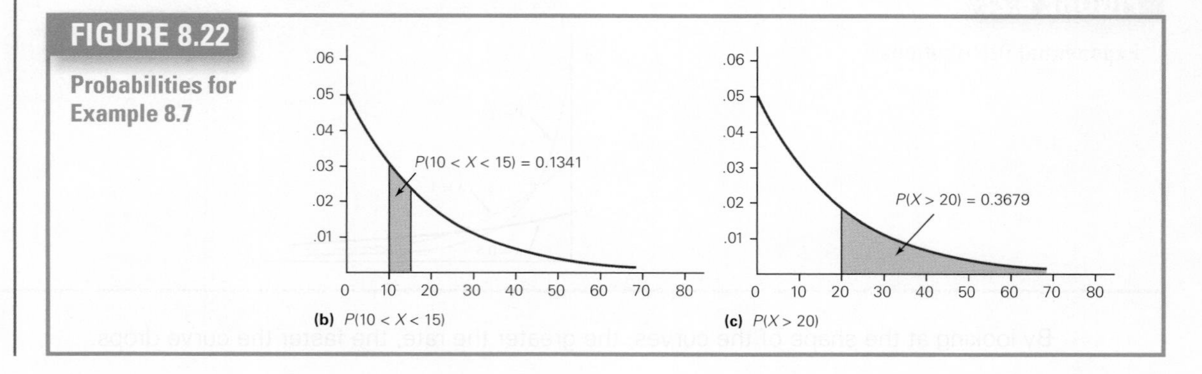

Using the Computer

EXCEL *Instructions*

Type (in any cell)

= **EXPON.DIST**(x, lambda, cumulative)

(or = **EXPONDIST** ([X], [λ], True) for Excel 2010 or below)

To produce the answer for Example 8.7c, we would find $P(X < 20)$ and subtract it from 1.

To find $P(X < 20)$, type

= **EXPONDIST**(20, 0.05, True)

which outputs 0.6321 and hence $P(X > 20) = 1 - 0.6321 = 0.3679$, which is exactly the number we produced manually.

MINITAB *Instructions*

Click **Calc, Probability Distributions** and **Exponential** ... and specify **Cumulative probability.** In the **Scale** box, type the mean, which is $1/\lambda$. In the **Threshold** box, type 0.

APPLICATIONS IN OPERATIONS MANAGEMENT

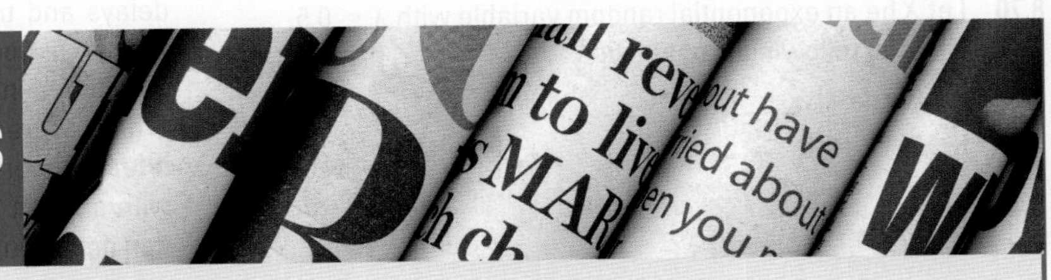

Queues

In Section 7-5, we described queue (or waiting-line) models and how the Poisson distribution is used to calculate the probabilities of the number of arrivals per time period. To calculate the operating characteristics of queues, management scientists often assume that the times to complete a service are exponentially distributed. In this application, the parameter λ is the service rate, which is defined as the mean number of service completions per time period. For example, if service times are exponentially distributed with $\lambda = 5$/hour, this tells us that the service rate is 5 units per hour or 5 per 60 minutes. Recall that the mean of an exponential distribution is $\mu = 1/\lambda$. In this case, the service facility can complete a service in an average of 12 minutes. This was calculated as

$$\mu = \frac{1}{\lambda} = \frac{1}{5\text{/hour}} = \frac{1}{5/60 \text{ minutes}} = \frac{60 \text{ minutes}}{5} = 12 \text{ minutes.}$$

We can use this distribution to make a variety of probability statements.

EXAMPLE 8.8

Supermarket Checkout Counter

A checkout counter at a supermarket completes the process according to an exponential distribution with a service rate of 6 per hour. A customer arrives at the checkout counter. Find the probability of the following events.

a. The service is completed in fewer than 5 minutes.

b. The customer leaves the checkout counter more than 10 minutes after arriving.

c. The service is completed in a time between 5 and 8 minutes.

SOLUTION:

One way to solve this problem is to convert the service rate so that the time period is 1 minute. (Alternatively, we can solve by converting the probability statements so that the time periods are measured in fractions of an hour.) Let the service rate = $X = 0.1$/minute.

a. $P(X < 5) = 1 - e^{-\lambda x} = 1 - e^{-0.1(5)} = 1 - e^{-0.5} = 1 - 0.6065 = 0.3935$

b. $P(X > 10) = e^{-\lambda x} = e^{-0.1(10)} = e^{-1} = 0.3679$

c. $P(5 < X < 8) = e^{-0.1(5)} - e^{-0.1(8)} = e^{-5} - e^{-8} = 0.6065 - 0.4493 = 0.1572$

EXERCISES

8.68 The random variable X is exponentially distributed with $\lambda = 3$. Sketch the graph of the distribution of X by plotting and connecting the points representing $f(x)$ for $x = 0, 0.5, 1, 1.5$ and 2.

8.69 X is an exponential random variable with $\lambda = 0.25$. Sketch the graph of the distribution of X by plotting and connecting the points representing $f(x)$ for $x = 0$, 2, 4, 6, 8, 10, 15, 20.

8.70 Let X be an exponential random variable with $\lambda = 0.5$. Find the following probabilities.

a. $P(X > 1)$

b. $P(X > 0.4)$

c. $P(X < 0.5)$

d. $P(X < 2)$

8.71 X is an exponential random variable with $\lambda = 0.3$. Find the following probabilities.

a. $P(X > 2)$

b. $P(X < 4)$

c. $P(1 < X < 2)$

d. $P(X = 3)$

8.72 The production of a complex chemical needed for anticancer drugs is exponentially distributed with $\lambda = 6$ kilograms per hour. What is the probability that the production process requires more than 15 minutes to produce the next kilogram of drugs?

8.73 The time between breakdowns of ageing machines is known to be exponentially distributed with a mean of 25 hours. The machine has just been repaired. Determine the probability that the next breakdown occurs more than 50 hours from now.

8.74 A bank wishing to increase its customer base advertises that it has the fastest service and that virtually all of its customers are served in less than 10 minutes. A management scientist has studied the service times and concluded that service times are exponentially distributed with a mean of 5 minutes. Determine what the bank means when it claims 'virtually all' its customers are served in less than 10 minutes.

8.75 Many European roads face traffic congestion and the use of toll roads can ease the traffic. Most toll booths use innovative technologies that help to eliminate the delays and traffic congestions. A consultant working for a company that produces technologies for toll booths in France concluded that if service times are measured from the time a car stops in line until it leaves, service times are exponentially distributed with a mean of 2.7 minutes. What proportion of cars can get through the toll booths in less than 3 minutes?

8.76 The manager of a petrol station has observed that the times required by drivers to fill their car's tank and pay are quite variable. In fact, the times are exponentially distributed with a mean of 7.5 minutes. What is the probability that a car can complete the transaction in less than 5 minutes?

8.77 Because automated teller machine (ATM) or cash machine customers can perform a number of transactions, the times to complete them can be quite variable. A banking consultant has noted that the times are exponentially distributed with a mean of 125 seconds. What proportion of the ATM customers take more than 3 minutes to do their banking?

8.78 The manager of a busy supermarket in Riyadh tracked the amount of time needed for customers to be served by the cashier. After checking with a local statistics lecturer, he concluded that the checkout times are exponentially distributed with a mean of 6 minutes. What proportion of customers requires more than 10 minutes to check out?

8-4 OTHER CONTINUOUS DISTRIBUTIONS

In this section, we introduce three more continuous distributions that are used extensively in statistical inference.

8-4a Student *t* Distribution

The Student *t* distribution was first derived by William S. Gosset in 1908. (Gosset published his findings under the pseudonym 'Student' and used the letter *t* to represent the random variable, hence the **Student *t* distribution** – also called the *Student's t distribution*). It is very commonly used in statistical inference, and we will employ it in Chapters 12, 13, 15 and 16.

Student *t* Density Function

The density function of the Student *t* distribution is as follows:

$$f(t) = \frac{\Gamma[(\nu + 1)/2]}{\sqrt{\nu\pi}\,\Gamma(\nu/2)}\left[1 + \frac{t^2}{\nu}\right]^{-(\nu+1)/2}$$

where ν (Greek letter *nu*) is the parameter of the Student *t* distribution called the **degrees of freedom**, $\pi = 3.14159$ (approximately), and Γ is the gamma function (its definition is not needed here).

The mean and variance of a Student *t* random variable are

$$E(t) = 0$$

and

$$V(t) = \frac{\nu}{\nu - 2} \quad \text{for } \nu > 2$$

Figure 8.23 depicts the Student *t* distribution. As you can see, it is similar to the standard normal distribution. Both are symmetrical about 0. (Both random variables have a mean of 0.) We describe the Student *t* distribution as mound shaped, whereas the normal distribution is bell shaped.

FIGURE 8.23

Student *t* Distribution

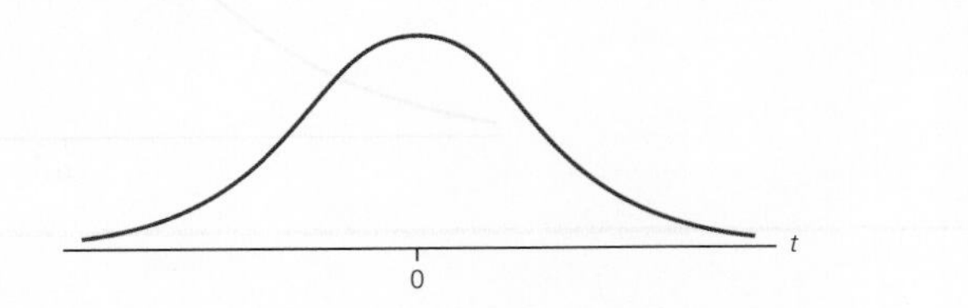

Figure 8.24 shows both the Student *t* and the standard normal distributions. The former is more widely spread out than the latter. [The variance of a standard normal random variable is 1, whereas the variance of a Student *t* random variable is $\nu/(\nu - 2)$, which is greater than 1 for all ν.]

FIGURE 8.24

Student *t* and Normal Distributions

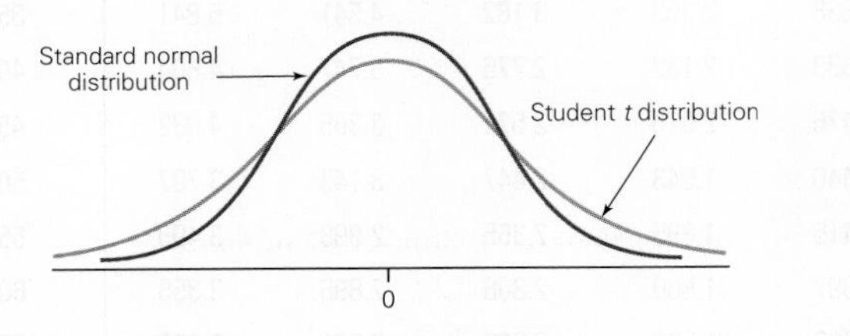

Figure 8.25 depicts Student *t* distributions with several different degrees of freedom. Notice that for larger degrees of freedom the Student *t* distribution's dispersion is smaller. For example, when $\nu = 10$, $V(t) = 1.25$; when $\nu = 50$, $V(t) = 1.042$; and when $\nu = 200$, $V(t) = 1.010$. As ν grows larger, the Student *t* distribution approaches the standard normal distribution.

FIGURE 8.25

Student *t* Distribution with $\nu = 2$, 10 and 30

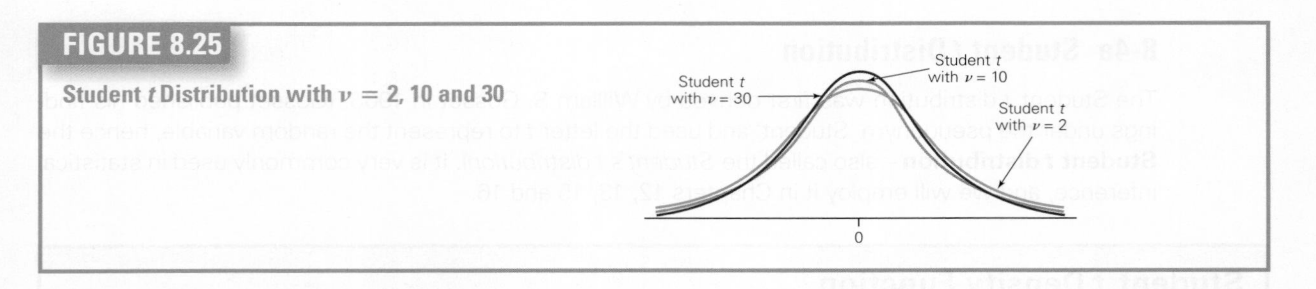

Student t Probabilities For each value of *v* (the number of degrees of freedom), there is a different Student *t* distribution. If we wanted to calculate probabilities of the Student *t* random variable manually as we did for the normal random variable, then we would need a different table for each *v*, which is not practical. Alternatively, we can use Microsoft Excel or Minitab. The instructions are given later in this section.

Determining Student t Values As you will discover later in this book, the Student *t* distribution is used extensively in statistical inference. And for inferential methods, we often need to find values of the random variable. To determine values of a normal random variable, we used Table 3 backwards. Finding values of a Student *t* random variable is considerably easier. Table 8.2 lists values of $t_{A,\nu}$, which are the values of a Student *t* random variable with *v* degrees of freedom such that

$$P(t > t_{A,\nu}) = A$$

Figure 8.26 depicts this notation.

FIGURE 8.26

Student *t* Distribution with t_A

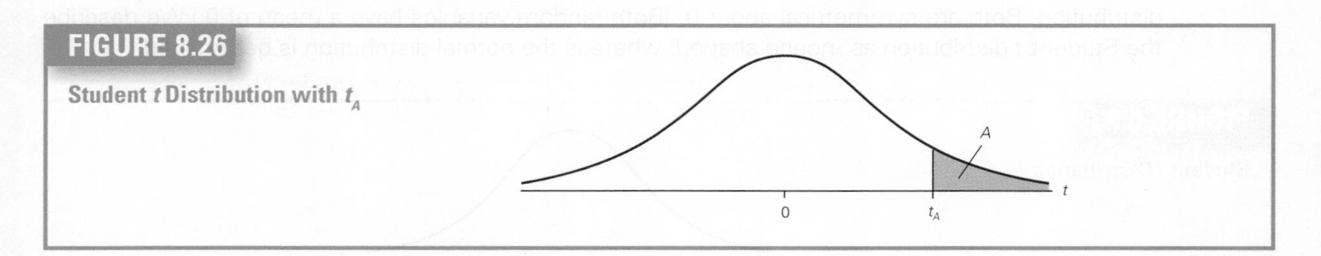

TABLE 8.2 **Critical Values of *t***

ν	$t_{0.100}$	$t_{0.050}$	$t_{0.025}$	$t_{0.010}$	$t_{0.005}$	ν	$t_{0.100}$	$t_{0.050}$	$t_{0.025}$	$t_{0.010}$	$t_{0.005}$
1	3.078	6.314	12.71	31.82	63.66	29	1.311	1.699	2.045	2.462	2.756
2	1.886	2.920	4.303	6.965	9.925	30	1.310	1.697	2.042	2.457	2.750
3	1.638	2.353	3.182	4.541	5.841	35	1.306	1.690	2.030	2.438	2.724
4	1.533	2.132	2.776	3.747	4.604	40	1.303	1.684	2.021	2.423	2.704
5	1.476	2.015	2.571	3.365	4.032	45	1.301	1.679	2.014	2.412	2.690
6	1.440	1.943	2.447	3.143	3.707	50	1.299	1.676	2.009	2.403	2.678
7	1.415	1.895	2.365	2.998	3.499	55	1.297	1.673	2.004	2.396	2.668
8	1.397	1.860	2.306	2.896	3.355	60	1.296	1.671	2.000	2.390	2.660
9	1.383	1.833	2.262	2.821	3.250	65	1.295	1.669	1.997	2.385	2.654

ν	$t_{0.100}$	$t_{0.050}$	$t_{0.025}$	$t_{0.010}$	$t_{0.005}$	ν	$t_{0.100}$	$t_{0.050}$	$t_{0.025}$	$t_{0.010}$	$t_{0.005}$
10	1.372	1.812	2.228	2.764	3.169	**70**	1.294	1.667	1.994	2.381	2.648
11	1.363	1.796	2.201	2.718	3.106	**75**	1.293	1.665	1.992	2.377	2.643
12	1.356	1.782	2.179	2.681	3.055	**80**	1.292	1.664	1.990	2.374	2.639
13	1.350	1.771	2.160	2.650	3.012	**85**	1.292	1.663	1.988	2.371	2.635
14	1.345	1.761	2.145	2.624	2.977	**90**	1.291	1.662	1.987	2.368	2.632
15	1.341	1.753	2.131	2.602	2.947	**95**	1.291	1.661	1.985	2.366	2.629
16	1.337	1.746	2.120	2.583	2.921	**100**	1.290	1.660	1.984	2.364	2.626
17	1.333	1.740	2.110	2.567	2.898	**110**	1.289	1.659	1.982	2.361	2.621
18	1.330	1.734	2.101	2.552	2.878	**120**	1.289	1.658	1.980	2.358	2.617
19	1.328	1.729	2.093	2.539	2.861	**130**	1.288	1.657	1.978	2.355	2.614
20	1.325	1.725	2.086	2.528	2.845	**140**	1.288	1.656	1.977	2.353	2.611
21	1.323	1.721	2.080	2.518	2.831	**150**	1.287	1.655	1.976	2.351	2.609
22	1.321	1.717	2.074	2.508	2.819	**160**	1.287	1.654	1.975	2.350	2.607
23	1.319	1.714	2.069	2.500	2.807	**170**	1.287	1.654	1.974	2.348	2.605
24	1.318	1.711	2.064	2.492	2.797	**180**	1.286	1.653	1.973	2.347	2.603
25	1.316	1.708	2.060	2.485	2.787	**190**	1.286	1.653	1.973	2.346	2.602
26	1.315	1.706	2.056	2.479	2.779	**200**	1.286	1.653	1.972	2.345	2.601
27	1.314	1.703	2.052	2.473	2.771	∞	1.282	1.645	1.960	2.326	2.576
28	1.313	1.701	2.048	2.467	2.763						

Observe that $t_{A,\nu}$ is provided for degrees of freedom ranging from 1 to 200 and ∞. To read this table, simply identify the degrees of freedom and find that value or the closest number to it if it is not listed. Then locate the column representing the t_A value you wish. For example, if we want the value of t with 10 degrees of freedom such that the area under the Student t curve is 0.05, we locate 10 in the first column and move across this row until we locate the number under the heading $t_{0.05}$. From Table 8.3, we find

$$t_{0.05,\,10} = 1.812$$

If the number of degrees of freedom is not shown, find its closest value. For example, suppose we wanted to find $t_{0.025,\,32}$. Because 32 degrees of freedom is not listed, we find the closest number of degrees of freedom, which is 30 and use $t_{0.025,\,30} = 2.042$ as an approximation.

TABLE 8.3 **Finding $t_{0.05,10}$**

DEGREES OF FREEDOM	$t_{0.10}$	$t_{0.05}$	$t_{0.025}$	$t_{0.01}$	$t_{0.005}$
1	3.078	6.314	12.706	31.821	63.657
2	1.886	2.920	4.303	6.965	9.925
3	1.638	2.353	3.182	4.541	5.841
4	1.533	2.132	2.776	3.747	4.604
5	1.476	2.015	2.571	3.365	4.032
6	1.440	1.943	2.447	3.143	3.707
7	1.415	1.895	2.365	2.998	3.499
8	1.397	1.860	2.306	2.896	3.355
9	1.383	1.833	2.262	2.821	3.250
10	1.372	1.812	2.228	2.764	3.169
11	1.363	1.796	2.201	2.718	3.106
12	1.356	1.782	2.179	2.681	3.055

Because the Student t distribution is symmetric about 0, the value of t such that the area to its *left is A* is $-t_{A,\nu}$. For example, the value of t with 10 degrees of freedom such that the area to its left is 0.05 is

$$-t_{0.05,10} = -1.812$$

Notice the last row in the Student t table. The number of degrees of freedom is infinite, and the t values are identical (except for the number of decimal places) to the values of z. For example,

$$t_{0.10,\infty} = 1.282$$
$$t_{0.05,\infty} = 1.645$$
$$t_{0.025,\infty} = 1.960$$
$$t_{0.01,\infty} = 2.326$$
$$t_{0.005,\infty} = 2.576$$

In the previous section, we showed (or showed how we determine) that

$$z_{0.10} = 1.28$$
$$z_{0.05} = 1.645$$
$$z_{0.025} = 1.96$$
$$z_{0.01} = 2.23$$
$$z_{0.005} = 2.575$$

Using the Computer

EXCEL *Instructions*

To compute Student t probabilities, type

= **T.DIST**(x, deg_freedom, cumulative)

which returns the left Student's t-distribution.

where x must be positive, *deg_freedom* is the number of degrees of freedom and *cumulative* is a logic value: for the cumulative distribution function, use TRUE; for the probability density function, use FALSE.

= **T. DIST.RT**(*x, deg_freedom*)

which returns the right-tailed Student's t-distribution and

= **T.DIST.2T**(*x, deg_freedom*)

which returns the two-tailed Student's t-distribution.

For example,

= **T.DIST.RT**(2, 50) = 0.02547

and

= **T.DIST.2T**(2, 50) = 0.05095

To determine the two-tailed inverse of the Student's t-distribution, t_A, type

= **T.INV.2T**(probability, deg_freedom)

For example, to find $t_{0.05,200}$ enter

= **T.INV.2T**(0.10, 200)

yielding 1.6525.

MINITAB *Instructions*

Click **Calc, Probability Distributions,** and **t** ... and type the **Degrees of freedom.**

Chi-Squared Distribution The density function of another very useful random variable is exhibited next.

Chi-Squared Density Function

The chi-squared density function is

$$f(\chi^2) = \frac{1}{\Gamma(\nu/2)} \frac{1}{2^{\nu/2}} (\chi^2)^{(\nu/2)-1} e^{-\chi^2/2} \quad \chi^2 > 0$$

The parameter ν is the number of degrees of freedom, which like the degrees of freedom of the Student *t* distribution affects the shape.

Figure 8.27 depicts a **chi-squared distribution**. As you can see, it is positively skewed ranging between 0 and ∞. Like that of the Student *t* distribution, its shape depends on its number of degrees of freedom. The effect of increasing the degrees of freedom is seen in Figure 8.28.

FIGURE 8.27
Chi-Squared Distribution

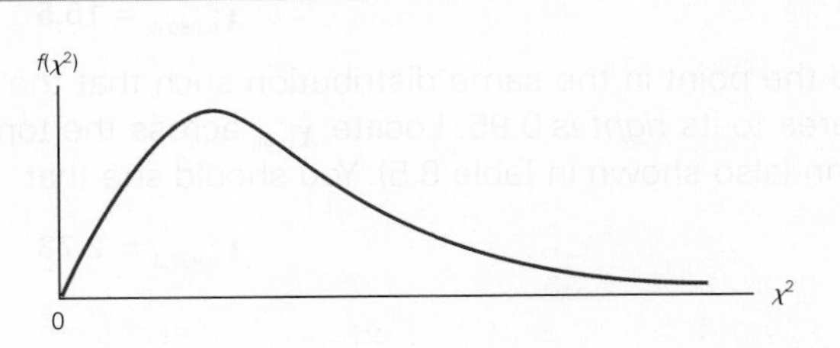

FIGURE 8.28
Chi-Squared Distribution with ν = 1, 5 and 10

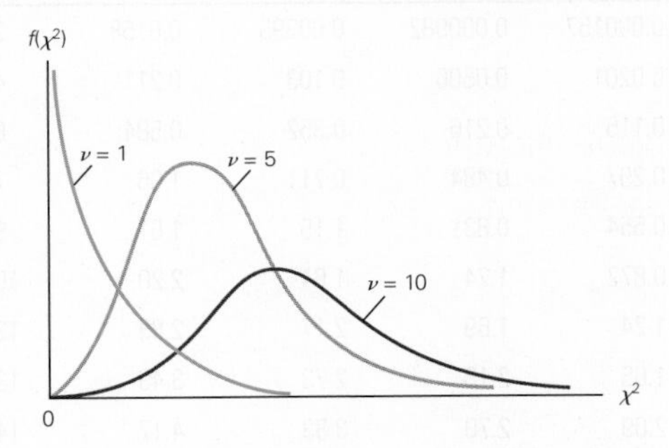

The mean and variance of a chi-squared random variable are

$$E(x^2) = v$$

and

$$V(x^2) = 2v$$

Determining Chi-Squared Values The value of χ^2 with v degrees of freedom such that the area to its right under the chi-squared curve is equal to A is denoted $\chi^2_{A,\nu}$. We cannot use $-\chi^2_{A,\nu}$ to represent the point such that the area to its *left* is A (as we did with the standard normal and Student t values) because χ^2 is always greater than 0. To represent left-tail critical values, we note that if the area to the left of a point is A, the area to its right must be $1 - A$ because the entire area under the chi-squared curve (as well as all continuous distributions) must equal 1. Thus, $\chi^2_{A,\nu}$ denotes the point such that the area to its left is A. See Figure 8.29.

FIGURE 8.29 χ^2_A and χ^2_{1-A}

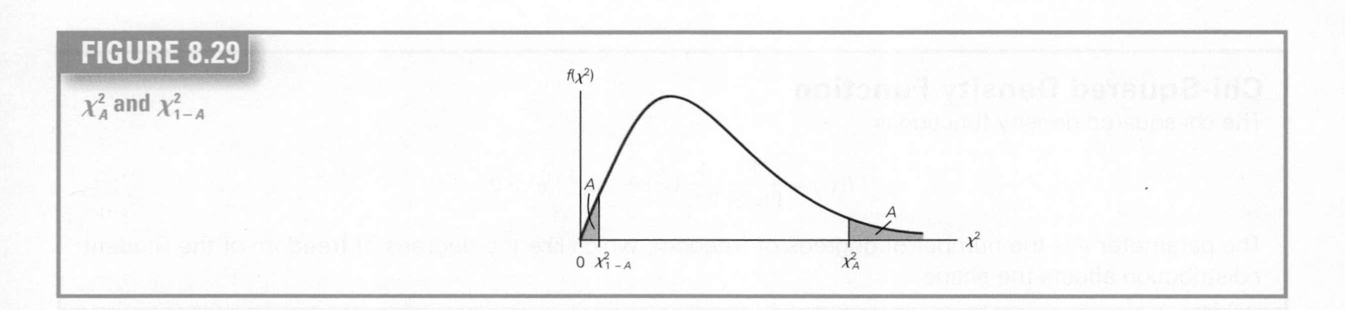

Table 8.4 lists critical values of the chi-squared distribution for degrees of freedom equal to 1 to 30, 40, 50, 60, 70, 80, 90 and 100. For example, to find the point in a chi-squared distribution with 8 degrees of freedom such that the area to its right is 0.05, locate 8 degrees of freedom in the left column and $\chi^2_{0.050}$ across the top. The intersection of the row and column contains the number we seek as shown in Table 8.5; that is,

$$\chi^2_{0.050,8} = 15.5$$

To find the point in the same distribution such that the area to its *left* is 0.05, find the point such that the area to its *right is* 0.95. Locate $\chi_{0.95}$ across the top row and 8 degrees of freedom down the left column (also shown in Table 8.5). You should see that

$$\chi^2_{0.950,8} = 2.73$$

TABLE 8.4 Critical Values of χ^2

ν	$\chi^2_{0.995}$	$\chi^2_{0.990}$	$\chi^2_{0.975}$	$\chi^2_{0.950}$	$\chi^2_{0.900}$	$\chi^2_{0.100}$	$\chi^2_{0.050}$	$\chi^2_{0.025}$	$\chi^2_{0.010}$	$\chi^2_{0.005}$
1	0.000039	0.000157	0.000982	0.00393	0.0158	2.71	3.84	5.02	6.63	7.88
2	0.0100	0.0201	0.0506	0.103	0.211	4.61	5.99	7.38	9.21	10.6
3	0.072	0.115	0.216	0.352	0.584	6.25	7.81	9.35	11.3	12.8
4	0.207	0.297	0.484	0.711	1.06	7.78	9.49	11.1	13.3	14.9
5	0.412	0.554	0.831	1.15	1.61	9.24	11.1	12.8	15.1	16.7
6	0.676	0.872	1.24	1.64	2.20	10.6	12.6	14.4	16.8	18.5
7	0.989	1.24	1.69	2.17	2.83	12.0	14.1	16.0	18.5	20.3
8	1.34	1.65	2.18	2.73	3.49	13.4	15.5	17.5	20.1	22.0
9	1.73	2.09	2.70	3.33	4.17	14.7	16.9	19.0	21.7	23.6
10	2.16	2.56	3.25	3.94	4.87	16.0	18.3	20.5	23.2	25.2
11	2.60	3.05	3.82	4.57	5.58	17.3	19.7	21.9	24.7	26.8
12	3.07	3.57	4.40	5.23	6.30	18.5	21.0	23.3	26.2	28.3
13	3.57	4.11	5.01	5.89	7.04	19.8	22.4	24.7	27.7	29.8
14	4.07	4.66	5.63	6.57	7.79	21.1	23.7	26.1	29.1	31.3
15	4.60	5.23	6.26	7.26	8.55	22.3	25.0	27.5	30.6	32.8

ν	$\chi^2_{0.995}$	$\chi^2_{0.990}$	$\chi^2_{0.975}$	$\chi^2_{0.950}$	$\chi^2_{0.900}$	$\chi^2_{0.100}$	$\chi^2_{0.050}$	$\chi^2_{0.025}$	$\chi^2_{0.010}$	$\chi^2_{0.005}$
16	5.14	5.81	6.91	7.96	9.31	23.5	26.3	28.8	32.0	34.3
17	5.70	6.41	7.56	8.67	10.09	24.8	27.6	30.2	33.4	35.7
18	6.26	7.01	8.23	9.39	10.86	26.0	28.9	31.5	34.8	37.2
19	6.84	7.63	8.91	10.12	11.65	27.2	30.1	32.9	36.2	38.6
20	7.43	8.26	9.59	10.85	12.44	28.4	31.4	34.2	37.6	40.0
21	8.03	8.90	10.28	11.59	13.24	29.6	32.7	35.5	38.9	41.4
22	8.64	9.54	10.98	12.34	14.04	30.8	33.9	36.8	40.3	42.8
23	9.26	10.20	11.69	13.09	14.85	32.0	35.2	38.1	41.6	44.2
24	9.89	10.86	12.40	13.85	15.66	33.2	36.4	39.4	43.0	45.6
25	10.52	11.52	13.12	14.61	16.47	34.4	37.7	40.6	44.3	46.9
26	11.16	12.20	13.84	15.38	17.29	35.6	38.9	41.9	45.6	48.3
27	11.81	12.88	14.57	16.15	18.11	36.7	40.1	43.2	47.0	49.6
28	12.46	13.56	15.31	16.93	18.94	37.9	41.3	44.5	48.3	51.0
29	13.12	14.26	16.05	17.71	19.77	39.1	42.6	45.7	49.6	52.3
30	13.79	14.95	16.79	18.49	20.60	40.3	43.8	47.0	50.9	53.7
40	20.71	22.16	24.43	26.51	29.05	51.8	55.8	59.3	63.7	66.8
50	27.99	29.71	32.36	34.76	37.69	63.2	67.5	71.4	76.2	79.5
60	35.53	37.48	40.48	43.19	46.46	74.4	79.1	83.3	88.4	92.0
70	43.28	45.44	48.76	51.74	55.33	85.5	90.5	95.0	100	104
80	51.17	53.54	57.15	60.39	64.28	96.6	102	107	112	116
90	59.20	61.75	65.65	69.13	73.29	108	113	118	124	128
100	67.33	70.06	74.22	77.93	82.36	118	124	130	136	140

TABLE 8.5 Critical Values of $\chi^2_{0.05,8}$ and $\chi^2_{0.950,8}$

DEGREES OF FREEDOM	$\chi^2_{0.995}$	$\chi^2_{0.990}$	$\chi^2_{0.975}$	$\chi^2_{0.950}$	$\chi^2_{0.900}$	$\chi^2_{0.100}$	$\chi^2_{0.050}$	$\chi^2_{0.025}$	$\chi^2_{0.010}$	$\chi^2_{0.005}$
1	0.000039	0.000157	0.000982	0.00393	0.0158	2.71	3.84	5.02	6.63	7.88
2	0.0100	0.0201	0.0506	0.103	0.211	4.61	5.99	7.38	9.21	10.6
3	0.072	0.115	0.216	0.352	0.584	6.25	7.81	9.35	11.3	12.8
4	0.207	0.297	0.484	0.711	1.06	7.78	9.49	11.1	13.3	14.9
5	0.412	0.554	0.831	1.15	1.61	9.24	11.1	12.8	15.1	16.7
6	0.676	0.872	1.24	1.64	2.20	10.6	12.6	14.4	16.8	18.5
7	0.989	1.24	1.69	2.17	2.83	12.0	14.1	16.0	18.5	20.3
8	1.34	1.65	2.18	2.73	3.49	13.4	15.5	17.5	20.1	22.0
9	1.73	2.09	2.70	3.33	4.17	14.7	16.9	19.0	21.7	23.6
10	2.16	2.56	3.25	3.94	4.87	16.0	18.3	20.5	23.2	25.2
11	2.60	3.05	3.82	4.57	5.58	17.3	19.7	21.9	24.7	26.8

For values of degrees of freedom greater than 100, the chi-squared distribution can be approximated by a normal distribution with $\mu = \nu$ and $\sigma = \sqrt{2\nu}$.

EXCEL *Instructions*

To calculate $P(\chi^2 > x)$, type into any cell

= **CHISQ.DIST**(*x, deg_freedom, cumulative*)

which returns the left-tailed probability of the chi-squared distribution

= **CHISQ.DIST.RT**(*x, deg_freedom*)

which returns the right tailed probability of the chi-squared distribution

For example, **CHISQ.DIST.RT**(6.25,3) = 0.100

To determine $\chi_{A,\nu}$, type

= **CHISQ.INV**(*probability, deg_freedom*)

For example, = **CHIINV**(0.10, 3) = 6.25

MINITAB *Instructions*

Click **Calc**, **Probability Distributions** and **Chi-square** .... Specify the **Degrees of freedom.**

8-4b *F* Distribution

The density function of the *F* distribution is given in the following box.

F Density Function

$$f(F) = \frac{\Gamma\left(\frac{\nu_1 + \nu_2}{2}\right)}{\Gamma\left(\frac{\nu_1}{2}\right)\Gamma\left(\frac{\nu_2}{2}\right)} \left(\frac{\nu_1}{\nu_2}\right)^{\frac{\nu_1}{2}} \frac{F^{\frac{\nu_1 - 2}{2}}}{\left(1 + \frac{\nu_1 F}{\nu_2}\right)^{\frac{\nu_1 + \nu_2}{2}}} \quad F > 0$$

where *F* ranges from 0 to ∞ and ν_1 and ν_2 are the parameters of the distribution called degrees of freedom. For reasons that are clearer in Chapter 12, we call ν_1 the *numerator degrees of freedom* and ν_2 the *denominator degrees of freedom.*

The mean and variance of an *F* random variable are

$$E(F) = \frac{\nu_2}{\nu_2 - 2} \quad \nu_2 > 2$$

and

$$V(F) = \frac{2\nu_2^2(\nu_1 + \nu_2 - 2)}{\nu_1(\nu_2 - 2)^2(\nu_2 - 4)} \quad \nu_2 > 4$$

Notice that the mean depends only on the denominator degrees of freedom and that for large ν_2 the mean of the *F* distribution is approximately 1. Figure 8.30 describes the density function when it is graphed. As you can see, the *F* distribution is positively skewed. Its actual shape depends on the two numbers of degrees of freedom.

FIGURE 8.30

F **Distribution**

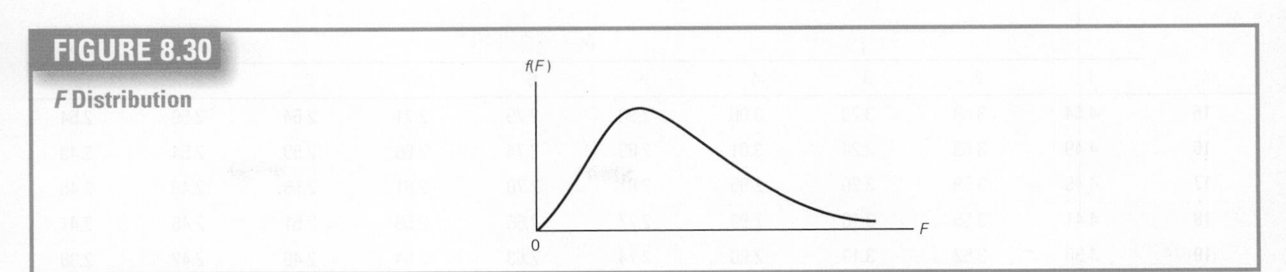

Determining Values of F We define F_{A,ν_1,ν_2} as the value of *F* with ν_1 and ν_2, degrees of freedom such that the area to its right under the curve is *A*; that is,

$$P(F > F_{A,\nu_1,\nu_2}) = A$$

Because the *F* random variable like the chi-squared can equal only positive values, we define F_{1-A,ν_1,ν_2} as the value such that the area to its left is *A*. Figure 8.31 depicts this notation. Table 6 in Appendix A provides values of F_{A,ν_1,ν_2} for $A = 0.05, 0.025, 0.01$ and 0.005. Part of Table 6 is reproduced here as Table 8.6.

FIGURE 8.31

F_{1-A} **and** F_A

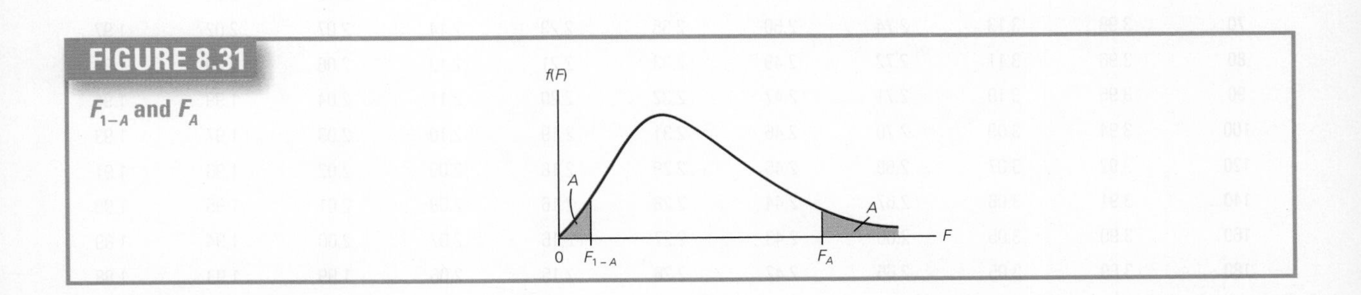

TABLE 8.6 **Critical Values of F_A for $A = 0.05$**

ν_2 \ ν_1	1	2	3	4	5	6	7	8	9	10
1	161	199	216	225	230	234	237	239	241	242
2	18.5	19.0	19.2	19.2	19.3	19.3	19.4	19.4	19.4	19.4
3	10.1	9.55	9.28	9.12	9.01	8.94	8.89	8.85	8.81	8.79
4	7.71	6.94	6.59	6.39	6.26	6.16	6.09	6.04	6.00	5.96
5	6.61	5.79	5.41	5.19	5.05	4.95	4.88	4.82	4.77	4.74
6	5.99	5.14	4.76	4.53	4.39	4.28	4.21	4.15	4.10	4.06
7	5.59	4.74	4.35	4.12	3.97	3.87	3.79	3.73	3.68	3.64
8	5.32	4.46	4.07	3.84	3.69	3.58	3.50	3.44	3.39	3.35
9	5.12	4.26	3.86	3.63	3.48	3.37	3.29	3.23	3.18	3.14
10	4.96	4.10	3.71	3.48	3.33	3.22	3.14	3.07	3.02	2.98
11	4.84	3.98	3.59	3.36	3.20	3.09	3.01	2.95	2.90	2.85
12	4.75	3.89	3.49	3.26	3.11	3.00	2.91	2.85	2.80	2.75
13	4.67	3.81	3.41	3.18	3.03	2.92	2.83	2.77	2.71	2.67
14	4.60	3.74	3.34	3.11	2.96	2.85	2.76	2.70	2.65	2.60

(continued)

	ν_1									
ν_2	1	2	3	4	5	6	7	8	9	10
15	4.54	3.68	3.29	3.06	2.90	2.79	2.71	2.64	2.59	2.54
16	4.49	3.63	3.24	3.01	2.85	2.74	2.66	2.59	2.54	2.49
17	4.45	3.59	3.20	2.96	2.81	2.70	2.61	2.55	2.49	2.45
18	4.41	3.55	3.16	2.93	2.77	2.66	2.58	2.51	2.46	2.41
19	4.38	3.52	3.13	2.90	2.74	2.63	2.54	2.48	2.42	2.38
20	4.35	3.49	3.10	2.87	2.71	2.60	2.51	2.45	2.39	2.35
22	4.30	3.44	3.05	2.82	2.66	2.55	2.46	2.40	2.34	2.30
24	4.26	3.40	3.01	2.78	2.62	2.51	2.42	2.36	2.30	2.25
26	4.23	3.37	2.98	2.74	2.59	2.47	2.39	2.32	2.27	2.22
28	4.20	3.34	2.95	2.71	2.56	2.45	2.36	2.29	2.24	2.19
30	4.17	3.32	2.92	2.69	2.53	2.42	2.33	2.27	2.21	2.16
35	4.12	3.27	2.87	2.64	2.49	2.37	2.29	2.22	2.16	2.11
40	4.08	3.23	2.84	2.61	2.45	2.34	2.25	2.18	2.12	2.08
45	4.06	3.20	2.81	2.58	2.42	2.31	2.22	2.15	2.10	2.05
50	4.03	3.18	2.79	2.56	2.40	2.29	2.20	2.13	2.07	2.03
60	4.00	3.15	2.76	2.53	2.37	2.25	2.17	2.10	2.04	1.99
70	3.98	3.13	2.74	2.50	2.35	2.23	2.14	2.07	2.02	1.97
80	3.96	3.11	2.72	2.49	2.33	2.21	2.13	2.06	2.00	1.95
90	3.95	3.10	2.71	2.47	2.32	2.20	2.11	2.04	1.99	1.94
100	3.94	3.09	2.70	2.46	2.31	2.19	2.10	2.03	1.97	1.93
120	3.92	3.07	2.68	2.45	2.29	2.18	2.09	2.02	1.96	1.91
140	3.91	3.06	2.67	2.44	2.28	2.16	2.08	2.01	1.95	1.90
160	3.90	3.05	2.66	2.43	2.27	2.16	2.07	2.00	1.94	1.89
180	3.89	3.05	2.65	2.42	2.26	2.15	2.06	1.99	1.93	1.88
200	3.89	3.04	2.65	2.42	2.26	2.14	2.06	1.98	1.93	1.88
∞	3.84	3.00	2.61	2.37	2.21	2.10	2.01	1.94	1.88	1.83

Values of F_{1-A,ν_1,ν_2} are unavailable. However, we do not need them because we can determine $F_{A,\,\nu_1,\nu_2}$ from F_A. Statisticians can show that

$$F_{1-A,\nu_1,\nu_2} = \frac{1}{F_{A,\nu_1,\nu_2}}$$

To determine any critical value, find the numerator degrees of freedom ν_1 across the top of Table 6 and the denominator degrees of freedom ν_2 down the left column. The intersection of the row and column contains the number we seek. To illustrate, suppose that we want to find $F_{0.05,\,5,7}$. Table 8.7 shows how this point is found. Locate the numerator degrees of freedom, 5, across the top and the denominator degrees of freedom, 7, down the left column. The intersection is 3.97. Thus, $F_{0.05,\,5,7} = 3.97$.

TABLE 8.7 $F_{0.05,\,5,7}$

DENOMINATOR DEGREES OF FREEDOM ν_2 \ NUMERATOR DEGREES OF FREEDOM ν_1	1	2	3	4	5	6	7	8	9
1	161	199	216	225	230	234	237	239	241
2	18.5	19.0	19.2	19.2	19.3	19.3	19.4	19.4	19.4
3	10.1	9.55	9.28	9.12	9.01	8.94	8.89	8.85	8.81
4	7.71	6.94	6.59	6.39	6.26	6.16	6.09	6.04	6.00
5	6.61	5.79	5.41	5.19	5.05	4.95	4.88	4.82	4.77
6	5.99	5.14	4.76	4.53	4.39	4.28	4.21	4.15	4.1
7	5.59	4.74	4.35	4.12	3.97	3.87	3.79	3.73	3.68
8	5.32	4.46	4.07	3.84	3.69	3.58	3.5	3.44	3.39
9	5.12	4.26	3.86	3.63	3.48	3.37	3.29	3.23	3.18
10	4.96	4.10	3.71	3.48	3.33	3.22	3.14	3.07	3.02

Note that the order in which the degrees of freedom appear is important. To find $F_{0.05,\,5,7}$ (numerator degrees of freedom = 7 and denominator degrees of freedom = 5), we locate 7 across the top and 5 down the side. The intersection is $F_{0.05,\,5,7} = 4.88$.

Suppose that we want to determine the point in an F distribution with $\nu_1 = 4$ and $\nu_2 = 8$ such that the area to its right is 0.95. Thus,

$$F_{0.95,4,8} = \frac{1}{F_{0.05,8,4}} = \frac{1}{6.04} = 0.166$$

Using the Computer

EXCEL *Instructions*

For probabilities, type

= **F.DIST.RT**([X], [ν_1], [ν_2])

which returns the right-tailed F probability distribution.

= **F.DIST**(*X, deg_freedom1, deg_freedom2, cumulative*)

returns the left-tailed F probability distribution.

For example, = **F.DIST.RT**(3.97, 5,7) = 0.05.

To determine $F_{A\nu_1,\nu_2}$, type

= **F.INV.RT**[A], [ν_1], [ν_2])

which returns the inverse of the (right-tailed) F probability distribution.

For example, = **F.INV.RT**(0.05, 5,7) = 3.97.

MINITAB *Instructions*

Click **Calc, Probability Distributions** and ***F*...**. Specify the **Numerator degrees of freedom** and the **Denominator degrees of freedom**.

EXERCISES

Some of the following exercises require the use of a computer and software.

8.79 Use the t table (Table 4) to find the following values of t.

a. $t_{0.10,15}$ **b.** $t_{0.10,23}$ **c.** $t_{0.025,83}$ **d.** $t_{0.05,195}$

8.80 Use a computer to find the following values of t.

a. $t_{0.10,15}$ **b.** $t_{0.10,23}$ **c.** $t_{0.025,83}$ **d.** $t_{0.05,195}$

8.81 Use a computer to find the following probabilities.

a. $P(t_{64} > 2.12)$ **c.** $P(t_{159} > 1.33)$
b. $P(t_{27} > 1.90)$ **d.** $P(t_{550} > 1.85)$

8.82 Use the χ^2 table (Table 5) to find the following values of χ^2.

a. $\chi^2_{0.10,5}$ **b.** $\chi^2_{0.01,100}$ **c.** $\chi^2_{0.95,18}$ **d.** $\chi^2_{0.99,60}$

8.83 Use a computer to find the following values of χ^2.

a. $\chi^2_{0.25,66}$ **b.** $\chi^2_{0.40,100}$ **c.** $\chi^2_{0.50,17}$ **d.** $\chi^2_{0.10,17}$

8.84 Use a computer to find the following probabilities.

a. $P(\chi^2_{73} > 80)$
b. $P(\chi^2_{200} > 125)$
c. $P(\chi^2_{88} > 60)$
d. $P(\chi^2_{1000} > 450)$

8.85 Use the F table (Table 6) to find the following values of F.

a. $F_{0.05,3,7}$ **b.** $F_{0.05,7,3}$ **c.** $F_{0.025,5,20}$ **d.** $F_{0.01,12,60}$

8.86 Use a computer to find the following values of F.

a. $F_{0.05,70,70}$ **c.** $F_{0.025,36,50}$
b. $F_{0.01,45,100}$ **d.** $F_{0.05,500,500}$

8.87 Use a computer to find the following probabilities.

a. $P(F_{7,20} > 2.5)$ **c.** $P(F_{34,62} > 1.8)$
b. $P(F_{18,63} > 1.4)$ **d.** $P(F_{200,400} > 1.1)$

CHAPTER SUMMARY

This chapter dealt with **continuous random variables** and their distributions. Because a continuous random variable can assume an infinite number of values, the probability that the random variable equals any single value is 0. Consequently, we addressed the problem of computing the probability of a range of values. We showed that the probability of any interval is the area in the interval under the curve representing the **density function**.

We introduced the most important distribution in statistics and showed how to compute the probability that a **normal random variable** falls into any interval. Additionally, we demonstrated how to use the normal table backwards to find values of a normal random variable given a probability. Next we introduced the **exponential distribution**, a distribution that is particularly useful in several management-science applications. Finally, we presented three more continuous random variables and their probability density functions. The **Student** t, **chi-squared** and F **distributions** will be used extensively in statistical inference.

SYMBOLS:

Symbol	Pronounced	Represents
π	pi	3.14159 . . .
z_A	*z-sub-A* or *z-A*	Value of Z such that area to its right is A
ν	nu	Degrees of freedom
t_A	*t-sub-A* or *t-A*	Value of t such that area to its right is A
χ^2_A	chi-squared-sub-*A* or chi-squared-*A*	Value of chi-squared such that area to its right is A
F_A	*F*-sub-*A* or *F-A*	Value of F such that area to its right is A
ν_1	nu-sub-one or nu-one	Numerator degrees of freedom
ν_2	nu-sub-two or nu-two	Denominator degrees of freedom

PART FOUR

SAMPLING THEORY AND INFERENTIAL STATISTICS

9 Sampling Distributions
10 Introduction to Estimation
11 Introduction to Hypothesis Testing
12 Inference about Populations
13 Analysis of Variance
14 Chi-Squared Tests
15 Simple Linear Regression and Correlation
16 Multiple Regression

9 SAMPLING DISTRIBUTIONS

9-1 SAMPLING DISTRIBUTION OF THE MEAN

9-2 SAMPLING DISTRIBUTION OF A PROPORTION

9-3 SAMPLING DISTRIBUTION OF THE DIFFERENCE BETWEEN TWO MEANS

9-4 FROM HERE TO INFERENCE

SALARIES OF A BUSINESS SCHOOL'S GRADUATES

Deans and other faculty members in professional schools often monitor how well the graduates of their programmes fare in the job market. Information about the types of jobs and their salaries may provide useful information about the success of a programme.

In the advertisements for a large university, the dean of the School of Business claims that the average salary of the school's graduates 1 year after graduation is £800 per week, with a standard deviation of £100. A second-year student in the business school who has just completed his statistics course would like to check whether the claim about the mean is correct. He does a survey of 25 people who graduated 1 year earlier and determines their weekly salary. He discovers the sample mean to be £750. To interpret his finding, he needs to calculate the probability that a sample of 25 graduates would have a mean of 750 or less when the population mean is £800 and the standard deviation is £100. After calculating the probability, he needs to draw some conclusion. **See Section 9-1b for the answer.**

INTRODUCTION

This chapter introduces the *sampling distribution*, a fundamental element in statistical inference that is completely different to sample distribution which is a collection of measurements. We remind you that statistical inference is the process of converting data into information. Here are the parts of the process we have thus far discussed:

1. Parameters describe populations.
2. Parameters are almost always unknown.
3. We take a random sample of a population to obtain the necessary data.
4. We calculate one or more statistics from the data.

For example, to estimate a population mean, we compute the sample mean. Although there is very little chance that the sample mean and the population mean are identical, we would expect them to be quite close. However, for the purposes of statistical inference, we need to be able to measure *how* close. The sampling distribution provides this service. It plays a crucial role in the process because the measure of proximity it provides is the key to statistical inference.

9-1 SAMPLING DISTRIBUTION OF THE MEAN

A **sampling distribution** is created by, as the name suggests, sampling. There are two ways to create a sampling distribution. The first is to actually draw samples of the same size from a population, calculate the statistic of interest and then use descriptive techniques to learn more about the sampling distribution. The second method relies on the rules of probability and the laws of expected value and variance to derive the sampling distribution. We will demonstrate the latter approach by developing the sampling distribution of the mean of two dice.

9-1a Sampling Distribution of the Mean of Two Dice

The population is created by throwing a fair die infinitely many times, with the random variable X indicating the number of spots showing on any one throw. The probability distribution of the random variable X is as follows:

x	1	2	3	4	5	6
$P(x)$	1/6	1/6	1/6	1/6	1/6	1/6

The population is infinitely large because we can throw the die infinitely many times (or at least imagine doing so). From the definitions of expected value and variance presented in Section 7-1, we calculate the population mean, variance and standard deviation.

Population mean:

$$\begin{aligned}\mu &= \sum xP(x)\\ &= 1(1/6) + 2(1/6) + 3(1/6) + 4(1/6) + 5(1/6) + 6(1/6)\\ &= 3.5\end{aligned}$$

Population variance:

$$\begin{aligned}\sigma^2 &= \sum (x-\mu)^2P(x)\\ &= (1-3.5)^2(1/6) + (2-3.5)^2(1/6) + (3-3.5)^2(1/6) + (4-3.5)^2(1/6)\\ &\quad + (5-3.5)^2(1/6) + (6-3.5)^2(1/6)\\ &= 2.92\end{aligned}$$

Population standard deviation:

$$\sigma = \sqrt{\sigma^2} = \sqrt{2.92} = 1.71$$

The sampling distribution is created by drawing samples of size 2 from the population. In other words, we toss two dice. Figure 9.1 depicts this process in which we compute the mean for each sample. Because the value of the sample mean varies randomly from sample to sample, we can regard $\overline{X}$ as a new random variable created by sampling. Table 9.1 lists all the possible samples and their corresponding values of $\overline{X}$.

FIGURE 9.1

Drawing Samples of Size 2 from a Population

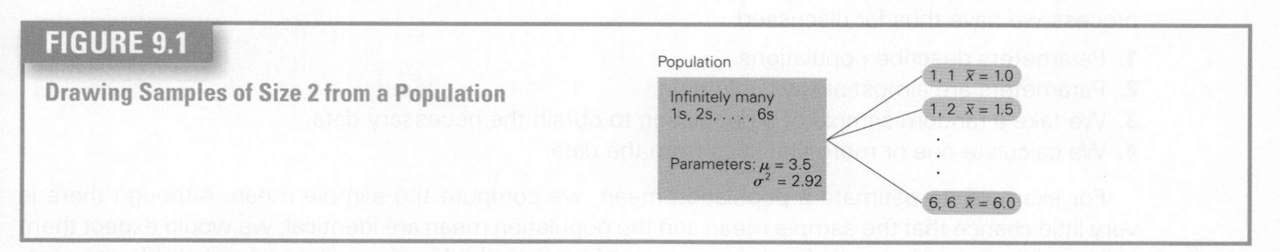

TABLE 9.1 **All Samples of Size 2 and Their Means**

Sample	$\overline{X}$	Sample	$\overline{X}$	Sample	$\overline{X}$
1, 1	1.0	3, 1	2.0	5, 1	3.0
1, 2	1.5	3, 2	2.5	5, 2	3.5
1, 3	2.0	3, 3	3.0	5, 3	4.0
1, 4	2.5	3, 4	3.5	5, 4	4.5
1, 5	3.0	3, 5	4.0	5, 5	5.0
1, 6	3.5	3, 6	4.5	5, 6	5.5
2, 1	1.5	4, 1	2.5	6, 1	3.5
2, 2	2.0	4, 2	3.0	6, 2	4.0
2, 3	2.5	4, 3	3.5	6, 3	4.5
2, 4	3.0	4, 4	4.0	6, 4	5.0
2, 5	3.5	4, 5	4.5	6, 5	5.5
2, 6	4.0	4, 6	5.0	6, 6	6.0

There are 36 different possible samples of size 2; because each sample is equally likely, the probability of any one sample being selected is 1/36. However, $\overline{X}$ can assume only 11 different possible values: 1.0,1.5, 2.0, . . . , 6.0, with certain values of $\overline{X}$ occurring more frequently than others. The value $\overline{X} = \mathbf{1.0}$ occurs only once, so its probability is 1/36. The value $\overline{X} = \mathbf{1.5}$ can occur in two ways–(1, 2) and (2, 1)–each having the same probability (1/36). Thus, $P(\overline{X} = 1.5) = 2/36$. The probabilities of the other values of $\overline{X}$ are determined in similar fashion, and the resulting **sampling distribution of the sample mean** is shown in Table 9.2.

TABLE 9.2 **Sampling Distribution of $\overline{X}$**

$\overline{X}$	$P(\overline{x})$
1.0	1/36
1.5	2/36
2.0	3/36
2.5	4/36
3.0	5/36
3.5	6/36
4.0	5/36
4.5	4/36
5.0	3/36
5.5	2/36
6.0	1/36

The most interesting aspect of the sampling distribution of $\overline{X}$ is how different it is from the distribution of X, as can be seen in Figure 9.2.

FIGURE 9.2

Distributions of X and $\overline{X}$

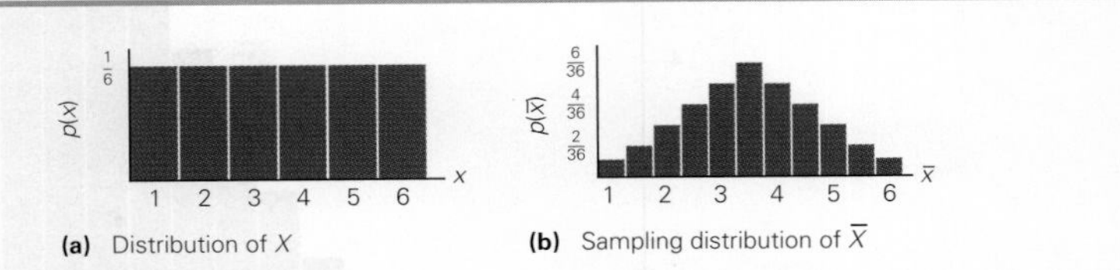

(a) Distribution of X

(b) Sampling distribution of $\overline{X}$

We can also compute the mean, variance and standard deviation of the sampling distribution. Once again using the definitions of expected value and variance, we determine the following parameters of the sampling distribution.

Mean of the sampling distribution of $\overline{X}$:

$$\mu_{\bar{x}} = \sum \bar{x}P(\bar{x})$$
$$= 1.0(1/36) + 1.5(2/36) + \cdots + 6.0(1/36)$$
$$= 3.5$$

Notice that the mean of the sampling distribution of $\overline{X}$ is equal to the mean of the population of the toss of a die computed previously.

Variance of the sampling distribution of $\overline{X}$:

$$\sigma^2_{\bar{x}} = \sum(\bar{x} - \mu_{\bar{x}})^2 P(\bar{x})$$
$$= (1.0 - 3.5)^2(1/36) + (1.5 - 3.5)^2(2/36) + \cdots + (6.0 - 3.5)^2(1/36)$$
$$= 1.46$$

It is no coincidence that the variance of the sampling distribution of $\overline{X}$ is exactly half of the variance of the population of the toss of a die (computed previously as $\sigma^2 = 2.92$).

Standard deviation of the sampling distribution of $\overline{X}$:

$$\sigma_{\bar{x}} = \sqrt{\sigma^2_{\bar{x}}} = \sqrt{1.46} = 1.21$$

It is important to recognise that the distribution of $\overline{X}$ is different from the distribution of X as depicted in Figure 9.2. However, the two random variables are related. Their means are the same ($\mu_{\bar{x}} = \mu = 3.5$) and their variances are related ($\sigma^2_{\bar{x}} = \sigma^2/2$).

Do not get lost in the terminology and notation. Remember that μ and σ^2 are the parameters of the population of X. To create the sampling distribution of $\overline{X}$, we repeatedly drew samples of size $n = 2$ from the population and calculated $\overline{X}$ for each sample. Thus, we treat $\overline{X}$ as a brand-new random variable, with its own distribution, mean and variance. The mean is denoted $\mu_{\bar{x}}$, and the variance is denoted $\sigma^2_{\bar{x}}$.

If we now repeat the sampling process with the same population but with other values of n, we produce somewhat different sampling distributions of $\overline{X}$. Figure 9.3 shows the sampling distributions of $\overline{X}$ when $n = 5$, 10 and 25.

For each value of n, the mean of the sampling distribution of $\overline{X}$ is the mean of the population from which we are sampling; that is,

$$\mu_{\bar{x}} = \mu = 3.5$$

The variance of the sampling distribution of the sample mean is the variance of the population divided by the sample size:

$$\sigma^2_{\bar{x}} = \frac{\sigma^2}{n}$$

FIGURE 9.3

Sampling Distributions of $\overline{X}$ for $n = 5$, 10 and 25

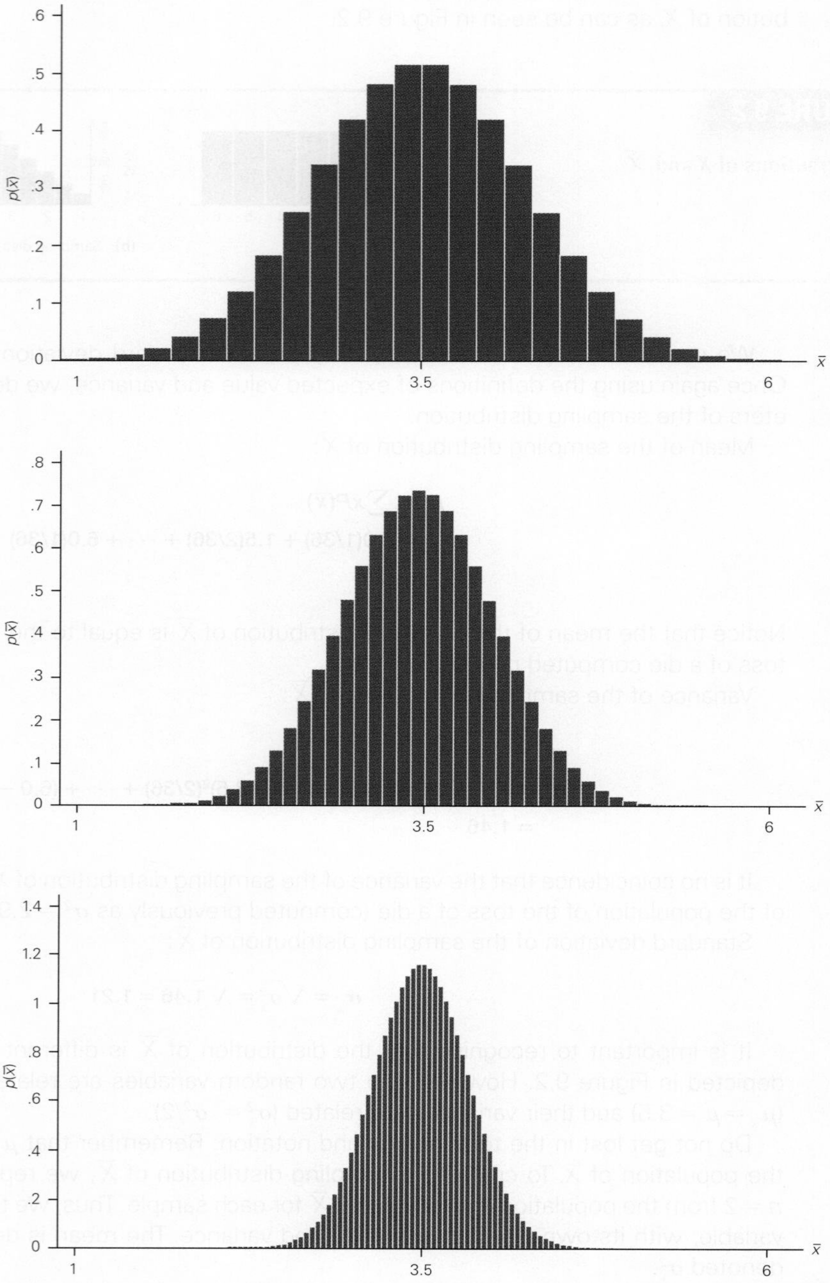

The standard deviation of the sampling distribution is called the **standard error of the mean**; that is,

$$\sigma_{\bar{x}} = \frac{\sigma}{\sqrt{n}}$$

As you can see, the variance of the sampling distribution of $\overline{X}$ is less than the variance of the population we are sampling from all sample sizes. Thus, a randomly selected value of $\overline{X}$ (the mean of the number of spots observed in, say five throws of the die) is likely to be closer to the mean value of 3.5 than is a randomly selected value of X (the number of spots observed in one throw). Indeed, this is what you would expect, because in five throws of the die you are likely to get some 5s and 6s and some 1s and 2s, which will tend to offset one another in the averaging process and produce a sample mean reasonably close to 3.5. As the number of throws of the die increases, the probability that the sample mean will be close to 3.5 also increases. Thus, we observe in Figure 9.3 that the sampling distribution of $\overline{X}$ becomes narrower (or more concentrated about the mean) as n increases.

Another thing that happens as n gets larger is that the sampling distribution of $\overline{X}$ becomes increasingly bell shaped. This phenomenon is summarised in the **central limit theorem**.

Central Limit Theorem

The sampling distribution of the mean of a random sample drawn from any population is approximately normal for a sufficiently large sample size. The larger the sample size, the more closely the sampling distribution of $\overline{X}$ will resemble a normal distribution.

The accuracy of the approximation alluded to in the central limit theorem depends on the probability distribution of the population and on the sample size. If the population is normal, then $\overline{X}$ is normally distributed for all values of n. If the population is non-normal, then $\overline{X}$ is approximately normal only for larger values of n. In many practical situations, a sample size of 30 may be sufficiently large to allow us to use the normal distribution as an approximation for the sampling distribution of $\overline{X}$. However, if the population is extremely non-normal (e.g., bimodal and highly skewed distributions), the sampling distribution will also be non-normal even for moderately large values of n.

Sampling Distribution of the Mean of Any Population We can extend the discoveries we have made to all infinitely large populations. Statisticians have shown that the mean of the sampling distribution is always equal to the mean of the population and that the standard error is equal to $\sigma/\sqrt{n}$ for infinitely large populations. (In the website Appendix Using the Laws of Expected Value and Variance to Derive the Parameters of Sampling Distributions we describe how to mathematically prove that $\mu_{\bar{x}} = \mu$ and $\sigma^2_{\bar{x}} = \sigma^2/n$.) However, if the population is finite the standard error is

$$\sigma_{\bar{x}} = \frac{\sigma}{\sqrt{n}}\sqrt{\frac{N-n}{N-1}}$$

where N is the population size and $\sqrt{\frac{N-n}{N-1}}$ is called the **finite population correction factor**. (The source of the correction factor is provided in the website Appendix Hypergeometric Distribution.) An analysis (see Exercises 9.13 and 9.14) reveals that if the population size is large relative to the sample size, then the finite population correction factor is close to 1 and can be ignored. As a rule of thumb, we will treat any population that is at least 20 times larger than the sample size as large. In practice, most applications involve populations that qualify as large because if the population is small, it may be possible to investigate each member of the population, and in so doing, calculate the parameters precisely. As a consequence, the finite population correction factor is usually omitted.

We can now summarise what we know about the sampling distribution of the sample mean for large populations.

Sampling Distribution of the Sample Mean

1. $\mu_{\bar{x}} = \mu$
2. $\sigma^2_{\bar{x}} = \sigma^2/n$ and $\sigma_{\bar{x}} = \sigma/\sqrt{n}$
3. If X is normal, then $\overline{X}$ is normal. If X is non-normal, then $\overline{X}$ is approximately normal for sufficiently large sample sizes. The definition of 'sufficiently large' depends on the extent of non-normality of X.

9-1b Creating the Sampling Distribution Empirically

In the previous analysis, we created the sampling distribution of the mean theoretically. We did so by listing all the possible samples of size 2 and their probabilities. (They were all equally likely with probability 1/36.) From this distribution, we produced the sampling distribution. We could also create the distribution empirically by actually tossing two fair dice repeatedly, calculating the sample mean for each sample, counting the number of times each value of $\overline{X}$ occurs, and computing the relative frequencies to estimate the theoretical probabilities. If we toss the two dice a large enough number of times, the relative frequencies and theoretical probabilities (computed previously) will be similar. Try it yourself. Toss two dice 500 times, calculate the mean of the two tosses, count the number of times each sample mean occurs and construct the histogram representing the sampling distribution. Obviously, this approach is far from ideal because of the excessive amount of time required to toss the dice enough times to make the relative frequencies good approximations for the theoretical probabilities. However, we can use the computer to quickly simulate tossing dice many times.

EXAMPLE 9.1

Contents of a 2 Litre Bottle

The foreman of a bottling plant has observed that the amount of soda in each 2 litre bottle is actually a normally distributed random variable, with a mean of 2.2 litres and a standard deviation of 0.3 litres.

a. If a customer buys one bottle, what is the probability that the bottle will contain more than 2 litres?

b. If a customer buys a carton of four bottles, what is the probability that the mean amount of the four bottles will be greater than 2 litres?

Solution:

a. Because the random variable is the amount of soda in one bottle, we want to find $P(X > 2)$, where X is normally distributed, $\mu = 2.2$ and $\sigma = 0.3$. Hence,

$$\begin{aligned} P(X > 2) &= P\left(\frac{X-\mu}{\rho} > \frac{2-2.2}{0.3}\right) \\ &= P(z > 0.67) \\ &= 1 - P(Z < -0.67) \\ &= 1 - 0.2514 \\ &= 0.7486 \end{aligned}$$

$$\begin{aligned} P(X > 32) &= P\left(\frac{X-\mu}{\sigma} > \frac{32-32.2}{0.3}\right) \\ &= P(Z > -0.67) \\ &= 1 - P(Z < -0.67) \\ &= 1 - 0.2514 = 0.7486 \end{aligned}$$

b. Now we want to find the probability that the mean amount of four filled bottles exceeds 2 litres; that is, we want $(\overline{X} > 2)$. From our previous analysis and from the central limit theorem, we know the following:

1. $\overline{X}$ is normally distributed.
2. $\mu_{\overline{x}} = \mu = 2.2$
3. $\sigma_{\overline{x}} = \dfrac{\sigma}{\sqrt{n}} = \dfrac{0.3}{\sqrt{4}} = 0.15$

Hence,

$$P(\overline{X} > 2) = P\left(\frac{\overline{X}-\mu_{\overline{x}}}{\sigma_{\overline{x}}} > \frac{2-2.2}{0.15}\right) = P(Z > -1.33) = 1 - P(Z < -1.33) = 1 - 0.918 = 0.9082$$

Figure 9.4 illustrates the distributions used in this example.

FIGURE 9.4

Distribution of X and Sampling Distribution of $\overline{X}$

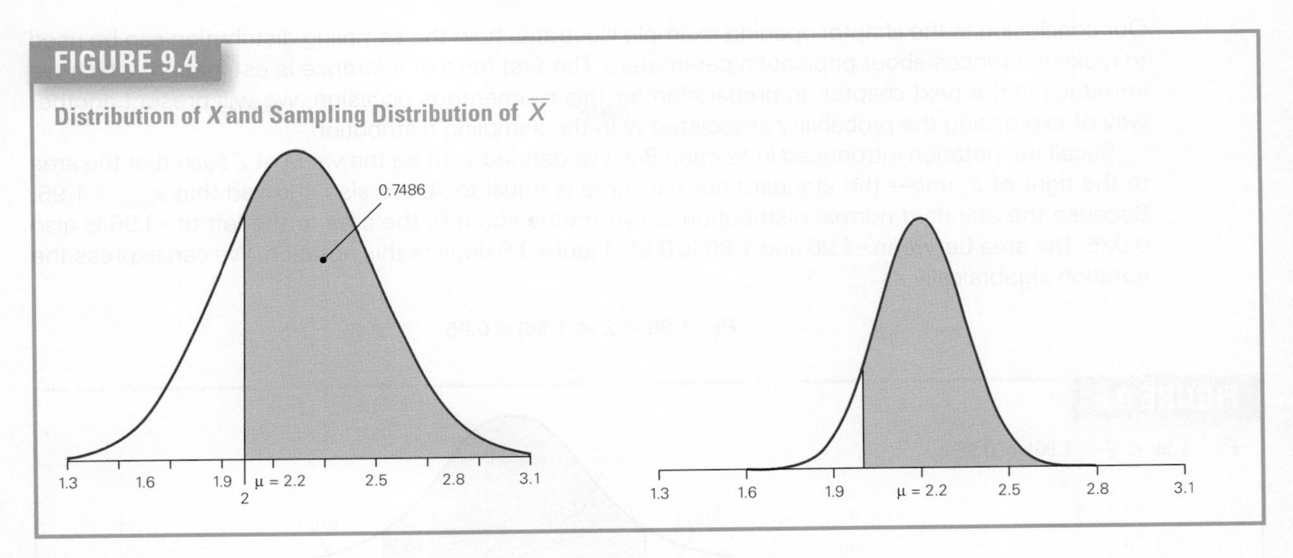

In Example 9.1(b), we began with the assumption that both μ and σ were known. Then, using the sampling distribution, we made a probability statement about $\overline{X}$. Unfortunately, the values of μ and σ are not usually known, so an analysis such as that in Example 9.1 cannot usually be conducted. However, we can use the sampling distribution to infer something about an unknown value of μ on the basis of a sample mean.

SALARIES OF A BUSINESS SCHOOL'S GRADUATES: SOLUTION

We want to find the probability that the sample mean is less than £750. Thus, we seek

$$P(\overline{X} < 750)$$

The distribution of X, the weekly income, is likely to be positively skewed, but not sufficiently so to make the distribution of $\overline{X}$ non-normal. As a result, we may assume that $\overline{X}$ is normal with mean $\mu_{\overline{x}} = \mu = 800$ and standard deviation $\sigma_{\overline{x}} = \sigma/\sqrt{n} = 100/\sqrt{25} = 20$. Thus,

$$P(\overline{X} < 750) = P\left(\frac{\overline{X} - \mu_{\overline{x}}}{\sigma_{\overline{x}}} < \frac{750 - 800}{20}\right) = P(Z < -2.5) = 0.0062$$

Figure 9.5 illustrates the distribution.

FIGURE 9.5

$P(\overline{X} < 750)$

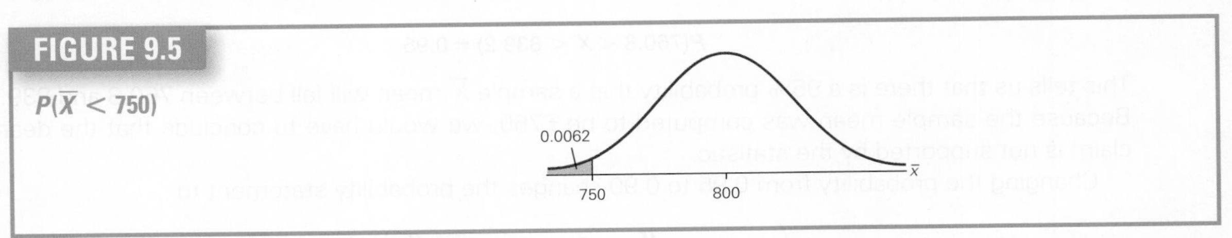

The probability of observing a sample mean as low as £750 when the population mean is £800 is extremely small. Because this event is quite unlikely, we would have to conclude that the dean's claim is not justified.

9-1c Using the Sampling Distribution for Inference

Our conclusion in the chapter-opening example illustrates how the sampling distribution can be used to make inferences about population parameters. The first form of inference is estimation, which we introduce in the next chapter. In preparation for this momentous occasion, we will present another way of expressing the probability associated with the sampling distribution.

Recall the notation introduced in Section 8-2. We defined z_A to be the value of Z such that the area to the right of z_A under the standard normal curve is equal to A. We also showed that $z_{0.025} = 1.96$. Because the standard normal distribution is symmetric about 0, the area to the left of −1.96 is also 0.025. The area between −1.96 and 1.96 is 0.95. Figure 9.6 depicts this notation. We can express the notation algebraically as

$$P(-1.96 < Z < 1.96) = 0.95$$

FIGURE 9.6

$P(-1.96 < Z < 1.96) = 0.95$

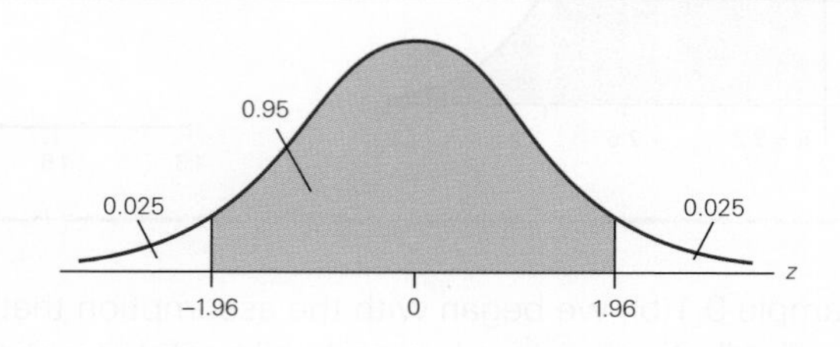

In this section, we established that

$$Z = \frac{\bar{X} - \mu}{\sigma/\sqrt{n}}$$

is standard normally distributed. Substituting this form of Z into the previous probability statement, we produce

$$P\left(-1.96 < \frac{\bar{X} - \mu}{\sigma/\sqrt{n}} < 1.96\right) = 0.95$$

With a little algebraic manipulation (multiply all three terms by $\sigma/\sqrt{n}$ and add μ to all three terms), we determine

$$P\left(\mu - 1.96\frac{\sigma}{\sqrt{n}} < \bar{X} < \mu + 1.96\frac{\sigma}{\sqrt{n}}\right) = 0.95$$

Returning to the chapter-opening example where $\mu = 800$, $\sigma = 100$ and $n = 25$, we compute

$$P\left(800 - 1.96\frac{100}{\sqrt{25}} < \bar{X} < 800 + 1.96\frac{100}{\sqrt{25}}\right) = 0.95$$

Thus, we can say that

$$P(760.8 < \bar{X} < 839.2) = 0.95$$

This tells us that there is a 95% probability that a sample $\bar{X}$ mean will fall between 760.8 and 839.2. Because the sample mean was computed to be £750, we would have to conclude that the dean's claim is not supported by the statistic.

Changing the probability from 0.95 to 0.90 changes the probability statement to

$$P\left(\mu - 1.645\frac{\sigma}{\sqrt{n}} < \bar{X} < \mu + 1.645\frac{\sigma}{\sqrt{n}}\right) = 0.90$$

We can also produce a general form of this statement:

$$P\left(\mu - z_{\alpha/2}\frac{\sigma}{\sqrt{n}} < \bar{X} < \mu + z_{\alpha/2}\frac{\sigma}{\sqrt{n}}\right) = 1 - \alpha$$

In this formula α (Greek letter *alpha*) is the probability that $\bar{X}$ does not fall into the interval. To apply this formula, all we need do is substitute the values for μ, σ, n and α. For example, with $\mu = 800$, $\sigma = 100$, $n = 25$ and $\alpha = 0.01$, we produce

$$P\left(\mu - z_{0.005}\frac{\sigma}{\sqrt{n}} < \bar{X} < \mu + z_{0.005}\frac{\sigma}{\sqrt{n}}\right) = 1 - 0.01$$

$$P\left(800 - 2.575\frac{100}{\sqrt{25}} < \bar{X} < 800 + 2.575\frac{100}{\sqrt{25}}\right) = 0.99$$

$$P(748.5 < \bar{X} < 851.5) = 0.99$$

which is another probability statement about $\bar{X}$. In Section 10-2, we will use a similar type of probability statement to derive the first statistical inference technique.

EXERCISES

9.1 Let X represent the result of the toss of a fair die. Find the following probabilities.

a. $P(X = 1)$

b. $P(X = 6)$

9.2 Let $\bar{X}$ represent the mean of the toss of two fair dice. Use the probabilities listed in Table 9.2 to determine the following probabilities.

a. $P(\bar{X} = 1)$

b. $P(\bar{X} = 6)$

9.3 An experiment consists of tossing five balanced dice. Find the following probabilities. (Determine the exact probabilities as we did in Tables 9.1 and 9.2 for two dice.)

a. $P(\bar{X} = 1)$

b. $P(\bar{X} = 6)$

9.4 Refer to Exercises 9.1 to 9.3. What do the probabilities tell you about the variances of X and $\bar{X}$?

9.5 A normally distributed population has a mean of 40 and a standard deviation of 12. What does the central limit theorem say about the sampling distribution of the mean if samples of size 100 are drawn from this population?

9.6 Refer to Exercise 9.5. Suppose that the population is not normally distributed. Does this change your answer? Explain.

9.7 A sample of $n = 16$ observations is drawn from a normal population with $\mu = 1000$ and $\sigma = 200$. Find the following.

a. $P(\bar{X} > 1050)$

b. $P(\bar{X} < 960)$

c. $P(\bar{X} > 1100)$

9.8 Repeat Exercise 9.7 with $n = 25$.

9.9 Repeat Exercise 9.7 with $n = 100$.

9.10 Given a normal population whose mean is 50 and whose standard deviation is 5, find the probability that a random sample of

a. four has a mean between 49 and 52.

b. 16 has a mean between 49 and 52.

c. 25 has a mean between 49 and 52.

9.11 Repeat Exercise 9.10 for a standard deviation of 10.

9.12 Repeat Exercise 9.10 for a standard deviation of 20.

9.13 a. Calculate the finite population correction factor when the population size is $N = 1000$ and the sample size is $n = 100$.

b. Repeat part (a) when $N = 3000$.

c. Repeat part (a) when $N = 5000$.

d. What have you learned about the finite population correction factor when N is large relative to n?

9.14 a. Suppose that the standard deviation of a population with $N = 10\,000$ members is 500. Determine the standard error of the sampling distribution of the mean when the sample size is 1000.

b. Repeat part (a) when $n = 500$.

c. Repeat part (a) when $n = 100$.

9.15 The heights of Spanish women are normally distributed with a mean of 161 cm and a standard deviation of 2 cm.

a. What is the probability that a randomly selected woman is taller than 163 cm?

b. A random sample of four women is selected. What is the probability that the sample mean height is greater than 163 cm?

c. What is the probability that the mean height of a random sample of 100 women is greater than 163 cm?

9.16 Refer to Exercise 9.15. If the population of women's heights is not normally distributed, which, if any, of the questions can you answer? Explain.

9.17 An automatic machine in a manufacturing process is operating properly if the lengths of an important subcomponent are normally distributed with mean = 117 cm and standard deviation = 5.2 cm.

a. Find the probability that one selected subcomponent is longer than 120 cm.

b. Find the probability that if four subcomponents are randomly selected, their mean length exceeds 120 cm.

c. Find the probability that if four subcomponents are randomly selected, all four have lengths that exceed 120 cm.

9.18 The amount of time the university lecturers devote to their jobs per week is normally distributed with a mean of 52 hours and a standard deviation of 6 hours.

a. What is the probability that a lecturer works for more than 60 hours per week?

b. Find the probability that the mean amount of work per week for three randomly selected lecturers is more than 60 hours.

c. Find the probability that if three lecturers are randomly selected all three work for more than 60 hours per week.

9.19 The number of pizzas consumed per month by university students is normally distributed with a mean of 10 and a standard deviation of 3.

a. What proportion of students consume more than 12 pizzas per month?

b. What is the probability that in a random sample of 25 students more than 275 pizzas are consumed? (*Hint:* What is the mean number of pizzas consumed by the sample of 25 students?)

9.20 The marks on a statistics midterm test are normally distributed with a mean of 78 and a standard deviation of 6.

a. What proportion of the class has a midterm mark of less than 75?

b. What is the probability that a class of 50 has an average midterm mark that is less than 75?

9.21 The amount of time spent by British consumers watching television per month is normally distributed with a mean of 43 hours and a standard deviation of 1.5 hours.

a. What is the probability that a randomly selected British consumer watches television for more than 44 hours per month?

b. What is the probability that the average time watching television by a random sample of five British consumers is more than 44 hours?

c. What is the probability that, in a random sample of five British consumers, all watch television for more than 44 hours per month?

9.22 The manufacturer of tinned salmon that is supposed to have a net weight of 100 g tells you that the net weight is actually a normal random variable with a mean of 100.5 g and a standard deviation of 0.18 g. Suppose that you draw a random sample of 36 tins.

a. Find the probability that the mean weight of the sample is less than 97 g.

b. Suppose your random sample of 36 tins of salmon produced a mean weight that is less than 97 g. Comment on the statement made by the manufacturer.

9.23 The number of customers who enter a supermarket each hour is normally distributed with a mean of 600 and a standard deviation of 200. The supermarket is open 16 hours per day. What is the probability that the total number of customers who enter the supermarket in 1 day is greater than 10 000? (*Hint:* Calculate the average hourly number of customers necessary to exceed 10 000 in one 16-hour day.)

9.24 The sign on the elevator in a supermarket states 'Maximum capacity 1140 kilograms or 16 persons'. A statistics lecturer wonders what the probability is that 16 persons would weigh more than 1140 kilograms. Discuss what the lecturer needs (besides the ability to perform the calculations) in order to satisfy her curiosity.

9.25 Refer to Exercise 9.24. Suppose that the lecturer discovers that the weights of people who use the elevator are normally distributed with an average of 75 kilograms and a standard deviation of 10 kilograms. Calculate the probability the lecturer seeks.

9.26 The time it takes for a statistics lecturer to mark his midterm test is normally distributed with a mean of 4.8 minutes and a standard deviation of 1.3 minutes. There are 60 students in the lecturer's class. What is the probability that he needs more than 5 hours to mark all the midterm tests? (The 60 midterm tests of the students in this year's class can be considered a random sample of the many thousands of midterm tests the professor has marked and will mark.)

9.27 Refer to Exercise 9.26. Does your answer change if you discover that the times needed to mark a midterm test are not normally distributed?

9.28 The restaurant in a large commercial building in Bordeaux provides coffee for the occupants in the building. The restaurateur has determined that the mean number of cups of coffee consumed in a day by all the occupants is 2.0 with a standard deviation of 6. A new tenant of the building intends to have a total of 125 new employees. What is the probability that the new employees will consume more than 240 cups per day?

9.29 The number of pages produced by a fax machine in a busy office is normally distributed with a mean of 275 and a standard deviation of 75. Determine the probability that in 1 week (5 days) more than 1500 faxes will be received.

9-2 SAMPLING DISTRIBUTION OF A PROPORTION

In Section 7-4 we introduced the binomial distribution whose parameter is *p*, the probability of success in any trial. In order to compute binomial probabilities, we assumed that *p* was known. However, in the real world *p* is unknown, requiring the statistics practitioner to estimate its value from a sample. The estimator of a population proportion of successes is the sample proportion; that is, we count the number of successes in a sample and compute

$$\hat{P} = \frac{X}{n}$$

($\hat{P}$ is read as *p hat*) where X is the number of successes and n is the sample size. When we take a sample of size n, we are actually conducting a binomial experiment; as a result, X is binomially distributed. Thus, the probability of any value of $\hat{P}$ can be calculated from its value of X. For example, suppose that we have a binomial experiment with $n = 10$ and $p = 0.4$. To find the probability that the sample proportion $\hat{P}$ is less than or equal to 0.50, we find the probability that X is less than or equal to 5 (because 5/10 = 0.50). From Table 1 in Appendix A we find $n = 10$ and $p = 0.4$

$$P(\hat{P} \leq 0.50) = P(X \leq 5) = 0.8338$$

We can calculate the probability associated with other values of $\hat{P}$ similarly.

Discrete distributions such as the binomial do not lend themselves easily to the kinds of calculation needed for inference. And inference is the reason we need sampling distributions. Fortunately, we can approximate the binomial distribution by a normal distribution.

What follows is an explanation of how and why the normal distribution can be used to approximate a binomial distribution. Disinterested readers can skip to Section 9-2c, where we present the approximate **sampling distribution of a sample proportion**.

9-2a Normal Approximation to the Binomial Distribution

Recall how we introduced continuous probability distributions in Chapter 8. We developed the density function by converting a histogram so that the total area in the rectangles equalled 1. We can do the same for a binomial distribution. To illustrate, let X be a binomial random variable with $n = 20$ and $p = 0.5$. We can easily determine the probability of each value of X, where $X = 0, 1, 2, \ldots, 19, 20$. A rectangle representing a value of x is drawn so that its area equals the probability. We accomplish this by letting the height of the rectangle equal the probability and the base of the rectangle equal 1. Thus, the base of each rectangle for x is the interval $x - 0.5$ to $x + 0.5$. Figure 9.7 depicts this graph. As you can see, the rectangle representing $x = 10$ is the rectangle whose base is the interval 9.5 to 10.5 and whose height is $P(X = 10) = 0.1762$.

If we now smooth the ends of the rectangles, we produce a bell-shaped curve as seen in Figure 9.8. Thus, to use the normal approximation, all we need do is find the area under the *normal* curve between 9.5 and 10.5.

To find normal probabilities it is required to first standardise x by subtracting the mean and dividing by the standard deviation. The values for μ and σ are derived from the binomial distribution being approximated. In Section 7.4 we pointed out that

$$\mu = np$$

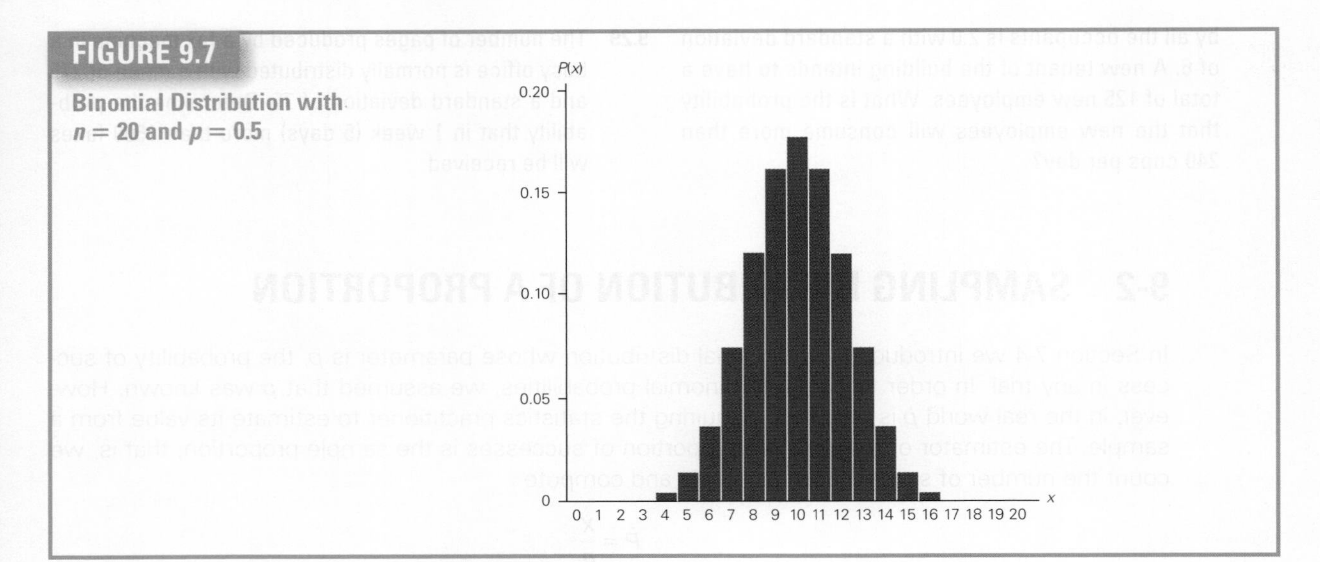

FIGURE 9.7 Binomial Distribution with $n = 20$ and $p = 0.5$

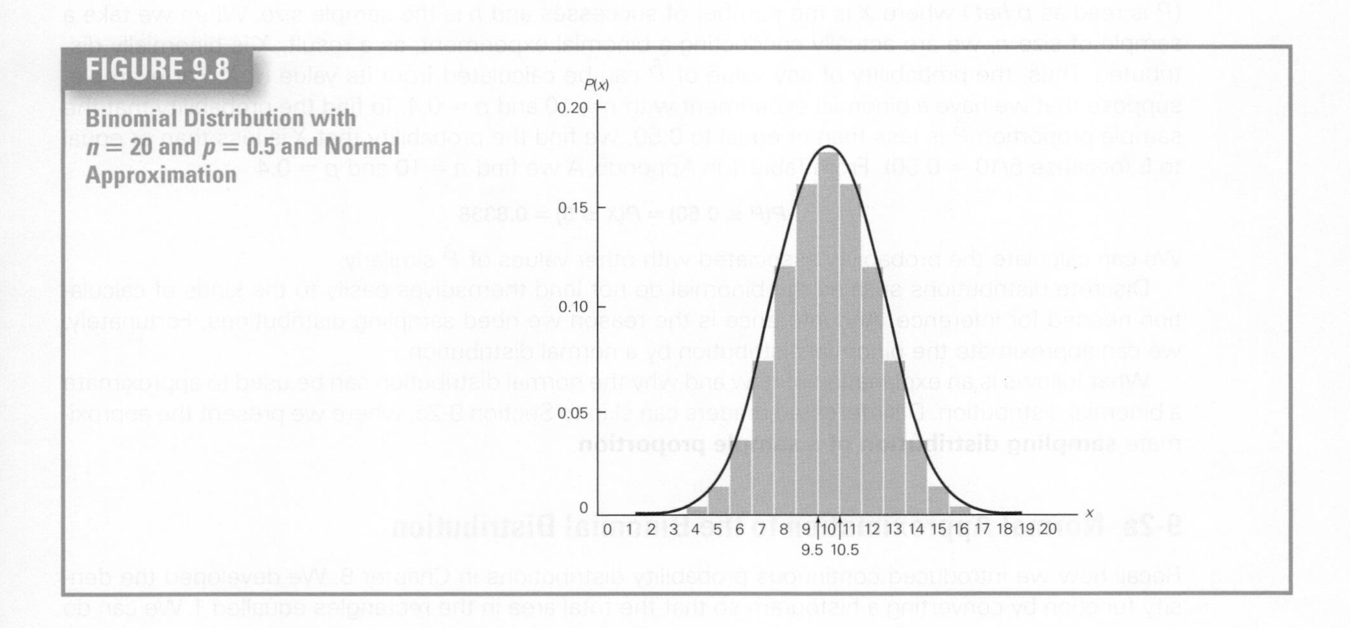

FIGURE 9.8 Binomial Distribution with $n = 20$ and $p = 0.5$ and Normal Approximation

and

$$\sigma = \sqrt{np(1-p)}$$

For $n = 20$ and $p = 0.5$, we have

$$\mu = np = 20(0.5) = 10$$

and

$$\sigma = \sqrt{np(1-p)} = \sqrt{20(0.5)(1-0.5)} = 2.24$$

To calculate the probability that $X = 10$ using the normal distribution requires that we find the area under the normal curve between 9.5 and 10.5; that is,

$$P(X = 10) \approx P(9.5 < Y < 10.5)$$

where Y is a normal random variable approximating the binomial random variable X. We standardise Y and use Table 3 of Appendix A to find

$$P(9.5 < Y < 10.5) = P\left(\frac{9.5 - 10}{2.24} < \frac{Y - \mu}{\sigma} < \frac{10.5 - 10}{2.24}\right)$$
$$= P(-0.22 < Z < 0.22) = (Z < 0.22) - P(Z < -0.22)$$
$$= 0.5871 - 0.4129 = 0.1742$$

The actual probability that X equals 10 is

$$P(X = 10) = 0.1762$$

As you can see, the approximation is quite good.

Notice that to draw a binomial distribution, which is discrete, it was necessary to draw rectangles whose bases were constructed by adding and subtracting 0.5 to the values of X. The 0.5 is called the **continuity correction factor**.

The approximation for any other value of X would proceed in the same manner. In general, the binomial probability $P(X = x)$ is approximated by the area under a normal curve between $x - 0.5$ and $x + 0.5$. To find the binomial probability $P(X \leq x)$, we calculate the area under the normal curve to the left of $x + 0.5$. For the same binomial random variable, the probability that its value is less than or equal to 8 is $P(X \leq 8) = 0.2517$. The normal approximation is

$$P(X \leq 8) \approx P(Y < 8.5) = P\left(\frac{Y - \mu}{\sigma} < \frac{8.5 - 10}{2.24}\right) = P(Z < -0.67) = 0.2514$$

We find the area under the normal curve to the right of $x - 0.5$ to determine the binomial probability $P(X \geq x)$. To illustrate, the probability that the binomial random variable (with $n = 20$ and $p = 0.5$) is greater than or equal to 14 is $P(X \geq 14) = 0.577$. The normal approximation is

$$P(X \geq 14) \approx P(Y > 13.5) = P\left(\frac{Y - \mu}{\sigma} > \frac{13.5 - 10}{2.24}\right) = P(Z > 1.56) = 0.0594$$

9-2b Omitting the Correction Factor for Continuity

When calculating the probability of *individual* values of X as we did when we computed the probability that X equals 10 earlier, the correction factor *must* be used. If we do not, we are left with finding the area in a line, which is 0. When computing the probability of a *range* of values of X, we can omit the correction factor. However, the omission of the correction factor will decrease the accuracy of the approximation. For example, if we approximate $P(X \leq 8)$ as we did previously except without the correction factor, we find

$$P(X \leq 8) \approx P(Y < 8) = P\left(\frac{Y - \mu}{\sigma} < \frac{8 - 10}{2.24}\right) = P(Z < -0.89) = 0.1867$$

The absolute size of the error between the actual cumulative binomial probability and its normal approximation is quite small when the values of x are in the tail regions of the distribution. For example, the probability that a binomial random variable with $n = 20$ and $p = 0.5$ is less than or equal to 3 is

$$P(X \leq 3) = 0.0013$$

The normal approximation with the correction factor is

$$P(X \leq 3) \approx P(Y < 3.5) = P\left(\frac{Y - \mu}{\sigma} < \frac{3.5 - 10}{2.24}\right) = P(Z < -2.90) = 0.0019$$

The normal approximation without the correction factor is (using Excel)

$$P(X \leq 3) \approx P(Y < 3) = P\left(\frac{Y - \mu}{\sigma} < \frac{3 - 10}{2.24}\right) = P(Z < -3.13) = 0.0009$$

For larger values of n, the differences between the normal approximation with and without the correction factor are small even for values of X near the centre of the distribution. For example, the probability that a binomial random variable with $n = 1000$ and $p = 0.3$ is less than or equal to 260 is

$$P(X \leq 260) = 0.0029 \text{ (using Excel)}$$

The normal approximation with the correction factor is

$$P(X \leq 260) \approx P(Y < 260.5) = P\left(\frac{Y - \mu}{\sigma} < \frac{260.5 - 300}{14.49}\right) = P(Z < -2.73) = 0.0032$$

The normal approximation without the correction factor is

$$P(X \leq 260) \approx P(Y < 260) = P\left(\frac{Y - \mu}{\sigma} < \frac{260 - 300}{14.49}\right) = P(Z < -2.76) = 0.0029$$

As we pointed out, the normal approximation of the binomial distribution is made necessary by the needs of statistical inference. As you will discover, statistical inference generally involves the use of large values of n, and the part of the sampling distribution that is of greatest interest lies in the tail regions. The correction factor was a temporary tool that allowed us to convince you that a binomial distribution can be approximated by a normal distribution. Now that we have done so, we will use the normal approximation of the binomial distribution to approximate the sampling distribution of a sample proportion, and in such applications the correction factor will be omitted.

9-2c Approximate Sampling Distribution of a Sample Proportion

Using the laws of expected value and variance (see the website Appendix Using the Laws of Expected Value and Variance to Derive the Parameters of Sampling Distributions), we can determine the mean, variance and standard deviation of $\hat{P}$. We will summarise what we have learned.

Sampling Distribution of a Sample Proportion

1. $\hat{P}$ is approximately normally distributed provided that np and $n(1 - p)$ are greater than or equal to 5.
2. The expected value: $E(\hat{P}) = p$
3. The variance: $V(\hat{P}) = \sigma^2_{\hat{p}} = \dfrac{p(1 - p)}{n}$
4. The standard deviation: $\sigma_{\hat{p}} = \sqrt{p(1 - p)/n}$*

(The standard deviation of $\hat{P}$ is called the **standard error of the proportion**.)

The sample size requirement is theoretical because, in practice, much larger sample sizes are needed for the normal approximation to be useful.

EXAMPLE 9.2

Political Survey

In the last election, a political leader representative received 52% of the votes cast. One year after the election, the political leader representative organised a survey that asked a random sample of 300 people whether they would vote for him in the next election. If we assume that his popularity has not changed, what is the probability that more than half of the sample would vote for him?

*As was the case with the standard error of the mean, the standard error of a proportion is $\sqrt{p(1 - p)/n}$ when sampling from infinitely large populations. When the population is finite, the standard error of the proportion must include the finite population correction factor, which can be omitted when the population is large relative to the sample size, a very common occurrence in practice.

SOLUTION:

The number of respondents who would vote for the political leader representative is a binomial random variable with $n = 300$ and $p = 0.52$. We want to determine the probability that the sample proportion is greater than 50%. In other words, we want to find $P(\hat{P} > 0.50)$.

We now know that the sample proportion $\hat{P}$ is approximately normally distributed with mean $p = 0.52$ and standard deviation $= \sqrt{p(1-p)/n} = \sqrt{(0.52)(0.48)/300} = 0.0288$.

Thus, we calculate

$$P(\hat{P} > 0.50) = P\left(\frac{\hat{P} - p}{\sqrt{p(1-p)/n}} > \frac{0.50 - 0.52}{0.0288}\right)$$
$$= P(Z > -0.69) = 1 - P(Z < -0.69) = 1 - 0.2451 = 0.7549$$

If we assume that the level of support remains at 52%, the probability that more than half the sample of 300 people would vote for the representative is 0.7549.

EXERCISES

Use the normal approximation without the correction factor to find the probabilities in the following exercises.

9.30 **a.** In a binomial experiment with $n = 300$ and $p = 0.5$, find the probability that $\hat{P}$ is greater than 60%.

b. Repeat part (a) with $p = 0.55$.

c. Repeat part (a) with $p = 0.6$.

9.31 **a.** The probability of success on any trial of a binomial experiment is 25%. Find the probability that the proportion of successes in a sample of 500 is less than 22%.

b. Repeat part (a) with $n = 800$.

c. Repeat part (a) with $n = 1000$.

9.32 Determine the probability that in a sample of 100 the sample proportion is less than 0.75 if $p = 0.80$.

9.33 A binomial experiment where $p = 0.4$ is conducted. Find the probability that in a sample of 60 the proportion of successes exceeds 0.35.

9.34 The proportion of eligible voters in the next election who will vote for the incumbent is assumed to be 55%. What is the probability that in a random sample of 500 voters less than 49% say they will vote for the incumbent?

9.35 The assembly line that produces an electronic component of a missile system has historically resulted in a 2% defective rate. A random sample of 800 components is drawn. What is the probability that the defective rate is greater than 4%? Suppose that in the random sample the defective rate is 4%. What does that suggest about the defective rate on the assembly line?

9.36 **a.** The manufacturer of aspirin claims that the proportion of headache sufferers who get relief with just two aspirin tablets is 53%. What is the probability that in a random sample of 400 headache sufferers, less than 50% obtain relief? If 50% of the sample actually obtained relief, what does this suggest about the manufacturer's claim?

b. Repeat part (a) using a sample of 1000.

9.37 The manager of a restaurant in a commercial building has determined that the proportion of customers who drink tea is 14%. What is the probability that in the next 100 customers at least 10% will be tea drinkers?

9.38 A commercial for a manufacturer of household appliances claims that 3% of all its products require a service call in the first year. A consumer protection association wants to check the claim by surveying 400 households that recently purchased one of the company's appliances. What is the probability that more than 5% require a service call within the first year? What would you say about the commercial's accuracy if in a random sample of 400 households 5% report at least one service call?

9.39 The Laurier Company's brand has a market share of 30%. Suppose that 1000 consumers of the product are asked in a survey which brand they prefer. What is the probability that more than 32% of the respondents say they prefer the Laurier brand?

9.40 A university bookshop claims that 50% of its customers are satisfied with the service and prices.

a. If this claim is true, what is the probability that in a random sample of 600 customers less than 45% are satisfied?

b. Suppose that in a random sample of 600 customers, 270 express satisfaction with the bookshop. What does this tell you about the bookshop's claim?

9.41 A psychologist believes that 80% of male drivers when lost continue to drive hoping to find the location they seek rather than ask directions. To examine this belief, he took a random sample of 350 male drivers and asked each what they did when lost. If the belief is true, determine the probability that less than 75% said they continue driving.

9.42 The Cape Town Grill restaurant chain regularly surveys its customers. On the basis of these surveys, the management of the chain claims that 75% of its customers rate the food as excellent. A consumer testing service wants to examine the claim by asking 460 customers to rate the food. What is the probability that less than 70% rate the food as excellent?

9.43 An accounting professor claims that no more than one-quarter of undergraduate business students will major in accounting. What is the probability that in a random sample of 1200 undergraduate business students, 336 or more will major in accounting?

9.44 Refer to Exercise 9.43. A survey of a random sample of 1200 undergraduate business students indicates that 336 students plan to major in accounting. What does this tell you about the professor's claim?

9-3 SAMPLING DISTRIBUTION OF THE DIFFERENCE BETWEEN TWO MEANS

Another sampling distribution that you will soon encounter is that of the **difference between two sample means**. The sampling plan calls for independent random samples drawn from each of two normal populations. The samples are said to be independent if the selection of the members of one sample is independent of the selection of the members of the second sample. We will expand upon this discussion in Chapter 13. We are interested in the sampling distribution of the difference between the two sample means.

In Section 9-1, we introduced the central limit theorem, which states that in repeated sampling from a normal population whose mean is μ and whose standard deviation is σ, the sampling distribution of the sample mean is normal with mean μ and standard deviation $\sigma/\sqrt{n}$. Statisticians have shown that the difference between two independent normal random variables is also normally distributed. Thus, the difference between two sample means $\bar{X}_1 - \bar{X}_2$ is normally distributed if both populations are normal.

By using the laws of expected value and variance we derive the expected value and variance of the **sampling distribution of** $\bar{X}_1 - \bar{X}_2$:

$$\mu_{\bar{x}_1-\bar{x}_2} = \mu_1 - \mu_2$$

and

$$\sigma^2_{\bar{x}-\bar{x}_2} = \frac{\sigma^2}{n_1} + \frac{\sigma_2^2}{n_2}$$

Thus, it follows that in repeated independent sampling from two populations with means μ_1 and μ_2 and standard deviations σ_1 and σ_2, respectively, the sampling distribution of $\bar{X}_1 - \bar{X}_2$ is normal with mean

$$\mu_{\bar{x}_1-\bar{x}_2} = \mu_1 - \mu_2$$

and standard deviation (which is the **standard error of the difference between two means**)

$$\sigma_{\bar{x}_1-\bar{x}_2} = \sqrt{\frac{\sigma_1^2}{n_1} + \frac{\sigma_2^2}{n_2}}$$

If the populations are non-normal, then the sampling distribution is only approximately normal for large sample sizes. The required sample sizes depend on the extent of non-normality. However, for most populations, sample sizes of 30 or more are sufficient.

Figure 9.9 depicts the sampling distribution of the difference between two means.

FIGURE 9.9

Sampling Distribution of $\overline{X}_1 - \overline{X}_2$

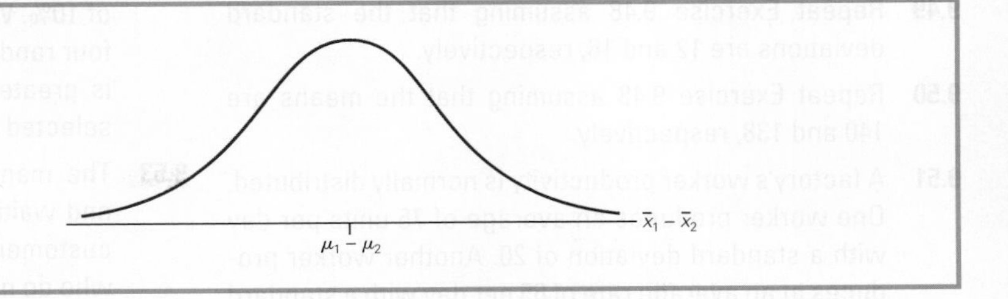

EXAMPLE 9.3

Starting Salaries of MBAs

Suppose that the starting salaries of MBA graduates from INSEAD (France) are normally distributed, with a mean of €92 000 and a standard deviation of €14 500. The starting salaries of MBA graduates from London Business School (LBS) are normally distributed, with a mean of €90 000 and a standard deviation of €18 300. If a random sample of 50 INSEAD MBAs and a random sample of 60 LBS MBAs are selected, what is the probability that the sample mean starting salary of INSEAD graduates will exceed that of the LBS graduates?

SOLUTION:

We want to determine $P(\overline{X}_1 - \overline{X}_2 > 0)$. We know that $\overline{X}_1 - \overline{X}_2$ is normally distributed with mean $\mu_1 - \mu_2 = 98\,000 - 90\,000 = 28\,000$ and standard deviation

$$\sqrt{\frac{\sigma_2^2}{n_1} + \frac{\sigma_2^2}{n_2}} = \sqrt{\frac{14\,500^2}{50} + \frac{18\,300^2}{60}} = 3128$$

We can standardise the variable and refer to Table 3 of Appendix A:

$$P(\overline{X}_1 - \overline{X}_2 > 0) = P\left(\frac{(\overline{X}_1 - \overline{X}_2) - (\mu_2 - \mu_1)}{\sqrt{\frac{\sigma_1^2}{n_1} + \frac{\sigma_2^2}{n_2}}} > \frac{0 - 2000}{3128}\right)$$

$$= P(Z > -0.64) = 1 - P(Z < -0.64) = 1 - 0.2611 = 0.7389$$

$$P(\overline{X}_1 - \overline{X}_2 > 0) = P\left(\frac{(\overline{X}_1 - \overline{X}_2) - (\mu_1 - \mu_2)}{\sqrt{\frac{\sigma_1^2}{n_1} = \frac{\sigma_2^2}{n_2}}} > \frac{0 - 2000}{3128}\right)$$

$$= P(Z > -64) = 1 - P(Z < -64) = 1 - 0.2611 = 0.7389$$

There is a 0.7389 probability that for a sample of size 50 from the INSEAD graduates and a sample of size 60 from the LBS graduates, the sample mean starting salary of INSEAD graduates will exceed the sample mean of LBS graduates.

EXERCISES

9.45 Independent random samples of ten observations each are drawn from normal populations. The parameters of these populations are

Population 1: $\mu = 280, \sigma = 25$
Population 2: $\mu = 270, \sigma = 30$

Find the probability that the mean of sample 1 is greater than the mean of sample 2 by more than 25.

9.46 Repeat Exercise 9.45 with samples of size 50.

9.47 Repeat Exercise 9.45 with samples of size 100.

9.48 Suppose that we have two normal populations with the means and standard deviations listed here. If random samples of size 25 are drawn from each population, what is the probability that the mean of sample 1 is greater than the mean of sample 2?

Population 1: $\mu = 40, \sigma = 6$
Population 2: $\mu = 38, \sigma = 8$

9.49 Repeat Exercise 9.48 assuming that the standard deviations are 12 and 16, respectively.

9.50 Repeat Exercise 9.48 assuming that the means are 140 and 138, respectively.

9.51 A factory's worker productivity is normally distributed. One worker produces an average of 75 units per day with a standard deviation of 20. Another worker produces at an average rate of 65 per day with a standard deviation of 21. What is the probability that in 1 week (5 working days), worker 1 will outperform worker 2?

9.52 A professor of statistics noticed that the marks in his course are normally distributed. He has also noticed that his morning classes average 73%, with a standard deviation of 12% on their final exams. His afternoon classes average 77%, with a standard deviation of 10%. What is the probability that the mean mark of four randomly selected students from a morning class is greater than the average mark of four randomly selected students from an afternoon class?

9.53 The manager of a restaurant believes that waiters and waitresses who introduce themselves by telling customers their names will get larger tips than those who do not. In fact, she claims that the average tip for the former group is 18%, whereas that of the latter is only 15%. If tips are normally distributed with a standard deviation of 3%, what is the probability that in a random sample of ten tips recorded from waiters and waitresses who introduce themselves, and ten tips from waiters and waitresses who do not, the mean of the former will exceed that of the latter?

9-4 FROM HERE TO INFERENCE

The primary function of the sampling distribution is statistical inference. To see how the sampling distribution contributes to the development of inferential methods, we need to briefly review how we got to this point.

In Chapters 7 and 8, we introduced probability distributions, which allowed us to make probability statements about values of the random variable. A prerequisite of this calculation is knowledge of the distribution and the relevant parameters. In Example 7.9, we needed to know that the probability that Pat Statsdud guesses the correct answer is 20% ($p = 0.2$) and that the number of correct answers (successes) in ten questions (trials) is a binomial random variable. We could then compute the probability of any number of successes. In Example 8.3, we needed to know that the return on investment is normally distributed with a mean of 10% and a standard deviation of 5%. These three bits of information allowed us to calculate the probability of various values of the random variable.

Figure 9.10 symbolically represents the use of probability distributions. Simply put, knowledge of the population and its parameter(s) allows us to use the probability distribution to make probability statements about individual members of the population. The direction of the arrows indicates the direction of the flow of information.

FIGURE 9.10
Probability Distribution

Population and parameter(s) —*Probability distribution*→ *Individual*

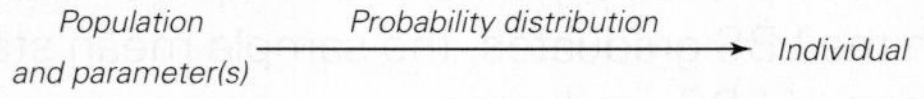

In this chapter, we developed the sampling distribution, wherein knowledge of the parameter(s) and some information about the distribution allow us to make probability statements about a sample statistic. In Example 9.1(b), knowing the population mean and standard deviation and assuming that the population is not extremely non-normal enabled us to calculate a probability statement about a sample mean. Figure 9.11 describes the application of sampling distributions.

FIGURE 9.11
Sampling Distribution

Population and parameter(s) —*Sampling distribution*→ *Statistic*

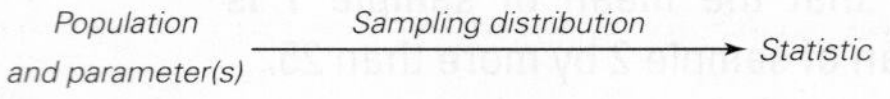

Notice that in applying both probability distributions and sampling distributions, we must know the value of the relevant parameters, a highly unlikely circumstance. In the real world, parameters are almost always unknown because they represent descriptive measurements about extremely large populations. Statistical inference addresses this problem. It does so by reversing the direction of the flow of knowledge in Figure 9.11. In Figure 9.12, we display the character of statistical inference. Starting in Chapter 10, we will assume that most population parameters are unknown. The statistics practitioner will sample from the population and compute the required statistic. The sampling distribution of that statistic will enable us to draw inferences about the parameter.

FIGURE 9.12

Sampling Distribution in Inference

Statistic —— *Sampling distribution* ——→ *Parameter*

You may be surprised to learn that, by and large, that is all we do in the remainder of this book. Why then do we need more chapters? They are necessary because there are many more parameter and sampling distribution combinations that define the inferential procedures to be presented in an introductory statistics course. However, they all work in the same way. If you understand how one procedure is developed, then you will likely understand all of them. Our task in the next two chapters is to ensure that you understand the first inferential method. Your job is identical.

CHAPTER SUMMARY

The sampling distribution of a statistic is created by repeated sampling from one population. In this chapter, we introduced the sampling distribution of the mean, the proportion and the difference between two means. We described how these distributions are created theoretically and empirically.

SYMBOLS:

Symbol	Pronounced	Represents
$\mu_{\bar{x}}$	mu *x* bar	Mean of the sampling distribution of the sample mean
$\sigma^2_{\bar{x}}$	sigma squared *x* bar	Variance of the sampling distribution of the sample mean
$\sigma_{\bar{x}}$	sigma *x* bar	Standard deviation (standard error) of the sampling distribution of the sample mean
α	alpha	Probability
$\hat{P}$	*p* hat	Sample proportion
$\sigma^2_{\hat{p}}$	sigma squared *p* hat	Variance of the sampling distribution of the sample proportion
$\sigma_{\hat{p}}$	sigma *p* hat	Standard deviation (standard error) of the sampling distribution of the sample proportion
$\mu_{\bar{x}_1-\bar{x}_2}$	mu *x* bar 1 minus *x* bar 2	Mean of the sampling distribution of the difference between two sample means
$\sigma^2_{\bar{x}_1-\bar{x}_2}$	sigma squared *x* bar 1 minus *x* bar 2	Variance of the sampling distribution of the difference between two sample means
$\sigma^2_{\bar{x}_1-\bar{x}_2}$	sigma *x* bar 1 minus *x* bar 2	Standard deviation (standard error) of the sampling distribution of the difference between two sample means

FORMULAS:

Expected value of the sample mean

$$E(\bar{X}) = \mu_{\bar{x}} = \mu$$

Variance of the sample mean

$$V(\bar{X}) = \sigma^2_{\bar{x}} = \frac{\sigma^2}{n}$$

Standard error of the sample mean

$$\sigma_{\bar{x}} = \frac{\sigma}{\sqrt{n}}$$

Standardizing the sample mean

$$Z = \frac{\bar{X} - \mu}{\sigma/\sqrt{n}}$$

Expected value of the sample proportion

$$E(\hat{P}) = \mu_p = p$$

Variance of the sample proportion

$$V(\hat{P}) = \sigma^2_{\hat{p}} = \frac{p(1-p)}{n}$$

Standard error of the sample proportion

$$\sigma_{\hat{p}} = \sqrt{\frac{p(1-p)}{n}}$$

Standardising the sample proportion

$$Z = \frac{\hat{P} - p}{\sqrt{p(1-p)/n}}$$

Expected value of the difference between two means

$$E(\bar{X}_1 - \bar{X}_2) = \mu_{\bar{x}_1 - \bar{x}_2} = \mu_1 - \mu_2$$

Variance of the difference between two means

$$V(\bar{X}_1 - \bar{X}_2) = \sigma^2_{\bar{x}_1 - \bar{x}_2} = \frac{\sigma_1^2}{n_1} + \frac{\sigma_2^2}{n_2}$$

Standard error of the difference between two means

$$\sigma_{\bar{x}_1 - \bar{x}_2} = \sqrt{\frac{\sigma_1^2}{n_1} + \frac{\sigma_2^2}{n_2}}$$

Standardising the difference between two sample means

$$Z = \frac{(\bar{X}_1 - \bar{X}_2) - (\mu_1 - \mu_2)}{\sqrt{\frac{\sigma_1^2}{n_1} + \frac{\sigma_2^2}{n_2}}}$$

10 INTRODUCTION TO ESTIMATION

10-1 CONCEPTS OF ESTIMATION

10-2 ESTIMATING THE POPULATION MEAN WHEN THE POPULATION STANDARD DEVIATION IS KNOWN

10-3 SELECTING THE SAMPLE SIZE

DETERMINING THE SAMPLE SIZE TO ESTIMATE THE MEAN TREE DIAMETER

A lumber company has just acquired the rights to a large tract of land containing thousands of trees.

A wood-felling company needs to be able to estimate the amount of timber it can harvest in a tract of land to determine whether the effort will be profitable. To do so, it must estimate the mean diameter of the trees. It decides to estimate that parameter to within 1 cm with 90% confidence. A forester familiar with the territory guesses that the diameters of the trees are normally distributed with a standard deviation of 6 cm. Using the formula on Section 10-3b, he determines that he should sample 98 trees. After sampling those 98 trees, the forester calculates the sample mean to be 25 cm. Suppose that after he has completed his sampling and calculations, he discovers that the actual standard deviation is 12 cm. Will he be satisfied with the result? **See Section 10-3b for the answer.**

INTRODUCTION

Having discussed descriptive statistics (Chapter 4), probability distributions (Chapters 7 and 8) and sampling distributions (Chapter 9), we are ready to tackle statistical inference. As we explained in Chapter 1, *statistical inference* is the process by which we acquire information and draw conclusions about populations from samples. There are two general procedures for making inferences about populations: *estimation* and *hypothesis testing*. In this chapter, we introduce the concepts and foundations of estimation and demonstrate them with simple examples. In Chapter 11, we describe the fundamentals of hypothesis testing. Because most of what we do in the remainder of this book applies the concepts of estimation and hypothesis testing, understanding Chapters 10 and 11 is vital to your development as a statistics practitioner.

10-1 CONCEPTS OF ESTIMATION

As its name suggests, the objective of estimation is to determine the approximate value of a population parameter on the basis of a sample statistic. For example, the sample mean is employed to estimate the population mean. We refer to the sample mean as the *estimator* of the population mean. Once the sample mean has been computed, its value is called the *estimate*. In this chapter, we will introduce the statistical process whereby we estimate a population mean using sample data. In the rest of the book, we use the concepts and techniques introduced here for other parameters.

10-1a Point and Interval Estimators

We can use sample data to estimate a population parameter in two ways. First, we can compute the value of the estimator and consider that value as the estimate of the parameter. Such an estimator is called a *point estimator*.

Point Estimator

A **point estimator** draws inferences about a population by estimating the value of an unknown parameter using a single value or point.

There are three drawbacks to using point estimators. First, it is virtually certain that the estimate will be wrong. (The probability that a continuous random variable will equal a specific value is 0; that is, the probability that $\bar{x}$ will exactly equal μ is 0.) Second, we often need to know how close the estimator is to the parameter. Third, in drawing inferences about a population, it is intuitively reasonable to expect that a large sample will produce more accurate results because it contains more information than a smaller sample does. But point estimators do not have the capacity to reflect the effects of larger sample sizes. As a consequence, we use the second method of estimating a population parameter, the *interval estimator*.

Interval Estimator

An **interval estimator** draws inferences about a population by estimating the value of an unknown parameter using an interval.

As you will see, the interval estimator is affected by the sample size; because it possesses this feature, we will deal mostly with interval estimators in this text.

To illustrate the difference between point and interval estimators, suppose that a statistics professor wants to estimate the mean summer income of his second-year business students. Selecting

25 students at random, he calculates the sample mean weekly income to be £400. The point estimate is the sample mean. In other words, he estimates the mean weekly summer income of all second-year business students to be 400. Using the technique described subsequently, he may instead use an interval estimate; he estimates that the mean weekly summer income of second-year business students to lie between 380 and 420.

Numerous applications of estimation occur in the real world. For example, television network executives want to know the proportion of television viewers who are tuned in to their networks; an economist wants to know the mean income of university graduates; and a medical researcher wishes to estimate the recovery rate of heart attack victims treated with a new drug. In each of these cases, to accomplish the objective exactly, the statistics practitioner would have to examine each member of the population and then calculate the parameter of interest. For instance, television network executives would have to ask each person in the country what he or she is watching to determine the proportion of people who are watching their shows. Because there are millions of television viewers, the task is both impractical and prohibitively expensive. An alternative would be to take a random sample from this population, calculate the sample proportion, and use that as an estimator of the population proportion. The use of the sample proportion to estimate the population proportion seems logical. The selection of the sample statistic to be used as an estimator, however, depends on the characteristics of that statistic. Naturally, we want to use the statistic with the most desirable qualities for our purposes.

One desirable quality of an estimator is *unbiasedness*.

Unbiased Estimator

An **unbiased estimator** of a population parameter is an estimator whose expected value is equal to that parameter.

This means that if you were to take an infinite number of samples and calculate the value of the estimator in each sample, the average value of the estimators would equal the parameter. This amounts to saying that, on average, the sample statistic is equal to the parameter.

We know that the sample mean $\overline{X}$ is an unbiased estimator of the population mean μ. In presenting the sampling distribution of $\overline{X}$ in Section 9-1, we stated that $E(\overline{X}) = \mu$. We also know that the sample proportion is an unbiased estimator of the population proportion because $E(\hat{P}) = p$ and that the difference between two sample means is an unbiased estimator of the difference between two population means because $E(\overline{X}_1 - \overline{X}_2) = \mu_1 - \mu_2$.

Recall that in Chapter 4 we defined the sample variance as

$$s^2 = \sum \frac{(x_i - \overline{x})^2}{n - 1}$$

At the time, it seemed odd that we divided by $n - 1$ rather than by n. The reason for choosing $n - 1$ was to make $E(s^2) = \sigma^2$ so that this definition makes the sample variance an unbiased estimator of the population variance. (The proof of this statement requires about a page of algebraic manipulation, which is more than we would be comfortable presenting here.) Had we defined the sample variance using n in the denominator, the resulting statistic would be a biased estimator of the population variance, one whose expected value is less than the parameter.

Knowing that an estimator is unbiased only assures us that its expected value equals the parameter; it does not tell us how close the estimator is to the parameter. Another desirable quality is that, as the sample size grows larger, the sample statistic should come closer to the population parameter. This quality is called *consistency*.

Consistency

An unbiased estimator is said to be **consistent** if the difference between the estimator and the parameter grows smaller as the sample size grows larger.

The measure we use to gauge closeness is the variance (or the standard deviation). Thus, $\overline{X}$ is a consistent estimator of μ because the variance of $\overline{X}$ is σ^2/n. This implies that as n grows larger, the variance of $\overline{X}$ grows smaller. As a consequence, an increasing proportion of sample means falls close to μ.

Figure 10.1 depicts two sampling distributions of $\overline{X}$. One sampling distribution is based on samples of size 25, and the other is based on samples of size 100. The former is more spread out than the latter.

FIGURE 10.1

Sampling Distribution of $\overline{X}$ with $n = 25$ and $n = 100$

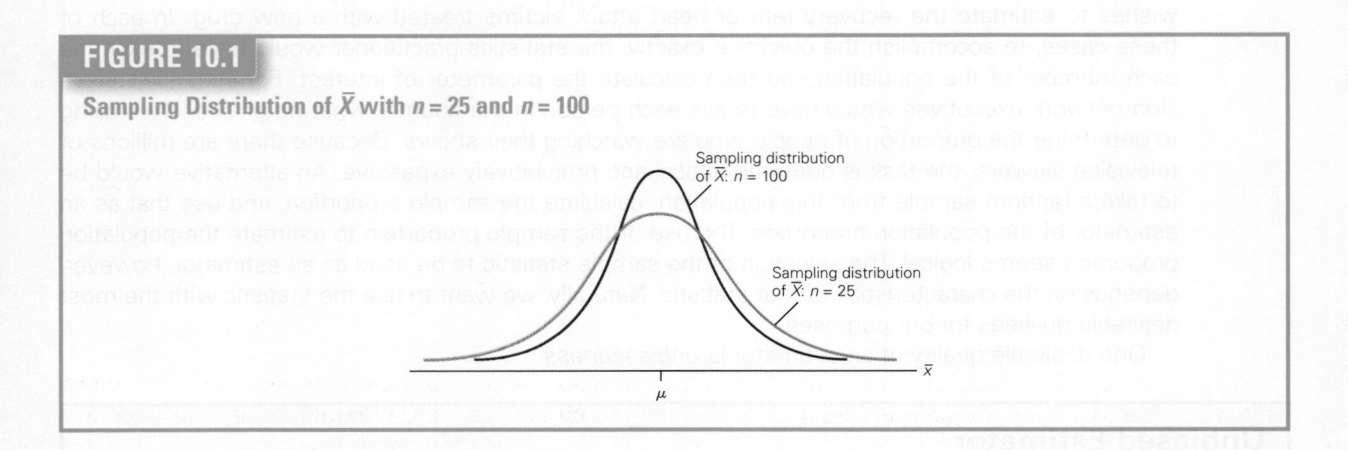

Similarly, $\hat{P}$ is a consistent estimator of p because it is unbiased and the variance of $\hat{P}$ is $p(1 - p)/n$, which grows smaller as n grows larger.

A third desirable quality is *relative efficiency*, which compares two unbiased estimators of a parameter.

Relative Efficiency

If there are two unbiased estimators of a parameter, the one whose variance is smaller is said to have **relative efficiency**.

We have already seen that the sample mean is an unbiased estimator of the population mean and that its variance is σ^2/n. In the next section, we will discuss the use of the sample median as an estimator of the population mean. Statisticians have established that the sample median is an unbiased estimator but that its variance is greater than that of the sample mean (when the population is normal). As a consequence, the sample mean is relatively more efficient than the sample median when estimating the population mean.

In the remaining chapters of this book we will present the statistical inference of a number of different population parameters. In each case, we will select a sample statistic that is unbiased and consistent. When there is more than one such statistic, we will choose the one that is relatively efficient to serve as the estimator.

10-1b Developing an Understanding of Statistical Concepts

In this section, we described three desirable characteristics of estimators: unbiasedness, consistency and relative efficiency. An understanding of statistics requires that you know that there are several potential estimators for each parameter, but that we choose the estimators used in this book because they possess these characteristics.

EXERCISES

10.1 How do point estimators and interval estimators differ?

10.2 Define unbiasedness.

10.3 Draw a sampling distribution of an unbiased estimator.

10.4 Draw a sampling distribution of a biased estimator.

10.5 Define consistency.

10.6 Draw diagrams representing what happens to the sampling distribution of a consistent estimator when the sample size increases.

10.7 Define relative efficiency.

10.8 Draw a diagram that shows the sampling distribution representing two unbiased estimators, one of which is relatively efficient.

10-2 ESTIMATING THE POPULATION MEAN WHEN THE POPULATION STANDARD DEVIATION IS KNOWN

We now describe how an interval estimator is produced from a sampling distribution. We choose to demonstrate estimation with an example that is unrealistic. However, this liability is offset by the example's simplicity. When you understand more about estimation, you will be able to apply the technique to more realistic situations.

Suppose we have a population with mean μ and standard deviation σ. The population mean is assumed to be unknown, and our task is to estimate its value. As we just discussed, the estimation procedure requires the statistics practitioner to draw a random sample of size n and calculate the sample mean $\bar{X}$.

The central limit theorem presented in Section 9-1 stated that $\bar{X}$ is normally distributed if X is normally distributed, or approximately normally distributed if X is non-normal and n is sufficiently large. This means that the variable

$$Z = \frac{\bar{X} - \mu}{\sigma/\sqrt{n}}$$

is standard normally distributed (or approximately so). In Section 9-1 we developed the following probability statement associated with the sampling distribution of the mean:

$$P\left(\mu - Z_{\alpha/2}\frac{\sigma}{\sqrt{n}} < \bar{X} < \mu + Z_{\alpha/2}\frac{\sigma}{\sqrt{n}}\right) = 1 - \alpha$$

which was derived from

$$P\left(-Z_{\alpha/2} < \frac{\bar{X} - \mu}{\sigma/\sqrt{n}} < Z_{\alpha/2}\right) = 1 - \alpha$$

Using a similar algebraic manipulation, we can express the probability in a slightly different form:

$$P\left(\bar{X} - Z_{\alpha/2}\frac{\sigma}{\sqrt{n}} < \mu < \bar{X} + Z_{\alpha/2}\frac{\sigma}{\sqrt{n}}\right) = 1 - \alpha$$

Notice that in this form the population mean is in the centre of the interval created by adding and subtracting $Z_{\alpha/2}$ standard errors to and from the sample mean. It is important for you to understand that this is merely another form of probability statement about the sample mean. This equation says that, with repeated sampling from this population, the proportion of values of $\bar{X}$ for which the interval

$$\bar{X} - Z_{\alpha/2}\frac{\sigma}{\sqrt{n}}, \bar{X} + Z_{\alpha/2}\frac{\sigma}{\sqrt{n}}$$

includes the population mean μ is equal to $1 - \sigma$. This form of probability statement is very useful to us because it is the **confidence interval estimator of μ.**

To apply this formula, we specify the confidence level $1 - \alpha$, from which we determine α, $\alpha/2$, $Z_{\alpha/2}$, (from Table 3 in Appendix A). Because the confidence level is the probability that the interval includes the actual value of μ, we generally set $1 - \alpha$ close to 1 (usually between 0.90 and 0.99).

Confidence Interval Estimator of μ*

$$\bar{X} - Z_{\alpha/2}\frac{\sigma}{\sqrt{n}},\ \bar{X} + Z_{\alpha/2}\frac{\sigma}{\sqrt{n}}$$

The probability $1 - \alpha$ is called the **confidence level**.

$\bar{X} - Z_{\alpha/2}\frac{\sigma}{\sqrt{n}}$ is called the **lower confidence limit (LCL)**.

$\bar{X} + Z_{\alpha/2}\frac{\sigma}{\sqrt{n}}$ is called the **upper confidence limit (UCL)**.

We often represent the confidence interval estimator as

$$\bar{X} \pm Z_{\alpha/2}\frac{\sigma}{\sqrt{n}}$$

where the minus sign defines the lower confidence limit and the plus sign defines the upper confidence limit.

In Table 10.1 we list four commonly used confidence levels and their associated values of $z_{\alpha/2}$. For example, if the confidence level is $1 - \alpha = 0.95$, $\alpha = 0.05$, $\alpha/2 = 0.025$, and $z_{\alpha/2} = z_{0.025} = 1.96$. The resulting confidence interval estimator is then called the **95% confidence interval estimator of μ.**

TABLE 10.1 Four Commonly Used Confidence Levels and $Z_{\alpha/2}$

$1-\alpha$	α	$\alpha/2$	$Z_{\alpha/2}$
0.90	0.10	0.05	$z_{0.05} = 1.645$
0.95	0.05	0.025	$z_{0.025} = 1.96$
0.98	0.02	0.01	$z_{0.01} = 2.33$
0.99	0.01	0.0005	$z_{0.005} = 2.575$

The following example illustrates how statistical techniques are applied. It also illustrates how we intend to solve problems in the rest of this book. The solution process that we advocate and use throughout this book is by and large the same one that statistics practitioners use to apply their skills in the real world. The process is divided into three stages. Simply stated, the stages are (1) the activities we perform before the calculations, (2) the calculations, and (3) the activities we perform after the calculations.

In stage 1, we determine the appropriate statistical technique to employ. Of course, for this example you will have no difficulty identifying the technique because you know only one at this point. (In practice, stage 1 also addresses the problem of *how* to gather the data. The methods used in the examples, exercises, and cases are described in the problem.)

In the second stage we calculate the statistics. We will do this in three ways.† To illustrate how the computations are completed, we will do the arithmetic manually with the assistance of a calculator. Solving problems by hand often provides insights into the statistical inference technique. Additionally, we will use the computer in two ways. First, in Excel we will use the Analysis ToolPak (**Data** menu item **Data Analysis**) or the add-ins we created for this book (**Add-Ins** menu item **Data Analysis Plus**). (Additionally, we will teach how to create do-it-yourself Excel spreadsheets that use built-in statistical functions.) Finally, we will use Minitab, one of the easiest software packages to use.

In the third and last stage of the solution, we intend to interpret the results and deal with the question presented in the problem. To be capable of properly interpreting statistical results, one needs to have an understanding of the fundamental principles underlying statistical inference.

*Since Chapter 7 we have been using the convention whereby an upper case letter (usually *X*) represents a random variable and a lower case letter (usually *x*) represents one of its values. However, in the formulas used in statistical inference, the distinction between the variable and its value becomes blurred. Accordingly, we will discontinue the notational convention and simply use lower case letters except when we wish to make a probability statement.

†We anticipate that students in most statistics modules will use only one of the three methods of computing statistics: the choice made by the instructor. If such is the case, readers are directed to ignore the other two.

APPLICATIONS IN OPERATIONS MANAGEMENT

Inventory Management

Operations managers use inventory models to determine the stock level that minimises total costs. In Section 8-2 we showed how the probabilistic model is used to make the inventory level decision. One component of that model is the mean demand during lead time. Recall that *lead time* refers to the interval between the time an order is made and when it is delivered. Demand during lead time is a random variable that is often assumed to be normally distributed. There are several ways to determine mean demand during lead time, but the simplest is to estimate that quantity from a sample.

EXAMPLE 10.1

Data Xm 10-01

Cape Computer Company

The Cape Computer Company makes its own computers and delivers them directly to customers who order them via the Internet. Cape competes primarily on price and speed of delivery. To achieve its objective of speed, Cape makes each of its five most popular computers and transports them to warehouses across the country. The computers are stored in the warehouses from which it generally takes 1 day to deliver a computer to the customer. This strategy requires high levels of inventory that add considerably to the cost. To lower these costs, the operations manager wants to use an inventory model. He notes that both daily demand and lead time are random variables. He concludes that demand during lead time is normally distributed, and he needs to know the mean to compute the optimum inventory level. He observes 25 lead time periods and records the demand during each period. These data are listed here. The manager would like a 95% confidence interval estimate of the mean demand during lead time. From long experience, the manager knows that the standard deviation is 75 computers.

Demand During Lead Time

235	374	309	499	253
421	361	514	462	369
394	439	348	344	330
261	374	302	466	535
386	316	296	332	334

SOLUTION:

IDENTIFY

To ultimately determine the optimum inventory level, the manager must know the mean demand during lead time. Thus, the parameter to be estimated is μ. At this point, we have described only one interval estimator. Thus, the confidence interval estimator that we intend to use is

$$\bar{x} \pm z_{\alpha/2}\frac{\sigma}{\sqrt{n}}$$

The next step is to perform the calculations. As we discussed previously, we will perform the calculations in three ways: manually, using Excel and using Minitab.

COMPUTE

Manually:

We need four values to construct the confidence interval estimate of μ. They are

$$\bar{X}, z_{\alpha/2}, \sigma, n$$

Using a calculator, we determine the summation $\sum x_i = 9254$. From this, we find

$$\bar{X} = \frac{\sum x_i}{n} = \frac{9254}{25} = 370.16$$

The confidence level is set at 95%; thus, $1 - \alpha = 0.95$, $\alpha = 1 - 0.95 = 0.05$, and $\alpha/2 = 0.025$. From Table 3 in Appendix A or from Table 10.1, we find

$$z_{\alpha/2} = z_{0.025} = 1.96$$

The population standard deviation is $\sigma = 75$, and the sample size is 25. Substituting $\bar{x}$, $z_{\alpha/2}$, σ and n into the confidence interval estimator, we find

$$\bar{x} \pm z_{\alpha/2}\frac{\sigma}{\sqrt{n}} = 370.16 \pm z_{0.025}\frac{75}{\sqrt{25}} = 370.06 \pm 1.96\frac{75}{\sqrt{25}} = 370.16 \pm 29.40$$

The lower and upper confidence limits are LCL = 340.76 and UCL = 399.56, respectively.

EXCEL

	A	B	C
1	*z*-Estimate: Mean		
2			
3			*Demand*
4	Mean		370.16
5	Standard deviation		80.783
6	Observations		25
7	SIGMA		75
8	LCL		340.76
9	UCL		399.56

Instructions

1. Type or import the data into one column. (**Open** Xm10-01.)
2. Click **Add-Ins**, **Data Analysis Plus** and **Z-Estimate: Mean**.
3. Fill in the dialog box: **Input Range** (A1:A26), type the value for the **Standard Deviation** (75), click **Labels** if the first row contains the name of the variable, and specify the confidence level by typing the value of α (0.05).

DO-IT-YOURSELF EXCEL

There is another way to produce the interval estimate for this problem. If you have already calculated the sample mean and know the sample size and population standard deviation, you need not employ the data set and **Data Analysis Plus** previously described. Instead you can create a spreadsheet that performs the calculations. Our suggested spreadsheet is shown next.

	A	B	C	D	E
1	**z-Estimate of a Mean**				
2					
3	**Sample mean**	370.16	**Confidence Interval Estimate**		
4	**Population standard deviation**	75	**370.16**	±	**29.40**
5	**Sample size**	25	**Lower confidence limit**		**340.76**
6	**Confidence level**	0.95	**Upper confidence limit**		**399.56**

Here are the tools (Excel functions) you will need to create this spreadsheet.

SQRT: Syntax: SQRT (*X*) Computes the square root of the quantity in parentheses. Use the **Insert** and **Ω Symbol** to input the ± sign.

NORM.S.INV: Syntax: NORM.S.INV (Probability) This function calculates the value of *z* such that $P(Z < z)$ = the probability in parentheses. For example, NORM.S.INV (0.95) determines the value of $z_{0.05}$ which is 1.645. You will need to figure out how to convert the confidence level specified in cell B6 into the value for $z_{\alpha/2}$.

We recommend that you save the spreadsheet. It can be used to solve some of the exercises at the end of this section.

In addition to providing another method of using Excel, this spreadsheet allows you to perform a 'what-if' analysis; that is, this worksheet provides you the opportunity to learn how changing some of the inputs affects the estimate. For example, type 0.99 in cell B6 to see what happens to the size of the interval when you increase the confidence level. Type 1000 in cell B5 to examine the effect of increasing the sample size. Type 10 in cell B4 to see what happens when the population standard deviation is smaller.

MINITAB

One-Sample *Z*: Demand

The assumed standard deviation = 75

Variable	N	Mean	StDev	SE Mean	95% CI
Demand	25	370.160	80.783	15.000	(340.761,399.559)

The output includes the sample standard deviation (**StDev** = 80.783), which is not needed for this interval estimate. Also printed is the standard error (**SE Mean** $= \sigma/\sqrt{n} =$ 15.0) and last, but not least, the 95% confidence interval estimate of the population mean.

Instructions

1. Type or import that data into one column. (**Open** Xm10-01.)
2. Click **Stat**, **Basic Statistics** and **1-Sample *Z***. . . .
3. Type or use the **Select** button to specify the name of the variable or the column it is stored in. In the **Samples in columns** box (Demand), type the value of the population standard deviation (75), and click **Options** . . . .
4. Type the value for the confidence level (0.95) and in the **Alternative** box select **not equal**.

INTERPRET

The operations manager estimates that the mean demand during lead time lies between 340.76 and 399.56. He can use this estimate as an input in developing an inventory policy. The model discussed in Section 8-2 computes the reorder point, assuming a particular value of the mean demand during lead time. In this example, he could have used the sample mean as a point estimator of the mean demand, from which the inventory policy could be determined. However, the use of the confidence interval estimator allows the manager to use both the lower and upper limits so that he can understand the possible outcomes.

10-2a Interpreting the Confidence Interval Estimate

Some people erroneously interpret the confidence interval estimate in Example 10.1 to mean that there is a 95% probability that the population mean lies between 340.76 and 399.56. This interpretation is wrong because it implies that the population mean is a variable about which we can make probability statements. In fact, the population mean is a fixed but unknown quantity. Consequently, we cannot interpret the confidence interval estimate of μ as a probability statement about μ. To translate the confidence interval estimate properly, we must remember that the confidence interval estimator was derived from the sampling distribution of the sample mean. In Section 9-1, we used the sampling distribution to make probability statements about the sample mean. Although the form has changed, the confidence interval estimator is also a probability statement about the sample mean. It states that there is $1 - \alpha$ probability that the sample mean will be equal to a value such that the interval $\bar{x} - z_{\alpha/2}\sigma/\sqrt{n}$ to $\bar{x} + z_{\alpha/2}\sigma/\sqrt{n}$ will include the population mean. Once the sample mean is computed, the interval acts as the lower and upper limits of the interval estimate of the population mean.

As an illustration, suppose we want to estimate the mean value of the distribution resulting from the throw of a fair die. Because we know the distribution, we also know that $\mu = 3.5$ and $\sigma = 1.71$. Pretend now that we know only that $\sigma = 1.71$, that μ is unknown, and that we want to estimate its value. To estimate μ, we draw a sample of size $n = 100$ (we throw the die 100 times) and calculate $\bar{x}$. The confidence interval estimator of μ is

$$\bar{x} \pm z_{\alpha/2}\frac{\sigma}{\sqrt{n}}$$

The 90% confidence interval estimator is

$$\bar{x} + z_{\alpha/2}\frac{\sigma}{\sqrt{n}} = \bar{x} \pm 1.645\frac{1.71}{\sqrt{100}} = \bar{x} \pm 0.281$$

This notation means that, if we repeatedly draw samples of size 100 from this population, 90% of the values of $\bar{x}$ will be such that μ would lie somewhere between $\bar{x} - 0.281$ and $\bar{x} + 0.281$, and 10% of the values of $\bar{x}$ will produce intervals that would not include μ. Now, imagine that we draw 40 samples of 100 observations each. The values of $\bar{x}$ and the resulting confidence interval estimates of μ are shown in Table 10.2. Notice that not all the intervals include the true value of the parameter. Samples 5, 16, 22 and 34 produce values of $\bar{x}$ that in turn produce intervals that exclude μ.

Students often react to this situation by asking 'What went wrong with samples 5, 16, 22 and 34?'. The answer is nothing. Statistics does not promise 100% certainty. In fact, in this illustration, we expected 90% of the intervals to include μ and 10% to exclude μ. Since we produced 40 intervals,

TABLE 10.2 90% Confidence Interval Estimates of μ

Sample	$\bar{x}$	$LCL = \bar{x} - 0.281$	$UCL = \bar{x} + 0.281$	Does interval include $\mu = 3.5$?
1	3.550	3.269	3.831	Yes
2	3.610	3.329	3.891	Yes
3	3.470	3.189	3.751	Yes
4	3.480	3.199	3.761	Yes
5	3.800	3.519	4.081	No
6	3.370	3.089	3.651	Yes
7	3.480	3.199	3.761	Yes
8	3.520	3.239	3.801	Yes
9	3.740	3.459	4.021	Yes
10	3.510	3.229	3.791	Yes
11	3.230	2.949	3.511	Yes
12	3.450	3.169	3.731	Yes
13	3.570	3.289	3.851	Yes
14	3.770	3.489	4.051	Yes
15	3.310	3.029	3.591	Yes
16	3.100	2.819	3.381	No
17	3.500	3.219	3.781	Yes

Sample	$\bar{x}$	LCL = $\bar{x}$ − 0.281	UCL = $\bar{x}$ + 0.281	Does interval include $\mu = 3.5$?
18	3.550	3.269	3.831	Yes
19	3.650	3.369	3.931	Yes
20	3.280	2.999	3.561	Yes
21	3.400	3.119	3.681	Yes
22	3.880	3.599	4.161	No
23	3.760	3.479	4.041	Yes
24	3.400	3.119	3.681	Yes
25	3.340	3.059	3.621	Yes
26	3.650	3.369	3.931	Yes
27	3.450	3.169	3.731	Yes
28	3.470	3.189	3.751	Yes
29	3.580	3.299	3.861	Yes
30	3.360	3.079	3.641	Yes
31	3.710	3.429	3.991	Yes
32	3.510	3.229	3.791	Yes
33	3.420	3.139	3.701	Yes
34	3.110	2.829	3.391	No
35	3.290	3.009	3.571	Yes
36	3.640	3.359	3.921	Yes
37	3.390	3.109	3.671	Yes
38	3.750	3.469	4.031	Yes
39	3.260	2.979	3.541	Yes
40	3.540	3.259	3.821	Yes

we expected that 4.0 (10% of 40) intervals would not contain $\mu = 3.5$.* It is important to understand that, even when the statistics practitioner performs experiments properly, a certain proportion (in this example, 10%) of the experiments will produce incorrect estimates by random chance.

We can improve the confidence associated with the interval estimate. If we let the confidence level $1 - \alpha$ equal 0.95, the 95% confidence interval estimator is

$$\bar{x} \pm z_{\alpha/2}\frac{\sigma}{\sqrt{n}} = \bar{x} \pm 1.96\frac{1.71}{\sqrt{100}} = \bar{x} \pm 0.335$$

Because this interval is wider, it is more likely to include the value of μ. If you re-do Table 10.2, this time using a 95% confidence interval estimator, only samples 16, 22 and 34 will produce intervals that do not include μ. (Notice that we expected 5% of the intervals to exclude μ and that we actually observed 3/40 = 7.5%.) The 99% confidence interval estimator is

$$\bar{x} \pm z_{\alpha/2}\frac{\sigma}{\sqrt{n}} = \bar{x} \pm 2.575\frac{1.71}{\sqrt{100}} = \bar{x} \pm 0.440$$

Applying this interval estimate to the sample means listed in Table 10.2 would result in having all 40 interval estimates include the population mean $\mu = 3.5$. (We expected 1% of the intervals to exclude μ; we observed 0/40 = 0%.)

In actual practice, only one sample will be drawn, and thus only one value of $\bar{x}$ will be calculated. The resulting interval estimate will either correctly include the parameter or incorrectly exclude it.

*In this illustration, exactly 10% of the sample means produced interval estimates that excluded the value of μ, but this will not always be the case. Remember, we expect 10% of the sample means in the long run to result in intervals excluding μ. This group of 40 sample means does not constitute 'the long run.'

Unfortunately, statistics practitioners do not know whether they are correct in each case; they know only that, in the long run, they will incorrectly estimate the parameter some of the time. Statistics practitioners accept that as a fact of life.

We summarise our calculations in Example 10.1 as follows. We estimate that the mean demand during lead time falls between 340.76 and 399.56, and this type of estimator is correct 95% of the time. Thus, the confidence level applies to our estimation procedure and not to any one interval. Incidentally, the media often refer to the 95% figure as '19 times out of 20', which emphasises the long-run aspect of the confidence level.

10-2b Information and the Width of the Interval

Interval estimation, like all other statistical techniques, is designed to convert data into information. However, a wide interval provides little information. For example, suppose that as a result of a statistical study we estimate with 95% confidence that the average starting salary of an accountant lies between £15 000 and £100 000. This interval is so wide that very little information was derived from the data. Suppose, however, that the interval estimate was £52 000 to £55 000. This interval is much narrower, providing accounting students more precise information about the mean starting salary.

The width of the confidence interval estimate is a function of the population standard deviation, the confidence level and the sample size. Consider Example 10.1, where σ was assumed to be 75. The interval estimate was 370.16 ± 29.40. If σ equalled 150, the 95% confidence interval estimate would become

$$\bar{x} \pm z_{\alpha/2}\frac{\sigma}{\sqrt{n}} = 370.16 \pm z_{0.025}\frac{150}{\sqrt{25}} = 370.16 \pm 1.96\frac{150}{\sqrt{25}} = 370.16 \pm 58.80$$

Thus, doubling the population standard deviation has the effect of doubling the width of the confidence interval estimate. This result is quite logical. If there is a great deal of variation in the random variable (measured by a large standard deviation), it is more difficult to accurately estimate the population mean. That difficulty is translated into a wider interval.

Although we have no control over the value of σ, we do have the power to select values for the other two elements. In Example 10.1, we chose a 95% confidence level. If we had chosen 90% instead, the interval estimate would have been

$$\bar{x} \pm z_{\alpha/2}\frac{\sigma}{\sqrt{n}} = 370.16 \pm z_{0.05}\frac{75}{\sqrt{25}} = 370.16 \pm 1.645\frac{75}{\sqrt{25}} = 370.16 \pm 24.68$$

A 99% confidence level results in this interval estimate:

$$\bar{x} \pm z_{\alpha/2}\frac{\sigma}{\sqrt{n}} = 370.16 \pm z_{0.005}\frac{75}{\sqrt{25}} = 370.16 \pm 2.575\frac{75}{\sqrt{25}} = 370.16 \pm 38.63$$

As you can see, decreasing the confidence level narrows the interval; increasing it widens the interval. However, a large confidence level is generally desirable because that means a larger proportion of confidence interval estimates that will be correct in the long run. There is a direct relationship between the width of the interval and the confidence level. This is because we need to widen the interval to be more confident in the estimate. (The analogy is that to be more likely to capture a butterfly, we need a larger butterfly net.) The trade-off between increased confidence and the resulting wider confidence interval estimates must be resolved by the statistics practitioner. As a general rule, however, 95% confidence is considered 'standard'.

The third element is the sample size. Had the sample size been 100 instead of 25, the confidence interval estimate would become

$$\bar{x} \pm z_{\alpha/2}\frac{\sigma}{\sqrt{n}} = 370.16 \pm z_{0.025}\frac{75}{\sqrt{100}} = 370.16 \pm 1.96\frac{75}{\sqrt{100}} = 370.16 \pm 14.70$$

Increasing the sample size fourfold decreases the width of the interval by half. A larger sample size provides more potential information. The increased amount of information is reflected in a narrower interval. However, there is another trade-off: Increasing the sample size increases the sampling cost. We will discuss these issues when we present sample size selection in Section 10-3.

10-2c Estimating the Population Mean Using the Sample Median

To understand why the sample mean is most often used to estimate a population mean, let's examine the properties of the sampling distribution of the sample median (denoted here as m). The sampling distribution of a sample median is normally distributed provided that the population is normal. Its mean and standard deviation are

$$\mu_m = \mu$$

and

$$\sigma_m = \frac{1.2533\sigma}{\sqrt{n}}$$

Using the same algebraic steps that we used above, we derive the confidence interval estimator of a population mean using the sample median

$$m \pm z_{\alpha/2}\frac{1.2533\sigma}{\sqrt{n}}$$

To illustrate, suppose that we have drawn the following random sample from a normal population whose standard deviation is 2.

1 1 1 3 4 5 6 7 8

The sample mean is $\bar{x} = 4$, and the sample median is $m = 4$.

The 95% confidence interval estimates using the sample mean and the sample median are

$$\bar{x} \pm z_{\alpha/2}\frac{\sigma}{\sqrt{n}} = 4.0 \pm 1.96\frac{2}{\sqrt{9}} = 4 \pm 1.307$$

$$m \pm z_{\alpha/2}\frac{1.2533\sigma}{\sqrt{n}} = 4.0 \pm 1.96\frac{(1.2533)(2)}{\sqrt{9}} = 4 \pm 1.638$$

As you can see, the interval based on the sample mean is narrower; as we pointed out previously, narrower intervals provide more precise information. To understand why the sample mean produces better estimators than the sample median, recall how the median is calculated. We simply put the data in order and select the observation that falls in the middle. Thus, as far as the median is concerned the data appear as

1 2 3 4 5 6 7 8 9

By ignoring the actual observations and using their ranks instead, we lose information. With less information, we have less precision in the interval estimators and so ultimately make poorer decisions.

EXERCISES

Developing an Understanding of Statistical Concepts

Exercises 10.9 to 10.16 are 'what-if' analyses designed to determine what happens to the interval estimate when the confidence level, sample size and standard deviation change. These problems can be solved manually, using the spreadsheet you created (that is, if you did create one) or Minitab.

10.9 **a.** A statistics practitioner took a random sample of 50 observations from a population with a standard deviation of 25 and computed the sample mean to be 100. Estimate the population mean with 90% confidence.

b. Repeat part (a) using a 95% confidence level.

c. Repeat part (a) using a 99% confidence level.

d. Describe the effect on the confidence interval estimate of increasing the confidence level.

10.10 **a.** The mean of a random sample of 25 observations from a normal population with a standard deviation of 50 is 200. Estimate the population mean with 95% confidence.

b. Repeat part (a) changing the population standard deviation to 25.

c. Repeat part (a) changing the population standard deviation to 10.

d. Describe what happens to the confidence interval estimate when the standard deviation is decreased.

10.11 **a.** A random sample of 25 was drawn from a normal distribution with a standard deviation of 5. The sample mean is 80. Determine the 95% confidence interval estimate of the population mean.

b. Repeat part (a) with a sample size of 100.

c. Repeat part (a) with a sample size of 400.

d. Describe what happens to the confidence interval estimate when the sample size increases.

10.12 **a.** Given the following information, determine the 98% confidence interval estimate of the population mean:

$$\bar{x} = 500 \quad \sigma = 12 \quad n = 50$$

b. Repeat part (a) using a 95% confidence level.

c. Repeat part (a) using a 90% confidence level.

d. Review parts (a)–(c) and discuss the effect on the confidence interval estimator of decreasing the confidence level.

10.13 **a.** The mean of a sample of 25 was calculated as $\bar{x} = 500$. The sample was randomly drawn from a population with a standard deviation of 15. Estimate the population mean with 99% confidence.

b. Repeat part (a) changing the population standard deviation to 30.

c. Repeat part (a) changing the population standard deviation to 60.

d. Describe what happens to the confidence interval estimate when the standard deviation is increased.

10.14 **a.** A statistics practitioner randomly sampled 100 observations from a population with a standard deviation of 5 and found that $\bar{x}$ is 10. Estimate the population mean with 90% confidence.

b. Repeat part (a) with a sample size of 25.

c. Repeat part (a) with a sample size of 10.

d. Describe what happens to the confidence interval estimate when the sample size decreases.

10.15 **a.** From the information given here determine the 95% confidence interval estimate of the population mean.

$$\bar{x} = 100 \quad \sigma = 20 \quad n = 25$$

b. Repeat part (a) with $\bar{x} = 200$.

c. Repeat part (a) with $\bar{x} = 500$.

d. Describe what happens to the width of the confidence interval estimate when the sample mean increases.

10.16 **a.** A random sample of 100 observations was randomly drawn from a population with a standard deviation of 5. The sample mean was calculated as $\bar{x} = 400$. Estimate the population mean with 99% confidence.

b. Repeat part (a) with $\bar{x} = 200$.

c. Repeat part (a) with $\bar{x} = 100$.

d. Describe what happens to the width of the confidence interval estimate when the sample mean decreases.

Exercises 10.17 to 10.20 are based on the optional subsection 'Estimating the Population Mean Using the Sample Median'. All exercises assume that the population is normal.

10.17 Is the sample median an unbiased estimator of the population mean? Explain.

10.18 Is the sample median a consistent estimator of the population mean? Explain.

10.19 Show that the sample mean is relatively more efficient than the sample median when estimating the population mean.

10.20 **a.** Given the following information, determine the 90% confidence interval estimate of the population mean using the sample median.

Sample median $= 500$, $\sigma = 12$, and $n = 50$

b. Compare your answer in part (a) to that produced in part (c) of Exercise 10.12. Why is the confidence interval estimate based on the sample median wider than that based on the sample mean?

Applications

The following exercises may be answered manually or with the assistance of a computer. The names of the files containing the data are shown.

10.20 Xr10-21 The following data represent a random sample of 9 marks (out of 10) on a statistics quiz. The marks are normally distributed with a standard deviation of 2. Estimate the population mean with 90% confidence.

7 9 7 5 4 8 3 10 9

10.22 Xr10-22 The following observations are the ages of a random sample of 8 men in a bar. It is known that the ages are normally distributed with a standard deviation of 10. Determine the 95% confidence interval estimate of the population mean. Interpret the interval estimate.

52 68 22 35 30 56 39 48

10.23 Xr10-23 How many rounds of golf do physicians (who play golf) play per year? A survey of 12 physicians revealed the following numbers:

3 41 17 1 33 37 18 15 17 12 29 51

Estimate with 95% confidence the mean number of rounds per year played by physicians, assuming that the number of rounds is normally distributed with a standard deviation of 12.

10.24 Xr10-24 Among the most exciting aspects of a university lecturer's life are the departmental meetings where such critical issues as the colour that the walls will be painted and who gets a new desk are decided. A sample of 20 lecturers was asked how many hours per year are devoted to these meetings. The responses are listed here. Assuming that the variable is normally distributed with a standard deviation of 8 hours, estimate the mean number of hours spent at departmental meetings by all lecturers. Use a confidence level of 90%.

14 17 3 6 17 3 8 4 20 15
7 9 0 5 11 15 18 13 8 4

10.25 Xr10-25 The number of cars sold annually by used car salespeople is normally distributed with a standard deviation of 15. A random sample of 15 salespeople was taken, and the number of cars each sold is listed here. Find the 95% confidence interval estimate of the population mean. Interpret the interval estimate.

79 43 58 66 101 63 79 33 58
71 60 101 74 55 88

10.26 Xr10-26 It is known that the amount of time needed to change the oil on a car is normally distributed with a standard deviation of 5 minutes. The amount of time to complete a random sample of ten oil changes was recorded and listed here. Compute the 99% confidence interval estimate of the mean of the population.

11 10 16 15 18 12 25 20 18 24

10.27 Xr10-27 Suppose that the amount of time teenagers spend weekly working at part-time jobs is normally distributed with a standard deviation of 40 minutes. A random sample of 15 teenagers was drawn, and each reported the amount of time spent at part-time jobs (in minutes). These are listed here. Determine the 95% confidence interval estimate of the population mean.

180 130 150 165 90 130 120 60 200
180 80 240 210 150 125

10.28 Xr10-28 One of the few negative side-effects of quitting smoking is weight gain. Suppose that the weight gain in the 12 months following a cessation in smoking is normally distributed with a standard deviation of 6 pounds. To estimate the mean weight gain, a random sample of 13 quitters was drawn; their recorded weights are listed here. Determine the 90% confidence interval estimate of the mean 12-month weight gain for all quitters.

16 23 8 2 14 22 18 11 10 19 5 8 15

10.29 Xr10-29 Because of different sales ability, experience and devotion, the incomes of real estate agents vary considerably. Suppose that in a large city the annual income is normally distributed with a standard deviation of £15 000. A random sample of 16 real estate agents was asked to report their annual income (in £1000). The responses are listed here. Determine the 99% confidence interval estimate of the mean annual income of all real estate agents in the city.

65 94 57 111 83 61 50 73 68 80
93 84 113 41 60 77

The following exercises require the use of a computer and software. The answers may be calculated manually.

10.30 Xr10-30 A survey of 400 statistics professors was undertaken. Each professor was asked how much time was devoted to teaching graphical techniques. We believe that the times are normally distributed with a standard deviation of 30 minutes. Estimate the population mean with 95% confidence.

10.31 Xr10-31 In a survey conducted to determine, among other things, the cost of holidays, 64 individuals were randomly sampled. Each person was asked to calculate the cost of his or her most recent holiday. Assuming that the standard deviation is €400, estimate with 95% confidence the average cost of all holidays.

10.32 Xr10-32 In an article about *disinflation*, various investments were examined. The investments included stocks, bonds and real estate. Suppose that a random sample of 200 rates of return on real estate investments was computed and recorded. Assuming that the standard deviation of all rates of return on real estate investments is 2.1%, estimate the mean rate of return on all real estate investments with 90% confidence. Interpret the estimate.

10.33 Xr10-33 A statistics lecturer is in the process of investigating how many tutorials university students miss each semester. To help answer this question, she took a random sample of 100 university students and asked each to report how many tutorials he or she had missed in the previous semester. Estimate the mean number of classes missed by all students at the university. Use a 99% confidence level and assume that the population standard deviation is known to be 2.2 classes.

10.34 Xr10-34 As part of a project to develop better lawn fertilizers, a research chemist wanted to determine the mean weekly growth rate of EverGreen lawn grass, a common type of grass. A sample of 250 blades of grass was measured, and the amount of growth in 1 week was recorded. Assuming that weekly growth is normally distributed with a standard deviation of 0.25 cm, estimate with 99% confidence the mean weekly growth of EverGreen lawn grass. Briefly describe what the interval estimate tells you about the growth of EverGreen lawn grass.

10.35 Xr10-35 A time study of a large production facility in Korea was undertaken to determine the mean time required to assemble a smartphone. A random sample of the times to assemble 50 smartphones was recorded. An analysis of the assembly times reveals that they are normally distributed with a standard deviation of 1.3 minutes. Estimate with 95% confidence the mean assembly time for all smartphones. What do your results tell you about the assembly times?

10.36 Xr10-36 One particular image of the Japanese manager is that of a workaholic with little or no leisure time. In a survey, a random sample of 250 Japanese middle managers was asked how many hours per week they spent in leisure activities (e.g. sports, movies, television). The results of the survey were recorded. Assuming that the population standard deviation is 6 hours, estimate with 90% confidence the mean leisure time per week for all Japanese middle managers. What do these results tell you?

10.37 Xr10-37 One measure of physical fitness is the amount of time it takes for the pulse rate to return to normal after exercise. A random sample of 100 women age 40 to 50 exercised on stationary bicycles for 30 minutes. The amount of time it took for their pulse rates to return to pre-exercise levels was measured and recorded. If the times are normally distributed with a standard deviation of 2.3 minutes, estimate with 99% confidence the true mean pulse-recovery time for all 40- to 50-year-old women. Interpret the results.

10.38 Xr10-38 A survey of 80 randomly selected companies asked them to report the annual income of their presidents. Assuming that incomes are normally distributed with a standard deviation of €30 000, determine the 90% confidence interval estimate of the mean annual income of all company presidents. Interpret the statistical results.

APPLICATIONS IN MARKETING

Advertising

One of the major tools in the promotion mix is advertising. An important decision to be made by the advertising manager is how to allocate the company's total advertising budget among the various competing media types, including television, radio and newspapers. Ultimately, the manager wants to know, for example, which television programmes are most watched by potential customers, and how effective it is to sponsor these programmes through advertising. But first the manager must assess the size of the audience, which involves estimating the amount of exposure potential customers have to the various media types, such as television.

10.39 **Xr10-39** To help make a decision about expansion plans, the president of a music company needs to know how many albums teenagers buy annually. Accordingly, he commissions a survey of 250 teenagers. Each is asked to report how many albums he or she purchased in the previous 12 months. Estimate with 90% confidence the mean annual number of albums purchased by all teenagers. Assume that the population standard deviation is three albums.

10.40 **Xr10-40** The sponsors of television shows targeted at the children's market wanted to know the amount of time children spend watching television because the types and number of programmes and commercials are greatly influenced by this information. As a result, it was decided to survey 100 children and ask them to keep track of the number of hours of television they watch each week. From past experience, it is known that the population standard deviation of the weekly amount of television watched is $\sigma = 8.0$ hours. The television sponsors want an estimate of the amount of television watched by the average child. A confidence level of 95% is judged to be appropriate.

10-3 SELECTING THE SAMPLE SIZE

As we discussed in the previous section, if the interval estimate is too wide, it provides little information. In Example 10.1 the interval estimate was 340.76 to 399.56. If the manager is to use this estimate as input for an inventory model, he needs greater precision. Fortunately, statistics practitioners can control the width of the interval by determining the sample size necessary to produce narrow intervals.

To understand how and why we can determine the sample size, we discuss the error of estimation.

10-3a Error of Estimation

In Chapter 5 we pointed out that sampling error is the difference between the sample and the population that exists only because of the observations that happened to be selected for the sample. Now that we have discussed estimation, we can define the sampling error as the difference between an estimator and a parameter. We can also define this difference as the **error of estimation**. In this chapter, this can be expressed as the difference between $\overline{X}$ and μ. In our derivation of the confidence interval estimator of μ, we expressed the following probability,

$$P\left(-Z_{\alpha/2} < \frac{\overline{X}-\mu}{\sigma/\sqrt{n}} < Z_{\alpha/2}\right) = 1-\alpha$$

which can also be expressed as

$$P\left(-Z_{\alpha/2}\frac{\sigma}{\sqrt{n}} < \overline{X}-\mu < +Z_{\alpha/2}\frac{\sigma}{\sqrt{n}}\right) = 1-\alpha$$

This tells us that the difference between $\overline{X}$ and μ lies between $-Z_{\alpha/2}\sigma/\sqrt{n}$ and $+Z_{\alpha/2}\sigma/\sqrt{n}$ with probability $1-\alpha$. Expressed another way, we have with probability $1-\alpha$,

$$|\overline{X}-\mu| < Z_{\alpha/2}\frac{\sigma}{\sqrt{n}}$$

In other words, the error of estimation is less than $Z_{\alpha/2}\sigma/\sqrt{n}$. We interpret this to mean that $Z_{\alpha/2}\sigma/\sqrt{n}$ is the maximum error of estimation that we are willing to tolerate. We label this value B, which stands for the **bound on the error of estimation**; that is,

$$B = Z_{\alpha/2}\frac{\sigma}{\sqrt{n}}$$

10-3b Determining the Sample Size

We can solve the equation for n if the population standard deviation σ, the confidence level $1-\alpha$ and the bound on the error of estimation B are known. Solving for n, we produce the following.

Sample Size to Estimate a Mean

$$n = \left(\frac{z_{\alpha/2}\sigma}{B}\right)^2$$

To illustrate, suppose that in Example 10.1 before gathering the data, the manager had decided that he needed to estimate the mean demand during lead time to within 16 units, which is the bound on the error of estimation. We also have $1 - \alpha = 0.95$ and $\sigma = 75$. We calculate

$$n = \left(\frac{z_{\alpha/2}\sigma}{B}\right)^2 = \left(\frac{(1.96)(75)}{16}\right)^2 = 84.41$$

Because n must be an integer and because we want the bound on the error of estimation to be *no more than* 16, any non-integer value must be rounded up. Thus, the value of n is rounded to 85, which means that to be 95% confident that the error of estimation will be no larger than 16, we need to randomly sample 85 lead time intervals.

DETERMINING THE SAMPLE SIZE TO ESTIMATE THE MEAN TREE DIAMETER: SOLUTION

Before the sample was taken, the forester determined the sample size as follows. The bound on the error of estimation is $B = 1$. The confidence level is 90% ($1 - \alpha = 0.90$). Thus $\alpha = 0.10$ and $\alpha/2 = 0.05$. It follows that $z_{\alpha/2} = 1.645$. The population standard deviation is assumed to be $\sigma = 6$. Thus,

$$n = \left(\frac{z_{\alpha/2}\sigma}{B}\right)^2 = \left(\frac{1.645 \times 6}{1}\right)^2 = 97.42$$

which is rounded to 98.

However, after the sample is taken the forester discovered that $\sigma = 12$. The 90% confidence interval estimate is

$$x \pm z_{\alpha/2}\frac{\sigma}{\sqrt{n}} = 25 \pm z_{0.05}\frac{12}{\sqrt{98}} = 25 \pm 1.645\frac{12}{\sqrt{98}} = 25 \pm 2$$

As you can see, the bound on the error of estimation is 2 and not 1. The interval is twice as wide as it was designed to be. The resulting estimate will not be as precise as needed.

In this chapter we have assumed that we know the value of the population standard deviation. In practice, this is seldom the case. (In Chapter 12 we introduce a more realistic confidence interval estimator of the population mean.) It is frequently necessary to 'guesstimate' the value of σ to calculate the sample size; that is, we must use our knowledge of the variable with which we are dealing to assign some value to σ.

Unfortunately, we cannot be very precise in this guess. However, in guesstimating the value of σ, we prefer to err on the high side. For the chapter-opening example, if the forester had determined the sample size using $\sigma = 12$, he would have computed

$$n = \left(\frac{z_{\alpha/2}\sigma}{B}\right)^2 = \left(\frac{(1.645)(12)}{1}\right)^2 = 389.67 \text{ (rounded to 390)}$$

Using $n = 390$ (assuming that the sample mean is again 25), the 90% confidence interval estimate is

$$\bar{x} \pm z_{\alpha/2} \frac{\sigma}{\sqrt{n}} = 25 \pm 1.645 \frac{12}{\sqrt{390}} = 25 \pm 1$$

This interval is as narrow as the forester wanted.

What happens if the standard deviation is *smaller* than assumed? If we discover that the standard deviation is less than we assumed when we determined the sample size, the confidence interval estimator will be narrower and therefore more precise. Suppose that after the sample of 98 trees was taken (assuming again that $\sigma = 6$), the forester discovers that $\sigma = 3$. The confidence interval estimate is

$$\bar{x} \pm z_{\alpha/2} \frac{\sigma}{\sqrt{n}} = 25 \pm 1.645 \frac{3}{\sqrt{98}} = 25 \pm 0.5$$

which is narrower than the forester wanted. Although this means that he would have sampled more trees than needed, the additional cost is relatively low when compared to the value of the information derived.

EXERCISES

Developing an Understanding of Statistical Concepts

10.41 a. Determine the sample size required to estimate a population mean to within 10 units given that the population standard deviation is 50. A confidence level of 90% is judged to be appropriate.

b. Repeat part (a) changing the standard deviation to 100.

c. Redo part (a) using a 95% confidence level.

d. Repeat part (a) wherein we wish to estimate the population mean to within 20 units.

10.42 Review Exercise 10.41. Describe what happens to the sample size when

a. the population standard deviation increases.

b. the confidence level increases.

c. the bound on the error of estimation increases.

10.43 a. A statistics practitioner would like to estimate a population mean to within 50 units with 99% confidence given that the population standard deviation is 250. What sample size should be used?

b. Redo part (a) changing the standard deviation to 50.

c. Redo part (a) using a 95% confidence level.

d. Redo part (a) wherein we wish to estimate the population mean to within 10 units.

10.44 Review the results of Exercise 10.43. Describe what happens to the sample size when

a. the population standard deviation decreases.

b. the confidence level decreases.

c. the bound on the error of estimation decreases.

10.45 a. Determine the sample size necessary to estimate a population mean to within 1 with 90% confidence given that the population standard deviation is 10.

b. Suppose that the sample mean was calculated as 150. Estimate the population mean with 90% confidence.

10.46 a. Repeat part (b) in Exercise 10.45 after discovering that the population standard deviation is actually 5.

b. Repeat part (b) in Exercise 10.45 after discovering that the population standard deviation is actually 20.

10.47 Review Exercises 10.45 and 10.46. Describe what happens to the confidence interval estimate when

a. the standard deviation is equal to the value used to determine the sample size.

b. the standard deviation is smaller than the one used to determine the sample size.

c. the standard deviation is larger than the one used to determine the sample size.

10.48 a. A statistics practitioner would like to estimate a population mean to within 10 units. The confidence level has been set at 95% and $\sigma = 200$. Determine the sample size.

b. Suppose that the sample mean was calculated as 500. Estimate the population mean with 95% confidence.

10.49 a. Repeat part (b) of Exercise 10.48 after discovering that the population standard deviation is actually 100.

b. Repeat part (b) of Exercise 10.48 after discovering that the population standard deviation is actually 400.

10.50 Review Exercises 10.48 and 10.49. Describe what happens to the confidence interval estimate when

a. the standard deviation is equal to the value used to determine the sample size.

b. the standard deviation is smaller than the one used to determine the sample size.

c. the standard deviation is larger than the one used to determine the sample size.

Applications

10.51 A medical statistician wants to estimate the average weight loss of people who are on a new diet plan. In a preliminary study, he guesses that the standard deviation of the population of weight losses is about 5 kg. How large a sample should he take to estimate the mean weight loss to within 1 kg, with 90% confidence?

10.52 The operations manager of a large production plant would like to estimate the average amount of time workers take to assemble a new electronic component. After observing a number of workers assembling similar devices, she guesses that the standard deviation is 6 minutes. How large a sample of workers should she take if she wishes to estimate the mean assembly time to within 20 seconds? Assume that the confidence level is to be 99%.

10.53 A statistics lecturer wants to compare today's students with those 25 years ago. All his current students' marks are stored on a computer so that he can easily determine the population mean. However, the marks 25 years ago reside only in his musty files. He does not want to retrieve all the marks and will be satisfied with a 95% confidence interval estimate of the mean mark 25 years ago. If he assumes that the population standard deviation is 12, how large a sample should he take to estimate the mean to within 2 marks?

10.54 A medical researcher wants to investigate the amount of time it takes for patients' headache pain to be relieved after taking a new prescription painkiller. She plans to use statistical methods to estimate the mean of the population of relief times. She believes that the population is normally distributed with a standard deviation of 20 minutes. How large a sample should she take to estimate the mean time to within 1 minute with 90% confidence?

10.55 The label on 4-litre cans of paint states that the amount of paint in the can is sufficient to paint about 35 square metres. However, this number is quite variable depending on the type of paint, dilution and thinner used. In fact, the amount of coverage is known to be approximately normally distributed with a standard deviation of 2.5 square metres. How large a sample should be taken to estimate the true mean coverage of all 4-litre cans to within 0.5 square metres with 95% confidence?

10.56 The operations manager of a plant making handheld tablet devices has proposed rearranging the production process to be more efficient. She wants to estimate the time to assemble the tablet using the new arrangement. She believes that the population standard deviation is 15 seconds. How large a sample of workers should she take to estimate the mean assembly time to within 2 seconds with 95% confidence?

CHAPTER SUMMARY

This chapter introduced the concepts of **estimation**, the **estimator** of a population mean when the population variance is known, and the error of estimation. Additional consistency, relative efficiency and confidence level are introduced. It also presented a formula to calculate the sample size necessary to estimate a population mean.

SYMBOLS:

Symbol	Pronounced	Represents
$1 - \alpha$	One minus alpha	Confidence level
B		Bound on the error of estimation
$z_{\alpha/2}$	z alpha by 2	Value of Z such that the area to its right is equal to $a/2$

FORMULAS:

Confidence interval estimator of μ with σ known

$$\bar{x} \pm z_{\alpha/2} \frac{\sigma}{\sqrt{n}}$$

Sample size to estimate μ

$$n = \left(\frac{z_{\alpha/2} \sigma}{B} \right)^2$$

11 INTRODUCTION TO HYPOTHESIS TESTING

11-1 CONCEPTS OF HYPOTHESIS TESTING

11-2 TESTING THE POPULATION MEAN WHEN THE POPULATION STANDARD DEVIATION IS KNOWN

11-3 CALCULATING THE PROBABILITY OF A TYPE II ERROR

11-4 THE ROAD AHEAD

SSA ENVELOPE PLAN

DATA Xm11-00

Federal Express (FedEx) sends invoices to customers requesting payment within 30 days. Each bill lists an address, and customers are expected to use their own envelopes to return their payments. Currently, the mean and standard deviation of the amount of time taken to pay bills are 24 days and 6 days, respectively. The chief financial officer (CFO) believes that including a stamped self-addressed (SSA) envelope would decrease the amount of time. She calculates that the improved cash flow from a 2-day decrease in the payment period would pay for the costs of the envelopes and stamps. Any further decrease in the payment period would generate a profit. To test her belief, she randomly selects 220 customers and includes an SSA envelope with their invoices. The numbers of days until payment is received were recorded. Can the CFO conclude that the plan will be profitable? **See Section 11-2h for the answer.**

INTRODUCTION

In Chapter 10 we introduced estimation and showed how it is used. Now we are going to present the second general procedure of making inferences about a population – hypothesis testing. The purpose of this type of inference is to determine whether enough statistical evidence exists to enable us to conclude that a belief or hypothesis about a parameter is supported by the data. You will discover that hypothesis testing has a wide variety of applications in business and economics, as well as many other fields. This chapter will lay the foundation upon which the rest of the book is based. As such it represents a critical contribution to your development as a statistics practitioner.

In the next section, we will introduce the concepts of hypothesis testing, and in Section 11-2 we will develop the method employed to test a hypothesis about a population mean when the population standard deviation is known. The rest of the chapter deals with related topics.

11-1 CONCEPTS OF HYPOTHESIS TESTING

A hypothesis is a supposition which is based on some basic information. The term **hypothesis testing** is probably new to most readers, but the concepts underlying hypothesis testing are quite familiar and this procedure follows a logical sequence of stages from proposing the hypothesis to deciding whether to accept or reject it. There are a variety of non-statistical applications of hypothesis testing, the best known of which is a criminal trial.

When a person is accused of a crime, he or she faces a trial. The prosecution presents its case, and a jury must make a decision on the basis of the evidence presented. In fact, the jury conducts a test of hypothesis. There are actually two hypotheses that are tested. The first is called the **null hypothesis** and is represented by H_0 (pronounced *H nought – nought* is a British term for zero). It is

H_0: **The defendant is innocent**

The second is called the **alternative hypothesis** (or **research hypothesis**) and is denoted H_1. In a criminal trial it is

H_1: **The defendant is guilty**

The null hypothesis acts as both a starting point and a benchmark against which the outcomes can be measured. Of course, the jury does not know which hypothesis is correct. The members must make a decision on the basis of the evidence presented by both the prosecution and the defence. There are only two possible decisions. Convict or acquit the defendant. In statistical parlance, convicting the defendant is equivalent to *rejecting the null hypothesis in favour of the alternative*; that is, the jury is saying that there was enough evidence to conclude that the defendant was guilty. Acquitting a defendant is phrased as *not rejecting the null hypothesis in favour of the alternative,* which means that the jury decided that there was not enough evidence to conclude that the defendant was guilty. Notice that we do not say that we accept the null hypothesis. In a criminal trial, that would be interpreted as finding the defendant *innocent.* Our justice system does not allow this decision.

There are two possible diametrically opposite types of errors. A **Type I error** occurs when we reject a true null hypothesis when it is in fact true. A **Type II error** is defined as not rejecting a false null hypothesis when it is in fact false. In the criminal trial, a Type I error is made when an innocent person is wrongly convicted. A Type II error occurs when a guilty defendant is acquitted. The probability of a Type I error is denoted by α which is also called the **significance level.** The probability of a Type II error is denoted by β (Greek letter *beta*). The error probabilities α and β are inversely related, meaning that any attempt to reduce one will increase the other. In practice, before starting a significance test we must decide on the importance of each type of error and then set the confidence level accordingly. Table 11.1 summarises the terminology and the concepts.

TABLE 11.1 Terminology of Hypothesis Testing

DECISION	H_0 IS TRUE (DEFENDANT IS INNOCENT)	H_0 IS FALSE (DEFENDANT IS GUILTY)
REJECT H_0 Convict defendant	Type I Error P(Type I Error) $= \alpha$	Correct decision
DO NOT REJECT H_0 Acquit defendant	Correct decision	Type II Error P(Type II Error) $= \beta$

In the UK justice system, as with many other countries (including the USA), Type I errors are regarded as more serious. As a consequence, the system is typically set up so that the probability of a Type I error is small. This is arranged by placing the burden of proof on the prosecution (the prosecution must prove guilt – the defence need not prove anything) and by having judges instruct the jury to find the defendant guilty only if there is 'evidence beyond a reasonable doubt'. In the absence of enough evidence, the jury must acquit even though there may be some evidence of guilt. The consequence of this arrangement is that the probability of acquitting guilty people is relatively large. A US Supreme Court justice once phrased the relationship between the probabilities of Type I and Type II errors in the following way: 'Better to acquit 100 guilty men than convict one innocent one'. In this opinion, the probability of a Type I error should be 1/100 of the probability of a Type II error.

The critical concepts in hypothesis testing follow.

1. There are two hypotheses. One is called the null hypothesis, and the other the alternative or research hypothesis.
2. The testing procedure begins with the assumption that the null hypothesis is true.
3. The goal of the process is to determine whether there is enough evidence to infer that the alternative hypothesis is true.
4. There are two possible decisions:

   Conclude that there is enough evidence to support the alternative hypothesis.
   Conclude that there is not enough evidence to support the alternative hypothesis.
5. Two possible errors can be made in any test. A Type I error occurs when we reject a true null hypothesis, and a Type II error occurs when we do not reject a false null hypothesis. The probabilities of Type I and Type II errors are

   $P(\text{Type I error}) = \alpha$
   $P(\text{Type II error}) = \beta$

Let's extend these concepts to statistical hypothesis testing.

In statistics we frequently test hypotheses about parameters. The hypotheses we test are generated by questions that managers need to answer. To illustrate, suppose that in Example 10.1 (Chapter 10) the operations manager did not want to estimate the mean demand during lead time but instead wanted to know whether the mean is different from 350, which may be the point at which the current inventory policy needs to be altered. In other words, the manager wants to determine whether he can infer that μ is not equal to 350. We can rephrase the question so that it now reads: 'Is there enough evidence to conclude that μ is not equal to 350?'. This wording is analogous to the criminal trial wherein the jury is asked to determine whether there is enough evidence to conclude that the defendant is guilty. Thus, the alternative (research) hypothesis is

$$H_1: \mu \neq 350$$

In a criminal trial, the process begins with the assumption that the defendant is innocent. In a similar fashion, we start with the assumption that the parameter equals the value we are testing. Consequently, the operations manager would assume that $\mu = 350$, and the null hypothesis is expressed as

$$H_0: \mu = 350$$

When we state the hypotheses, we list the null first followed by the alternative hypothesis. To determine whether the mean is different from 350, we test

$$H_0: \mu = 350$$
$$H_1: \mu \neq 350$$

Now suppose that in this illustration the current inventory policy is based on an analysis that revealed that the actual mean demand during lead time is 350. After a vigorous advertising campaign, the manager suspects that there has been an increase in demand and thus an increase in mean demand during lead time. To test whether there is evidence of an increase, the manager would specify the alternative hypothesis as

$$H_1: \mu > 350$$

Because the manager knew that the mean was (and maybe still is) 350, the null hypothesis would state

$$H_0: \mu = 350$$

Further suppose that the manager does not know the actual mean demand during lead time, but the current inventory policy is based on the assumption that the mean is *less than or equal to* 350. If the advertising campaign increases the mean to a quantity larger than 350, a new inventory plan will have to be instituted. In this scenario, the hypotheses become

$$H_0: \mu \leq 350$$
$$H_1: \mu > 350$$

Notice that in both illustrations the alternative hypothesis is designed to determine whether there is enough evidence to conclude that the mean is greater than 350. Although the two null hypotheses are different (one states that the mean is equal to 350, and the other states that the mean is less than or equal to 350), when the test is conducted, the process begins by assuming that the mean is *equal to* 350. In other words, no matter the form of the null hypothesis, we use the equal sign in the null hypothesis. Here is the reason. If there is enough evidence to conclude that the alternative hypothesis (the mean is greater than 350) is true when we assume that the mean is *equal to* 350, we would certainly draw the same conclusion when we assume that the mean is a value that is *less than* 350. As a result, the null hypothesis will always state that the parameter equals the value specified in the alternative hypothesis.

To emphasise this point, suppose the manager now wanted to determine whether there has been a decrease in the mean demand during lead time. We express the null and alternative hypotheses as

$$H_0: \mu = 350$$
$$H_1: \mu < 350$$

The hypotheses are often set up to reflect a manager's decision problem wherein the null hypothesis represents the *status quo*. Often this takes the form of some course of action such as maintaining a particular inventory policy. If there is evidence of an increase or decrease in the value of the parameter, a new course of action will be taken. Examples include deciding to produce a new product, switching to a better drug to treat an illness or sentencing a defendant to prison.

The next element in the procedure is to randomly sample the population and calculate the sample mean. This is called the test statistic. The test statistic is the criterion on which we base our decision about the hypotheses. (In the criminal trial analogy, this is equivalent to the evidence presented in the case.) The test statistic is based on the best estimator of the parameter. In Chapter 10 we stated that the best estimator of a population mean is the sample mean.

If the test statistic's value is inconsistent with the null hypothesis, we reject the null hypothesis and infer that the alternative hypothesis is true. For example, if we are trying to decide whether the mean is greater than 350, a large value of $\overline{x}$ (say, 600) would provide enough evidence. If $\overline{x}$ is close to 350 (say, 355), we would say that this does not provide much evidence to infer that the mean is greater than 350. In the absence of sufficient evidence, we do not reject the null hypothesis in favour of the alternative. (In the absence of sufficient evidence of guilt, a jury finds the defendant not guilty.)

In a criminal trial, 'sufficient evidence' is defined as 'evidence beyond a reasonable doubt'. In statistics, we need to use the test statistic's sampling distribution to define 'sufficient evidence'. We will do so in the next section.

EXERCISES

Exercises 11.1–11.5 feature non-statistical applications of hypothesis testing. For each, identify the hypotheses, define Type I and Type II errors, and discuss the consequences of each error. In setting up the hypotheses, you will have to consider where to place the 'burden of proof'.

11.1 It is the responsibility of the European Medicines Evaluation Agency, set up in January 1995, to judge the safety and effectiveness of the most important drugs, including anti-cancer medicines. There are two possible decisions: approve the drug or disapprove the drug.

11.2 You are contemplating a PhD in business or economics. If you succeed, a life of fame, fortune and happiness awaits you. If you fail, you have wasted 3 years of your life. Should you go for it?

11.3 You are faced with two investments. One is very risky, but the potential returns are high. The other is safe, but the potential is quite limited. Pick one.

11.4 You are the pilot of a jumbo jet. You smell smoke in the cockpit. The nearest airport is less than 5 minutes away. Should you land the plane immediately?

11.5 Several years ago in a high-profile case, a defendant was acquitted in a double-murder trial but was subsequently found responsible for the deaths in a civil trial. In a civil trial the plaintiff (the victims' relatives) are required only to show that the preponderance of evidence points to the guilt of the defendant. Aside from the other issues in the cases, discuss why these results are logical.

11-2 TESTING THE POPULATION MEAN WHEN THE POPULATION STANDARD DEVIATION IS KNOWN

To illustrate the process, consider the following example.

EXAMPLE 11.1

Data Xm11-01

Department Store's New Billing System

The manager of a department store is thinking about establishing a new billing system for the store's credit customers. After a thorough financial analysis, she determines that the new system will be cost-effective only if the mean monthly account is more than €170. A random sample of 400 monthly accounts is drawn, for which the sample mean is €178. The manager knows that the accounts are approximately normally distributed with a standard deviation of €65. Can the manager conclude from this that the new system will be cost-effective?

SOLUTION:

IDENTIFY

This example deals with the population of the credit accounts at the store. To conclude that the system will be cost-effective requires the manager to show that the mean account for all customers is greater than €170. Consequently, we set up the alternative hypothesis to express this circumstance:

$$H_1: \mu > 170 \text{ (Install new system)}$$

If the mean is less than or equal to 170, then the system will not be cost-effective. The null hypothesis can be expressed as

$$H_0: \mu \leq 170 \text{ (Do not install new system)}$$

However, as was discussed in Section 11-1, we will actually test $\mu = 170$, which is how we specify the null hypothesis:

$$H_0: \mu = 170$$

As we previously pointed out, the test statistic is the best estimator of the parameter. In Chapter 10, we used the sample mean to estimate the population mean. To conduct this test, we ask and answer the following question: Is a sample mean of 178 sufficiently greater than 170 to allow us to confidently infer that the population mean is greater than 170?

There are two approaches to answering this question. The first is called the *rejection region method.* It can be used in conjunction with the computer, but it is mandatory for those computing statistics manually and taking into consideration the critical values and rejection region. The second is the *p-value approach*, which in general can be employed only in conjunction with a computer and statistical software. The conclusions from both approaches are exactly the same and we recommend, however, that users of statistical software be familiar with both approaches.

11-2a Rejection Region

It seems reasonable to reject the null hypothesis in favour of the alternative if the value of the sample mean is large relative to 170. If we had calculated the sample mean to be say, 500, it would be quite apparent that the null hypothesis is false and we would reject it. On the other hand, values of $\bar{x}$ close to 170, such as 171, do not allow us to reject the null hypothesis because it is entirely possible to observe a sample mean of 171 from a population whose mean is 170. Unfortunately, the decision is not always so obvious. In this example, the sample mean was calculated to be 178, a value apparently neither very far away from nor very close to 170. To make a decision about this sample mean, we set up the *rejection region.*

Rejection Region

The rejection region is a range of values such that if the test statistic falls into that range, we decide to reject the null hypothesis in favour of the alternative hypothesis.

Suppose we define the value of the sample mean that is just large enough to reject the null hypothesis as $\bar{x}_L$. The rejection region is

$$\bar{x} > \bar{x}_L$$

Because a Type I error is defined as rejecting a true null hypothesis, and the probability of committing a Type I error is α, it follows that

$$\begin{aligned}\alpha &= P(\text{rejecting } H_0 \text{ given that } H_0 \text{ is true})\\ &= P(\bar{x} > \bar{x}_L \text{ given that } H_0 \text{ is true})\end{aligned}$$

Figure 11.1 depicts the sampling distribution and the rejection region.

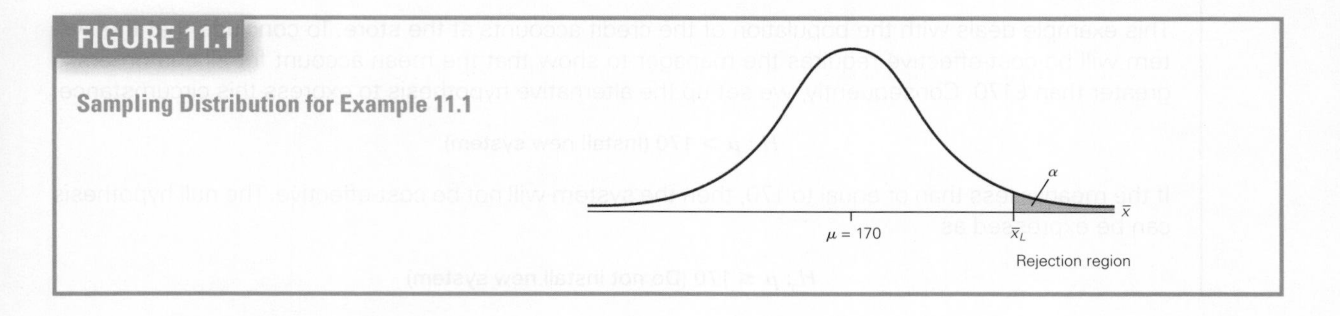

FIGURE 11.1 Sampling Distribution for Example 11.1

From Section 9-1 we know that the sampling distribution of $\overline{x}$ is normal or approximately normal, with mean μ and standard deviation $\sigma/\sqrt{n}$. As a result, we can standardise $\overline{x}$ and obtain the following probability:

$$P\left(\frac{\overline{x}-\mu}{\sigma/\sqrt{n}} > \frac{\overline{x}_L-\mu}{\sigma/\sqrt{n}}\right) = P\left(Z > \frac{\overline{x}_L-\mu}{\sigma/\sqrt{n}}\right) = \alpha$$

From Section 8-2, we defined z_α to be the value of a standard normal random variable such that

$$P(Z > z_\alpha) = \alpha$$

Because both probability statements involve the same distribution (standard normal) and the same probability (α), it follows that the limits are identical. Thus,

$$\frac{\overline{x}_L-\mu}{\sigma/\sqrt{n}} = z_\alpha$$

We know that $\sigma = 65$ and $n = 400$. Because the probabilities defined earlier are conditional on the null hypothesis being true, we have $\mu = 170$. To calculate the rejection region, we need a value of α at the significance level. Suppose that the manager chose α to be 5%. It follows that $z_\alpha = z_{0.5} = 1.645$. We can now calculate the value of $\overline{x}_L$:

$$\frac{\overline{x}_L-\mu}{\sigma/\sqrt{n}} = z_\alpha$$

$$\frac{\overline{x}_L-170}{65/\sqrt{400}} = 1.645$$

$$\overline{x}_L = 175.34$$

Therefore, the rejection region is

$$\overline{x} > 175.34$$

The sample mean was computed to be 178. Because the test statistic (sample mean) is in the rejection region (it is greater than 175.34), we reject the null hypothesis. Thus, there is sufficient evidence to infer that the mean monthly account is greater than €170.

Our calculations determined that any value of $\overline{x}$ above 175.34 represents an event that is quite unlikely when sampling (with $n = 400$) from a population whose mean is 170 (and whose standard deviation is 65). This suggests that the assumption that the null hypothesis is true is incorrect, and consequently we reject the null hypothesis in favour of the alternative hypothesis.

11-2b Standardised Test Statistic

The preceding test used the test statistic $\overline{x}$; as a result, the rejection region had to be set up in terms of $\overline{x}$. An easier method specifies that the test statistic be the standardised value of $\overline{x}$; that is, we use the **standardised test statistic**.

$$z = \frac{\overline{x}-\mu}{\sigma/\sqrt{n}}$$

and the rejection region consists of all values of z that are greater than z_α. Algebraically, the rejection region is

$$z > z_\alpha$$

We can redo Example 11.1 using the standardised test statistic.

The rejection region is

$$z > z_\alpha = z_{0.05} = 1.645$$

The value of the test statistic is calculated next:

$$z = \frac{\bar{x} - \mu}{\sigma/\sqrt{n}} = \frac{178 - 170}{65/\sqrt{400}} = 2.46$$

Because 2.46 is greater than 1.645, reject the null hypothesis and conclude that there is enough evidence to infer that the mean monthly account is greater than €170.

As you can see, the conclusions we draw from using the test statistic $\bar{x}$ and the standardised test statistic z are identical. Figures 11.2 and 11.3 depict the two sampling distributions, highlighting the equivalence of the two tests.

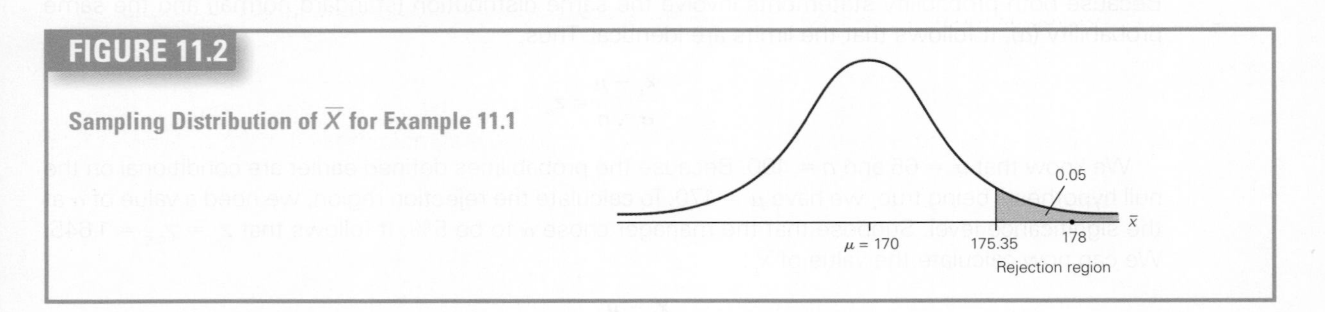

FIGURE 11.2 Sampling Distribution of $\bar{X}$ for Example 11.1

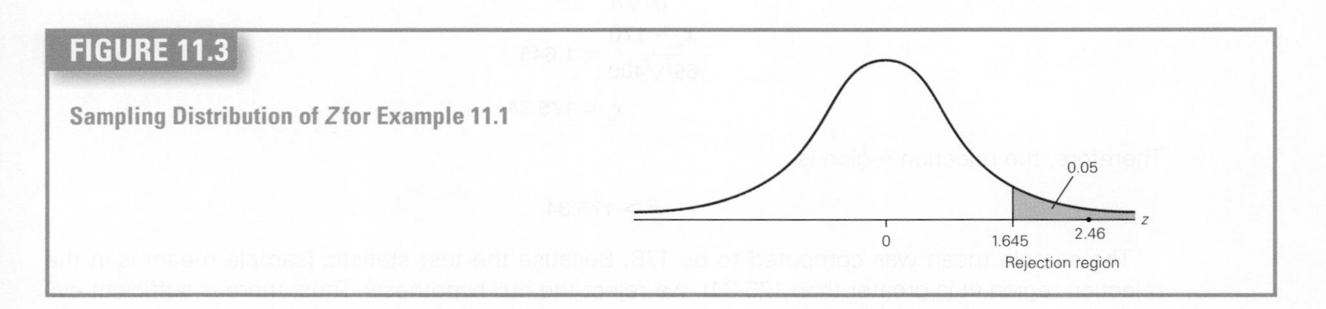

FIGURE 11.3 Sampling Distribution of Z for Example 11.1

Because it is convenient and because statistical software packages employ it, the standardised test statistic will be used throughout this book. For simplicity, we will refer to the *standardised test statistic* simply as the *test statistic.*

Incidentally, when a null hypothesis is rejected, the test is said to be **statistically significant** at whatever significance level the test was conducted. Summarizing Example 11.1, we would say that the test was significant at the 5% significance level.

11-2c *p*-Value

There are several drawbacks to the rejection region method. Foremost among them is the type of information provided by the result of the test. The rejection region method produces a yes or no response to the question: Is there sufficient statistical evidence to infer that the alternative hypothesis is true? The implication is that the result of the test of hypothesis will be converted automatically into one of two possible courses of action: one action as a result of rejecting the null hypothesis in favour of the alternative and another as a result of not rejecting the null hypothesis in favour of the alternative. In Example 11.1, the rejection of the null hypothesis seems to imply that the new billing system will be installed.

In fact, this is not the way in which the result of a statistical analysis is utilised. The statistical procedure is only one of several factors considered by a manager when making a decision. In Example 11.1 the manager discovered that there was enough statistical evidence to conclude that the mean monthly account is greater than €170. However, before taking any action, the manager would like to consider a number of factors including the cost and feasibility of restructuring the billing system and the possibility of making an error, in this case a Type I error.

What is needed to take full advantage of the information available from the test result and make a better decision is a measure of the amount of statistical evidence supporting the alternative hypothesis so that it can be weighed in relation to the other factors, especially the financial ones. The *p-value of a test* provides this measure and also give us some idea of the strength of the evidence against null hypothesis.

p-Value

The ***p*-value** of a test is the probability of observing a test statistic at least as extreme as the one computed given that the null hypothesis is true.

In Example 11.1 the *p*-value is the probability of observing a sample mean at least as large as 178 when the population mean is 170. Thus,

$$p\text{-value} = P(\bar{X} > 178) = P\left(\frac{\bar{X} - \mu}{\sigma/\sqrt{n}} > \frac{178 - 170}{65/\sqrt{400}}\right) = P(Z > 2.46)$$

$$= 1 - P(Z < 2.46) = 1 - 0.9931 = 0.0069$$

Figure 11.4 describes this calculation.

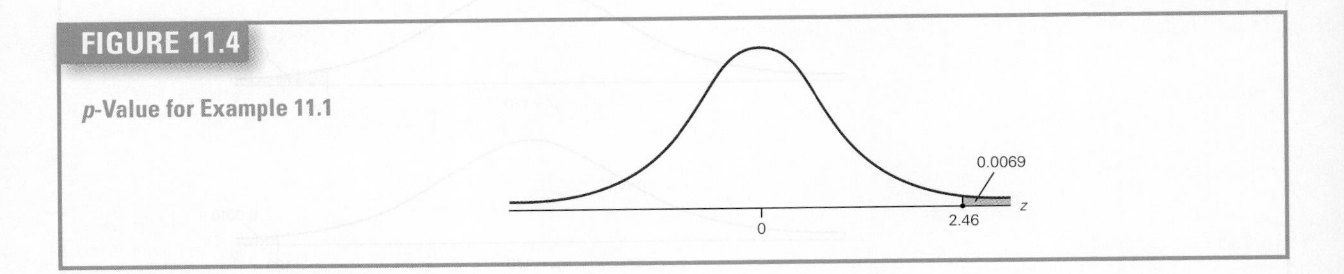

FIGURE 11.4

p-Value for Example 11.1

11-2d Interpreting the *p*-Value

To properly interpret the results of an inferential procedure, you must remember that the technique is based on the sampling distribution. The sampling distribution allows us to make probability statements about a sample statistic assuming knowledge of the population parameter. Thus, the probability of observing a sample mean at least as large as 178 from a population whose mean is 170, is 0.0069, which is very small. In other words, we have just observed an unlikely event, an event so unlikely that we seriously doubt the assumption that began the process – that the null hypothesis is true. Consequently, we have reason to reject the null hypothesis and support the alternative.

Students may be tempted to simplify the interpretation by stating that the *p*-value is the probability that the null hypothesis is true. Do not! As was the case with interpreting the confidence interval estimator, you cannot make a probability statement about a parameter. It is not a random variable.

The *p*-value of a test provides valuable information because it is a measure of the amount of statistical evidence that supports the alternative hypothesis. To understand this interpretation fully, refer to Table 11.2 where we list several values of *x*, their *z*-statistics, and *p*-values for Example 11.1. Notice that the closer $\bar{x}$ is to the hypothesised mean, 170, the larger the *p*-value is. The further $\bar{x}$ is above 170, the smaller the *p*-value is. Values of $\bar{x}$ far above 170 tend to indicate that the alternative hypothesis is true. Thus, the smaller the *p*-value, the more the statistical evidence supports the alternative hypothesis. Figure 11.5 graphically depicts the information in Table 11.2.

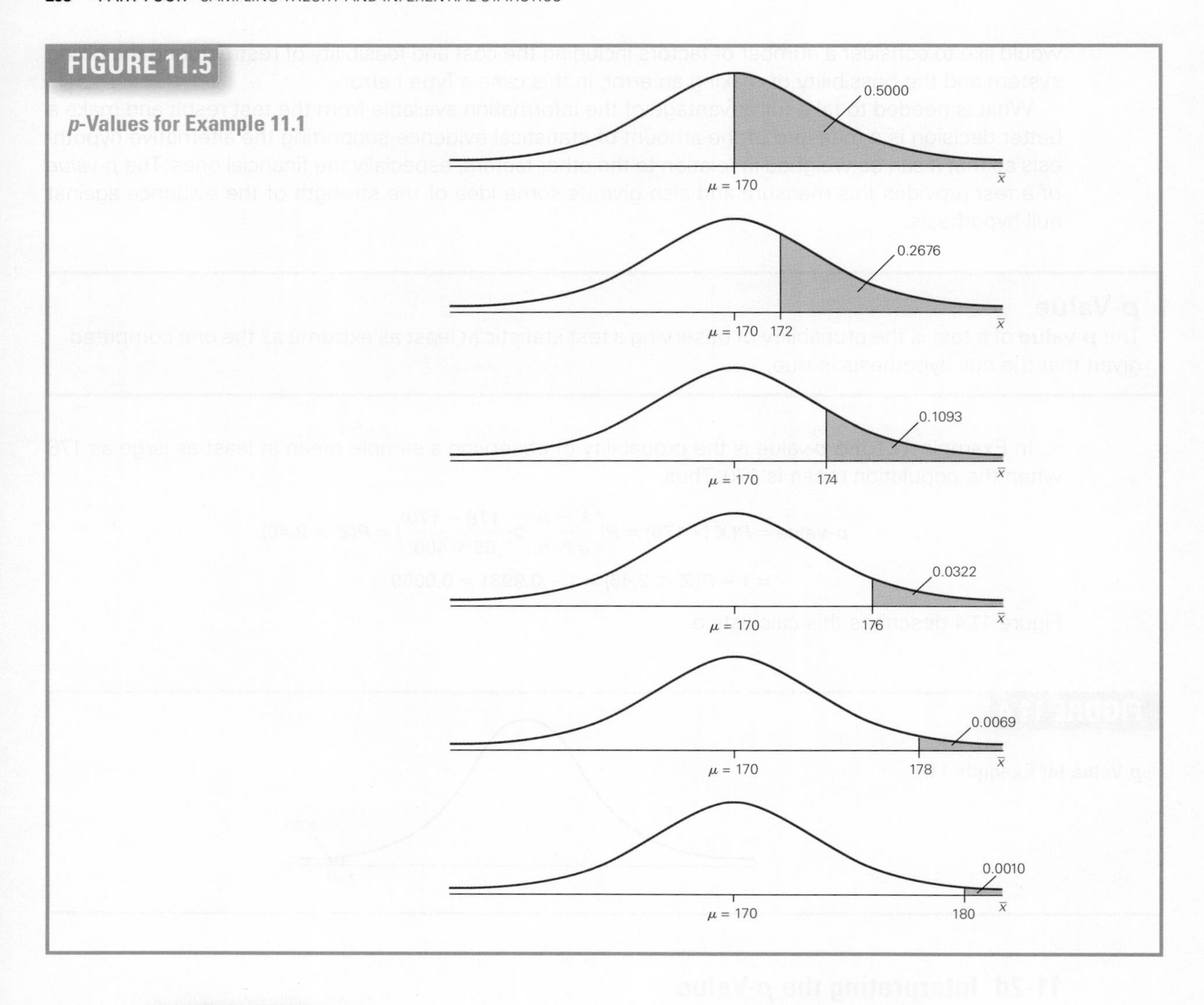

FIGURE 11.5 *p*-Values for Example 11.1

TABLE 11.2 Test Statistics and *p*-Values for Example 11.1

SAMPLE MEAN $\bar{x}$	TEST STATISTIC $z = \frac{\bar{x} - \mu}{\sigma/\sqrt{n}} = \frac{\bar{x} - 170}{65/\sqrt{400}}$	*p*-VALUE
170	0	0.5000
172	0.62	0.2676
174	1.23	0.1093
176	1.85	0.0322
178	2.46	0.0069
180	3.08	0.0010

This raises the question, 'How small does the *p*-value have to be to infer that the alternative hypothesis is true?'. In general, the answer depends on a number of factors, including the costs of making Type I and Type II errors. In Example 11.1, a Type I error would occur if the manager adopts the new billing system when it is not cost-effective. If the cost of this error is high, we attempt to minimise its probability. In the rejection region method, we do so by setting the significance level quite low – say, 1%. Using the *p*-value method, we would insist that the *p*-value be quite small, providing sufficient evidence to infer that the mean monthly account is greater than €170 before proceeding with the new billing system.

11-2e Describing the *p*-Value

Statistics practitioners can translate *p*-values using the following descriptive terms:

If the *p*-value is less than 0.01, we say that there is *overwhelming* evidence to infer that the alternative hypothesis is true. We also say that the test is **highly significant**.

If the *p*-value lies between 0.01 and 0.05, there is *strong* evidence to infer that the alternative hypothesis is true. The result is deemed to be **significant**.

If the *p*-value is between 0.05 and 0.10, we say that there is *weak* evidence to indicate that the alternative hypothesis is true. When the *p*-value is greater than 5%, we say that the result is **not statistically significant**.

When the *p*-value exceeds 0.10, we say that there is little to no evidence to infer that the alternative hypothesis is true.

Figure 11.6 summarises these terms.

FIGURE 11.6 Describing *p*-Values

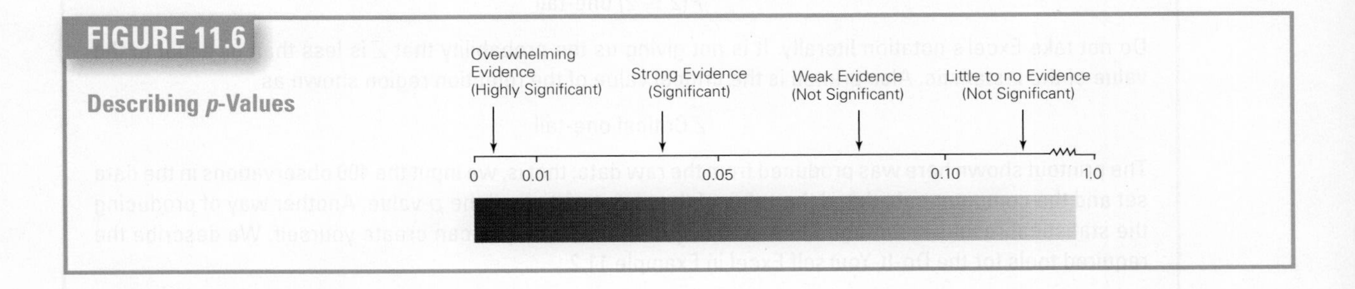

11-2f The *p*-Value and Rejection Region Methods

If we so choose, we can use the *p*-value to make the same type of decisions we make in the rejection region method. The rejection region method requires the decision-maker to select a significance level from which the rejection region is constructed. We then decide to reject or not reject the null hypothesis. Another way of making that type of decision is to compare the *p*-value with the selected value of the significance level. If the *p*-value is less than α, we judge the *p*-value to be small enough to reject the null hypothesis. If the *p*-value is greater than α, we do not reject the null hypothesis.

11-2g Solving Manually, Using Excel and Using Minitab

As you have already seen, we offer three ways to solve statistical problems. When we perform the calculations manually, we will use the rejection region approach. We will set up the rejection region using the test statistic's sampling distribution and associated table (in Appendix A). The calculations will be performed manually and a reject–do not reject decision will be made. In this chapter, it is possible to compute the *p*-value of the test manually. However, in later chapters we will be using test statistics that are not normally distributed, making it impossible to calculate the *p*-values manually. In these instances, manual calculations require the decision to be made via the rejection region method only.

Most software packages that compute statistics, including Excel and Minitab, print the *p*-value of the test. When we employ the computer, we will not set up the rejection region. Instead we will focus on the interpretation of the *p*-value.

EXCEL

	A	B	C	D
1	**Z-Test: Mean**			
2				
3				*Accounts*
4	Mean			178.00
5	Standard Deviation			68.37
6	Observations			400
7	Hypothesized Mean			170
8	SIGMA			65
9	z Stat			2.46
10	P(Z<=z) one-tail			0.0069
11	z Critical one-tail			1.6449
12	P(Z<=z) two-tail			0.0138
13	z Critical two-tail			1.96

Instructions

1. Type or import the data into one column. (**Open** Xm11-01.)
2. Click **Add-Ins, Data Analysis Plus** and **Z-Test: Mean.**
3. Fill in the dialog box: **Input Range** (A1:A401), type the **Hypothesised Mean** (170), type a positive value for the **Standard Deviation** (65), click **Labels** if the first row contains the name of the variable, and type the significance level α (0.05).

The first part of the printout reports the statistics and the details of the test. As you can see, the test statistic is $z = 2.46$. The *p*-value* of the test is $P(Z > 2.46) = 0.0069$. Excel reports this probability as

$$P(Z \leq z) \text{ one-tail}$$

Do not take Excel's notation literally. It is not giving us the probability that Z is less than or equal to the value of the *z*-statistic. Also printed is the critical value of the rejection region shown as

$$Z \text{ Critical one-tail}$$

The printout shown here was produced from the raw data; that is, we input the 400 observations in the data set and the computer calculated the value of the test statistic and the *p*-value. Another way of producing the statistical results is through the use of a spreadsheet that you can create yourself. We describe the required tools for the Do-It-Yourself Excel in Example 11.2.

*Excel provides two probabilities in its printout. The way in which we determine the *p*-value of the test from the printout is somewhat more complicated. Interested students are advised to read the website Appendix Converting Excel's Probabilities to *p*-Values.

MINITAB

One-Sample Z: Accounts

Test of mu = 170 vs > 170
The assumed standard deviation = 65

Variable	N	Mean	StDev	SE Mean	95% Lower Bound	Z	P
Accounts	400	177.997	68.367	3.250	172.651	2.46	0.007

Instructions

1. Type or import the data into one column. (**Open** Xm11-01.)
2. Click **Stat, Basic Statistics** and **1-Sample Z** . . . .
3. Type or use the **Select** button to specify the name of the variable or the column in the **Samples in Columns** box (Accounts). Type the value of the Standard deviation (65), check the **Perform hypothesis test** box and type the value of μ under the null hypothesis in the **Hypothesised mean** box (170).
4. Click **Options** . . . and specify the form of the alternative hypothesis in the **Alternative** box (greater than).

11-2h Interpreting the Results of a Test

In Example 11.1, we rejected the null hypothesis. Does this prove that the alternative hypothesis is true? The answer is no; because our conclusion is based on sample data (and not on the entire population), we can never *prove* anything by using statistical inference. Consequently, we summarise the test by stating that there is enough statistical evidence to infer that the null hypothesis is false and that the alternative hypothesis is true.

Now suppose that $\bar{x}$ had equalled 174 instead of 178. We would then have calculated $z = 1.23$ (p-value $= 0.1093$), which is not in the rejection region. Could we conclude on this basis that there is enough statistical evidence to infer that the null hypothesis is true and hence that $\mu = 170$? Again the answer is 'no' because it is absurd to suggest that a sample mean of 174 provides enough evidence to infer that the population mean is 170. (If it proved anything, it would prove that the population mean is 174.) Because we are testing a single value of the parameter under the null hypothesis, we can never have enough statistical evidence to establish that the null hypothesis is true (unless we sample the entire population). (The same argument is valid if you set up the null hypothesis as H_0: $\mu \leq 170$. It would be illogical to conclude that a sample mean of 174 provides enough evidence to conclude that the population mean is *less than or equal to 170*.)

Consequently, if the value of the test statistic does not fall into the rejection region (or the p-value is large), rather than say we accept the null hypothesis (which implies that we are stating that the null hypothesis is true), we state that we do not reject the null hypothesis, and we conclude that not enough evidence exists to show that the alternative hypothesis is true. Although it may appear to be the case, we are not being overly technical. Your ability to set up tests of hypotheses properly and to interpret their results correctly very much depends on your understanding of this point. The point is that the conclusion is based on the alternative hypothesis. In the final analysis, there are only two possible conclusions of a test of hypothesis.

Conclusions of a Test of Hypothesis

If we reject the null hypothesis, we conclude that there is enough statistical evidence to infer that the alternative hypothesis is true. If we do *not* reject the null hypothesis, we conclude that there is *not* enough statistical evidence to infer that the alternative hypothesis is true.

Observe that the alternative hypothesis is the focus of the conclusion. It represents what we are investigating, which is why it is also called the *research hypothesis*. Whatever you are trying to show statistically must be represented by the alternative hypothesis (bearing in mind that you have only three choices for the alternative hypothesis – the parameter is greater than, less than, or not equal to the value specified in the null hypothesis).

When we introduced statistical inference in Chapter 10, we pointed out that the first step in the solution is to identify the technique. When the problem involves hypothesis testing, part of this process is the specification of the hypotheses. Because the alternative hypothesis represents the condition we are researching, we will identify it first. The null hypothesis automatically follows because the null hypothesis must specify equality. However, by tradition, when we list the two hypotheses, the null hypothesis comes first, followed by the alternative hypothesis. All examples in this book will follow that format.

SSA ENVELOPE PLAN: SOLUTION

IDENTIFY

The objective of the study is to draw a conclusion about the mean payment period. Thus, the parameter to be tested is the population mean μ. We want to know whether there is enough statistical evidence to show that the population mean is less than 22 days. Thus, the alternative hypothesis is

$$H_1: \mu < 22$$

The null hypothesis is

$$H_0: \mu = 22$$

The test statistic is the only one we have presented thus far. It is

$$z = \frac{\bar{x} - \mu}{\sigma/\sqrt{n}}$$

COMPUTE MANUALLY:

To solve this problem manually, we need to define the rejection region, which requires us to specify a significance level. A 10% significance level is deemed to be appropriate. (We will discuss our choice later.)

We wish to reject the null hypothesis in favour of the alternative only if the sample mean and hence the value of the test statistic is small enough. As a result, we locate the rejection region in the left tail of the sampling distribution. To understand why, remember that we are trying to decide whether there is enough statistical evidence to infer that the mean is less than 22 (which is the alternative hypothesis). If we observe a large sample mean (and hence a large value of *z*), do we want to reject the null hypothesis in favour of the alternative? The answer is an emphatic 'no'. It is illogical to think that if the sample mean is, say, 30, there is enough evidence to conclude that the mean payment period for all customers would be less than 22.

Consequently, we want to reject the null hypothesis only if the sample mean (and hence the value of the test statistic *z*) is small. How small is small enough? The answer is determined by the significance level and the rejection region. Thus, we set up the rejection region as

$$z < -z_\alpha = -z_{0.10} = -1.28$$

Note that the direction of the inequality in the rejection region ($z < -z_\alpha$) matches the direction of the inequality in the alternative hypothesis ($\mu < 22$). Also note that we use the negative sign, because the rejection region is in the left tail (containing values of *z* less than 0) of the sampling distribution.

From the data, we compute the sum and the sample mean. They are

$$\sum x_i = 4759$$

$$\bar{x} = \frac{\sum x_i}{220} = \frac{4759}{220} = 21.63$$

We will assume that the standard deviation of the payment periods for the SSA plan is unchanged from its current value of $\sigma = 6$. The sample size is n = 220, and the value of μ is hypothesised to be 22. We compute the value of the test statistic as

$$z = \frac{\bar{x} - \mu}{\sigma/\sqrt{n}} = \frac{21.63 - 22}{6/\sqrt{220}} = -0.91$$

Because the value of the test statistic, $z = -0.91$, is not less than –1.28, we do not reject the null hypothesis and we do not conclude that the alternative hypothesis is true. There is insufficient evidence to infer that the mean is less than 22 days.

We can determine the *p*-value of the test as follows:

$$p\text{-}value = P(Z < -0.91) = 0.1814$$

In this type of one-tail (left-tail) test of hypothesis, we calculate the *p*-value as $P(Z < z)$, where z is the actual value of the test statistic. Figure 11.7 depicts the sampling distribution, rejection region and *p*-value.

EXCEL

	A	B	C	D
1	**Z-Test: Mean**			
2				
3				Payment
4	Mean			21.63
5	Standard Deviation			5.84
6	Observations			220
7	Hypothesized Mean			22
8	SIGMA			6
9	z Stat			–0.91
10	P(Z<=z) one-tail			0.1814
11	z Critical one-tail			1.6449
12	P(Z<=z) two-tail			0.3628
13	z Critical two-tail			1.96

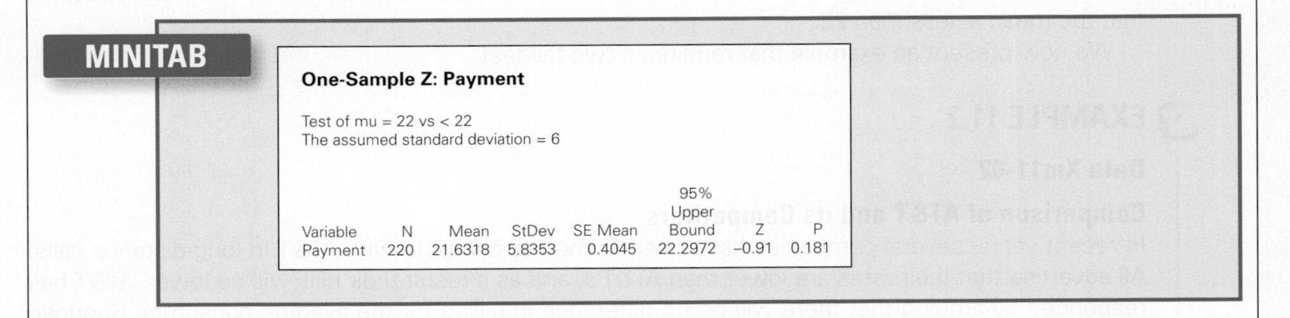

MINITAB

One-Sample Z: Payment

Test of mu = 22 vs < 22
The assumed standard deviation = 6

Variable	N	Mean	StDev	SE Mean	95% Upper Bound	Z	P
Payment	220	21.6318	5.8353	0.4045	22.2972	–0.91	0.181

INTERPRET

The value of the test statistic is –0.91, and its *p*-value is 0.1814, a figure that does not allow us to reject the null hypothesis. Because we were not able to reject the null hypothesis, we say that there is not enough evidence to infer that the mean payment period is less than 22 days. Note that there was some evidence to indicate that the mean of the entire population of payment periods is less than 22 days. We did calculate the sample mean to be 21.63. However, to reject the null hypothesis we need *enough* statistical evidence – and in this case we simply did not have enough reason to reject the null hypothesis in favour

(*continued*)

of the alternative. In the absence of evidence to show that the mean payment period for all customers sent a SSA envelope would be less than 22 days, we cannot infer that the plan would be profitable.

A Type I error occurs when we conclude that the plan works when it actually does not. The cost of this mistake is not high. A Type II error occurs when we do not adopt the SSA envelope plan when it would reduce costs. The cost of this mistake can be high. As a consequence, we would like to minimise the probability of a Type II error. Thus, we chose a large value for the probability of a Type I error; we set

$$\alpha = 0.10$$

Figure 11.7 exhibits the sampling distribution for this example.

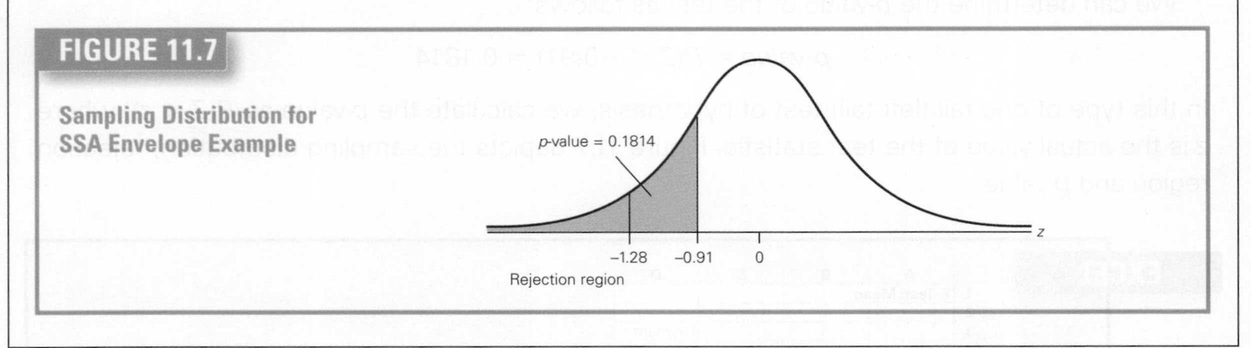

FIGURE 11.7 Sampling Distribution for SSA Envelope Example

11-2i One- and Two-Tail Tests

The statistical tests conducted in Example 11.1 and the SSA envelope example are called one-tail tests because the rejection region is located in only one tail of the sampling distribution. The *p*-value is also computed by finding the area in one tail of the sampling distribution. The right tail in Example 11.1 is the important one because the alternative hypothesis specifies that the mean is *greater than* 170. In the SSA envelope example, the left tail is emphasised because the alternative hypothesis specifies that the mean is *less than* 22.

We now present an example that requires a two-tail test.

EXAMPLE 11.2

Data Xm11-02

Comparison of AT&T and its Competitors

In recent years, several companies have been formed to compete with AT&T in long-distance calls. All advertise that their rates are lower than AT&T's, and as a result their bills will be lower. AT&T has responded by arguing that there will be no difference in billing for the average consumer. Suppose that a statistics practitioner working for AT&T determines that the mean and standard deviation of monthly long-distance bills for all its residential customers are €17.09 and €3.87, respectively. He then takes a random sample of 100 customers and recalculates their last month's bill using the rates quoted by a leading competitor. Assuming that the standard deviation of this population is the same as for AT&T, can we conclude at the 5% significance level that there is a difference between the average AT&T bill and that of the leading competitor?

SOLUTION:

IDENTIFY

In this problem, we want to know whether the mean monthly long-distance bill is different from €17.09. Consequently, we set up the alternative hypothesis to express this condition:

$$H_1: \mu \neq 17.09$$

The null hypothesis specifies that the mean is equal to the value specified under the alternative hypothesis. Hence

$$H_0: \mu = 17.09$$

COMPUTE MANUALLY:

To set up the rejection region, we need to realise that we can reject the null hypothesis when the test statistic is large or when it is small. In other words, we must set up a *two-tail rejection region.* Because the total area in the rejection region must be α, we divide this probability by 2. Thus, the rejection region* is

$$z < -z_{\alpha/2} \quad \text{or} \quad z > z_{\alpha/2}$$

For $\alpha = 0.05$, $\alpha/2 = 0.025$, and $z_{\alpha/2} = z_{0.025} = 1.96$.

$$z < -1.96 \quad \text{or} \quad z > 1.96$$

From the data, we compute

$$\sum x_i = 1754.99$$

$$\bar{x} = \frac{\sum x_i}{n} = \frac{1754.99}{100} = 17.55$$

The value of the test statistic is

$$z = \frac{\bar{x} - \mu}{\sigma/\sqrt{n}} = \frac{17.55 - 17.09}{3.87/\sqrt{100}} = 1.19$$

Because 1.19 is neither greater than 1.96 nor less than −1.96, we cannot reject the null hypothesis.

We can also calculate the *p*-value of the test. Because it is a two-tail test, we determine the *p*-value by finding the area in both tails; that is,

$$p\text{-value} = P(Z < -1.19) + P(Z > 1.19) = 0.1170 + 0.1170 = 0.2340$$

Or, more simply, multiply the probability in one tail by 2.

In general, the *p*-value in a two-tail test is determined by

$$p\text{-value} = 2P(Z > |z|)$$

where z is the actual value of the test statistic and $|z|$ is its absolute value.

EXCEL

	A	B	C	D
1	**Z-Test: Mean**			
2				
3				*Bills*
4	Mean			17.55
5	Standard Deviation			3.94
6	Observations			100
7	Hypothesized Mean			17.09
8	SIGMA			3.87
9	z Stat			1.19
10	P(Z<=z) one-tail			0.1173
11	z Critical one-tail			1.6449
12	P(Z<=z) two-tail			0.2346
13	z Critical two-tail			1.96

*Statistics practitioners often represent this rejection region as $|z| > z_{\alpha/2}$, which reads, 'the absolute value of z is greater than $z_{\alpha/2}$'. We prefer our method because it is clear that we are performing a two-tail test.

DO-IT-YOURSELF EXCEL

As was the case with the spreadsheet you created in Chapter 10, to estimate a population mean you can produce a spreadsheet that does the same for testing a population mean.

Tools: **SQRT** and **NORM.S.INV** are functions described in Example 10.1.

NORM.S.DIST: Syntax: NORM.S.DIST(*z*, cumulative) This function computes the probability that a standard normal random variable is less than the quantity in parentheses. For example, NORM.S.DIST(1.19, True) $= P(z < 1.19)$. However, the quantity we show as $P(Z \leq z)$ **one-tail,** which is computed by both Data Analysis and Data Analysis Plus is actually calculated in the following way. Find the probability to the left of the ***z* Stat** and the probability to its right. $P(Z \leq z)$ **one-tail** is the smaller of the two probabilities. $P(Z \leq z)$ **two-tail** is twice $P(Z \leq z)$ **one-tail.** For more details, we suggest you read the website Appendix Converting Excel's Probabilities to *p*-Values.

MINITAB

One-Sample Z: Bills

Test of mu = 17.09 vs not = 17.09
The assumed standard deviation = 3.87

Variable	N	Mean	StDev	SE Mean	95% CI	Z	P
Bills	100	17.5499	3.9382	0.3870	(16.7914, 18.3084)	1.19	0.235

INTERPRET

There is not enough evidence to infer that the mean long-distance bill is different from AT&T's mean of €17.09. Figure 11.8 depicts the sampling distribution for this example.

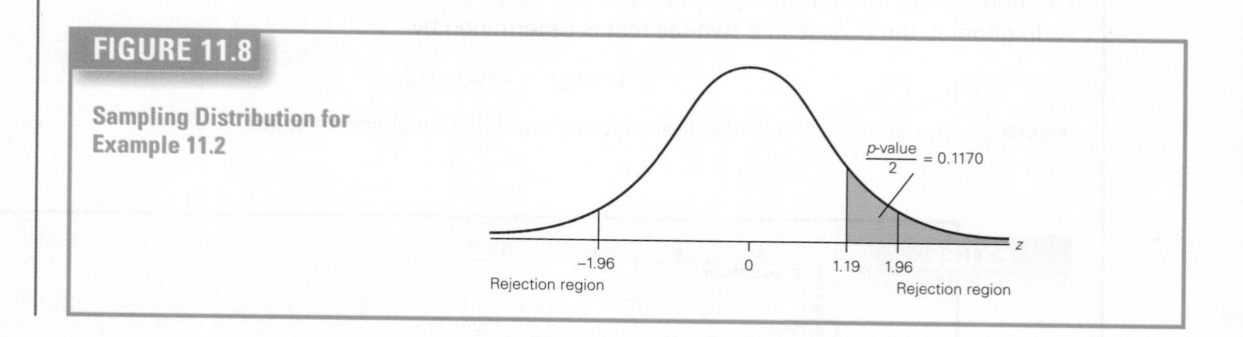

FIGURE 11.8 Sampling Distribution for Example 11.2

11-2j When Do We Conduct One- and Two-Tail Tests?

A two-tail test is conducted whenever the alternative hypothesis specifies that the mean is *not equal* to the value stated in the null hypothesis – that is, when the hypotheses assume the following form:

$$H_0: \mu = \mu_0$$
$$H_1: \mu \neq \mu_0$$

There are two one-tail tests. We conduct a one-tail test that focuses on the right tail of the sampling distribution whenever we want to know whether there is enough evidence to infer that the mean is greater than the quantity specified by the null hypothesis – that is, when the hypotheses are

$$H_0: \mu = \mu_0$$
$$H_1: \mu > \mu_0$$

The second one-tail test involves the left tail of the sampling distribution. It is used when the statistics practitioner wants to determine whether there is enough evidence to infer that the mean is less than the value of the mean stated in the null hypothesis. The resulting hypotheses appear in this form:

$$H_0: \mu = \mu_0$$
$$H_1: \mu < \mu_0$$

The techniques introduced in Chapters 12, 15, 16 and 17 require you to decide which of the three forms of the test to employ. Make your decision in the same way as we described the process.

11-2k Testing Hypotheses and Confidence Interval Estimators

As you have seen, the test statistic and the confidence interval estimator are both derived from the sampling distribution. It should not be a surprise then that we can use the confidence interval estimator to test hypotheses. To illustrate, consider Example 11.2. The 95% confidence interval estimate of the population mean is

$$\bar{x} \pm z_{\alpha/2}\frac{\sigma}{\sqrt{n}} = 17.55 \pm 1.96\frac{3.87}{\sqrt{100}} = 17.55 \pm 0.76$$

$$\text{LCL} = 16.79 \text{ and UCL} = 18.31$$

We estimate that μ lies between 16.79 and 18.31. Because this interval includes 17.09, we cannot conclude that there is sufficient evidence to infer that the population mean differs from 17.09.

In Example 11.1, the 95% confidence interval estimate is LCL = 171.63 and UCL = 184.37. The interval estimate excludes 170, allowing us to conclude that the population mean account is not equal to €170.

As you can see, the confidence interval estimator can be used to conduct tests of hypotheses. This process is equivalent to the rejection region approach. However, instead of finding the critical values of the rejection region and determining whether the test statistic falls into the rejection region, we compute the interval estimate and determine whether the hypothesised value of the mean falls into the interval.

Using the interval estimator to test hypotheses has the advantage of simplicity. Apparently, we do not need the formula for the test statistic; we need only the interval estimator. However, there are two serious drawbacks.

First, when conducting a one-tail test, our conclusion may not answer the original question. In Example 11.1, we wanted to know whether there was enough evidence to infer that the mean is *greater than* 170. The estimate concludes that the mean *differs from* 170. You may be tempted to say that because the entire interval is greater than 170, there is enough statistical evidence to infer that the population mean is greater than 170. However, in attempting to draw this conclusion, we run into the problem of determining the procedure's significance level. Is it 5% or is it 2.5%? We may be able to overcome this problem through the use of **one-sided confidence interval estimators**. However, if the purpose of using confidence interval estimators instead of test statistics is simplicity, one-sided estimators are a contradiction.

Second, the confidence interval estimator does not yield a *p*-value, which we have argued is the better way to draw inferences about a parameter. Using the confidence interval estimator to test hypotheses forces the decision-maker into making a reject–do not reject decision rather than providing information about how much statistical evidence exists to be judged with other factors in the decision process. Furthermore, we only postpone the point in time when a test of hypothesis must be used. In later chapters, we will present problems where only a test produces the information we need to make decisions.

11-2l Developing an Understanding of Statistical Concepts 1

As is the case with the confidence interval estimator, the test of hypothesis is based on the sampling distribution of the sample statistic. The result of a test of hypothesis is a probability statement about the sample statistic. We assume that the population mean is specified by the null hypothesis. We then compute the test statistic and determine how likely it is to observe this large (or small) value when the null hypothesis is true. If the probability is small, we conclude that the assumption that the null hypothesis is true is unfounded and we reject it.

11-2m Developing an Understanding of Statistical Concepts 2

When we (or the computer) calculate the value of the test statistic

$$z = \frac{\bar{x} - \mu}{\sigma/\sqrt{n}}$$

we are also measuring the difference between the sample statistic $\bar{x}$ and the hypothesised value of the parameter μ in terms of the standard error $\sigma/\sqrt{n}$. In Example 11.2, we found that the value of the test statistic was $z = 1.19$. This means that the sample mean was 1.19 standard errors above the hypothesised value of μ. The standard normal probability table told us that this value is not considered unlikely. As a result, we did not reject the null hypothesis.

The concept of measuring the difference between the sample statistic and the hypothesised value of the parameter in terms of the standard errors is one that will be used throughout this book.

EXERCISES

Developing an Understanding of Statistical Concepts

In Exercises 11.6–11.10, calculate the value of the test statistic, set up the rejection region, determine the p-value, interpret the result and draw the sampling distribution.

11.6 $H_0: \mu = 1000$

$H_1: \mu \neq 1000$

$\sigma = 200, n = 100, \bar{x} = 980, \alpha = 0.01$

11.7 $H_0: \mu = 50$

$H_1: \mu > 50$

$\sigma = 5, n = 9, \bar{x} = 51, \alpha = 0.03$

11.8 $H_0: \mu = 15$

$H_1: \mu < 15$

$\sigma = 2, n = 25, \bar{x} = 14.3, \alpha = 0.10$

11.9 $H_0: \mu = 70$

$H_1: \mu > 70$

$\sigma = 20, n = 100, \bar{x} = 80, \alpha = 0.01$

11.10 $H_0: \mu = 50$

$H_1: \mu < 50$

$\sigma = 15, n = 100, \bar{x} = 48, \alpha = 0.05$

Exercises 11.11 to 11.23 are 'what-if analyses' designed to determine what happens to the test statistic and p-value when the sample size, standard deviation and sample mean change. These problems can be solved manually, using the spreadsheet you created in this section or by using Minitab.

11.11 a. Compute the *p*-value in order to test the following hypotheses given that $\bar{x} = 52$, $n = 9$ and $\sigma = 5$.

$H_0: \mu = 50$

$H_1: \mu > 50$

b. Repeat part (a) with $n = 25$.

c. Repeat part (a) with $n = 100$.

d. Describe what happens to the value of the test statistic and its *p*-value when the sample size increases.

11.12 a. A statistics practitioner formulated the following hypotheses.

$H_0: \mu = 200$

$H_1: \mu < 200$

and learned that $\bar{x} = 190$, $n = 9$ and $\sigma = 50$

Compute the *p*-value of the test.

b. Repeat part (a) with $\sigma = 30$.

c. Repeat part (a) with $\sigma = 10$.

d. Discuss what happens to the value of the test statistic and its *p*-value when the standard deviation decreases.

11.13 a. Given the following hypotheses, determine the *p*-value when $\bar{x} = 21$, $n = 25$ and $\sigma = 5$.

$H_0: \mu = 20$

$H_1: \mu \neq 20$

b. Repeat part (a) with $\bar{x} = 22$.

c. Repeat part (a) with $\bar{x} = 23$.

d. Describe what happens to the value of the test statistic and its *p*-value when the value of $\bar{x}$ increases.

11.14 a. Test these hypotheses by calculating the *p*-value given that $\bar{x} = 99$, $n = 100$ and $\sigma = 8$

$H_0: \mu = 100$

$H_1: \mu \neq 100$

b. Repeat part (a) with $n = 50$.

c. Repeat part (a) with $n = 20$.

d. What is the effect on the value of the test statistic and the *p*-value of the test when the sample size decreases?

11.15 a. Find the *p*-value of the following test given that $\bar{x} = 990$, $n = 100$ and $\sigma = 25$.

$H_0: \mu = 1000$

$H_1: \mu < 1000$

b. Repeat part (a) with $\sigma = 50$.

c. Repeat part (a) with $\sigma = 100$.

d. Describe what happens to the value of the test statistic and its *p*-value when the standard deviation increases.

11.16 a. Calculate the *p*-value of the test described here.

$H_0: \mu = 60$

$H_1: \mu > 60$

$\bar{x} = 72$, $n = 25$, $\sigma = 20$

b. Repeat part (a) with $\bar{x} = 68$.

c. Repeat part (a) with $\bar{x} = 64$.

d. Describe the effect on the test statistic and the *p*-value of the test when the value of $\bar{x}$ decreases.

11.17 Redo Example 11.1 with

a. $n = 200$

b. $n = 100$

c. Describe the effect on the test statistic and the *p*-value when *n* increases.

11.18 Perform a what-if analysis to calculate the *p*-values in Table 11.2.

11.19 Redo the SSA example with

a. $n = 100$

b. $n = 500$

11.20 For the SSA example, create a table that shows the effect on the test statistic and the *p*-value of decreasing the value of the sample mean. Use $\bar{x} = 22.0$, 21.8, 21.6, 21.4, 21.2, 21.0, 20.8, 20.6 and 20.4.

11.21 Redo Example 11.2 with

a. $n = 50$

b. $n = 400$

c. Briefly describe the effect on the test statistic and the *p*-value when *n* increases.

11.22 Redo Example 11.2 with

a. $\sigma = 2$

b. $\sigma = 10$

c. What happens to the test statistic and the *p*-value when σ increases?

11.23 Refer to Example 11.2. Create a table that shows the effect on the test statistic and the *p*-value of changing the value of the sample mean. Use $\bar{x} = 15.0$, 15.5, 16.0, 16.5, 17.0, 17.5, 18.0, 18.5 and 19.0.

Applications

The following exercises may be answered manually or with the assistance of a computer. The files containing the data are given.

11.24 **Xr11-24** A business student claims that, on average, an MBA student is required to prepare more than five cases per week. To examine the claim, a statistics professor asks a random sample of ten MBA students to report the number of cases they prepare weekly. The results are exhibited here. Can the professor conclude at the 5% significance level that the claim is true, assuming that the number of cases is normally distributed with a standard deviation of 1.5?

2 7 4 8 9 5 11 3 7 4

11.25 **Xr11-25** A random sample of 18 young adult men (20–30 years old) was sampled. Each person was asked how many minutes of sports he watched on television daily. The responses are listed here. It is known that $\sigma = 10$. Test to determine at the 5% significance level whether there is enough statistical evidence to infer that the mean amount of television watched daily by all young adult men is greater than 50 minutes.

50	48	65	74	66	37	45	68	64
65	58	55	52	63	59	57	74	65

11.26 **Xr11-26** A random sample of 12 second-year university students enrolled in a business statistics course was drawn. At the course's completion, each student was asked how many hours he or she spent doing homework in statistics. The data are listed here. It is known that the population standard deviation is $\sigma = 8.0$.

The instructor has recommended that students devote 3 hours per week for the duration of the 12-week semester, for a total of 36 hours. Test to determine whether there is evidence that the average student spent less than the recommended amount of time. Compute the *p*-value of the test.

31 40 26 30 36 38 29 40 38 30 35 38

11.27 Xr11-27 A machine that produces ball bearings is set so that the average diameter is 0.50 cm. A sample of ten ball bearings was measured, with the results shown here. Assuming that the standard deviation is 0.05 cm, can we conclude at the 5% significance level that the mean diameter is not 0.50 cm?

0.48 0.50 0.49 0.52 0.53 0.48 0.49 0.47 0.46 0.51

11.28 Xr11-28 Spam e-mail has become a serious and costly nuisance. An office manager in Slough believes that the average amount of time spent by office workers reading and deleting spam exceeds 25 minutes per day. To test this belief, he takes a random sample of 18 workers and measures the amount of time each spends reading and deleting spam. The results are listed here. If the population of times is normal with a standard deviation of 12 minutes, can the manager infer at the 1% significance level that he is correct?

35 48 29 44 17 21 32 28 34
23 13 9 11 30 42 37 43 48

The following exercises require the use of a computer and software. The answers may be calculated manually.

11.29 Xr11-29 A manufacturer of light bulbs advertises that, on average, its long-life bulb will last more than 5000 hours. To test the claim, a statistician took a random sample of 100 bulbs and measured the amount of time until each bulb burned out. If we assume that the lifetime of this type of bulb has a standard deviation of 400 hours, can we conclude at the 5% significance level that the claim is true?

11.30 Xr11-30 In the midst of labour–management negotiations, the president of a company argues that the company's working class workers, who are paid an average of £30 000 per year, are well paid because the mean annual income of all working class workers in the country is less than £30 000. That figure is disputed by the union, which does not believe that the mean working class income is less than £30 000. To test the company president's belief, an arbitrator draws a random sample of 350 working class workers from across the country and asks each to report his or her annual income. If the arbitrator assumes that the working class workers' incomes are normally distributed with a standard deviation of £8000, can it be inferred at the 5% significance level that the company president is correct?

11.31 Xr11-31 A dean of a business school claims that the Graduate Management Admission Test (GMAT) scores of applicants to the school's MBA programme have increased during the past 5 years. Five years ago, the mean and standard deviation of GMAT scores of MBA applicants were 560 and 50, respectively. Twenty applications for this year's programme were randomly selected and the GMAT scores recorded. If we assume that the distribution of GMAT scores of this year's applicants is the same as that of 5 years ago, with the possible exception of the mean, can we conclude at the 5% significance level that the dean's claim is true?

11.32 Xr11-32 Past experience indicates that the monthly long-distance telephone bill is normally distributed with a mean of €17.85 and a standard deviation of €3.87. After an advertising campaign aimed at increasing long-distance telephone usage, a random sample of 25 household bills was taken.

a. Do the data allow us to infer at the 10% significance level that the campaign was successful?

b. What assumption must you make to answer part (a)?

11.33 Xr11-33 In an attempt to reduce the number of person-hours lost as a result of industrial accidents, a large production plant installed new safety equipment. In a test of the effectiveness of the equipment, a random sample of 50 departments was chosen. The number of person-hours lost in the month before and the month after the installation of the safety equipment was recorded. The percentage change was calculated and recorded. Assume that the population standard deviation is $\sigma = 6$. Can we infer at the 10% significance level that the new safety equipment is effective?

11.34 Xr11-34 A traffic officer believes that the average speed of cars travelling over a certain stretch of the Paris Ring Road exceeds the posted limit of 80 kmph. The speeds of a random sample of 200 cars were recorded. Do these data provide sufficient evidence at the 1% significance level to support the officer's belief? What is the *p*-value of the test? (Assume that the standard deviation is known to be 8.)

11.35 Xr11-35 An automotive expert claims that the large number of self-serve petrol stations has resulted in poor automobile maintenance, and that the average tyre pressure is more than 0.28 bar below its manufacturer's specification. As a quick test, 50 tyres are examined, and the number of psi each tyre is below specification is recorded. If we assume that tyre pressure is normally distributed with $\sigma = 0.10$ bar, can we infer at the 10% significance level that the expert is correct? What is the *p*-value?

11.36 Xr11-36 For the past few years, the number of customers of a takeaway on a motorway has averaged 20 per hour, with a standard deviation of 3 per hour. This year, another takeaway 2 km away opened. The manager of the takeaway believes that this will result in a decrease in the number of customers. The number of customers who arrived during 36 randomly selected hours was recorded. Can we conclude at the 5% significance level that the manager is correct?

11.37 Xr11-37 A fast-food franchise is considering building a restaurant at a certain location. Based on financial analyses, a site is acceptable only if the number of pedestrians passing the location averages more than 100 per hour. The number of pedestrians observed for each of 40 hours was recorded. Assuming that the population standard deviation is known to be 16, can we conclude at the 1% significance level that the site is acceptable?

11.38 Xr11-38 Many Alpine ski centres base their projections of revenues and profits on the assumption that the average Alpine skier skis four times per year. To investigate the validity of this assumption, a random sample of 63 skiers is drawn and each is asked to report the number of times he or she skied the previous year. If we assume that the standard deviation is 2, can we infer at the 10% significance level that the assumption is wrong?

11.39 Xr11-39 The current no-smoking regulations in office buildings require workers who smoke to take breaks and leave the building in order to satisfy their habits. A study indicates that such workers average 32 minutes per day taking smoking breaks. The standard deviation is 8 minutes. To help reduce the average break, rooms with powerful extractors were installed in the buildings. To see whether these rooms serve their designed purpose, a random sample of 110 smokers was taken. The total amount of time away from their desks was measured for 1 day. Test to determine whether there has been a decrease in the mean time away from their desks. Compute the *p*-value and interpret it relative to the costs of Type I and Type II errors.

11.40 Xr11-40 A low-handicap golfer who uses Titleist brand golf balls observed that his average drive is 27 metres and the standard deviation is 10 metres. Nike has just introduced a new ball, which has been endorsed by Tiger Woods. Nike claims that the ball will travel farther than Titleist. To test the claim, the golfer hits 100 drives with a Nike ball and measures the distances. Conduct a test to determine whether Nike is correct. Use a 5% significance level.

11-3 CALCULATING THE PROBABILITY OF A TYPE II ERROR

To properly interpret the results of a test of hypothesis, you must be able to specify an appropriate significance level or to judge the *p*-value of a test. However, you also must understand the relationship between Type I and Type II errors. In this section, we describe how the probability of a Type II error is computed and interpreted.

Recall Example 11.1, where we conducted the test using the sample mean as the test statistic and we computed the rejection region (with $\alpha = 0.05$) as

$$\bar{x} > 175.34$$

A Type II error occurs when a false null hypothesis is not rejected. In Example 11.1, if $\bar{x}$ is less than 175.34, we will not reject the null hypothesis. If we do not reject the null hypothesis, we will not install the new billing system. Thus, the consequence of a Type II error in this example is that we will not install the new system when it would be cost-effective. The probability of this occurring is the probability of a Type II error. It is defined as

$$\beta = P(\bar{X} < 175.34, \text{ given that the null hypothesis is false})$$

The condition that the null hypothesis is false tells us only that the mean is not equal to 170. If we want to compute β, we need to specify a value for μ. Suppose that when the mean account is at least €180, the new billing system's savings become so attractive that the manager would hate to make the mistake of not installing the system. As a result, she would like to determine the probability of not installing the new system when it would produce large cost savings. Because calculating probability from an approximately normal sampling distribution requires a value of μ (as well as σ and n), we will calculate the probability of not installing the new system when μ is *equal* to 180:

$$\beta = P(\bar{X} < 175.34, \text{ given that } \mu = 180)$$

We know that $\bar{x}$ is approximately normally distributed with mean μ and standard deviation $\sigma/\sqrt{n}$. To proceed, we standardise $\bar{x}$ and use the standard normal table (Table 3 in Appendix A):

$$\beta = P\left(\frac{\bar{X}-\mu}{\sigma/\sqrt{n}} < \frac{175.34-180}{65/\sqrt{400}}\right) = P(Z < -1.43) = 0.0764$$

This tells us that when the mean account is actually €180, the probability of incorrectly not rejecting the null hypothesis is 0.0764. Figure 11.9 graphically depicts how the calculation was performed. Notice that to calculate the probability of a Type II error, we had to express the rejection region in terms of the unstandardised test statistic x, and we had to specify a value for μ other than the one shown in the null hypothesis. In this illustration, the value of μ used was based on a financial analysis indicating that when μ is at least €180 the cost savings would be very attractive.

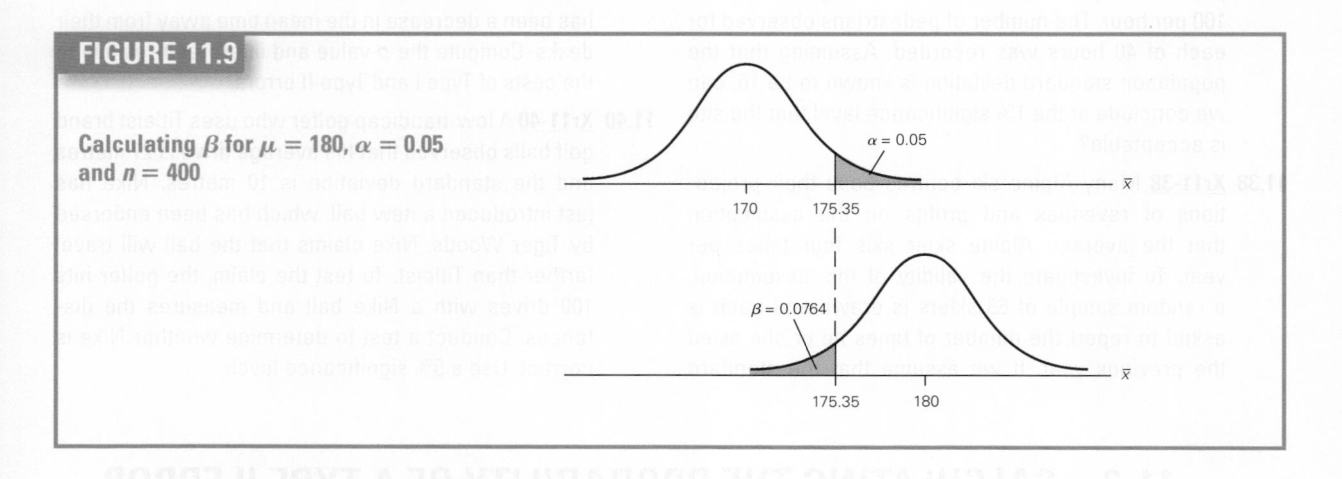

FIGURE 11.9

Calculating β for $\mu = 180$, $\alpha = 0.05$ and $n = 400$

11-3a Effect on β of Changing α

Suppose that in the previous illustration we had used a significance level of 1% instead of 5%. The rejection region expressed in terms of the standardised test statistic would be

$$z > z_{0.01} = 2.33$$

or

$$\frac{\bar{X}-170}{65/\sqrt{400}} > 2.33$$

Solving for $\bar{x}$, we find the rejection region in terms of the unstandardised test statistic:

$$\bar{X} > 177.57$$

The probability of a Type II error when $\mu = 180$ is

$$\beta = P\left(\frac{\bar{X}-\mu}{\sigma/\sqrt{n}} < \frac{177.57-180}{65/\sqrt{400}}\right) = P(Z < -0.75) = 0.2266$$

Figure 11.10 depicts this calculation. Compare this figure with Figure 11.9. As you can see, by decreasing the significance level from 5% to 1%, we have shifted the critical value of the rejection region to the right and thus enlarged the area where the null hypothesis is not rejected. The probability of a Type II error increases from 0.0764 to 0.2266.

This calculation illustrates the inverse relationship between the probabilities of Type I and Type II errors alluded to in Section 11-1. It is important to understand this relationship. From a practical point of view, it tells us that if you want to decrease the probability of a Type I error (by specifying a small

FIGURE 11.10

Calculating β for $\mu = 180$, $\alpha = 0.01$ and $n = 400$

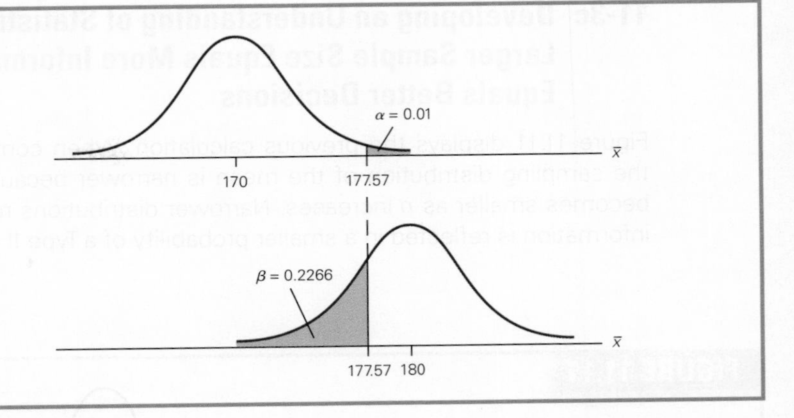

value of α), you increase the probability of a Type II error. In applications where the cost of a Type I error is considerably larger than the cost of a Type II error, this is appropriate. In fact, a significance level of 1% or less is probably justified. However, when the cost of a Type II error is relatively large, a significance level of 5% or more may be appropriate.

Unfortunately, there is no simple formula to determine what the significance level should be. The manager must consider the costs of both mistakes in deciding what to do. Judgement and knowledge of the factors in the decision are crucial.

11-3b Judging the Test

There is another important concept to be derived from this section. A statistical test of hypothesis is effectively defined by the significance level and the sample size, both of which are selected by the statistics practitioner. We can judge how well the test functions by calculating the probability of a Type II error at some value of the parameter. To illustrate, in Example 11.1 the manager chose a sample size of 400 and a 5% significance level on which to base her decision. With those selections, we found β to be 0.0764 when the actual mean is 180. If we believe that the cost of a Type II error is high and thus that the probability is too large, we have two ways to reduce the probability. We can increase the value of α; however, this would result in an increase in the chance of making a Type I error, which is very costly.

Alternatively, we can increase the sample size. Suppose that the manager chose a sample size of 1000. We will now recalculate β with $n = 1000$ (and $\alpha = 0.05$). The rejection region is

$$z > z_{0.05} = 1.645$$

or

$$\frac{\bar{X} - 170}{65/\sqrt{1000}} > 1.645$$

which yields

$$\bar{X} > 173.38$$

The probability of a Type II error is

$$\beta = P\left(\frac{\bar{X} - \mu}{\sigma/\sqrt{n}} < \frac{173.38 - 180}{65/\sqrt{1000}}\right) = P(Z < -3.22) = 0 \text{ (approximately)}$$

In this case, we maintained the same value of α (0.05), but we reduced the probability of not installing the system when the actual mean account is €180 to virtually 0.

11-3c Developing an Understanding of Statistical Concepts: Larger Sample Size Equals More Information Equals Better Decisions

Figure 11.11 displays the previous calculation. When compared with Figure 11.9, we can see that the sampling distribution of the mean is narrower because the standard error of the mean $\sigma/\sqrt{n}$ becomes smaller as n increases. Narrower distributions represent more information. The increased information is reflected in a smaller probability of a Type II error.

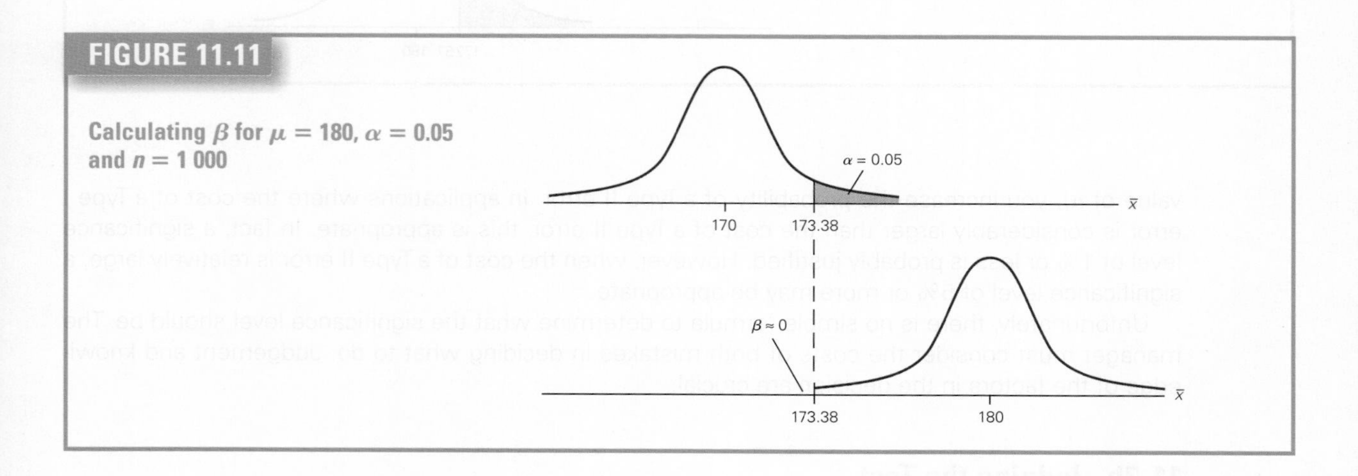

FIGURE 11.11

Calculating β for $\mu = 180$, $\alpha = 0.05$ and $n = 1\,000$

The calculation of the probability of a Type II error for $n = 400$ and for $n = 1000$ illustrates a concept whose importance cannot be overstated. By increasing the sample size, we reduce the probability of a Type II error. By reducing the probability of a Type II error, we make this type of error less frequently. Hence, larger sample sizes allow us to make better decisions in the long run. This finding lies at the heart of applied statistical analysis and reinforces the book's first sentence: 'Statistics is a way to get information from data'.

Throughout this book we introduce a variety of applications in accounting, finance, marketing, operations management, human resources management and economics. In all such applications, the statistics practitioner must make a decision, which involves converting data into information. The more information, the better the decision. Without such information, decisions must be based on guesswork, instinct and luck. W. Edwards Deming, a famous statistician, said it best: 'Without data you are just another person with an opinion'.

11-3d Power of a Test

Another way of expressing how well a test performs is to report its *power*: the probability of it leading us to reject the null hypothesis when it is false. Thus, the power of a test is $1 - \beta$.

When more than one test can be performed in a given situation, we would naturally prefer to use the test that is correct more frequently. If (given the same alternative hypothesis, sample size and significance level) one test has a higher power than a second test, the first test is said to be more powerful.

Using the Computer

DO-IT-YOURSELF EXCEL

You will need to create three spreadsheets, one for a left-tail, one for a right-tail and one for a two-tail test.

Here is our spreadsheet for the right-tail test for Example 11.1.

	A	B	C	D
1	Right-Tail Test			
2				
3	H0: MU	170	Critical value	175.35
4	SIGMA	65	Prob(Type II error)	0.0761
5	Sample size	400	Power of the test	0.9239
6	ALPHA	0.05		
7	H1: MU	180		

Tools: **NORM.S.INV**: Use this function to help compute the critical value in cell D3.
NORM.S.DIST: This function is needed to calculate the probability in cell D4.

MINITAB

Minitab computes the power of the test.

```
Power and Sample Size

1-Sample Z Test

Testing mean = null (versus > null)
Calculating power for mean = null + difference
Alpha = 0.05  Assumed standard deviation = 65


              Sample
Difference    Size      Power
        10     400   0.923938
```

Instructions

1. Click **Stat, Power and Sample Size** and **1-Sample Z** . . . .
2. Specify the sample size in the **Sample sizes** box. (You can specify more than one value of n. Minitab will compute the power for each value.) Type the difference between the actual value of β and the value of β under the null hypothesis. (You can specify more than one value.) Type the value of the standard deviation in the **Standard deviation box.**
3. **Click Options** . . . and specify the **Alternative Hypothesis** and the **Significance level.**

For Example 11.1, we typed **400** to select the **Sample sizes,** the **Differences** was **10 (= 180 − 170), Standard deviation** was **65,** the **Alternative Hypothesis** was **Greater than** and the **Significance level** was **0.05.**

11-3e Operating Characteristic Curve

To compute the probability of a Type II error, we must specify the significance level, the sample size and an alternative value of the population mean. One way to keep track of all these components is to draw the operating characteristic (OC) curve, which plots the values of β versus the values of μ. Because of the time-consuming nature of these calculations, the computer is a virtual necessity. To illustrate, we will draw the OC curve for Example 11.1. We used Excel (we could have used Minitab instead) to compute the probability of a Type II error in Example 11.1 for μ = 170, 171, . . . , 185, with

$n = 400$. Figure 11.12 depicts this curve. Notice as the alternative value of μ increases the value of β decreases. This tells us that as the alternative value of μ moves farther from the value of μ under the null hypothesis, the probability of a Type II error decreases. In other words, it becomes easier to distinguish between $\mu = 170$ and other values of μ when μ is farther from 170. Note that when $\mu = 170$ (the hypothesised value of μ), $\beta = 1 - \alpha$.

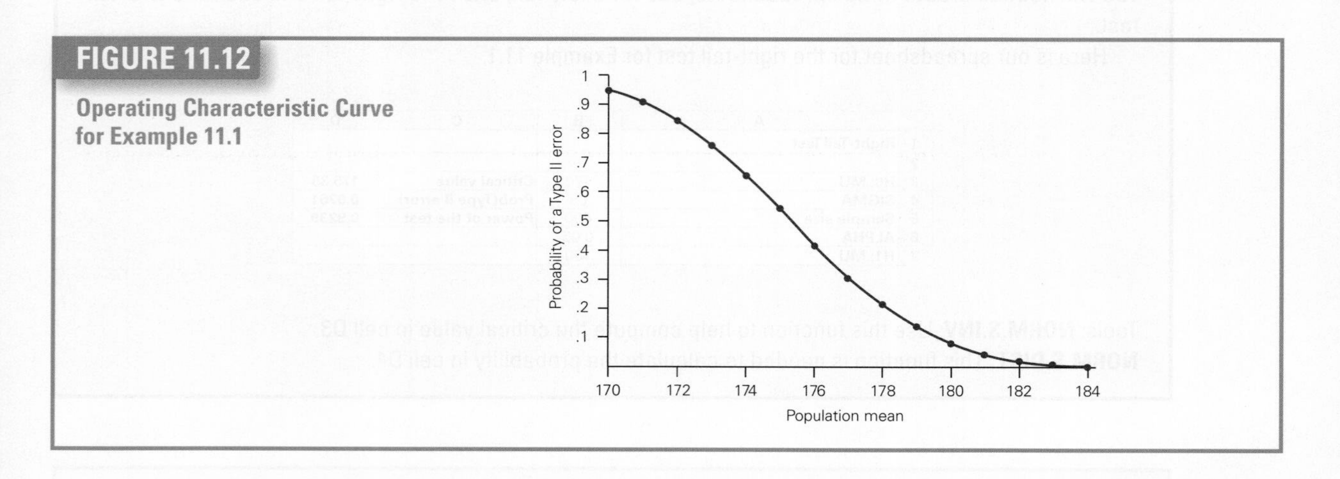

FIGURE 11.12

Operating Characteristic Curve for Example 11.1

The OC curve can also be useful in selecting a sample size. Figure 11.13 shows the OC curve for Example 11.1 with $n = 100$, 400, 1000 and 2000. An examination of this chart sheds some light concerning the effect increasing the sample size has on how well the test performs at different values of μ. For example, we can see that smaller sample sizes will work well to distinguish between 170 and values of μ larger than 180. However, to distinguish between 170 and smaller values of μ requires larger sample sizes. Although the information is imprecise, it does allow us to select a sample size that is suitable for our purposes.

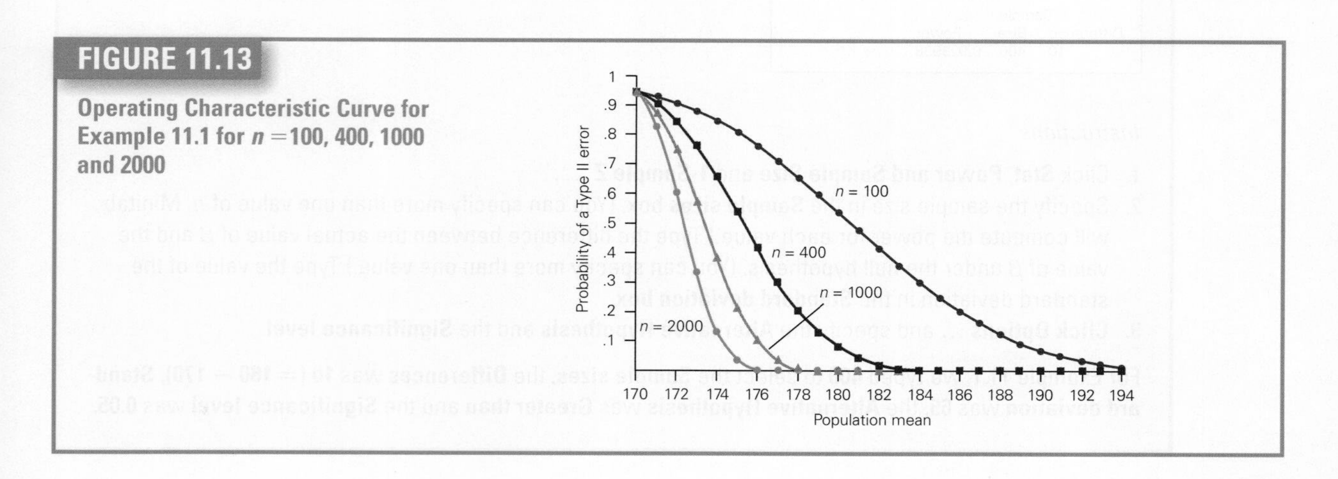

FIGURE 11.13

Operating Characteristic Curve for Example 11.1 for $n = 100$, 400, 1000 and 2000

11-3f Determining the Alternative Hypothesis to Define Type I and Type II Errors

We have already discussed how the alternative hypothesis is determined. It represents the condition we are investigating. In Example 11.1 we wanted to know whether there was sufficient statistical evidence to infer that the new billing system would be cost-effective – that is, whether the mean monthly account is greater than €170. In this textbook, you will encounter many problems using similar phraseology. Your job will be to conduct the test that answers the question.

In real life, however, the manager (that is you 5 years from now) will be asking and answering the question. In general, you will find that the question can be posed in two ways. In Example 11.1, we asked whether there was evidence to conclude that the new system would be cost-effective. Another way of investigating the issue is to determine whether there is sufficient evidence to infer that the new system would *not* be cost-effective. We remind you of the criminal trial analogy. In a criminal trial, the burden of proof falls on the prosecution to prove that the defendant is guilty. In other countries with less emphasis on individual rights, the defendant is required to prove his or her innocence. The UK, the USA, Canada, and many other countries chose the former because the conviction of an innocent defendant is considered to be the greater error. Thus, the test is set up with the null and alternative hypotheses as described in Section 11-1.

In a statistical test where we are responsible for both asking and answering a question, we must ask the question so that we directly control the error that is more costly. As you have already seen, we control the probability of a Type I error by specifying its value (the significance level). Consider Example 11.1 once again. There are two possible errors: (1) conclude that the billing system is cost-effective when it is not and (2) conclude that the system is not cost-effective when it is. If the manager concludes that the billing plan is cost-effective, the company will install the new system. If, in reality, the system is not cost-effective, the company will incur a loss. On the other hand, if the manager concludes that the billing plan is not going to be cost-effective, the company will not install the system. However, if the system is actually cost-effective, the company will lose the potential gain from installing it. Which cost is greater?

Suppose we believe that the cost of installing a system that is not cost-effective is higher than the potential loss of not installing an effective system. The error we wish to avoid is the erroneous conclusion that the system is cost-effective. We define this as a Type I error. As a result, the burden of proof is placed on the system to deliver sufficient statistical evidence that the mean account is greater than €170. The null and alternative hypotheses are as formulated previously:

$$H_0: \mu = 170$$
$$H_1: \mu > 170$$

However, if we believe that the potential loss of not installing the new system when it would be cost-effective is the larger cost, we would place the burden of proof on the manager to infer that the mean monthly account is less than €170. Consequently, the hypotheses would be

$$H_0: \mu = 170$$
$$H_1: \mu < 170$$

This discussion emphasises the need in practice to examine the costs of making both types of error before setting up the hypotheses. However, it is important for readers to understand that the questions posed in exercises throughout this book have already taken these costs into consideration. Accordingly, your task is to set up the hypotheses to answer the questions.

EXERCISES

Developing an Understanding of Statistical Concepts

11.41 Calculate the probability of a Type II error for the following test of hypothesis, given that $\mu = 203$.

$H_0: \mu = 200$

$H_1: \mu \neq 200$

$\alpha = 0.05, \sigma = 10, n = 100$

11.42 Find the probability of a Type II error for the following test of hypothesis, given that $\mu = 1\,050$.

$H_0: \mu = 1\,000$

$H_1: \mu > 1\,000$

$\alpha = 0.01, \sigma = 50, n = 25$

11.43 Determine β for the following test of hypothesis, given that $\mu = 48$.

$H_0: \mu = 50$

$H_1: \mu < 50$

$\alpha = 0.05, \sigma = 10, n = 40$

11.44 For each of Exercises 11.41–11.43, draw the sampling distributions similar to Figure 11.9.

11.45 A statistics practitioner wants to test the following hypotheses with $\sigma = 20$ and $n = 100$:

$H_0: \mu = 100$

$H_1: \mu > 100$

a. Using $\alpha = 0.10$ find the probability of a Type II error when $\mu = 102$.

b. Repeat part (a) with $\alpha = 0.02$.

c. Describe the effect on β of decreasing α.

11.46 a. Calculate the probability of a Type II error for the following hypotheses when $\mu = 37$:

H_0: $\mu = 40$

H_1: $\mu < 40$

The significance level is 5%, the population standard deviation is 5 and the sample size is 25.

b. Repeat part (a) with $\alpha = 15\%$.

c. Describe the effect on β of increasing α.

11.47 Draw the figures of the sampling distributions for Exercises 11.45 and 11.46.

11.48 For the test of hypothesis

H_0: $\mu = 1000$

H_1: $\mu \neq 1000$

$\alpha = 0.05, \sigma = 200$

draw the operating characteristic curve for $n = 25$, 100 and 200.

11.49 Suppose that in Example 11.1 we wanted to determine whether there was sufficient evidence to conclude that the new system would *not* be cost-effective. Set up the null and alternative hypotheses and discuss the consequences of Type I and Type II errors. Conduct the test. Is your conclusion the same as the one reached in Example 11.1? Explain.

Applications

11.50 In Exercise 11.33, we tested to determine whether the installation of safety equipment was effective in reducing person-hours lost to industrial accidents. The null and alternative hypotheses were

H_0: $\mu = 0$

H_1: $\mu < 0$

with $\sigma = 6$, $\alpha = 0.10$, $n = 50$ and $\mu =$ the mean percentage change. The test failed to indicate that the new safety equipment is effective. The manager is concerned that the test was not sensitive enough to detect small but important changes. In particular, he worries that if the true reduction in time lost to accidents is actually 2% (i.e. $\mu = -2$), then the firm may miss the opportunity to install very effective equipment. Find the probability that the test with $\sigma = 6$, $\alpha = 0.10$ and $n = 50$ will fail to conclude that such equipment is effective. Discuss ways to decrease this probability.

11.51 The test of hypothesis in the SSA example concluded that there was not enough evidence to infer that the plan would be profitable. The company would hate to not institute the plan if the actual reduction was as little as 3 days (i.e., $\mu = 21$). Calculate the relevant probability and describe how the company should use this information.

11.52 The fast-food franchiser in Exercise 11.37 was unable to provide enough evidence that the site is acceptable. She is concerned that she may be missing an opportunity to locate the restaurant in a profitable location. She feels that if the actual mean is 104, the restaurant is likely to be very successful. Determine the probability of a Type II error when the mean is 104. Suggest ways to improve this probability.

11.53 Refer to Exercise 11.39. A financial analyst has determined that a 2-minute reduction in the average break would increase productivity. As a result the company would hate to lose this opportunity. Calculate the probability of erroneously concluding that the renovation would not be successful when the average break is 30 minutes. If this probability is high, describe how it can be reduced.

11.54 A school-board administrator believes that the average number of days absent per year among students is less than 10 days. From past experience, he knows that the population standard deviation is 3 days. In testing to determine whether his belief is true, he could use one of the following plans:

i. $n = 100, \quad \alpha = 0.01$

ii. $n = 75, \quad \alpha = 0.05$

iii. $n = 50, \quad \alpha = 0.10$

Which plan has the lowest probability of a Type II error, given that the true population average is 9 days?

11.55 The feasibility of constructing a profitable electricity-producing windmill depends on the mean velocity of the wind and density of air. For a certain type of windmill, the mean would have to exceed 20 metres per second to warrant its construction. The determination of a site's feasibility is a two-stage process. In the first stage, readings of the wind velocity are taken and the mean is calculated. The test is designed to answer the question, 'Is the site feasible?' In other words, is there sufficient evidence to conclude that the mean wind velocity exceeds 20 m/s? If there is enough evidence, further testing is conducted. If there is not enough evidence, the site is removed from consideration. Discuss the consequences and potential costs of Type I and Type II errors.

11.56 The number of potential sites for the first-stage test in Exercise 11.55 is quite large and the readings can be expensive. Accordingly, the test is conducted with a sample of 25 observations. Because the second-stage cost is high, the significance level is set at 1%. A financial analysis of the potential profits and costs reveals that if the mean wind velocity is as high as 25 m/s, the windmill would be extremely profitable. Calculate the probability that the first-stage test will not conclude that the site is feasible when the actual mean wind velocity is 25 m/s. (Assume that σ is 8.) Discuss how the process can be improved.

11-4 THE ROAD AHEAD

We had two principal goals to accomplish in Chapters 10 and 11. First, we wanted to present the concepts of estimation and hypothesis testing. Second, we wanted to show how to produce confidence interval estimates and conduct tests of hypotheses. The importance of both goals should not be underestimated. Almost everything that follows this chapter will involve either estimating a parameter or testing a set of hypotheses. Consequently, Sections 10-2 and 11-2 set the pattern for the way in which statistical techniques are applied. It is no exaggeration to state that if you understand how to produce and use confidence interval estimates and how to conduct and interpret hypothesis tests, then you are well on your way to the ultimate goal of being competent at analysing, interpreting and presenting data. It is fair for you to ask what more you must accomplish to achieve this goal. The answer, simply put, is much more of the same.

In the chapters that follow, we plan to present about three dozen different statistical techniques that can be (and frequently are) employed by statistics practitioners. To calculate the value of test statistics or confidence interval estimates requires nothing more than the ability to add, subtract, multiply, divide and compute square roots. If you intend to use the computer, all you need to know are the commands. The key, then, to applying statistics is knowing which formula to calculate or which set of commands to issue. Thus, the real challenge of the subject lies in being able to define the problem and identify which statistical method is the most appropriate one to use.

Most students have some difficulty recognising the particular kind of statistical problem they are addressing unless, of course, the problem appears among the exercises at the end of a section that just introduced the technique needed. Unfortunately, in practice, statistical problems do not appear already so identified. Consequently, we have adopted an approach to teaching statistics that is designed to help identify the statistical technique.

A number of factors determine which statistical method should be used, but two are especially important: the type of data and the purpose of the statistical inference. In Chapter 2 we pointed out that there are effectively three types of data – interval, ordinal and nominal. Recall that nominal data represent categories such as marital status, occupation and gender. Statistics practitioners often record nominal data by assigning numbers to the responses (e.g.1 = Single, 2 = Married, 3 = Divorced, 4 = Widowed). Because these numbers are assigned completely arbitrarily, any calculations performed on them are meaningless. All that we can do with nominal data is count the number of times each category is observed. Ordinal data are obtained from questions whose answers represent a rating or ranking system. For example, if students are asked to rate a university professor, the responses may be excellent, good, fair or poor. To draw inferences about such data, we convert the responses to numbers. Any numbering system is valid as long as the order of the responses is preserved. Thus '4 = Excellent, 3 = Good, 2 = Fair, 1 = Poor' is just as valid as '15 = Excellent, 8 = Good, 5 = Fair, 2 = Poor' Because of this feature, the most appropriate statistical procedures for ordinal data are ones based on a ranking process.

Interval data are real numbers, such as those representing income, age, height, weight and volume. Computation of means and variances is permissible.

The second key factor in determining the statistical technique is the purpose of doing the work. Every statistical method has some specific objective. We address five such objectives in this book.

11-4a Problem Objectives

1. **Describe a population.** Our objective here is to describe some property of a population of interest. The decision about which property to describe is generally dictated by the type of data. For example, suppose the population of interest consists of all purchasers of computers. If we are interested in the purchasers' incomes (for which the data are interval), we may calculate the mean or the variance to describe that aspect of the population. But if we are interested in the brand of computer that has been bought (for which the data are nominal), all we can do is compute the proportion of the population that purchases each brand.
2. **Compare two populations.** In this case, our goal is to compare a property of one population with a corresponding property of a second population. For example, suppose the populations of interest are male and female purchasers of computers. We could compare the means of their incomes, or we could compare the proportion of each population that purchases a certain brand. Once again, the data type generally determines what kinds of properties we compare.
3. **Compare two or more populations.** We might want to compare the average income in each of several locations in order (for example) to decide where to build a new shopping centre. Or we might want to compare the proportions of defective items in a number of production lines in order to determine which line is the best. In each case, the problem objective involves comparing two or more populations.
4. **Analyze the relationship between two variables.** There are numerous situations in which we want to know how one variable is related to another. Governments need to know what effect rising interest rates have on the unemployment rate. Companies want to investigate how the sizes of their advertising budgets influence sales volume. In most of the problems in this introductory text, the two variables to be analysed will be of the same type; we will not attempt to cover the fairly large body of statistical techniques that has been developed to deal with two variables of different types.
5. **Analyse the relationship among two or more variables.** Our objective here is usually to forecast one variable (called the *dependent variable*) on the basis of several other variables (called *independent variables*). We will deal with this problem only in situations in which all variables are interval.

11-4b Derivations

Because this book is about statistical applications, we assume that our readers have little interest in the mathematical derivations of the techniques described. However, it might be helpful for you to have some understanding about the process that produces the formulas.

As described previously, factors such as the problem objective and the type of data determine the parameter to be estimated and tested. For each parameter, statisticians have determined which statistic to use. That statistic has a sampling distribution that can usually be expressed as a formula. For example, in this chapter the parameter of interest was the population mean μ, whose best estimator is the sample mean $\bar{x}$. Assuming that the population standard deviation σ is known, the sampling distribution of $\overline{X}$ is normal (or approximately so) with mean μ and standard deviation $\sigma/\sqrt{n}$. The sampling distribution can be described by the formula

$$Z = \frac{\overline{X} - \mu}{\sigma/\sqrt{n}}$$

This formula also describes the test statistic for μ with σ known. With a little algebra, we were able to derive (in Section 10-2) the confidence interval estimator of μ.

In future chapters, we will repeat this process, which in several cases involves the introduction of a new sampling distribution (introduced in Chapter 8). Although its shape and formula will differ from the sampling distribution used in this chapter, the pattern will be the same. In general, the formula that expresses the sampling distribution will describe the test statistic. Then some algebraic manipulation (which we will not show) produces the interval estimator. Consequently, we will reverse the order of presentation of the two techniques. In other words, we will present the test of hypothesis first, followed by the confidence interval estimator.

CHAPTER SUMMARY

In this chapter, we introduced the concepts of hypothesis testing and applied them to testing hypotheses about a population mean. We showed how to specify the null and alternative hypotheses, set up the rejection region, compute the value of the test statistic and, finally, to make a decision. Equally as important, we discussed how to interpret the test results. This chapter also demonstrated another way to make decisions; by calculating and using the *p*-value of the test. To help interpret test results, we showed how to calculate the probability of a Type II error. Finally, we provided a road map of how we plan to present statistical techniques.

SYMBOLS:

Symbol	Pronounced	Represents
H_0	*H* nought	Null hypothesis
H_1	*H* one	Alternative (research) hypothesis
α	alpha	Probability of a Type I error
β	beta	Probability of a Type II error
$\bar{x}_L$	*X* bar sub *L* or *X* bar *L*	Value of $\bar{x}$ large enough to reject H_0
$\lvert z \rvert$	Absolute *z*	Absolute value of *z*

FORMULA:

Test statistic for μ

$$Z = \frac{\bar{x} - \mu}{\sigma/\sqrt{n}}$$

12 INFERENCE ABOUT A POPULATION

12-1 INFERENCE ABOUT A POPULATION MEAN WHEN THE STANDARD DEVIATION IS UNKNOWN

12-2 INFERENCE ABOUT A POPULATION VARIANCE

12-3 INFERENCE ABOUT A POPULATION PROPORTION

12-4 INFERENCE ABOUT THE DIFFERENCE BETWEEN TWO MEANS: INDEPENDENT SAMPLES

12-5 INFERENCE ABOUT THE DIFFERENCE BETWEEN TWO MEANS: MATCHED PAIRS EXPERIMENT

12-6 INFERENCE ABOUT THE RATIO OF TWO VARIANCES

12-7 INFERENCE ABOUT THE DIFFERENCE BETWEEN TWO POPULATION PROPORTIONS

TV RATINGS

DATA Xm12-00

Statistical techniques play a vital role in helping advertisers determine how many viewers watch the shows that they sponsor. Several companies sample television viewers to determine what shows they watch, for example, TNS Gallup (in Austria, Norway, Russia), AGB Nielsen firm (in the USA), BARB (in the UK), Auditel (in Italy) and SAARF in South Africa. Satellite households are measured via software on the decoder, which outputs the channel viewed to the metre. South Africa is one of the countries to have used this system since 1989. The television audience measurement (TAMS) service provides daily ratings of all programmes and commercial breaks, and individual spot data is sent 4 days after the week in question. The ratings are issued on a weekly basis.

A group of students in their final year is conducting a face-to-to face questionnaire survey to establish a new TV rating in South Africa. The weekly analysis of the top 10 regular shows (excluding News bulletins) ratings has been produced and the results are displayed below. **See Section 12-3e for the answer.**

Network	Show	Rank
SABC1	Generations	1
SABC1	Amaza	2
SABC1	Loxion Lyric	3
SABC1	Task Force	4
SABC2	Muvhango	5
Television turned off or watched some other channel		6

We would like to use the data to estimate how many South African viewers age 15 to 49 were watching 'Amaza', a TV show for teenagers and young adults based in the coastal Cape Town suburb, Muizenherg.

INTRODUCTION

In the previous two chapters, we introduced the concepts of statistical inference and showed how to estimate and test a population mean. However, the illustration we chose is unrealistic because the techniques require us to use the population standard deviation σ, which, in general, is unknown. The purpose, then, of Chapters 10 and 11 was to set the pattern for the way in which we plan to present other statistical techniques. In other words, we will begin by identifying the parameter to be estimated or tested. We will then specify the parameter's estimator (each parameter has an estimator chosen because of the characteristics we discussed at the beginning of Chapter 10) and its sampling distribution. Using simple mathematics, statisticians have derived the interval estimator and the test statistic. This pattern will be used repeatedly as we introduce new techniques.

In Section 11-4 we described the five problem objectives addressed in this book, and we laid out the order of presentation of the statistical methods. In this chapter we will present techniques employed when the problem objective is to describe a population. When the data are interval, the parameters of interest are the population mean μ and the population variance σ^2. In Section 12-1, we describe how to make inferences about the population mean under the more realistic assumption that the population standard deviation is unknown. In Section 12-2 we continue to deal with interval data, but our parameter of interest becomes the population variance.

In Chapter 2 and Section 11-4 we pointed out that when the data are nominal, the only computation that makes sense is determining the proportion of times each value occurs. Section 12-3 discusses inference about the proportion σ^2. In Sections 12-4 and 12-6 we deal with interval variables; the parameter of interest is the difference between two means. The difference between these two sections introduces yet another factor that determines the correct statistical method – the design of the experiment used to gather the data. In Section 12-4 the samples are independently drawn, whereas in Section 12-5 the samples are taken from a matched pairs experiment.

Section 12-6 presents the procedures employed to infer whether two population variances differ. The parameter is the ratio σ_1^2/σ_2^2. (When comparing two variances, we use the ratio rather than the difference because of the nature of the sampling distribution.)

Section 12-7 addresses the problem of comparing two populations of nominal data. The parameter to be tested and estimated is the difference between two proportions.

12-1 INFERENCE ABOUT A POPULATION MEAN WHEN THE STANDARD DEVIATION IS UNKNOWN

In Sections 10-2 and 11-2, we demonstrated how to estimate and test the population mean when the population standard deviation is known. The confidence interval estimator and the test statistic were derived from the sampling distribution of the sample mean with σ known, expressed as

$$z = \frac{\bar{x} - \mu}{\sigma/\sqrt{n}}$$

In this section, we take a more realistic approach by acknowledging that if the population mean is unknown, then so is the population standard deviation. Consequently, the previous sampling distribution cannot be used. Instead, we substitute the sample standard deviation s in place of the unknown population standard deviation σ. The result is called a *t*-statistic because that is what mathematician William S. Gosset called it. In 1908, Gosset showed that the *t*-statistic defined as

$$t = \frac{\bar{x} - \mu}{s/\sqrt{n}}$$

is Student *t*-distributed when the sampled population is normal. (Gosset published his findings under the pseudonym 'Student', hence the **Student *t*-distribution**.) Recall that we introduced the Student *t*-distribution in Section 8-4.

With exactly the same logic used to develop the test statistic in Section 11-2 and the confidence interval estimator in Section 10-2, we derive the following inferential methods.

Test statistic for μ when σ is unknown

When the population standard deviation is unknown and the population is normal, the test statistic for testing hypotheses about μ is

$$t = \frac{\bar{x} - \mu}{s/\sqrt{n}}$$

which is Student *t*-distributed with $\nu = n - 1$ degrees of freedom.

Confidence interval estimator of μ when σ is unknown

$$\bar{x} \pm t_{\alpha/2}\frac{s}{\sqrt{n}} \qquad \nu = n - 1$$

These formulas now make obsolete the test statistic and interval estimator employed in Chapters 10 and 11 to estimate and test a population mean. Although we continue to use the concepts developed in Chapters 10 and 11 (as well as all the other chapters), we will no longer use the *z*-statistic and the *z*-estimator of μ. All future inferential problems involving a population mean will be solved using the *t*-statistic and μ-estimator of μ shown in the preceding boxes.

EXAMPLE 12.1

DATA Xm12-01

Newspaper Recycling Plant

In the near future, nations will probably have to do more to save the environment. Possible actions include reducing energy use and recycling. Currently, most products manufactured from recycled material are considerably more expensive than those manufactured from material found in the Earth. For example, it is approximately three times as expensive to produce glass bottles from recycled glass than from silica sand, soda ash and limestone, all plentiful materials mined in numerous countries. It is more expensive to manufacture aluminium cans from recycled cans than from bauxite. Newspapers are an exception. It can be profitable to recycle newspaper. A major expense is the collection from homes. In recent years, many companies have gone into the business of collecting used newspapers from households and recycling them. A financial analyst for one such company has recently computed that the firm would make a profit if the mean weekly newspaper collection from each household exceeded 2.0 kg. In a study to determine the feasibility of a recycling plant, a random sample of 148 households was drawn from a large community, and the weekly weight of newspapers discarded for recycling for each household was recorded. Do these data provide sufficient evidence to allow the analyst to conclude that a recycling plant would be profitable?

Weights of Discarded Newspapers

2.5	0.7	3.4	1.8	1.9	2.0	1.3	1.2	2.2	0.9	2.7	2.9	1.5	1.5	2.2
3.2	0.7	2.3	3.1	1.3	4.2	3.4	1.5	2.1	1.0	2.4	1.8	0.9	1.3	2.6
3.6	0.8	3.0	2.8	3.6	3.1	2.4	3.2	4.4	4.1	1.5	1.9	3.2	1.9	1.6
3.0	3.7	1.7	3.1	2.4	3.0	1.5	3.1	2.4	2.1	2.1	2.3	0.7	0.9	2.7
1.2	2.2	1.3	3.0	3.0	2.2	1.5	2.7	0.9	2.5	3.2	3.7	1.9	2.0	3.7
2.3	0.6	0.0	1.0	1.4	0.9	2.6	2.1	3.4	0.5	4.1	2.2	3.4	3.3	0.0
2.2	4.2	1.1	2.3	3.1	1.7	2.8	2.5	1.8	1.7	0.6	3.6	1.4	2.2	2.2
1.3	1.7	3.0	0.8	1.6	1.8	1.4	3.0	1.9	2.7	0.8	3.3	2.5	1.5	2.2
2.6	3.2	1.0	3.2	1.6	3.4	1.7	2.3	2.6	1.4	3.3	1.3	2.4	2.0	
1.3	1.8	3.3	2.2	1.4	3.2	4.3	0.0	2.0	1.8	0.0	1.7	2.6	3.1	

SOLUTION:

IDENTIFY

The problem objective is to describe the population of the amounts of newspaper discarded by each household in the population. The data are interval, indicating that the parameter to be tested is the population mean. Because the financial analyst needs to determine whether the mean is greater than 2.0 kg, the alternative hypothesis is

$$H_1: \mu > 2.0$$

As usual, the null hypothesis states that the mean is equal to the value listed in the alternative hypothesis:

$$H_0: \mu > 2.0$$

The test statistic is

$$t = \frac{\bar{x} - \mu}{s/\sqrt{n}} \quad \nu = n - 1$$

COMPUTE

MANUALLY:

The manager believes that the cost of a Type I error (concluding that the mean is greater than 2 when it is not) is quite high. Consequently, he sets the significance level at 1%. The rejection region is

$$t > t_{\alpha,n-1} = t_{0.01,147} \approx t_{0.01,150} = 2.351$$

To calculate the value of the test statistic, we need to calculate the sample mean $\bar{X}$ and the sample standard deviation s. From the data, we determine

$$\sum x_i = 322.7 \textbf{ and } \sum x_i^2 = 845.1$$

Thus,

$$\bar{x} = \frac{\sum x_i}{n} = \frac{322.7}{148} = 2.18$$

$$s^2 = \frac{\sum x_i^2 - \frac{\left(\sum x_i\right)^2}{n}}{n-1} = \frac{845.1 - \frac{(322.7)^2}{148}}{148-1} = 0.962$$

and

$$s = \sqrt{s^2} = \sqrt{0.962} = 0.981$$

The value of μ is to be found in the null hypothesis. It is 2.0. The value of the test statistic is

$$t = \frac{\bar{x} - \mu}{s/\sqrt{n}} = \frac{2.18 - 2.0}{0.981/\sqrt{148}} = 2.23$$

Because 2.23 is not greater than 2.351, we cannot reject the null hypothesis in favour of the alternative. (Students performing the calculations manually can approximate the *p*-value. The website Appendix Approximating the *p*-Value from the Student *t* Table describes how.)

EXCEL

	A	B	C	D
1	t-Test: Mean			
2				
3				Newspaper
4	Mean			2.18
5	Standard Deviation			0.98
6	Hypothesised Mean			2
7	df			147
8	t Stat			2.24
9	P(T<=t) one-tail			0.0134
10	t Critical one-tail			2.3520
11	P(T<=t) two-tail			0.0268
12	t Critical two-tail			2.6097

Instructions

1. Type or import the data into one column*. (**Open** Xm12-01.)
2. Click **Add-Ins, Data Analysis Plus** and ***t*-Test: Mean.**
3. Specify the **Input Range** (A1:A149), the **Hypothesised Mean** (2) and α (0.01).

MINITAB

One-Sample T: Newspaper

Test of mu = 2 vs > 2

Variable	N	Mean	StDev	SE Mean	95% Lower Bound	T	P
Newspaper	148	2.1804	0.9812	0.0807	2.0469	2.24	0.013

Instructions

1. Type or import the data into one column. (**Open** Xm12-01.)
2. Click **Stat**, **Basic Statistics** and **1-Sample *t* ....**
3. Type or use the **Select** button to specify the name of the variable or the column in the **Samples in columns** box (Newspaper), choose **Perform hypothesis test** and type the value of σ in the **Hypothesised mean** box (2) and click **Options . . . .**
4. Select one of **less than, not equal** or **greater than** in the **Alternative** box (greater than).

*If there is an empty cell (representing missing data), it must be removed. See the website Appendix Removing Empty Cells in Excel.

INTERPRET

The value of the test statistic is $t = 2.24$ and its p-value is 0.0134. There is not enough evidence to infer that the mean weight of discarded newspapers is greater than 2.0. Note that there is some evidence: The p-value is 0.0134. However, because we wanted the probability of a Type I error to be small, we insisted on a 1% significance level. Thus, we cannot conclude that the recycling plant would be profitable.

Figure 12.1 exhibits the sampling distribution for this example.

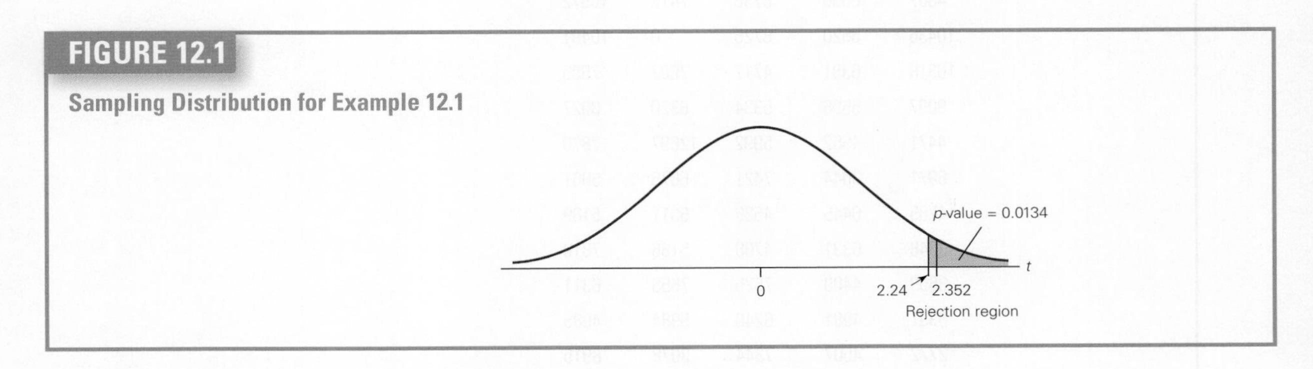

FIGURE 12.1

Sampling Distribution for Example 12.1

EXAMPLE 12.2

DATA Xm12-02

Income Tax

Every year HM Revenue and Customs (HMRC) publishes the Income Tax Liabilities Statistics (ITLS). These statistics are based on HMRC's annual Survey of Personal Incomes (SPI), a representative sample survey of the tax records of individuals on HMRC's Pay as You Earn, Self-Assessment and Payment Claims administrative systems. Let's assume that in 2013–14 the number of taxpayers in the UK was going to be 30 million and HMRC's survey examined 1.107% or 332 100 taxpayers. To determine how well the survey was performed, a random sample of income tax returns was drawn. Estimate with 95% confidence the mean additional income tax collected from the 332 100 files audited.

Income Tax

7973	4219	9824	665	1767
1443	5376	6025	4155	8380
1418	3032	5149	9732	3967
11608	3676	10828	4561	6910
7501	4308	0	7462	5133
10760	4701	1296	5375	5528
1521	9701	4743	5070	4818
3699	5644	5645	6648	1659
4162	1375	8888	2968	6188
4397	7765	5062	11755	5995
6779	1939	6950	6222	3333
5940	2638	7218	6902	5646
2072	3549	2456	2249	4083
12822	6522	2306	4103	6885

Income Tax

5607	0	7489	7460	4771
7101	940	12728	2633	6631
9582	6087	2629	3289	8774
4091	5373	10330	8953	5326
4101	6129	8413	4973	10556
7007	9940	5262	4785	2125
4807	6590	8736	7412	10972
10436	5520	8725	0	10491
10318	6381	4717	7607	7585
9097	5506	5334	6320	8027
4471	4962	5992	12697	7870
6871	3044	7421	6875	6061
4785	6445	4536	5511	5189
3348	6331	4708	5186	7810
3609	4409	7525	7555	6311
5391	4981	6240	5984	4985
2772	4507	7344	3078	8915
11400	6556	9025	7904	6659
976	5110	6603	322	5589

SOLUTION:

IDENTIFY

The problem objective is to describe the population of additional income tax. The data are interval and hence, the parameter is the population mean μ. The question asks us to estimate this parameter. The confidence interval estimator is:

$$\bar{x} \pm t_{\alpha/2} \frac{s}{\sqrt{n}}$$

COMPUTE

MANUALLY:

From the data we determine:

$$\sum x_i = 1\,115\,933 \text{ and } \sum x_i^2 = 7\,924\,385\,209$$

Thus,

$$x = \frac{\sum x_i}{n} = \frac{1\,115\,933}{192} = 5812$$

and

$$s^2 = \frac{\sum x_i^2 - \frac{\left(\sum x_i\right)^2}{n}}{n-1} = \frac{7\,924\,385\,209 - \frac{(1\,112\,933)^2}{192}}{192-1} = 7\,530\,964$$

Thus,

$$s = \sqrt{s^2} = \sqrt{7\,530\,964} = 2744$$

Because we want a 95% confidence interval estimate, $1 - \alpha = 0.95$, $\alpha = 0.05$, $\alpha/2 = 0.025$ and $t_{\alpha/2,n-1} = t_{0.025,191} \approx t_{0.025,190} = 1.973$. Thus, the 95% confidence interval estimate of μ is:

$$x \pm t_{\alpha/2}\frac{s}{\sqrt{n}} = 5812 \pm 1.973\frac{2744}{\sqrt{192}} = 5812 \pm 391$$

or

LCL = £5421 and UCL = £6203

EXCEL

	A	B	C	D
1	**t-Estimate: Mean**			
2				
3			*Taxes*	
4	Mean		5812	
5	Standard Deviation		2744	
6	Observations		192	
7	Standard Error		198	
8	LCL		5422	
9	UCL		6203	

Instructions

1. Type or import the data into one column*. (**Open** Xm12-02.)
2. Click **Add-Ins, Data Analysis Plus** and ***t*-Estimate: Mean.**
3. Specify the **Input Range** (A1:A193) and α (0.0.5).

MINITAB

One-Sample T: Taxes

Variable	N	Mean	StDev	SE Mean	95% CI
Taxes	192	5812	2744	198	(5422, 6203)

Instructions

1. Type or import the data into one column. (**Open** Xm12-02.)
2. Click **Stat, Basic Statistics and 1-Sample t...**
3. Select or type the variable name in the **Samples in columns** box (Taxes) and click **Options . . . .**
4. Specify the **Confidence level** (95) and **not equal** for the **Alternative.**

INTERPRET

We estimate that the mean additional tax collected lies between £5421 and £6203. We can use this estimate to help decide whether the HMRC is surveying the right individuals.

12-1a Checking the Required Conditions

When we introduced the Student *t*-distribution, we pointed out that the *t*-statistic is Student *t*-distributed if the population from which we have sampled is normal. However, statisticians have shown that the mathematical process that derived the Student *t*-distribution is **robust**, which means that if the population is non-normal, the results of the *t*-test and confidence interval estimate are still valid provided that the population is not *extremely* non-normal.** To check this requirement, we draw the histogram and determine whether it is far from bell shaped. Figures 12.2 and 12.3 depict

*If the column contains a blank (representing missing data) the row will have to be deleted.

**Statisticians have shown that when the sample size is large, the results of a *t*-test and estimator of a mean are valid even when the population is extremely non-normal. The sample size required depends on the extent of non-normality.

the Excel histograms for Examples 12.1 and 12.2, respectively. (The Minitab histograms are similar.) Both histograms suggest that the variables are not extremely non-normal and, in fact, may be normal.

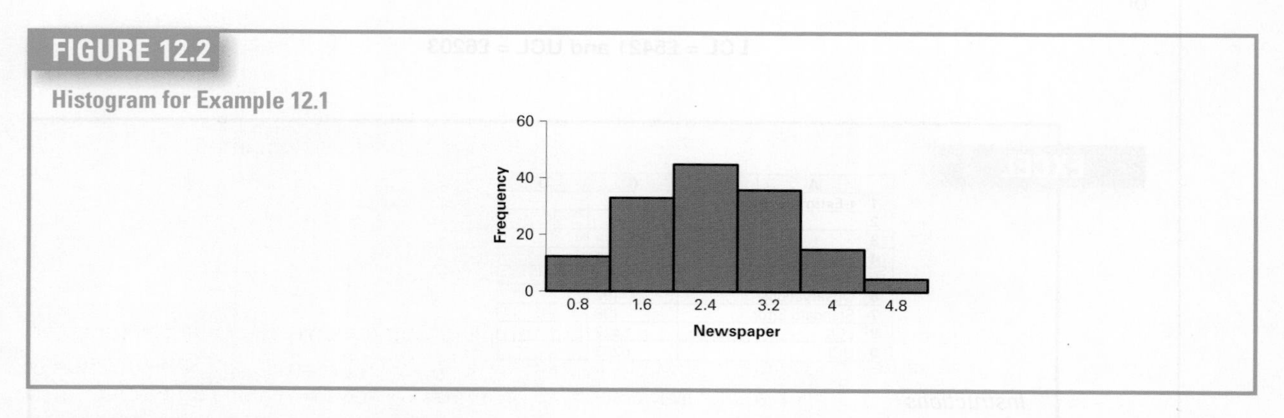

FIGURE 12.2 Histogram for Example 12.1

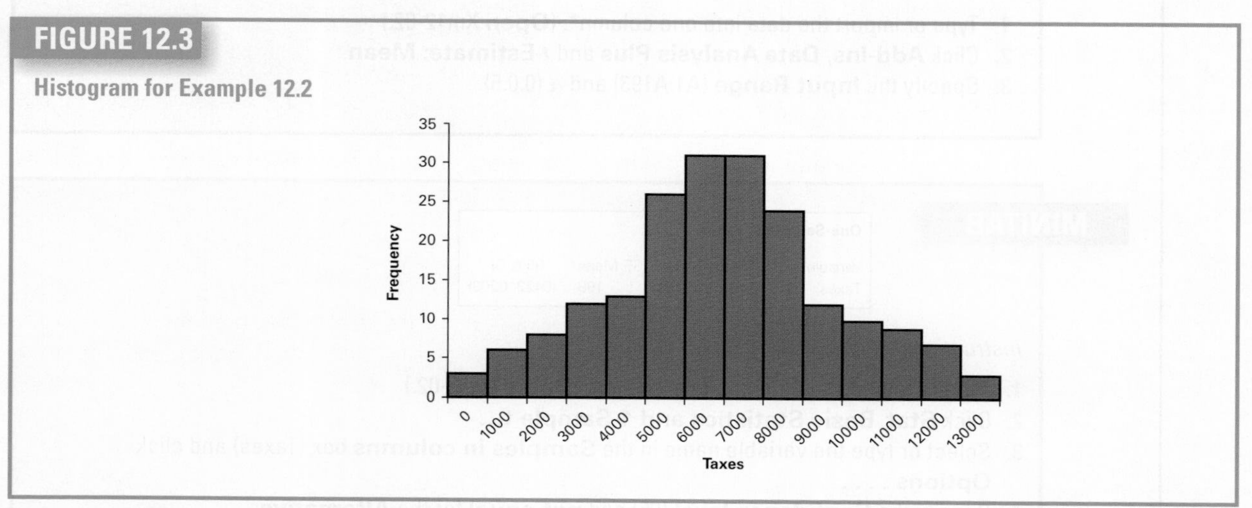

FIGURE 12.3 Histogram for Example 12.2

12-1b Estimating the Totals of Finite Populations

The inferential techniques introduced thus far were derived by assuming infinitely large populations. In practice, however, most populations are finite. (Infinite populations are usually the result of some endlessly repeatable process, such as flipping a coin or selecting items with replacement.) When the population is small, we must adjust the test statistic and interval estimator using the finite population correction factor introduced in Chapter 9. However, in populations that are large relative to the sample size, we can ignore the correction factor. Large populations are defined as populations that are at least 20 times the sample size.

Finite populations allow us to use the confidence interval estimator of a mean to produce a confidence interval estimator of the population total. To estimate the total, we multiply the lower and upper confidence limits of the estimate of the mean by the population size. Thus, the confidence interval estimator of the total is:

$$N\left[\bar{x} \pm t_{\alpha/2}\frac{s}{\sqrt{n}}\right]$$

For example, suppose that we wish to estimate the total amount of additional income tax collected from the 332 100 returns that were examined. The 95% confidence interval estimate of the total is:

$$N\left[\bar{x} \pm t_{\alpha/2}\frac{s}{\sqrt{n}}\right] = 332\,100\,(5812 \pm 391)$$

Which is

LCL = 1 800 314 100 and UCL = 2 060 016 300

12-1c Developing an Understanding of Statistical Concepts 1

This section introduces the term *degrees of freedom*. This is a value that is different for different statistical tests. We will encounter this term many times in this book, so a brief discussion of its meaning is warranted. The Student *t*-distribution is based on using the sample variance to estimate the unknown population variance. The sample variance is defined as

$$s^2 = \frac{\sum (x_i - \bar{x})^2}{n - 1}$$

To compute s^2, we must first determine $\bar{x}$. Recall that sampling distributions are derived by repeated sampling from the same population. To repeatedly take samples to compute s^2, we can choose any numbers for the first $n - 1$ observations in the sample. However, we have no choice on the nth value because the sample mean must be calculated first. To illustrate, suppose that $n = 3$ and we find $\bar{x} = 10$. We can have x_1 and x_2 assume any values without restriction. However, x_3 must be such that $\bar{x} = 10$. For example, if $x_1 = 6$ and $x_2 = 8$, then x_3 must equal 16. Therefore, there are only two degrees of freedom in our selection of the sample. We say that we lose one degree of freedom because we had to calculate $\bar{x}$.

Notice that the denominator in the calculation of s^2 is equal to the number of degrees of freedom. This is not a coincidence and will be repeated throughout this book.

12-1d Developing an Understanding of Statistical Concepts 2

The *t*-statistic like the *z*-statistic measures the difference between the sample mean $\bar{X}$ and the hypothesised value of μ in terms of the number of standard errors. However, when the population standard deviation σ is unknown, we estimate the standard error by

$$s/\sqrt{n}$$

12-1e Developing an Understanding of Statistical Concepts 3

When we introduced the Student *t*-distribution in Section 8-4, we pointed out that it is more widely spread out than the standard normal. This circumstance is logical. The only variable in the *z*-statistic is the sample mean $\bar{x}$, which will vary from sample to sample. The *t*-statistic has two variables: the sample mean $\bar{x}$ and the sample standard deviation s, both of which will vary from sample to sample. Because of the greater uncertainty, the *t*-statistic will display greater variability.

We complete this section with a review of how we identify the techniques introduced in this section.

Factors that Identify the *t*-Test and Estimator of μ

1. **Problem objective**: Describe a population.
2. **Data type**: Interval.
3. **Type of descriptive measurement**: Central location.

EXERCISES

DO-IT-YOURSELF EXCEL

12.1 Construct an Excel spreadsheet that performs the t-test of μ. Inputs: sample mean, sample standard deviation, sample size, hypothesised mean. Outputs: Test statistic, critical values, and one- and two-tail p-values. Tools: **T.INV, T.DIST.**

12.2 Create a spreadsheet that computes the t-estimate of μ. Inputs: sample mean, sample standard deviation, sample size and confidence level. Outputs: Upper and lower confidence limits. Tools: **T.INV.**

Developing an Understanding of Statistical Concepts

The following exercises are 'what-if' analyses designed to determine what happens to the test statistics and interval estimates when elements of the statistical inference change. These problems can be solved manually or using the Do-It-Yourself Excel spreadsheets you created.

12.3 **a.** A random sample of 25 was drawn from a population. The sample mean and standard deviation are $\bar{x}$ and $s = 125$. Estimate μ with 95% confidence.

b. Repeat part (a) with $n = 50$.

c. Repeat part (a) with $n = 100$.

d. Describe what happens to the confidence interval estimate when the sample size increases.

12.4 **a.** A statistics practitioner drew a random sample of 400 observations and found that $\bar{x} = 700$ and $s = 100$. Estimate the population mean with 90% confidence.

b. Repeat part (a) with a 95% confidence level.

c. Repeat part (a) with a 99% confidence level.

d. What is the effect on the confidence interval estimate of increasing the confidence level?

12.5 **a.** The mean and standard deviation of a sample of 100 are $\bar{x} = 10$ and $s = 1$.
Estimate the population mean with 95% confidence.

b. Repeat part (a) with $s = 4$.

c. Repeat part (a) with $s = 10$.

d. Discuss the effect on the confidence interval estimate of increasing the standard deviation s.

12.6 **a.** The sample mean and standard deviation from a random sample of ten observations from a normal population were computed as $\bar{x} = 23$ and $s = 9$. Calculate the value of the test statistic (and for Excel users, the p-value) of the test required to determine whether there is enough evidence to infer at the 5% significance level that the population mean is greater than 20.

b. Repeat part (a) with $n = 30$.

c. Repeat part (a) with $n = 50$.

d. Describe the effect on the t-statistic (and for Excel users, the p-value) of increasing the sample size.

12.7 **a.** Calculate the test statistic (and for Excel users, the p-value) when $\bar{x} = 145$, $s = 50$ and $n = 100$,

H_0: $\mu = 150$

H_1: $\mu < 150$

b. Repeat part (a) with $\bar{x} = 140$.

c. Repeat part (a) with $\bar{x} = 135$.

d. What happens to the t-statistic (and for Excel users, the p-value) when the sample mean decreases?

12.8 **a.** Estimate the population mean with 90% confidence given the following: $\bar{x} = 175$, $s = 30$, and $n = 5$.

b. Repeat part (a) assuming that you know that the population standard deviation is $\sigma = 30$.

c. Explain why the interval estimate produced in part (b) is narrower than that in part (a).

12.9 **a.** A random sample of 11 observations was taken from a normal population. The sample mean and standard deviation are $s = 9$. Can we infer at the 5% significance level that the population mean is greater than 70?

b. Repeat part (a) assuming that you know that the population standard deviation is $\sigma = 90$.

c. Explain why the conclusions produced in parts (a) and (b) differ.

Applications

The following exercises may be answered manually or with the assistance of a computer. The data are stored in files. Assume that the random variable is normally distributed.

12.10 Xr12-10 A courier service in Andover advertises that its average delivery time is less than 6 hours for local deliveries. A random sample of times for 12 deliveries to an address across town was recorded. These data are shown here. Is this sufficient evidence to support the courier's advertisement, at the 5% level of significance?

3.03 6.33 6.50 5.22 3.56 6.76
7.98 4.82 7.96 4.54 5.09 6.46

12.11 Xr12-11 A Trading Standards Officer responsible for enforcing laws governing weights and measures routinely inspects packages to determine whether the weight of the contents is at least as great as that advertised on the package. A random sample of 18 containers whose packaging states that the contents weigh 225 g was drawn. The contents were weighed and the results follow. Can we conclude at the 1% significance level that on average the containers are mislabelled?

221.13 224.25 224.81 226.51 225.10 219.71
225.95 225.38 220.84 228.50 221.70 223.70
224.53 223.11 224.53 226.23 228.21 224.25

12.12 Xr12-12 Part of a PhD student's job is to publish his or her research. This task often entails reading a variety of journal articles to keep up-to-date. To help determine faculty standards, a dean of a business school surveyed a random sample of 12 PhD students across China and asked them to count the number of journal articles they read in a typical month. These data are listed here. Estimate with 90% confidence the mean number of journal articles read monthly by research students.

9 17 4 23 56 30 41 45 21 10 44 20

12.13 Xr12-13 University bookshops order books that instructors adopt for their courses. The number of copies ordered matches the projected demand. However, at the end of the semester, the bookshop has too many copies on hand and must return them to the publisher. A bookshop has a policy that the proportion of books returned should be kept as small as possible. The average is supposed to be less than 10%. To see whether the policy is working, a random sample of book titles was drawn, and the fraction of the total originally ordered that are returned is recorded and listed here. Can we infer at the 10% significance level that the mean proportion of returns is less than 10%?

4 15 11 7 5 9 4 3 5 8

The following exercises require the use of a computer and software. Use a 5% significance level for all tests.

12.14 Xr12-14 A company that produces universal remote controls wanted to determine the number of remote control devices Saudi homes contain. The company hired a statistician to survey 240 randomly selected homes and determine the number of remote controls. If there are 5 million households, estimate with 99% confidence the total number of remote controls in Saudi Arabia.

12.15 Xr12-15 The UK chain WHSmith, a high street retailer that sells a wide variety of books, office supplies and stationery, often features sales of products whose prices are reduced because of rebates. Some rebates are so large that the effective price becomes £0. The goal is to lure customers into the store to buy other non-sale items. A secondary objective is to acquire addresses and telephone numbers to sell to telemarketers and other mass marketers. During 1 week in January, WHSmith offered a 100-pack of CD-ROMs (regular price £29.99 minus £10 instant rebate, £12 manufacturer's rebate and £8 WHSmith mail-in rebate). The number of packages was limited, and no rain checks were issued. In all WHSmith stores, 2800 packages were in stock. All were sold. A random sample of 122 buyers was undertaken. Each was asked to report the total value of the other purchases made that day. Estimate with 95% confidence the total spent on other products purchased by those who bought the CD-ROMs.

12.16 Xr12-16 Traffic congestion seems to worsen each year in all EU countries. Additionally, in 2004 and 2007, several Eastern European countries joined the EU and their economies have a considerable impact on the transport system. This raises the question: 'How much does traffic congestion add to the external costs of transport?' UIC studies on external cost of transport can help to obtain an up-to-date and comprehensive picture of transport in the EU. Assume that the UIC has recently conducted an analysis to produce an estimate of the total cost. Drivers in the 99 most congested areas in the EU were sampled and for each driver the congestion cost in time and petrol was recorded. The total number of drivers in these 99 areas was 171 000 000. Estimate with 95% confidence the total cost of congestion in the 99 areas.

12.17 Xr12-17 Companies that sell groceries over the Internet have to compete in an increasingly crowded market, particularly in the UK, with all the graduate supermarket chains now offering delivery services to most regions across the country. Online grocery selling in South Africa, however, is still relatively undeveloped, despite a strong consumer marketplace, perhaps in part due to poorer Internet coverage and less homes online. Customers must enter their orders, pay by credit card and receive delivery by lorry. A potential e-grocer analyzed the market and determined that the average order would have to exceed ZAR850 for the business to be profitable. To determine whether an e-grocery would be profitable in one large city, she offered the service and recorded the size of the order for a random sample of customers. Can we infer from this data that an e-grocery will be profitable in this city?

12-2 INFERENCE ABOUT A POPULATION VARIANCE

In Section 12-1, where we presented the inferential methods about a population mean, we were interested in acquiring information about the central location of the population. As a result, we tested and estimated the population mean. If we are interested instead in drawing inferences about a population's variability, the parameter we need to investigate is the population variance σ^2. Inference about the variance can be used to make decisions in a variety of problems. In an example illustrating the use of the normal distribution, in Section 8-2, we showed why variance is a measure of risk. In Section 7-3 we described an important application in finance wherein stock diversification was shown to reduce the variance of a portfolio and, in so doing, reduce the risk associated with that portfolio. In both sections, we assumed that the population variances were known. In this section, we take a more realistic approach and acknowledge that we need to use statistical techniques to draw inferences about a population variance.

Another application of the use of variance comes from operations management. Quality technicians attempt to ensure that their company's products consistently meet specifications. One way of judging the consistency of a production process is to compute the variance of a product's size, weight or volume; that is, if the variation in size, weight or volume is large, it is likely that an unsatisfactorily large number of products will lie outside the specifications for that product. We will return to this subject later in this book. In Section 14-6, we discuss how operations managers search for and reduce the variation in production processes.

The task of deriving the test statistic and the interval estimator provides us with another opportunity to show how statistical techniques in general are developed. We begin by identifying the best estimator. That estimator has a sampling distribution, from which we produce the test statistic and the interval estimator.

12-2a Statistic and Sampling Distribution

The estimator of σ^2 is the sample variance introduced in Section 4-2. The statistic s^2 has the desirable characteristics presented in Section 10-1; that is, s^2 is an unbiased, consistent estimator of σ^2.

Statisticians have shown that the sum of squared deviations from the mean $\sum(x_i - \bar{x})^2$ [which is equal to $(n - 1)s^2$] divided by the population variance is chi-squared distributed with $v = n - 1$ degrees of freedom provided that the sampled population is normal. The statistic

$$\chi^2 = \frac{(n-1)s^2}{\sigma^2}$$

is called the **chi-squared statistic** (χ^2-statistic). The chi-squared distribution was introduced in Section 8-4.

12-2b Testing and Estimating a Population Variance

As we discussed in Section 11-4, the formula that describes the sampling distribution is the formula of the test statistic.

Test Statistic for σ^2

The test statistic used to test hypotheses about σ^2 is

$$\chi^2 = \frac{(n-1)s^2}{\sigma^2}$$

which is chi-squared distributed with $v = n-1$ degrees of freedom when the population random variable is normally distributed with variance equal to σ^2.

Using the notation introduced in Section 8-4, we can make the following probability statement:

$$P(\chi^2_{1-\alpha/2} < \chi^2 < \chi^2_{\alpha/2}) = 1 - \alpha$$

Substituting

$$\chi^2 = \frac{(n-1)s^2}{\sigma^2}$$

and with some algebraic manipulation, we derive the confidence interval estimator of a population variance.

Confidence Interval Estimator of σ^2

$$\text{Lower confidence limit (LCL)} = \frac{(n-1)s^2}{\chi^2_{\alpha/2}}$$

$$\text{Upper confidence limit (UCL)} = \frac{(n-1)s^2}{\chi^2_{1-\alpha/2}}$$

APPLICATIONS IN OPERATIONS MANAGEMENT

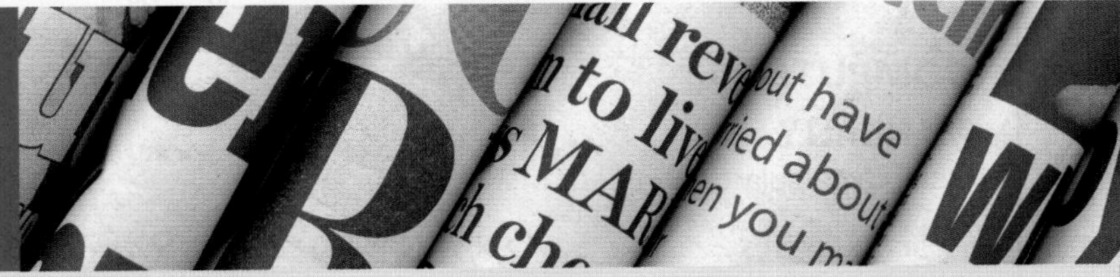

Quality

A critical aspect of production is quality. The quality of a final product is a function of the quality of the product's components. If the components do not fit, the product will not function as planned and likely cease functioning before its customers expect it to. For example, if a car door is not made to its specifications, it will not fit. As a result, the door will leak both water and air.

Operations managers attempt to maintain and improve the quality of products by ensuring that all components are made so that there is as little variation as possible. As you have already seen, statisticians measure variation by computing the variance.

Incidentally, an entire chapter (Chapter 19) is devoted to the topic of quality.

EXAMPLE 12.3

DATA Xm12-03

Consistency of a Container-Filling Machine, Part 1

Container-filling machines are used to package a variety of liquids, including milk, soft drinks and paint. Ideally, the amount of liquid should vary only slightly because large variations will cause some containers to be underfilled (cheating the customer) and some to be overfilled (resulting in costly waste). The president of a company that developed a new type of machine boasts that this machine can fill 1-litre (1000 cubic centimetres) containers so consistently that the variance of the fills will be less than 1 cubic centimetre. To examine the veracity of the claim, a random sample of 25 1-litre fills was taken and the results (cubic centimetres) recorded. These data are listed here. Do these data allow the president to make this claim at the 5% significance level?

Fills

999.6	1000.7	999.3	1000.1	999.5
1000.5	999.7	999.6	999.1	997.8
1001.3	1000.7	999.4	1000.0	998.3
999.5	1000.1	998.3	999.2	999.2
1000.4	1000.1	1000.1	999.6	999.9

SOLUTION:

IDENTIFY

The problem objective is to describe the population of 1-litre fills from this machine. The data are interval, and we are interested in the variability of the fills. It follows that the parameter of interest is the population variance. Because we want to determine whether there is enough evidence to support the claim, the alternative hypothesis is

$$H_1: \sigma^2 < 1$$

The null hypothesis is

$$H_0: \sigma^2 = 1$$

and the test statistic we will use is

$$\chi^2 = \frac{(n-1)s^2}{\sigma^2}$$

COMPUTE

MANUALLY:

Using a calculator, we find

$$\sum x_i = 24\,992.0 \quad \text{and} \quad \sum x_i^2 = 24\,984\,017.76$$

Thus,

$$s^2 = \frac{\sum x_i^2 - \frac{\left(\sum x_i\right)^2}{n}}{n-1} = \frac{24\,984\,017.76 - \frac{(24\,992.0)^2}{25}}{25-1} = 0.6333$$

The value of the test statistic is

$$\chi^2 = \frac{(n-1)s^2}{\sigma^2} = \frac{(25-1)(0.6333)}{1} = 15.20$$

The rejection region is

$$\chi^2 < \chi^2_{1-\alpha,n-1} = \chi^2_{1-0.05,25-1} = \chi^2_{0.95,24} = 13.85$$

Because 15.20 is not less than 13.85, we cannot reject the null hypothesis in favour of the alternative.

EXCEL

	A	B	C	D
1	**Chi-Squared Test: Variance**			
2				
3				*Fills*
4	Sample Variance			0.6333
5	Hypothesised Variance			1
6	df			24
7	Chi-Squared Stat			15.20
8	P (CHI<=chi) one-tail			0.0852
9	Chi-Squared Critical one tail	Left-tail		13.85
10		Right-tail		36.42
11	P (CHI<=chi) two-tail			0.1705
12	Chi-Squared Critical two tail	Left-tail		12.40
13		Right-tail		39.36

The value of the test statistic is 15.20. $P(\text{CHI} \leq \text{chi})$ one-tail is the probability $P(X^2 < 15.20)$, which is equal to 0.0852. Because this is a one-tail test, the *p*-value is 0.0852.

Instructions

1. Type or import the data into one column*. (Open Xm12-03.)
2. Click **Add-Ins, Data Analysis Plus** and **Chi-squared Test: Variance.**

Specify the **Input Range** (A1:A26), type the **Hypothesised Variance** (1) and the value of α (0.05).

MINITAB

Test and CI for One Standard Deviation: Fills

```
Null hypothesis          sigma = 1
Alternative hypothesis   sigma = < 1

Statistics

Variable   N  StDev    Variance
Fills     25  0.796       0.633

Tests

Variable  Method    Chi-Square     DF  P-Value
Fills     Standard       15.20  24.00    0.085
```

Instructions

Some of the output has been deleted.

Besides computing the Chi-Squared statistic and *p*-value, and because we are conducting a one-tail test, Minitab calculates a one-sided confidence interval estimate. (See Section 11-2k for a discussion of one-sided confidence interval estimators.)

1. Type or import the data into one column. (**Open** Xm12-03.)
2. Click **Stat, Basic Statistics** and **1 Variance . . . .**
3. Type or use the **Select** button to specify the name of the variable or the column in the **Samples in columns** box (Fills), check **Perform hypothesis test**, and type the value in the **Hypothesised standard deviation** box (1).
4. Click **Options . . .** and select one of **less than, not equal** or **greater than** in the **Alternative** box (less than).

INTERPRET

There is not enough evidence to infer that the claim is true. As we discussed before, the result does not say that the variance is equal to 1; it merely states that we are unable to show that the variance is less than 1. Figure 12.4 depicts the sampling distribution of the test statistic.

FIGURE 12.4

Sampling Distribution for Example 12.3

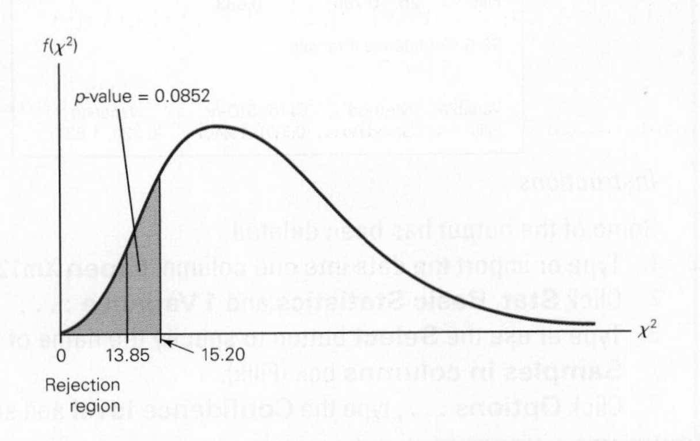

*If the column contains a blank (representing missing data) the row will have to be deleted.

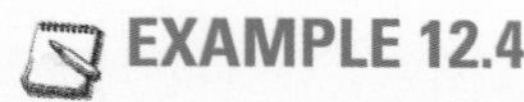

EXAMPLE 12.4

Consistency of a Container-Filling Machine, Part 2

Estimate with 99% confidence the variance of fills in Example 12.3.

SOLUTION:

MANUALLY:

In the solution to Example 12.3, we found $(n-1)s^2$ to be 15.20. From Table 5 in Appendix A, we find

$$\chi^2_{\alpha/2,n-1} = \chi^2_{0.005,24} = 45.6$$

$$\chi^2_{1-\alpha/2,n-1} = \chi^2_{0.995,24} = 9.89$$

Thus,

$$\textbf{LCL} = \frac{(n-1)s^2}{\chi^2_{\alpha/2}} = \frac{15.20}{45.6} = 0.3333$$

$$\textbf{UCL} = \frac{(n-1)s^2}{\chi^2_{1-\alpha/2}} = \frac{15.20}{9.89} = 1.537$$

We estimate that the variance of fills is a number that lies between 0.3333 and 1.537.

EXCEL

	A	B
1	**Chi-Squared Estimate: Variance**	
2		
3		*Fills*
4	Sample Variance	0.6333
5	df	24
6	LCL	0.3336
7	UCL	1.5375

Instructions

1. Type or import the data into one column*. (**Open** Xm12-03.)
2. Click **Add-Ins, Data Analysis Plus** and **Chi-Squared Estimate: Variance**.
3. Specify the **Input Range** (A1:A26) and σ (0.01).

MINITAB

Test and CI for One Standard Deviation: Fills

Statistics

Variable	N	StDev	Variance
Fills	25	0.796	0.633

99% Confidence Intervals

Variable	Method	CI for StDev	CI for Variance
Fills	Standard	(0.578, 1.240)	(0.334, 1.537)

Instructions

Some of the output has been deleted.

1. Type or import the data into one column. (**Open** Xm12-03.)
2. Click **Stat**, **Basic Statistics** and **1 Variance . . . .**
3. Type or use the **Select** button to specify the name of the variable or the column in the **Samples in columns** box (Fills).
4. Click **Options . . .**, type the **Confidence level** and select **not equal** in the **Alternative** box.

*If the column contains a blank (representing missing data) the row will have to be deleted.

INTERPRET

In Example 12.3, we saw that there was not sufficient evidence to infer that the population variance is less than 1. Here we see that σ^2 estimated to lie between 0.3336 and 1.5375. Part of this interval is above 1, which tells us that the variance may be larger than 1, confirming the conclusion we reached in Example 12.3. We may be able to use the estimate to predict the percentage of overfilled and underfilled bottles. This may allow us to choose among competing machines.

12-2c Checking the Required Condition

Like the t-test and estimator of σ introduced in Section 12-1, the chi-squared test and estimator of σ^2 theoretically require that the sample population be normal. In practice, however, the technique is valid so long as the population is not extremely non-normal. We can gauge the extent of non-normality by drawing the histogram. Figure 12.5 depicts Excel's version of this histogram. As you can see, the fills appear to be somewhat asymmetric. However the variable does not appear to be very non-normal. We conclude that the normality requirement is not seriously violated.

FIGURE 12.5

Histogram for Examples 12.3 and 12.4

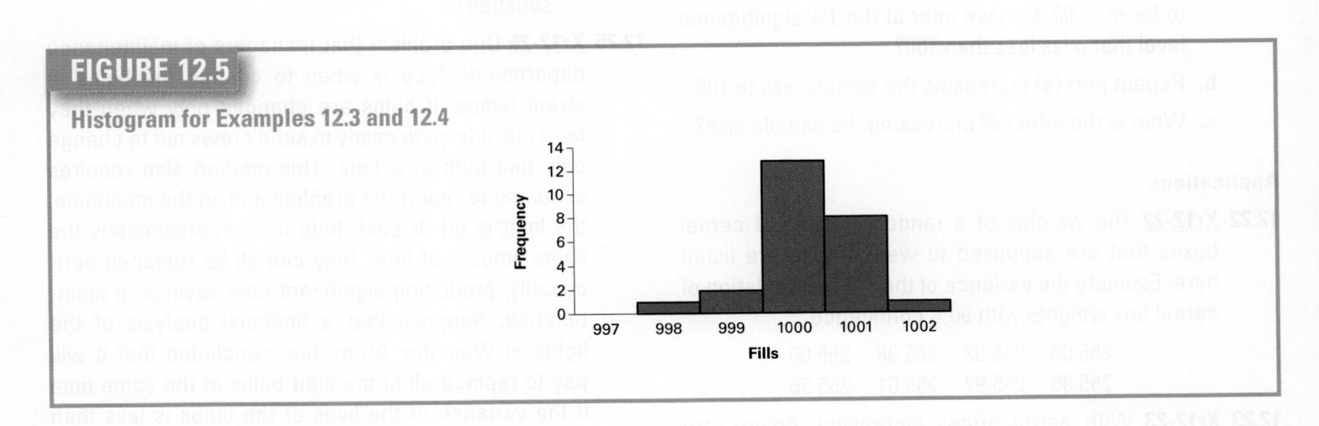

Here is how we recognise when to use the techniques introduced in this section.

Factors that Identify the Chi-Squared t-Test and Estimator of μ

1. **Problem objective**: Describe a population.
2. **Data type**: Interval.
3. **Type of descriptive measurement**: Variability.

EXERCISES

DO-IT-YOURSELF EXCEL

Construct Excel spreadsheets that perform the following techniques

12.18 χ^2-test of σ^2. Inputs: sample proportion, sample size and hypothesised proportion. Outputs: Test statistic, critical values, and one- and two-tail p-values. Tools: **CHI.INV, CHI.TEST**.

12.19 χ^2-estimate of σ^2. Inputs: sample proportion, sample size and confidence level. Outputs: Upper and lower confidence limits. Tools: **CHI. INV**.

Developing an Understanding of Statistical Concepts

The following three exercises are 'what-if analyses' designed to determine what happens to the test statistics and interval estimates when elements of the statistical inference change. These problems can be solved manually or using the Do-It-Yourself Excel spreadsheets you created.

12.20 a. A random sample of 100 observations was drawn from a normal population. The sample variance was calculated to be $s^2 = 220$. Test with $\alpha = 0.05$ to determine whether we can infer that the population variance differs from 300.

b. Repeat part (a) changing the sample size to 50.

c. What is the effect of decreasing the sample size?

12.21 a. The sample variance of a random sample of 50 observations from a normal population was found to be $s^2 = 80$. Can we infer at the 1% significance level that σ^2 is less than 100?

b. Repeat part (a) increasing the sample size to 100.

c. What is the effect of increasing the sample size?

Applications

12.22 Xr12-22 The weights of a random sample of cereal boxes that are supposed to weigh 255 g are listed here. Estimate the variance of the entire population of cereal box weights with 90% confidence.

255.05 255.03 255.98 255.00
255.99 255.97 255.01 255.96

12.23 Xr12-23 With petrol prices increasing, drivers are more concerned with their cars' petrol consumption. For the past 5 years a driver has tracked the fuel efficiency of his car and found that the variance from fill-up to fill-up was $\sigma^2 = 12$ 1/100 km^2. Now that his car is 5 years old, he would like to know whether the variability of fuel efficiency has changed. He recorded the fuel consumption (1/100 km) from his last eight fill-ups; these are listed here. Conduct a test at a 10% significance level to infer whether the variability has changed.

17 15 18 14 21 25 16 13

The following exercises require the use of a computer and software.

12.24 Xr12-24 One important factor in inventory control is the variance of the daily demand for the product. A management scientist has developed the optimal order quantity and reorder point, assuming that the variance is equal to 250. Recently, the company has experienced some inventory problems, which induced the operations manager to doubt the assumption. To examine the problem, the manager took a sample of 25 days and recorded the demand.

a. Do these data provide sufficient evidence at the 5% significance level to infer that the management scientist's assumption about the variance is wrong?

b. What is the required condition for the statistical procedure in part (a)?

c. Does it appear that the required condition is not satisfied?

12.25 Xr12-25 One problem that managers of maintenance departments face is when to change the bulbs in street lamps. If bulbs are changed only when they burn out, it is quite costly to send crews out to change only one bulb at a time. This method also requires someone to report the problem and, in the meantime, the light is off. If each bulb lasts approximately the same amount of time, they can all be replaced periodically, producing significant cost savings in maintenance. Suppose that a financial analysis of the lights at Wembley Arena has concluded that it will pay to replace all of the light bulbs at the same time if the variance of the lives of the bulbs is less than 200 hours^2. The lengths of life of the last 100 bulbs were recorded. What conclusion can be drawn from these data? Use a 5% significance level.

12.26 Xr12-26 Home blood-pressure monitors have been on the market for several years. This device allows people with high blood pressure to measure their own and determine whether additional medication is necessary. Concern has been expressed about inaccurate readings. To judge the severity of the problem a laboratory technician measured his own blood pressure 25 times using the leading brand of monitors. Estimate the population variance with 95% confidence.

12-3 INFERENCE ABOUT A POPULATION PROPORTION

In this section, we continue to address the problem of describing a population. However, we shift our attention to populations of nominal data, which means that the population consists of nominal or categorical values. For example, in a brand-preference survey in which the statistics practitioner asks consumers of a particular product which brand they purchase, the values of the random variable are the brands. If there are five brands, the values could be represented by their names, by letters

(A, B, C, D and E) or by numbers (1, 2, 3, 4 and 5). When numbers are used, it should be understood that the numbers only represent the name of the brand, are completely arbitrarily assigned and cannot be treated as real numbers – that is, we cannot calculate means and variances.

12-3a Parameter

Recall the discussion of types of data in Chapter 2. When the data are nominal, all that we are permitted to do to describe the population or sample is count the number of occurrences of each value. From the counts, we calculate proportions. Thus, the parameter of interest in describing a population of nominal data is the population proportion p. In Section 7-4, this parameter was used to calculate probabilities based on the binomial experiment. One of the characteristics of the binomial experiment is that there are only two possible outcomes per trial. Most practical applications of inference about p involve more than two outcomes. However, in many cases we are interested in only one outcome, which we label a 'success'. All other outcomes are labelled as 'failures'. For example, in brand-preference surveys we are interested in our company's brand. In political surveys, we wish to estimate or test the proportion of voters who will vote for one particular candidate – probably the one who has paid for the survey.

12-3b Statistic and Sampling Distribution

The logical statistic used to estimate and test the population proportion is the sample proportion defined as

$$\hat{p} = \frac{x}{n}$$

where x is the number of successes in the sample and n is the sample size. In Section 9-2 we presented the approximate sampling distribution of $\hat{P}$. (The actual distribution is based on the binomial distribution, which does not lend itself to statistical inference.) The sampling distribution of $\hat{P}$ is approximately normal with mean p and standard deviation $\sqrt{p(1-p)/n}$ [provided that np and $n(1-p)$ are greater than 5]. We express this sampling distribution as

$$z = \frac{\hat{P} - p}{\sqrt{p(1-p)/n}}$$

12-3c Testing and Estimating a Proportion

As you have already seen, the formula that summarises the sampling distribution also represents the test statistic.

Test Statistic for p

$$z = \frac{\hat{P} - p}{\sqrt{p(1-p)/n}}$$

which is approximately normal when np and $n(1-p)$ are greater than 5.

Using the same algebra employed in Sections 10-2 and 12-1, we attempt to derive the confidence interval estimator of p from the sampling distribution. The result is

$$\hat{p} \pm z_{\alpha/2}\sqrt{p(1-p)/n}$$

This formula, although technically correct, is useless. To understand why, examine the standard error of the sampling distribution $\sqrt{p(1-p)/n}$. To produce the interval estimate, we must compute the standard error, which requires us to know the value of p, the parameter we wish to estimate. This is the first of several statistical techniques where we face the same problem: how to determine the value of the standard error. In this application, the problem is easily and logically solved: simply estimate the value of p with $\hat{p}$.

Thus, we estimate the standard error with $\sqrt{\hat{p}(1-\hat{p})/n}$.

Confidence Interval Estimator of p

$$\hat{p} \pm z_{\alpha/2}\sqrt{\hat{p}(1-\hat{p})/n}$$

which is valid provided that np and $n(1-p)$ are greater than 5.

EXAMPLE 12.5

DATA Xm12-05

Election Day Exit Poll

When an election for political office takes place, the television networks cancel regular programming and instead provide election coverage. When the ballots are counted, the results are reported. However, for important offices such as president, the networks actively compete to see which will be the first to predict a winner. This is done through exit polls, wherein a random sample of voters who exit the polling booth is asked for whom they voted. From the data, the sample proportion of voters supporting the candidates is computed. A statistical technique is applied to determine whether there is enough evidence to infer that the leading candidate will garner enough votes to win. Suppose that in the exit poll from the state X, the pollsters recorded only the votes of the two candidates who had any chance of winning, the first candidate was from Party A and the second one was from Party B. Can the networks conclude from these exit polls' data the winner of the election?

SOLUTION:

IDENTIFY

The problem objective is to describe the population of votes in the state. The data are nominal because the values are 'Party A' (code = 1) and 'Party B' (code = 2). Thus the parameter to be tested is the proportion of votes in the entire country that are for the Party B candidate. Because we want to determine whether the network can declare the Party B candidate to be the winner at 8:01 pm, the alternative hypothesis is

$$H_1: p > 0.5$$

which makes the null hypothesis

$$H_0: p = 0.5$$

The test statistic is

$$z = \frac{\hat{p}-p}{\sqrt{p(1-p)/n}}$$

COMPUTE

MANUALLY:

It appears that this is a 'standard' problem that requires a 5% significance level. Thus, the rejection region is

$$z > z_{\alpha} = z_{.05} = 1.645$$

From the file, we count the number of 'successes', which is the number of votes cast for the Party B candidate, and find $x = 407$. The sample size is 765. Hence, the sample proportion is

$$\hat{p} = \frac{x}{n} = \frac{407}{765} = 0.532$$

The value of the test statistic is

$$z = \frac{\hat{p}-p}{\sqrt{p(1-p)/n}} = \frac{0.532-0.5}{\sqrt{0.5(1-0.5)/765}} = 1.77$$

Because the test statistic is (approximately) normally distributed, we can determine the *p*-value. It is

$$p - value = P(Z > 1.77) = 1 - p(z < 1.77) = 1 - 0.9616 = 0.0384$$

There is enough evidence at the 5% significance level that the Party B candidate has won.

EXCEL

	A	B	C	D
1	**z-Test: Proportion**			
2				
3				*Votes*
4	Sample Proportion			0.532
5	Observations			765
6	Hypothesised Proportion			0.5
7	z Stat			1.7716
8	P(Z<=z) one-tail			0.0382
9	z Critical one-tail			1.6449
10	P(Z<=z) two-tail			0.0764
11	z Critical two-tail			1.9600

Instructions

1. Type or import the data into one column*. (Open Xm12-05.)
2. Click **Add-Ins, Data Analysis Plus** and **Z-Test: Proportion.**
3. Specify the **Input Range** (A1:A766), type the **Code for Success** (2), the **Hypothesised Proportion** (0.5) and a value of α (0.05).

MINITAB

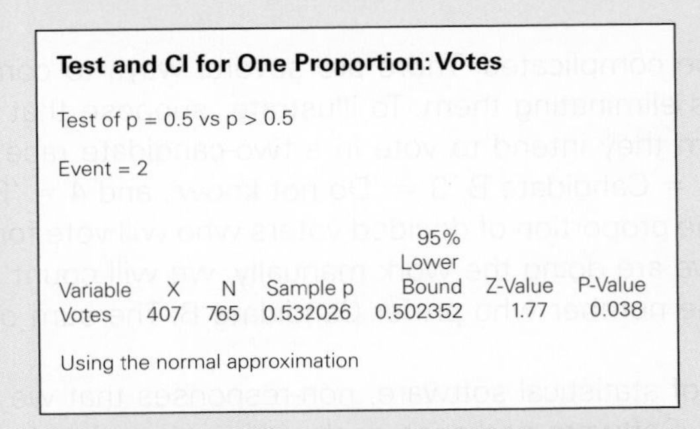

Test and CI for One Proportion: Votes

Test of p = 0.5 vs p > 0.5

Event = 2

Variable	X	N	Sample p	95% Lower Bound	Z-Value	P-Value
Votes	407	765	0.532026	0.502352	1.77	0.038

Using the normal approximation

As was the case with the test of a variance, Minitab calculates a one-sided confidence interval estimate when we are conducting a one-tail test.

Instructions

The data must represent successes and failures. The codes can be numbers or text. There can be only two kinds of entries: one representing success and the other representing failure. If numbers are used, Minitab will interpret the larger one as a success.

1. Type or import the data into one column. (**Open** Xm12-05.)
2. Click **Stat, Basic Statistics** and **1 Proportion . . . .**
3. Use the **Select** button or type the name of the variable or its column in the **Samples in columns** box (Votes) and check **Perform hypothesis test** and type the **Hypothesised proportion** (0.5).
4. Click **Options . . .** and specify the **Alternative** hypothesis (greater than). To use the normal approximation of the binomial, click **Use test and interval based on normal approximation.**

*If the column contains a blank (representing missing data) the row will have to be deleted.

The value of the test statistic is $z = 1.77$ and the one-tail p-value = 0.382. Using a 5% significance level, we reject the null hypothesis and conclude that there is enough evidence to infer that Party B candidate won the presidential election.

One of the key issues to consider here is the cost of Type I and Type II errors. A Type I error occurs if we conclude that the Party B will win when in fact he has lost. Such an error would mean that a network would announce at 8:01 pm, that the Party B candidate has won and then later in the evening would have to admit to a mistake. If a particular network were the only one that made this error, it would cast doubt on their integrity and possibly affect the number of viewers.

12-3d Missing Data

In real statistical applications, we occasionally find that the data set is incomplete. In some instances, the statistics practitioner may have failed to properly record some observations or some data may have been lost. In other cases, respondents may refuse to answer. For example, in political surveys where the statistics practitioner asks voters for whom they intend to vote in the next election, some people will answer that they have not decided or that their vote is confidential and refuse to answer. In surveys where respondents are asked to report their income, people often refuse to divulge this information. This is a troublesome issue for statistics practitioners. We cannot force people to answer our questions. However, if the number of non-responses is high, the results of our analysis may be invalid because the sample is no longer truly random. To understand why, suppose that people who are in the top quarter of household incomes regularly refuse to answer questions about their incomes. The resulting estimate of the population household income mean will be lower than the actual value.

The issue can be complicated. There are several ways to compensate for non-responses. The simplest method is eliminating them. To illustrate, suppose that in a political survey respondents are asked for whom they intend to vote in a two-candidate race. Surveyors record the results as 1 = Candidate A, 2 = Candidate B, 3 = 'Do not know', and 4 = 'Refuse to say'. If we wish to infer something about the proportion of decided voters who will vote for Candidate A, we can simply omit codes 3 and 4. If we are doing the work manually, we will count the number of voters who prefer Candidate A and the number who prefer Candidate B. The sum of these two numbers is the total sample size.

In the language of statistical software, non-responses that we wish to eliminate are collectively called *missing data*. Software packages deal with missing data in different ways. In Excel, the non-responses appear as blanks; in Minitab, they appear as asterisks.

12-3e Estimating the Total Number of Successes in a Large Finite Population

As was the case with the inference about a mean, the techniques in this section assume infinitely large populations. When the populations are small, it is necessary to include the finite population correction factor. In our definition a population is small when it is less than 20 times the sample size. When the population is large and finite, we can estimate the total number of successes in the population.

To produce the confidence interval estimator of the total, we multiply the lower and upper confidence limits of the interval estimator of the proportion of successes by the population size. The confidence interval estimator of the total number of successes in a large finite population is

$$N\left(\hat{p} \pm z_{\alpha/2}\sqrt{\frac{\hat{p}(1-\hat{p})}{n}}\right)$$

We will use this estimator in the chapter-opening example and several of this section's exercises.

TV RATINGS: SOLUTION

IDENTIFY

The problem objective is to describe the population of television shows watched by viewers across the country. The data are nominal. The combination of problem objective and data type make the parameter to be estimated the proportion of the entire population of 14- to 49-year-olds that watched Amaza (code = 2). The confidence interval estimator of the proportion is:

$$\hat{p} \pm z_{\alpha/2}\sqrt{\frac{\hat{p}(1-\hat{p})}{n}}$$

$$\hat{p} = \frac{x}{n} = \frac{275}{5000} = 0.0550$$

COMPUTE

MANUALLY:

To solve manually, we count the number of 2s in the file. We find this value to be 77. Thus,

$$\hat{p} = \frac{x}{n} = \frac{77}{1356} = 0.0567$$

The confidence level is $1 - \alpha = 0.95$. It follows that $\alpha = 0.05$, $\alpha/2 = 0.025$ and $z_{\alpha/2} = z_{0.025} = 1.96$. The 95% confidence interval estimate of p is:

$$\hat{p} \pm z_{\alpha/2}\sqrt{\frac{\hat{p}(1-\hat{p})}{n}} = 0.0567 \pm 1.96\sqrt{\frac{0.0567(1-0.0567)}{1356}} = 0.0567 \pm 0.0123$$

LCL = 0.0445 and UCL = 0.0691

EXCEL

	A	B	C
1	**z-Estimate: Proportion**		
2			
3			*TV Programme*
4	Sample Proportion		0.0568
5	Observations		1356
6	LCL		0.0445
7	UCL		0.0691

Instructions

1. Type or import the data into one column*. (**Open** Xm12-00.)
2. Click **Add-Ins, Data Analysis Plus** and **Z-Estimate: Proportion.**
3. Specify the **Input Range** (B1:B1357), the **Code for Success** (2) and the value of α (0.05).

MINITAB

Minitab requires that the data set contain only two values, the larger of which would be considered a success. In this example there are six values. If there are more than two codes or if the code for success is smaller than that for failure, we must recode.

(continued)

*If a cell is empty (representing missing data) it must be removed. See the website Appendix Removing Empty Cells in Excel.

Results for: Xm12-00.MTW

Test and CI for One Proportion: Recorded Programme

Event = 2

Variable	X	N	Sample p	95% CI
Recorded Programme	77	1356	0.056785	(0.044467, 0.069103)

Using the normal approximation.

Instructions

Recode data

1. Click **Data, Code** and **Numeric to Numeric . . .**
2. In the **Code data from columns** box, type or **Select** the data you wish to recode.
3. In the **Store coded data in columns** box, type the column where the recoded data are to be stored. (We named the column 'Recoded Programme'.)
4. Specify the **Original values:** you wish to recode (e.g. 1 3:6) and their **New:** values (e.g. 0).

Estimate the proportion

1. Click **Stat, Basic Statistics** and 1 **Proportion . . . .**
2. In the **Samples in columns** box, type or **Select** the data ('Recoded Programme').
3. Click **Options . . . .**
4. Specify the **Confidence level:** (0.95), select **Alternative: not equal,** and **Use test and interval based on normal distribution.**

INTERPRET

We estimate that between 4.44% and 6.91% of all South Africans who were between 15 and 49 years old were watching the TV drama 'Amaza'. If we multiply these figures by the total number of South Africans who were between 15 and 49 years old, 25.26 million, we produce an interval estimate of the number of South African adults 15–49 watching *Amaza*. Thus,

$$\text{LCL} = 0.0445 \times 25.26 \text{ million} = 1.12 \text{ million}$$

and

$$\text{UCL} = 0.0691 \times 25.26 \text{ million} = 1.75 \text{ million}$$

Sponsoring companies can then determine the value of any commercials that appeared on the show.

12-3f Selecting the Sample Size to Estimate the Proportion

When we introduced the sample size selection method to estimate a mean in Section 10-3, we pointed out that the sample size depends on the confidence level and the bound on the error of estimation that the statistics practitioner is willing to tolerate. When the parameter to be estimated is a proportion, the bound on the error of estimation is

$$B = z_{\alpha/2}\sqrt{\frac{\hat{p}(1-\hat{p})}{n}}$$

Solving for n, we produce the required sample size as indicated in the box.

Sample Size to Estimate a Proportion

$$n = \left(\frac{z_{\alpha/2}\sqrt{\hat{p}(1-\hat{p})}}{B}\right)^2$$

To illustrate the use of this formula, suppose that in a brand-preference survey we want to estimate the proportion of consumers who prefer our company's brand to within 0.03 with 95% confidence. This means that the bound on the error of estimation is $B = 0.03$. Because $1 - \alpha = 0.95$, $\alpha = 0.05$, $\alpha/2 = 0.025$ and $z_{\alpha/2} = z_{0.025} = 1.96$,

$$n = \left(\frac{1.96\sqrt{\hat{p}(1-\hat{p})}}{0.03}\right)^2$$

To solve for n, we need to know $\hat{p}$. Unfortunately, this value is unknown, because the sample has not yet been taken. At this point, we can use either of two methods to solve for n.

Method 1 If we have no knowledge of even the approximate value of $\hat{p}$, we let $\hat{p} = 0.5$. We choose $\hat{p} = 0.5$ because the product $\hat{p}(1-\hat{p})$ equals its maximum value at $\hat{p} = 0.5$. (Figure 12.6 illustrates this point.) This, in turn, results in a conservative value of n; as a result, the confidence interval will be no wider than the interval $\hat{p} \pm 0.03$. If, when the sample is drawn, $\hat{p}$ does not equal 0.5, the confidence interval estimate will be better (that is, narrower) than planned. Thus,

$$n = \left(\frac{1.96\sqrt{(0.5)(0.5)}}{0.03}\right)^2 = (32.67)^2 = 1068$$

If it turns out that $\hat{p} = 0.5$, the interval estimate is $\hat{p} \pm 0.03$. If not, the interval estimate will be narrower. For instance, if it turns out that $\hat{p} = 0.2$, then the estimate is $\hat{p} \pm 0.024$, which is better than we had planned.

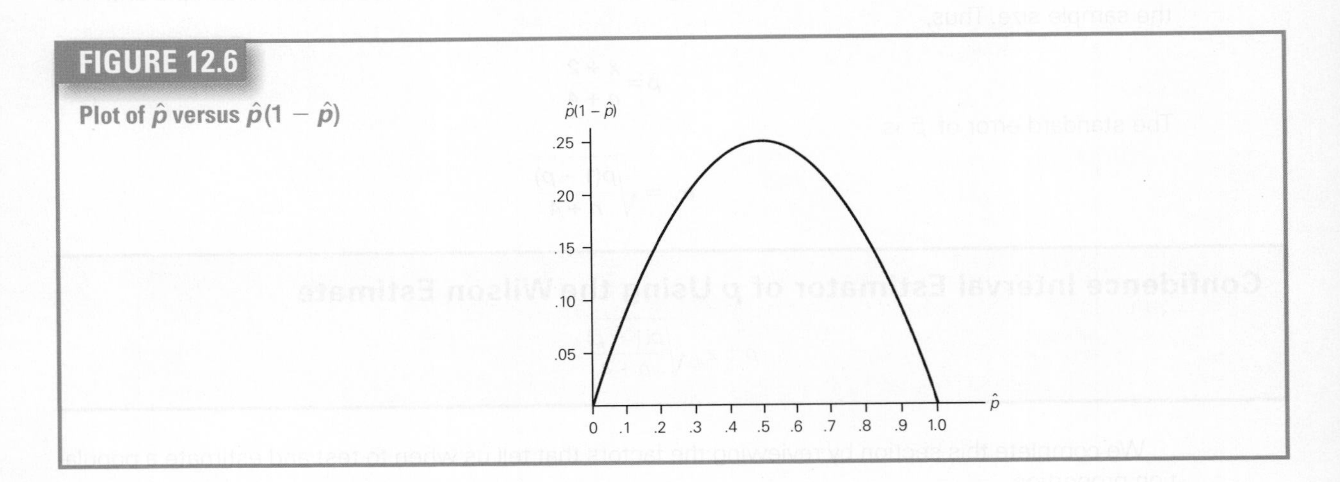

FIGURE 12.6

Plot of $\hat{p}$ versus $\hat{p}(1-\hat{p})$

Method 2 If we have some idea about the value of $\hat{p}$, we can use that quantity to determine n. For example, if we believe that $\hat{p}$ will turn out to be approximately 0.2, we can solve for n as follows:

$$n = \left(\frac{1.96\sqrt{(0.2)(0.8)}}{0.03}\right)^2 = (26.13)^2 = 683$$

Notice that this produces a smaller value of n (thus reducing sampling costs) than does method 1. If $\hat{p}$ actually lies between 0.2 and 0.8, however, the estimate will not be as good as we wanted, because the interval will be wider than desired.

Method 1 is often used to determine the sample size used in public opinion surveys reported by newspapers, magazines, television and radio. These polls usually estimate proportions to within 3%, with 95% confidence. (The media often state the confidence level as '19 times out of 20'.) If you have ever wondered why opinion polls almost always estimate proportions to within 3%, consider the sample size required to estimate a proportion to within 1%:

$$n = \left(\frac{1.96\sqrt{(0.5)(0.5)}}{0.01}\right)^2 = (98)^2 = 9604$$

The sample size 9604 is 9 times the sample size needed to estimate a proportion to within 3%. Thus, to divide the width of the interval by 3 requires multiplying the sample size by 9. The cost would also increase considerably. For most applications, the increase in accuracy (created by decreasing the width of the confidence interval estimate) does not overcome the increased cost. Confidence interval estimates with 5% or 10% bounds (sample sizes 385 and 97, respectively) are generally considered too wide to be useful. Thus, the 3% bound provides a reasonable compromise between cost and accuracy.

12-3g Wilson Estimators

When using the confidence interval estimator of a proportion when success is a relatively rare event, it is possible to find no successes, especially if the sample size is small. To illustrate, suppose that a sample of 100 produced $x = 0$, which means that $\hat{p} = 0$. The 95% confidence interval estimator of the proportion of successes in the population becomes

$$\hat{p} \pm z_{\alpha/2}\sqrt{\frac{\hat{p}(1-\hat{p})}{n}} = 0 \pm 1.96\sqrt{\frac{0(1-0)}{100}} = 0 \pm 0$$

This implies that if we find no successes in the sample, then there is no chance of finding a success in the population. Drawing such a conclusion from virtually any sample size is unacceptable. The remedy may be a suggestion made by Edwin Wilson in 1927. The Wilson estimate denoted $\tilde{p}$ (pronounced '*p* tilde') is computed by adding 2 to the number of successes in the sample and 4 to the sample size. Thus,

$$\tilde{p} = \frac{x+2}{n+4}$$

The standard error of $\tilde{p}$ is

$$\sigma_{\tilde{p}} = \sqrt{\frac{\tilde{p}(1-\tilde{p})}{n+4}}$$

Confidence Interval Estimator of *p* Using the Wilson Estimate

$$\tilde{p} \pm z_{\alpha/2}\sqrt{\frac{\tilde{p}(1-\tilde{p})}{n+4}}$$

We complete this section by reviewing the factors that tell us when to test and estimate a population proportion.

Factors that Identify the *z*-Test and Interval Estimator of *p*

1. **Problem objective**: Describe a population.
2. **Data type**: Nominal.

EXERCISES

DO-IT-YOURSELF EXCEL

Construct Excel spreadsheets that perform the following techniques

12.27 *z*-test of *p*. Inputs: sample variance, sample size and hypothesised variance. Outputs: Test statistic, critical values, and one- and two-tail *p*-values. Tools: **NORM.INV, NORM.S.DIST**.

12.28 *z*-estimate of *p*. Inputs: sample variance, sample size and confidence level. Outputs: Upper and lower confidence limits. Tools: **NORM.INV**.

Developing an Understanding of Statistical Concepts

Next exercises are 'what-if analyses' designed to determine what happens to the test statistics and interval estimates when elements of the statistical inference change. These problems can be solved manually or using your Do-It-Yourself Excel spreadsheets.

12.29 a. In a random sample of 200 observations, we found the proportion of successes to be 48%. Estimate with 95% confidence the population proportion of successes.

b. Repeat part (a) with $n = 500$.

c. Repeat part (a) with $n = 1000$.

d. Describe the effect on the confidence interval estimate of increasing the sample size.

12.30 a. Calculate the *p*-value of the test of the following hypotheses given that $\hat{p} = 0.63$ and $n = 100$:

$H_0: p = 0.60$

$H_0: p > 0.60$

b. Repeat part (a) with $n = 200$.

c. Repeat part (a) with $n = 400$.

d. Describe the effect on the *p*-value of increasing the sample size.

12.31 a. A statistics practitioner wants to test the following hypotheses:

$H_0: p = 0.70$

$H_1: p > 0.70$

A random sample of 100 produced $\hat{p} = 0.73$. Calculate the *p*-value of the test.

b. Repeat part (a) with $\hat{p} = 0.72$.

c. Repeat part (a) with $\hat{p} = 0.71$.

d. Describe the effect on the *z*-statistic and its *p*-value of decreasing the sample proportion.

12.32 Determine the sample size necessary to estimate a population proportion to within 0.03 with 90% confidence assuming you have no knowledge of the approximate value of the sample proportion.

12.33 Suppose that you used the sample size calculated in Exercise 12.40 and found $\hat{p} = 0.5$.

a. Estimate the population proportion with 90% confidence.

b. Is this the result you expected? Explain.

Applications

The following three exercises require the use of the Wilson Estimator.

12.34 In Chapter 6 we discussed how an understanding of probability allows one to properly interpret the results of medical screening tests. The use of Bayes's Law requires a set of prior probabilities, which are based on historical records. Suppose that a physician wanted to estimate the probability that a woman under 35 years of age would give birth to a Down syndrome baby. She randomly sampled 200 births and discovered only one such case. Use the Wilson estimator to produce a 95% confidence interval estimate of the proportion of women under 35 who will have a Down syndrome baby.

12.35 Spam is of concern to anyone with an email address. Several companies offer protection by eliminating spam emails as soon as they hit an inbox. To examine one such product, a manager randomly sampled his daily emails for 50 days after installing spam software. A total of 374 emails were received, of which three were spam. Use the Wilson estimator to estimate with 90% confidence the proportion of spam emails that get through.

12.36 A management lecturer was in the process of investigating the relationship between education and managerial level achieved. The source of his data was a survey of 385 CEOs of medium and large companies. He discovered that there was only 1 CEO who did not have at least one university degree. Estimate (using a Wilson estimator) with 99% confidence the proportion of CEOs of medium and large companies with no university degrees.

The following exercises require the use of a computer and software. Use a 5% significance level for all tests.

12.37 Xr12-37 An increasing number of people are giving gift certificates as birthday presents. To measure the extent of this practice, a random sample of people was asked (survey conducted 26–29 December) whether they had received a gift certificate for their birthday. The responses are recorded as 1 = No and 2 = Yes. Estimate with 95% confidence the proportion of people who received a gift certificate for their birthday.

12.38 Xr12-38 Because television audiences of newscasts tend to be older (and because older people suffer from a variety of medical ailments) pharmaceutical companies' advertising often appears on national news in the three Italian networks (RAI, RET 4 and CANALE 5). The ads concern prescription drugs such as those to treat heartburn. To determine how effective the ads are, a survey was undertaken. Adults over 50 who regularly watch network newscasts were asked whether they had contacted their physician to ask about one of the prescription drugs advertised during the newscast. The responses (1 = No and 2 = Yes) were recorded. Estimate with 95% confidence the fraction of adults over 50 who have contacted their physician to enquire about a prescription drug.

12.39 Xr12-39 A lecturer of business statistics recently adopted a new textbook. At the completion of the course, 100 randomly selected students were asked to assess the book. The responses are as follows:

Excellent (1), Good (2), Adequate (3), Poor (4)

The results are stored using the codes in parentheses. Do the data allow us to conclude that more than 50% of all business students would rate the book as excellent?

12.40 Refer to Exercise 12.39. Do the data allow us to conclude that more than 90% of all business students would rate it as at least adequate?

12.41 Xm12-00 Refer to the chapter-opening example. Estimate with 95% confidence the number of South Africans 15 to 49 years old who were tuned to the *Task Force*.

12-4 INFERENCE ABOUT THE DIFFERENCE BETWEEN TWO MEANS: INDEPENDENT SAMPLES

In order to test and estimate the difference between two population means, the statistics practitioner draws random samples from each of two populations. In this section, we discuss independent samples. In Section 12-5, where we present the matched pairs experiment, the distinction between independent samples and matched pairs will be made clear. For now, we define independent samples as samples completely unrelated to one another.

Figure 12.7 depicts the sampling process. Observe that we draw a sample of size n_1 from population 1 and a sample of size n_2 from population 2. For each sample, we compute the sample means and sample variances.

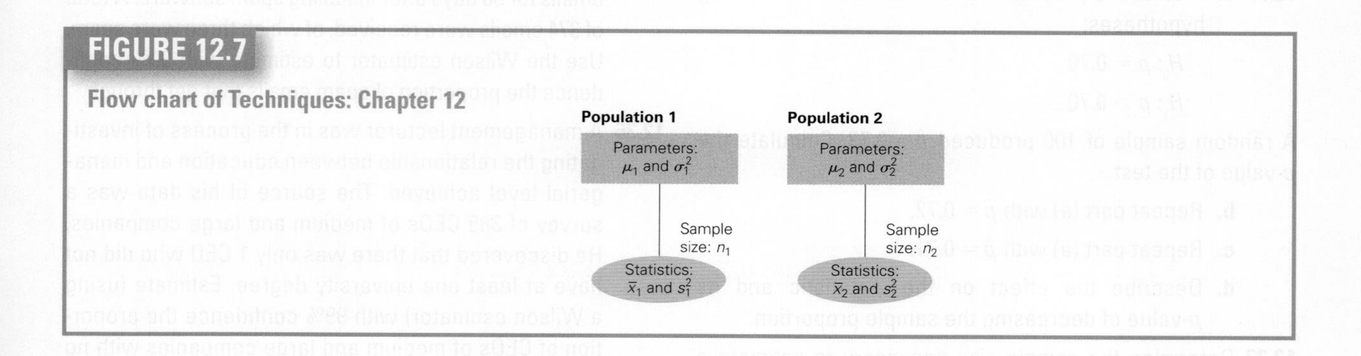

The best estimator of the difference between two population means, $\mu_1 - \mu_2$, is the difference between two sample means, $\bar{x}_1 - \bar{x}_2$. In Section 9-3 we presented the sampling distribution of $\bar{x}_1 - \bar{x}_2$.

Sampling Distribution of $\bar{x}_1 - \bar{x}_2$

1. $\bar{x}_1 - \bar{x}_2$ is normally distributed if the populations are normal and approximately normal if the populations are non-normal and the sample sizes are large.
2. The expected value of $\bar{x}_1 - \bar{x}_2$ is

$$E(\bar{x}_1 - \bar{x}_2) = \mu_1 - \mu_2$$

3. The variance of $\bar{x}_1 - \bar{x}_2$ is

$$V(\bar{x}_1 - \bar{x}_2) = \frac{\sigma_1^2}{n_1} + \frac{\sigma_2^2}{n_2}$$

The standard error of

$$\sqrt{\frac{\sigma_1^2}{n_1} + \frac{\sigma_2^2}{n_2}}$$

Thus,

$$z = \frac{(\bar{x}_1 - \bar{x}_2) - (\mu_1 - \mu_2)}{\sqrt{\frac{\sigma_1^2}{n_1} + \frac{\sigma_2^2}{n_2}}}$$

is a standard normal (or approximately normal) random variable. It follows that the test statistic is

$$z = \frac{(\bar{x}_1 - \bar{x}_2) - (\mu_1 - \mu_2)}{\sqrt{\frac{\sigma_1^2}{n_1} + \frac{\sigma_2^2}{n_2}}}$$

The interval estimator is

$$(\bar{x}_1 - \bar{x}_2) \pm z_{\alpha/2}\sqrt{\frac{\sigma_1^2}{n_1} + \frac{\sigma_2^2}{n_2}}$$

However, these formulas are rarely used because the population variances σ_1^2 and σ_2^2 are virtually always unknown. Consequently, it is necessary to estimate the standard error of the sampling distribution. The way to do this depends on whether the two unknown population variances are equal. When they are equal, the test statistic is defined in the following way.

Test Statistic for $\mu_1 - \mu_2$ when $\sigma_1^2 = \sigma_2^2$

$$t = \frac{(\bar{x}_1 - \bar{x}_2) - (\mu_1 - \mu_2)}{\sqrt{s_p^2\left(\frac{1}{n_1} + \frac{1}{n_2}\right)}} \qquad \nu = n_1 + n_2 - 2$$

where

$$s_p^2 = \frac{(n_1 - 1)s_1^2 + (n_2 - 1)s_2^2}{n_1 + n_2 - 2}$$

The quantity s_p^2 is called the **pooled variance estimator**. It is the weighted average of the two sample variances with the number of degrees of freedom in each sample used as weights. The requirement that the population variances be equal makes this calculation feasible because we need only one estimate of the common value of σ_1^2 and σ_2^2. It makes sense for us to use the pooled variance estimator because, in combining both samples, we produce a better estimate.

The test statistic is Student t distributed with $n_1 + n_2 - 2$ degrees of freedom, provided that the two populations are normal. The confidence interval estimator is derived by mathematics that by now has become routine.

Confidence Interval Estimator of $\mu_1 - \mu_2$ When $\sigma_1^2 = \sigma_2^2$

$$(\bar{x}_1 - \bar{x}_2) \pm t_{\alpha/2}\sqrt{s_p^2\left(\frac{1}{n_1} + \frac{1}{n_2}\right)} \qquad \nu = n_1 + n_2 - 2$$

We will refer to these formulas as the **equal-variances test statistic** and **confidence interval estimator**, respectively.

When the population variances are unequal, we cannot use the pooled variance estimate. Instead, we estimate each population variance with its sample variance. Unfortunately, the sampling distribution of the resulting statistic

$$\frac{(\bar{x}_1 - \bar{x}_2) - (\mu_1 - \mu_2)}{\sqrt{\frac{s_1^2}{n_1} + \frac{s_2^2}{n_2}}}$$

is neither normally nor Student t distributed. However, it can be approximated by a Student t distribution with degrees of freedom equal to

$$\nu = \frac{(s_1^2/n_1 + s_2^2/n_2)^2}{\frac{(s_1^2/n_1)^2}{n_1 - 1} + \frac{(s_2^2/n_2)^2}{n_2 - 1}}$$

(It is usually necessary to round this number to the nearest integer.) The test statistic and confidence interval estimator are easily derived from the sampling distribution.

Test Statistic for $\mu_1 - \mu_2$ When $\sigma_1^2 \neq \sigma_2^2$

$$t = \frac{(\bar{x}_1 - \bar{x}_2) - (\mu_1 - \mu_2)}{\sqrt{\left(\frac{s_1^2}{n_1} + \frac{s_2^2}{n_2}\right)}} \qquad \nu = \frac{(s_1^2/n_1 + s_2^2/n_2)^2}{\frac{(s_1^2/n_1)^2}{n_1 - 1} + \frac{(s_2^2/n_2)^2}{n_2 - 1}}$$

Confidence Interval Estimator of $\mu_1 - \mu_2$ When $\sigma_1^2 \neq \sigma_2^2$

$$(\bar{x}_1 - \bar{x}_2) \pm t_{\alpha/2}\sqrt{\left(\frac{s_1^2}{n_1} + \frac{s_2^2}{n_2}\right)} \qquad \nu = \frac{(s_1^2/n_1 + s_2^2/n_2)^2}{\frac{(s_1^2/n_1)^2}{n_1 - 1} + \frac{(s_2^2/n_2)^2}{n_2 - 1}}$$

We will refer to these formulas as the **unequal-variances test statistic** and **confidence interval estimator**, respectively.

The question naturally arises, How do we know when the population variances are equal? The answer is that because σ_1^2 and σ_2^2 are unknown, we cannot know for certain whether they are equal. However, we can perform a statistical test to determine whether there is evidence to infer that the population variances differ. We conduct the F-test of the ratio of two variances, which we briefly present here and save the details for Section 13-4.

Testing the Population Variances

The hypotheses to be tested are

$$H_0: \sigma_1^2/\sigma_2^2 = 1$$

$$H_1: \sigma_1^2/\sigma_2^2 \neq 1$$

The test statistic is the ratio of the sample variances s_1^2/s_2^2, which is *F*-distributed with degrees of freedom $\nu_1 = n_1 - 1$ and $\nu_2 = n_2 - 1$. Recall that we introduced the *F*-distribution in Section 8-4. The required condition is the same as that for the *t*-test of $\mu_1 - \mu_2$, which is that both populations are normally distributed. This is a two-tail test so that the rejection region is

$$F > F_{\alpha/2,\nu_1,\nu_2} \quad \text{or} \quad F < F_{1-\alpha/2,\nu_1,\nu_2}$$

Put simply, we will reject the null hypothesis that states that the population variances are equal when the ratio of the sample variances is large or if it is small. Table 6 in Appendix A, which lists the critical values of the *F*-distribution, defines 'large' and 'small'.

12-4a Decision Rule: Equal-Variances or Unequal-Variances *t*-Tests and Estimators

Recall that we can never have enough statistical evidence to conclude that the null hypothesis is true. This means that we can only determine whether there is enough evidence to infer that the population variances *differ*. Accordingly, we adopt the following rule: We will use the equal-variances test statistic and confidence interval estimator unless there is evidence (based on the *F*-test of the population variances) to indicate that the population variances are unequal, in which case we will apply the unequal-variances test statistic and confidence interval estimator.

EXAMPLE 12.6

DATA Xm12-06

Direct and Broker-Purchased Mutual Funds

Millions of investors buy mutual funds (see Example 6.1 for a description of mutual funds), choosing from thousands of possibilities. Some funds can be purchased directly from banks or other financial institutions whereas others must be purchased through brokers, who charge a fee for this service. This raises the question: Can investors do better by buying mutual funds directly than by purchasing mutual funds through brokers? To help answer this question, a group of researchers randomly sampled the annual returns from mutual funds that can be acquired directly and mutual funds that are bought through brokers and recorded the net annual returns, which are the returns on investment after deducting all relevant fees. These are listed next.

Direct					Broker				
9.33	4.68	4.23	14.69	10.29	3.24	3.71	16.4	4.36	9.43
6.94	3.09	10.28	−2.97	4.39	−6.76	13.15	6.39	−11.07	8.31
16.17	7.26	7.1	10.37	−2.06	12.8	11.05	−1.9	9.24	−3.99
16.97	2.05	−3.09	−0.63	7.66	11.1	−3.12	9.49	−2.67	−4.44
5.94	13.07	5.6	−0.15	10.83	2.73	8.94	6.7	8.97	8.63
12.61	0.59	5.27	0.27	14.48	−0.13	2.74	0.19	1.87	7.06
3.33	13.57	8.09	4.59	4.8	18.22	4.07	12.39	−1.53	1.57
16.13	0.35	15.05	6.38	13.12	−0.8	5.6	6.54	5.23	−8.44
11.2	2.69	13.21	−0.24	−6.54	−5.75	−0.85	10.92	6.87	−5.72
1.14	18.45	1.72	10.32	−1.06	2.59	−0.28	−2.15	−1.69	6.95

Can we conclude at the 5% significance level that directly purchased mutual funds outperform mutual funds bought through brokers?

SOLUTION:

IDENTIFY

To answer the question, we need to compare the population of returns from direct and returns from broker-bought mutual funds. The data are obviously interval (we have recorded real numbers).This problem objective-data type combination tells us that the parameter to be tested is the difference between two means, $\mu_1 - \mu_2$ The hypothesis to be tested is that the mean net annual return from directly purchased mutual funds (μ_1) is larger than the mean of broker-purchased funds (μ_2). Hence, the alternative hypothesis is

$$H_1: (\mu_1 - \mu_2) > 0$$

As usual, the null hypothesis automatically follows:

$$H_0: (\mu_1 - \mu_2) = 0$$

To decide which of the t-tests of $\mu_1 - \mu_2$ to apply, we conduct the F-test of σ_1^2/σ_2^2.

$$H_0: \sigma_1^2/\sigma_2^2 = 1$$
$$H_1: \sigma_1^2/\sigma_2^2 \neq 1$$

COMPUTE

MANUALLY:

From the data, we calculated the following statistics:

$$s_1^2 = 37.49 \quad \text{and} \quad s_2^2 = 43.34$$

Test statistic: $F = s_1^2/s_2^2 = 37.49/43.34 = 0.86$

Rejection region: $F > F_{\alpha/2,\nu_1,\nu_2} = F_{0.025,49,49} \approx F_{0.025,50,50} = 1.75$

or

$$F < F_{1-\alpha/2,\nu_1,\nu_2} = F_{0.975,49,49} = 1/F_{0.025,49,49} \approx 1/F_{0.025,50,50} = 1/1.75 = 0.57$$

Because $F = 0.86$ is not greater than 1.75 or smaller than 0.57, we cannot reject the null hypothesis.

EXCEL

	A	B	C
1	F-Test: Two-Sample for Variances		
2			
3		*Direct*	*Broker*
4	Mean	6.63	3.72
5	Variance	37.49	43.34
6	Observations	50	50
7	df	49	49
8	F	0.8650	
9	P(F<=f) one-tail	0.3068	
10	F Critical one-tail	0.6222	

The value of the test statistic is $F = 0.8650$. Excel outputs the one-tail p-value. Because we are conducting a two-tail test, we double that value. Thus, the p-value of the test we are conducting is $2 \times 0.3068 = 0.6136$.

Instructions

1. Type or import the data into two columns*. (**Open** Xm12-06.)
2. Click **Data, Data Analysis** and ***F*-test Two-Sample for Variances.**
3. Specify the **Variable 1 Range** (A1:A51) and the **Variable 2 Range** (B1:B51). Type a value for α (0.05).

*If a cell is empty (representing missing data), it must be removed.

MINITAB

Test for Equal Variances: Direct, Broker

F-Test (Normal Distribution)
Test statistic = 0.86, *p*-value = 0.614

Instructions

(*Note:* Some of the printout has been omitted.)

1. Type or import the data into two columns. (**Open** Xm12-06.)
2. Click **Stat**, **Basic Statistics** and **2 Variances ...**
3. **In** the **Samples in different columns** box, select the **First** (Direct) and **Second** (Broker) variables.

INTERPRET

There is not enough evidence to infer that the population variances differ. It follows that we must apply the equal-variances *t*-test of $\mu_1 - \mu_2$ The hypotheses are

$$H_0: (\mu_1 - \mu_2) = 0$$
$$H_1: (\mu_1 - \mu_2) > 0$$

COMPUTE MANUALLY:

From the data, we calculated the following statistics:

$$\bar{x}_1 = 6.63$$
$$\bar{x}_2 = 3.72$$
$$s_1^2 = 37.49$$
$$s_2^2 = 43.34$$

The pooled variance estimator is

$$s_p^2 = \frac{(n_1 - 1)s_1^2 + (n_2 - 1)s_2^2}{n_1 + n_2 - 2}$$
$$= \frac{(50 - 1)37.49 + (50 - 1)43.34}{50 + 50 - 2}$$
$$= 40.42$$

The number of degrees of freedom of the test statistic is

$$\nu = n_1 + n_2 - 2 = 50 + 50 - 2 = 98$$

The rejection region is

$$t > t_{\alpha,\nu} = t_{0.05,98} \approx t_{0.05,100} = 1.660$$

We determine that the value of the test statistic is

$$t = \frac{(\bar{x}_1 - \bar{x}_2) - (\mu_1 - \mu_2)}{\sqrt{s_p^2\left(\frac{1}{n_1} + \frac{1}{n_2}\right)}}$$
$$= \frac{(6.63 - 3.72) - 0}{\sqrt{40.42\left(\frac{1}{50} + \frac{1}{50}\right)}}$$
$$= 2.29$$

EXCEL

	A	B	C
1	t-Test: Two-Sample Assuming Equal Variances		
2			
3		*Direct*	*Broker*
4	Mean	6.63	3.72
5	Variance	37.49	43.34
6	Observations	50	50
7	Pooled Variance	40.41	
8	Hypothesised Mean Difference	0	
9	df	98	
10	t Stat	2.29	
11	P(T<=t) one-tail	0.0122	
12	t Critical one-tail	1.6606	
13	P(T<=t) two-tail	0.0243	
14	t Critical two-tail	1.9845	

Instructions

1. Type or import the data into two columns. (**Open** Xm12-06.)
2. Click **Data, Data Analysis** and ***t*-Test: Two-Sample Assuming Equal Variances.**
3. Specify the **Variable 1 Range** (A1:A51) and the **Variable 2 Range** (B1:B51). Type the value of the **Hypothesised Mean Difference*** (0) and type a value for α (0.05).

MINITAB

Two-Sample T-Test and CI: Direct, Broker

Two-sample T for Direct vs Broker

	N	Mean	StDev	SE Mean
Direct	50	6.63	6.12	0.87
Broker	50	3.72	6.58	0.93

Difference = mu (Direct) – mu (Broker)
Estimate for difference: 2.91
95% lower bond for difference: 0.80
T-Test of difference = 0 (vs >): T-Value = 2.29 P-Value = 0.012 DF = 98
Both use Pooled StDev = 6.3572

Instructions

1. Type or import the data into two columns. (**Open** Xm12-06.)
2. Click **Stat, Basic Statistics** and **2-Sample *t*. . .**
3. If the data are stacked, use the **Samples in one column** box to specify the names of the variables. If the data are unstacked (as in Example 12.6), specify the **First** and **Second** variables in the **Samples in different columns** box (Direct, Broker). (See the discussion on Data Formats on Section 12-6d for a discussion of stacked and unstacked data.) Click **Assume equal variances.** Click **Options ....**
4. **In** the **Test difference** box, type the value of the parameter under the null hypothesis (0) and select one of **less than, not equal** or **greater than** for the **Alternative** hypothesis (greater than).

INTERPRET

The value of the test statistic is 2.29. The one-tail *p*-value is 0.0122. We observe that the *p*-value of the test is small (and the test statistic falls into the rejection region). As a result, we conclude that there is sufficient evidence to infer that on average directly purchased mutual funds outperform broker-purchased mutual funds.

*This term is technically incorrect. Because we are testing $\mu_1 - \mu_2$, Excel should ask for and output the 'Hypothesised Difference between Means'.

Estimating $\mu_1 - \mu_2$: Equal-Variances

In addition to testing a value of the difference between two population means, we can also estimate the difference between means. Next we compute the 95% confidence interval estimate of the difference between the mean return for direct and broker mutual funds.

COMPUTE

MANUALLY:

The confidence interval estimator of the difference between two means with equal population variances is

$$(\bar{x}_1 - \bar{x}_2) \pm t_{\alpha/2}\sqrt{s_p^2\left(\frac{1}{n_1} + \frac{1}{n_2}\right)}$$

The 95% confidence interval estimate of the difference between the return for directly purchased mutual funds and the mean return for broker-purchased mutual funds is

$$(\bar{x}_1 - \bar{x}_2) \pm t_{\alpha/2}\sqrt{s_p^2\left(\frac{1}{n_1} + \frac{1}{n_2}\right)} = (6.63 - 3.72) \pm 1.984\sqrt{40.42\left(\frac{1}{50} + \frac{1}{50}\right)}$$

$$= 2.91 \pm 2.52$$

The lower and upper limits are 0.39 and 5.43.

EXCEL

	A	B	C	D	E	F
1	t-Estimate : Two Means (Equal Variances)					
2						
3				Direct	Broker	
4	Mean			6.63	3.72	
5	Variance			37.49	43.34	
6	Observations			50	50	
7						
8	Pooled Variance			40.41		
9	Degrees of Freedom			98		
10	Confidence Level			0.95		
11	Confidence Interval Estimate			2.91	±	2.52
12	LCL			0.38		
13	UCL			5.43		

Instructions

1. Type or import the data into two columns*. (**Open** Xm12-06.)
2. Click **Add-Ins, Data Analysis Plus** and ***t*-Estimate: Two Means.**
3. Specify the **Variable 1 Range** (A1:A51) and the **Variable 2 Range** (B1:B51). Click **Independent Samples with Equal Variances** and the value for α (0.05).

MINITAB

Two-Sample T-Test and CI: Direct, Broker

Two-sample T for Direct vs Broker

	N	Mean	StDev	SE Mean
Direct	50	6.63	6.12	0.87
Broker	50	3.72	6.58	0.93

Difference = mu (Direct) – mu (Broker)
Estimate for difference: 2.91
95% CI for difference: (0.38, 5.43)
T-Test of difference = 0 (vs not =): T-Value = 2.29 P-Value = 0.024 DF = 98
Both use Pooled StDev = 6.3572

(*continued*)

*If a cell is empty (representing missing data), it must be removed.

Instructions

To produce a confidence interval estimate, follow the instructions for the test, but specify **not equal** for the **Alternative.** Minitab will conduct a two-tail test and produce the confidence interval estimate.

We estimate that the return on directly purchased mutual funds is on average between 0.38 and 5.43 percentage points larger than broker-purchased mutual funds.

EXAMPLE 12.7

DATA Xm12-07

Effect of New CEO in Family-Run Businesses

What happens to the family-run business when the boss's son or daughter takes over? Does the business do better after the change if the new boss is the offspring of the owner or does the business do better when an outsider is made chief executive officer (CEO)? In pursuit of an answer, researchers randomly selected 140 firms between 1994 and 2002, 30% of which passed ownership to an offspring and 70% of which appointed an outsider as CEO. For each company, the researchers calculated the operating income as a proportion of assets in the year before and the year after the new CEO took over. The change (operating income after−operating income before) in this variable was recorded and is listed next. Do these data allow us to infer that the effect of making an offspring CEO is different from the effect of hiring an outsider as CEO?

Offspring			Outsider						
−1.95	0.91	−3.15	0.69	−1.05	1.58	−2.46	3.33	−1.32	−0.51
0	−216	3.27	−0.95	−4.23	−1.98	1.59	3.2	5.93	8.68
0.56	1.22	−0.67	−2.2	−0.16	4.41	−2.03	0.55	−0.45	1.43
1.44	0.67	2.61	2.65	2.77	4.62	−1.69	−1.4	−3.2	−0.37
1.5	−0.39	1.55	5.39	−0.96	4.5	0.55	2.79	5.08	−0.49
1.41	−1.43	−2.67	4.15	1.01	2.37	0.95	5.62	0.23	−0.08
−0.32	−0.48	−1.91	4.28	0.09	2.44	3.06	−2.69	−2.69	−1.16
−1.7	0.24	1.01	2.97	6.79	1.07	4.83	−2.59	3.76	1.04
−1.66	0.79	−1.62	4.11	1.72	−1.11	5.67	2.45	1.05	1.28
−1.87	−1.19	−5.25	2.66	6.64	0.44	−0.8	3.39	0.53	1.74
−1.38	1.89	0.14	6.31	4.75	1.36	1.37	5.89	3.2	−0.14
0.57	−3.7	2.12	−3.04	2.84	0.88	0.72	−0.71	−3.07	−0.82
3.05	−0.31	2.75	−0.42	−2.1	0.33	4.14	4.22	−4.34	0
2.98	−1.37	0.3	−0.89	2.07	−5.96	3.04	0.46	−1.16	2.68

SOLUTION:

IDENTIFY

The objective is to compare two populations, and the data are interval. It follows that the parameter of interest is the difference between two population means $\mu_1 - \mu_2$, where μ_1 is the mean difference for companies where the owner's son or daughter became CEO and μ_2 is the mean difference for companies who appointed an outsider as CEO.

To determine whether to apply the equal or unequal variances *t*-test, we use the *F*-test of two variances.

$$H_0: \sigma_1^2/\sigma_2^2 = 1$$
$$H_1: \sigma_1^2/\sigma_2^2 \neq 1$$

COMPUTE MANUALLY:

From the data, we calculated the following statistics:

$$s_1^2 = 3.79 \quad \text{and} \quad s_2^2 = 8.03$$

Test statistic: $F = s_1^2/s_2^2 = 3.79/8.03 = 0.47$

The degrees of freedom are $\nu_1 = n_1 - 1 = 42 - 1 = 41$ and $\nu_2 = n_2 - 1 = 98 - 1 = 97$

Rejection region: $F > F_{\alpha/2,\nu_1,\nu_2} = F_{0.025,41,97} \approx F_{0.025,40,100} = 1.64$

or

$$F < F_{1-\alpha/2,\nu_1,\nu_2} = F_{0.975,41,97} \approx 1/F_{0.025,97,41} \approx 1/F_{0.025,100,40} = 1/1.74 = 0.57$$

Because $F = 0.47$ is less than 0.57, we reject the null hypothesis.

EXCEL

	A	B	C
1	F-Test: Two-Sample for Variances		
2			
3		*Offspring*	*Outsider*
4	Mean	−0.10	1.24
5	Variance	3.79	8.03
6	Observations	42	98
7	df	41	97
8	F	0.47	
9	P(F<=f) one-tail	0.0040	
10	F Critical one-tail	0.6314	

The value of the test statistic is $F = 0.47$, and the *p*-value $= 2 \times 0.0040 = 0.0080$.

MINITAB

Test for Equal Variances: Offspring, Outsider

F-Test (Normal Distribution)
Test statistic = 0.47, p-value = 0.008

INTERPRET

There is enough evidence to infer that the population variances differ. The appropriate technique is the unequal-variances *t*-test of $\mu_1 - \mu_2$.

Because we want to determine whether there is a *difference* between means, the alternative hypothesis is

$$H_1: (\mu_1 - \mu_2) \neq 0$$

and the null hypothesis is

$$H_0: (\mu_1 - \mu_2) = 0$$

COMPUTE MANUALLY:

From the data, we calculated the following statistics:

$$\bar{x}_1 = -0.10$$

$$\bar{x}_2 = 1.24$$

$$s_1^2 = 3.79$$

$$s_2^2 = 8.03$$

The number of degrees of freedom of the test statistic is

$$\nu = \frac{(s_1^2/n_1 + s_2^2/n_2)^2}{\frac{(s_1^2/n_1)^2}{n_1 - 1} + \frac{(s_2^2/n_2)^2}{n_2 - 1}}$$

$$= \frac{(3.79/42 + 8.03/98)^2}{\frac{(3.79/42)^2}{42 - 1} + \frac{(8.03/98)^2}{98 - 1}}$$

$$= 110.69 \text{ rounded to } 111$$

The rejection region is

$$t < -t_{\alpha/2,\nu} = -t_{0.025,111} \approx -t_{0.025,110} = -1.982 \text{ or } t > t_{\alpha/2,\nu} = t_{0.025,111} \approx 1.982$$

The value of the test statistic is computed next:

$$t = \frac{(\bar{x}_1 - \bar{x}_2) - (\mu_1 - \mu_2)}{\sqrt{\left(\frac{s_1^2}{n_1} + \frac{s_2^2}{n_2}\right)}}$$

$$= \frac{(-0.10 - 1.24) - (0)}{\sqrt{\left(\frac{3.79}{42} + \frac{8.03}{98}\right)}} = -3.22$$

EXCEL

	A	B	C
1	*t-Test: Two-Sample Assuming Unequal Variances*		
2			
3		*Offspring*	*Outsider*
4	Mean	–0.10	1.24
5	Variance	3.79	8.03
6	Observations	42	98
7	Hypothesised Mean Difference	0	
8	df	111	
9	t Stat	–3.22	
10	P(T<=t) one-tail	0.0008	
11	t Critical one-tail	1.6587	
12	P(T<=t) two-tail	0.0017	
13	t Critical two-tail	1.9816	

Instructions

Follow the instructions for Example 12.6 except at step 2 click **Data, Data Analysis and t-Test: Two-Sample Assuming Unequal Variances.**

MINITAB

Two-Sample *t*-Test and CI: Offspring, Outsider

Two-sample *t*-Test for Offspring vs Outsider

	N	Mean	StDev	SE Mean
Offspring	42	−0.10	1.95	0.30
Outsider	98	1.24	2.83	0.29

Difference = mu (Offspring) − mu (Outsider)
Estimate for difference: −1.336
95% CI for difference: (−2.158, −0.514)
t-Test of difference = 0 (vs not =): T-Value = −3.22 P-Value = 0.002 DF = 110

Instructions

Follow the instructions for Example 12.6 except at step 3 do not click **Assume equal variances**.

INTERPRET

The *t*-statistic is −3.22, and its *p*-value is 0.0017. Accordingly, we conclude there is sufficient evidence to infer that the mean changes in operating income differ.

Estimating $\mu_1 - \mu_2$: Unequal-Variances

We can also draw inferences about the difference between the two population means by calculating the confidence interval estimator. We use the unequal-variances confidence interval estimator of $\mu_1 - \mu_2$ and a 95% confidence level.

COMPUTE

MANUALLY:

$$(x_1 - x_2) \pm t_{\alpha/2}\sqrt{\left(\frac{s_1^2}{n_1} + \frac{s_2^2}{n_2}\right)}$$

$$= (-0.10 - 1.24) \pm 1.982\sqrt{\left(\frac{3.79}{42} + \frac{8.03}{98}\right)}$$

$$= -1.34 \pm 0.82$$

$$\text{LCL} = -2.16 \text{ and UCL} = -0.52$$

EXCEL

	A	B	C	D
1	*t*-Estimate : Two Means (Unequal Variances)			
2				
3		*Offspring*	*Outsider*	
4	Mean	−0.10	1.24	
5	Variance	3.79	8.03	
6	Observations	42	98	
7				
8	Degrees of Freedom	110.75		
9	Confidence Level	0.95		
10	Confidence Interval Estimate	−1.34	±	0.82
11	LCL	−2.16		
12	UCL	−0.51		

(*continued*)

Instructions

1. Type or import the data into two columns*. (**Open** Xm12-07.)
2. Click **Add-Ins, Data Analysis Plus** and ***t*-Estimate: Two Means.**
3. Specify the **Variable 1 Range** (A1:A43) and the **Variable 2 Range** (B1:B99). Click **Independent Samples with Unequal Variances** and the value for α (0.05).

Minitab prints the confidence interval estimate as part of the output of the test statistic. However, you must specify the **Alternative** hypothesis as **not equal** to produce a two-sided interval.

INTERPRET

We estimate that the mean change in operating incomes for outsiders exceeds the mean change in the operating income for offspring and lies between 0.51 and 2.16 percentage points.

12-4b Checking the Required Condition

Both the equal-variances and unequal-variances techniques require that the populations be normally distributed.† As before, we can check to see whether the requirement is satisfied by drawing the histograms of the data.

To illustrate, we used Excel (Minitab histograms are almost identical) to create the histograms for Example 12.6 (Figures 12.8 and 12.9) and Example 12.7 (Figures 12.10 and 12.11).

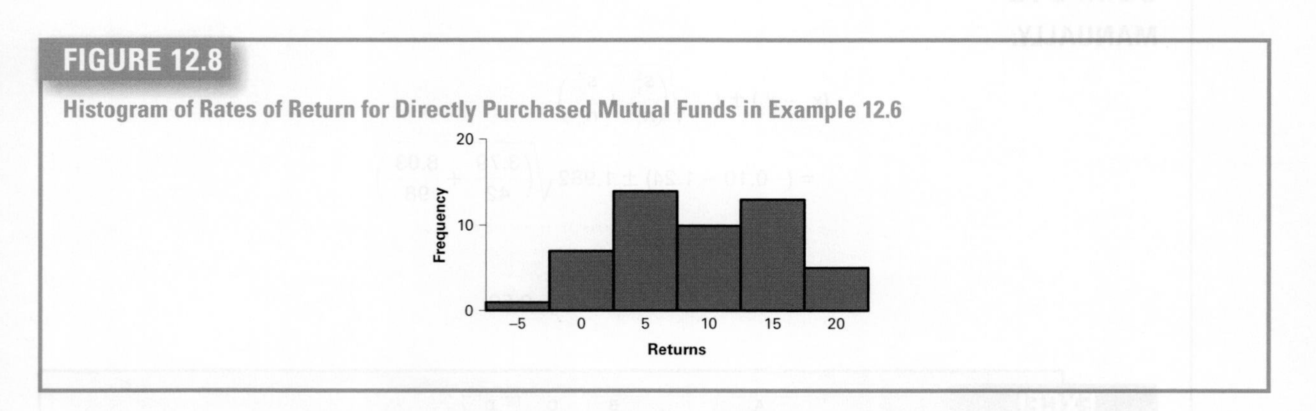

Although the histograms are not perfectly bell shaped, it appears that in both examples the data are at least approximately normal. Because this technique is robust, we can be confident in the validity of the results.

---

*If a cell is empty (representing missing data), it must be removed. See the website Appendix Removing Empty Cells in Excel.

†Large sample sizes can overcome the effects of extreme abnormality.

FIGURE 12.9

Histogram of Rates of Return for Broker-Purchased Mutual Funds in Example 12.6

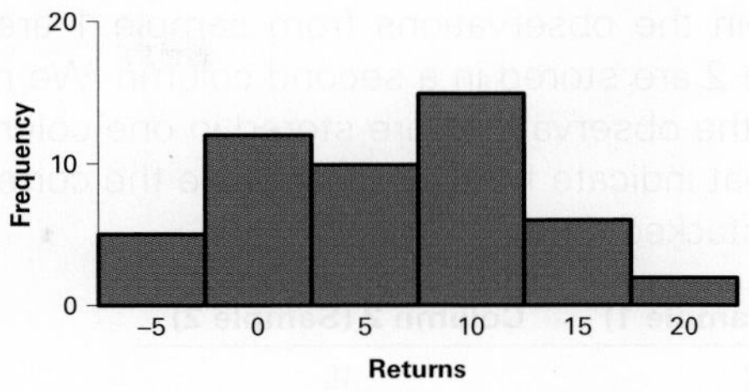

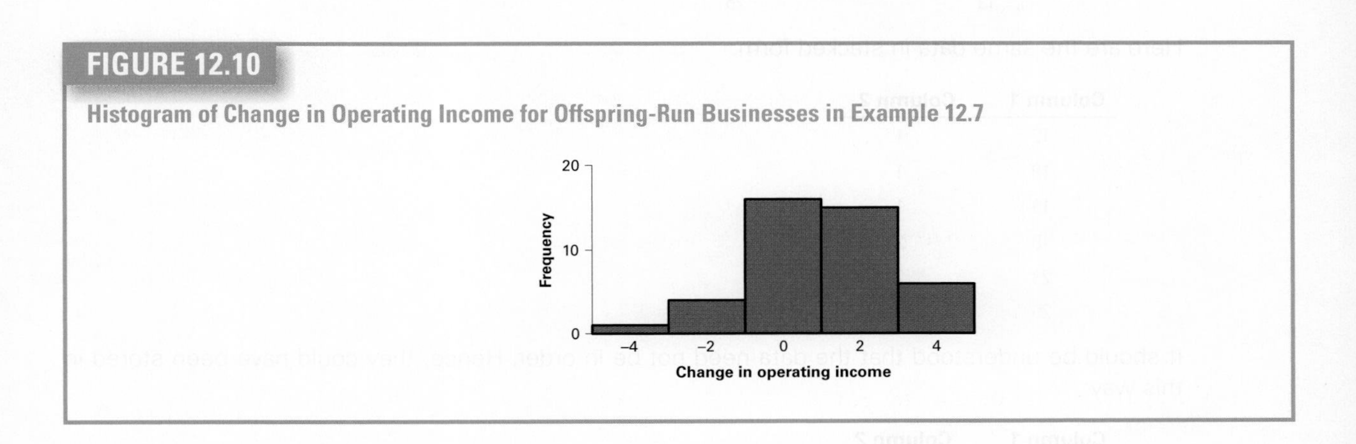

FIGURE 12.10

Histogram of Change in Operating Income for Offspring-Run Businesses in Example 12.7

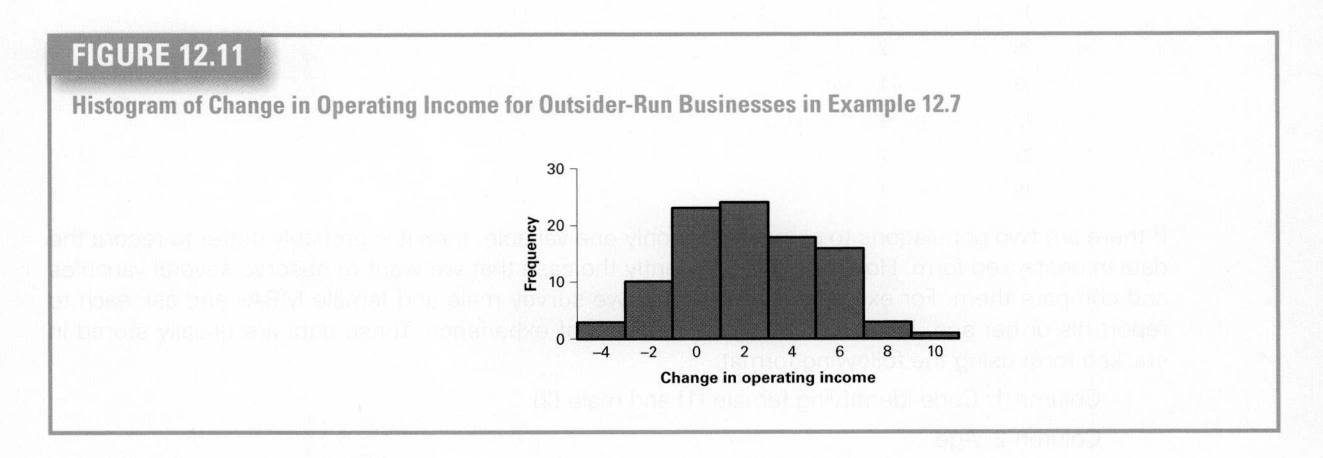

FIGURE 12.11

Histogram of Change in Operating Income for Outsider-Run Businesses in Example 12.7

12-4c Violation of the Required Condition

When the normality requirement is unsatisfied, we can use a nonparametric technique: the Wilcoxon Rank Sum Test (Chapter 17*) to replace the equal-variances test of $\mu_1 - \mu_2$. We have no alternative to the unequal-variances test of $\mu_1 - \mu_2$ when the populations are very non-normal.

*Instructors who wish to teach the use of nonparametric techniques for testing the difference between two means when the normality requirement is not satisfied should use the website Appendix Introduction to Nonparametric Techniques and the website Appendix Wilcoxon Rank Sum Test and Wilcoxon Signed Rank Sum Test.

12-4d Data Formats

There are two formats for storing the data when drawing inferences about the difference between two means. The first, which you have seen demonstrated in both Examples 12.6 and 12.7, is called *unstacked,* wherein the observations from sample 1 are stored in one column and the observations from sample 2 are stored in a second column. We may also store the data in stacked format. In this format, all the observations are stored in one column. A second column contains the codes, usually 1 and 2, that indicate from which sample the corresponding observation was drawn. Here is an example of unstacked data.

Column 1 (Sample 1)	Column 2 (Sample 2)
12	18
19	23
13	25

Here are the same data in stacked form.

Column 1	Column 2
12	1
19	1
13	1
18	2
23	2
25	2

It should be understood that the data need not be in order. Hence, they could have been stored in this way:

Column 1	Column 2
18	2
25	2
13	1
12	1
23	2
19	1

If there are two populations to compare and only one variable, then it is probably better to record the data in unstacked form. However, it is frequently the case that we want to observe several variables and compare them. For example, suppose that we survey male and female MBAs and ask each to report his or her age, income and number of years of experience. These data are usually stored in stacked form using the following format.

Column 1: Code identifying female (1) and male (2)

Column 2: Age

Column 3: Income

Column 4: Years of experience

To compare ages, we would use columns 1 and 2. Columns 1 and 3 are used to compare incomes and columns 1 and 4 are used to compare experience levels.

Most statistical software requires one format or the other. Some but not all of Excel's techniques require unstacked data. Some of Minitab's procedures allow either format, whereas others specify only one. Fortunately, both of our software packages allow the statistics practitioner to alter the format. We say 'fortunately' because this allowed us to store the data in either form on our website. In fact, we have used both forms to allow you to practise your ability to manipulate the data as necessary. You will need this ability to perform statistical techniques in this and other chapters in this book.

12-4e Developing an Understanding of Statistical Concepts 1

The formulas in this section are relatively complicated. However, conceptually both test statistics are based on the techniques we introduced earlier The value of the test statistic is the difference between the statistic $\bar{x}_1 - \bar{x}_2$ and the hypothesised value of the parameter $\mu_1 - \mu_2$ measured in terms of the standard error.

12-4f Developing an Understanding of Statistical Concepts 2

The standard error must be estimated from the data for all inferential procedures introduced here. The method we use to compute the standard error of $x_1 - x_2$ depends on whether the population variances are equal. When they are equal we calculate and use the pooled variance estimator s_p^2. We are applying an important principle here, and we will do so again in Section 12-8 and in later chapters. The principle can be loosely stated as follows: Where possible, it is advantageous to pool sample data to estimate the standard error. In Example 12.6 we are able to pool because we assume that the two samples were drawn from populations with a common variance. Combining both samples increases the accuracy of the estimate. Thus, s_p^2 is a better estimator of the common variance than either s_1^2 or s_2^2 separately. When the two population variances are unequal, we cannot pool the data and produce a common estimator. We must compute s_1^2 and s_2^2 and use them to estimate σ_1^2 and σ_2^2, respectively.

Here is a summary of how we recognise the techniques presented in this section.

Factors that Identify the Equal–Variances *t*-Test and Estimator of $\mu_1 - \mu_2$

1. **Problem objective:** Compare two populations.
2. **Data type:** Interval.
3. **Descriptive measurement:** Central location.
4. **Experimental design:** Independent samples.
5. **Population variances:** Equal.

Factors that Identify the Unequal–Variances *t*-Test and Estimator of $\mu_1 - \mu_2$

1. **Problem objective:** Compare two populations.
2. **Data type:** Interval.
3. **Descriptive measurement:** Central location.
4. **Experimental design:** Independent samples.
5. **Population variances:** Unequal.

EXERCISES

DO-IT-YOURSELF EXCEL

Construct Excel spreadsheets for each of the following:

12.42 Equal-variance *t*-test of $\mu_1 - \mu_2$. Inputs: Sample means, sample standard deviations, sample sizes, hypothesised difference between means. Outputs: Test statistic, critical values and one- and two-tail *p*-values. Tools: **T.INV,T.DIST.**

12.43 Equal-variance *t*-estimator of $\mu_1 - \mu_2$. Inputs: Sample means, sample standard deviations, sample sizes and confidence level. Outputs: Upper and lower confidence limits. Tools: **T.INV.**

12.44 Unequal-variance *t*-test of $\mu_1 - \mu_2$. Inputs: Sample means, sample standard deviations, sample sizes and hypothesised difference between means. Outputs: Test statistic, critical values, and one- and two-tail *p*-values. Tools: **T.INV,T.DIST.**

12.45 Unequal-variance *t*-estimator of $\mu_1 - \mu_2$. Inputs: Sample means, sample standard deviations, sample sizes and confidence level. Outputs: Upper and lower confidence limits. Tools: **T.INV.**

Developing an Understanding of Statistical Concepts

The next exercises are 'what-if' analyses designed to determine what happens to the test statistics and interval estimates when elements of the statistical inference change. These problems can be solved manually, using the Excel spreadsheets you created or Minitab.

12.46 In random samples of 25 from each of two normal populations, we found the following statistics:

$\bar{x}_1 = 524 \quad s_1 = 129$

$\bar{x}_2 = 469 \quad s_2 = 141$

a. Estimate the difference between the two population means with 95% confidence.

b. Repeat part (a) increasing the standard deviations to $s_1 = 255$ and $s_2 = 260$.

c. Describe what happens when the sample standard deviations get larger.

d. Repeat part (a) with samples of size 100.

e. Discuss the effects of increasing the sample size.

12.47 Random sampling from two normal populations produced the following results:

$\bar{x}_1 = 63 \quad s_1 = 18 \quad n_1 = 50$

$\bar{x}_2 = 60 \quad s_2 = 7 \quad n_2 = 45$

a. Estimate with 90% confidence the difference between the two population means.

b. Repeat part (a) changing the sample standard deviations to 41 and 15, respectively.

c. What happens when the sample standard deviations increase?

d. Repeat part (a) doubling the sample sizes.

e. Describe the effects of increasing the sample sizes.

12.48 For each of the following, determine the number of degrees of freedom assuming equal population variances and unequal population variances.

a. $n_1 = 15, n_2 = 15, s_1^2 = 25, s_2^2 = 15$

b. $n_1 = 10, n_2 = 16, s_1^2 = 100, s_2^2 = 15$

c. $n_1 = 50, n_2 = 50, s_1^2 = 8, s_2^2 = 14$

d. $n_1 = 60, n_2 = 45, s_1^2 = 75, s_2^2 = 10$

12.49 Refer to Exercise 12.48.

a. Confirm that in each case the number of degrees of freedom for the equal-variances test statistic and confidence interval estimator is larger than that for the unequal-variances test statistic and confidence interval estimator.

b. Try various combinations of sample sizes and sample variances to illustrate that the number of degrees of freedom for the equal-variances test statistic and confidence interval estimator is larger than that for the unequal-variances test statistic and confidence interval estimator.

Applications

For the next exercises use a 10% significance level.

12.50 Xr12-50 A number of restaurants feature a device that allows credit card users to swipe their cards at the table. It allows the user to specify a percentage or the amount of money to leave as a tip. In an experiment to see how it works, a random sample of credit card users was drawn. Some paid the usual way, and some used the new device. The percent left as a tip was recorded and listed below. Can we infer that users of the device leave larger tips?

Usual	10.3	15.2	13.0	9.9	12.1	13.4	12.2	14.9	13.2	12.0	
Device	13.6	15.7	12.9	13.2	12.9	13.4	12.1	13.9	15.7	15.4	17.4

12.51 Xr12-51 Who spends more on their holidays: golfers or skiers? To help answer this question, a travel agency surveyed 15 customers who regularly take their spouses on either a skiing or a golfing holiday. The amounts spent on holidays last year are shown here. Can we infer that golfers and skiers differ in their holiday expenses?

Golfer	2450	3860	4528	1944	3166	3275
	4490	3685	2950			
Skier	3805	3725	2990	4357	5550	4130

For the next exercises use a 5% significance level unless specified otherwise.

12.52 Xr12-52 Is eating oat bran an effective way to reduce cholesterol? Early studies indicated that eating oat bran daily reduces cholesterol levels by 5% to 10%. Reports of this study resulted in the introduction of many new breakfast cereals with various percentages of oat bran as an ingredient. However, an experiment performed by medical researchers in Boston cast doubt on the effectiveness of oat bran. In that study, 120 volunteers ate oat bran for breakfast, and another 120 volunteers ate another grain cereal for breakfast. At the end of 6 weeks, the percentage of cholesterol reduction was computed for both groups. Can we infer that oat bran is different from other cereals in terms of cholesterol reduction?

12.53 Xr13-53 In assessing the value of radio advertisements, sponsors consider not only the total number of listeners but also their ages. The 18 to 34 age group is considered to spend the most money. To examine the issue, the manager of an FM station commissioned a survey. One objective was to measure the difference in listening habits between the 18 to 34 age and 35 to 50 age groups. The survey asked 250 people in each age category how much time they spent listening to FM radio per day. The results (in minutes) were recorded and stored in stacked format (column 1 = Age group and column 2 = Listening times).

a. Can we conclude that a difference exists between the two groups?

b. Estimate with 95% confidence the difference in mean time listening to FM radio between the two age groups.

c. Are the required conditions satisfied for the techniques you used in parts (a) and (b)?

12.54 Xr12-54 The president of a company that manufactures automobile air conditioners is considering switching his supplier of condensers. Supplier A, the current producer of condensers for the manufacturer, prices its product 5 % higher than supplier B. Because the president wants to maintain his company's reputation for quality, he wants to be sure that supplier B's condensers last at least as long as supplier A's. After a careful analysis, the president decided to retain supplier A if there is sufficient statistical evidence that supplier A's condensers last longer on average than supplier B's. In an experiment, 30 midsize cars were equipped with air conditioners using type A condensers while another 30 midsize cars were equipped with type B condensers. The number of miles (in thousands) driven by each car before the condenser broke down was recorded. Should the president retain supplier A?

12.55 Xr12-55 It is often useful for companies to know who their customers are and how they became customers. In a study of credit card use, random samples were drawn of cardholders who applied for the credit card and credit cardholders who were contacted by telemarketers or by mail. The total purchases made by each in the last month were recorded. Can we conclude from these data that differences exist on average between the two types of customers?

12.56 Xr12-56 Traditionally, wine has been sold in glass bottles with cork stoppers. The stoppers are supposed to keep air out of the bottle because oxygen is the enemy of wine, particularly red wine. Recent research appears to indicate that metal screw caps are more effective in keeping air out of the bottle. However, metal caps are perceived to be inferior and usually associated with cheaper brands of wine. To determine if this perception is wrong, a random sample of 130 people who drink at least one bottle per week on average was asked to participate in an experiment. All were given the same wine in two types of bottles. One group was given a corked bottle, and the other was given a bottle with a metal cap and asked to taste the wine and indicate what they think the retail price of the wine should be. Determine whether there is enough evidence to conclude that bottles of wine with metal caps are perceived to be cheaper.

12-5 INFERENCE ABOUT THE DIFFERENCE BETWEEN TWO MEANS: MATCHED PAIRS EXPERIMENT

We continue our presentation of statistical techniques that address the problem of comparing two populations of interval data. In Section 12-4, the parameter of interest was the difference between two population means, where the data were generated from independent samples. In this section, the data are gathered from a matched pairs experiment. To illustrate why matched pairs experiments are needed and how we deal with data produced in this way, consider the following example.

EXAMPLE 12.8

DATA Xm12-08
Comparing Salary Offers for Finance and Marketing MBA Graduates, Part 1

In the last few years, a number of web-based companies that offer job placement services have been created. The manager of one such company wanted to investigate the job offers recent MBAs were obtaining. In particular, she wanted to know whether finance students were being offered higher salaries than marketing students. In a preliminary study, she randomly sampled 50 recently graduated MBAs, half of whom studied finance and half studied marketing. From each she obtained the highest salary offer (including benefits). These data are listed here. Can we infer that finance graduates obtain higher salary offers than do those studying marketing among MBAs?

Highest salary offer made to finance graduates

61 228	51 836	20 620	73 356	84 186	79 782	29 523	80 645	76 125
62 531	77 073	86 705	70 286	63 196	64 358	47 915	86 792	75 155
65 948	29 392	96 382	80 644	51 389	61 955	63 573		

Highest salary offer made to marketing graduates

73 361	36 956	63 627	71 069	40 203	97 097	49 442	75 188	59 854
79 816	51 943	35 272	60 631	63 567	69 423	68 421	56 276	47 510
58 925	78 704	62 553	81 931	30 867	49 091	48 843		

SOLUTION:

IDENTIFY

The objective is to compare two populations of interval data. The parameter is the difference between two means $\mu_1 - \mu_2$ (where μ_1 = mean highest salary offer to finance graduates and μ_2 = mean highest salary offer to marketing graduates). Because we want to determine whether finance graduates are offered higher salaries, the alternative hypothesis will specify that μ_1 is greater than μ_2. The *F*-test for variances was conducted, and the results indicate that there is not enough evidence to infer that the population variances differ. Hence we use the equal-variances test statistic:

$$H_0: (\mu_1 - \mu_2) = 0$$

$$H_1: (\mu_1 - \mu_2) > 0$$

$$\textbf{Test statistic: } t = \frac{(\bar{x}_1 - \bar{x}_2) - (\mu_1 - \mu_2)}{\sqrt{s_p^2\left(\frac{1}{n_1} + \frac{1}{n_2}\right)}}$$

COMPUTE

MANUALLY:

From the data, we calculated the following statistics:

$$\bar{x}_1 = 65\,624$$

$$\bar{x}_2 = 60\,423$$

$$s_1^2 = 360\,433\,294$$

$$s_2^2 = 262\,228\,559$$

$$s_p^2 = \frac{(n_1 - 1)s_1^2 + (n_2 - 1)s_2^2}{n_1 + n_2 - 2}$$

$$= \frac{(25 - 1)(360\,433\,294) + (25 - 1)(262\,228\,559)}{25 + 25 - 2}$$

$$= 311\,330\,926$$

The value of the test statistic is computed next:

$$t = \frac{(\bar{x}_1 - \bar{x}_2) - (\mu_1 - \mu_2)}{\sqrt{s_p^2\left(\frac{1}{n_1} + \frac{1}{n_2}\right)}}$$

$$= \frac{(65\,624 - 60\,423) - (0)}{\sqrt{311\,330\,926\left(\frac{1}{25} + \frac{1}{25}\right)}}$$

$$= 1.04$$

The number of degrees of freedom of the test statistic is

$$\nu = n_1 + n_2 - 2 = 25 + 25 - 2 = 48$$

The rejection region is

$$t > t_{\alpha,\nu} = t_{0.05,48} \approx 1.676$$

EXCEL

	A	B	C
1	*t*-Test: Two-Sample Assuming Equal Variances		
2			
3		*Finance*	*Marketing*
4	Mean	65,624	60,423
5	Variance	360,433,294	262,228,559
6	Observations	25	25
7	Pooled Variance	311,330,926	
8	Hypothesised Mean Difference	0	
9	df	48	
10	t Stat	1.04	
11	P(T<=t) one-tail	0.1513	
12	t Critical one-tail	1.6772	
13	P(T<=t) two-tail	0.3026	
14	t Critical two-tail	2.0106	

MINITAB

Two-Sample T-Test and CI: Finance, Marketing

```
Two-sample T for Finance vs Marketing

            N   Mean  StDev   SE Mean
Finance    25  65624  18985      3797
Marketing  25  60423  16193      3239


Difference = mu (Finance) – mu (Marketing)
Estimate for difference: 5201.00
95% lower bound for difference: –3169.42
T-Test of difference = 0 (vs >): T-Value = 1.04  P-Value = 0.151  DF = 48
Both use Pooled StDev = 17644.5722
```

INTERPRET

The value of the test statistic (t = 1.04) and its *p*-value (0.1513) indicate that there is very little evidence to support the hypothesis that finance graduates receive higher salary offers than marketing graduates.

Notice that we have some evidence to support the alternative hypothesis. The difference in sample means is

$$(\bar{x}_1 - \bar{x}_2) = (65\,624 - 60\,423) = 5201$$

However, we judge the difference between sample means in relation to the standard error of $\bar{X}_1 - \bar{X}_2$. As we have already calculated,

$$s_p^2 = 311\,330\,926$$

and

$$\sqrt{s_p^2\left(\frac{1}{n_1} + \frac{1}{n_2}\right)} = 4991$$

Consequently, the value of the test statistic is t = 5201/4991 = 1.04, a value that does not allow us to infer that finance graduates attract higher salary offers. We can see that although the difference between the sample means was quite large, the variability of the data as measured by s_p^2 was also large, resulting in a small test statistic value.

EXAMPLE 12.9

DATA Xm12-09

Comparing Salary Offers for Finance and Marketing MBA Graduates, Part 2

Suppose now that we redo the experiment in the following way. We examine the transcripts of finance and marketing MBA graduates. We randomly select a finance and a marketing graduate whose grade point average (GPA) falls between 3.92 and 4 (based on a maximum of 4). We then randomly select a finance and a marketing graduate whose GPA is between 3.84 and 3.92. We continue this process until the 25th pair of finance and marketing graduates is selected whose GPA fell between 2.0 and 2.08. (The minimum GPA required for graduation is 2.0.) We recorded the highest salary offer. These data, together with the GPA group, are listed here. Can we conclude from these data that finance graduates draw larger salary offers than do marketing graduates?

Group	Finance	Marketing
1	95 171	89 329
2	88 009	92 705
3	98 089	99 205
4	106 322	99 003
5	74 566	74 825
6	87 089	77 038
7	88 664	78 272
8	71 200	59 462
9	69 367	51 555
10	82 618	81 591
11	69 131	68 110
12	58 187	54 970
13	64 718	68 675
14	67 716	54 110
15	49 296	46 467
16	56 625	53 559
17	63 728	46 793
18	55 425	39 984
19	37 898	30 137
20	56 244	61 965
21	51 071	47 438
22	31 235	29 662
23	32 477	33 710
24	35 274	31 989
25	45 835	38 788

SOLUTION:

The experiment described in Example 12.8 is one in which the samples are independent. In other words, there is no relationship between the observations in one sample and the observations in the second sample. However, in this example the experiment was designed in such a way that each observation in one sample is matched with an observation in the other sample. The matching is conducted by selecting finance and marketing graduates with similar GPAs. Thus, it is logical to compare the salary offers for finance and marketing graduates in each group. This type of experiment is called matched pairs. We now describe how we conduct the test.

For each GPA group, we calculate the matched pair difference between the salary offers for finance and marketing graduates.

Group	Finance	Marketing	Difference
1	95 171	89 329	5 842
2	88 009	92 705	–4 696
3	98 089	99 205	–1 116
4	106 322	99 003	7 319
5	74 566	74 825	–259
6	87 089	77 038	10 051
7	88 664	78 272	10 392
8	71 200	59 462	11 738
9	69 367	51 555	17 812
10	82 618	81 591	1 027
11	69 131	68 110	1 021
12	58 187	54 970	3 217
13	64 718	68 675	–3 957
14	67 716	54 110	13 606
15	49 296	46 467	2 829
16	56 625	53 559	3 066
17	63 728	46 793	16 935
18	55 425	39 984	15 441
19	37 898	30 137	7 761
20	56 244	61 965	–5 721
21	51 071	47 438	3 633
22	31 235	29 662	1 573
23	32 477	33 710	–1 233
24	35 274	31 989	3 285
25	45 835	38 788	7 047

In this experimental design, the parameter of interest is the **mean of the population of differences**, which we label μ_D. Note that μ_D does in fact equal $\mu_1 - \mu_2$, but we test μ_D because of the way the experiment was designed. Hence, the hypotheses to be tested are

$$H_0: \mu_D = 0$$
$$H_1: \mu_D > 0$$

We have already presented inferential techniques about a population mean. Recall that in previous Chapter 12 sections we introduced the *t*-test of μ. Thus, to test hypotheses about μ_D, we use the following test statistic.

Test Statistic for μ_D

$$t = \frac{\bar{x}_D - \mu_D}{s_D/\sqrt{n_D}}$$

which is Student *t* distributed with $\nu = n_D - 1$ degrees of freedom, provided that the differences are normally distributed.

Aside from the subscript *D*, this test statistic is identical to the one presented in previous Chapter 12 sections. We conduct the test in the usual way.

COMPUTE MANUALLY:

Using the differences computed above, we find the following statistics:

$$\bar{x}_D = 5065$$

$$s_D = 6647$$

from which we calculate the value of the test statistic:

$$t = \frac{\bar{x}_D - \mu_D}{s_D/\sqrt{n_D}} = \frac{5065 - 0}{6647/\sqrt{25}} = 3.81$$

The rejection region is

$$t > t_{\alpha,\nu} = t_{0.05,24} = 1.711$$

EXCEL

	A	B	C
1	*t*-Test: Paired Two Sample for Means		
2			
3		*Finance*	*Marketing*
4	Mean	65,438	60,374
5	Variance	444,981,810	469,441,785
6	Observations	25	25
7	Pearson Correlation	0.9520	
8	Hypothesised Mean Difference	0	
9	df	24	
10	t Stat	3.81	
11	P(T<=t) one-tail	0.0004	
12	t Critical one-tail	1.7109	
13	P(T<=t) two-tail	0.0009	
14	t Critical two-tail	2.0639	

Excel prints the sample means, variances and sample sizes for each sample (as well as the coefficient of correlation), which implies that the procedure uses these statistics. It does not. The technique is based on computing the paired differences from which the mean, variance and sample size are determined. Excel should have printed these statistics.

Instructions

1. Type or import the data into two columns*. (**Open** Xm12-09.)
2. Click **Data, Data Analysis** and ***t*-Test: Paired Two-Sample for Means.**
3. Specify the **Variable 1 Range** (B1:B26) and the **Variable 2 Range** (C1:C26). Type the value of **the Hypothesised Mean Difference** (0) and specify a value for α (0.05).

MINITAB

Paired T-Test and CI: Finance, Marketing

Paired T for Finance vs Marketing

	N	Mean	StDev	SE Mean
Finance	25	65438.2	21094.6	4218.9
Marketing	25	60373.7	21666.6	4333.3
Difference	25	5064.52	6646.90	1329.38

95% lower bound for mean difference: 2790.11

T-Test of mean difference = 0 (vs > 0): T-Value = 3.81 P-Value = 0.000

*If one or both columns contain an empty cell (representing missing data) the entire row must be removed.

Instructions

1. Type or import the data into two columns. (**Open** Xm12-09.)
2. **Click Stat, Basic Statistics** and **Paired** *t*. ...
3. Select the variable names of the **First sample** (Finance) and **Second sample** (Marketing). Click **Options ...** .
4. In the **Test Mean** box, type the hypothesised mean of the paired difference (0), and specify the **Alternative** (greater than).

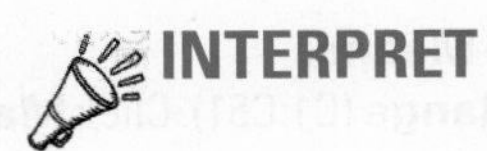

The value of the test statistic is $t = 3.81$ with a *p*-value of 0.0004. There is now overwhelming evidence to infer that finance graduates obtain higher salary offers than marketing graduates. By redoing the experiment as matched pairs, we were able to extract this information from the data.

12-5a Estimating the Mean Difference

We derive the confidence interval estimator of μ_D using the usual form for the confidence interval.

Confidence Interval Estimator of μ_D

$$\bar{x}_D \pm t_{\alpha/2}\frac{s_D}{\sqrt{n_D}}$$

DATA Xm12-09

Comparing Salary Offers for Finance and Marketing MBA Graduates, Part 3

Compute the 95% confidence interval estimate of the mean difference in salary offers between finance and marketing graduates in Example 12.9.

SOLUTION:

COMPUTE MANUALLY:

The 95% confidence interval estimate of the mean difference is

$$\bar{x}_D \pm t_{\alpha/2}\frac{s_D}{\sqrt{n_D}} = 5065 \pm 2.064\frac{6647}{\sqrt{25}} = 5065 \pm 2744$$

$$LCL = 2321 \quad \text{and} \quad UCL = 7809$$

EXCEL

	A	B	C	D
1	*t*-Estimate : Two Means (Matched Pairs)			
2				
3		*Difference*		
4	Mean	5065		
5	Variance	44181217		
6	Observations	25		
7	Degrees of Freedom	24		
8	Confidence Level	0.95		
9	Confidence Interval Estimate	5065	±	2744
10	LCL	2321		
11	UCL	7808		

Instructions

1. Type or import the data into two columns*. (**Open** Xm12-09.)
2. Click **Add-Ins**, **Data Analysis Plus** and ***t*-Estimate: Two Means.**
3. Specify the **Variable 1 Range** (B1:B51) and the **Variable 2 Range** (C1:C51). Click **Matched Pairs** and the value for α (0.05).

MINITAB

Paired T-Test and CI: Finance, Marketing

Paired T for Finance vs Marketing

	N	Mean	StDev	SE Mean
Finance	25	65438.2	21094.6	4218.9
Marketing	25	60373.7	21666.6	4333.3
Difference	25	5064.52	6646.90	1329.38

95% CI for mean difference: (2320.82, 7808.22)
T-Test of mean difference = 0 (vs not = 0): T-Value = 3.81 P-Value = 0.001

Instructions

Follow the instructions to test the paired difference. However, you must specify **not equal** for the **Alternative** hypothesis to produce the two-sided confidence interval estimate of the mean difference.

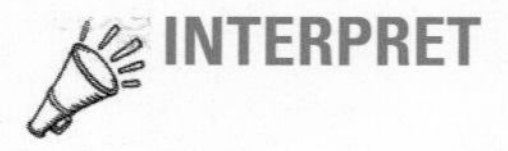

We estimate that the mean salary offer to finance graduates exceeds the mean salary offer to marketing graduates by an amount that lies between €2321 and €7808 (using the computer output).

12-5b Independent Samples or Matched Pairs: Which Experimental Design Is Better?

Examples 12.8 and 12.9 demonstrated that the experimental design is an important factor in statistical inference. However, these two examples raise several questions about experimental designs.

1. Why does the matched pairs experiment result in concluding that finance graduates receive higher salary offers than do marketing graduates, whereas the independent samples experiment could not?
2. Should we always use the matched pairs experiment? In particular, are there disadvantages to its use?
3. How do we recognise when a matched pairs experiment has been performed?

---

*If one or both columns contain an empty cell (representing missing data) the entire row must be removed.

Here are our answers.

1. The matched pairs experiment worked in Example 12.9 by reducing the variation in the data. To understand this point, examine the statistics from both examples. In Example 12.8 we found $\bar{x}_1 - \bar{x}_2 = 5201$. In Example 12.9 we computed $\bar{x}_D = 5065$. Thus, the numerators of the two test statistics were quite similar. However, the test statistic in Example 12.9 was much larger than the test statistic in Example 12.8 because of the standard errors. In Example 12.8 we calculated

$$s_p^2 = 311\,330\,926 \quad \text{and} \quad \sqrt{s_p^2\left(\frac{1}{n_1}+\frac{1}{n_2}\right)} = 4991$$

Example 12.9 produced

$$s_D = 6647 \quad \text{and} \quad \frac{s_D}{\sqrt{n_D}} = 1329$$

As you can see, the difference in the test statistics was caused not by the numerator, but by the denominator. This raises another question: Why was the variation in the data of Example 12.8 so much greater than the variation in the data of Example 12.9? If you examine the data and statistics from Example 12.8, you will find that there was a great deal of variation *between* the salary offers in each sample. In other words, some MBA graduates received high salary offers and others relatively low ones. This high level of variation, as expressed by s_p^2, made the difference between the samples means appear to be small. As a result, we could not conclude that finance graduates attract higher salary offers.

Looking at the data from Example 12.9, we see that there is very little variation between the observations of the paired differences. The variation caused by different GPAs has been decreased markedly. The smaller variation causes the value of the test statistic to be larger. Consequently, we conclude that finance students obtain higher salary offers.

2. Will the matched pairs experiment always produce a larger test statistic than the independent samples experiment? The answer is, not necessarily. Suppose that in our example we found that companies did not consider GPAs when making decisions about how much to offer the MBA graduates. In such circumstances, the matched pairs experiment would result in no significant decrease in variation when compared to independent samples. It is possible that the matched pairs experiment may be less likely to reject the null hypothesis than the independent samples experiment. The reason can be seen by calculating the degrees of freedom. In Example 12.8, the number of degrees of freedom was 48, whereas in Example 12.9 it was 24. Even though we had the same number of observations (25 in each sample), the matched pairs experiment had half the number of degrees of freedom as the equivalent independent samples experiment. For exactly the same value of the test statistic, a smaller number of degrees of freedom in a Student *t* distributed test statistic yields a larger *p*-value. What this means is that if there is little reduction in variation to be achieved by the matched pairs experiment, the statistics practitioner should choose instead to conduct the experiment with independent samples.
3. As you have seen, in this book we deal with questions arising from experiments that have already been conducted. Consequently, one of your tasks is to determine the appropriate test statistic. In the case of comparing two populations of interval data, you must decide whether the samples are independent (in which case the parameter is $\mu_1 - \mu_2$) or matched pairs (in which case the parameter is μ_D) to select the correct test statistic. To help you do so, we suggest you ask and answer the following question: Does some natural relationship exist between each pair of observations that provides a logical reason to compare the first observation of sample 1 with the first observation of sample 2, the second observation of sample 1 with the second observation of sample 2, and so on? If so, the experiment was conducted by matched pairs. If not, it was conducted using independent samples.

12-5c Observational and Experimental Data

We can design a matched pairs experiment where the data are gathered using a controlled experiment or by observation. The data in Examples 12.8 and 12.9 are observational. As a consequence, when the statistical result provided evidence that finance graduates attracted higher salary offers, it did not necessarily mean that students educated in finance are more attractive to prospective employers. It may be, for example, that better students graduate in finance and better students achieve higher starting salaries.

12-5d Checking the Required Condition

The validity of the results of the *t*-test and estimator of μ_D depends on the normality of the differences (or large enough sample sizes). The histogram of the differences (Figure 12.12) is positively skewed but not enough so that the normality requirement is violated.

FIGURE 12.12

Histogram of Differences in Example 13.5

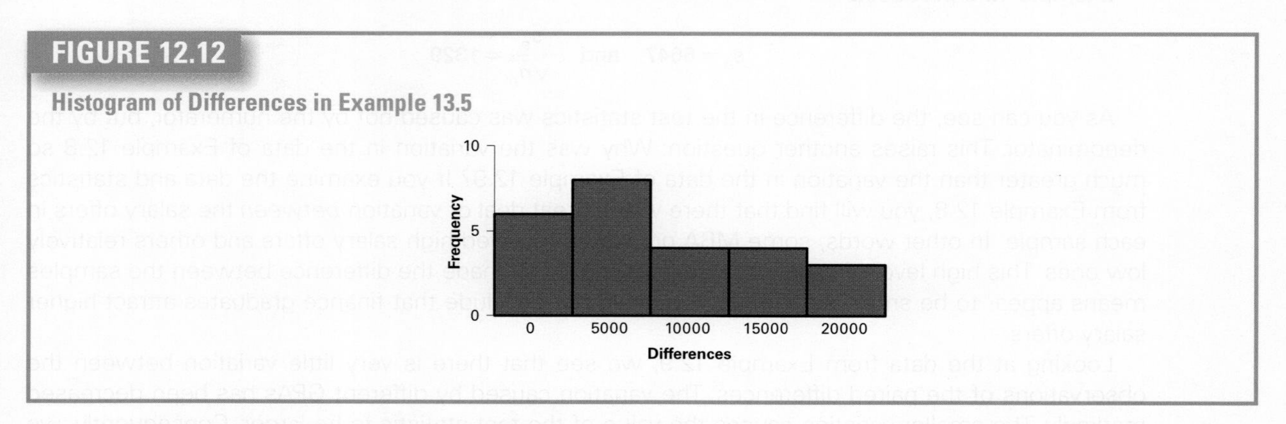

12-5e Violation of Required Condition

If the differences are very non-normal, we cannot use the *t*-test of μ_D. We can, however, employ a nonparametric technique – the Wilcoxon Signed Rank Sum Test for matched pairs, which we present in Chapter 17.*

12-5f Developing an Understanding of Statistical Concepts 1

Two of the most important principles in statistics were applied in this section. The first is the concept of analysing sources of variation. In Examples 12.8 and 12.9, we showed that by reducing the variation between salary offers in each sample we were able to detect a real difference between the two graduates. This was an application of the more general procedure of analysing data and attributing some fraction of the variation to several sources. In Example 12.9, the two sources of variation were the GPA and the MBA graduate. However, we were not interested in the variation between graduates with differing GPAs. Instead, we only wanted to eliminate that source of variation, making it easier to determine whether finance graduates draw larger salary offers.

In Chapter 13 we will introduce a technique called the *analysis of variance* that does what its name suggests: It analyses sources of variation in an attempt to detect real differences. In most applications of this procedure, we will be interested in each source of variation and not simply in reducing one source. We refer to the process as *explaining the variation.* The concept of explained variation is also applied in Chapters 15–17, where we introduce regression analysis.

12-5g Developing an Understanding of Statistical Concepts 2

The second principle demonstrated in this section is that statistics practitioners can design data-gathering procedures in such a way that they can analyse sources of variation. Before conducting the

*Instructors who wish to teach the use of nonparametric techniques for testing the mean difference when the normality requirement is not satisfied should use the website Appendix Introduction to Nonparametric Techniques and the website Appendix Wilcoxon Rank Sum Test and Wilcoxon Signed Rank Sum Test.

experiment in Example 12.9, the statistics practitioner suspected that there were large differences between graduates with different GPAs. Consequently, the experiment was organised so that the effects of those differences were mostly eliminated. It is also possible to design experiments that allow for easy detection of real differences and minimise the costs of data gathering. Unfortunately, we will not present this topic. However, you should understand that the entire subject of the design of experiments is an important one, because statistics practitioners often need to be able to analyse data to detect differences, and the cost is almost always a factor. Here is a summary of how we determine when to use these techniques.

Factors that Identify the *t*-Test and Estimator of μ_D

1. **Problem objective:** Compare two populations.
2. **Data type:** Interval.
3. **Descriptive measurement:** Central location.
4. **Experimental design:** Matched pairs.

EXERCISES

Applications

Conduct all tests at the 5% significance level.

12.57 Xr12-57 Many people use scanners to read documents and store them in a Word (or some other software) file. To help determine which brand of scanner to buy, a student conducts an experiment wherein eight documents are scanned by each of the two scanners in which he is interested. He records the number of errors made by each. These data are listed here. Can he infer that Brand A (the more expensive scanner) is better than Brand B?

Document	1	2	3	4	5	6	7	8
Brand A	17	29	18	14	21	25	22	29
Brand B	21	38	15	19	22	30	31	37

12.58 Xr12-58 The president of a large company in Durban is in the process of deciding whether to adopt a lunch-time exercise programme. The purpose of such programmes is to improve the health of workers and, in so doing, reduce medical expenses. To get more information, he institutes an exercise programme for the employees in one office. The president knows that during the winter months medical expenses are relatively high because of the incidence of colds and flu. Consequently, he decides to use a matched pairs design by recording medical expenses for the 12 months before the programme and for 12 months after the programme. The 'before' and 'after' expenses (in thousands of euros) are compared on a month-to-month basis and shown here.

a. Do the data indicate that exercise programmes reduce medical expenses? (Test with $\alpha = 0.05$.)

b. Estimate with 95% confidence the mean savings produced by exercise programmes.

c. Was it appropriate to conduct a matched pairs experiment? Explain.

Month	Jan	Feb	Mar	Apr	May	Jun
Before programme	68	44	30	58	35	33
After programme	59	42	20	62	25	30
Month	**Jul**	**Aug**	**Sep**	**Oct**	**Nov**	**Dec**
Before programme	52	69	23	69	48	30
After programme	56	62	25	75	40	26

The following exercises require the use of a computer and software.

Use a 5% significance level unless specified otherwise.

12.59 Xr12-59 One measure of the state of the economy is the amount of money homeowners pay on their mortgage each month. To determine the extent of change between this year and 5 years ago, a random sample of 150 homeowners was drawn. The monthly mortgage payments for each homeowner for this year and for 5 years ago were recorded. (The amounts have been adjusted so that we are comparing constant currency.) Can we infer that mortgage payments have risen over the past 5 years?

12.60 Xr12-60 To determine the effect of advertising in the Yellow Pages, Bell Telephone took a sample of 40 retail stores that did not advertise in the Yellow Pages last year but did so this year. The annual sales (in thousands of euros) for each store in both years were recorded.

a. Estimate with 90% confidence the improvement in sales between the 2 years.

b. Can we infer that advertising in the Yellow Pages improves sales?

c. Check to ensure that the required condition(s) of the techniques used in parts (a) and (b) is satisfied.

d. Would it be advantageous to perform this experiment with independent samples? Explain why or why not.

12.61 Xr12-61 Because of the high cost of energy, homeowners in northern climates need to find ways to cut their heating costs. A building contractor wanted to investigate the effect on heating costs of increasing the insulation. As an experiment, he located a large sub-development built around 1970 with minimal insulation. His plan was to insulate some of the houses and compare the heating costs in the insulated homes with those that remained uninsulated. However, it was clear to him that the size of the house was a critical factor in determining heating costs. Consequently, he found 16 pairs of identical-sized houses ranging from about 111 to 260 square metres. He insulated one house in each pair (levels of R20 in the walls and R32 in the attic) and left the other house unchanged. The heating cost for the following winter season was recorded for each house.

a. Do these data allow the contractor to infer at the 10% significance level that the heating cost for insulated houses is less than that for the uninsulated houses?

b. Estimate with 95% confidence the mean savings due to insulating the house.

c. What is the required condition for the use of the techniques in parts (a) and (b)?

12.62 Xr12-62 The fluctuations in the stock market induce some investors to sell and move their money into more stable investments. To determine the degree to which recent fluctuations affected ownership, a random sample of 170 people who confirmed that they owned some stock was surveyed. The values of the holdings were recorded at the end of last year and at the end of the year before. Can we infer that the value of the stock holdings has decreased?

12.63 Xr12-63 Suppose that another experiment is conducted and the finance and marketing MBA graduates were matched according to their undergraduate GPA. As in the previous examples, the highest starting salary offers were recorded. Can we infer from these data that finance graduates attract higher salary offers than marketing graduates?

12.64 Discuss whether the experiment in Exercise 12.63 produced a significant test result or not.

12-6 INFERENCE ABOUT THE RATIO OF TWO VARIANCES

In Sections 12-4 and 12-5, we dealt with statistical inference concerning the difference between two population means. The problem objective in each case was to compare two populations of interval data, and our interest was in comparing measures of central location. This section discusses the statistical technique to use when the problem objective and the data type are the same as in Sections 12-4 and 12-5, but our interest is in comparing variability. Here we will study the ratio of two population variances. We make inferences about the ratio because the sampling distribution is based on ratios rather than differences.

We have already encountered this technique when we used the *F*-test of two variances to determine which *t*-test and estimator of the difference between two means to use. In this section, we apply the technique to other problems where our interest is in comparing the variability in two populations.

In the previous chapter, we presented the procedures used to draw inferences about a single population variance and we pointed out that variance can be used to address problems where we need to judge the consistency of a production process. Also the variance was used to measure the risk associated with a portfolio of investments. In this section, we compare two variances, enabling us to compare the consistency of two production processes. We can also compare the relative risks of two sets of investments.

We will proceed in a manner that is probably becoming quite familiar.

12-6a Parameter

As you will see shortly, we compare two population variances by determining the ratio. Consequently, the parameter is σ_1^2/σ_2^2.

12-6b Statistic and Sampling Distribution

We have previously noted that the sample variance (defined in Chapter 4) is an unbiased and consistent estimator of the population variance. Not surprisingly, the estimator of the parameter σ_1^2/σ_2^2 is the ratio of the two sample variances drawn from their respective populations s_1^2/s_2^2.

The sampling distribution of s_1^2/s_2^2 is said to be *F*-distributed provided that we have independently sampled from two normal populations. (The *F*-distribution was introduced in Section 8-4.)

Statisticians have shown that the ratio of two independent chi-squared variables divided by their degrees of freedom is *F*-distributed. The degrees of freedom of the *F*-distribution are identical to the degrees of freedom for the two chi-squared distributions. In Section 12-2, we pointed out that $(n-1)s^2/\sigma^2$ is chi-squared distributed, provided that the sampled population is normal. If we have independent samples drawn from two normal populations, then both $(n_1 - 1)s_1^2/\sigma_1^2$ are chi-squared distributed. If we divide each by their respective number of degrees of freedom and take the ratio, we produce

$$\frac{\dfrac{(n_1-1)s_1^2/\sigma_1^2}{(n_1-1)}}{\dfrac{(n_2-1)s_2^2/\sigma_2^2}{(n_2-1)}}$$

which simplifies to

$$\frac{s_1^2/\sigma_1^2}{s_2^2/\sigma_2^2}$$

This statistic is *F*-distributed with $\nu_1 = n_1 - 1$ and $\nu_2 = n_2 - 1$ degrees of freedom. Recall that ν_1 is called the **numerator degrees of freedom** and ν_2 is called the **denominator degrees of freedom.**

12-6c Testing and Estimating a Ratio of Two Variances

In this book, our null hypothesis will always specify that the two variances are equal. As a result, the ratio will equal 1. Thus, the null hypothesis will always be expressed as

$$H_0: \sigma_1^2/\sigma_2^2 = 1$$

The alternative hypothesis can state that the ratio σ_1^2/σ_2^2 is either not equal to 1, greater than 1 or less than 1. Technically, the test statistic is

$$F = \frac{s_1^2/\sigma_1^2}{s_2^2/\sigma_2^2}$$

However, under the null hypothesis, which states that $\sigma_1^2/\sigma_2^2 = 1$, the test statistic becomes as follows.

Test Statistic for σ_1^2/σ_2^2

The test statistic employed to test that σ_1^2/σ_2^2 is equal to 1 is

$$F = \frac{s_1^2}{s_2^2}$$

which is *F*-distributed with $\nu_1 = n_1 - 1$ and $\nu_2 = n_2 - 1$ degrees of freedom provided that the populations are normal.

With the usual algebraic manipulation, we can derive the confidence interval estimator of the ratio of two population variances.

Confidence Interval Estimator of σ_1^2/σ_2^2

$$LCL = \left(\frac{s_1^2}{s_2^2}\right)\frac{1}{F_{\alpha/2,\nu_1,\nu_2}}$$

$$UCL = \left(\frac{s_1^2}{s_2^2}\right)F_{\alpha/2,\nu_2,\nu_1}$$

where $\nu_1 = n_1 - 1$ and $\nu_2 = n_2 - 1$.

EXAMPLE 12.11

DATA Xm12-11

Testing the Quality of Two-Bottle Filling Machines

In Example 12.3, we applied the chi-squared test of a variance to determine whether there was sufficient evidence to conclude that the population variance was less than 1.0. Suppose that the statistics practitioner also collected data from another container-filling machine and recorded the fills of a randomly selected sample. Can we infer at the 5% significance level that the second machine is superior in its consistency?

SOLUTION:

IDENTIFY

The problem objective is to compare two populations where the data are interval. Because we want information about the consistency of the two machines, the parameter we wish to test is σ_1^2/σ_2^2, where σ_1^2 is the variance of machine 1 and σ_2^2 is the variance for machine 2. We need to conduct the *F*-test of σ_1^2/σ_2^2 to determine whether the variance of population 2 is less than that of population 1. Expressed differently, we wish to determine whether there is enough evidence to infer that σ_1^2 is larger than σ_2^2. Hence, the hypotheses we test are

$$H_0: \sigma_1^2/\sigma_2^2 = 1$$

$$H_1: \sigma_1^2/\sigma_2^2 > 1$$

COMPUTE

MANUALLY:

The sample variances are $s_1^2 = 0.6333$ and $s_2^2 = 0.4528$. The value of the test statistic is

$$F = \frac{s_1^2}{s_2^2} = \frac{0.6333}{0.4528} = 1.40$$

The rejection region is

$$F > F_{\alpha,\nu_1,\nu_2} = F_{0.05,24,24} = 1.98$$

Because the value of the test statistic is not greater than 1.98, we cannot reject the null hypothesis.

EXCEL

	A	B	C
1	*F-Test Two-Sample for Variances*		
2			
3		*Machine 1*	*Machine 2*
4	Mean	999.7	999.8
5	Variance	0.6333	0.4528
6	Observations	25	25
7	df	24	24
8	F	1.3988	
9	P(F<=f) one-tail	0.2085	
10	F Critical one-tail	1.9838	

The value of the test statistic is $F = 1.3988$. Excel outputs the one-tail *p*-value, which is 0.2085.

Instructions

1. Type or import the data into two columns. (**Open** Xm12-11.)
2. Click **Data, Data Analysis** and ***F*-test Two-Sample for Variances.**
3. Specify the **Variable 1 Range** (A1:A26) and the **Variable 2 Range** (B1:B26). Type a value for α (0.05).

MINITAB

Test for Equal Variances: Machine 1, Machine 2

F-Test (Normal Distribution)
Test statistic = 1.40, p-value = 0.417

Note that Minitab conducts a two-tail test only. Thus, the *p*-value = 0.417/2 = 0.2085.

Instructions

1. Type or import the data into two columns. (**Open** Xm12-11.)
2. Click **Stat, Basic Statistics** and **2 Variances ....**
3. In the **Samples in different columns** box, select the **First** (Machine 1) and **Second** (Machine 2) variables.

INTERPRET

There is not enough evidence to infer that the variance of machine 2 is less than the variance of machine 1.

The histograms (not shown) appear to be sufficiently bell shaped to satisfy the normality requirement.

EXAMPLE 12.12

DATA Xm12-11

Estimating the Ratio of the Variances in Example 12.11

Determine the 95% confidence interval estimate of the ratio of the two population variances in Example 12.11.

SOLUTION:

COMPUTE

MANUALLY:

We find

$$F_{\alpha/2,\nu_1,\nu_2} = F_{0.025,24,24} = 2.27$$

Thus,

$$\text{LCL} = \left(\frac{s_1^2}{s_2^2}\right)\frac{1}{F_{\alpha/2,\nu_1,\nu_2}} = \left(\frac{0.6333}{0.4528}\right)\frac{1}{2.27} = 0.616$$

$$\text{UCL} = \left(\frac{s_1^2}{s_2^2}\right)F_{\alpha/2,\nu_2,\nu_1} = \left(\frac{0.6333}{0.4528}\right)2.27 = 3.17$$

We estimate that σ_1^2/σ_2^2 lies between 0.616 and 3.17.

EXCEL

	A	B	C
1	*F*-Estimate : Two Variances		
2			
3		*Machine 1*	*Machine 2*
4	Mean	999.7	999.8
5	Variance	0.6333	0.4528
6	Observations	25	25
7	df	24	24
8	LCL	0.6164	
9	UCL	3.1743	

Instructions

1. Type or import the data into two columns*. (**Open** Xm12-11.)
2. Click **Add-ins, Data Analysis Plus** and ***F* Estimate 2 Variances.**
3. Specify the **Variable 1 Range** (A1:A26) and the **Variable 2 Range** (B1:B26). Type a value for α (0.05).

MINITAB

Minitab does not compute the estimate of the ratio of two variances.

INTERPRET

As we pointed out in Chapter 11, we can often use a confidence interval estimator to test hypotheses. In this example, the interval estimate excludes the value of 1. Consequently, we can draw the same conclusion as we did in Example 12.11.

Factors that Identify the *F*-Test and Estimator of σ_1^2/σ_2^2

1. **Problem objective:** Compare two populations.
2. **Data type:** Interval.
3. **Descriptive measurement:** Variability.

EXERCISES

DO-IT-YOURSELF EXCEL

Construct Excel spreadsheets for each of the following:

12.65 *F*-test of σ_1^2/σ_2^2. Inputs: sample variances, sample sizes and hypothesised ratio of population variances. Outputs: Test statistic, critical values and one-and two-tail *p*-values. Tools: **F.INV, F.DIST.**

12.66 *F*-estimate of σ_1^2/σ_2^2. Inputs: sample variances, sample sizes and confidence level. Outputs: Upper and lower confidence limits, Tools: **F.INV.**

*If one or both columns contain(s) an empty cell (representing missing data) it must be removed. See the website Appendix Removing Empty Cells in Excel.

Developing an Understanding of Statistical Concepts

The next exercises are 'what-if analyses' designed to determine what happens to the test statistics and interval estimates when elements of the statistical inference change. These problems can be solved manually, using Excel's Do-It-Yourself Excel spreadsheets you just created, or Minitab.

12.67 Random samples from two normal populations produced the following statistics:

$s_1^2 = 350 \quad n_1 = 30 \quad s_2^2 = 700 \quad n_2 = 30$

a. Can we infer at the 10% significance level that the two population variances differ?

b. Repeat part (a) changing the sample sizes to $n_1 = 15$ and $n_2 = 15$.

c. Describe what happens to the test statistic and the conclusion when the sample sizes decrease.

12.68 Random samples from two normal populations produced the following statistics:

$s_1^2 = 28 \quad n_1 = 10 \quad s_2^2 = 19 \quad n_2 = 10$

a. Estimate with 95% confidence the ratio of the two population variances.

b. Repeat part (a) changing the sample sizes to $n_1 = 25$ and $n_2 = 25$.

c. Describe what happens to the width of the confidence interval estimate when the sample sizes increase.

Applications

Use a 5% significance level in all tests, unless specified otherwise.

12.69 Xr12-69 The manager of a dairy is in the process of deciding which of two new carton-filling machines to use. The most important attribute is the consistency of the fills. In a preliminary study she measured the fills in the 1-litre carton and listed them here. Can the manager infer that the two machines differ in their consistency of fills?

Machine 1 0.998 0.997 1.003 1.000 0.999
1.000 0.998 1.003 1.004 1.000

Machine 2 1.003 1.004 0.997 0.996 0.999 1.003
1.000 1.005 1.002 1.004 0.996

12.70 Xr12-70 An operations manager who supervises an assembly line has been experiencing problems with the sequencing of jobs. The problem is that bottlenecks are occurring because of the inconsistency of sequential operations. He decides to conduct an experiment wherein two different methods are used to complete the same task. He measures the times (in seconds). The data are listed here. Can he infer that the second method is more consistent than the first method?

Method 1 8.8 9.6 8.4 9.0 8.3 9.2 9.0 8.7 8.5 9.4

Method 2 9.2 9.4 8.9 9.6 9.7 8.4 8.8 8.9 9.0 9.7

12.71 Xr12-71 A new motorway in the UK has just been completed and the government must decide on speed limits. There are several possible choices. However, on advice from police who monitor traffic, the objective was to reduce the variation in speeds, which it is thought contribute to the number of collisions. It has been acknowledged that speed contributes to the severity of collisions. It is decided to conduct an experiment to acquire more information. Signs are posted for 1 week indicating that the speed limit is 70 mph. A random sample of cars' speeds is measured. During the second week, signs are posted indicating that the maximum speed is 70 mph and that the minimum speed is 60 mph. Once again a random sample of speeds is measured. Can we infer that limiting the minimum and maximum speeds reduces the variation in speeds?

12.72 Xr12-72 In Exercise 12.25 we described the problem of whether to change all the light bulbs at Wembley Stadium or change them one by one as they burn out. There are two brands of bulbs that can be used. Because both the mean and the variance of the lengths of life are important, it was decided to test the two brands. A random sample of both brands was drawn and left on until they burned out. The times were recorded. Can the Wembley Stadium management conclude that the variances differ?

12.73 Xr12-73 In deciding where to invest her retirement fund, an investor recorded the weekly returns of two portfolios for 1 year. Can we conclude that portfolio 2 is riskier than portfolio 1?

12-7 INFERENCE ABOUT THE DIFFERENCE BETWEEN TWO POPULATION PROPORTIONS

In this section, we present the procedures for drawing inferences about the difference between populations whose data are nominal. The number of applications of these techniques is almost limitless. For example, pharmaceutical companies test new drugs by comparing the new and old or the new versus a placebo. Marketing managers compare market shares before and after advertising campaigns. Operations managers compare defective rates between two machines. Political pollsters measure the difference in popularity before and after an election.

12-7a Parameter

When data are nominal, the only meaningful computation is to count the number of occurrences of each type of outcome and calculate proportions. Consequently, the parameter to be tested and estimated in this section is the difference between two population proportions $p_1 - p_2$.

12-7b Statistic and Sampling Distribution

To draw inferences about $p_1 - p_2$, we take a sample of size n_1 from population 1 and a sample of size n_2 from population 2 (Figure 12.13 depicts the sampling process).

FIGURE 12.13

Sampling From Two Populations of Nominal Data

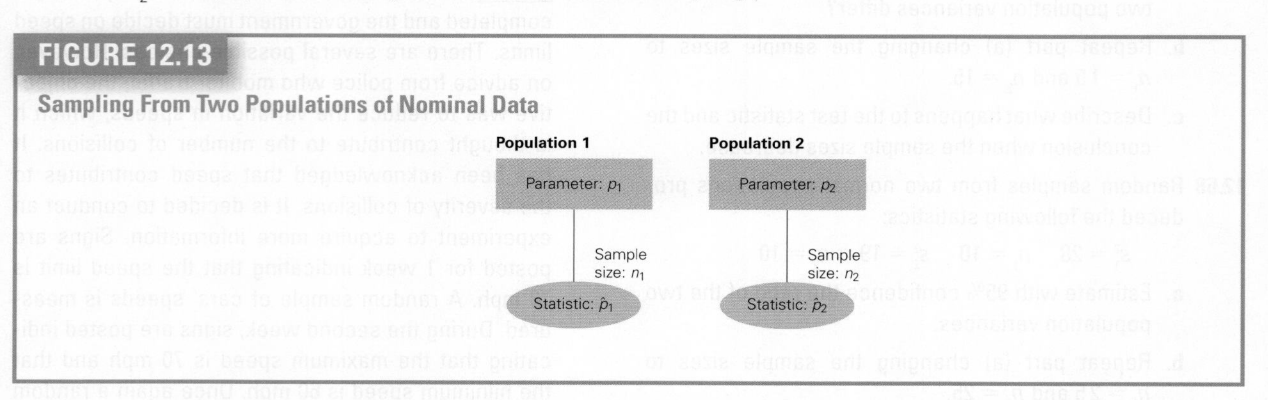

For each sample, we count the number of successes (recall that we call anything we are looking for a success), which we label x_1 and x_2, respectively. The sample proportions are then computed:

$$\hat{p}_1 = \frac{x_1}{n_1} \quad \text{and} \quad \hat{p}_2 = \frac{x_2}{n_2}$$

Statisticians have proven that the statistic $\hat{p}_1 - \hat{p}_2$ is an unbiased consistent estimator of the parameter $p_1 - p_2$. Using the same mathematics as we did in Chapter 9 to derive the sampling distribution of the sample proportion $\hat{p}$, we determine the sampling distribution of the difference between two sample proportions.

Sampling Distribution of $\hat{p}_1 - \hat{p}_2$

1. The statistic $\hat{p}_1 - \hat{p}_2$ is approximately normally distributed provided that the sample sizes are large enough so that n_1p_1, $n_1(1 - p_1)$, n_2p_2 and $n_2(1 - p_2)$ are all greater than or equal to 5. [Because p_1 and p_2 are unknown, we express the sample size requirement as $n_1\hat{p}_1$, $n_1(1 - \hat{p}_1)$, $n_2\hat{p}_2$ and $n_2(1 - \hat{p}_2)$ are greater than or equal to 5.]
2. The mean of $\hat{p}_1 - \hat{p}_2$ is

$$E(\hat{p}_1 - \hat{p}_2) = p_1 - p_2$$

3. The variance of $\hat{p}_1 - \hat{p}_2$ is

$$V(\hat{p}_1 - \hat{p}_2) = \frac{p_1(1 - p_1)}{n_1} + \frac{p_2(1 - p_2)}{n_2}$$

The standard error is

$$\sigma_{\hat{p}_1 - \hat{p}_2} = \sqrt{\frac{p_1(1 - p_1)}{n_1} + \frac{p_2(1 - p_2)}{n_2}}$$

Thus, the variable

$$z = \frac{(\hat{p}_1 - \hat{p}_2) - (p_1 - p_2)}{\sqrt{\frac{p_1(1-p_1)}{n_1} + \frac{p_2(1-p_2)}{n_2}}}$$

is approximately standard normally distributed.

12-7c Testing and Estimating the Difference between Two Proportions

We would like to use the z-statistic just described as our test statistic; however, the standard error of $\hat{p}_1 - \hat{p}_2$, which is

$$\sigma_{\hat{p}_1 - \hat{p}_2} = \sqrt{\frac{p_1(1-p_1)}{n_1} + \frac{p_2(1-p_2)}{n_2}}$$

is unknown because both p_1 and p_2 are unknown. As a result, the standard error of $\hat{p}_1 - \hat{p}_2$ must be estimated from the sample data. There are two different estimators of this quantity, and the determination of which one to use depends on the null hypothesis. If the null hypothesis states that $p_1 - p_2 = 0$, the hypothesised equality of the two population proportions allows us to pool the data from the two samples to produce an estimate of the common value of the two proportions p_1 and p_2. The **pooled proportion estimate** is defined as

$$\hat{p} = \frac{x_1 + x_2}{n_1 + n_2}$$

Thus, the estimated standard error of $\hat{p}_1 - \hat{p}_2$ is

$$\sqrt{\frac{\hat{p}(1-\hat{p})}{n_1} + \frac{\hat{p}(1-\hat{p})}{n_2}} = \sqrt{\hat{p}(1-\hat{p})\left(\frac{1}{n_1} + \frac{1}{n_2}\right)}$$

The principle used in estimating the standard error of $\hat{p}_1 - \hat{p}_2$ is analogous to that applied in Section 12-4 to produce the pooled variance estimate s_p^2, which is used to test $\mu_1 - \mu_2$ with σ_1^2 and σ_2^2 unknown but equal. The principle roughly states that, where possible, pooling data from two samples produces a better estimate of the standard error. Here, pooling is made possible by hypothesising (under the null hypothesis) that $p_1 = p_2$. (In Section 12-4, we used the pooled variance estimate because we assumed that $\sigma_1^2 = \sigma_2^2$.) We will call this application Case 1.

The second case applies when, under the null hypothesis, we state that $p_1 - p_2 = D$, where D is some value other than 0. Under such circumstances, we cannot pool the sample data to estimate the standard error of $\hat{p}_1 - \hat{p}_2$. The appropriate test statistic is described next as Case 2.

Test Statistic for $p_1 - p_2$: Case 1

If the null hypothesis specifies

$$H_0: (p_1 - p_2) = 0$$

the test statistic is

$$z = \frac{(\hat{p}_1 - \hat{p}_2) - (p_1 - p_2)}{\sqrt{\hat{p}(1-\hat{p})\left(\frac{1}{n_1} + \frac{1}{n_2}\right)}}$$

Because we hypothesise that $p_1 - p_2 = 0$, we simplify the test statistic to

$$z = \frac{(\hat{p}_1 - \hat{p}_2)}{\sqrt{\hat{p}(1-\hat{p})\left(\frac{1}{n_1} + \frac{1}{n_2}\right)}}$$

Test Statistic for $p_1 - p_2$: Case 2

If the null hypothesis specifies

$$H_0: (p_1 - p_2) = D \quad (D \neq 0)$$

the test statistic is

$$z = \frac{(\hat{p}_1 - \hat{p}_2) - (p_1 - p_2)}{\sqrt{\dfrac{\hat{p}_1(1 - \hat{p}_1)}{n_1} + \dfrac{\hat{p}_2(1 - \hat{p}_2)}{n_2}}}$$

which can also be expressed as

$$z = \frac{(\hat{p}_1 - \hat{p}_2) - D}{\sqrt{\dfrac{\hat{p}_1(1 - \hat{p}_1)}{n_1} + \dfrac{\hat{p}_2(1 - \hat{p}_2)}{n_2}}}$$

Notice that this test statistic is determined by simply substituting the sample statistics $\hat{p}_1$ and $\hat{p}_2$ in the standard error of $\hat{p}_1 - \hat{p}_2$.

You will find that, in most practical applications (including the exercises in this book), Case 1 applies – in most problems, we want to know whether the two population proportions differ: that is,

$$H_1: (p_1 - p_2) \neq 0$$

or if one proportion exceeds the other; that is,

$$H_1: (p_1 - p_2) > 0 \quad \text{or} \quad H_1: (p_1 - p_2) < 0$$

In some other problems, however, the objective is to determine whether one proportion exceeds the other by a specific non-zero quantity. In such situations, Case 2 applies.

We derive the interval estimator of $p_1 - p_2$ in the same manner we have been using since Chapter 10.

Confidence Interval Estimator of $p_1 - p_2$

$$(\hat{p}_1 - \hat{p}_2) \pm z_{\alpha/2}\sqrt{\frac{\hat{p}_1(1 - \hat{p}_1)}{n_1} + \frac{\hat{p}_2(1 - \hat{p}_2)}{n_2}}$$

This formula is valid when $n_1\hat{p}_1$, $n_1(1 - \hat{p}_1)$, $n_2\hat{p}_2$ and $n_2(1 - \hat{p}_2)$ are greater than or equal to 5.

Notice that the standard error is estimated using the individual sample proportions rather than the pooled proportion. In this procedure we cannot assume that the population proportions are equal as we did in the Case 1 test statistic.

APPLICATIONS IN MARKETING

Test Marketing

Marketing managers frequently make use of test marketing to assess consumer reaction to a change in a characteristic (such as price or packaging) of an existing product, or to assess consumers' preferences regarding a proposed new product. *Test marketing* involves experimenting with changes to the marketing mix in a small, limited test market and assessing consumers' reaction in the test market before undertaking costly changes in production and distribution for the entire market.

EXAMPLE 12.13

DATA Xm12-13

Test Marketing of Package Designs, Part 1

The General Products Company produces and sells a variety of household products. Because of stiff competition, one of its products, a bath soap, is not selling well. Hoping to improve sales, General Products decided to introduce more attractive packaging. The company's advertising agency developed two new designs. The first design features several bright colours to distinguish it from other brands. The second design is light green in colour with just the company's logo on it. As a test to determine which design is better, the marketing manager selected two supermarkets. In one supermarket, the soap was packaged in a box using the first design; in the second supermarket, the second design was used. The product scanner at each supermarket tracked every buyer of soap over a 1-week period. The supermarkets recorded the last four digits of the scanner code for each of the five brands of soap the supermarket sold. The code for the General Products brand of soap is 9077 (the other codes are 4255, 3745, 7118 and 8855. After the trial period, the scanner data were transferred to a computer file. Because the first design is more expensive, management has decided to use this design only if there is sufficient evidence to allow it to conclude that design is better. Should management switch to the brightly coloured design or the simple green one?

SOLUTION:

IDENTIFY

The problem objective is to compare two populations. The first is the population of soap sales in supermarket 1, and the second is the population of soap sales in supermarket 2. The data are nominal because the values are 'buy General Products soap' and 'buy other companies' soap'. These two factors tell us that the parameter to be tested is the difference between two population proportions $p_1 - p_2$ (where p_1 and p_2 are the proportions of soap sales that are a General Products brand in supermarkets 1 and 2, respectively). Because we want to know whether there is enough evidence to adopt the brightly coloured design, the alternative hypothesis is

$$H_1: (p_1 - p_2) > 0$$

The null hypothesis must be

$$H_0: (p_1 - p_2) = 0$$

which tells us that this is an application of Case 1. Thus, the test statistic is

$$z = \frac{(\hat{p}_1 - \hat{p}_2)}{\sqrt{\hat{p}(1-\hat{p})\left(\frac{1}{n_1} + \frac{1}{n_2}\right)}}$$

COMPUTE

MANUALLY:

To compute the test statistic manually requires the statistics practitioner to tally the number of successes in each sample, where success is represented by the code 9077. Reviewing all the sales reveals that

$$x_1 = 180 \quad n_1 = 904 \quad x_2 = 155 \quad n_2 = 1038$$

The sample proportions are

$$\hat{p}_1 = \frac{180}{904} = 0.1991$$

and

$$\hat{p}_2 = \frac{155}{1038} = 0.1493$$

The pooled proportion is

$$\hat{p} = \frac{180 + 155}{904 + 1038} = \frac{335}{1942} = 0.1725$$

The value of the test statistic is

$$z = \frac{(\hat{p}_1 - \hat{p}_2)}{\sqrt{\hat{p}(1-\hat{p})\left(\frac{1}{n_1} + \frac{1}{n_2}\right)}} = \frac{(0.1991 - 0.1493)}{\sqrt{(0.1725)(1-0.1725)\left(\frac{1}{904} + \frac{1}{1038}\right)}} = 2.90$$

A 5% significance level seems to be appropriate. Thus, the rejection region is

$$z > z_{\alpha} = z_{0.05} = 1.645$$

EXCEL

	A	B	C	D
1	**z-Test: Two Proportions**			
2				
3			*Supermarket 1*	*Supermarket 2*
4	Sample Proportions		0.1991	0.1493
5	Observations		904	1038
6	Hypothesised Difference		0	
7	z Stat		2.90	
8	P(Z<=z) one tail		0.0019	
9	z Critical one-tail		1.6449	
10	P(Z<=z) two-tail		0.0038	
11	z Critical two-tail		1.96	

Instructions

1. Type or import the data into two adjacent columns*. (**Open** Xm12-13.)
2. Click **Add-Ins, Data Analysis Plus** and **Z-Test: 2 Proportions.**
3. Specify the **Variable 1 Range** (A1:A905) and the **Variable 2 Range** (B1:B1039). Type the **Code for Success** (9077), the **Hypothesised Difference** (0) and a value for α (0.05).

MINITAB

Test and CI for Two Proportions: Supermarket 1, Supermarket 2

Event = 9077

Variable	X	N	Sample p
Supermarket 1	180	904	0.199115
Supermarket 2	155	1038	0.149326

Difference = p (Supermarket 1) – p (Supermarket 2)
Estimate for difference: 0.0497894
95% lower bound for difference: 0.0213577
Test for difference = 0 (vs > 0): Z = 2.90 P-Value = 0.002

Instructions

1. Type or import the data into two adjacent columns. (**Open** Xm12-13.) Recode the data if necessary. (Minitab requires that there be only two codes and the higher value is deemed to be a success.)
2. Click **Stat, Basic Statistics** and **2 Proportions ....**
3. In the **Samples in different columns** specify the **First** (Supermarket 1) and **Second** (Supermarket 2) samples. Click **Options ....**
4. Type the value of the **Test difference** (00), specify the **Alternative** hypothesis (greater than) and click **Use pooled estimate of *p* for test.**

*If one or both columns contain an empty cell (representing missing data) the row will have to be removed.

Warning: If there are asterisks representing missing data, Minitab will be unable to conduct either the test or the estimate of the difference between two proportions. Click **Data** and **Sort,** which will eliminate the asterisks.

INTERPRET

The value of the test statistic is $z = 2.90$; its p-value is 0.0019. There is enough evidence to infer that the brightly coloured design is more popular than the simple design. As a result, it is recommended that management switch to the first design.

EXAMPLE 12.14

DATA Xm12-13

Test Marketing of Package Designs, Part 2

Suppose that in Example 12.13 the additional cost of the brightly coloured design requires that it outsell the simple design by more than 3%. Should management switch to the brightly coloured design?

SOLUTION:

IDENTIFY

The alternative hypothesis is

$$H_1: (p_1 - p_2) > 0.03$$

and the null hypothesis follows as

$$H_0: (p_1 - p_2) = 0.03$$

Because the null hypothesis specifies a non-zero difference, we would apply the Case 2 test statistic.

COMPUTE

MANUALLY:

The value of the test statistic is

$$z = \frac{(\hat{p}_1 - \hat{p}_2) - (p_1 - p_2)}{\sqrt{\frac{\hat{p}_1(1-\hat{p}_1)}{n_1} + \frac{\hat{p}_2(1-\hat{p}_2)}{n_2}}} = \frac{(0.1991 - 0.1493) - (0.03)}{\sqrt{\frac{0.1991(1-0.1991)}{904} + \frac{0.1493(1-0.1493)}{1038}}} = 1.15$$

EXCEL

	A	B	C	D
1	z-Test: Two Proportions			
2				
3			*Supermarket 1*	*Supermarket 2*
4	Sample Proportions		0.1991	0.1493
5	Observations		904	1038
6	Hypothesised Difference		0.03	
7	z Stat		1.14	
8	P(Z<=z) one tail		0.1261	
9	z Critical one-tail		1.6449	
10	P(Z<=z) two-tail		0.2522	
11	z Critical two-tail		1.96	

Instructions

Use the same commands we used previously, except specify that the **Hypothesised Difference** is 0.03. Excel will apply the Case 2 test statistic when a non-zero value is typed.

MINITAB

Test and CI for Two Proportions: Supermarket 1, Supermarket 2

Event = 9077

Variable	X	N	Sample p
Supermarket 1	180	904	0.199115
Supermarket 2	155	1038	0.149326

Difference = p (Supermarket 1) – p (Supermarket 2)
Estimate for difference: 0.0497894
95% lower bound for difference: 0.0213577
Test for difference = 0.03 (vs > 0.03): Z = 1.14 P-Value = 0.126

Instructions

Use the same commands detailed previously except at step 4, specify that the **Test difference** is 0.03 and do not click **Use pooled estimate of *p* for test.**

INTERPRET

There is not enough evidence to infer that the proportion of soap customers who buy the product with the brightly coloured design is more than 3% higher than the proportion of soap customers who buy the product with the simple design. In the absence of sufficient evidence, the analysis suggests that the product should be packaged using the simple design.

EXAMPLE 12.15

DATA Xm12-13

Test Marketing of Package Designs, Part 3

To help estimate the difference in profitability, the marketing manager in Examples 12.13 and 12.14 would like to estimate the difference between the two proportions. A confidence level of 95% is suggested.

SOLUTION:

IDENTIFY

The parameter is $p_1 - p_2$, which is estimated by the following confidence interval estimator:

$$(\hat{p}_1 - \hat{p}_2) \pm z_{\alpha/2}\sqrt{\frac{\hat{p}_1(1-\hat{p}_1)}{n_1} + \frac{\hat{p}_2(1-\hat{p}_2)}{n_2}}$$

COMPUTE

MANUALLY:

The sample proportions have already been computed. They are

$$\hat{p}_1 = \frac{180}{904} = 0.1991$$

and

$$\hat{p}_2 = \frac{155}{1038} = 0.1493$$

The 95% confidence interval estimate of $p_1 - p_2$ is

$$(\hat{p}_1 - \hat{p}_2) \pm z_{\alpha/2}\sqrt{\frac{\hat{p}_1(1-\hat{p}_1)}{n_1} + \frac{\hat{p}_2(1-\hat{p}_2)}{n_2}}$$

$$= (0.1991 - 0.1493) \pm 1.96\sqrt{\frac{0.1991(1-0.1991)}{904} + \frac{0.1493(1-0.1493)}{1038}}$$

$$= 0.0498 \pm 0.0339$$

$$LCL = 0.0159 \quad \text{and} \quad UCL = 0.0837$$

EXCEL

	A	B	C	D
1	**z-Estimate: Two Proportions**			
2				
3			*Supermarket 1*	*Supermarket 2*
4	Sample Proportions		0.1991	0.1493
5	Observations		904	1038
6				
7	LCL		0.0159	
8	UCL		0.0837	

Instructions

1. Type or import the data into two adjacent columns*. (**Open** Xm12-13.)
2. Click **Add-Ins**, **Data Analysis Plus** and **Z-Estimate: 2 Proportions.**
3. Specify the **Variable 1 Range** (A1:A905) and the **Variable 2 Range** (B1:B1039). Specify the **Code for Success** (9077) and a value for α (0.05).

MINITAB

Test and CI for Two Proportions: Supermarket 1, Supermarket 2

Event = 9077

Variable	X	N	Sample p
Supermarket 1	180	904	0.199115
Supermarket 2	155	1038	0.149326

Difference = p (Supermarket 1) – p (Supermarket 2)
Estimate for difference: 0.0497894
95% CI for difference: (0.0159109, 0.0836679)
Test for difference = 0 (vs not = 0): Z = 2.88 P-Value = 0.004

Instructions

Follow the commands to test hypotheses about two proportions. Specify the alternative hypothesis as **not equal** and do not click **Use pooled estimate of *p* for test.**

INTERPRET

We estimate that the market share for the brightly coloured design is between 1.59% and 8.37% larger than the market share for the simple design.

*If one or both columns contain an empty cell (representing missing data) the row must be removed.

EXERCISE

DO-IT-YOURSELF EXCEL

Construct Excel spreadsheets for each of the following:

12.74 z-test of $p_1 - p_2$. Inputs: sample proportions, sample sizes and hypothesised difference between two populations. Outputs: Test statistic, critical values and one- and two-tail p-values. Tools, **NORM.S.INV, NORM.S.DIST.**

12.75 z-estimate of $p_1 - p_2$. Inputs: sample proportions, sample sizes and confidence level. Outputs: Test statistic, one- and two-tail p-values. Tools, **NORM.S.INV.**

Developing an Understanding of Statistical Concepts

The next exercises are 'what-if analyses' designed to determine what happens to the test statistics and interval estimates when elements of the statistical inference change. These problems can be solved manually, using Do-It-Yourself Excel spreadsheets you created, or using Minitab.

12.76 Random samples from two binomial populations yielded the following statistics:

$\hat{p}_1 = 0.45 \quad n_1 = 100 \quad \hat{p}_2 = 0.40 \quad n_2 = 100$

a. Calculate the p-value of a test to determine whether we can infer that the population proportions differ.

b. Repeat part (a) increasing the sample sizes to 400.

c. Describe what happens to the p-value when the sample sizes increase.

12.77 These statistics were calculated from two random samples:

$\hat{p}_1 = 0.60 \quad n_1 = 225 \quad \hat{p}_2 = 0.55 \quad n_2 = 225$

a. Calculate the p-value of a test to determine whether there is evidence to infer that the population proportions differ.

b. Repeat part (a) with $\hat{p}_1 = 0.95$ and $\hat{p}_2 = 0.90$.

c. Describe the effect on the p-value of increasing the sample proportions.

d. Repeat part (a) with $\hat{p}_1 = 0.10$ and $\hat{p}_2 = 0.05$.

e. Describe the effect on the p-value of decreasing the sample proportions.

12.78 After sampling from two binomial populations we found the following.

$\hat{p}_1 = 0.18 \quad n_1 = 100 \quad \hat{p}_2 = 0.22 \quad n_2 = 100$

a. Estimate with 9 0 % confidence the difference in population proportions.

b. Repeat part (a) increasing the sample proportions to 0.48 and 0.52, respectively.

c. Describe the effects of increasing the sample proportions.

Applications

12.79 Many stores sell extended warranties for products they sell. These are very lucrative for store owners. To learn more about who buys these warranties, a random sample of a store's customers who recently purchased a product for which an extended warranty was available, was drawn. Among other variables each respondent reported whether they paid the regular price or a sale price and whether they purchased an extended warranty.

	Regular price	Sale price
Sample size	229	178
Number who bought extended warranty	47	25

Can we conclude at the 10% significance level that those who paid the regular price are more likely to buy an extended warranty?

12.80 A firm has classified its customers in two ways: (1) according to whether the account is overdue and (2) whether the account is new (less than 12 months) or old. To acquire information about which customers are paying on time and which are overdue, a random sample of 292 customer accounts was drawn. Each was categorised as a new account (less than 12 months) and old, and whether the customer has paid or is overdue. The results are summarised next.

	New Account	Old Account
Sample size	83	209
Overdue account	12	49

Is there enough evidence at the 5% significance level to infer that new and old accounts are different with respect to overdue accounts?

APPLICATIONS IN OPERATIONS MANAGEMENT

Pharmaceutical and Medical Experiments

When new products are developed, they are tested in several ways. First, does the new product work? Second, is it better than the existing product? Third, will customers buy it at a price that is profitable? Performing a customer survey or some other experiment that yields the information needed often tests the last question. This experiment is usually the domain of the marketing manager.

The other two questions are dealt with by the developers of the new product, which usually means the research department or the operations manager. When the product is a new drug, there are particular ways in which the data are gathered. The sample is divided into two groups. One group is assigned the new drug and the other is assigned a placebo, a pill that contains no medication. The experiment is often called 'double-blind' because neither the subjects who take the drug nor the physician/scientist who provides the drug knows whether any individual is taking the drug or the placebo. At the end of the experiment the data that are compiled allow statistics practitioners to do their work. Exercises 12.81-12.83 are examples of this type of statistical application.

12.81 Cold and allergy medicines have been available for a number of years. One serious side-effect of these medications is that they cause drowsiness, which makes them dangerous for industrial workers. In recent years, a non-drowsy cold and allergy medicine has been developed. One such product, Hismanal, is claimed by its manufacturer to be the first once-a-day non-drowsy allergy medicine. The non-drowsy part of the claim is based on a clinical experiment in which 1604 patients were given Hismanal and 1109 patients were given a placebo. Of the first group 7.1% reported drowsiness: of the second group, 6.4% reported drowsiness. Do these results allow us to infer at the 5% significance level that Hismanal's claim is false?

12.82 Plavix is a drug that is given to angioplasty patients to help prevent blood clots. A researcher at a medical university organized a study that involved 12 562 patients in 482 hospitals in 28 countries. All the patients had acute coronary syndrome, which produces mild heart attacks or unstable angina, chest pain that may precede a heart attack. The patients were divided into two equal groups. Group 1 received daily Plavix pills, while group 2 received a placebo. After 1 year, 9.3 % of patients on Plavix suffered a stroke or new heart attack, or had died of cardiovascular disease, compared with 11.5% of those who took the placebo.

a. Can we infer that Plavix is effective?

b. Describe your statistical analysis in a report to the marketing manager of the pharmaceutical company.

12.83 A study described in the *British Medical Journal* (January 2004) sought to determine whether exercise would help extend the lives of patients with heart failure. A sample of 801 patients with heart failure was recruited; 395 received exercise training and 406 did not. There were 88 deaths among the exercise group and 105 among those who did not exercise. Can researchers infer that exercise training reduces mortality?

CHAPTER SUMMARY

The inferential methods presented in this chapter address the problem of describing a single population. When the data are interval, the parameters of interest are the population mean σ and the population variance σ^2. The Student t-distribution is used to test and estimate the mean when the population standard deviation is unknown. The chi-squared distribution is used to make inferences about a population variance. When the data are nominal, the parameter to be tested and estimated is the population proportion p. The sample proportion follows an approximate normal distribution, which produces the test statistic and the interval estimator. We also discussed how to determine the sample size required to estimate a population proportion. We introduced market segmentation and described how statistical techniques presented in this chapter can be used to estimate the size of a segment.

In this chapter we also presented a variety of techniques that allow statistics practitioners to compare two populations. When the data are interval and we are interested in measures of central location, we encountered two more factors that must be considered when choosing the appropriate technique. When the samples are independent, we can use either the **equal-variances** or **unequal-variances formulas.** When the samples are **matched pairs**, we have only one set of formulas. We introduced the ***F*-statistic**, which is used to make inferences about two population variances. When the data are nominal, the parameter of interest is the difference between two proportions. For this parameter, we had two test statistics and one interval estimator.

SYMBOLS:

Symbol	Pronounced	Represents
ν	Nu	Degrees of freedom
X^2	chi squared	Chi-squared statistic
$\hat{p}$	*p* hat	Sample proportion
$\tilde{p}$	*p* tilde	Wilson estimator
s_p^2	*s* sub *p* squared	Pooled variance estimator
μ_D	mu sub *D* or mu *D*	Mean of the paired differences
$\bar{x}_D$	*x* bar sub *D* or *x* bar *D*	Sample mean of the paired differences
s_D	*s* sub *D* or *s D*	Sample standard deviation of the paired differences
n_D	*n* sub *D* or *n D*	Sample size of the paired differences

FORMULAS:

Test statistic for μ

$$t = \frac{\bar{x} - \mu}{s/\sqrt{n}}$$

Confidence interval estimator of σ

$$\bar{x} \pm t_{\alpha/2}\frac{s}{\sqrt{n}}$$

Test statistic for μ

$$\chi^2 = \frac{(n-1)s^2}{\sigma^2}$$

Confidence interval estimator of μ

$$\text{LCL} = \frac{(n-1)s^2}{\chi^2_{\alpha/2}}$$

$$\text{UCL} = \frac{(n-1)s^2}{\chi^2_1 - \alpha/2}$$

Test statistic for σ^2

$$z = \frac{\hat{p} - p}{\sqrt{p(1-p)/n}}$$

Confidence interval estimator of p

$$\hat{p} \pm z_{\alpha/2}\sqrt{\hat{p}(1-\hat{p})/n}$$

Sample size to estimate p

$$n = \left(\frac{z_{\alpha/2}\sqrt{\hat{p}(1-\hat{p})}}{B}\right)^2$$

Wilson estimator

$$\tilde{p} = \frac{x+2}{n+4}$$

Confidence interval estimator of p using the Wilson estimator

$$\tilde{p} \pm z_{\alpha/2}\sqrt{\tilde{p}(1-\tilde{p})/(n+4)}$$

Confidence interval estimator of the total of a large finite population

$$N\left[\bar{x} \pm t_{\alpha/2}\frac{s}{\sqrt{n}}\right]$$

Confidence interval estimator of the total number of successes in a large finite population

$$N\left[\hat{p} \pm z_{\alpha/2}\sqrt{\frac{\hat{p}(1-\hat{p})}{n}}\right]$$

Equal-variances *t*-test of $\mu_1 - \mu_2$

$$t = \frac{(\bar{x}_1 - \bar{x}_2) - (\mu_1 - \mu_2)}{\sqrt{s_p^2\left(\frac{1}{n_1} + \frac{1}{n_2}\right)}} \qquad \nu = n_1 + n_2 - 2$$

Equal-variances interval estimator of $(\mu_1 - \mu_2)$

$$(\bar{x}_1 - \bar{x}_2) \pm t_{\alpha/2}\sqrt{s_p^2\left(\frac{1}{n_1} + \frac{1}{n_2}\right)} \qquad \nu = n_1 + n_2 - 2$$

Unequal-variances *t*-test of $\mu_1 - \mu_2$

$$t = \frac{(\bar{x}_1 - \bar{x}_2) - (\mu_1 - \mu_2)}{\sqrt{\left(\frac{s_1^2}{n_1} + \frac{s_2^2}{n_2}\right)}} \qquad \nu = \frac{\left(\frac{s_1^2}{n_1} + \frac{s_2^2}{n_2}\right)^2}{\frac{\left(\frac{s_1^2}{n_1}\right)^2}{n_1 - 1} + \frac{\left(\frac{s_2^2}{n_2}\right)^2}{n_2 - 1}}$$

Unequal-variances interval estimator of $\mu_1 - \mu_2$

$$(\bar{x}_1 - \bar{x}_2) \pm t_{\alpha/2}\sqrt{\frac{s_1^2}{n_1} + \frac{s_2^2}{n_2}} \qquad \nu = \frac{\left(\frac{s_1^2}{n_1} + \frac{s_2^2}{n_2}\right)^2}{\frac{\left(\frac{s_1^2}{n_1}\right)^2}{n_1 - 1} + \frac{\left(\frac{s_2^2}{n_2}\right)^2}{n_2 - 1}}$$

t-test of μ_D

$$t = \frac{\bar{x}_D - \mu_D}{\frac{s_D}{\sqrt{n_D}}} \qquad \nu = n_D - 1$$

t-estimator of μ_D

$$\bar{x}_D \pm t_{\alpha/2}\frac{s_D}{\sqrt{n_D}} \qquad \nu = n_D - 1$$

F-test of σ_1^2/σ_2^2

$$F = \frac{s_1^2}{s_2^2} \qquad \nu_1 = n_1 - 1 \text{ and } \nu_2 = n_2 - 1$$

F-estimator of σ_1^2/σ_2^2

$$\text{LCL} = \left(\frac{s_1^2}{s_2^2}\right)\frac{1}{F_{\alpha/2,\nu_1,\nu_2}}$$

$$\text{UCL} = \left(\frac{s_1^2}{s_2^2}\right)F_{\alpha/2,\nu_2,\nu_1}$$

z-test and estimator of $p_1 - p_2$

$$\text{Case 1:} \quad z = \frac{(\hat{p}_1 - \hat{p}_2)}{\sqrt{\hat{p}(1 - \hat{p})\left(\frac{1}{n_1} + \frac{1}{n_2}\right)}}$$

$$\text{Case 2:} \quad z = \frac{(\hat{p}_1 - \hat{p}_2) - (p_1 - p_2)}{\sqrt{\frac{\hat{p}_1(1 - \hat{p}_1)}{n_1} + \frac{\hat{p}_2(1 - \hat{p}_2)}{n_2}}}$$

z-estimator of $p_1 - p_2$

$$(\hat{p}_1 - \hat{p}_2) \pm z_{\alpha/2}\sqrt{\frac{\hat{p}_1(1 - \hat{p}_1)}{n_1} + \frac{\hat{p}_2(1 - \hat{p}_2)}{n_2}}$$

We present the flow chart in Figure 12.14 as part of our ongoing effort to help you identify the appropriate statistical technique. This flow chart shows the techniques introduced in this chapter only. As we add new techniques in the upcoming chapters, we will expand this flow chart until it contains all the statistical inference techniques covered in this book. Use the flow chart to select the correct method in the chapter exercises that follow.

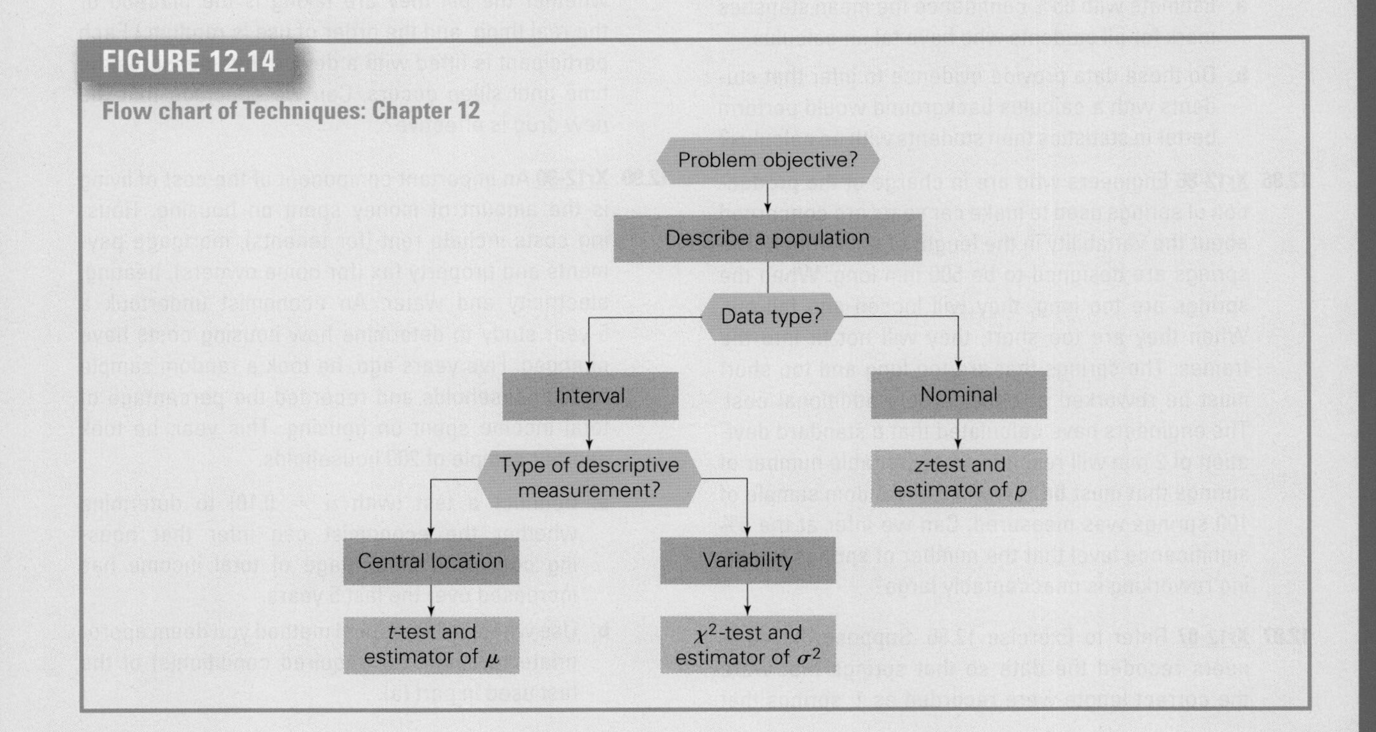

FIGURE 12.14

Flow chart of Techniques: Chapter 12

CHAPTER EXERCISES

The following exercises require the use of a computer and software. Use a 5% significance level unless specified otherwise.

12.84 Xr12-84 Robots are being used with increasing frequency on production lines to perform monotonous tasks. To determine whether a robot welder should replace human welders in producing automobiles, an experiment was performed. The time for the robot to complete a series of welds was found to be 38 seconds. A random sample of 20 workers was taken, and the time for each worker to complete the welds was measured. The mean was calculated to be 38 seconds, the same as the robot's time. However, the robot's time did not vary, whereas there was variation among the workers' times. An analysis of the production line revealed that if the variance exceeds 17 seconds^2, there will be problems. Perform an analysis of the data, and determine whether problems using human welders are likely.

12.85 Xr12-85 In a large university (with numerous campuses), the marks in an introductory statistics course are normally distributed with a mean of 68%. To determine the effect of requiring students to pass a calculus test (which at present is not a prerequisite), a random sample of 50 students who have taken calculus is given a statistics course. The marks out of 100 were recorded.

a. Estimate with 95% confidence the mean statistics mark for all students who have taken calculus.

b. Do these data provide evidence to infer that students with a calculus background would perform better in statistics than students with no calculus?

12.86 Xr12-86 Engineers who are in charge of the production of springs used to make car seats are concerned about the variability in the length of the springs. The springs are designed to be 500 mm long. When the springs are too long, they will loosen and fall out. When they are too short, they will not fit into the frames. The springs that are too long and too short must be reworked at considerable additional cost. The engineers have calculated that a standard deviation of 2 mm will result in an acceptable number of springs that must be reworked. A random sample of 100 springs was measured. Can we infer at the 5% significance level that the number of springs requiring reworking is unacceptably large?

12.87 Xr12-87 Refer to Exercise 12.86. Suppose the engineers recoded the data so that springs that were the correct length were recorded as 1, springs that were too long were recorded as 2 and springs that were too short were recorded as 3. Can we infer at the 10% significance level that less than 90% of the springs are the correct length?

12.88 Xr12-88 How important to your health are regular holidays? In a study a random sample of men and women were asked how frequently they take holidays. The men and women were divided into two groups each. The members of group 1 had suffered a heart attack; the members of group 2 had not. The number of days of holiday last year was recorded for each person. Can we infer that men and women who suffer heart attacks holiday less than those who did not suffer a heart attack?

12.89 Xr13-89 Research scientists at a pharmaceutical company have recently developed a new non-prescription sleeping pill. They decide to test its effectiveness by measuring the time it takes for people to fall asleep after taking the pill. Preliminary analysis indicates that the time to fall asleep varies considerably from one person to another. Consequently, they organise the experiment in the following way. A random sample of 100 volunteers who regularly suffer from insomnia is chosen. Each person is given one pill containing the newly developed drug and one placebo. (They do not know whether the pill they are taking is the placebo or the real thing, and the order of use is random.) Each participant is fitted with a device that measures the time until sleep occurs. Can we conclude that the new drug is effective?

12.90 Xr12-90 An important component of the cost of living is the amount of money spent on housing. Housing costs include rent (for tenants), mortgage payments and property tax (for home owners), heating, electricity and water. An economist undertook a 5-year study to determine how housing costs have changed. Five years ago, he took a random sample of 200 households and recorded the percentage of total income spent on housing. This year, he took another sample of 200 households.

a. Conduct a test (with $\alpha = 0.10$) to determine whether the economist can infer that housing cost as a percentage of total income has increased over the last 5 years.

b. Use whatever statistical method you deem appropriate to check the required condition(s) of the test used in part (a).

12.91 Xr12-91 In designing advertising campaigns to sell magazines, it is important to know how much time each of a number of demographic groups spends reading magazines. In a preliminary study, 40 people were randomly selected. Each was asked how much time per week he or she spends reading magazines; additionally, each was categorised by gender and by income level (high or low). The data are stored in the following way: column 1 = Time spent reading magazines per week in minutes for all respondents; column 2 = Gender (1 = Male, 2 = Female); column 3 = Income level (1 = Low, 2 = High).

a. Is there sufficient evidence at the 10% significance level to conclude that men and women differ in the amount of time spent reading magazines?

b. Is there sufficient evidence at the 10% significance level to conclude that high-income individuals devote more time to reading magazines than low-income people?

13 ANALYSIS OF VARIANCE

13-1 ONE-WAY ANALYSIS OF VARIANCE (ONE-WAY ANOVA)

13-2 MULTIPLE COMPARISONS

13-3 ANALYSIS OF VARIANCE EXPERIMENTAL DESIGNS

13-4 RANDOMISED BLOCK (TWO-WAY) ANALYSIS OF VARIANCE

13-5 TWO-FACTOR ANALYSIS OF VARIANCE

CAR ACCELERATION AND TEST DRIVER

DATA Xm13-00

Suppose that a car manufacturer wants to examine if the car acceleration is affected by the test drivers. A closed course is set up for four test drivers (Driver 1, Driver 2, Driver 3 and Driver 4) and the time it takes to accelerate from 0 to 100 km/h is recorded. For each test driver a sample of five is collected. **See Section 13-1f for the answer.**

INTRODUCTION

The technique presented in this chapter allows statistics practitioners to compare two or more populations of interval data. The technique is called the **analysis of variance (ANOVA),** and it is an extremely powerful and commonly used procedure. The analysis of variance technique determines whether differences exist between the means of more than two groups, which is actually a more general form of the t-test that is appropriate to use with three or more groups, but it can be also used with two groups. Ironically, the procedure works by analysing the sample variance, hence the name. We will examine several different forms of the technique.

One of the first applications of the analysis of variance was conducted in the 1920s to determine whether different treatments of fertiliser produced different crop yields. The terminology of that original experiment is still used. No matter what the experiment, the procedure is designed to determine whether there are significant differences between the **treatment means.**

13-1 ONE-WAY ANALYSIS OF VARIANCE (ONE-WAY ANOVA)

The analysis of variance is a procedure that tests to determine whether differences exist between two or more population means. The name of the technique derives from the way in which the calculations are performed; that is, the technique analyses the variance of the data to determine whether we can infer that the population means differ. As in Chapter 12, the experimental design is a determinant in identifying the proper method to use. In this section, we describe the procedure to apply when the samples are independently drawn and when there is only one grouping dimension. The technique is called the **one-way analysis of variance**. Figure 13.1 depicts the sampling process for drawing independent samples. The mean and variance of population j ($j = 1, 2, \dots, k$) are labelled μ_j and σ_j^2, respectively. Both parameters are unknown. For each population, we draw independent random samples. For each sample, we can compute the mean $\bar{x}_j$ and the variance s_j^2.

FIGURE 13.1

Sampling Scheme for Independent Samples

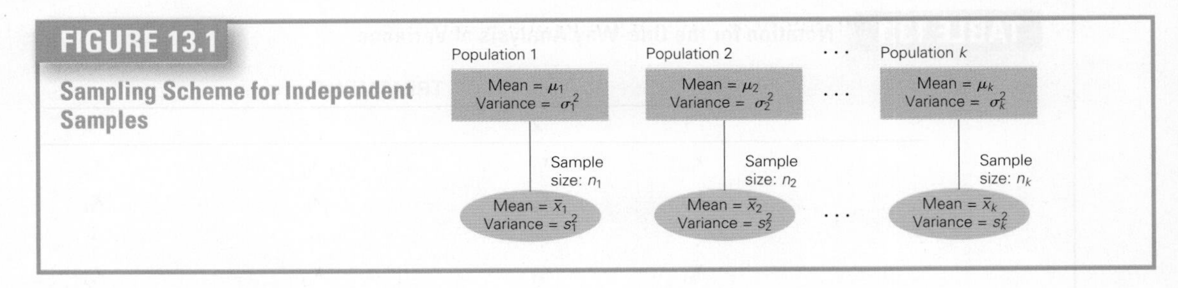

EXAMPLE 13.1

DATA Xm13-01

Proportion of Total Assets Invested in Stocks

In the last decade, stockbrokers have drastically changed the way they do business. Internet trading has become quite common, and online trades can cost as little as £6. It is now easier and cheaper to invest in the stock market than ever before. What are the effects of these changes? To help answer this question, assume that a financial analyst randomly sampled 366 British households and asked each to report the age category of the head of the household and the proportion of its financial assets that are invested in the stock market. The age categories are

Young (less than 35)
Early middle age (35 to 49)
Late middle age (50 to 65)
Senior (older than 65)

The analyst was particularly interested in determining whether the ownership of stocks varied by age. Some of the data are listed next. Do these data allow the analyst to determine that there are differences in stock ownership between the four age groups?

Young	Early Middle Age	Late Middle Age	Senior
24.8	28.9	81.5	66.8
35.5	7.3	0.0	77.4
68.7	61.8	61.3	32.9
42.2	53.6	0.0	74.0
⋮	⋮	⋮	⋮

SOLUTION:

You should confirm that the data are interval (percentage of total assets invested in the stock market) and that the problem objective is to compare four populations (age categories). The parameters are the four population means: μ_1, μ_2, μ_3 and μ_4. The null hypothesis will state that there are no differences between the population means. Hence,

$$H_0: \quad \mu_1 = \mu_2 = \mu_3 = \mu_4$$

The analysis of variance determines whether there is enough statistical evidence to show that the null hypothesis is false. Consequently, the alternative hypothesis will always specify the following:

H_1: At least two means differ

The next step is to determine the test statistic, which is somewhat more involved than the test statistics we have introduced thus far. The process of performing the analysis of variance is facilitated by the notation in Table 13.1.

TABLE 13.1 Notation for the One-Way Analysis of Variance

	TREATMENT					
	1	2		j		k
	x_{11}	x_{12}	...	x_{1j}	...	x_{1k}
	x_{21}	x_{22}	...	x_{2j}	...	x_{2k}
	⋮	⋮		⋮		⋮
	$x_{n_1 1}$	$x_{n_2 2}$		$x_{n_j j}$		$x_{n_k k}$
Sample size	n_1	n_2		n_j		n_k
Sample mean	$\bar{x}_1$	$\bar{x}_2$		$\bar{x}_j$		$\bar{x}_k$

x_{ij} = ith observation of the jth sample

n_j = number of observations in the sample taken from the jth population

$$\bar{x}_j = \text{mean of the } j\text{th sample} = \frac{\sum_{i=1}^{n_j} x_{ij}}{n_j}$$

$$\bar{\bar{x}} = \text{grand mean of all the observations} = \frac{\sum_{j=1}^{k}\sum_{i=1}^{n_j} x_{ij}}{n} \text{ where } n = n_1 + n_2 + \dots + n_k \text{ and } k \text{ is the number of populations}$$

The variable X is called the **response** (or **dependent**) **variable** and its values are called **responses.** The unit that we measure is called an **experimental** (or **independent** variable) **unit.** In this example, the response variable is the percentage of assets invested in stocks, and the experimental units are the heads of households sampled. The criterion by which we classify the populations is called a **factor**. Each population is called a factor **level**. The factor in Example 13.1 is the age category of the head of the household and there are four levels. Later in this chapter, we will discuss an experiment where the populations are classified using two factors. In this section, we deal with single-factor experiments only.

Test Statistic

The test statistic is computed in accordance with the following rationale. If the null hypothesis is true, the population means would all be equal. We would then expect that the sample means would be close to one another. If the alternative hypothesis is true, however, there would be large differences between some of the sample means. The statistic that measures the proximity of the sample means to each other is called the **between-treatments variation**; it is denoted **SST**, which stands for **sum of squares for treatments**.

Sum of Squares for Treatments

$$SST = \sum_{j=1}^{k} n_j(\bar{x}_j - \bar{\bar{x}})^2$$

As you can deduce from this formula, if the sample means are close to each other, all of the sample means would be close to the **grand mean** ($\bar{\bar{x}}$ is calculated across all individuals and all groups); as a result, SST would be small. In fact, SST achieves its smallest value (zero) when all the sample means are equal. In other words, if

$$\bar{x}_1 = \bar{x}_2 = \cdots = \bar{x}_k$$

then

$$SST = 0$$

It follows that a small value of SST supports the null hypothesis. In this example, we compute the sample means and the grand mean as

$$\bar{x}_1 = 44.40$$
$$\bar{x}_2 = 52.47$$
$$\bar{x}_3 = 51.14$$
$$\bar{x}_4 = 51.84$$
$$\bar{\bar{x}} = 49.96$$

The sample sizes are

$$n_1 = 84$$
$$n_2 = 131$$
$$n_3 = 93$$
$$n_4 = 58$$
$$n = n_1 + n_2 + n_3 + n_4 = 84 + 131 + 93 + 58 = 366$$

Then

$$\begin{aligned} SST &= \sum_{j=1}^{k} n_j(\bar{x}_j - \bar{\bar{x}})^2 \\ &= 84(44.40 - 49.96)^2 + 131(52.47 - 49.96)^2 \\ &\quad + 93(51.14 - 49.96)^2 + 58(51.84 - 49.96)^2 \\ &= 3758.8 \end{aligned}$$

If large differences exist between the sample means, at least some sample means differ considerably from the grand mean, producing a large value of SST. It is then reasonable to reject the null hypothesis in favour of the alternative hypothesis. The key question to be answered in this test (as in all other statistical tests) is: How large does the statistic have to be for us to justify rejecting the null hypothesis? In our example, SST = 3758.8. Is this value large enough to indicate that the population means differ? To answer this question, we need to know how much variation exists in the percentage of assets, which is measured by the **within-treatments variation**, which is denoted by **SSE (sum of squares for error)**. The within-treatments variation provides a measure of the amount of variation in the response variable that is not caused by the treatments. In this example, we are trying to determine whether the percentages of total assets invested in stocks vary by the age of the head of the household. However, other variables also affect the responses variable. We would expect that variables such as household income, occupation, and the size of the family would play a role in determining how much money families invest in stocks. All of these (as well as others we may not even be able to identify) are sources of variation, which we would group together and call the error. This source of variation is measured by the sum of squares for error.

Sum of Squares for Error

$$SSE = \sum_{j=1}^{k}\sum_{i=1}^{n_j}(x_{ij} - \bar{x}_j)^2$$

When SSE is partially expanded, we get

$$SSE = \sum_{i=1}^{n_1}(x_{i1} - \bar{x}_1)^2 + \sum_{i=1}^{n_2}(x_{i2} - \bar{x}_2)^2 + \cdots + \sum_{i=1}^{n_k}(x_{ik} - \bar{x}_k)^2$$

If you examine each of the k components of SSE, you will see that each is a measure of the variability of that sample. If we divide each component by $n_j - 1$, we obtain the sample variances. We can express this by rewriting SSE as

$$SSE = (n_1 - 1)s_1^2 + (n_2 - 1)s_2^2 + \cdots + (n_k - 1)s_k^2$$

where s_j^2 is the sample variance of sample j. SSE is thus the combined or pooled variation of the k samples. This is an extension of a calculation we made in Section 12-4, where we tested and estimated the difference between two means using the pooled estimate of the common population variance (denoted s_p^2). One of the required conditions for that statistical technique is that the population variances are equal. That same condition is now necessary for us to use SSE; that is, we require that

$$\sigma_1^2 = \sigma_2^2 = \cdots = \sigma_k^2$$

Returning to our example, we calculate the sample variances as follows:

$$s_1^2 = 386.55$$

$$s_2^2 = 469.44$$

$$s_3^2 = 471.82$$

$$s_4^2 = 444.79$$

Thus,

$$\begin{aligned} SSE &= (n_1 - 1)s_1^2 + (n_2 - 1)s_2^2 + (n_3 - 1)s_3^2 + (n_4 - 1)s_4^2 \\ &= (84 - 1)(386.55) + (131 - 1)(469.44) \\ &\quad + (93 - 1)(471.82) + (58 - 1)(444.79) \\ &= 161\,871.3 \end{aligned}$$

The next step is to compute quantities called the **mean squares**. The **mean square for treatments** is computed by dividing SST by the number of treatments minus 1.

Mean Square for Treatments

$$MST = \frac{SST}{k-1}$$

The **mean square for error** is determined by dividing SSE by the total sample size (labelled n) minus the number of treatments.

Mean Square for Error

$$MSE = \frac{SSE}{n-k}$$

Finally, the test statistic is defined as the ratio of the two mean squares.

Test Statistic

$$F = \frac{MST}{MSE}$$

Sampling Distribution of the Test Statistic

The test statistic is F-distributed with $k - 1$ and $n - k$ degrees of freedom, provided that the response variable is normally distributed. In Section 8-4, we introduced the F-distribution, and in Chapter 12 we used it to test and estimate the ratio of two population variances. The test statistic in that application was the ratio of two sample variances s_1^2 and s_2^2. If you examine the definitions of SST and SSE, you will see that both measure variation similar to the numerator in the formula used to calculate the sample variance s^2 used throughout this book. When we divide SST by $k - 1$ and SSE by $n - k$ to calculate MST and MSE, respectively, we are actually computing unbiased estimators of the common population variance, assuming (as we do) that the null hypothesis is true. Thus, the ratio $F =$ MST/MSE is the ratio of two sample variances. The degrees of freedom for this application are the denominators in the mean squares; that is, $\nu_1 = k - 1$ and $\nu_2 = n - k$. For Example 13.1, the degrees of freedom are

$$\nu_1 = k - 1 = 4 - 1 = 3$$

$$\nu_2 = n - k = 366 - 4 = 362$$

In our example, we found

$$MST = \frac{SST}{k-1} = \frac{3\,758.8}{3} = 1\,252.92$$

$$MSE = \frac{SSE}{n-k} = \frac{161\,871.3}{362} = 447.16$$

$$F = \frac{MST}{k - MSE} = \frac{1\,252.92}{447.16} = 2.80$$

Rejection Region and *p*-Value

The purpose of calculating the *F*-statistic is to determine whether the value of SST is large enough to reject the null hypothesis. As you can see, if SST is large, *F* will be large. Hence, we reject the null hypothesis only if

$$F > F_{\alpha,k-1,n-k}$$

If we let $\alpha = 0.05$, the rejection region for Example 13.1 is

$$F > F_{\alpha,k-1,n-k} = F_{0.05,3,362} \approx F_{0.05,3,\infty} = 2.61$$

We found the value of the test statistic to be $F = 2.80$. Thus, there is enough evidence to infer that the mean percentage of total assets invested in the stock market differs between the four age groups.

The *p*-value of this test is

$$P(F > 2.80)$$

A computer is required to calculate this value, which is 0.0405.

Figure 13.2 depicts the sampling distribution for Example 13.1.

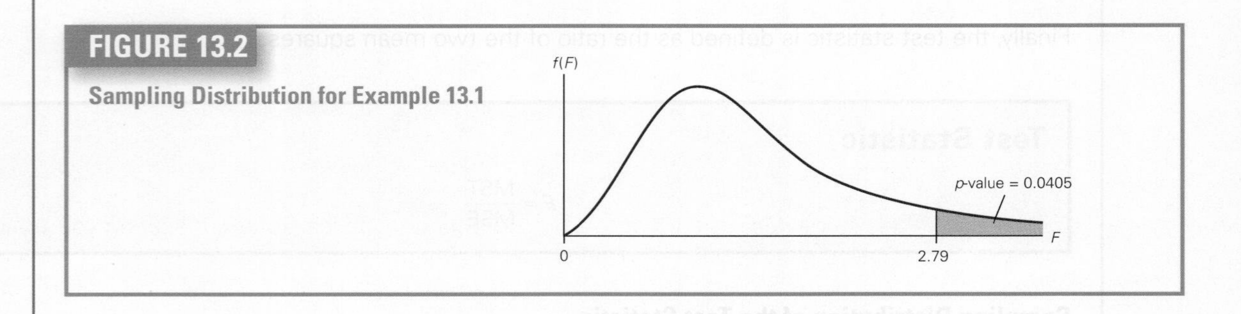

FIGURE 13.2 Sampling Distribution for Example 13.1

The results of the analysis of variance are usually reported in an **analysis of variance (ANOVA) table**. Table 13.2 shows the general organisation of the ANOVA table, and Table 13.3 shows the ANOVA table for Example 13.1.

TABLE 13.2 ANOVA Table for the One-Way Analysis of Variance

SOURCE OF VARIATION	DEGREES OF FREEDOM	SUMS OF SQUARES	MEAN SQUARES	*F*-STATISTIC
Treatments	$k-1$	SST	MST = SST/($k-1$)	F = MST/MSE
Error	$n-k$	SSE	MSE = SSE/($n-k$)	
Total	$n-1$	SS (Total)		

TABLE 13.3 ANOVA Table for Example 13.1

SOURCE OF VARIATION	DEGREES OF FREEDOM	SUMS OF SQUARES	MEAN SQUARES	*F*-STATISTIC
Treatments	3	3 758.8	1 252.92	2.80
Error	362	161 871.3	447.16	
Total	365	165 630.1		

The terminology used in the ANOVA table (and for that matter, in the test itself) is based on the partitioning of the sum of squares. Such partitioning is derived from the following equation (whose validity can be demonstrated by using the rules of summation):

$$\sum_{j=1}^{k}\sum_{i=1}^{n_j}(x_{ij}-\bar{\bar{x}})^2 = \sum_{j=1}^{k}n_j(\bar{x}_j-\bar{\bar{x}})^2 + \sum_{j=1}^{k}\sum_{i=1}^{n_j}(x_{ij}-\bar{x}_j)^2$$

The term on the left represents the **total variation** of all the data. This expression is denoted **SS(Total)**. If we divide SS(Total) by the total sample size minus 1 (that is, by $n - 1$), we would obtain the sample variance (assuming that the null hypothesis is true). The first term on the right of the equal sign is SST, and the second term is SSE. As you can see, the total variation SS(Total) is partitioned into two sources of variation. The sum of squares for treatments (SST) is the variation attributed to the differences between the treatment means, whereas the sum of squares for error (SSE) measures the variation within the samples. The preceding equation can be restated as

$$\text{SS(Total)} = \text{SST} + \text{SSE}$$

The test is then based on the comparison of the mean squares of SST and SSE.

Recall that in discussing the advantages and disadvantages of the matched pairs experiment in Section 12-5, we pointed out that statistics practitioners frequently seek ways to reduce or explain the variation in a random variable. In the analysis of variance introduced in this section, the sum of squares for treatments explains the variation attributed to the treatments (age categories). The sum of squares for error measures the amount of variation that is unexplained by the different treatments. If SST explains a significant portion of the total variation, we conclude that the population means differ. In Sections 13-4 and 13-5 we will introduce other experimental designs of the analysis of variance – designs that attempt to reduce or explain even more of the variation.

If you have felt some appreciation of the computer and statistical software sparing you the need to manually perform the statistical techniques in earlier chapters, your appreciation should now grow, because the computer will allow you to avoid the incredibly time-consuming and boring task of performing the analysis of variance by hand. As usual, we have solved Example 13.1 using Excel and Minitab, whose outputs are shown here.

COMPUTE

EXCEL

	A	B	C	D	E	F	G
1	Anova: Single Factor						
2							
3	SUMMARY						
4	*Groups*	*Count*	*Sum*	*Average*	*Variance*		
5	Young	84	3729.5	44.40	386.55		
6	Early Middle Age	131	6873.9	52.47	469.44		
7	Late Middle Age	93	4755.9	51.14	471.82		
8	Senior	58	3006.6	51.84	444.79		
9							
10							
11	ANOVA						
12	*Source of Variation*	*SS*	*df*	*MS*	*F*	*P-value*	*F crit*
13	Between Groups	3741.4	3	1247.12	2.79	0.0405	2.6296
14	Within Groups	161871.0	362	447.16			
15							
16	Total	165612.3	365				

Instructions

1. Type or import the data into adjacent columns. (**Open** Xm13-01.)
2. Click **Data, Data Analysis** and **Anova: Single Factor.**
3. Specify the **Input Range** (A1:D132) and a value for α (0.05).

MINITAB

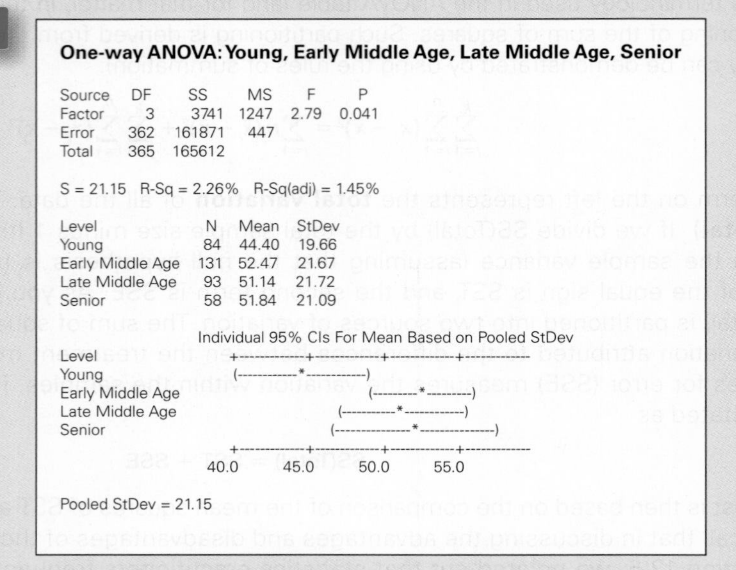

One-way ANOVA: Young, Early Middle Age, Late Middle Age, Senior

Source	DF	SS	MS	F	P
Factor	3	3741	1247	2.79	0.041
Error	362	161871	447		
Total	365	165612			

S = 21.15 R-Sq = 2.26% R-Sq(adj) = 1.45%

Level	N	Mean	StDev
Young	84	44.40	19.66
Early Middle Age	131	52.47	21.67
Late Middle Age	93	51.14	21.72
Senior	58	51.84	21.09

```
                   Individual 95% CIs For Mean Based on Pooled StDev
Level              --+---------+---------+---------+-------
Young               (-----------*-----------)
Early Middle Age                      (---------*---------)
Late Middle Age                    (-----------*-----------)
Senior                            (---------------*---------------)
                   --+---------+---------+---------+-------
                   40.0      45.0      50.0      55.0

Pooled StDev = 21.15
```

Instructions

If the data are unstacked:

1. Type or import the data. (**Open** Xm13-01.)
2. Click **Stat**, **ANOVA** and **One-way (Unstacked)**
3. **In** the **Responses (in separate columns)** box, type or select the variable names of the treatments (Young, Early Middle Age, Late Middle Age, Senior).

If the data are stacked:

1. Type or import the data in two columns.
2. Click **Stat**, **ANOVA** and **One-way**
3. Type the variable name of the response variable and the name of the factor variable.

INTERPRET

The value of the test statistic is $F = 2.80$ and its *p*-value is 0.0405, which means there is evidence to infer that the percentage of total assets invested in stocks are different in at least two of the age categories.

Note that in this example the data are observational. We cannot conduct a controlled experiment. To do so would require the financial analyst to randomly assign households to each of the four age groups, which is impossible.

Incidentally, when the data are obtained through a controlled experiment in the one-way analysis of variance, we call the experimental design the **completely randomised design** of the analysis of variance.

13-1a Checking the Required Conditions

The *F-test* of the analysis of variance requires that the random variable be normally distributed with equal variances. The normality requirement is easily checked graphically by producing the histograms for each sample. From the Excel histograms in Figure 13.3 we can see that there is no reason to believe that the requirement is not satisfied.

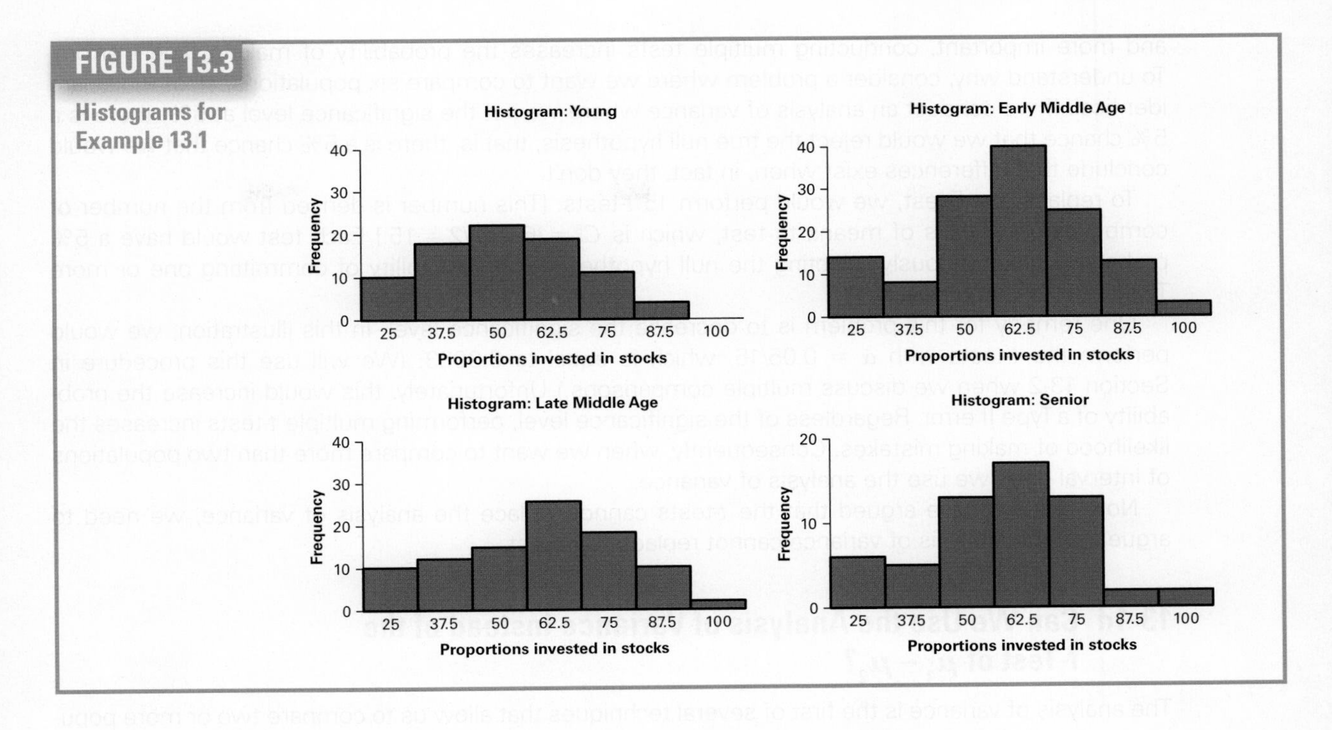

FIGURE 13.3 Histograms for Example 13.1

The equality of variances is examined by printing the sample standard deviations or variances. Excel output includes the variances, and Minitab calculates the standard deviations. The similarity of sample variances allows us to assume that the population variances are equal. In the website Appendix Bartlett's Test, we present a statistical procedure designed to test for the equality of variances.

13-1b Violation of the Required Conditions

If the data are not normally distributed, we can replace the one-way analysis of variance with its nonparametric counterpart, which is the Kruskal–Wallis Test.*) If the population variances are unequal, we can use several methods to correct the problem. However, these corrective measures are beyond the level of this book.

13-1c Can We Use the *t*-Test of the Difference between Two Means Instead of the Analysis of Variance?

The analysis of variance tests to determine whether there is evidence of differences between two or more population means. The *t*-test of $\mu_1 - \mu_2$ determines whether there is evidence of a difference between two population means. The question arises, can we use *t*-tests instead of the analysis of variance? In other words, instead of testing all the means in one test as in the analysis of variance, why not test each pair of means? In Example 13.1 we would test $(\mu_1 - \mu_2)$, $(\mu_1 - \mu_3)$, $(\mu_1 - \mu_4)$, $(\mu_2 - \mu_3)$, $(\mu_2 - \mu_4)$, If we find no evidence of a difference in each test, we would conclude that none of the means differ. If there was evidence of a difference in at least one test, we would conclude that some of the means differ.

There are two reasons why we don't use multiple *t*-tests instead of one *F*-test. First, we would have to perform many more calculations. Even with a computer, this extra work is tedious. Second,

*Instructors who wish to teach the use of nonparametric techniques for testing the difference between two or more means when the normality requirement is not satisfied should use the website Appendix Kruskal–Wallis Test and Friedman Test.

and more important, conducting multiple tests increases the probability of making Type I errors. To understand why, consider a problem where we want to compare six populations, all of which are identical. If we conduct an analysis of variance where we set the significance level at 5%, there is a 5% chance that we would reject the true null hypothesis; that is, there is a 5% chance that we would conclude that differences exist when, in fact, they don't.

To replace the *F*-test, we would perform 15 *t*-tests. [This number is derived from the number of combinations of pairs of means to test, which is $C_2^6 = (6 \times 5)/2 = 15$.] Each test would have a 5% probability of erroneously rejecting the null hypothesis. The probability of committing one or more Type I errors is about 54%.*

One remedy for this problem is to decrease the significance level. In this illustration, we would perform the *t*-tests with $\alpha = 0.05/15$, which is equal to 0.0033. (We will use this procedure in Section 13-2 when we discuss multiple comparisons.) Unfortunately, this would increase the probability of a Type II error. Regardless of the significance level, performing multiple *t*-tests increases the likelihood of making mistakes. Consequently, when we want to compare more than two populations of interval data, we use the analysis of variance.

Now that we have argued that the *t*-tests cannot replace the analysis of variance, we need to argue that the analysis of variance cannot replace the *t*-test.

13-1d Can We Use the Analysis of Variance Instead of the *t*-Test of $\mu_1 - \mu_2$?

The analysis of variance is the first of several techniques that allow us to compare two or more populations. Most of the examples and exercises deal with more than two populations. However, it should be noted that, like all other techniques whose objective is to compare two or more populations, the analysis of variance can be used to compare only two populations. If that is the case, then why do we need techniques to compare exactly two populations? Specifically, why do we need the *t*-test of $\mu_1 - \mu_2$ when the analysis of variance can be used to test two population means?

To understand why, we still need the *t*-test to make inferences about $\mu_1 - \mu_2$. Suppose that we plan to use the analysis of variance to test two population means. The null and alternative hypotheses are

$$H_0: \mu_1 = \mu_2$$

$$H_1: \text{At least two means differ}$$

Of course, the alternative hypothesis specifies that $\mu_1 \neq \mu_2$. However, if we want to determine whether μ_1 is greater than μ_2 (or vice versa), we cannot use the analysis of variance because this technique allows us to test for a difference only. Thus, if we want to test to determine whether one population mean exceeds the other, we must use the *t*-test of $\mu_1 - \mu_2$ (with $\sigma_1^2 = \sigma_2^2$). Moreover, the analysis of variance requires that the population variances are equal. If they are not, we must use the unequal variances test statistic.

13-1e Relationship between the *F*-Statistic and the *t*-Statistic

It is probably useful for you to understand the relationship between the *t*-statistic and the *F*-statistic. The test statistic for testing hypotheses about $\mu_1 - \mu_2$ with equal variances is

$$t = \frac{(\bar{x}_1 - \bar{x}_2) - (\mu_1 - \mu_2)}{\sqrt{s_p^2\left(\frac{1}{n_1} + \frac{1}{n_2}\right)}}$$

*The probability of committing at least one Type I error is computed from a binomial distribution with $n = 15$ and $p = 0.05$. Thus, $P(X \geq 1) = 1 - P(X = 0) = 1 - 0.463 = 0.537$.

If we square this quantity, the result is the F-statistic: $F = t^2$. To illustrate this point, we will redo the calculation of the test statistic in Example 12.6 using the analysis of variance. Recall that because we were able to assume that the population variances were equal, the test statistic was as follows:

$$t = \frac{(6.63 - 3.72) - 0}{\sqrt{40.42\left(\frac{1}{50} + \frac{1}{50}\right)}} = 2.29$$

Using the analysis of variance (the Excel output is shown here; Minitab's is similar), we find that the value of the test statistic is $F = 5.23$, which is $(2.29)^2$. Notice though that the analysis of variance p-value is 0.0243, which is twice the t-test p-value, which is 0.0122. The reason: the analysis of variance is conducting a test to determine whether the population means *differ*. If Example 12.6 had asked to determine whether the means differ, we would have conducted a two-tail test and the p-value would be 0.0243, the same as the analysis of variance p-value.

Excel Analysis of Variance Output for Example 12.6

	A	B	C	D	E	F	G
1	Anova: Single Factor						
2							
3	SUMMARY						
4	*Groups*	*Count*	*Sum*	*Average*	*Variance*		
5	Direct	50	331.6	6.63	37.49		
6	Broker	50	186.2	3.72	43.34		
7							
8							
9	ANOVA						
10	*Source of Variation*	*SS*	*df*	*MS*	*F*	*P-value*	*F crit*
11	Between Groups	211.4	1	211.41	5.23	0.0243	3.9381
12	Within Groups	3960.5	98	40.41			
13							
14	Total	4172.0	99				

13-1f Developing an Understanding of Statistical Concepts

Conceptually and mathematically, the F-test of the independent samples' single-factor analysis of variance is an extension of the t-test of $\mu_1 - \mu_2$. Moreover, if we simply want to determine whether a difference between two means exists, we can use the analysis of variance. The advantage of using the analysis of variance is that we can partition the total sum of squares, which enables us to measure how much variation is attributable to differences between populations and how much variation is attributable to differences within populations. As we pointed out in Section 12-6, explaining the variation is an extremely important topic, one that we will see again in other experimental designs of the analysis of variance and in regression analysis (Chapters 15 and 16).

CAR ACCELERATION AND TEST DRIVER: SOLUTION

IDENTIFY

The variable is the time the car takes to accelerate from 0 to 100 km/h. The problem objective is to compare four populations (test drivers) and the experimental design is independent samples. Thus, we apply the one-way analysis of variance.

(continued)

COMPUTE

EXCEL

	A	B	C	D	E	F	G
1	Anova: Single Factor						
2							
3	SUMMARY						
4	*Groups*	*Count*	*Sum*	*Average*	*Variance*		
5	Driver 1	5	29.4	5.88	0.037		
6	Driver 2	5	27.9	5.58	0.027		
7	Driver 3	5	29.6	5.92	0.017		
8	Driver 4	5	27.8	5.56	0.023		
9							
10							
11	ANOVA						
12	*Source of Variation*	*SS*	*df*	*MS*	*F*	*P-value*	*F crit*
13	Between Groups	0.5495	3	0.183	7.045	0.0031	3.239
14	Within Groups	0.416	16	0.026			
15							
16	Total	0.9655	19				

MINITAB

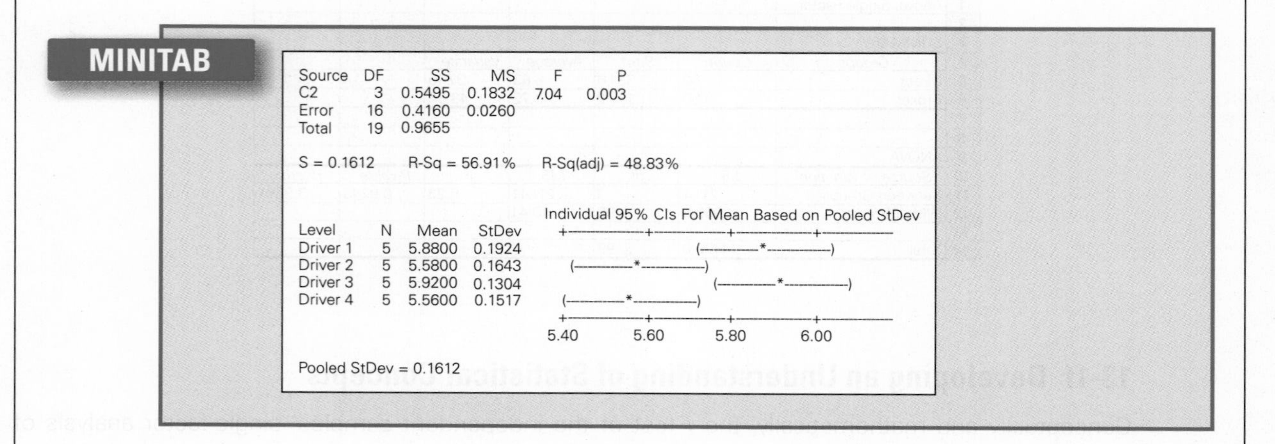

Source	DF	SS	MS	F	P
C2	3	0.5495	0.1832	7.04	0.003
Error	16	0.4160	0.0260		
Total	19	0.9655			

S = 0.1612 R-Sq = 56.91% R-Sq(adj) = 48.83%

```
                                Individual 95% CIs For Mean Based on Pooled StDev
Level     N    Mean    StDev   +---------+---------+---------+---------
Driver 1  5  5.8800   0.1924             (--------*--------)
Driver 2  5  5.5800   0.1643   (--------*--------)
Driver 3  5  5.9200   0.1304               (--------*--------)
Driver 4  5  5.5600   0.1517  (--------*--------)
                               +---------+---------+---------+---------
                              5.40      5.60      5.80      6.00

Pooled StDev = 0.1612
```

INTERPRET

The *p*-value is 0.0031. There is sufficient evidence to infer that the times differ between the four test drivers

Factors That Identify the One-Way Analysis of Variance

1. **Problem objective:** Compare two or more populations.
2. **Data type:** Interval.
3. **Experimental design:** Independent samples.

EXERCISES

Developing an Understanding of Statistical Concepts

Exercises 13.1–13.3 are 'what-if' analyses designed to determine what happens to the test statistic when the means, variances and sample sizes change. These problems can be solved manually or by creating an Excel worksheet.

13.1 A statistics practitioner calculated the following statistics:

	Treatment		
Statistic	1	2	3
n	5	5	5
$\bar{x}$	10	15	20
s^2	50	50	50

a. Complete the ANOVA table.

b. Repeat part (a) changing the sample sizes to 10 each.

c. Describe what happens to the F-statistic when the sample sizes increase.

13.2 You are given the following statistics:

	Treatment		
Statistic	1	2	3
n	4	4	4
$\bar{x}$	20	22	25
s^2	10	10	10

a. Complete the ANOVA table.

b. Repeat part (a) changing the variances to 25 each.

c. Describe the effect on the F-statistic of increasing the sample variances.

13.3 The following statistics were calculated:

	Treatment			
Statistic	1	2	3	4
n	10	14	11	18
$\bar{x}$	30	35	33	40
s^2	10	10	10	10

a. Complete the ANOVA table.

b. Repeat part (a) changing the sample means to 130, 135, 133 and 140.

c. Describe the effect on the F-statistic of increasing the sample means by 100.

Applications

13.4 Xr13-04 How does an MBA degree specialism affect the number of job offers received? An MBA student randomly sampled four recent graduates, one each in finance, marketing and management, and asked them to report the number of job offers. Can we conclude at the 5% significance level that there are differences in the number of job offers between the three MBA specialisms?

Finance	Marketing	Management
3	1	8
1	5	5
4	3	4
1	4	6

13.5 Xr13-05 A consumer organisation was concerned about the differences between the advertised sizes of containers and the actual amount of product. In a preliminary study, six packages of three different brands of margarine that are supposed to contain 500 ml were measured. The differences from 500 ml are listed here. Do these data provide sufficient evidence to conclude that differences exist between the three brands? Use $\alpha = 0.01$.

Brand 1	Brand 2	Brand 3
1	2	1
3	2	2
3	4	4
0	3	2
1	0	3
0	4	4

13.6 Xr13-06 Many college and university students obtain summer jobs. A statistics lecturer wanted to determine whether students in different degree programmes earn different amounts. A random sample of five students in the BA, BSc and BBA programmes were asked to report what they earned the previous summer. The results (in €1000s) are listed here.

Can the lecturer infer at the 5% significance level that students in different degree programmes differ in their summer earnings?

BA	BSc	BBA
3.3	3.9	4.0
2.5	5.1	6.2
4.6	3.9	6.3
5.4	6.2	5.9
3.9	4.8	6.4

13.7 <u>**Xr13-07**</u> Spam is the price we pay for being able to easily communicate by email. Does spam affect everyone equally? In a preliminary study, university lecturers, administrators and students were randomly sampled. Each person was asked to count the number of spam messages received that day. The results follow. Can we infer at the 2.5% significance level that the differing university communities differ in the amount of spam they receive in their emails?

Lecturers	Administrators	Students
7	5	12
4	9	4
0	12	5
3	16	18
18	10	15

13.8 <u>**Xr13-08**</u> A management scientist believes that one way of judging whether a computer comes equipped with enough memory is to determine the age of the computer. In a preliminary study, random samples of computer users were asked to identify the brand of computer and its age (in months). The categorised responses are shown here. Do these data provide sufficient evidence to conclude that there are differences in age between the computer brands? (Use $\alpha = 0.05$.)

IBM	Dell	Hewlett-Packard	Other
17	8	6	24
10	4	15	12
13	21	8	15

Exercises 13.9–13.15 require the use of a computer and software. Use a 5% significance level unless specified otherwise.

13.09 <u>**Xr13-09**</u> Are proficiency test scores affected by the education of the child's parents? For example, TOEFL iBT tests are administered to a sample of students and pupils. Test scores can range from 0 to 120. To answer this question, a random sample of 9-year-old children was drawn. Each child's test score and the educational level of the parent with the higher level were recorded. The education categories are less than high (or secondary) school, high school graduate, some college and college graduate. Can we infer that there are differences in test scores between children whose parents have different educational levels?

13.10 <u>**Xr13-10**</u> A manufacturer of outdoor brass lamps and mailboxes has received numerous complaints about premature corrosion. The manufacturer has identified the cause of the problem as the low-quality lacquer used to coat the brass. He decides to replace his current lacquer supplier with one of five possible alternatives. To judge which is best, he uses each of the five lacquers to coat 25 brass mailboxes and puts all 125 mailboxes outside. He records, for each, the number of days until the first sign of corrosion is observed.

a. Is there sufficient evidence at the 1% significance level to allow the manufacturer to conclude that differences exist between the five lacquers?

b. What are the required conditions for the test conducted in part (a)?

c. Does it appear that the required conditions of the test in part (a) are satisfied?

13.11 <u>**Xr13-11**</u> In early 2014, various companies (i.e. Honda) and even supermarkets (i.e. Asda) have announced redundancies. A statistician asked a random sample of workers how long it would be before they had significant financial hardships if they lost their jobs and could not find new ones. They also classified their income. The classifications are

More than £50 000

£30 000 to £50 000

£20 000 to £30 000

Less than £20 000

Can we infer that differences exist between the four groups?

13.12 <u>**Xr13-12**</u> In the introduction to this chapter, we mentioned that the first use of the analysis of variance was in the 1920s. It was employed to determine whether different amounts of fertiliser yielded different amounts of crop. Suppose that a scientist at an agricultural college wanted to redo the original experiment using three different types of fertiliser. Accordingly, she applied fertiliser A to 20 one-acre plots of land, fertiliser B to another 20 plots and fertiliser C to yet another 20 plots of land. At the end of the growing season, the crop yields were recorded. Can the scientist infer that differences exist between the crop yields?

13.13 <u>**Xr13-13**</u> There is a huge number of breakfast cereals on the market. Each company produces several different products in the belief that there are distinct markets. For example, there is a market composed primarily of children, another for diet-conscious adults and another for health-conscious adults. Each cereal the companies produce has at least one market as its target. However, consumers make their own decisions, which may or may not match the target predicted by the cereal maker. In an attempt to distinguish between consumers, a survey of adults between the ages of 25 and 65 was undertaken.

Each was asked several questions, including age, income and years of education, as well as which brand of cereal they consumed most frequently. The cereal choices are

1. Sugar Smacks, a children's cereal
2. Special K, a cereal aimed at dieters
3. Fibre One, a cereal that is designed and advertised as healthy
4. Cheerios, a combination of healthy and tasty

The results of the survey were recorded using the following format:

Column 1: Cereal choice

Column 2: Age of respondent

Column 3: Annual household income

Column 4: Years of education

a. Determine whether there are differences between the ages of the consumers of the four cereals.

b. Determine whether there are differences between the incomes of the consumers of the four cereals.

c. Determine whether there are differences between the educational levels of the consumers of the four cereals.

d. Summarise your findings in parts (a) through (c) and prepare a report describing the differences between the four groups of cereal consumers.

13.14 Xr13-14 According to the 2007 survey by GulfTalent.com, Dubai was officially the most congested region in the Middle East. As large cities grow larger, traffic congestion also increases. To measure how commuting time differs between Dubai, Kuwait and Jeddah, random samples of commuters in each region were drawn. Is there sufficient evidence to infer that differences in commuting time exists between the three regions?

13.15 Xr13-15 The Programme for International Student Assessment (PISA) conducts tests of 15-year-olds. The tests jointly developed by the participating countries are tests for reading literary test, mathematical literacy test and scientific literary test. Random samples from the USA, Canada and the UK were recorded. For each test, determine whether there are differences between the three countries.

13-2 MULTIPLE COMPARISONS

When we conclude from the one-way analysis of variance that at least two treatment means differ, we often need to know which treatment means are responsible for these differences. For example, if an experiment is undertaken to determine whether different locations within a store produce different mean sales, the manager would be keenly interested in determining which locations result in significantly higher sales and which locations result in lower sales. Similarly, a stockbroker would like to know which one of several mutual funds outperforms the others, and a television executive would like to know which television commercials hold the viewers' attention and which are ignored.

Although it may appear that all we need to do is examine the sample means and identify the largest or the smallest to determine which population means are largest or smallest, this is not the case. To illustrate, suppose that in a five-treatment analysis of variance, we discover that differences exist and that the sample means are as follows:

$$\bar{x}_1 = 20 \quad \bar{x}_2 = 19 \quad \bar{x}_3 = 25 \quad \bar{x}_4 = 22 \quad \bar{x}_5 = 17$$

The statistics practitioner wants to know which of the following conclusions are valid:

1. μ_3 is larger than the other means.
2. μ_3 and μ_4, are larger than the other means.
3. μ_5, is smaller than the other means.
4. μ_5 and μ_2 are smaller than the other means.
5. μ_3 is larger than the other means, and μ_5 is smaller than the other means.

From the information we have, it is impossible to determine which, if any, of the statements are true. We need a statistical method to make this determination. The technique is called **multiple comparisons**.

EXAMPLE 13.2

DATA Xm13-02

Comparing the Costs of Repairing Car Bumpers

Because of foreign competition, North American automobile manufacturers have become more concerned with quality. One aspect of quality is the cost of repairing damage caused by accidents. A manufacturer is considering several new types of bumpers. To test how well they react to low-speed collisions, ten bumpers of each of four different types were installed on mid-size cars, which were then driven into a wall at 5 mph. The cost of repairing the damage in each case was assessed. The data are shown below.

a. Is there sufficient evidence at the 5% significance level to infer that the bumpers differ in their reactions to low-speed collisions?

b. If differences exist, which bumpers differ?

Bumper 1	Bumper 2	Bumper 3	Bumper 4
610	404	599	272
354	663	426	405
234	521	429	197
399	518	621	363
278	499	426	297
358	374	414	538
379	562	332	181
548	505	460	318
196	375	494	412
444	438	637	499

SOLUTION:

IDENTIFY

The problem objective is to compare four populations. The data are interval, and the samples are independent. The correct statistical method is the one-way analysis of variance, which we perform using Excel and Minitab.

COMPUTE

EXCEL

	A	B	C	D	E	F	G
1	Anova: Single Factor						
2							
3	SUMMARY						
4	*Groups*	*Count*	*Sum*	*Average*	*Variance*		
5	Bumper 1	10	3800	380.0	16,924		
6	Bumper 2	10	4859	485.9	8,197		
7	Bumper 3	10	4838	483.8	10,426		
8	Bumper 4	10	3482	348.2	14,049		
9							
10							
11	ANOVA						
12	*Source of Variation*	*SS*	*df*	*MS*	*F*	*P-value*	*F crit*
13	Between Groups	150,884	3	50,295	4.06	0.0139	2.8663
14	Within Groups	446,368	36	12,399			
15							
16	Total	597,252	39				

MINITAB

One-way ANOVA: Bumper 1, Bumper 2, Bumper 3, Bumper 4

```
Source  DF      SS     MS     F      P
Factor   3  150884  50295  4.06  0.014
Error   36  446368  12399
Total   39  597252

S = 111.4   R-Sq = 25.26%   R-Sq(adj) = 19.03%

                                   Individual 95% CIs For Mean Based on Pooled StDev
Level      N   Mean  StDev   -------+---------+---------+---------+--
Bumper 1  10  380.0  130.1      (------*------)
Bumper 2  10  485.9   90.5                      (------*------)
Bumper 3  10  483.8  102.1                      (------*------)
Bumper 4  10  348.2  118.5   (------*------)
                             -------+---------+---------+---------+--
                                  320       400       480       560
```

INTERPRET

The test statistic is $F = 4.06$ and the p-value $= 0.0139$. There is enough statistical evidence to infer that there are differences between some of the bumpers. The question is now, 'Which bumpers differ?'.

There are several statistical inference procedures that deal with this problem. We will present three methods that allow us to determine which population means differ. All three methods apply to the one-way experiment only.

13-2a Fisher's Least Significant Difference Method

The **least significant difference (LSD)** method was briefly introduced in Section 13-1. To determine which population means differ, we could perform a series of t-tests of the difference between two means on all pairs of population means to determine which are significantly different. In Chapter 12 we introduced the equal-variances t-test of the difference between two means. The test statistic and confidence interval estimator are, respectively,

$$t = \frac{(\bar{x}_1 - \bar{x}_2) - (\mu_1 - \mu_2)}{\sqrt{s_p^2\left(\frac{1}{n_1} + \frac{1}{n_2}\right)}}$$

$$(\bar{x}_1 - \bar{x}_2) \pm t_{\alpha/2}\sqrt{s_p^2\left(\frac{1}{n_1} + \frac{1}{n_2}\right)}$$

with degrees of freedom $\nu = n_1 + n_2 - 2$.

Recall that s_p^2 is the pooled variance estimate, which is an unbiased estimator of the variance of the two populations. (Recall that the use of these techniques requires that the population variances be equal.) In this section, we modify the test statistic and interval estimator.

Earlier in this chapter, we pointed out that MSE is an unbiased estimator of the common variance of the populations we are testing. Because MSE is based on all the observations in the k samples, it will be a better estimator than s_p^2 (which is based on only two samples). Thus, we could draw inferences about every pair of means by substituting MSE for s_p^2 in the formulas for test statistic and confidence interval estimator shown previously. The number of degrees of freedom would also change to $\nu = n - k$ (where n is the total sample size). The test statistic to determine whether μ_i and μ_j differ is

$$t = \frac{(\bar{x}_i - \bar{x}_j) - (\mu_i - \mu_j)}{\sqrt{MSE\left(\frac{1}{n_i} + \frac{1}{n_j}\right)}}$$

The confidence interval estimator is

$$(\bar{x}_i - \bar{x}_j) \pm t_{\alpha/2}\sqrt{\text{MSE}\left(\frac{1}{n_i} + \frac{1}{n_j}\right)}$$

with degrees of freedom $\nu = n - k$.

We define the least significant difference LSD as

$$\text{LSD} = t_{\alpha/2}\sqrt{\text{MSE}\left(\frac{1}{n_i} + \frac{1}{n_j}\right)}$$

A simple way of determining whether differences exist between each pair of population means is to compare the absolute value of the difference between their two sample means and LSD. In other words, we will conclude that μ_i and μ_j differ if

$$\bar{x}_i - \bar{x}_j > \text{LSD}$$

LSD will be the same for all pairs of means if all k sample sizes are equal. If some sample sizes differ, LSD must be calculated for each combination.

In Section 13-1 we argued that this method is flawed because it will increase the probability of committing a Type I error. That is, it is more likely than the analysis of variance to conclude that a difference exists in some of the population means when in fact none differ. In Section 13-1c, we calculated that if $k = 6$ and all population means are equal, the probability of erroneously inferring at the 5% significance level that at least two means differ is about 54%. The 5% figure is now referred to as the *comparison-wise Type I error rate.* The true probability of making at least one Type I error is called the *experiment-wise Type I error rate,* denoted α_E. The experimentwise Type I error rate can be calculated as

$$\alpha_E = 1 - (1 - \alpha)^C$$

Here C is the number of pairwise comparisons, which can be calculated by $C = k(k - 1)/2$. Mathematicians have proven that

$$\alpha_E \leq C\alpha$$

which means that if we want the probability of making at least one Type I error to be no more than α_E, we simply specify $\alpha = \alpha_E/C$. The resulting procedure is called the **Bonferroni adjustment or correction**.

13-2b Bonferroni Adjustment to LSD Method

The adjustment is made by dividing the specified experimentwise Type I error rate by the number of combinations of pairs of population means. For example, if $k = 6$, then

$$C = \frac{k(k-1)}{2} = \frac{6(5)}{2} = 15$$

If we want the true probability of a Type I error to be no more than 5%, we divide this probability by C. Thus, for each test we would use a value of α equal to

$$\alpha = \frac{\alpha_E}{C} = \frac{0.05}{15} = 0.0033$$

We use Example 13.2 to illustrate Fisher's LSD method and the Bonferroni adjustment. The four sample means are

$$\bar{x}_1 = 380.0$$

$$\bar{x}_2 = 485.9$$

$$\bar{x}_3 = 483.8$$

$$\bar{x}_4 = 348.2$$

The pairwise absolute differences are

$$|\bar{x}_1 - \bar{x}_2| = |380.0 - 485.9| = |-105.9| = 105.9$$

$$|\bar{x}_1 - \bar{x}_3| = |380.0 - 483.8| = |-103.8| = 103.8$$

$$|\bar{x}_1 - \bar{x}_4| = |380.0 - 348.2| = |31.8| = 31.8$$

$$|\bar{x}_2 - \bar{x}_3| = |485.9 - 483.8| = |2.1| = 2.1$$

$$|\bar{x}_2 - \bar{x}_4| = |485.9 - 348.2| = |137.7| = 137.7$$

$$|\bar{x}_3 - \bar{x}_4| = |483.8 - 348.2| = |135.6| = 135.6$$

From the computer output, we learn that MSE = 12 399 and $\nu = n - k = 40 - 4 = 36$. If we conduct the LSD procedure with $\alpha = 0.05$ we find $t_{\alpha/2,n-k} = t_{0.025,36} \approx t_{0.025,35} = 2.030$. Thus,

$$t_{\alpha/2}\sqrt{\text{MSE}\left(\frac{1}{n_i} + \frac{1}{n_j}\right)} = 2.030\sqrt{12\,399\left(\frac{1}{10} + \frac{1}{10}\right)} = 101.09$$

We can see that four pairs of sample means differ by more than 101.09. In other words, $|\bar{x}_1 - \bar{x}_2| = 105.9$, $|\bar{x}_1 - \bar{x}_3| = 103.8$, $|\bar{x}_2 - \bar{x}_4| = 137.7$ and $|\bar{x}_3 - \bar{x}_4| = 135.6$. Hence, μ_1 and μ_2, μ_1 and μ_3, μ_2 and μ_4, and μ_3 and μ_4 differ. The other two pairs – μ_1 and μ_4, and μ_2 and μ_3 – do not differ.

If we perform the LSD procedure with the Bonferroni adjustment, the number of pairwise comparisons is 6 (calculated as $C = k(k - 1)/2 = 4(3)/2$). We set $\alpha = 0.05/6 = 0.0083$. Thus $t_{\alpha/2,36} = t_{0.0042,36} = 2.794$ (available from Excel and difficult to approximate manually) and

$$\text{LSD} = t_{\alpha/2}\sqrt{\text{MSE}\left(\frac{1}{n_i} + \frac{1}{n_j}\right)} = 2.794\sqrt{12\,399\left(\frac{1}{10} + \frac{1}{10}\right)} = 139.13$$

Now no pair of means differ because all the absolute values of the differences between sample means are less than 139.19.

The drawback to the LSD procedure is that we increase the probability of at least one Type I error. The Bonferroni adjustment corrects this problem. However, recall that the probabilities of Type I and Type II errors are inversely related. The Bonferroni adjustment uses a smaller value of α, which results in an increased probability of a Type II error. A Type II error occurs when a difference between population means exists, yet we cannot detect it. This may be the case in this example. The next multiple comparison method addresses this problem.

13-2c Tukey's Multiple Comparison Method

A more powerful test is **Tukey's multiple comparison method**. This technique determines a critical number similar to LSD for Fisher's test, denoted by ω (Greek letter *omega*) such that, if any pair of sample means has a difference greater than ω, we conclude that the pair's two corresponding population means are different.

The test is based on the Studentised range, which is defined as the variable

$$q = \frac{\bar{x}_{max} - \bar{x}_{min}}{s/\sqrt{n}}$$

where $\bar{x}_{max}$ and $\bar{x}_{min}$ are the largest and smallest sample means, respectively, assuming that there are no differences between the population means. We define ω as follows.

Critical Number ω

$$\omega = q_\alpha(k,\nu)\sqrt{\frac{\text{MSE}}{n_g}}$$

where

k = Number of treatments
n = Number of observations ($n = n_1 + n_2 + \cdots + n_k$)
ν = Number of degrees of freedom associated with MSE ($\nu = n - k$)
n_g = Number of observations in each of k samples
α = Significance level
$q_\alpha(k,\nu)$ = Critical value of the Studentised range

Theoretically, this procedure requires that all sample sizes be equal. However, if the sample sizes are different, we can still use this technique provided that the sample sizes are at least similar. The value of n_g used previously is the *harmonic mean* of the sample sizes; that is,

$$n_g = \frac{k}{\frac{1}{n_1} + \frac{1}{n_2} + \cdots + \frac{1}{n_k}}$$

Table 7 in Appendix A provides values of $q_\alpha(k,\nu)$ for a variety of values of k and ν, and for $\alpha = 0.01$ and 0.05. Applying Tukey's method to Example 13.2, we find

$$k = 4$$

$$n_1 = n_2 = n_3 = n_4 = n_g = 10$$

$$\nu = n - k = 40 - 4 = 36$$

$$\text{MSE} = 12{,}399$$

$$q_{0.05}(4,37) \approx q_{0.05}(4,40) = 3.79$$

Thus,

$$\omega = q_\alpha(k,\nu)\sqrt{\frac{\text{MSE}}{n_g}} = (3.79)\sqrt{\frac{12\,399}{10}} = 133.45$$

There are two absolute values larger than 133.45. Hence, we conclude that μ_2 and μ_4, and μ_3 and μ_4 differ. The other four pairs do not differ.

EXCEL

	A	B	C	D	E
1	**Multiple Comparisons**				
2					
3				LSD	Omega
4	Treatment	Treatment	Difference	Alpha = 0.05	Alpha = 0.05
5	*Bumper 1*	*Bumper 2*	−105.9	100.99	133.45
6		*Bumper 3*	−103.8	100.99	133.45
7		*Bumper 4*	31.8	100.99	133.45
8	*Bumper 2*	*Bumper 3*	2.1	100.99	133.45
9		*Bumper 4*	137.7	100.99	133.45
10	*Bumper 3*	*Bumper 4*	135.6	100.99	133.45

(*continued*)

Tukey and Fisher's LSD with the Bonferroni Adjustment ($\alpha = 0.05/6 = 0.083$)

	A	B	C	D	E
1	**Multiple Comparisons**				
2					
3				LSD	Omega
4	Treatment	Treatment	Difference	Alpha = 0.0083	Alpha = 0.05
5	*Bumper 1*	*Bumper 2*	−105.9	139.11	133.45
6		*Bumper 3*	−103.8	139.11	133.45
7		*Bumper 4*	31.8	139.11	133.45
8	*Bumper 2*	*Bumper 3*	2.1	139.11	133.45
9		*Bumper 4*	137.7	139.11	133.45
10	*Bumper 3*	*Bumper 4*	135.6	139.11	133.45

The printout includes ω (Tukey's method), the differences between sample means for each combination of populations, and Fisher's LSD. (The Bonferroni adjustment is made by specifying another value for α.)

Instructions

1. Type or import the data into adjacent columns. (**Open** Xm13-02.)
2. Click **Add-Ins, Data Analysis Plus** and **Multiple Comparisons.**
3. Specify the **Input Range** (A1:D11).Type the value of α. To use the Bonferroni adjustment divide α by $C = k(k - l)/2$. For Tukey, Excel computes ω only for $\alpha = 0.05$.

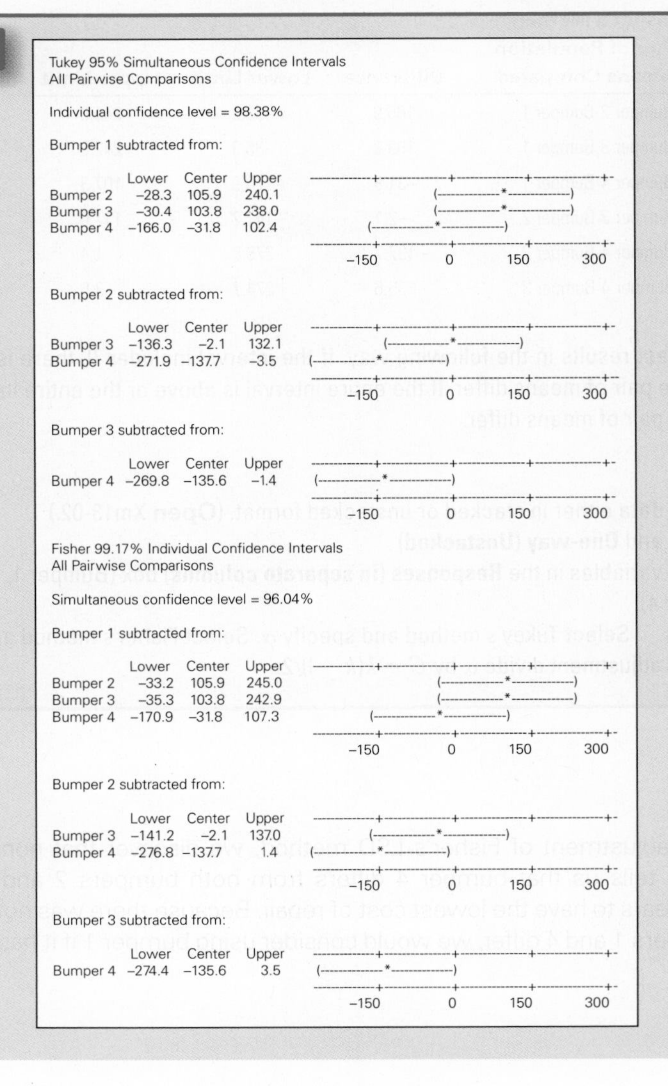

```
Tukey 95% Simultaneous Confidence Intervals
All Pairwise Comparisons

Individual confidence level = 98.38%

Bumper 1 subtracted from:

          Lower  Center  Upper   ---------+---------+---------+---------+-
Bumper 2  -28.3   105.9  240.1                  (---------*---------)
Bumper 3  -30.4   103.8  238.0                  (---------*---------)
Bumper 4 -166.0   -31.8  102.4          (---------*---------)
                                 ---------+---------+---------+---------+-
                                       -150         0       150       300

Bumper 2 subtracted from:

          Lower  Center  Upper   ---------+---------+---------+---------+-
Bumper 3 -136.3    -2.1  132.1             (---------*---------)
Bumper 4 -271.9  -137.7   -3.5   (---------*---------)
                                 ---------+---------+---------+---------+-
                                       -150         0       150       300

Bumper 3 subtracted from:

          Lower  Center  Upper   ---------+---------+---------+---------+-
Bumper 4 -269.8  -135.6   -1.4   (---------*---------)
                                 ---------+---------+---------+---------+-
                                       -150         0       150       300

Fisher 99.17% Individual Confidence Intervals
All Pairwise Comparisons

Simultaneous confidence level = 96.04%

Bumper 1 subtracted from:

          Lower  Center  Upper   ---------+---------+---------+---------+-
Bumper 2  -33.2   105.9  245.0                  (---------*---------)
Bumper 3  -35.3   103.8  242.9                  (---------*---------)
Bumper 4 -170.9   -31.8  107.3          (---------*---------)
                                 ---------+---------+---------+---------+-
                                       -150         0       150       300

Bumper 2 subtracted from:

          Lower  Center  Upper   ---------+---------+---------+---------+-
Bumper 3 -141.2    -2.1  137.0             (---------*---------)
Bumper 4 -276.8  -137.7    1.4   (---------*---------)
                                 ---------+---------+---------+---------+-
                                       -150         0       150       300

Bumper 3 subtracted from:

          Lower  Center  Upper   ---------+---------+---------+---------+-
Bumper 4 -274.4  -135.6    3.5   (---------*---------)
                                 ---------+---------+---------+---------+-
                                       -150         0       150       300
```

Minitab reports the results of Tukey's multiple comparisons by printing interval estimates of the differences between each pair of means. The estimates are computed by calculating the pairwise difference between sample means minus ω for the lower limit and plus ω for the upper limit. The calculations are described in the following table.

Tukey's Method Pair of Population Means Compared	Difference	Lower Limit	Upper Limit
Bumper 2-Bumper 1	105.9	−28.3	240.1
Bumper 3-Bumper 1	103.8	−30.4	238.0
Bumper 4-Bumper 1	−31.8	−166.0	102.4
Bumper 3-Bumper 2	−2.1	−136.3	132.1
Bumper 4-Bumper 2	−137.7	−271.9	−3.5
Bumper 4-Bumper 3	−135.6	−269.8	−1.4

A similar calculation is performed for Fisher's method replacing ω by LSD.

Fisher's Method Pair of Population Means Compared	Difference	Lower Limit	Upper Limit
Bumper 2-Bumper 1	105.9	−33.2	245.0
Bumper 3-Bumper 1	103.8	−35.3	242.9
Bumper 4-Bumper 1	−31.8	−170.9	107.3
Bumper 3-Bumper 2	−2.1	−141.2	137.0
Bumper 4-Bumper 2	−137.7	−276.8	1.4
Bumper 4-Bumper 3	−135.6	274.7	3.5

We interpret the test results in the following way. If the interval includes 0, there is not enough evidence to infer that the pair of means differ. If the entire interval is above or the entire interval is below 0, we conclude that the pair of means differ.

Instructions

1. Type or import the data either in stacked or unstacked format. (**Open** Xm13-02.)
2. Click **Stat, ANOVA** and **One-way (Unstacked)**
3. Type or **Select** the variables in the **Responses (in separate columns)** box (Bumper 1, Bumper 2, Bumper 3, Bumper 4).
4. Click **Comparisons**. . . Select Tukey's method and specify α. Select Fisher's method and specify α. For the Bonferroni adjustment divide α by $C = k(k - 1)/2$.

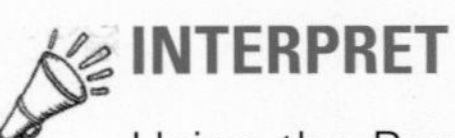

INTERPRET

Using the Bonferroni adjustment of Fisher's LSD method, we discover that none of the bumpers differ. Tukey's method tells us that bumper 4 differs from both bumpers 2 and 3. Based on this sample, bumper 4 appears to have the lowest cost of repair. Because there was not enough evidence to conclude that bumpers 1 and 4 differ, we would consider using bumper 1 if it has other advantages over bumper 4.

13-2d Which Multiple Comparison Method to Use

Unfortunately, no one procedure works best in all types of problems. Most statisticians agree with the following guidelines:

If you have identified two or three pairwise comparisons that you wish to make before conducting the analysis of variance, use the Bonferroni method. This means that if there are ten populations in a problem but you are particularly interested in comparing, say, populations 3 and 7 and populations 5 and 9, use Bonferroni with $C = 2$.

If you plan to compare all possible combinations, use Tukey.

When do we use Fisher's LSD? If the purpose of the analysis is to point to areas that should be investigated further, Fisher's LSD method is indicated.

Incidentally, to employ Fisher's LSD or the Bonferroni adjustment, you must perform the analysis of variance first. Tukey's method can be employed instead of the analysis of variance.

EXERCISES

Developing an Understanding of Statistical Concepts

13.16 a. Use Fisher's LSD method with $\alpha = 0.05$ to determine which population means differ in the following problem.

$k = 3$ $n_1 = 10$ $n_2 = 10$ $n_3 = 10$

$\text{MSE} = 700$ $\bar{x}_1 = 128.7$ $\bar{x}_2 = 101.4$ $\bar{x}_3 = 133.7$

b. Repeat part (a) using the Bonferroni adjustment.

c. Repeat part (a) using Tukey's multiple comparison method.

13.17 a. Use Fisher's LSD procedure with $a = 0.05$ to determine which population means differ given the following statistics:

$k = 5$ $n_1 = 5$ $n_2 = 5$ $n_3 = 5$

$\text{MSE} = 125$ $\bar{x}_1 = 227$ $\bar{x}_2 = 205$ $\bar{x}_3 = 219$

$n_4 = 5$ $n_5 = 5$

$\bar{x}_4 = 248$ $\bar{x}_5 = 202$

b. Repeat part (a) using the Bonferroni adjustment.

c. Repeat part (a) using Tukey's multiple comparison method

Applications

Unless specified otherwise, use a 5% significance level.

13.18 Apply Tukey's method to determine which brands differ in Exercise 13.5.

13.19 Refer to Exercise 13.6.

a. Employ Fisher's LSD method to determine which degrees differ (use $\alpha = 0.10$).

b. Repeat part (a) using the Bonferroni adjustment.

The following exercises require the use of a computer and software.

13.20 Xr13-20 Current regulations require recreational craft that are operating commercially to carry flares. One of the most important features of flares is their burning times. To help decide which of four brands on the market to use, a vessel technician measured the burning time for a random sample of ten flares of each brand. The results were recorded to the nearest minute.

a. Can we conclude that differences exist between the burning times of the four brands of flares?

b. Apply Fisher's LSD method with the Bonferroni adjustment to determine which flares are better.

c. Repeat part (b) using Tukey's method.

13.21 Xr13-10 Refer to Exercise 13.10.

a. Apply Fisher's LSD method with the Bonferroni adjustment to determine which lacquers differ.

b. Repeat part (a) applying Tukey's method instead.

13.22 Xr13-22 An engineering student who is about to graduate decided to survey various firms in the motor industry in the Thames Valley to see which offered the best chance for early promotion and career advancement. He surveyed 30 small firms (size level is based on gross revenues), 30 medium-sized firms and 30 large firms, and determined how much time must elapse before an average engineer can receive a promotion.

a. Can the engineering student conclude that speed of promotion varies between the three sizes of engineering firms?

b. If differences exist, which of the following is true? Use Tukey's method.

i. Small firms differ from the other two.

ii. Medium-sized firms differ from the other two.

iii. Large firms differ from the other two.

iv. All three firms differ from one another.

v. Small firms differ from large firms.

13.23 Xr13-12 **a.** Apply Tukey's multiple comparison method to determine which fertilisers differ in Exercise 13.12.

b. Repeat part (a) applying the Bonferroni adjustment.

13-3 ANALYSIS OF VARIANCE EXPERIMENTAL DESIGNS

Since we introduced the matched pairs experiment in Section 12-5, the experimental design has been one of the factors that determine which technique we use. Statistics practitioners often design experiments to help extract the information they need to assist them in making decisions. The one-way analysis of variance introduced in Section 13-1 is only one of many different experimental designs of the analysis of variance. For each type of experiment, we can describe the behaviour of the response variable using a mathematical expression or model. Although we will not exhibit the mathematical expressions in this chapter (we introduce models in Chapter 15), we think it is useful for you to be aware of the elements that distinguish one experimental design or model from another. In this section, we present some of these elements; in so doing, we introduce two of the experimental designs that will be presented later in this chapter.

13-3a Single-Factor and Multifactor Experimental Designs

As we pointed out in Section 13-1, the criterion by which we identify populations is called a *factor*. The experiment described in Section 13-1 is a single-factor analysis of variance because it addresses the problem of comparing two or more populations defined on the basis of only one factor. A **multifactor experiment** is one in which two or more factors define the treatments. The experiment described in Example 13.1 is a single-factor design because we had one treatment: age of the head of the household. In other words, the factor is the age, and the four age categories were the levels of this factor.

Suppose that we can also look at the gender of the household head in another study. We would then develop a two-factor analysis of variance in which the first factor, age, has four levels, and the second factor, gender, has two levels. We will discuss two-factor experiments in Section 13-5.

13-3b Independent Samples and Blocks

In Section 12-5, we introduced statistical techniques where the data were gathered from a matched pairs experiment. This type of experimental design reduces the variation within the samples, making it easier to detect differences between the two populations. When the problem objective is to compare more than two populations, the experimental design that is the counterpart of the matched pairs experiment is called the **randomised block design**. The term *block* refers to a matched group of observations from each population. Suppose that in Examples 12.7 and 12.8 we had wanted to compare the salary offers for finance, marketing, accounting and operations management majors. To redo Example 12.10 we would conduct a randomised block experiment where the blocks are the 25 GPA groups and the treatments are the four MBA majors.

Once again, the experimental design should reduce the variation in each treatment to make it easier to detect differences.

We can also perform a blocked experiment by using the same subject (person, plant and store) for each treatment. For example, we can determine whether sleeping pills are effective by giving three brands of pills to the same group of people to measure the effects. Such experiments are called **repeated measures** designs. Technically, this is a different design than the randomised block. However, the data are analysed in the same way for both designs. Hence, we will treat repeated measures designs as randomised block designs.

The randomised block experiment is also called the **two-way analysis of variance.** In Section 13-4, we introduce the technique used to calculate the test statistic for this type of experiment.

13-3c Fixed and Random Effects

If our analysis includes all possible levels of a factor, the technique is called a **fixed-effects analysis of variance**. If the levels included in the study represent a random sample of all the levels that exist, the technique is called a **random-effects analysis of variance.** In Example 13.2, there were only four possible bumpers. Consequently, the study is a fixed-effects experiment. However, if there were other bumpers besides the four described in the example, and we wanted to know whether there were differences in repair costs between all bumpers, the application would be a random-effects experiment. Here is another example.

To determine whether there is a difference in the number of units produced by the machines in a large factory, four machines out of 50 in the plant are randomly selected for study. The number of units each produces per day for 10 days will be recorded. This experiment is a random-effects experiment because we selected a random sample of four machines and the statistical results thus allow us to determine whether there are differences between the 50 machines.

In some experimental designs, there are no differences in calculations of the test statistic between fixed and random effects. However, in others, including the two-factor experiment presented in Section 13-5, the calculations are different.

13-4 RANDOMISED BLOCK (TWO-WAY) ANALYSIS OF VARIANCE

The purpose of designing a randomised block experiment is to reduce the within-treatments variation to more easily detect differences between the treatment means. In the one-way analysis of variance, we partitioned the total variation into the between-treatments and the within-treatments variation; that is,

$$\text{SS(Total)} = \text{SST} + \text{SSE}$$

In the randomised block design of the analysis of variance, we partition the total variation into three sources of variation,

$$\text{SS(Total)} = \text{SST} + \text{SSB} + \text{SSE}$$

where **SSB**, the **sum of squares for blocks**, measures the variation between the blocks. When the variation associated with the blocks is removed, SSE is reduced, making it easier to determine whether differences exist between the treatment means.

At this point in our presentation of statistical inference, we will deviate from our usual procedure of solving examples in three ways: manually, using Excel and using Minitab. The calculations for this experimental design and for the experiment presented in the next section are so time-consuming that solving them by hand adds little to your understanding of the technique. Consequently, although we will continue to present the concepts by discussing how the statistics are calculated, we will solve the problems only by computer.

To help you understand the formulas, we will use the following notation:

$\bar{x}[T]_j$ = Mean of the observations in the jth treatment ($j = 1, 2, \ldots, k$)
$\bar{x}[B]_i$ = Mean of the observations in the ith block ($i = 1, 2, \ldots, b$)
b = Number of blocks

Table 13.4 summarises the notation we use in this experimental design.

TABLE 13.4 Notation for the Randomised Block Analysis of Variance

	TREATMENTS				
BLOCK	1	2		k	BLOCK MEAN
1	x_{11}	x_{12}	...	x_{1k}	$\bar{x}[B]_1$
2	x_{21}	x_{22}	...	x_{2k}	$\bar{x}[B]_2$
⋮	⋮	⋮		⋮	⋮
b	x_{b1}	x_{b2}	...	x_{bk}	$\bar{x}[B]_b$
Treatment mean	$\bar{x}[T]_1$	$\bar{x}[T]_2$	...	$\bar{x}[T]_k$	

The definitions of SS(Total) and SST in the randomised block design are identical to those in the independent samples design. SSE in the independent samples design is equal to the sum of SSB and SSE in the randomised block design.

Sums of Squares in the Randomised Block Experiment

$$\text{SS(Total)} = \sum_{j=1}^{k}\sum_{i=1}^{b}(x_{ij} - \bar{\bar{x}})^2$$

$$\text{SST} = \sum_{j=1}^{k} b(\bar{x}[T]_j - \bar{\bar{x}})^2$$

$$\text{SSB} = \sum_{i=1}^{b} k(\bar{x}[B]_i - \bar{\bar{x}})^2$$

$$\text{SSE} = \sum_{j=1}^{k}\sum_{i=1}^{b}(x_{ij} - \bar{x}[T]_j - \bar{x}[B]_i + \bar{\bar{x}})^2$$

The test is conducted by determining the mean squares, which are computed by dividing the sums of squares by their respective degrees of freedom.

Mean Squares for the Randomised Block Experiment

$$\text{MST} = \frac{\text{SST}}{k-1}$$

$$\text{MSB} = \frac{\text{SSB}}{b-1}$$

$$\text{MSE} = \frac{\text{SSE}}{n-k-b+1}$$

Finally, the test statistic is the ratio of mean squares, as described in the box.

Test Statistic for the Randomised Block Experiment

$$F = \frac{\text{MST}}{\text{MSE}}$$

which is F-distributed with $\nu_1 = k - 1$ and $\nu_2 = n - k - b + 1$ degrees of freedom.

An interesting, and sometimes useful, by-product of the test of the treatment means is that we can also test to determine whether the block means differ. This will allow us to determine whether the experiment should have been conducted as a randomised block design. (If there are no differences between the blocks, the randomised block design is less likely to detect real differences between the treatment means.) Such a discovery could be useful in future similar experiments. The test of the block means is almost identical to that of the treatment means except the test statistic is

$$F = \frac{\text{MSB}}{\text{MSE}}$$

which is *F*-distributed with $\nu_1 = b - 1$ and $\nu_2 = n - k - b + 1$ degrees of freedom.

As with the one-way experiment, the statistics generated in the randomised block experiment are summarised in an ANOVA table, whose general form is exhibited in Table 13.5.

TABLE 13.5 **ANOVA Table for the Randomised Block Analysis of Variance**

SOURCE OF VARIATION	DEGREES OF FREEDOM	SUMS OF SQUARES	MEAN SQUARES	*F*-STATISTIC
Treatments	$k - 1$	SST	MST = SST/(k − 1)	F = MST/MSE
Blocks	$b - 1$	SSB	MSB=SSB/(b − 1)	F = MSB/MSE
Error	$n - k - b + 1$	SSE	MSE = SSE/(n − k-b + 1)	
Total	$n - 1$	SS(Total)		

EXAMPLE 13.3

DATA Xm13-03

Comparing Cholesterol-Lowering Drugs

One in six people under 35 suffers from high levels of cholesterol, which can lead to heart attacks. For those with very high levels (above 280), doctors prescribe drugs to reduce cholesterol levels. A pharmaceutical company has recently developed four such drugs. To determine whether any differences exist in their benefits, an experiment was organised. The company selected 25 groups of four men, each of whom had cholesterol levels in excess of 280. In each group, the men were matched according to age and weight. The drugs were administered over a 2-month period, and the reduction in cholesterol was recorded. Do these results allow the company to conclude that differences exist between the four new drugs?

Group	Drug 1	Drug 2	Drug 3	Drug 4
1	6.6	12.6	2.7	8.7
2	7.1	3.5	2.4	9.3
3	7.5	4.4	6.5	10.0
4	9.9	7.5	16.2	12.6
5	13.8	6.4	8.3	10.6
6	13.9	13.5	5.4	15.4
7	15.9	16.9	15.4	16.3
6	14.3	11.4	17.1	18.9
9	16.0	16.9	7.7	13.7
10	16.3	14.8	16.1	19.4
11	14.6	18.6	9.0	18.5
12	18.7	21.2	24.3	21.1
13	17.3	10.0	9.3	19.3
14	19.6	17.0	19.2	21.9

Group	Drug 1	Drug 2	Drug 3	Drug 4
15	20.7	21.0	18.7	22.1
16	18.4	27.2	18.9	19.4
17	21.5	26.8	7.9	25.4
18	20.4	28.0	23.8	26.5
19	21.9	31.7	8.8	22.2
20	22.5	11.9	26.7	23.5
21	21.5	28.7	25.2	19.6
22	25.2	29.5	27.3	30.1
23	23.0	22.2	17.6	26.6
24	23.7	19.5	25.6	24.5
25	28.4	31.2	26.1	27.4

SOLUTION:

IDENTIFY

The problem objective is to compare four populations, and the data are interval. Because the researchers recorded the cholesterol reduction for each drug for each member of the similar groups of men, we identify the experimental design as randomised block. The response variable is the cholesterol reduction, the treatments are the drugs, and the blocks are the 25 similar groups of men. The hypotheses to be tested are as follows.

$$H_0: \mu_1 = \mu_2 = \mu_3 = \mu_4$$

H_1: At least two means differ

COMPUTE

EXCEL

	A	B	C	D	E	F	G
36	ANOVA						
37	*Source of Variation*	*SS*	*df*	*MS*	*F*	*P-value*	*F crit*
38	Rows	3848.66	24	160.36	10.11	9.70E-15	1.6695
39	Columns	195.95	3	65.32	4.12	0.0094	2.7318
40	Error	1142.56	72	15.87			
41							
42	Total	5187.17	99				

Note the use of scientific notation for one of the *p*-values. The number 9.70E−15 (E stands for *exponent*) is 9.70 multiplied by 10 raised to the power −15, that is, 9.70×10^{-15}. You can increase or decrease the number of decimal places, and you can convert the number into a regular number, but you would need many decimal places, which is why Excel uses scientific notation when the number is very small. (Excel also uses scientific notation for very large numbers.)

The output includes block and treatment statistics (sums, averages and variances, which are not shown here) and the ANOVA table. The *F*-statistic to determine whether differences exist between the four drugs **(Columns)** is 4.12. Its *p*-value is 0.0094. The other *F*-statistic, 10.11 (*p*-value $= 9.70 \times 10^{-15}$ = virtually 0), indicates that there are differences between the groups of men **(Rows)**.

Instructions

1. Type or import the data into adjacent columns*. (**Open** Xm13-03.)
2. Click **Data, Data Analysis . . .** and **Anova: Two-Factor Without Replication**.
3. Specify the **Input Range** (A1:E26). Click **Labels** if applicable. If you do, both the treatments and blocks must be labelled (as in Xm13-03). Specify the value of α (0.05).

*If one or more columns contain an empty cell (representing missing data) the entire row must be removed. See the website Appendix Removing Empty Cells in Excel.

MINITAB

Two-way ANOVA: Reduction versus Group, Drug

Analysis of Variance for Reduction

Source	DF	SS	MS	F	P
Group	24	3848.7	160.4	10.11	0.000
Drug	3	196.0	65.3	4.12	0.009
Error	72	1142.6	15.9		
Total	99	5187.2			

The *F*-statistic for **Drug** is 4.12 with a *p*-value of 0.009. The *F*-statistic for the blocks **(Group)** is 10.11, with a *p*-value of 0.

Instructions

The data must be in stacked format in three columns. One column contains the responses, another contains codes for the levels of the blocks, and a third column contains codes for the levels of the treatments.

1. Click **Stat, ANOVA** and **Two-way** ....
2. Specify the **Responses, Row factor** and **Column factor.**

INTERPRET

A Type I error occurs when you conclude that differences exist when, in fact, they do not. A Type II error is committed when the test reveals no difference when at least two means differ. It would appear that both errors are equally costly. Accordingly, we judge the *p*-value against a standard of 5%. Because the *p*-value = 0.0094, we conclude that there is sufficient evidence to infer that at least two of the drugs differ. An examination reveals that cholesterol reduction is greatest using drugs 2 and 4. Further testing is recommended to determine which is better.

13-4a Checking the Required Conditions

The *F*-test of the randomised block design of the analysis of variance has the same requirements as the independent samples design. That is, the random variable must be normally distributed and the population variances must be equal. The histograms (not shown) appear to support the validity of our results; the reductions appear to be normal. The equality of variances requirement also appears to be met.

13-4b Violation of the Required Conditions

When the response is not normally distributed, we can replace the randomised block analysis of variance with the Friedman test, which is introduced in Section 18-4.

13-4c Criteria for Blocking

In Section 12-5 we listed the advantages and disadvantages of performing a matched pairs experiment. The same comments are valid when we discuss performing a blocked experiment. The purpose of blocking is to reduce the variation caused by differences between the experimental units. By grouping the experimental units into homogeneous blocks with respect to the response variable, the statistics practitioner increases the chances of detecting actual differences between the treatment means. Hence, we need to find criteria for blocking that significantly affect the response variable. For example, suppose that a statistics professor wants to determine which of four methods of teaching statistics is best. In a one-way experiment, he might take four samples of ten students, teach each sample by a different method, grade the students at the end of the course and perform an *F*-test to determine whether differences exist. However, it is likely that there are very large differences between the students within each class that may hide differences between classes. To reduce this variation, the statistics professor must identify variables that are linked to a student's grade in statistics. For example, overall ability of the student, completion of mathematics courses and exposure to other statistics courses are all related to performance in a statistics course.

The experiment could be performed in the following way. The statistics professor selects four students at random whose average grade before statistics is 95–100. He then randomly assigns the students to one of the four classes. He repeats the process with students whose average is 90–95, 85–90, … and 50–55. The final grades would be used to test for differences between the classes.

Any characteristics that are related to the experimental units are potential blocking criteria. For example, if the experimental units are people, we may block according to age, gender, income, work experience, intelligence, residence (country, county or city), weight or height. If the experimental unit is a factory and we are measuring number of units produced hourly, blocking criteria include work-force experience, age of the plant and quality of suppliers.

13-4d Developing an Understanding of Statistical Concepts

As we explained previously, the randomised block experiment is an extension of the matched pairs experiment discussed in Section 12-5. In the matched pairs experiment, we simply remove the effect of the variation caused by differences between the experimental units. The effect of this removal is seen in the decrease in the value of the standard error (compared to the standard error in the test statistic produced from independent samples) and the increase in the value of the *t*-statistic. In the randomised block experiment of the analysis of variance, we actually measure the variation between the blocks by computing SSB. The sum of squares for error is reduced by SSB, making it easier to detect differences between the treatments. In addition, we can test to determine whether the blocks differ – a procedure we were unable to perform in the matched pairs experiment.

To illustrate, let's return to Examples 12.9 and 12.10, which were experiments to determine whether there was a difference in starting salaries offered to finance and marketing MBA majors. (In fact, we tested to determine whether finance majors draw higher salary offers than do marketing majors. However, the analysis of variance can test only for differences.) In Example 12.9 (independent samples), there was insufficient evidence to infer a difference between the two types of majors. In Example 12.10 (matched pairs experiment), there was enough evidence to infer a difference. As we pointed out in Section 12-6, matching by grade point average allowed the statistics practitioner to more easily discern a difference between the two types of majors. If we repeat Examples 12.9 and 12.10 using the analysis of variance, we come to the same conclusion. The Excel outputs are shown here. (Minitab's printouts are similar.)

Excel Analysis of Variance Output for Example 12.9

	A	B	C	D	E	F	G
9	ANOVA						
10	*Source of Variation*	*SS*	*df*	*MS*	*F*	*P-value*	*F crit*
11	Between Groups	338,130,013	1	338,130,013	1.09	0.3026	4.0427
12	Within Groups	14,943,884,470	48	311,330,926			
13							
14	Total	15,282,014,483	49				

Excel Analysis of Variance Output for Example 12.10

	A	B	C	D	E	F	G
34	ANOVA						
35	*Source of Variation*	*SS*	*df*	*MS*	*F*	*P-value*	*F crit*
36	Rows	21,415,991,654	24	892,332,986	40.39	4.17E-14	1.9838
37	Columns	320,617,035	1	320,617,035	14.51	0.0009	4.2597
38	Error	530,174,605	24	22,090,609			
39							
40	Total	22,266,783,295	49				

In Example 12.9, we partition the total sum of squares [SS(Total) = 15 282 014 483] into two sources of variation: SST = 338 130 013 and SSE = 14 943 884 470. In Example 12.10, the total sum of squares is SS(Total) = 22 266 783 295 SST (sum of squares for majors) = 320 617 035 SSB (sum of squares for GPA) = 21 415 991 654 and SSE = 530 174 605. As you can see, the sums of squares for treatments are approximately equal (338 130 013 and 320 617 035). However, the two calculations differ in the sums of squares for error. SSE in Example 12.10 is much smaller than SSE in Example 12.9 because the randomised block experiment allows us to measure and remove the effect of the variation between MBA students with the same majors. The sum of squares for blocks (sum of squares for GPA groups) is 21 415 991 654, a statistic that measures how much variation exists between the salary offers within majors. As a result of removing this variation, SSE is small. Thus, we conclude in Example 12.10 that the salary offers differ between majors whereas there was not enough evidence in Example 12.9 to draw the same conclusion.

Notice that in both examples the *t*-statistic squared equals the *F*-statistic in Example 13.4, $t = 1.04$, which when squared equals 1.09, which is the *F*-statistic (rounded). In Example 13.5, $t = 3.81$, which when squared equals 14.51, the *F*-statistic for the test of the treatment means. Moreover, the *p*-values are also the same.

We now complete this section by listing the factors that we need to recognise to use this experiment of the analysis of variance.

Factors That Identify the Randomised Block of the Analysis of Variance

1. **Problem objective:** Compare two or more populations.
2. **Data type:** Interval.
3. **Experimental design:** Blocked samples.

EXERCISES

Developing an Understanding of Statistical Concepts

13.24 The following statistics were generated from a randomised block experiment with $k = 3$ and $b = 7$:

SST = 100   SSB = 50   SSE = 25

a. Test to determine whether the treatment means differ. (Use $\alpha = 0.05$.)

b. Test to determine whether the block means differ. (Use $\alpha = 0.05$.)

13.25 A randomised block experiment produced the following statistics:

k = 5   b = 12   SST = 1500   SSB = 1000   SS(Total) = 3500

a. Test to determine whether the treatment means differ. (Use $\alpha = 0.01$.)

b. Test to determine whether the block means differ. (Use $\alpha = 0.01$.)

13.26 Suppose the following statistics were calculated from data gathered from a randomised block experiment with $k = 4$ and $b = 10$:

SS(Total) = 1210   SST = 275   SSB = 625

a. Can we conclude from these statistics that the treatment means differ? (Use $\alpha = 0.01$.)

b. Can we conclude from these statistics that the block means differ? (Use $\alpha = 0.01$.)

13.27 A randomised block experiment produced the following statistics.

k = 3   b = 8   SST = 1500   SS(Total) = 3500

a. Test at the 5% significance level to determine whether the treatment means differ given that SSB = 500.

b. Repeat part (a) with SSB = 1000.

c. Repeat part (a) with SSB = 1500.

d. Describe what happens to the test statistic as SSB increases.

13.28 Xr13-28 **a.** Assuming that the data shown here were generated from a randomised block experiment calculate SS(Total), SST, SSB and SSE.

b. Assuming that the data below were generated from a one-way (independent samples) experiment calculate SS(Total), SST and SSE.

c. Why does SS(Total) remain the same for both experimental designs?

d. Why does SST remain the same for both experimental designs?

e. Why does SSB + SSE in part (a) equal SSE in part (b)?

Treatment		
1	2	3
7	12	8
10	6	9
12	16	13
9	13	6
12	10	11

13.29 Xr13-29 a. Calculate SS(Total), SST, SSB and SSE, assuming that the accompanying data were generated from a randomised block experiment.

b. Calculate SS(Total), SST and SSE, assuming that the data below were generated from a one-way (independent samples) experiment.

c. Explain why SS(Total) remains the same for both experimental designs.

d. Explain why SST remains the same for both experimental designs.

e. Explain why SSB + SSE in part (a) equals SSE in part (b).

Treatment			
1	2	3	4
6	5	4	4
8	5	5	6
7	6	5	6

Applications

13.30 Xr13-30 As an experiment to understand measurement error, a statistics lecturer asks four students to measure the height of the lecturer, a male student and a female student. The differences (in centimetres) between the correct dimension and the ones produced by the students are listed here. Can we infer that there are differences in the errors between the subjects being measured? (Use $\alpha = 0.05$.)

	Errors in Measuring Heights of		
Student	Lecturer	Male Student	Female Student
1	1.4	1.5	1.3
2	3.1	2.6	2.4
3	2.8	2.1	1.5
4	3.4	3.6	2.9

The following exercises require the use of a computer and software. Use a 5% significance level, unless specified otherwise.

13.31 Xr13-31 A large company is starting an e-commerce site and it is in the process of selecting one of three possible couriers to act as its sole delivery method. To help in making the decision, an experiment was performed whereby letters were sent using each of the three couriers at 12 different times of the day to a delivery point across town. The number of minutes required for delivery was recorded.

a. Can we conclude that there are differences in delivery times between the three couriers?

b. Did the statistics practitioner choose the correct design? Explain.

13.32 Xr13-32 Refer to Exercise 13.12. Despite failing to show that differences in the three types of fertiliser exist, the scientist continued to believe that there were differences, and that the differences were masked by the variation between the plots of land. Accordingly, he conducted another experiment. In the second experiment he found 20 three-acre plots of land scattered across the county. He divided each into three plots and applied the three types of fertiliser on each of the one-acre plots. The crop yields were recorded.

a. Can the scientist infer that there are differences between the three types of fertiliser?

b. What do these test results reveal about the variation between the plots?

13.33 Xr13-33 A recruiter for a computer company would like to determine whether there are differences in sales ability between business, arts and science graduates. She takes a random sample of 20 business graduates who have been working for the company for the past 2 years. Each is then matched with an arts graduate and a science graduate with similar educational and working experience. The commission earned by each (in £1000s) in the last year was recorded.

a. Is there sufficient evidence to allow the recruiter to conclude that there are differences in sales ability between the holders of the three types of degrees?

b. Conduct a test to determine whether an independent samples design would have been a better choice.

c. What are the required conditions for the test in part (a)?

d. Are the required conditions satisfied?

13.34 Xr13-34 The advertising revenues commanded by a radio station depend on the number of listeners it has. The manager of a station that plays mostly hard rock music wants to learn more about its listeners – mostly teenagers and young adults. In particular, he wants to know whether the amount of time they spend listening to radio music varies by the day of the week.

If the manager discovers that the mean time per day is about the same, he will schedule the most popular music evenly throughout the week. Otherwise, the top hits will be played mostly on the days that attract the greatest audience. An opinion survey company is hired, and it randomly selects 200 teenagers and asks them to record the amount of time spent listening to music on the radio for each day of the previous week. What can the manager conclude from these data?

13-5 TWO-FACTOR ANALYSIS OF VARIANCE

In Section 13-1, we addressed problems where the data were generated from single-factor experiments. In Example 13.1 the treatments were the four age categories. Thus, there were four levels of a single factor. In this section, we address the problem where the experiment features two factors. The general term for such data-gathering procedures is **factorial experiment (or ANOVA)**. In factorial experiments, we can examine the effect on the response variable of two or more factors, although in this book we address the problem of only two factors. We can use the analysis of variance to determine whether the levels of each factor are different from one another. In addition, the ANOVA design permits very detailed analysis of both individual independent variables as well as interactions among independent variables.

We will present the technique for fixed effects only. That means we will address problems where all the levels of the factors are included in the experiment. As was the case with the randomised block design, calculating the test statistic in this type of experiment is quite time-consuming. As a result, we will use Excel and Minitab to produce our statistics.

EXAMPLE 13.4

DATA Xm13-04

Comparing the Lifetime Number of Jobs by Educational Level

One measure of the health of a nation's economy is how quickly it creates jobs. One aspect of this issue is the number of jobs individuals hold. As part of a study on job tenure, you are supposed to conduct a survey in which British men and women aged between 37 and 45 were asked how many jobs they have held in their lifetimes. Also recorded was educational attainment. The categories are

Less than high school (E1)

High school (E2)

Some college/university but no degree (E3)

At least one university degree (E4)

The data are shown for each of the eight categories of gender and education. Can we infer that differences exist between genders and educational levels?

Male E1	Male E2	Male E3	Male E4	Female E1	Female E2	Female E3	Female E4
10	12	15	8	7	7	5	7
9	11	8	9	13	12	13	9
12	9	7	5	14	6	12	3
16	14	7	11	6	15	3	7
14	12	7	13	11	10	13	9
17	16	9	8	14	13	11	6
13	10	14	7	13	9	15	10
9	10	15	11	11	15	5	15
11	5	11	10	14	12	9	4
15	11	13	8	12	13	8	11

SOLUTION:

IDENTIFY

We begin by treating this example as a one-way analysis of variance. Notice that there are eight treatments. However, the treatments are defined by two different factors. One factor is gender, which has two levels. The second factor is educational attainment, which has four levels.

We can proceed to solve this problem in the same way we did in Section 13-1: we test the following hypotheses.

$$H_0: \quad \mu_1 = \mu_2 = \mu_3 = \mu_4 = \mu_5 = \mu_6 = \mu_7 = \mu_8$$

$$H_1: \quad \text{At least two means differ}$$

COMPUTE

EXCEL

	A	B	C	D	E	F	G
1	Anova: Single Factor						
2							
3	SUMMARY						
4	*Groups*	*Count*	*Sum*	*Average*	*Variance*		
5	Male E1	10	126	12.60	8.27		
6	Male E2	10	110	11.00	8.67		
7	Male E3	10	106	10.60	11.60		
8	Male E4	10	90	9.00	5.33		
9	Female E1	10	115	11.50	8.28		
10	Female E2	10	112	11.20	9.73		
11	Female E3	10	94	9.40	16.49		
12	Female E4	10	81	8.10	12.32		
13							
14							
15	ANOVA						
16	*Source of Variation*	*SS*	*df*	*MS*	*F*	*P-value*	*F crit*
17	Between Groups	153.35	7	21.91	2.17	0.0467	2.1397
18	Within Groups	726.20	72	10.09			
19							
20	Total	879.55	79				

MINITAB

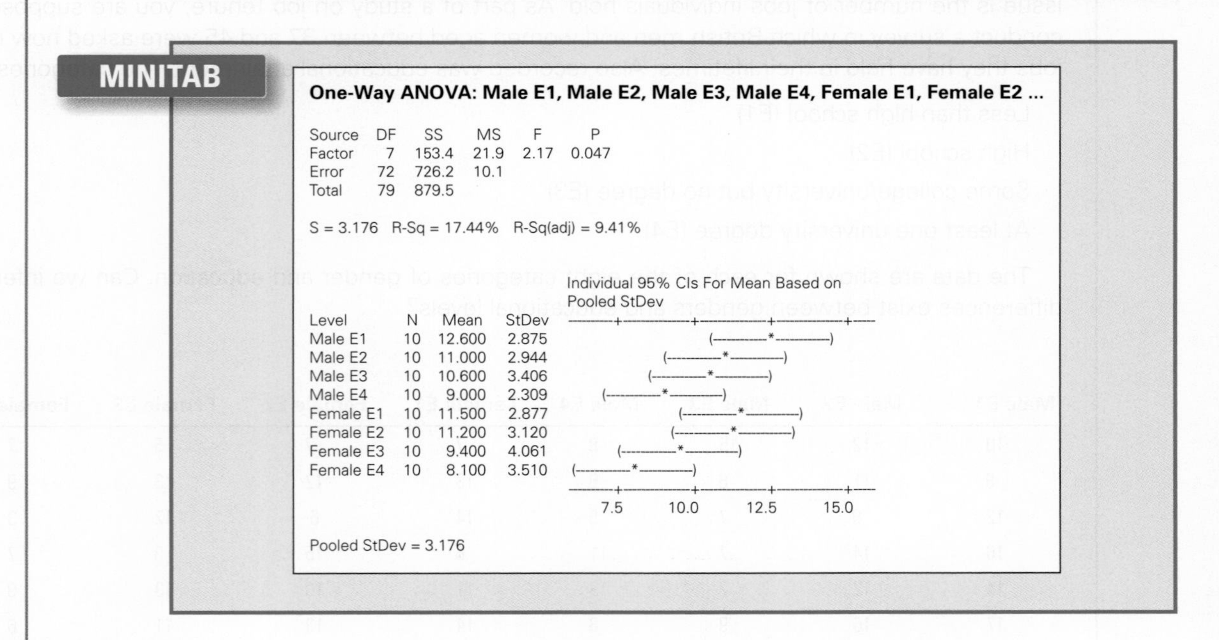

One-Way ANOVA: Male E1, Male E2, Male E3, Male E4, Female E1, Female E2 ...

Source	DF	SS	MS	F	P
Factor	7	153.4	21.9	2.17	0.047
Error	72	726.2	10.1		
Total	79	879.5			

S = 3.176 R-Sq = 17.44% R-Sq(adj) = 9.41%

Individual 95% CIs For Mean Based on Pooled StDev

Level	N	Mean	StDev
Male E1	10	12.600	2.875
Male E2	10	11.000	2.944
Male E3	10	10.600	3.406
Male E4	10	9.000	2.309
Female E1	10	11.500	2.877
Female E2	10	11.200	3.120
Female E3	10	9.400	4.061
Female E4	10	8.100	3.510

7.5 10.0 12.5 15.0

Pooled StDev = 3.176

INTERPRET

The value of the test statistic is $F = 2.17$ with a p-value of 0.0467. We conclude that there are differences in the number of jobs between the eight treatments.

This statistical result raises more questions – namely, can we conclude that the differences in the mean number of jobs are caused by differences between males and females? Or are they caused by differences between educational levels? Or, perhaps, are there combinations, called **interactions**, of gender and education that result in especially high or low numbers? To show how we test for each type of difference, we need to develop some terminology.

A **complete factorial experiment** is an experiment in which the data for all possible combinations of the levels of the factors are gathered. That means that in Example 13.4 we measured the number of jobs for all eight combinations. This experiment is called a complete 2 × 4 factorial experiment, where the actual numbers represent both the number of factors and the number of levels within each factor.

In general, we will refer to one of the factors as factor A (arbitrarily chosen). The number of levels of this factor will be denoted by *a*. The other factor is called factor B, and its number of levels is denoted by *b*. This terminology becomes clearer when we present the data from Example 13.4 in another format. Table 13.6 depicts the layout for a *two-way classification,* which is another name for the complete factorial experiment. The number of observations for each combination is called a **replicate**. The number of replicates is denoted by *r*. In this book, we address only problems in which the number of replicates is the same for each treatment. Such a design is called **balanced**.

TABLE 13.6 **Two-Way Classification for Example 13.4**

	MALE	FEMALE
Less than high school	10	7
	9	13
	12	14
	16	6
	14	11
	17	14
	13	13
	9	11
	11	14
	15	12
High school	12	7
	11	12
	9	6
	14	15
	12	10
	16	13
	10	9
	10	15
	5	12
	11	13
Less than bachelor's degree	15	5
	8	13
	7	12
	7	3
	7	13
	9	11
	14	15

	MALE	FEMALE
	15	5
	11	9
	13	8
At least one bachelor's degree	8	7
	9	9
	5	3
	11	7
	13	9
	8	6
	7	10
	11	15
	10	4
	8	11

Thus, we use a complete factorial experiment where the number of treatments is *ab* with *r* replicates per treatment. In Example 13.4, $a = 2$, $b = 4$ and $r = 10$. As a result, we have ten observations for each of the eight treatments.

If you examine the ANOVA table, you can see that the total variation is SS(Total) = 879.55, the sum of squares for treatments is SST = 153.35, and the sum of squares for error is SSE = 726.20. The variation caused by the treatments is measured by SST. To determine whether the differences result from factor A, factor B or some interaction between the two factors, we need to partition SST into three sources. These are SS(A), SS(B) and SS(AB).

For those whose mathematical confidence is high, we have provided an explanation of the notation as well as the definitions of the sums of squares. Learning how the sums of squares are calculated is useful but hardly essential to your ability to conduct the tests.

13-5a How the Sums of Squares for Factors A and B and Interaction are Computed

To help you understand the formulas, we will use the following notation:

x_{ijk} = *k*th observation in the *ij*th treatment

$\bar{x}[AB]_{ij}$ = Mean of the response variable in the *ij*th treatment (mean of the treatment when the factor A level is *i* and the factor B level is *j*)

$\bar{x}[A]_i$ = Mean of the observations when the factor A level is *i*

$\bar{x}[B]_j$ = Mean of the observations when the factor B level is *j*

$\bar{\bar{x}}$ = Mean of all the observations

a = Number of factor A levels

b = Number of factor B levels

r = Number of replicates

In this notation, $\bar{x}[AB]_{11}$ is the mean of the responses for factor A level 1 and factor B level 1. The mean of the responses for factor A level 1 is $\bar{x}[A]_1$. The mean of the responses for factor B level 1 is $\bar{x}[B]_1$.

Table 13.7 describes the notation for the two-factor analysis of variance.

TABLE 13.7 Notation for Two-Factor Analysis of Variance

	FACTOR A							
FACTOR B	**1**		**2**		**…**	***a***		
1	x_{111} x_{112} ⋮ x_{11r}	$\bar{x}[AB]_{11}$	x_{211} x_{212} ⋮ x_{21r}	$\bar{x}[AB]_{21}$		x_{a11} x_{a12} ⋮ x_{a1r}	$\bar{x}[AB]_{a1}$	$\bar{x}[B]_1$
2	x_{121} x_{122} ⋮ x_{12r}	$\bar{x}[AB]_{12}$	x_{221} x_{222} ⋮ x_{22r}	$\bar{x}[AB]_{22}$		x_{a21} x_{a22} ⋮ x_{a2r}	$\bar{x}[AB]_{a2}$	$\bar{x}[B]_2$
⋮								
b	x_{1b1} x_{1b2} ⋮ x_{1br}	$\bar{x}[AB]_{1b}$	x_{2b1} x_{2b2} ⋮ x_{2br}	$\bar{x}[AB]_{2b}$		x_{ab1} x_{ab2} ⋮ x_{abr}	$\bar{x}[AB]_{ab}$	$\bar{x}[B]_b$
	$\bar{x}[A]_1$		$\bar{x}[A]_2$			$\bar{x}[A]_a$		$\bar{\bar{x}}$

The sums of squares are defined as follows.

Sums of Squares in the Two-Factor Analysis of Variance

$$\text{SS(Total)} = \sum_{i=1}^{a}\sum_{j=1}^{b}\sum_{k=1}^{r}(x_{ijk} - \bar{\bar{x}})^2$$

$$\text{SS(A)} = rb\sum_{i=1}^{a}(\bar{x}[A]_i - \bar{\bar{x}})^2$$

$$\text{SS(B)} = ra\sum_{i=1}^{b}(\bar{x}[B]_j - \bar{\bar{x}})^2$$

$$\text{SS(AB)} = r\sum_{i=1}^{a}\sum_{j=1}^{b}(\bar{x}[AB]_{ij} - \bar{x}[A]_i - \bar{x}[B]_j + \bar{\bar{x}})^2$$

$$\text{SSE} = \sum_{i=1}^{a}\sum_{j=1}^{b}\sum_{k=1}^{r}(x_{ijk} - \bar{x}[AB]_{ij})^2$$

To compute SS(A), we calculate the sum of the squared differences between the factor A level means, which are denoted $\bar{x}[A]_i$ and the grand mean $\bar{\bar{x}}$. The sum of squares for factor B, SS(B), is defined similarly. The interaction sum of squares, SS(AB), is calculated by taking each treatment mean (a treatment consists of a combination of a level of factor A and a level of factor B), subtracting the factor A level mean, subtracting the factor B level mean, adding the grand mean, squaring this quantity and adding. The sum of squares for error, SSE, is calculated by subtracting the treatment means from the observations, squaring and adding.

To test for each possibility, we conduct several *F*-tests similar to the one performed in Section 13-1. Figure 13.4 illustrates the partitioning of the total sum of squares that leads to the *F*-tests. We have included in this figure the partitioning used in the one-way study. When the one-way analysis of variance allows us to infer that differences between the treatment means exist, we continue our analysis by partitioning the treatment sum of squares into three sources of variation. The first is sum of squares for factor A, which we label SS(A), which measures the variation between the levels of factor A. Its degrees of freedom are $a - 1$. The second is the sum of squares for factor B, whose degrees of freedom are $b - 1$. SS(B) is the variation between the levels of factor B. The interaction sum of squares is labelled SS(AB), which is a measure of the amount of variation between the combinations of factors A and B; its degrees of freedom are $(a - 1) \times (b - 1)$. The sum of squares for error is SSE, and its degrees of freedom are $n - ab$. (Recall that n is the total sample size, which in this experiment is $n = abr$.) Notice that SSE and its number of degrees of freedom are identical in both partitions. As in the previous experiment, SSE is the variation within the treatments.

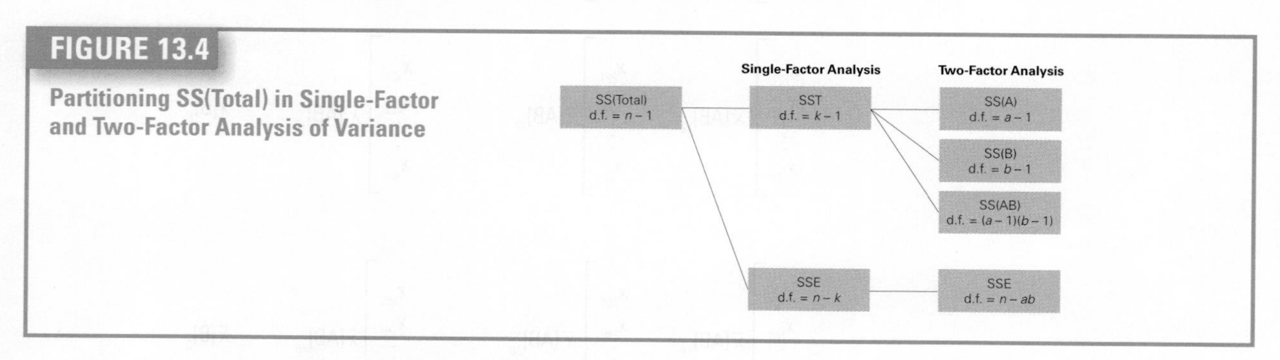

FIGURE 13.4

Partitioning SS(Total) in Single-Factor and Two-Factor Analysis of Variance

F-Tests Conducted in Two-Factor Analysis of Variance

Test for Differences between the Levels of Factor A

H_0: The means of the a levels of factor A are equal

H_1: At least two means differ

$$\text{Test statistic: } F = \frac{\text{MS(A)}}{\text{MSE}}$$

Test for Differences between the Levels of Factor B

H_0: The means of the b levels of factor B are equal

H_1: At least two means differ

$$\text{Test statistic: } F = \frac{\text{MS(B)}}{\text{MSE}}$$

Test for Interaction between Factors A and B

H_0: Factors A and B do not interact to affect the mean responses

H_1: Factors A and B do interact to affect the mean responses

$$\text{Test statistic: } F = \frac{\text{MS(AB)}}{\text{MSE}}$$

Required Conditions

1. The distribution of the response is normally distributed.
2. The variance for each treatment is identical.
3. The samples are independent.

As in the two previous experimental designs of the analysis of variance, we summarise the results in an ANOVA table. Table 13.8 depicts the general form of the table for the complete factorial experiment.

TABLE 13.8 **ANOVA Table for the Two-Factor Experiment**

SOURCE OF VARIATION	DEGREES OF FREEDOM	SUMS OF SQUARES	MEAN SQUARES	F-STATISTIC
Factor A	$a-1$	SS(A)	$MS(A) = SS(A)/(a-1)$	$F = MS(A)/MSE$
Factor B	$b-1$	SS(B)	$MS(B) = SS(B)/(b-1)$	$F = MS(B)/MSE$
Interaction	$(a-1)(b-1)$	SS(AB)	$MS(AB) = SS(AB)/[(a-1)(b-1)]$	$F = MS(AB)/MSE$
Error	$n-ab$	SSE	$MSE = SSE/(n-ab)$	
Total	$n-1$	SS(Total)		

We will illustrate the techniques using the data in Example 13.4. All calculations will be performed by Excel and Minitab.

EXCEL

	A	B	C	D	E	F	G
1	Anova: Two-Factor with Replication						
2							
3	SUMMARY	Male	Female	Total			
4	*Less than High School*						
5	Count	10	10	20			
6	Sum	126	115	241			
7	Average	12.6	11.5	12.1			
8	Variance	8.27	8.28	8.16			
9							
10	*High School*						
11	Count	10	10	20			
12	Sum	110	112	222			
13	Average	11.0	11.2	11.1			
14	Variance	8.67	9.73	8.73			
15							
16	*Less than Bachelor's*						
17	Count	10	10	20			
18	Sum	106	94	200			
19	Average	10.6	9.4	10.0			
20	Variance	11.6	16.49	13.68			
21							
22	*Bachelor's or more*						
23	Count	10	10	20			
24	Sum	90	81	171			
25	Average	9.0	8.1	8.6			
26	Variance	5.33	12.32	8.58			
27							
28	*Total*						
29	Count	40	40				
30	Sum	432	402				
31	Average	10.8	10.1				
32	Variance	9.50	12.77				
33							
34	ANOVA						
35	*Source of Variation*	*SS*	*df*	*MS*	*F*	*P-value*	*F crit*
36	Sample	135.85	3	45.28	4.49	0.0060	2.7318
37	Columns	11.25	1	11.25	1.12	0.2944	3.9739
38	Interaction	6.25	3	2.08	0.21	0.8915	2.7318
39	Within	726.20	72	10.09			
40							
41	Total	879.55	79				

In the ANOVA table, **Sample** refers to factor B (educational level) and **Columns** refers to factor A (gender). Thus, MS(B) = 45.28, MS(A) = 11.25, MS(AB) = 2.08 and MSE = 10.09. The *F*-statistics are 4.49 (educational level), 1.12 (gender) and 0.21 (interaction).

Instructions

1. Type or import the data using the same format as Xm13-04a. (*Note:* You must label the rows and columns as we did.)
2. Click **Data, Data Analysis** and **Anova: Two-Factor with Replication.**
3. Specify the **Input Range** (A1:C41). Type the number of replications in the **Rows per sample** box (10).
4. Specify a value for α (0.05).

MINITAB

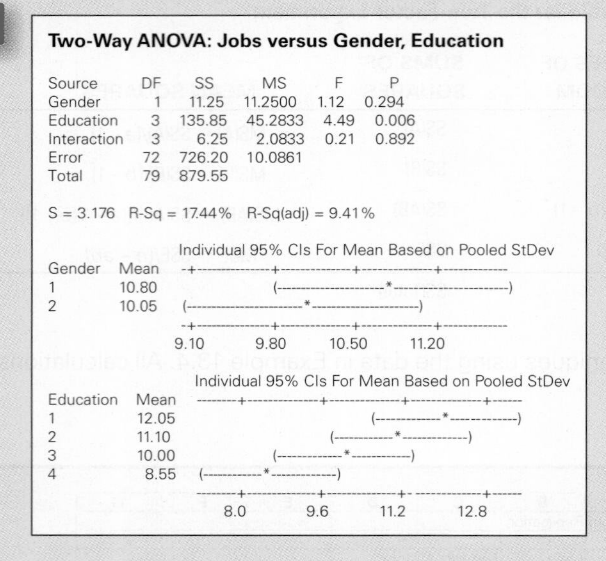

```
Two-Way ANOVA: Jobs versus Gender, Education

Source        DF      SS       MS       F      P
Gender         1   11.25  11.2500    1.12  0.294
Education      3  135.85  45.2833    4.49  0.006
Interaction    3    6.25   2.0833    0.21  0.892
Error         72  726.20  10.0861
Total         79  879.55

S = 3.176   R-Sq = 17.44%   R-Sq(adj) = 9.41%

                 Individual 95% CIs For Mean Based on Pooled StDev
Gender   Mean    -+-------------+-------------+-------------+------------
1        10.80              (--------------------*--------------------)
2        10.05   (--------------------*--------------------)
                 -+-------------+-------------+-------------+------------
                 9.10        9.80         10.50         11.20

                     Individual 95% CIs For Mean Based on Pooled StDev
Education   Mean     -------+-------------+-------------+-------------+------
1           12.05                                 (-------------*-------------)
2           11.10                           (------------*------------)
3           10.00                 (-------------*------------)
4            8.55    (------------*------------)
                     -------+-------------+-------------+-------------+------
                           8.0           9.6          11.2          12.8
```

Instructions

1. Type or import the data in stacked format in three columns. One column contains the responses, another contains codes for the levels of factor A, and a third column contains codes for the levels of factor B. (**Open** Xm13-04b.)
2. Click **Stat, ANOVA** and **Two-way** ....
3. Specify the **Responses** (Jobs), **Row factor** (Gender) and **Column factor** (Education).
4. To produce the graphics check **Display means.**

13-5b Test for Differences in Number of Jobs between Men and Women

H_0: The means of the two levels of factor A are equal

H_1: At least two means differ

$$\text{Test statistic: } F = \frac{\text{MS(A)}}{\text{MSE}}$$

Value of the test statistic: From the computer output, we have

MS(A) = 11.25, MSE = 10.09 and F = 11.25/10.09 = 1.12 (p-value = 0.2944)

There is no evidence at the 5% significance level to infer that differences in the number of jobs exist between men and women.

13-5c Test for Differences in Number of Jobs between Education Levels

H_0: The means of the four levels of factor B are equal

H_1: At least two means differ

$$\text{Test statistic: } F = \frac{\text{MS(B)}}{\text{MSE}}$$

Value of the test statistic: From the computer output, we find

MS(B) = 45.28 and MSE = 10.09. Thus, F = 45.28/10.09 = 4.49 (p-value = 0.0060).

There is sufficient evidence at the 5% significance level to infer that differences in the number of jobs exist between educational levels.

13-5d Test for Interaction between Factors A and B

H_0: Factors A and B do not interact to affect the mean number of jobs

H_1: Factors A and B do interact to affect the mean number of jobs

$$\text{Test statistic: } F = \frac{MS(AB)}{MSE}$$

Value of the test statistic: From the printouts,

$MS(AB) = 2.08$, $MSE = 10.09$ and $F = 2.08/10.09 = 0.21$ (p-value $= 0.8915$).

There is not enough evidence to conclude that there is an interaction between gender and education.

Figure 13.5 is a graph of the mean responses for each of the eight treatments. As you can see, there are small (not significant) differences between males and females. There are significant differences between men and women with different educational backgrounds. Finally, there is no interaction.

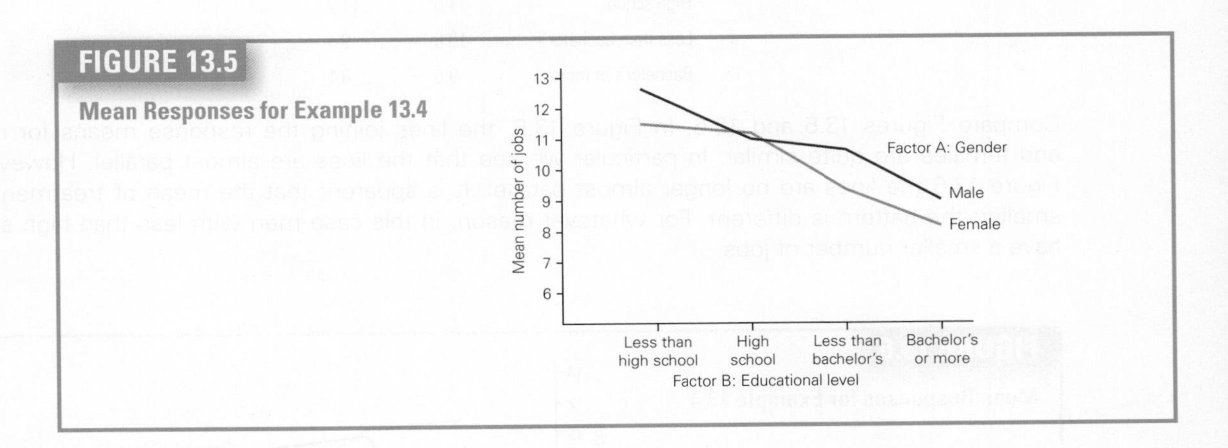

FIGURE 13.5
Mean Responses for Example 13.4

13-5e What Is Interaction?

To more fully understand interaction we have changed the sample associated with men who have not finished high school (Treatment 1). We subtracted 6 from the original numbers so that the sample in treatment 1 is

4, 3, 6, 10, 8, 11, 7, 3, 5, 9

The new data are stored in Xml3-04c (Excel format) and Xml3-04d (Minitab format). The mean is 6.6. Here are the Excel and Minitab ANOVA tables.

EXCEL

	A	B	C	D	E	F	G
35	ANOVA						
36	*Source of Variation*	*SS*	*df*	*MS*	*F*	*P-value*	*F crit*
37	Sample	75.85	3	25.28	2.51	0.0657	2.7318
38	Columns	11.25	1	11.25	1.12	0.2944	3.9739
39	Interaction	120.25	3	40.08	3.97	0.0112	2.7318
40	Within	726.20	72	10.09			
41							
42	Total	933.55	79				

MINITAB

Two-Way ANOVA: Jobs versus Gender, Education

Source	DF	SS	MS	F	P
Gender	1	11.25	11.2500	1.12	0.294
Education	3	75.85	25.2833	2.51	0.066
Interaction	3	120.25	40.0833	3.97	0.011
Error	72	726.20	10.0861		
Total	79	933.55			

INTERPRET

In this example there is not enough evidence (at the 5% significance level) to infer that there are differences between men and women and between the educational levels. However, there is sufficient evidence to conclude that there is interaction between gender and education.

	Male	Female
Less than high school	6.6	11.5
High school	11.0	11.2
Less than bachelor's	10.6	9.4
Bachelor's or more	9.0	8.1

Compare Figures 13.5 and 13.6. In Figure 13.5, the lines joining the response means for males and females are quite similar. In particular we see that the lines are almost parallel. However, in Figure 13.6 the lines are no longer almost parallel. It is apparent that the mean of treatment 1 is smaller; the pattern is different. For whatever reason, in this case men with less than high school have a smaller number of jobs.

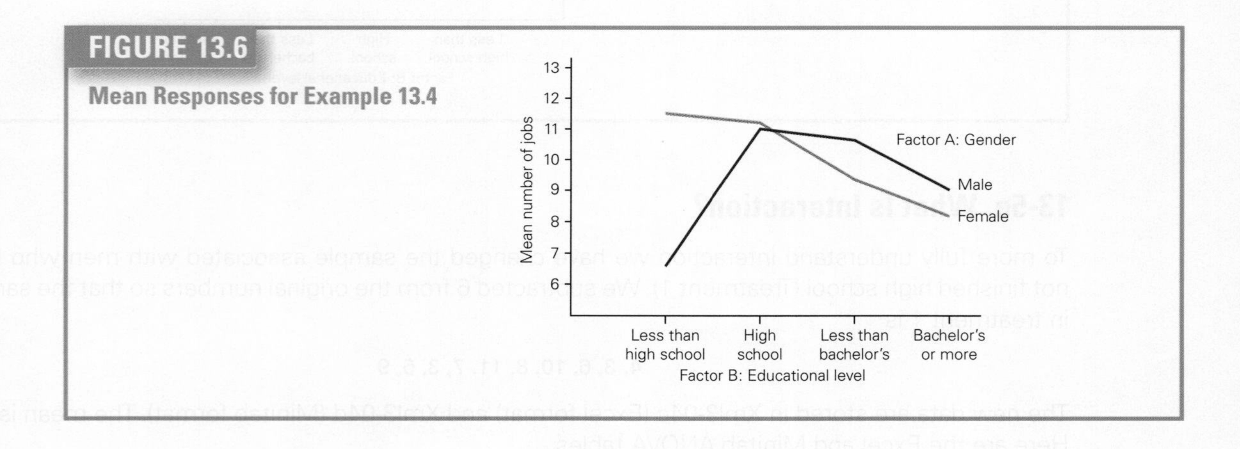

FIGURE 13.6
Mean Responses for Example 13.4

13-5f Conducting the Analysis of Variance for the Complete Factorial Experiment

In addressing the problem outlined in Example 13.4, we began by conducting a one-way analysis of variance to determine whether differences existed between the eight treatment means. This was done primarily for pedagogical reasons to enable you to see that when the treatment means differ; we need to analyse the reasons for the differences. However, in practice, we generally do not conduct this test in the complete factorial experiment (although it should be noted that some statistics practitioners prefer this 'two-stage' strategy). We recommend that you proceed directly to the two-factor analysis of variance.

In the two versions of Example 13.4 we conducted the tests of each factor and then the test for interaction.

However, if there is evidence of interaction, the tests of the factors are irrelevant. There may or may not be differences between the levels of factor A and the levels of factor B. Accordingly, we change the order of conducting the *F*-tests.

Order of Testing in the Two-Factor Analysis of Variance

Test for interaction first. If there is enough evidence to infer that there is interaction, do not conduct the other tests.

If there is not enough evidence to conclude that there is interaction, proceed to conduct the *F*-tests for factors A and B.

13-5g Developing an Understanding of Statistical Concepts

You may have noticed that there are similarities between the two-factor experiment and the randomised block experiment. In fact, when the number of replicates is one, the calculations are identical. (Minitab uses the same command.) This raises the question, What is the difference between a factor in a multifactor study and a block in a randomised block experiment? In general, the difference between the two experimental designs is that in the randomised block experiment, blocking is performed specifically to reduce variation, whereas in the two-factor model the effect of the factors on the response variable is of interest to the statistics practitioner. The criteria that define the blocks are always characteristics of the experimental units. Consequently, factors that are characteristics of the experimental units will be treated not as factors in a multifactor study, but as blocks in a randomised block experiment.

Let's review how we recognise the need to use the procedure described in this section.

Factors that Identify the Independent Samples in a Two-Factor Analysis of Variance

1. **Problem objective:** Compare two or more populations (populations are defined as the combinations of the levels of two factors).
2. **Data type:** Interval.
3. **Experimental design:** Independent samples.

EXERCISES

13.35 A two-factor analysis of variance experiment was performed with $a = 3$, $b = 4$ and $r = 20$. The following sums of squares were computed:

SS(Total) = 42 450 SS(A) = 1560

SS(B) = 2 880 SS(AB) = 7605

Conduct whatever test you deem necessary at the 1% significance level to determine whether there are differences between the levels of factor A, the levels of factor B or interaction between factors A and B.

13.36 A statistics practitioner conducted a two-factor analysis of variance experiment with $a = 4$, $b = 3$ and $r = 8$. The sums of squares are listed here:

SS(Total) = 9420 SS(A) = 203 SS(B) = 859

SS(AB) = 513

a. Test at the 5% significance level to determine whether factors A and B interact.

b. Test at the 5% significance level to determine whether differences exist between the levels of factor A.

c. Test at the 5% significance level to determine whether differences exist between the levels of factor B.

13.37 Xr13-37 The following data were generated from a 2 × 2 factorial experiment with three replicates:

Factor A	Factor B 1	2
1	6	12
	9	10
	7	11
2	9	15
	10	14
	5	10

a. Test at the 5% significance level to determine whether factors A and B interact.

b. Test at the 5% significance level to determine whether differences exist between the levels of factor A.

c. Test at the 5% significance level to determine whether differences exist between the levels of factor B.

13.38 Xr13-38 The data shown here were taken from a 2 × 3 factorial experiment with four replicates:

Factor A	Factor B 1	2
1	23	20
	18	17
	17	16
	20	19
2	27	29
	23	23
	20	27
	28	25
3	23	27
	21	19
	24	20
	16	22

a. Test at the 5% significance level to determine whether factors A and B interact.

b. Test at the 5% significance level to determine whether differences exist between the levels of factor A.

c. Test at the 5% significance level to determine whether differences exist between the levels of factor B.

13.39 Xr13-39 Refer to Example 13.4. We have revised the data by adding 2 to each of the numbers of the men. What do these data tell you?

13.40 Xr13-40 Refer to Example 13.4. We have altered the data by subtracting 4 from the numbers of treatment. What do these data tell you?

Applications

The following exercises require the use of a computer and software.

13.41 Xr13-41 Detergent manufacturers frequently make claims about the effectiveness of their products. A consumer-protection service decided to test the five best selling brands of detergent, where each manufacturer claims that its product produces the 'whitest whites' in all water temperatures. The experiment was conducted in the following way. One hundred and fifty white sheets were equally soiled. Thirty sheets were washed in each brand – ten with cold water, ten with warm water and ten with hot water. After washing, the 'whiteness' scores for each sheet were measured with laser equipment.

Column 1: Water temperature code

Column 2: Scores for detergent 1 (first 10 rows = cold water, middle 10 rows = warm and last 10 rows = hot)

Column 2: Scores for detergent 2 (same format as column 2) Column 3: Scores for detergent 3 (same format as column 2)

Column 4: Scores for detergent 4 (same format as column 2)

Column 5: Scores for detergent 5 (same format as column 2)

a. What are the factors in this experiment?

b. What is the response variable?

c. Identify the levels of each factor.

d. Perform a statistic analysis using a 5% significance level to determine whether there is sufficient statistical evidence to infer that there are differences in whiteness scores between the five detergents, differences in whiteness scores between the three water temperatures, or interaction between detergents and temperatures.

13.42 Xr13-42 Headaches are one of the most common, but least understood, ailments. Most people get headaches several times per month; over-the-counter medication is usually sufficient to eliminate their pain. However, for a significant proportion of people, headaches are debilitating and make their lives almost unbearable. Many such people have investigated a wide spectrum of possible treatments, including narcotic drugs, hypnosis, biofeedback and acupuncture, with little or no success. In the last few years, a promising new treatment has been developed. Simply described, the treatment involves a series of injections of a local anesthetic to the occipital nerve (located in the back of the neck). The current treatment procedure is to schedule the injections once a week for 4 weeks. However, it has been suggested that another procedure may be better – one that features one injection every other day for a total of four injections. Additionally, some physicians recommend other combinations of drugs that may increase the effectiveness of the injections. To analyse the problem, an experiment was organised. It was decided to test for a difference between the two schedules of injection and to determine whether there are differences between four drug mixtures. Because of the possibility of an interaction between the schedule and the drug, a complete factorial experiment was chosen. Five headache patients were randomly selected for each combination of schedule and drug. Forty patients were treated and each was asked to report the frequency, duration and severity of his or her headache prior to treatment and for the 30 days following the last injection. An index ranging from 0 to 100 was constructed for each patient, where 0 indicates no headache pain and 100 specifies the worst headache pain. The improvement in the headache index for each patient was recorded and reproduced in the accompanying table. (A negative value indicates a worsening condition.) (The author is grateful to Dr Lome Greenspan for his help in writing this example.)

a. What are the factors in this experiment?

b. What is the response variable?

c. Identify the levels of each factor.

d. Analyse the data and conduct whichever tests you deem necessary at the 5% significance level to determine whether there is sufficient statistical evidence to infer that there are differences in the improvement in the headache index between the two schedules, differences in the improvement in the headache index between the four drug mixtures, or interaction between schedules and drug mixtures.

Improvement in Headache Index

	Drug Mixture			
Schedule	**1**	**2**	**3**	**4**
One Injection	17	24	14	10
Every Week	6	15	9	−1
(4 Weeks)	10	10	12	0
	12	16	0	3
	14	14	6	−1
One Injection	18	−2	20	−2
Every 2 Days	9	0	16	7
(4 Days)	17	17	12	10
	21	2	17	6
	15	6	18	7

13.43 Xr13-43 Most college lecturers prefer to have their students participate actively in seminars. Ideally, students will ask their lecturers questions and answer their lecturers' questions, making the seminar experience more interesting and useful. Many lecturers seek ways to encourage their students to participate in seminars. A statistics lecturer believes that there are a number of external factors that affect student participation. He believes that the time of day and the configuration of seats are two such factors. Consequently, he organised the following experiment. Six groups of about 60 students each were scheduled for one semester. Two groups were scheduled at 9:00 am, two at 1:00 pm and two at 4:00 pm. At each of the three times, one of the groups was assigned to a room where the seats were arranged in rows of ten seats. The other group was a U-shaped, tiered room, where students not only face the instructor, but face their fellow students as well. In each of the six classrooms, over 5 days, student participation was measured by counting the number of times students asked and answered questions. These data are displayed in the accompanying table.

a. How many factors are there in this experiment? What are they?

b. What is the response variable?

c. Identify the levels of each factor.

d. What conclusions can the professor draw from these data?

	Time		
Group Configuration	**9:00 am**	**1:00 pm**	**4:00 pm**
Rows	10	9	7
	7	12	12
	9	12	9
	6	14	20
	6	8	7
U-Shape	15	4	7
	18	4	4
	11	7	9
	13	4	8
	13	6	7

CHAPTER SUMMARY

The analysis of variance allows us to test for differences between populations when the data are interval. The analyses of the results of three different experimental designs were presented in this chapter. The one-way analysis of variance defines the populations on the basis of one factor. The second experimental design also defines the treatments on the basis of one factor. However, the randomised block design uses data gathered by observing the results of a matched or blocked experiment (two-way analysis of variance). The third design is the two-factor experiment wherein the treatments are defined as the combinations of the levels of two factors. All the analyses of variance are based on partitioning the total sum of squares into sources of variation from which the mean squares and *F*-statistics are computed.

In addition, we introduced three multiple comparison methods that allow us to determine which means differ in the one-way analysis of variance.

Finally, we described an important application in operations management that employs the analysis of variance.

SYMBOLS:

Symbol	Pronounced	Represents
$\bar{\bar{x}}$	*x* double bar	Overall or grand mean
q		Studentised range
ω	Omega	Critical value of Tukey's multiple comparison method
$q_\alpha(k,\nu)$	*q* sub alpha *k v*	Critical value of the Studentised range
n_g		Number of observations in each of *k* samples
$\bar{x}[T]_j$	*x* bar *T* sub *j*	Mean of the *j*th treatment
$\bar{x}[B]_i$	*x* bar *B* sub *i*	Mean of the *i*th block
$\bar{x}[AB]_{ij}$	*x* bar *A B* sub *ij*	Mean of the *ij*th treatment
$\bar{x}[A]_i$	*x* bar *A* sub *i*	Mean of the observations when the factor A level is *i*
$\bar{x}[B]_j$	*x* bar *B* sub *j*	Mean of the observations when the factor B level is *j*

FORMULAS:

One-way analysis of variance

$$SST = \sum_{j=1}^{k} n_j(x_j - x)^2$$

$$SSE = \sum_{j=1}^{k}\sum_{i=1}^{k}(x_{ij} - \bar{x}_j)^2$$

$$MST = \frac{SST}{k-1}$$

$$MST = \frac{SSE}{n-k}$$

$$F = \frac{MST}{MSE}$$

Least significant difference comparison method

$$LSD = t_{\alpha/2}\sqrt{MSE\left(\frac{1}{n_i} + \frac{1}{n_j}\right)}$$

Tukey's multiple comparison method

$$\omega = q_\alpha(k,\nu)\sqrt{\frac{MSE}{n_g}}$$

Two-way analysis of variance (randomised block design of experiment)

$$SS(\text{Total}) = \sum_{j=1}^{k}\sum_{i=1}^{b}(x_{ij} - \overline{\overline{x}})^2$$

$$SST = \sum_{j=1}^{k} b(\overline{x}[T]_j - \overline{\overline{x}})^2$$

$$SSB = \sum_{i=1}^{k} b(\overline{x}[B]_i - \overline{\overline{x}})^2$$

$$SSE = \sum_{j=1}^{k}\sum_{i=1}^{k}(x_{ij} - \overline{x}[T]_j - \overline{x}[B]_i + \overline{\overline{x}})^2$$

$$MST = \frac{SST}{k-1}$$

$$MSB = \frac{SSB}{b-1}$$

$$MSE = \frac{SSE}{n-k-b+1}$$

$$F = \frac{MST}{MSE}$$

$$F = \frac{MSB}{MSE}$$

$$SS(\text{Total}) = \sum_{i=1}^{a}\sum_{j=1}^{b}\sum_{k=1}^{r}(x_{ijk} - \overline{\overline{x}})^2$$

$$SS(A) = rb\sum_{j=1}^{a}(\overline{x}[A]_i - \overline{\overline{x}})^2$$

$$SS(B) = ra\sum_{j=1}^{b}(x[B]_j - x)^2$$

$$SS(AB) = r\sum_{i=1}^{a}\sum_{j=1}^{b}(\overline{x}[AB]_{ij} - \overline{x}[A]_i - \overline{x}[B]_j + \overline{\overline{x}})^2$$

$$SSE = \sum_{i=1}^{a}\sum_{j=1}^{b}\sum_{k=1}^{r}(x_{ijk} - \overline{x}[AB]_{ij})^2$$

$$MS(A) = \frac{SS(A)}{a-1}$$

$$MS(B) = \frac{SS(B)}{b-1}$$

$$MS(AB) = \frac{SS(AB)}{(a-1)(b-1)}$$

$$F = \frac{MS(A)}{MSE}$$

$$F = \frac{MS(B)}{MSE}$$

$$F = \frac{MS(AB)}{MSE}$$

CHAPTER EXERCISES

The following exercises require the use of a computer and software. Use a 5% significance level.

13.44 Xr13-44 Each year billions of pounds are lost because of injuries at work. Costs can be decreased if injured workers can be rehabilitated quickly. As part of an analysis of the amount of time taken for workers to return to work, a sample of male working class workers aged 35 to 45, who suffered a common wrist fracture, was taken. The researchers believed that the mental and physical condition of the individual affects recovery time. Each man was given a questionnaire to complete, which measured whether he tended to be optimistic or pessimistic. Their physical condition was also evaluated and categorised as very physically fit, average, or in poor condition. The number of days until the wrist returned to full function was measured for each individual. These data were recorded in the following way:

Column 1: Time to recover for optimists ((columns 1–10) = very fit, rows 11–20 = in average condition, rows 21–30 = poor condition)

Column 2: Time to recover for pessimists (same format as column 1)

a. What are the factors in this experiment? What are the levels of each factor?

b. Can we conclude that pessimists and optimists differ in their recovery times?

c. Can we conclude that physical condition affects recovery times?

13.45 Xr13-45 To help high school students pick a major, a company called PayScale surveys graduates of a variety of programmes. In one such survey, graduates of the following degree programmes were asked what their annual salaries were after working at least 10 years in the field.

Elementary Education

Human Development

Social Work

Special Education

Can we infer that there are differences in salary between the four college degrees?

13.46 Xr13-46 The editor of the student newspaper was in the process of making some major changes in the newspaper's layout. He was also contemplating changing the typeface of the print used. To help himself make a decision, he set up an experiment in which 20 individuals were asked to read four newspaper pages, with each page printed in a different typeface. If the reading speed differed, then the typeface that was read fastest would be used. However, if there was not enough evidence to allow the editor to conclude that such differences existed, the current typeface would be continued. The times (in seconds) to completely read one page were recorded. What should the editor do?

13.47 Xr13-47 In marketing children's products, it is extremely important to produce television commercials that hold the attention of the children who view them. A psychologist hired by a marketing research firm wants to determine whether differences in attention span exist between children watching advertisements for different types of products. One hundred and fifty children under 10 years of age were recruited for an experiment. One-third watched a 60-second commercial for a new computer game, one-third watched a commercial for a breakfast cereal and one-third watched a commercial for children's clothes. Their attention spans (in seconds) were measured and recorded. Do these data provide enough evidence to conclude that there are differences in attention span between the three products advertised?

13.48 Xr13-48 On reconsidering the experiment in Exercise 13.47, the psychologist decides that the age of the child may influence the attention span. Consequently, the experiment is redone in the following way. Three 10-year-olds, three 9-year-olds, three 8-year-olds, three 7-year-olds, three 6-year-olds, three 5-year-olds and three 4-year-olds are randomly assigned to watch one of the commercials, and their attention spans are measured. Do the results indicate that there are differences in the abilities of the products advertised to hold children's attention?

13.49 Xr13-49 Increasing tuition has resulted in some students being saddled with large debts on graduation. To examine this issue, a random sample of recent graduates was asked to report whether they had student loans, and if so, how much was the debt at graduation. Each person who reported that they owed money was also asked whether their degree was a BA, BSc, BBA, or other. Can we conclude that debt levels differ between the four types of degree?

13.50 Xr13-50 Studies indicate that single male investors tend to take the most risk, whereas married female investors tend to be conservative. This raises the question, which does best? The risk-adjusted returns for single and married men, and for single and married women were recorded. Can we infer that differences exist between the four groups of investors?

13.51 Xr13-51 Virtually all restaurants attempt to have three 'seatings' on weekend nights. Three seatings means that each table gets three different sets of customers. Obviously, any group that lingers over dessert and coffee may result in the loss of one seating and profit for the restaurant. In an effort to determine which types of groups tend to linger, a random sample of 150 groups was drawn. For each group, the number of members and the length of time that the group stayed were recorded in the following way:

Column A: Length of time for 2 people
Column B: Length of time for 3 people
Column C: Length of time for 4 people
Column D: Length of time for more than 4 people

Do these data allow us to infer that the length of time in the restaurant depends on the size of the party?

13.52 Xr14-52 Stock market investors are always seeking the 'Holy Grail', a sign that tells them the market has bottomed out or achieved its highest level. There are several indicators. One is the buy signal developed by Gerald Appel, who believed that a bottom has been reached when the difference between the weekly close of the New York Stock Exchange (NYSE) index and the 10-week moving average is –4.0 points or more. Another bottom indicator is based on identifying a certain pattern in the line chart of the stock market index. As an experiment, a financial analyst randomly selected 100 weeks. For each week he determined whether there was an Appel buy, a chart buy, or no indication. For each type of week he recorded the percentage change over the next 4 weeks. Can we infer that the two buy indicators are not useful?

13.53 Xr13-53 Thousands of South Africans often spend several hours a day commuting to and from work. Other than the wasted time, are there other negative effects associated with fighting traffic in largest cities, such as Soweto? A statistics study may shed light on the issue. A random sample of adults was surveyed. Among other questions each was asked how much time he or she slept and how much time was spent commuting. The categories for commuting time are 1 to 30 minutes, 31 to 60 minutes and over 60 minutes. Is there sufficient evidence to conclude that the amount of sleep differs between commuting categories?

13.54 Xr13-54 A random sample of 500 teenagers were grouped in the following way: ages 13–14, 15–17 and 18–19. Each teenager was asked to record the number of Facebook friends each had. Is there sufficient evidence to infer that there are differences in the number of Facebook friends between the three teenage groups?

CASE 13.1

DATA C13-01

Comparing Three Methods of Treating Childhood Ear Infections

Acute otitis media, an infection of the middle ear, is a common childhood illness. There are various ways to treat the problem. To help determine the best way, researchers conducted an experiment. One hundred and eighty children between 10 months and 2 years with recurrent acute otitis media were divided into three equal groups. Group 1 was treated by surgically removing the adenoids (adenoidectomy), the second was treated with the drug Sulfafurazole, and the third with a placebo.

Each child was tracked for 2 years, during which time all symptoms and episodes of acute otitis media were recorded. The data were recorded in the following way:

Column 1: ID number
Column 2: Group number
Column 3: Number of episodes of the illness
Column 4: Number of visits to a physician because of any infection
Column 5: Number of prescriptions
Column 6: Number of days with symptoms of respiratory infection

a. Are there differences between the three groups with respect to the number of episodes, number of physician visits, number of prescriptions and number of days with symptoms of respiratory infection?

b. Assume that you are working for the company that makes the drug Sulfafurazole. Write a report to the company's executives discussing your results.

14 CHI-SQUARED TESTS

14-1 CHI-SQUARED GOODNESS-OF-FIT TEST

14-2 CHI-SQUARED TEST OF A CONTINGENCY TABLE

14-3 SUMMARY OF TESTS ON NOMINAL DATA

14-4 CHI-SQUARED TEST FOR NORMALITY

CONTINUOUS HOUSEHOLD SURVEYS: HAS LEVEL OF CONCERN FOR THE ENVIRONMENT CHANGED SINCE 2006?

DATA Xm14-00

The Continuous Household Survey (CHS) began in 1983, and samples approximately 1% of households in Northern Ireland each year. It is carried out by the Northern Ireland Statistics and Research Agency (NISRA). Northern Ireland households were asked to provide their views on environmental issues. Conduct a test to determine whether public level of concern for the environment varies from year to year. **See Section 14-2e for the answer.**

INTRODUCTION

We have seen a variety of statistical techniques that are used when the data are nominal. In Chapter 2, we introduced bar and pie charts, both graphical techniques to describe a set of nominal data. Later in Chapter 2 we showed how to describe the relationship between two sets of nominal data by producing a frequency table and a bar chart. However, these techniques simply describe the data, which may represent a sample or a population. In this chapter, we deal with similar problems, but the goal is to use statistical techniques to make inferences about populations from sample data.

This chapter develops two statistical techniques that involve nominal data. The first is a *goodness-of-fit test* applied to data produced by a *multinomial experiment,* a generalisation of a binomial experiment. The second uses data arranged in a table (called a *contingency table*) to determine whether two classifications of a population of nominal data are statistically independent; this test can also be interpreted as a comparison of two or more populations. The sampling distribution of the test statistics in both tests is the chi-squared distribution introduced in Chapter 8.

14-1 CHI-SQUARED GOODNESS-OF-FIT TEST

This section presents another test designed to describe a population of nominal data. The first such test was introduced in Section 12-3, where we discussed the statistical procedure employed to test hypotheses about a population proportion. In that case, the nominal variable could assume one of only two possible values: success or failure. Our tests dealt with hypotheses about the proportion of successes in the entire population. Recall that the experiment that produces the data is called a *binomial experiment*. In this section, we introduce the **multinomial experiment**, which is an extension of the binomial experiment, wherein there are two or more possible outcomes per trial.

Multinomial Experiment

A multinomial experiment is one that possesses the following properties.

1. The experiment consists of a fixed number n of trials.
2. The outcome of each trial can be classified into one of k categories, called *cells*.
3. The probability p_i that the outcome will fall into cell i remains constant for each trial. Moreover, $p_1 + p_2 + \cdots + p_k = 1$.
4. Each trial of the experiment is independent of the other trials.

When $k = 2$, the multinomial experiment is identical to the binomial experiment. Just as we count the number of successes (recall that we label the number of successes x) and failures in a binomial experiment, we count the number of outcomes falling into each of the k cells in a multinomial experiment. In this way, we obtain a set of observed frequencies $f_1, f_2, \ldots f_k$ where f_i is the observed frequency of outcomes falling into cell i, for $i = 1, 2, \ldots, k$. Because the experiment consists of n trials and an outcome must fall into some cell,

$$f_1 + f_2 + \ldots + f_k = n$$

Just as we used the number of successes x (by calculating the sample proportion $\hat{p}$, which is equal to x/n) to draw inferences about p, so we use the observed frequencies to draw inferences about the cell probabilities. We will proceed in what by now has become a standard procedure. Next, we will set up the hypotheses and develop the test statistic and its sampling distribution. The following example will demonstrate the process.

EXAMPLE 14.1

Testing Market Shares

Company A has recently conducted aggressive advertising campaigns to maintain and possibly increase its share of the market (currently 45%) for fabric softener. Its main competitor, company B, has 40% of the market, and a number of other competitors account for the remaining 15%. To determine whether the market shares changed after the advertising campaign, the marketing manager for company A solicited the preferences of a random sample of 200 customers of fabric softener. Of the 200 customers, 102 indicated a preference for company A's product, 82 preferred company B's fabric softener and the remaining 16 preferred the products of one of the competitors. Can the analyst infer at the 5% significance level that customer preferences have changed from their levels before the advertising campaigns were launched?

SOLUTION:

The population in question is composed of the brand preferences of the fabric softener customers. The data are nominal because each respondent will choose one of three possible answers: product A, product B or other. If there were only two categories, or if we were interested only in the proportion of one company's customers (which we would label as successes and label the others as failures), we would identify the technique as the *z*-test of *p*. However, in this problem we are interested in the proportions of all three categories. We recognise this experiment as a multinomial experiment, and we identify the technique as the **chi-squared goodness-of-fit test.**

Because we want to know whether the market shares have changed, we specify those pre-campaign market shares in the null hypothesis.

$$H_0\text{: } p_1 = 0.45, p_2 = 0.40, p_3 = 0.15$$

The alternative hypothesis attempts to answer our question, 'Have the proportions changed?' Thus,

$$H_1\text{: At least one } p_i \text{ is not equal to its specified value}$$

14-1a Test Statistic

If the null hypothesis is true, we would expect the number of customers selecting brand A, brand B and other to be 200 times the proportions specified under the null hypothesis; that is,

$$e_1 = 200(0.45) = 90$$
$$e_2 = 200(0.40) = 80$$
$$e_3 = 200(0.15) = 30$$

In general, the **expected frequency** for each cell is given by

$$e_i = np_i$$

This expression is derived from the formula for the expected value of a binomial random variable, introduced in Section 7-4.

Figure 14.1 is a bar chart (created by Excel) showing the comparison of actual and expected frequencies.

If the expected frequencies e_i and the **observed frequencies** f_i are quite different, we would conclude that the null hypothesis is false, and we would reject it. However, if the expected and observed frequencies are similar, we would not reject the null hypothesis. The test statistic defined in the box measures the similarity of the expected and observed frequencies.

Chi-Squared Goodness-of-Fit Test Statistic

$$\chi^2 = \sum_{i=1}^{k} \frac{(f_i - e_i)^2}{e_i}$$

FIGURE 14.1

Bar Chart for Example 14.1

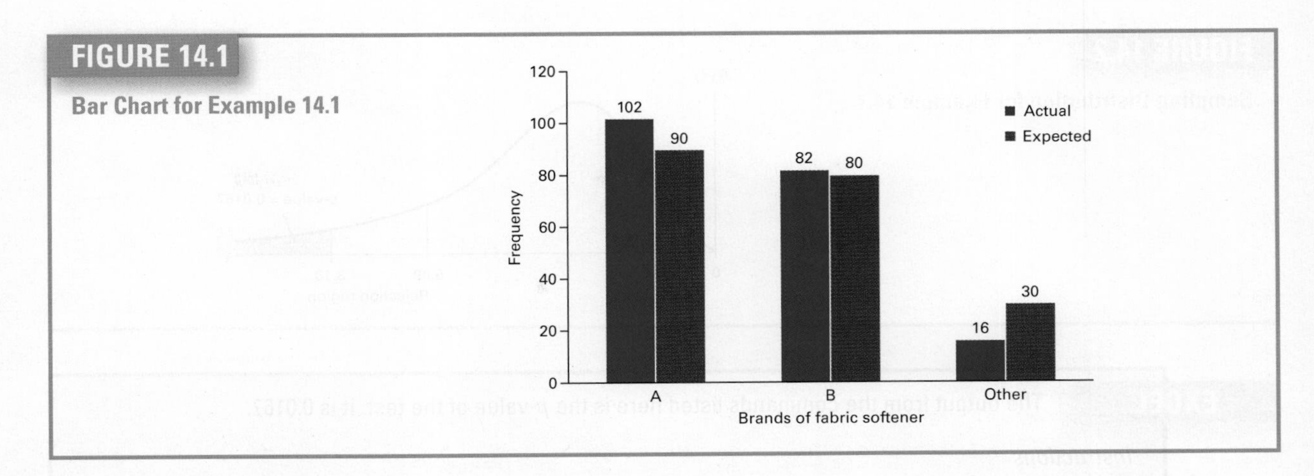

The sampling distribution of the test statistic is approximately chi-squared distributed with $\nu = k - 1$ degrees of freedom, provided that the sample size is large. We will discuss this required condition later. (The chi-squared distribution was introduced in Section 8-4.)

The following table demonstrates the calculation of the test statistic. Thus, the value $\chi^2 = 8.18$. As usual, we judge the size of this test statistic by specifying the rejection region or by determining the *p*-value.

Company	Observed Frequency f_i	Expected Frequency e_i	$(f_i - e_i)$	$\frac{(f_i - e_i)^2}{e_i}$
A	102	90	12	1.60
B	82	80	2	0.05
Other	16	30	−14	6.53
Total	200	200		$\chi^2 = 8 - 18$

When the null hypothesis is true, the observed and expected frequencies should be similar, in which case the test statistic will be small. Thus, a small test statistic supports the null hypothesis. If the null hypothesis is untrue, some of the observed and expected frequencies will differ and the test statistic will be large. Consequently, we want to reject the null hypothesis when χ^2 is greater than $\chi^2_{\alpha,k-1}$. In other words, the rejection region is

$$\chi^2 > \chi^2_{\alpha,k-1}$$

In Example 14.1, $k = 3$; the rejection region is

$$\chi^2 > \chi^2_{\alpha,k-1} = \chi^2_{0.05,2} = 5.99$$

Because the test statistic is $\chi^2 = 8.18$, we reject the null hypothesis. The *p*-value of the test is

$$p\text{-value} = P(\chi^2 > 8.18)$$

Unfortunately, Table 5 in Appendix A does not allow us to perform this calculation (except for approximation by interpolation). The *p*-value must be produced by computer. Figure 14.2 depicts the sampling distribution, rejection region and *p*-value. The *p*-value for the χ^2 test is the area under the χ^2 curve and to the right of the χ^2 value. The number of degrees of freedom is equal to the number of terms in $\chi^2 - 1$, in our case is 7.

FIGURE 14.2

Sampling Distribution for Example 14.1

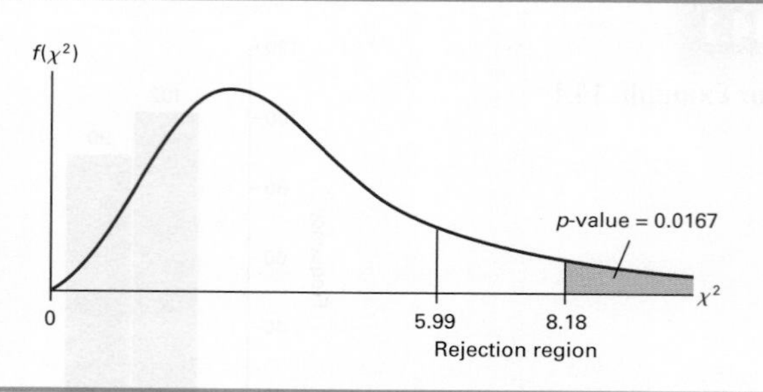

EXCEL

The output from the commands listed here is the *p*-value of the test. It is 0.0167.

Instructions

1. Type the observed values into one column and the expected values into another column. (If you wish, you can type the cell probabilities specified in the null hypothesis and let Excel convert these into expected values by multiplying by the sample size.)
2. Activate an empty cell and type

= **CHIQ.TEST([Actual _ range], [Expected _ range])**

where the ranges are the cells containing the actual observations and the expected values.

You can also perform 'what-if' analyses to determine for yourself the effect of changing some of the observed values and the sample size.

If we have the raw data representing the nominal responses we must first determine the frequency of each category (the observed values) using the **COUNTIF** function described in Section 2.

MINITAB

Chi-Square Goodness-of-Fit Test for Observed Counts in Variable: C1

Category	Observed	Test Proportion	Expected	Contribution to Chi-Sq
1	102	0.45	90	1.60000
2	82	0.40	80	0.05000
3	16	0.15	30	6.53333

N	DF	Chi-Sq	P-Value
200	2	8.18333	0.017

Instructions

1. Click **Stat, Tables** and **Chi-Square Goodness-of-Fit Test (One Variable)**. . . .
2. Type the observed values into the **Observed counts:** box (102 82 16). If you have a column of data click **Categorical data:** and specify the column or variable name.
3. Click **Proportions specified by historical counts** and **Input constants.** Type the values of the proportions under the null hypothesis (0.45 0.40 0.15).

INTERPRET

There is sufficient evidence at the 5% significance level to infer that the proportions have changed since the advertising campaigns were implemented. If the sampling was conducted properly, we can be quite confident in our conclusion. This technique has only one required condition, which is satisfied. (See the next subsection.) It is probably a worthwhile exercise to determine the nature and causes of the changes. The results of this analysis will determine the design and timing of other advertising campaigns.

14-1b Required Condition

The actual sampling distribution of the test statistic defined previously is discrete, but it can be approximated by the chi-squared distribution provided that the sample size is large. This requirement is similar to the one we imposed when we used the normal approximation to the binomial in the sampling distribution of a proportion. In that approximation we needed np and $n(1-p)$ to be 5 or more. A similar rule is imposed for the chi-squared test statistic. It is called the *rule of five,* which states that the sample size must be large enough so that the expected value for each cell must be 5 or more. Where necessary, cells should be combined to satisfy this condition. We discuss this required condition and provide more details on its application in the website Appendix Rule of Five.

Factors that Identify the Chi-Squared Goodness-of-Fit Test

1. **Problem objective:** Describe a single population.
2. **Data type:** Nominal.
3. **Number of categories:** 2 or more.

EXERCISES

Developing an Understanding of Statistical Concepts

Exercises 14.1–14.6 are 'what-if' analyses designed to determine what happens to the test statistic of the goodness-of-fit test when elements of the statistical inference change. These problems can be solved manually or using Excel's **CHIQ.TEST.**

14.1 Consider a multinomial experiment involving $n = 300$ trials and $k = 5$ cells. The observed frequencies resulting from the experiment are shown in the accompanying table, and the null hypothesis to be tested is as follows:

$$H_0: \ p_1 = 0.1, p_2 = 0.2, p_3 = 0.3, p_4 = 0.2, p_5 = 0.2$$

Test the hypothesis at the 1% significance level.

Cell	1	2	3	4	5
Frequency	24	64	84	72	56

14.2 Repeat Exercise 14.1 with the following frequencies:

Cell	1	2	3	4	5
Frequency	12	32	42	36	28

14.3 Repeat Exercise 14.1 with the following frequencies:

Cell	1	2	3	4	5
Frequency	6	16	21	18	14

14.4 Review the results of Exercises 14.1–14.3. What is the effect of decreasing the sample size?

14.5 Consider a multinomial experiment involving $n = 150$ trials and $k = 4$ cells. The observed frequencies resulting from the experiment are shown in the accompanying table, and the null hypothesis to be tested is as follows:

$$H_0: \ p_1 = 0.3, p_2 = 0.3, p_3 = 0.2, p_4 = 0.2$$

Cell	1	2	3	4
Frequency	38	50	38	24

Test the hypotheses, using $\alpha = 0.05$.

14.6 For Exercise 14.5, retest the hypotheses, assuming that the experiment involved twice as many trials ($n = 300$) and that the observed frequencies were twice as high as before, as shown here.

Cell	1	2	3	4
Frequency	76	100	76	48

Exercises 14.7–14.18 require the use of a computer and software. Use a 5% significance level unless specified otherwise. The answers to Exercises 14.7–14.15 may be calculated manually.

14.7 Xr14-07 The results of a multinomial experiment with $k = 5$ were recorded. Each outcome is identified by the numbers 1 to 5. Test to determine whether there is enough evidence to infer that the proportions of outcomes differ.

14.8 **Xr14-08** A multinomial experiment was conducted with $k = 4$. Each outcome is stored as an integer from 1 to 4 and the results of a survey were recorded. Test the following hypotheses.

H_0: $p_1 = 0.15, p_2 = 0.40, p_3 = 0.35, p_4 = 0.10$

H_1: At least one p_i is not equal to its specified value

14.9 **Xr14-09** To determine whether a single die is balanced, or fair, the die was rolled 600 times. Is there sufficient evidence to allow you to conclude that the die is not fair?

Applications

14.10 **Xr14-10** Grades assigned by an economics instructor have historically followed a symmetrical distribution: 5% A's, 25% B's, 40% C's, 25% D's and 5% F's. This year, a sample of 150 grades was drawn and the grades (1 = A, 2 = B, 3 = C, 4 = D and 5 = F) were recorded. Can you conclude, at the 10% level of significance, that this year's grades are distributed differently from grades in the past?

14.11 **Xr14-11** Pat Statsdud is about to write a multiple-choice exam but as usual knows absolutely nothing. Pat plans to guess one of the five choices. Pat has been given one of the lecturer's previous exams with the correct answers marked. The correct choices were recorded where 1 = (a), 2 = (b), 3 = (c), 4 = (d) and 5 = (e). Help Pat determine whether this lecturer does not randomly distribute the correct answer over the five choices? If this is true, how does it affect Pat's strategy?

14.12 **Xr14-12** European accounts managers are interested in the speed with which customers who make purchases on credit pay their bills. In addition to calculating the average number of days that unpaid bills (called *accounts receivable*) remain outstanding, they often prepare an aging schedule. An ageing schedule classifies outstanding accounts receivable according to the time that has elapsed since billing and records the proportion of accounts receivable belonging to each classification. A large firm has determined its ageing schedule for the past 5 years. These results are shown in the accompanying table. During the past few months, however, the economy has taken a downturn. The company would like to know whether the recession has affected the ageing schedule. A random sample of 250 accounts receivable was drawn and each account was classified as follows:

1 = 0–14 days outstanding

2 = 15–29 days outstanding

3 = 30–59 days outstanding

4 = 60 or more days outstanding

Number of Days Outstanding	Proportion of Accounts Receivable Past 5 Years
0–14	0.72
15–29	0.15
30–59	0.10
60 and more	0.03

Determine whether the ageing schedule has changed.

14.13 **Xr14-13** Licence records in a county reveal that 15% of cars are city car (1), 25% are small family (2), 40% are large family (3) and the rest are an executive car and models (4). A random sample of accidents involving cars licenced in the county was drawn. The type of car was recorded using the codes in parentheses. Can we infer that certain sizes of cars are involved in a higher than expected percentage of accidents?

14.14 **Xr14-14** In an election held last year that was contested by three parties, Party A captured 31% of the vote, party B garnered 51% and party C received the remaining votes. A survey of 1200 voters asked each to identify the party that they would vote for in the next election. These results were recorded where 1 = party A, 2 = party B and 3 = party C. Can we infer at the 10% significance level that voter support has changed since the election?

14.15 **Xr14-15** In a number of pharmaceutical studies volunteers who take placebos (but are told they have taken a cold remedy) report the following side-effects:

Headache (1)	5%
Drowsiness (2)	7%
Stomach upset (3)	4%
No side-effect (4)	84%

A random sample of 250 people who were given a placebo (but who thought they had taken an anti-inflammatory) reported whether they had experienced each of the side-effects. These responses were recorded using the codes in parentheses. Do these data provide enough evidence to infer that the reported side-effects of the placebo for an anti-inflammatory differ from that of a cold remedy?

CENSUS SURVEY EXERCISE

14.16 Census2011
According to the Office of National Statistics 2011 Census: Key Statistics for England and Wales, March 2011, the proportions for each category of marital status in 2011 was:

Single (never married) 35%
Married (including separated, but not divorced) 47%
Separated but still legally married 2.6%
Widowed 7%
Divorced 9%

Can we infer that the Census in 2011 over-represented at least one category of marital status (MARITAL STATUS)?

14.17 Census2011
According to the Office of National Statistics 2011 Census: Key Statistics for England and Wales, March 2011, most residents of England and Wales in 2011 were:

White 86%
Black 3.4%
Other 10.6%

Test to determine whether there is sufficient evidence that the Census 2011 over-represented at least one race (Ethnic Group).

14.18 Census2011 Can we infer that the Census in 2011 over-represented at least one category of marital status (MARITAL STATUS)?

14-2 CHI-SQUARED TEST OF A CONTINGENCY TABLE

In Chapter 2 we developed the **cross-classification table** as a first step in graphing the relationship between two nominal variables. Our goal was to determine whether the two variables were related. In this section we extend the technique to statistical inference. We introduce another chi-squared test, this one designed to satisfy two different problem objectives. The **chi-squared test of a contingency table** is used to determine whether there is enough evidence to infer that two nominal variables are related and to infer that differences exist between two or more populations of nominal variables. Completing both objectives entails classifying items according to two different criteria. To see how this is done, consider the following example.

Example 14.2

DATA Xm14-02

Relationship between Undergraduate Degree and MBA Specialism

The MBA programme was experiencing problems scheduling its courses. The demand for the programme's optional courses and specialisms was quite variable from 1 year to the next. In 1 year, students seem to want marketing courses; in other years, accounting or finance are the rage. In desperation, the dean of the business school turned to a statistics lecturer for assistance. The statistics lecturer believed that the problem may be the variability in the academic background of the students and that the undergraduate degree affects the choice of specialism. As a start, he took a random sample of last year's MBA students and recorded the undergraduate degree and the specialism selected in the graduate programme. The undergraduate degrees were BA, BEng, BBA and several others. There are three possible specialisms for the MBA students: accounting, finance and marketing. The results were summarised in a cross-classification table, which is shown here. Can the statistician conclude that the undergraduate degree affects the choice of specialism?

	MBA Specialism			
Undergraduate Degree	Accounting	Finance	Marketing	Total
BA	31	13	16	60
BEng	8	16	7	31
BBA	12	10	17	39
Other	10	5	7	22
Total	61	44	47	152

SOLUTION:

One way to solve the problem is to consider that there are two variables: undergraduate degree and MBA specialism. Both are nominal. The values of the undergraduate degree are BA, BEng, BBA and other. The values of MBA specialism are accounting finance and marketing. The problem objective is to analyse the relationship between the two variables. Specifically, we want to know whether one variable is related to the other.

Another way of addressing the problem is to determine whether differences exist between BAs, BEngs, BBAs and others. In other words, we treat the holders of each undergraduate degree as a separate population. Each population has three possible values represented by the MBA specialism. The problem objective is to compare four populations. (We can also answer the question by treating the MBA specialisms as populations and the undergraduate degrees as the values of the random variable.)

As you will shortly discover, both objectives lead to the same test. Consequently, we address both objectives at the same time.

The null hypothesis will specify that there is no relationship between the two variables. We state this in the following way:

H_0: The two variables are independent

The alternative hypothesis specifies one variable affects the other, expressed as

H_1: The two variables are dependent

14-2a Graphical Technique

Figure 14.3 depicts the graphical technique introduced in Chapter 2 to show the relationship (if any) between the two nominal variables.

The bar chart displays the data from the sample. It does appear that there is a relationship between the two nominal variables in the sample. However, to draw inferences about the population of MBA students we need to apply an inferential technique.

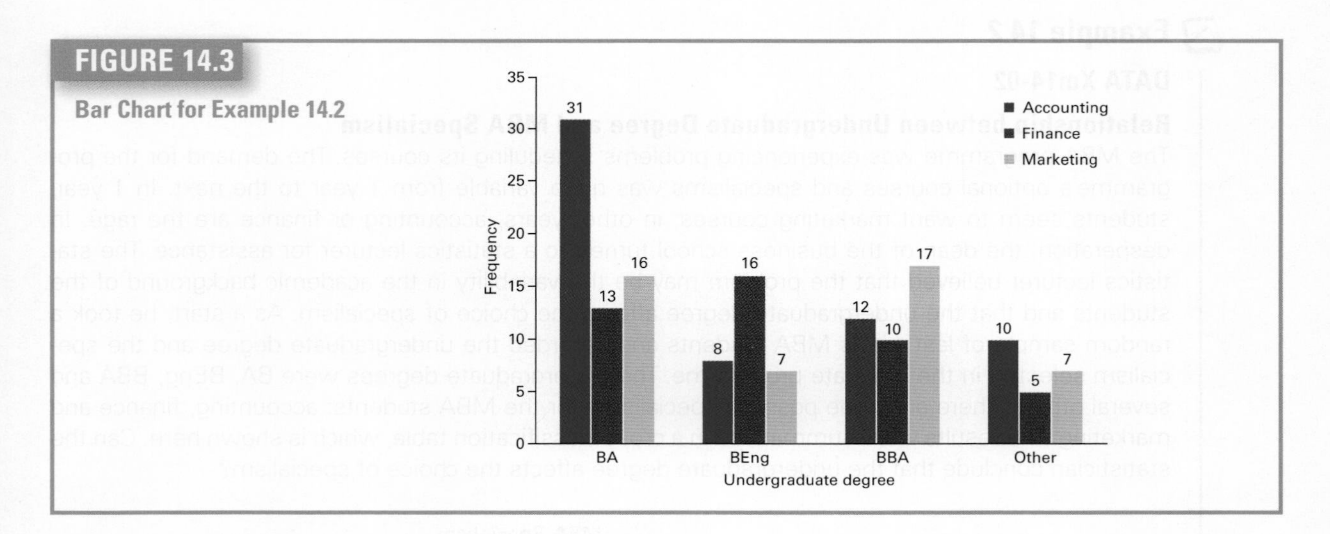

FIGURE 14.3
Bar Chart for Example 14.2

14-2b Test Statistic

The test statistic is the same as the one used to test proportions in the goodness-of-fit test; that is, the test statistic is

$$\chi^2 = \sum_{i=1}^{k} \frac{(f_i - e_i)^2}{e_i}$$

where k is the number of cells in the cross-classification table. If you examine the null hypothesis described in the goodness-of-fit test and the one described above, you will discover a major difference. In the goodness-of-fit test, the null hypothesis lists values for the probabilities p_i. The null hypothesis for the chi-squared test of a contingency table only states that the two variables are independent. However, we need the probabilities to compute the expected values e_i, which in turn are needed to calculate the value of the test statistic. (The entries in the table are the observed values f_i.) The question immediately arises, from where do we get the probabilities? The answer is that they must come from the data after we assume that the null hypothesis is true.

In Chapter 6 we introduced independent events and showed that if two events A and B are independent, the joint probability P(A and B) is equal to the product of P(A) and P(B). That is,

$$P(\text{A and B}) = P(\text{A}) \times P(\text{B})$$

The events in this example are the values each of the two nominal variables can assume. Unfortunately, we do not have the probabilities of A and B. However, these probabilities can be estimated from the data. Using relative frequencies, we calculate the estimated probabilities for the MBA specialism.

$$P(\text{Accounting}) = \frac{61}{152} = 0.401$$

$$P(\text{Finance}) = \frac{44}{152} = 0.289$$

$$P(\text{Marketing}) = \frac{47}{152} = 0.309$$

We calculate the estimated probabilities for the undergraduate degree.

$$P(\text{BA}) = \frac{60}{152} = 0.395$$

$$P(\text{BEng}) = \frac{31}{152} = 0.204$$

$$P(\text{BBA}) = \frac{39}{152} = 0.257$$

$$P(\text{Other}) = \frac{22}{152} = 0.145$$

Assuming that the null hypothesis is true, we can compute the estimated joint probabilities. To produce the expected values, we multiply the estimated joint probabilities by the sample size, $n = 152$. The results are listed in a **contingency table**, the word *contingency* derived by calculating the expected values contingent on the assumption that the null hypothesis is true (the two variables are independent).

Undergraduate Degree	MBA Specialism			Total
	Accounting	Finance	Marketing	
BA	$152 \times \frac{60}{152} \times \frac{61}{152} = 24.08$	$152 \times \frac{60}{152} \times \frac{44}{152} = 17.37$	$152 \times \frac{60}{152} \times \frac{47}{152} = 18.55$	60
BEng	$152 \times \frac{31}{152} \times \frac{61}{152} = 12.44$	$152 \times \frac{31}{152} \times \frac{44}{152} = 8.97$	$152 \times \frac{31}{152} \times \frac{47}{152} = 9.59$	31
BBA	$152 \times \frac{39}{152} \times \frac{61}{152} = 15.65$	$152 \times \frac{39}{152} \times \frac{44}{152} = 11.29$	$152 \times \frac{39}{152} \times \frac{47}{152} = 12.06$	39
Other	$152 \times \frac{22}{152} \times \frac{61}{152} = 8.83$	$152 \times \frac{22}{152} \times \frac{44}{152} = 6.37$	$152 \times \frac{22}{152} \times \frac{47}{152} = 6.80$	22
Total	61	44	47	152

As you can see, the expected value for each cell is computed by multiplying the row total by the column total and dividing by the sample size. For example, the BA and Accounting cell expected value is

$$152 \times \frac{60}{152} \times \frac{61}{152} = \frac{60 \times 61}{152} = 24.08$$

All the other expected values would be determined similarly.

Expected Frequencies for a Contingency Table

The expected frequency of the cell in row *i* and column *j* is

$$e_{ij} = \frac{\text{row } i \text{ total} \times \text{column } j \text{ total}}{\text{sample size}}$$

The expected cell frequencies are shown in parentheses in the following table. As in the case of the goodness-of-fit test, the expected cell frequencies should satisfy the rule of five.

	MBA Specialism		
Undergraduate Degree	**Accounting**	**Finance**	**Marketing**
BA	31 (24.08)	13 (17.37)	16 (18.55)
BEng	8 (12.44)	16 (8.97)	7 (9.59)
BBA	12 (15.65)	10 (11.29)	17 (12.06)
Other	10 (8.83)	5 (6.37)	7 (6.80)

We can now calculate the value of the test statistic:

$$\begin{aligned}\chi^2 = \sum_{i=1}^{k} \frac{(f_i - e_i)^2}{e_i} &= \frac{(31-24.08)^2}{24.08} + \frac{(13-17.37)^2}{17.37} + \frac{(16-18.55)^2}{18.55}\\ &+ \frac{(8-12.44)^2}{12.44} + \frac{(16-8.97)^2}{8.97} + \frac{(7-9.59)^2}{9.59} + \frac{(12-15.65)^2}{15.65}\\ &+ \frac{(10-11.29)^2}{11.29} + \frac{(17-12.06)^2}{12.06} + \frac{(10-8.83)^2}{8.83}\\ &+ \frac{(5-6.37)^2}{6.37} + \frac{(7-6.80)^2}{6.80}\\ &= 14.70\end{aligned}$$

Notice that we continue to use a single subscript in the formula of the test statistic when we should use two subscripts, one for the rows and one for the columns. We believe that it is clear, that for each cell we must calculate the squared difference between the observed and expected frequencies divided by the expected frequency. We don't believe that the satisfaction of using the mathematically correct notation overcomes the unnecessary complication.

14-2c Rejection Region and *p*-Value

To determine the rejection region we must know the number of degrees of freedom associated with the chi-squared statistic. The number of degrees of freedom for a contingency table with *r* rows and *c* columns is $\nu = (r - 1)(c - 1)$. For this example, the number of degrees of freedom is $\nu = (r - 1)(c - 1) = (4 - 1)(3 - 1) = 6$. If we employ a 5% significance level, the rejection region is

$$\chi^2 > \chi^2_{\alpha,\nu} = \chi^2_{0.05,6} = 12.6$$

Because $\chi^2 = 14.70$, we reject the null hypothesis and conclude that there is evidence of a relationship between undergraduate degree and MBA specialism. The p-value of the test statistic is

$$P(\chi^2 > 14.70)$$

Unfortunately, we cannot determine the p-value manually.

Using the Computer

Excel and Minitab can produce the chi-squared statistic either from a cross-classification table whose frequencies have already been calculated or from raw data. The respective printouts are almost identical.

File Xm14-02 contains the raw data using the following codes:

Column 1 (Undergraduate Degree)	Column 2 (MBA Major)
1 = BA	1 = Accounting
2 = BEng	2 = Finance
3 = BBA	3 = Marketing
4 = Other	

EXCEL

	A	B	C	D	E	F
1	Contingency Table					
2						
3		*Degree*				
4	*MBA Major*		1	2	3	TOTAL
5		1	31	13	16	60
6		2	8	16	7	31
7		3	12	10	17	39
8		4	10	5	7	22
9		TOTAL	61	44	47	152
10						
11						
12		chi-squared Stat			14.70	
13		df			6	
14		p-value			0.0227	
15		chi-squared Critical			12.5916	

Instructions (Raw Data)

1. Type or import the data into two adjacent columns*. (**Open** Xm14-02.) The codes must be positive integers greater than 0.
2. Click **Add-Ins, Data Analysis Plus** and **Contingency Table (Raw Data).**
3. Specify the **Input Range** (A1:B153) and specify the value of α (0.05).

Instructions (Completed Table)

1. Type the frequencies into adjacent columns.
2. Click **Add-Ins, Data Analysis Plus** and **Contingency Table.**
3. Specify the **Input Range.** Click **Labels** if the first row and first column of the input range contain the names of the categories. Specify the value for α.

*If one or both columns contain an empty cell (representing missing data), the entire row must be removed. See the website Appendix Removing Empty Cells in Excel.

MINITAB

Tabulated statistics: Degree, MBA Major

Rows: Degree Columns: MBA Major

	1	2	3	All
1	31	13	16	60
2	8	16	7	31
3	12	10	17	39
4	10	5	7	22
All	61	44	47	152

Cell Contents: Count

Pearson Chi-Square = 14.702, DF = 6, P-Value = 0.023
Likelihood Ratio Chi-Square = 13.781, DF = 6, P-Value = 0.032

Instructions (Raw Data)

1. Type or import the data into two columns. (**Open** Xm14-02.)
2. Click **Stat, Tables** and **Cross Tabulation and Chi-Square** . . . .
3. In the **Categorical variables** box, select or type the variables **For rows** (Degree) and **For columns** (MBA Major). Click **Chi-Square** . . . .
4. Under **Display** click **Chi-Square analysis.** Specify **Chi-Square analysis.**

Instructions (Completed Table)

1. Type the observed frequencies into adjacent columns.
2. Click **Stat, Tables** and **Chi-Square Test (Table in Worksheet)** . . . .
3. Select or type the names of the variables representing the columns.

INTERPRET

There is strong evidence to infer that the undergraduate degree and MBA specialism are related. This suggests that the dean can predict the number of optional courses by counting the number of MBA students with each type of undergraduate degree. We can see that BAs favour accounting courses, BEngs prefer finance, BBAs are partial to marketing, and others show no particular preference.

If the null hypothesis is true, undergraduate degree and MBA major are independent of one another. This means that whether an MBA student earned a BA, BEng, BBA or other degree does not affect his or her choice of programme specialism in the MBA. Consequently, there is no difference in specialism choice among the graduates of the undergraduate programmes. If the alternative hypothesis is true, undergraduate degree does affect the choice of MBA specialism. Thus, there are differences between the four undergraduate degree categories.

14-2d Rule of Five

In the previous section, we pointed out that the expected values should be at least 5 to ensure that the chi-squared distribution provides an adequate approximation of the sampling distribution. In a contingency table where one or more cells have expected values of less than 5, we need to combine rows or columns to satisfy the rule of five. This subject is discussed in the website Appendix Rule of Five.

14-2e Data Formats

In Example 14.2, the data were stored in two columns, one column containing the values of one nominal variable and the second column storing the values of the second nominal variable. The data can be stored in another way. In Example 14.2, we could have recorded the data in three columns, one column for each MBA specialism. The columns would contain the codes representing the undergraduate degree.

Alternatively, we could have stored the data in four columns, one column for each undergraduate degree. The columns would contain the codes for the MBA specialisms. In either case, we have to count the number of each value and construct the cross-tabulation table using the counts. Both Excel and Minitab can calculate the chi-squared statistic and its *p*-value from the cross-tabulation table. We will illustrate this approach with the solution to the chapter-opening example.

CONTINUOUS HOUSEHOLD SURVEYS

Has Level of Concern for the Environment Changed Since 2006?: Solution

IDENTIFY

The problem objective is to compare public opinion in four different years. The variable is nominal because its values are Concerned and Not concerned, represented by 1 and 2, respectively. The appropriate technique is the chi-squared test of a contingency table. The hypotheses are:

H_0: The two variables are independent
H_1: The two variables are dependent

In this application, the two variables are year (2006/07, 2007/08, . . . , 2012/13) and the answer to the question posed by the Continuous Household Survey (Concerned and Not concerned). Unlike Example 13.2, the data are not stored in two columns. To produce the statistical result, we will need to create a contingency table. The following table was determined by counting the numbers of 1s and 2s for each year.

	YEAR							
	2006	2007	2008	2009	2010	2011	2012	2013
Concerned	79	81	81	82	76	75	72	72
Not concerned	21	19	19	18	24	25	28	28

COMPUTE

	A	B	C	D	E	F	G	H	I	J	K
1	**Contingency Table**										
2											
3											
4			*2005/06*	*2006/07*	*2007/08*	*2008/09*	*2009/10*	*2010/11*	*2011/12*	*2012/13*	*TOTAL*
5		*Concerned*	79	81	81	82	76	75	72	72	618
6		*Not concerned*	21	19	19	18	24	25	28	28	182
7		*TOTAL*	100	100	100	100	100	100	100	100	800
8											
9											
10		Chi-Squared Statistic		6.5721							
11		Degrees of Freedom		7							
12		P-Value		0.4748							
13		Chi-Squared Critical		14.0671							

INTERPRET

The *p*-value is 0.4748 and the chi squared value is lower than the critical value. There is not enough evidence to infer that the two variables are independent. Thus, there is not enough evidence to conclude that the level of concern for the environment varies from year to year.

Here is a summary of the factors that tell us when to apply the chi-squared test of a contingency table. Note that there are two problem objectives satisfied by this statistical procedure.

Factors that Identify the Chi-Squared Test of a Contingency Table

1. **Problem objectives:** Analyse the relationship between two variables and compare two or more populations.
2. **Data type:** Nominal.

EXERCISES

Developing an Understanding of Statistical Concepts

14.19 Conduct a test to determine whether the two classifications L and M are independent, using the data in the accompanying cross-classification table. (Use $\alpha = 0.05$.)

	M_1	M_2
L_1	28	68
L_2	56	36

14.20 Repeat Exercise 14.19 using the following table:

	M_1	M_2
L_1	14	34
L_2	28	18

14.21 Repeat Exercise 14.19 using the following table:

	M_1	M_2
L_1	7	17
L_2	14	9

14.22 Review the results of Exercises 14.19–14.21. What is the effect of decreasing the sample size?

14.23 Conduct a test to determine whether the two classifications R and C are independent, using the data in the accompanying cross-classification table. (Use $\alpha = 0.10$.)

	C_1	C_1	C_1
R_1	40	32	48
R_2	30	48	52

Applications

Use a 5% significance level unless specified otherwise.

14.24 The trustee of a company's pension plan has solicited the opinions of a sample of the company's employees about a proposed revision of the plan. A breakdown of the responses is shown in the accompanying table. Is there enough evidence to infer that the responses differ between the three groups of employees?

Responses	Working Class Workers	Middle Class Workers	Managers
For	67	32	11
Against	63	18	9

14.25 The operations manager of a company that manufactures shirts wants to determine whether there are differences in the quality of workmanship among the three daily shifts. She randomly selects 600 recently made shirts and carefully inspects them. Each shirt is classified as either perfect or flawed, and the shift that produced it is also recorded. The accompanying table summarises the number of shirts that fell into each cell. Do these data provide sufficient evidence to infer that there are differences in quality between the three shifts?

	Shift		
Shirt Condition	1	2	3
Perfect	240	191	139
Flawed	10	9	11

14.26 McKinsey & Company, a well-known global management consulting firm with offices in America, Europe, Africa, the Middle East and Asia, wants to test how it can influence the proportion of questionnaires returned from surveys. In the belief that the inclusion of an inducement to respond may be important, the firm sends out 1000 questionnaires: 200 promise to send respondents a summary of the survey results, 300 indicate that 20 respondents (selected by lottery) will be awarded gifts and 500 are accompanied by no inducements. Of these, 80 questionnaires promising a summary, 100 questionnaires offering gifts and 120 questionnaires offering no inducements are returned. What can you conclude from these results?

Exercises 14.27-14.37 require the use of a computer and software. The answers may be calculated manually. Use a 5% significance level unless specified otherwise.

14.27 Xm02-04 (Example 2.4 revisited) A major UK city has four competing newspapers: *The Times, The Independent, The Star* and *The Sun*. To help design advertising campaigns, the advertising managers of the newspapers need to know which segments of the newspaper market are reading their papers. A survey was conducted to analyse the relationship between newspapers read and occupation. A sample of newspaper readers was asked to report which newspaper they read: *The Times* (1), *The Independent* (2), *The Star* (3), *The Sun* (4), and to indicate whether they were working class (1), middle-class workers (2) or upper class (3). Can we infer that occupation and newspaper are related?

14.28 Xrl4-28 An investor who can correctly forecast the direction and size of changes in foreign currency exchange rates is able to reap huge profits in the international currency markets. A knowledgeable reader of the *Financial Times* (in particular, of the currency futures market quotations) can determine the direction of change in various exchange rates that is predicted by all investors, viewed collectively. Predictions from 216 investors, together with the subsequent actual directions of change, were recorded in the following way: Column 1: predicted change where 1 = positive and 2 = negative; column 2: actual change where 1 = positive and 2 = negative.

a. Can we infer at the 10% significance level that a relationship exists between the predicted and actual directions of change?

b. To what extent would you make use of these predictions in formulating your forecasts of future exchange rate changes?

14.29 Xr14-29 Because television audiences of news broadcasting tend to be older (and because older people suffer from a variety of medical ailments), pharmaceutical companies' advertising often appears on national news on the three networks in South Africa (SABC, DSTV and MNet). To determine how effective the ads are a survey was undertaken. Adults over 50 were asked about their primary sources of news. The responses are

1. SABC News 2. DSRV News 3. MNet News
4. Newspapers 5. Radio 6. None of the above

Each person was also asked whether they suffer from heartburn, and if so, what remedy they take. The answers were recorded as follows:

1. Do not suffer from heartburn
2. Suffer from heartburn but take no remedy
3. Suffer from heartburn and take an over-the-counter remedy (e.g. Rennie, Gaviscon)
4. Suffer from heartburn and take a prescription pill

Is there a relationship between an adult's source of news and his or her heartburn condition?

14.30 Xr02-27(Exercise 2.27 revisited.) The associate dean of a business school was looking for ways to improve the quality of the applicants to its MBA programme. In particular, she wanted to know whether the undergraduate degree of applicants differed among her school and the three nearby universities with MBA programmes. She sampled 100 applicants of her programme and an equal number from each of the other universities. She recorded their undergraduate degrees (1 = BA, 2 = BEng, 3 = BBA, 4 = other) as well as universities (codes 1, 2, 3 and 4). Do these data provide sufficient evidence to infer that undergraduate degree and the university each person applied to are related?

14.31 Xr14-31 The relationship between drug companies and medical researchers is under scrutiny because of possible conflict of interest. The issue that started the controversy was a 1995 case control study that suggested that the use of calcium-channel blockers to treat hypertension led to an increased risk of heart disease. This led to an intense debate both in technical journals and in the press. Researchers writing in the *New England Journal of Medicine* ('Conflict of Interest in the Debate over Calcium Channel Antagonists', 8 January, 1998, p. 101) looked at the 70 reports that appeared during 1996–97, classifying them as favourable, neutral or critical towards the drugs. The researchers then contacted the authors of the reports and questioned them about financial ties to drug companies. The results were recorded in the following way:

Column 1: Results of the scientific study;
1 = favourable, 2 = neutral, 3 = critical

Column 2:1= financial ties to drug companies,
2 = no ties to drug companies

Do these data allow us to infer that the research findings for calcium-channel blockers are affected by whether the research is funded by drug companies?

14.32 Xr14-32 After a thorough analysis of the market, a publisher of business and economics statistics books has divided the market into three general approaches to teach applied statistics. These are (1) use of a computer and statistical software with no manual calculations, (2) traditional teaching of concepts and solution of problems by hand and (3) mathematical approach with emphasis on derivations and proofs. The publisher wanted to know whether this market could be segmented on the basis of the educational background of the instructor. As a result, the statistics editor organised a survey that asked 195 professors of business and economics statistics to report their approach to teaching and which one of the following categories represents their highest degree:

1. Business (MBA or PhD in business)
2. Economics
3. Mathematics or engineering
4. Other

a. Can the editor infer that there are differences in type of degree among the three teaching approaches? If so, how can the editor use this information?

b. Suppose that you work in the marketing department of a textbook publisher. Prepare a report for the editor that describes this analysis.

14.33 Xr14-33 Household types are categorised in the following way: 1. Married couple with children, 2. Married couple without children, 3. Single parent, 4. One person, 5. Other. Random samples of families in the USA, Canada and the UK were drawn and the household types recorded. Is there sufficient evidence to infer that there are differences in household types between the three countries?

14.34 Xr14-34 Refer to Exercise 14.33. Random samples from Denmark, Ireland, the Netherlands and Sweden were drawn. Is there sufficient evidence to infer that there are differences in household types between the four countries?

14.35 Xr14-35 A statistics practitioner took random samples from Canada, Australia, New Zealand and the UK, and classified each person as either obese (2) or not obese (1). Can we conclude from these data that there are differences in obesity rates between the four Commonwealth nations?

14.36 Xr14-36 To measure the extent of cigarette smoking around the world, random samples of adults in Denmark, Finland, Norway and Sweden were drawn. Each was asked whether he or she smoked (2 = Yes, 1 = No). Can we conclude that there are differences in smoking between the four Scandinavian countries?

14.37 Xr14-37 Refer to Exercise 14.36. The survey was performed in Canada, Australia, New Zealand and the UK. Is there enough evidence to infer that there are differences in adult cigarette smoking between the four Commonwealth countries?

14-3 SUMMARY OF TESTS ON NOMINAL DATA

At this point in the textbook, we have described four tests that are used when the data are nominal:

z-test of p (Section 12-3)

z-test of $p_1 - p_2$ (Section 12-7)

Chi-squared goodness-of-fit test (Section 14-1)

Chi-squared test of a contingency table (Section 14-2)

In the process of presenting these techniques, it was necessary to concentrate on one technique at a time and focus on the kinds of problems each addresses. However, this approach tends to conflict somewhat with our promised goal of emphasising the 'when' of statistical inference. In this section, we summarise the statistical tests on nominal data to ensure that you are capable of selecting the correct method.

There are two critical factors in identifying the technique used when the data are nominal. The first, of course, is the problem objective. The second is the number of categories that the nominal variable can assume. Table 14.1 provides a guide to help select the correct technique.

Notice that when we describe a population of nominal data with exactly two categories, we can use either of two techniques. We can employ the z-test of p or the chi-squared goodness-of-fit test. These two tests are equivalent because if there are only two categories, the multinomial

TABLE 14.1 Statistical Techniques for Nominal Data

PROBLEM OBJECTIVE	NUMBER OF CATEGORIES	STATISTICAL TECHNIQUE
Describe a population	2	*z*-test of *p* or the chi-squared goodness-of-fit test
Describe a population	More than 2	Chi-squared goodness-of-fit test
Compare two populations	2	*z*-test of $p_1 - p_2$ or chi-squared test of a contingency table
Compare two populations	More than 2	Chi-squared test of a contingency table
Compare two or more populations	2 or more	Chi-squared test of a contingency table
Analyse the relationship between two variables	2 or more	Chi-squared test of a contingency table

experiment is actually a binomial experiment (one of the categorical outcomes is labelled *success*, and the other is labelled *failure*). Mathematical statisticians have established that if we square the value of *z*, the test statistic for the test of *p*, we produce the χ^2-statistic; that is, $z^2 = \chi^2$. Thus, if we want to conduct a two-tail test of a population proportion, we can employ either technique. However, the chi-squared goodness-of-fit test can test only to determine whether the hypothesised values of p_1 (which we can label *p*) and p_2 (which we call $1 - p$) are not equal to their specified values. Consequently, to perform a one-tail test of a population proportion, we must use the z-test of *p*. (This issue was discussed in Chapter 13 when we pointed out that we can use either the *t*-test of $\mu_1 - \mu_2$ or the analysis of variance to conduct a test to determine whether two population means differ.)

When we test for differences between two populations of nominal data with two categories, we can also use either of two techniques: the *z*-test of $p_1 - p_2$ (Case 1) or the chi-squared test of a contingency table. Once again, we can use either technique to perform a two-tail test about $p_1 - p_2$. (Squaring the value of the z-statistic yields the value of the χ^2-statistic.) However, one-tail tests must be conducted by the *z*-test of $p_1 - p_2$.The rest of the table is quite straightforward. Notice that when we want to compare two populations when there are more than two categories, we use the chi-squared test of a contingency table.

Figure 14.4 offers another summary of the tests that deal with nominal data introduced in this book. There are two groups of tests: those that test hypotheses about single populations and those that test either for differences or for independence. In the first set, we have the *z*-test of *p*, which can be replaced by the chi-squared test of a multinomial experiment.

The latter test is employed when there are more than two categories. To test for differences between two **populations**, we apply the *z*-test of $p_1 - p_2$. Instead, we can use the chi-squared test of a contingency table, which can be applied to a variety of other problems.

14-3a Developing an Understanding of Statistical Concepts

Table 14.1 and Figure 14.4 summarise how we deal with nominal data. We determine the frequency of each category and use these frequencies to compute test statistics. We can then compute proportions to calculate *z*-statistics or use the frequencies to calculate χ^2-statistics. Because squaring a standard normal random variable produces a chi-squared variable, we can employ either statistic to test for differences. As a consequence, when you encounter nominal data in the problems described in this book (and other introductory applied statistics books), the most logical starting point in selecting the appropriate technique will be either a *z*-statistic or a χ^2-statistic. However, you should know that there are other statistical procedures that can be applied to nominal data, techniques that are not included in this book.

FIGURE 14.4

Tests on Nominal Data

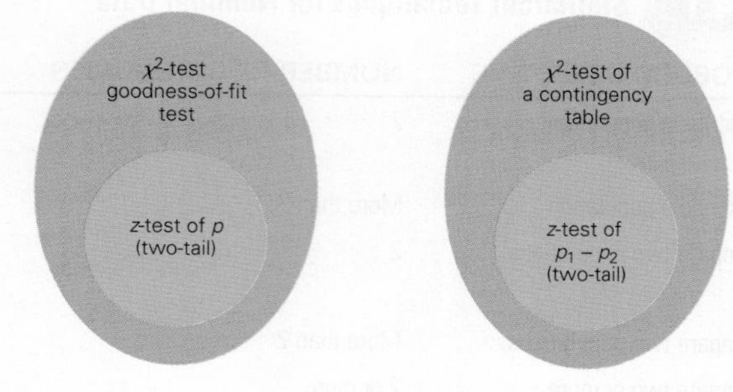

χ^2-test goodness-of-fit test

z-test of p (two-tail)

χ^2-test of a contingency table

z-test of $p_1 - p_2$ (two-tail)

14-4 CHI-SQUARED TEST FOR NORMALITY

We can use the goodness-of-fit test presented in Section 14-1 in another way. We can test to determine whether data were drawn from any distribution. The most common application of this procedure is a test of normality.

In the examples and exercises shown in Section 14-1, the probabilities specified in the null hypothesis were derived from the question. In Example 14.1, the probabilities p_1, p_2 and p_3 were the market shares before the advertising campaign. To test for normality (or any other distribution), the probabilities must first be calculated using the hypothesised distribution. To illustrate, consider Example 12.1, where we tested the mean amount of discarded newspaper using the Student *t* distribution. The required condition for this procedure is that the data must be normally distributed. To determine whether the 148 observations in our sample were indeed taken from a normal distribution, we must calculate the theoretical probabilities assuming a normal distribution. To do so, we must first calculate the sample mean and standard deviation: $\bar{x} = 2.18$ and $s = 0.981$. Next, we find the probabilities of an arbitrary number of intervals. For example, we can find the probabilities of the following intervals:

Interval 1: $X \leq 0.709$
Interval 2: $0.709 < X \leq 1.69$
Interval 3: $1.69 < X \leq 2.67$
Interval 4: $2.67 < X \leq 3.65$
Interval 5: $X > 3.65$

We will discuss the reasons for our choices of intervals later.

The probabilities are computed using the normal distribution and the values of $\bar{x}$ and s as estimators of μ and σ. We calculated the sample mean and standard deviation as $\bar{x} = 2.18$ and $s = 0.981$. Thus,

$$P(X \leq 0.709) = P\left(\frac{X-\mu}{\sigma} \leq \frac{0.709 - 2.18}{0.981}\right) = P(Z \leq -1.5) = 0.0668$$

$$P(0.709 < X \leq 1.69) = P\left(\frac{0.709 - 2.18}{0.981} < \frac{X-\mu}{\sigma} \leq \frac{1.69 - 2.18}{0.981}\right)$$
$$= P(-1.5 < Z \leq -0.5) = 2417$$

$$P(1.69 < X \leq 2.67) = P\left(\frac{1.69 - 2.18}{0.981} < \frac{X-\mu}{\sigma} \leq \frac{2.67 - 2.18}{0.981}\right)$$
$$= P(-0.5 < Z \leq 0.5) = 0.3829$$

$$P(2.67 < X \leq 3.65) = P\left(\frac{2.67 - 2.18}{0.981} < \frac{X-\mu}{\sigma} \leq \frac{3.65 - 2.18}{0.981}\right)$$
$$= P(0.5 < Z \leq 1.5) = 0.2417$$

$$P(X > 3.65) = P\left(\frac{X-\mu}{\sigma} > \frac{3.65 - 2.18}{0.981}\right) = P(Z > 1.5) = 0.0668$$

To test for normality is to test the following hypotheses:

H_0: $p_1 = 0.0668, p_2 = 0.2417, p_3 = 0.3829, p_4 = 0.2417, p_5 = 0.0668$

H_1: **At least two proportions differ from their specified values**

We complete the test as we did in Section 14-1, except that the number of degrees of freedom associated with the chi-squared statistic is the number of intervals minus 1 minus the number of parameters estimated, which in this illustration is two. (We estimated the population mean μ and the population standard deviation σ.) Thus, in this case, the number of degrees of freedom is $k - 1 - 2 = 5 - 1 - 2 = 2$.

The expected values are

$$e_1 = np_1 = 148(0.0668) = 9.89$$
$$e_2 = np_2 = 148(0.2417) = 35.78$$
$$e_3 = np_3 = 148(0.3829) = 56.67$$
$$e_4 = np_4 = 148(0.2417) = 35.78$$
$$e_5 = np_5 = 148(0.0668) = 9.89$$

The observed values are determined manually by counting the number of values in each interval. Thus,

$$f_1 = 10$$
$$f_2 = 36$$
$$f_3 = 54$$
$$f_4 = 39$$
$$f_5 = 9$$

The chi-squared statistic is

$$\chi^2 = \sum_{i=1}^{k} \frac{(f_i - e_i)^2}{e_i} = \frac{(10 - 9.89)^2}{9.89} + \frac{(36 - 35.78)^2}{35.78} + \frac{(54 - 56.67)^2}{56.67} + \frac{(39 - 35.78)^2}{35.78} + \frac{(9 - 9.89)^2}{9.89} = 0.50$$

The rejection region is

$$\chi^2 > \chi^2_{a,k-3}\ \chi^2 = \chi^2_{0.05,2} = 5.99$$

There is not enough evidence to conclude that these data are not normally distributed.

14-4a Class Intervals

In practice you can use any intervals you like. We chose the intervals we did to facilitate the calculation of the normal probabilities. The number of intervals was chosen to comply with the rule of five, which requires that all expected values be at least equal to 5. Because the number of degrees of freedom is $k - 3$, the minimum number of intervals is $k = 4$.

Using the Computer

EXCEL

	A	B	C	D
1	Chi-Squared Test of Normality			
2				
3		*Newspaper*		
4	Mean	2.18		
5	Standard deviation	0.981		
6	Observations	148		
7				
8	Intervals	Probability	Expected	Observed
9	(z <= −1.5)	0.0668	9.89	10
10	(−1.5 < z <= −0.5)	0.2417	35.78	36
11	(−0.5 < z <= 1.5)	0.3829	56.67	54
12	(0.5 < z <=1.5)	0.2417	35.78	39
13	(z > 1.5)	0.0668	9.89	9
14				
15				
16	chi-squared Stat	0.50		
17	df	2		
18	p-value	0.7792		
19	chi-squared Critical	5.9915		

We programmed Excel to calculate the value of the test statistic so that the expected values are at least 5 (where possible) and the minimum number of intervals is 4. Hence, if the number of observations is more than 220, the intervals and probabilities are

Interval	Probability
$Z \leq -2$	0.0228
$-2 < Z \leq - {}^{-}1$	0.1359
$-1 < Z \leq 0$	0.3413
$0 < Z \leq 1$	0.3413
$1 < Z \leq 2$	0.1359
$Z > 2$	0.0228

If the sample size is less than or equal to 220 and greater than 80, the intervals are

Interval	Probability
$Z \leq -1.5$	0.0668
$-1.5 < Z \leq -0.5$	0.2417
$-0.5 < Z \leq 0.5$	0.3829
$0.5 < Z \leq 1.5$	0.2417
$Z > 1.5$	0.0668

If the sample size is less than or equal to 80, we employ the minimum number of intervals, 4. When the sample size is less than 32, at least one expected value will be less than 5. The intervals are

Interval	Probability
$Z \leq -1$	0.1587
$-1 < Z \leq 0$	0.3413
$0 < Z \leq 1$	0.3413
$Z > 1$	0.1587

Instructions

1. Type or import the data into one column. (**Open** Xm12-01.)
2. Click **Add-Ins, Data Analysis Plus** and **Chi-Squared Test of Normality**.
3. Specify the **Input Range** (A1:A149) and the value of α (0.05).

MINITAB Minitab does not conduct this procedure.

14-4b Interpreting the Results of a Chi-Squared Test for Normality

In the example above, we found that there was little evidence to conclude that the weight of discarded newspaper is not normally distributed. However, had we found evidence of non-normality, this would not necessarily invalidate the *z*-test we conducted in Example 12.1. As we pointed out in Chapter 12, the *t*-test of a mean is a robust procedure, which means that only if the variable is extremely non-normal and the sample size is small can we conclude that the technique is suspect. The problem here is that if the sample size is large and the variable is only slightly non-normal, the chi-squared test for normality will, in many cases, conclude that the variable is not normally distributed. However, if the variable is even quite non-normal and the sample size is large, the *t*-test will still be valid. Although there are situations in which we need to know whether a variable is non-normal, we continue to advocate that the way to decide if the normality requirement for almost all statistical techniques applied to interval data is satisfied is to draw histograms and look for shapes that are far from bell shaped (e.g., highly skewed or bimodal). We will use this approach in Chapter 17 when we introduce nonparametric techniques that are used when interval data are non-normal.

EXERCISES

14.38 Suppose that a random sample of 100 observations was drawn from a population. After calculating the mean and standard deviation, each observation was standardised and the number of observations in each of the following intervals was counted. Can we infer at the 5% significance level that the data were not drawn from a normal population?

Interval	Frequency
$Z \leq 1.5$	10
$-1.5 < Z \leq -0.5$	18
$-0.5 < Z \leq 0.5$	48
$0.5 < Z \leq 1.5$	16
$Z > 1.5$	8

14.39 A random sample of 50 observations yielded the following frequencies for the standardised intervals:

Interval	Frequency
$Z \leq -1$	6
$-1 < Z \leq 0$	27
$0 < Z \leq 1$	14
$Z > 1$	3

Can we infer that the data are not normal? (Use $\alpha = 0.10$.)

The following exercises require the use of a computer and software.

14.40 Xr13-12 Exercise 13.12 asked you to conduct a *t*-test of the difference between two means (reaction times). Test to determine whether there is enough evidence to infer that the reaction times are not normally distributed. A 5% significance level is judged to be suitable.

CHAPTER SUMMARY

This chapter introduced three statistical techniques. The first is the chi-squared goodness-of-fit test, which is applied when the problem objective is to describe a single population of nominal data with two or more categories. The second is the chi-squared test of a contingency table. This test has two objectives: to analyse the relationship between two nominal variables and to compare two or more populations of nominal data. The last procedure is designed to test for normality.

SYMBOLS:

Symbol	Pronounced	Represents
f_i	*f* sub *i*	Frequency of the *i*th category
e_t	*e* sub *i*	Expected value of the *i*th category
χ^2	Chi squared	Test statistic

FORMULA:

Test statistic for all procedures

$$\chi^2 = \sum_{i=1}^{k} \frac{(f_i - e_i)^2}{e_i}$$

CHAPTER EXERCISES

Use a 5% significance level, unless specified otherwise.

14.41 An organisation dedicated to ensuring fairness in television game shows is investigating *Wheel of Fortune*. In this show, three contestants are required to solve puzzles by selecting letters. Each contestant gets to select the first letter and continues selecting until he or she chooses a letter that is not in the hidden word, phrase or name. The order of contestants is random. However, contestant 1 gets to start game 1, contestant 2 starts game 2, and so on. The contestant who wins the most money is declared the winner and he or she is given an opportunity to win a grand prize. Usually, more than three games are played per show, and as a result it appears that contestant 1 has an advantage: contestant 1 will start two games, whereas contestant 3 will usually start only one game. To see whether this is the case, a random sample of 30 shows was taken and the starting position of the winning contestant for each show was recorded. These are shown in the following table:

Starting position	1	2	3
Number of wins	14	10	6

Do the tabulated results allow us to conclude that the game is unfair?

14.42 Recent studies suggest that UK workers have an average 9 sick days each year, nearly four times more than Asia-Pacific and 1.25 times more than Western Europe. It has been estimated that employee absenteeism costs UK employers more than 30 billion per year. As a first step in addressing the rising cost of absenteeism, the personnel department of a large corporation recorded the weekdays during which individuals in a sample of 362 absentees were away over the past several months. Do these data suggest that absenteeism is higher on some days of the week than on others?

Day of the Week	Monday	Tuesday	Wednesday	Thursday	Friday
Number absent	87	62	71	68	74

14.43 Suppose that the personnel department in Exercise 14.34 continued its investigation by categorising absentees according to the shift on which they worked, as shown in the accompanying table. Is there sufficient evidence at the 10% significance level of a relationship between the days on which employees are absent and the shift on which the employees work?

Shift	Monday	Tuesday	Wednesday	Thursday	Friday
Day	52	28	37	31	33
Evening	35	34	34	37	41

14.44 A management behaviour analyst has been studying the relationship between male/female supervisory structures in the workplace and the level of employees' job satisfaction. The results of a recent survey are shown in the accompanying table. Is there sufficient evidence to infer that the level of job satisfaction depends on the boss/employee gender relationship?

Level of Satisfaction	BOSS/EMPLOYEE Female/ Male	Female/ Female	Male/ Male	Male/ Female
Satisfied	21	25	54	71
Neutral	39	49	50	38
Dissatisfied	31	48	10	11

The following exercises require the use of a computer and software. Use a 5% significance level, unless specified otherwise.

14.45 Xr14-45 Stress is a serious medical problem that costs businesses and government billions of money annually and is common throughout Europe. As a result, it is important to determine the causes and possible cures. It would be helpful to know whether the causes are universal or do they vary from country to country. In surveys carried out every 5 years by the European Foundation for the Improvement of Living and Working Conditions, respondents were asked to report their primary source of stress in their lives. The responses are:

1 = Job, 2 = Finances, 3 = Health
4 = Family life, 5 = Other

The data were recorded using these codes plus 1 = Sweden and 2 = Netherlands. Do these data provide sufficient evidence to conclude that Sweden and Netherlands differ in their sources of stress?

14.46 Xr14-46 In the 2012 survey carried out by the TNS Opinion and Social Network, one in five European Union citizens declare that they have stopped smoking. Because nicotine is one of the most addictive drugs, quitting smoking is a difficult and frustrating task. It usually takes several tries before success is achieved. There are various methods, including cold turkey, nicotine patch, hypnosis and group therapy sessions. In an experiment to determine how these methods differ, a random sample of smokers who have decided to quit is selected. Each smoker has chosen one of the methods listed above. After 1 year, the respondents report whether they have quit (1 = Yes, 2 = No) and which method they used (1 = Cold turkey, 2 = Nicotine patch, 3 = Hypnosis, 4 = Group therapy sessions). Is there sufficient evidence to conclude that the four methods differ in their success?

14.47 Xr14-47 A newspaper publisher trying to pinpoint his market's characteristics wondered whether the way people read a newspaper is related to the reader's educational level. A survey asked adult readers which section of the paper they read first and asked to report their highest educational level. These data were recorded (column 1 = First section read where 1 = Front page, 2 = Sports, 3 = Editorial and 4 = Other) and column 2 = Educational level where 1 = Did not complete high school, 2 = High school graduate, 3 = University or college graduate and 4 = Postgraduate degree). What do these data tell the publisher about how educational level affects the way adults read the newspaper?

14.48 Xr14-48 Every week, the UK Lottery draws six numbers between 1 and 49. Lottery ticket buyers are naturally interested in whether certain numbers are drawn more frequently than others. To assist players, the official UK National Lottery website publishes the number of times each of the 49 numbers has been drawn in the past 52 weeks. The numbers and the frequency with which each occurred were recorded.

a. If the numbers are drawn from a uniform distribution, what is the expected frequency for each number?

b. Can we infer that the data were not generated from a uniform distribution?

In Section 14-4, we showed how to test for normality. However, we can use the same process to test for any other distribution.

14.49 Xr14-49 A scientist believes that the gender of a child is a binomial random variable with probability = 0.5 for a boy and 0.5 for a girl. To help test her belief, she randomly samples 100 families with five children. She records the number of boys. Can the scientist infer that the number of boys in families with five children is not a binomial random variable with $p = 0.5$? (*Hint:* Find the probability of $X = 0, 1, 2, 3, 4$ and 5 from a binomial distribution with $n = 5$ and $p = 0.5$.)

14.50 Xr14-505 Given the high cost of medical care, research that points the way to avoid illness is welcome. Previously performed research tells us that stress affects the immune system. Two scientists at Clalit Health Services asked 114 healthy adults about their social circles; they were asked to list every group they had contact with at least once every 2 weeks – family, coworkers, neighbours, friends, religious groups and community groups. Participants also reported negative life events over the past year, events such as death of a friend or relative, divorce or job-related problems. The participants were divided into four groups:

Group 1: Highly social and highly stressed
Group 2: Not highly social and highly stressed
Group 3: Highly social and not highly stressed
Group 4: Not highly social and not highly stressed

Each individual was classified in this way. In addition, whether each person contracted a cold over the next 12 weeks was recorded (1 = Cold, 2 = No cold). Can we infer that there are differences between the four groups in terms of contracting a cold?

The following exercises employ data files associated with examples and exercises seen earlier in this book.

14.51 **Xr12-49** Exercise 12.49 described a study to determine whether viewers (older than 50) of the network news had contacted their physician to ask about one of the prescription drugs advertised during the newscast. The responses (1 = No, 2 = Yes) were recorded. Also recorded were which of the three networks they normally watch (1 = RAI, 2 = RETE 4, 3 = CANALE 5). Can we conclude that there are differences in responses between the three network news shows?

14.52 **Xm12-05** Example 12.5 described exit polls wherein people are asked whether they voted for the Party A or Party B candidate for president. The surveyors also record gender (1 = Female, 2 = Male), educational attainment (1 = Did not finish high school, 2 = completed high school, 3 = completed college or university, 4 = postgraduate degree), and income level (1 = Under €25 000, 2 = €25 000 to €49 999, 3 = €50 000 to €75 000, 4 = over €75 000).

a. Is there sufficient evidence to infer that voting and gender are related?

b. Do the data allow the conclusion that voting and educational level are related?

c. Can we infer that voting and income are related?

TABLE A14.1 **Summary of Statistical Techniques in Chapters 12 to 14**

t-test of μ

Estimator of μ (including estimator of $N\mu$)

χ^2-test of σ^2

Estimator of σ^2

z-test of p

Estimator of p (including estimator of Np)

Equal-variances t-test of $\mu_1 - \mu_2$

Equal-variances estimator of $\mu_1 - \mu_2$

Unequal-variances t-test of $\mu_1 - \mu_2$

Unequal-variances estimator of $\mu_1 - \mu_2$

t-test of μ_D

Estimator of μ_D

F-test of σ^2_1/σ^2_2

Estimator of σ^2_1/σ^2_2

z-test of $p_1 - p_2$ (Case 1)

z-test of $p_1 - p_2$ (Case 2)

Estimator of $p_1 - p_2$

One-way analysis of variance (including multiple comparisons)

Two-way (randomised blocks) analysis of variance

Two-factor analysis of variance

χ^2-goodness-of-fit test

χ^2-test of a contingency table

15 SIMPLE LINEAR REGRESSION AND CORRELATION

15-1 MODEL

15-2 ESTIMATING THE COEFFICIENTS

15-3 ERROR VARIABLE: REQUIRED CONDITIONS

15-4 ASSESSING THE MODEL

15-5 USING THE REGRESSION EQUATION

15-6 REGRESSION DIAGNOSTICS-I

EDUCATION AND INCOME: HOW ARE THEY RELATED?

DATA Xm15-00

You are probably a student in an undergraduate or graduate business or economics programme. Your plan is to graduate, get a good job and draw a high salary. You have probably assumed that more education equals better job equals higher income. Is this true? A Social Survey recorded two variables that will help determine whether education and income are related and, if so, what the value of an additional year of education might be. The next statistical procedures will help you to answer these questions. **See Section 15-4j for the answer.**

INTRODUCTION

Correlation and regression are like two sides of a coin. **Regression analysis** is used to predict the value of one variable on the basis of other variables. This technique may be the most commonly used statistical procedure because, as you can easily appreciate, almost all companies and government institutions forecast variables such as product demand, interest rates, inflation rates, prices of raw materials and labour costs.

The technique involves developing a mathematical equation or model that describes the relationship between the variable to be forecast, which is called the **dependent variable**, and variables that the statistics practitioner believes are related to the dependent variable. The dependent variable is denoted as Y, whereas the related variables are called **independent variables** and are denoted as $X_1, X_2, \ldots, X_k$ (where k is the number of independent variables).

If we are interested only in determining whether a relationship exists we employ correlation analysis, a technique that we have already introduced. In Chapter 3 we presented the graphical method to describe the association between two interval variables – the scatter diagram. We introduced the coefficient of correlation and covariance in Chapter 4.

Because regression analysis involves many new techniques and concepts, we divided the presentation into three chapters. In this chapter, we present techniques that allow us to determine the relationship between only two variables. In Chapter 16 we expand our discussion to more than two variables.

Here are three illustrations of the use of regression analysis.

Illustration 1 The product manager in charge of a particular brand of children's breakfast cereal would like to predict the demand for the cereal during the next year. To use regression analysis, she and her staff list the following variables as likely to affect sales:

Price of the product

Number of children 5 to 12 years of age (the target market)

Price of competitors' products

Effectiveness of advertising (as measured by advertising exposure)

Annual sales this year

Annual sales in previous years

Illustration 2 A gold speculator is considering a major purchase of gold bullion. He would like to forecast the price of gold 2 years from now (his planning horizon), using regression analysis. In preparation he produces the following list of independent variables:

Interest rates

Inflation rate

Price of oil

Demand for gold jewellery

Demand for industrial and commercial gold

Dow Jones Industrial Average

Illustration 3 An estate agent wants to predict the selling price of houses more accurately. She believes that the following variables affect the price of a house:

Size of the house (number of square feet)

Number of bedrooms

Frontage of the plot

Condition

Location

In each of these illustrations, the primary motive for using regression analysis is forecasting. Nonetheless, analysing the relationship among variables can also be quite useful in managerial decision-making. For instance, in the first application, the product manager may want to know how price is related to product demand so that a decision about a prospective change in pricing can be made.

Regardless of why regression analysis is performed, the next step in the technique is to develop a mathematical equation or model that accurately describes the nature of the relationship that exists between the dependent variable and the independent variables. This stage – which is only a small part of the total process – is described in the next section. In the ensuing sections of this chapter (and in Chapter 16), we will spend considerable time assessing and testing how well the model fits the actual data. Only when we are satisfied with the model do we use it to estimate and forecast.

15-1 MODEL

The job of developing a mathematical equation can be quite complex, because we need to have some idea about the nature of the relationship between each of the independent variables and the dependent variable. The number of different mathematical models that could be proposed is virtually infinite. Here is an example from Chapter 4.

Profit = (Price per unit − Variable cost per unit) × Number of units sold − Fixed costs

You may encounter the next example in a finance course:

$$F = P(1 + i)^n$$

where

F = **future value of an investment**
P = **principle or present value**
i = **interest rate per period**
n = **number of periods**

These are all examples of **deterministic models**, so named because such equations allow us to determine the value of the dependent variable (on the left side of the equation) from the values of the independent variables. In many practical applications of interest to us, deterministic models are unrealistic. For example, is it reasonable to believe that we can determine the selling price of a house solely on the basis of its size? Unquestionably, the size of a house affects its price, but many other variables (some of which may not be measurable) also influence price. What must be included in most practical models is a method to represent the randomness that is part of a real-life process. Such a model is called a **probabilistic model**.

To create a probabilistic model, we start with a deterministic model that approximates the relationship we want to model. We then add a term that measures the random error of the deterministic component.

Suppose that in illustration 3, the estate agent knows that the cost of building a new house is about £1000 per square metre and that most plots sell for about £100 000. The approximate selling price would be

$$y = 100\,000 + 1000x$$

where y = selling price and x = size of the house in square feet. A house of 200 square metres would therefore be estimated to sell for

$$y = 100\,000 + 1000(200) = 300\,000$$

We know, however, that the selling price is not likely to be exactly £300 000. Prices may actually range from £200 000 to £400 000. In other words, the deterministic model is not really suitable. To represent this situation properly, we should use the probabilistic model

$$y = 100\,000 + 1000x + \varepsilon$$

where ε (the Greek letter epsilon) represents the **error variable** – the difference between the actual selling price and the estimated price based on the size of the house. The error thus accounts for all the variables, measurable and immeasurable, that are not part of the model. The value of ε will vary from one sale to the next, even if x remains constant. In other words, houses of exactly the same size will sell for different prices because of differences in location and number of bedrooms and bathrooms, as well as other variables.

In the three chapters devoted to regression analysis, we will present only probabilistic models. In this chapter, we describe only the straight-line model with one independent variable. This model is called the **first-order linear model** – sometimes called the **simple linear regression model.***

First-Order Linear Model

$$y = \beta_0 + \beta_1 x + \varepsilon \quad (y = \textit{Regression (on X)} + \textit{error})$$

where

y = dependent variable
x = independent variable
β_0 = y-intercept
β_1 = slope of the line (or gradient of the line)
ε = error variable

The problem objective addressed by the model is to analyse the relationship between two variables, x and y, both of which must be interval. To define the relationship between x and y, we need to know the value of the coefficients β_0 and β_1. However, these coefficients are population parameters, which are almost always unknown. In the next section, we discuss how these parameters are estimated.

15-2 ESTIMATING THE COEFFICIENTS

We estimate the parameters β_0 and β_1 in a way similar to the methods used to estimate all the other parameters discussed in this book. We draw a random sample from the population of interest and calculate the sample statistics we need. However, because β_0 and β_1 represent the coefficients of a straight line, their estimators are based on drawing a straight line through the sample data. The straight line that we wish to use to estimate β_0 and β_1 is the 'best' straight line that fits the data – best in the sense that it comes closest to the sample data points. This best straight line, called the *least squares line* or *regression line*, is derived from calculus and is represented by the following equation:

$$\hat{y} = b_0 + b_1 x$$

Here b_0 is the y-intercept, b_1 is the slope and $\hat{y}$ is the predicted or fitted value of y. In Chapter 4, we introduced the **least squares method**, which produces a straight line that minimises the sum of the squared differences between the points and the line. The coefficients b_0 and b_1 are calculated so that the sum of squared deviations is minimised.

$$\sum_{i=1}^{n} (y_i - \hat{y}_i)^2$$

*We use the term *linear* in two ways. The 'linear' in linear regression refers to the form of the model wherein the terms form a linear combination of the coefficients β_0 and β_1. Thus, for example, the model $y = \beta_0 + \beta_0 x^2 + \varepsilon$ is a linear combination whereas $y = \beta_0 + \beta_1 x + \varepsilon$ is not. The simple linear regression model $y = \beta_0 + \beta_0 x^2 + \varepsilon$ describes a straight-line or linear relationship between the dependent variable and one independent variable. In this book, we use the linear regression technique only. Hence, when we use the word *linear* we will be referring to the straight-line relationship between the variables.

In other words, the values of $\hat{y}$ on average come closest to the observed values of y. The calculus derivation is available in the website Appendix, Deriving the Normal Equations, which shows how the following formulas, first shown in Chapter 4, were produced.

Least Squares Line Coefficients

$$b_1 = \frac{s_{xy}}{s_x^2}$$

$$b_0 = \bar{y} - b_1\bar{x}$$

where

$$s_{xy} = \frac{\sum_{i=1}^{n}(x_i - \bar{x})(y_i - \bar{y})}{n-1}$$

$$s_x^2 = \frac{\sum_{i=1}^{n}(x_i - \bar{x})^2}{n-1}$$

$$\bar{x} = \frac{\sum_{i=1}^{n} x_i}{n}$$

$$\bar{y} = \frac{\sum_{i=1}^{n} y_i}{n}$$

In Chapter 4, we provided shortcut formulas for the sample variance and the sample covariance. Combining them provides a shortcut method to manually calculate the slope coefficient.

Shortcut Formula for b_1

$$b_1 = \frac{s_{xy}}{s_x^2}$$

$$s_{xy} = \frac{1}{n-1}\left[\sum_{i=1}^{n} x_i y_i - \frac{\sum_{i=1}^{n} x_i \sum_{i=1}^{n} y_i}{n}\right]$$

$$s_x^2 = \frac{1}{n-1}\left[\sum_{i=1}^{n} x_i^2 - \frac{\left(\sum_{i=1}^{n} x_i\right)^2}{n}\right]$$

Statisticians have shown that b_0 and b_1 are unbiased estimators of β_0 and β_1, respectively. Although the calculations are straightforward, we would rarely compute the regression line manually because the work is time-consuming. However, we illustrate the manual calculations for a very small sample.

EXAMPLE 15.1

DATA Xm15-01

Annual Bonus and Years of Experience

The annual bonuses (€1000s) of six employees with different years of experience were recorded as follows. We wish to determine the straight-line relationship between annual bonus and years of experience.

Years of experience *x*	1	2	3	4	5	6
Annual bonus *y*	6	1	9	5	17	12

SOLUTION:

To apply the shortcut formula, we need to compute four summations. Using a calculator, we find.

$$\sum_{i=1}^{n} x_i = 21$$

$$\sum_{i=1}^{n} y_i = 50$$

$$\sum_{i=1}^{n} x_i y_i = 212$$

$$\sum_{i=1}^{n} x_i^2 = 91$$

The covariance and the variance of x can now be computed:

$$s_{xy} = \frac{1}{n-1}\left[\sum_{i=1}^{n} x_i y_i - \frac{\sum_{i=1}^{n} x_i \sum_{i=1}^{n} y_i}{n}\right] = \frac{1}{6-1}\left[212 - \frac{(21)(50)}{6}\right] = 7.4$$

$$s_x^2 = \frac{1}{n-1}\left[\sum_{i=1}^{n} x_i^2 - \frac{\left(\sum_{i=1}^{n} x_i\right)^2}{n}\right] = \frac{1}{6-1}\left[91 - \frac{(21)^2}{6}\right] = 3.5$$

The sample slope coefficient is calculated next:

$$b_1 = \frac{s_{xy}}{s_x^2} = \frac{7.4}{3.5} = 2.114$$

The y-intercept is computed as follows:

$$\bar{x} = \frac{\sum x_i}{n} = \frac{21}{6} = 3.5$$

$$y = \frac{\sum y_i}{n} = \frac{50}{6} = 8.333$$

$$b_0 = \bar{y} - b_1\bar{x} = 8.333 - (2.114)(3.5) = 0.934$$

Thus, the least squares line is

$$\hat{y} = 0.934 + 2.114x$$

Figure 15.1 depicts the least squares (or regression) line. As you can see, the line fits the data reasonably well. We can measure how well by computing the value of the minimised sum of squared deviations. The deviations between the actual data points and the line are called **residuals**, denoted e_i; that is,

$$e_i = y_i - \hat{y}_i$$

FIGURE 15.1

Scatter Diagram with Regression Line for Example 15.1

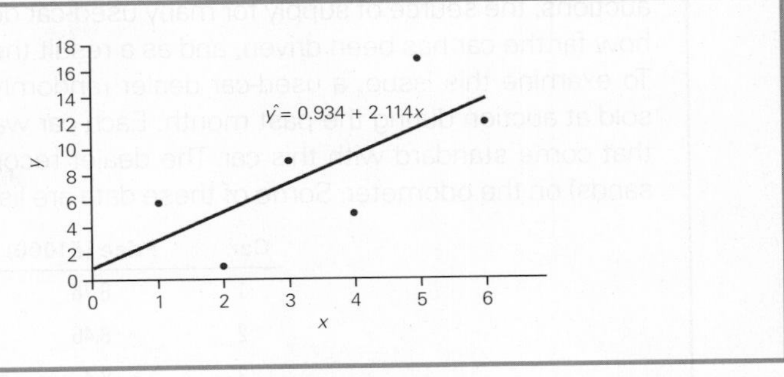

The residuals are observations of the error variable. Consequently, the minimised sum of squared deviations is called the **sum of squares for error**, denoted SSE.

The calculation of the residuals in this example is shown in Figure 15.2. Notice that we compute *by* $\hat{y}_i$ substituting x_i into the formula of the regression line. The residuals are the differences between the observed values of y_i and the fitted or predicted values of $\hat{y}_i$. Table 15.1 describes these calculations.

Thus, SSE = 81.104. No other straight line will produce a sum of squared deviations as small as 81.104. In that sense, the regression line fits the data best. The sum of squares for error is an important statistic because it is the basis for other statistics that assess how well the linear model fits the data. We will introduce these statistics in Section 15-4.

FIGURE 15.2

Calculation of Residuals in Example 15.1

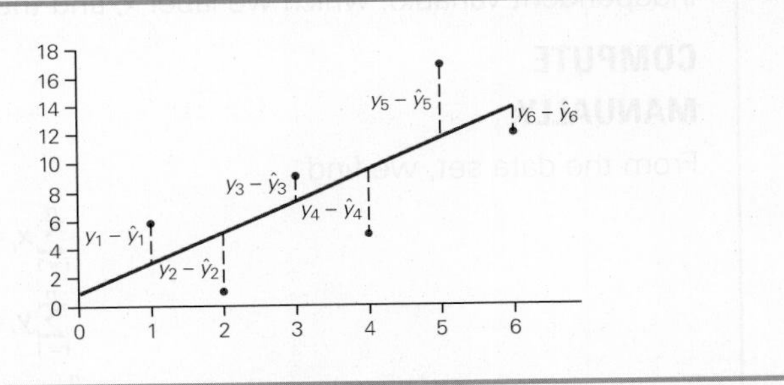

TABLE 15.1 **Calculation of Residuals in Example 15.1**

x_i	y_i	$\hat{y} = 0.934 = 2.114x_i$	$y_i - \hat{y}_i$	$(y_i - \hat{y}_i)^2$
1	6	3.048	2.952	8.714
2	1	5.162	−4.162	17.322
3	9	7.276	1.724	2.972
4	5	9.390	−4.390	19.272
5	17	11.504	5.496	30.206
6	12	13.618	−1.618	2.618
				$\sum(y_i - \hat{y}_i)^2 = 81.104$

EXAMPLE 15.2

DATA Xm15-02

Odometer Reading and Prices of Used Toyota Auris, Part 1

Whether you want to buy or sell a used car, you want to find its real value. Car dealers use the Black Book to help them determine the market value of used cars that their customers trade in when purchasing new cars. The Black Book is updated daily according to the market and lists the trade-in values for all basic models of cars. It provides alternative values for each car model according to its condition and optional features. The values are determined on the basis of the average paid at recent used-car auctions, the source of supply for many used-car dealers. However, a critical factor for used-car buyers is how far the car has been driven, and as a result the odometer reading is also a useful indicator of value. To examine this issue, a used-car dealer randomly selected 100 three-year old Toyota Auris that were sold at auction during the past month. Each car was in top condition and equipped with all the features that come standard with this car. The dealer recorded the price (£1000) and the number of km (thousands) on the odometer. Some of these data are listed here. The dealer wants to find the regression line.

Car	Price (£1000)	Odometer (1000 km)
1	8.76	59.84
2	8.46	71.68
3	8.4	73.28
⋮	⋮	⋮
98	8.7	53.12
99	8.82	62.72
100	8.58	58.24

SOLUTION:

IDENTIFY

Notice that the problem objective is to analyse the relationship between two interval variables. Because we believe that the odometer reading affects the selling price, we identify the former as the independent variable, which we label *x*, and the latter as the dependent variable, which we label *y*.

COMPUTE

MANUALLY:

From the data set, we find

$$\sum_{i=1}^{n} x_i = 5761.8$$

$$\sum_{i=1}^{n} y_i = 890.46$$

$$\sum_{i=1}^{n} x_i y_i = 51\,029.69$$

$$\sum_{i=1}^{n} x_i^2 = 343\,005.67$$

Next we calculate the covariance and the variance of the independent variable *x*:

$$s_{xy} = \frac{1}{n-1}\left[\sum_{i=1}^{n} x_i y_i - \frac{\sum_{i=1}^{n} x_i \sum_{i=1}^{n} y_i}{n}\right] = \frac{1}{100-1}\left[51\,029.69 - \frac{(5761.9)(890.46)}{100}\right] = -2.79268$$

$$s_x^2 = \frac{1}{n-1}\left[\sum_{i=1}^{n} x_i^2 - \frac{\left(\sum_{i=1}^{n} x_i\right)^2}{n}\right] = \frac{1}{100-1}\left[343\,005.67 - \frac{33197878.3}{100}\right] = 111.382$$

The sample slope coefficient is calculated next:

$$b_1 = \frac{s_{xy}}{s_x^2} = \frac{-2.79268}{111.382} = -0.0251$$

The *y*-intercept is computed as follows:

$$\bar{x} = \frac{\sum x_i}{n} = \frac{5761.8}{100} = 57.618$$

$$\bar{y} = \frac{\sum y_i}{n} = \frac{890.46}{100} = 8.9046$$

$$b_0 = \bar{y} - b_1\bar{x} = 8.9046 - (-0.0251)(57.618) = 10.349$$

The sample regression line is

$$\hat{y} = 10.349 - 0.0251x$$

EXCEL

	A	B	C	D	E	F
1	SUMMARY OUTPUT					
2						
3	*Regression Statistics*					
4	Multiple R	0.8052				
5	R Square	0.6483				
6	Adjusted R Square	0.6447				
7	Standard Error	0.1959				
8	Observations	100				
9						
10	ANOVA					
11		*df*	*SS*	*MS*	*F*	*Significance F*
12	Regression	1	6.93	6.93	180.64	5.75E-24
13	Residual	98	3.76	0.04		
14	Total	99	10.69			
15						
16		*Coefficients*	*Standard Error*	*t Stat*	*P-value*	
17	Intercept	10.35	0.109	94.73	3.57E-98	
18	Odometer	−0.0251	0.002	−13.44	5.75E-24	

Instructions

1. Type or import data into two columns*, one storing the dependent variable and the other the independent variable. (**Open** Xm15-02.)
2. Click **Data, Data Analysis** and **Regression.**
3. Specify the **Input *Y* Range** (A1:A101) and the **Input *X* Range** (B1:B101).

To draw the scatter diagram follow the instructions provided in Chapter 3.

*If one or both columns contain an empty cell (representing missing data), the row must be removed.

MINITAB

Regression Analysis: Price versus Odometer

```
The regression equation is
Price = 10.3 – 0.0251 Odometer

Predictor        Coef     SE Coef       T      P
Constant       10.3492     0.1093   94.73  0.000
Odometer     –0.025073   0.001865  –13.44  0.000

S = 0.195893   R-Sq = 64.8%   R-Sq(adj) = 64.5%

Analysis of Variance

Source          DF       SS       MS       F      P
Regression       1   6.9320   6.9320  180.64  0.000
Residual Error  98   3.7607   0.0384
Total           99  10.6927
```

Instructions

1. Type or import the data into two columns. (**Open** Xm15-02.)
2. Click **Stat, Regression** and **Regression** ....
3. Type the name of the dependent variable in the **Response** box (Price) and the name of the independent variable in the **Predictors** box (Odometer).

To draw the scatter diagram click **Stat, Regression** and **Fitted Line Plot**. Alternatively, follow the instructions provide in Chapter 3.

The printouts include more statistics than we need right now. However, we will be discussing the rest of the printouts later.

INTERPRET

The slope coefficient b_1 is −0.0251, which means that for each additional 1000 km on the odometer, the price decreases by an average of £0.0251 thousand. Expressed more simply, the slope tells us that for each additional km on the odometer, the price decreases on average by £0.0251 or 2.51 pence.

The intercept is $b_0 = 10.349$. Technically the intercept is the point at which the regression line and the *y-axis* intersect. This means that when $x = 0$ (i.e. the car was not driven at all) the selling price is £10.349 thousand or £10 349. We might be tempted to interpret this number as the price of cars that have not been driven. However, in this case, the intercept is probably meaningless. Because our sample did not include any cars with zero miles on the odometer, we have no basis for interpreting b_0. As a general rule, we cannot determine the value of $\hat{y}$ for a value of x that is far outside the range of the sample values of x. In this example, the smallest and largest values of x are 30.6 and 78.7, respectively. Because $x = 0$ is not in this interval, we cannot safely interpret the value of y when $x = 0$.

It is important to bear in mind that the interpretation of the coefficients pertains only to the sample, which consists of 100 observations. To infer information about the population, we need statistical inference techniques, which are described subsequently.

In the sections that follow, we will return to this problem and the computer output to introduce other statistics associated with regression analysis.

EXERCISES

15.1 The term *regression* was originally used in 1885 by Sir Francis Galton in his analysis of the relationship between the heights of children and parents. He formulated the 'law of universal regression', which specifies that 'each peculiarity in a man is shared by his kinsmen, but on average in a less degree'. (Evidently, people spoke this way in 1885.) In 1903, two statisticians, K. Pearson and A. Lee, took a random sample of 1078 father–son pairs to examine Galton's law ('On the Laws of Inheritance in Man, I. Inheritance of Physical Characteristics', *Biometrika* 2:457–462). Their sample regression line was

$$\text{Son's height} = 33.73 + 0.516 \times \text{Father's height}$$

a. Interpret the coefficients.

b. What does the regression line tell you about the heights of sons of tall fathers?

c. What does the regression line tell you about the heights of sons of short fathers?

15.2 Xr15-02 Attempting to analyse the relationship between advertising and sales, the owner of a furniture store recorded the monthly advertising budget (€ thousands) and the sales (€ millions) for a sample of 12 months. The data are listed here.

Advertising	23	46	60	54	28	33
Sales	9.6	11.3	12.8	9.8	8.9	12.5

Advertising	25.0	31.0	36.0	88.0	90.0	99.0
Sales	12.0	11.4	12.6	13.7	14.4	15.9

a. Draw a scatter diagram. Does it appear that advertising and sales are linearly related?

b. Calculate the least squares line and interpret the coefficients.

15.3 Xr15-03 To determine how the number of housing starts is affected by mortgage rates an economist recorded the average mortgage rate and the number of housing starts in a large county for the past 10 years. These data are listed here.

Rate	8.5	7.8	7.6	7.5	8.0
Starts	115	111	185	201	206

Rate	8.4	8.8	8.9	8.5	8.0
Starts	167	155	117	133	150

a. Determine the regression line.

b. What do the coefficients of the regression line tell you about the relationship between mortgage rates and housing starts?

15.4 Xr15-04 Critics of television often refer to the detrimental effects that all the violence shown on television has on children. However, there may be another problem. It may be that watching television also reduces the amount of physical exercise, causing weight gains. A sample of fifteen 10-year-old children was taken. The number of pounds each child was overweight was recorded (a negative number indicates the child is underweight). In addition, the number of hours of television viewing per week was also recorded. These data are listed here.

Television	42	34	25	35	37	38	31	33
Overweight	18	6	0	−1	13	14	7	7

Television	19	29	38	28	29	36	18
Overweight	−9	8	8	5	3	14	−7

a. Draw the scatter diagram.

b. Calculate the sample regression line and describe what the coefficients tell you about the relationship between the two variables.

15.5 Xr15-05 To help determine how many beers to stock the concession manager at Camp Nou Stadium wanted to know how the temperature affected beer sales. Accordingly, she took a sample of ten games and recorded the number of beers sold and the temperature in the middle of the game.

Temperature	26.6	20.0	25.6	26.1	30.6
Number of beers	20 533	1 439	13 829	21 286	30 985

Temperature	23.3	30.0	33.3	25.0	28.9
Number of beers	17 187	30 240	37 596	9 610	28 742

a. Compute the coefficients of the regression line.

b. Interpret the coefficients.

The exercises that follow were created to allow you to see how regression analysis is used to solve realistic problems. As a result, most feature a large number of observations. We anticipate that most students will solve these problems using a computer and statistical software.

15.6 Xr15-06 In television's early years, most commercials were 60 seconds long. Now, however, commercials can be any length. The objective of commercials remains the same: to have as many viewers as possible remember the product in a favourable way and eventually buy it. In an experiment to determine how the length of a commercial is related to people's memory of it, 60 randomly selected people were asked to watch a 1-hour television programme. In the middle of the show, a commercial advertising a brand of toothpaste appeared. Some viewers watched a commercial that lasted for 20 seconds, others watched one that lasted for 24 seconds, 28 seconds,..., 60 seconds. The essential content of the commercials was the same. After the show, each person was given a test to measure how much he or she remembered about the product. The commercial times and test scores (on a 30-point test) were recorded.

a. Draw a scatter diagram of the data to determine whether a linear model appears to be appropriate.

b. Determine the least squares line.

c. Interpret the coefficients.

APPLICATIONS IN HUMAN RESOURCES MANAGEMENT

Retaining Workers

Human resource managers are responsible for a variety of tasks within organisations. As we pointed out in the introduction in Chapter 1, personnel or human resource managers are involved with recruiting new workers, determining which applicants are most suitable to hire, and helping with various aspects of monitoring the workforce, including absenteeism and worker turnover. For many firms, worker turnover is a costly problem. First, there is the cost of recruiting and attracting qualified workers. The firm must advertise vacant positions and make certain that applicants are judged properly. Second, the cost of training hirees can be high, particularly in technical areas. Third, new employees are often not as productive and efficient as experienced employees. Consequently, it is in the interests of the firm to attract and keep the best workers. Any information that the personnel manager can obtain is likely to be useful.

15.7 Xr15-07 The human resource manager of a telemarketing firm is concerned about the rapid turnover of the firm's telemarketers. It appears that many telemarketers do not work very long before quitting. There may be a number of reasons, including relatively low pay, personal unsuitability for the work, and the low probability of advancement. Because of the high cost of hiring and training new workers, the manager decided to examine the factors that influence workers to quit. He reviewed the work history of a random sample of workers who have quit in the last year and recorded the number of weeks on the job before quitting and the age of each worker when originally hired.

a. Use regression analysis to describe how the work period and age are related.

b. Briefly discuss what the coefficients tell you.

15.8 Xr15-08 Besides their known long-term effects, do cigarettes also cause short-term illnesses such as colds? To help answer this question, a sample of smokers was drawn. Each person was asked to report the average number of cigarettes smoked per day and the number of days absent from work due to colds last year.

a. Determine the regression line.

b. What do the coefficients tell you about the relationship between smoking cigarettes and sick days because of colds?

15.9 Xr15-09 An estate agent specialising in commercial property wanted a more precise method of judging the likely selling price (in £1000s) of apartment buildings. As a first effort, she recorded the price of a number of apartment buildings sold recently and the number of square metres (in 1000s) in the building.

a. Calculate the regression line.

b. What do the coefficients tell you about the relationship between price and square metre?

15.10 Xr03-24 (Exercise 3.24 revisited) In an attempt to determine the factors that affect the amount of energy used, 200 households were analysed. In each, the number of occupants and the amount of electricity used were measured. Determine the regression line and interpret the results.

15.11 Xr15-11 An economist for the government is attempting to produce a better measure of poverty than is currently in use. To help acquire information, she recorded the annual household income (in €1000s) and the amount of money spent on food during one week for a random sample of households. Determine the regression line and interpret the coefficients.

15.12 Xr15-12 An economist wanted to investigate the relationship between office rents (the dependent variable) and vacancy rates. Accordingly, he took a random sample of monthly office rents and the percentage of vacant office space in 30 different cities.

a. Determine the regression line.

b. Interpret the coefficients.

APPLICATIONS IN HUMAN RESOURCES MANAGEMENT

Testing Job Applicants

The recruitment process at many firms involves tests to determine the suitability of candidates. The tests may be written to determine whether the applicant has sufficient knowledge in his or her area of expertise to perform well on the job. There may be oral tests to determine whether the applicant's personality matches the needs of the job. Manual or technical skills can be tested through a variety of physical tests. The test results contribute to the decision to hire. In some cases, the test result is the only criterion to hire. Consequently, it is vital to ensure that the test is a reliable predictor of job performance. If the tests are poor predictors, they should be discontinued. Statistical analyses allow personnel managers to examine the link between the test results and job performance.

15.13 Xr15-13 Although a large number of tasks in the computer industry are robotic, many operations require human workers. Some jobs require a great deal of dexterity to properly position components into place. A large computer maker routinely tests applicants for these jobs by giving a dexterity test that involves a number of intricate finger and hand movements. The tests are scored on a 100-point scale. Only those who have scored above 70 are hired. To determine whether the tests are valid predictors of job performance, the personnel manager drew a random sample of 45 workers who were hired 2 months ago. He recorded their test scores and the percentage of non-defective computers they produced in the last week. Determine the regression line and interpret the coefficients.

15-3 ERROR VARIABLE: REQUIRED CONDITIONS

In the previous section, we used the least squares method to estimate the coefficients of the linear regression model. A critical part of this model is the error variable ε. In the next section, we will present an inferential method that determines whether there is a relationship between the dependent and independent variables. Later we will show how we use the regression equation to estimate and predict. For these methods to be valid, however, four requirements involving the probability distribution of the error variable must be satisfied.

Required Conditions for the Error Variable

1. The probability distribution of ε is normal.
2. The mean of the distribution is 0; that is, $E(\varepsilon) = 0$.
3. The standard deviation of ε is σ_e, which is a constant regardless of the value of x.
4. The value of ε associated with any particular value of y is independent of ε associated with any other value of y.

Requirements 1, 2 and 3 can be interpreted in another way: For each value of x, y is a normally distributed random variable whose mean is

$$E(y) = \beta_1 + \beta_1 x$$

and whose standard deviation is σ_ε. Notice that the mean depends on x. The standard deviation, however, is not influenced by x because it is a constant over all values of x. Figure 15.3 depicts this interpretation. Notice that for each value of x, $E(y)$ changes, but the shape of the distribution of y remains the same. In other words, for each x, y is normally distributed with the same standard deviation.

FIGURE 15.3

Distribution of *y* Given *x*

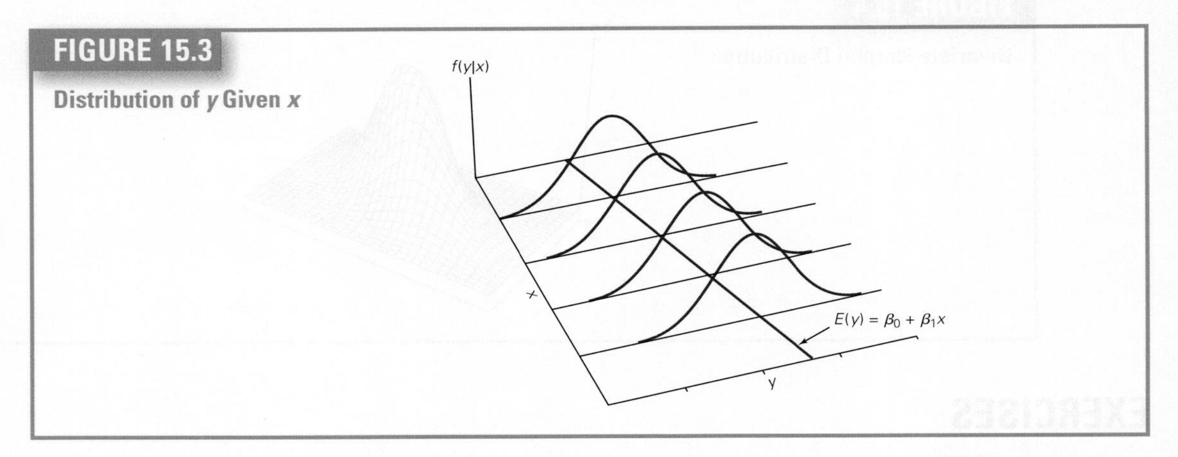

In Section 15-6, we will discuss how departures from these required conditions affect the regression analysis and how they are identified.

15-3a Observational and Experimental Data

In Chapter 5 and again in Chapter 12, we described the difference between observational and experimental data. We pointed out that statistics practitioners often design controlled experiments to enable them to interpret the results of their analyses more clearly than would be the case after conducting an observational study. Example 15.2 is an illustration of observational data. In that example, we merely observed the odometer reading and auction selling price of 100 randomly selected cars.

If you examine Exercise 15.6, you will see experimental data gathered through a controlled experiment. To determine the effect of the length of a television commercial on its viewers' memories of the product advertised, the statistics practitioner arranged for 60 television viewers to watch a commercial of differing lengths and then tested their memories of that commercial. Each viewer was randomly assigned a commercial length. The values of x ranged from 20 to 60 and were set by the statistics practitioner as part of the experiment. For each value of x, the distribution of the memory test scores is assumed to be normally distributed with a constant variance.

We can summarise the difference between the experiments described in Example 15.2 and the one described in Exercise 15.6. In Example 15.2, both the odometer reading and the auction selling price are random variables. We hypothesise that for each possible odometer reading, there is a theoretical population of auction selling prices that are normally distributed with a mean that is a linear function of the odometer reading and a variance that is constant. In Exercise 15.6, the length of the commercial is not a random variable but a series of values selected by the statistics practitioner. For each commercial length, the memory test scores are required to be normally distributed with a constant variance.

Regression analysis can be applied to data generated from either observational or controlled experiments. In both cases, our objective is to determine how the independent variable is related to the dependent variable. However, observational data can be analysed in another way. When the data are observational, both variables are random variables. We need not specify that one variable is independent and the other is dependent. We can simply determine *whether* the two variables are related. The equivalent of the required conditions described in the previous box is that the two variables are bivariate normally distributed. (Recall that in Section 7-2 we introduced the bivariate distribution, which describes the joint probability of two variables.) A bivariate normal distribution is described in Figure 15.4. As you can see, it is a three-dimensional bell-shaped curve. The dimensions are the variables x, y and the joint density function $f(x,y)$.

In Section 15-4, we will discuss the statistical technique that is used when both x and y are random variables and they are bivariate normally distributed. In Chapter 17, we will introduce a procedure applied when the normality requirement is not satisfied.

FIGURE 15.4
Bivariate Normal Distribution

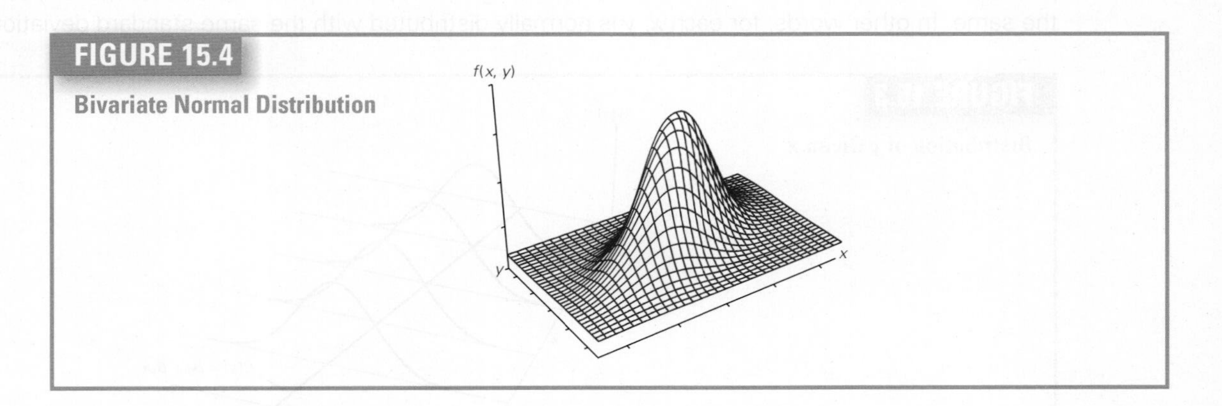

EXERCISES

15.14 Describe what the required conditions mean in Exercise 15.6. If the conditions are satisfied, what can you say about the distribution of memory test scores?

15-4 ASSESSING THE MODEL

The least squares method produces the best straight line. However, there may, in fact, be no relationship or perhaps a non-linear relationship between the two variables. If so, a straight-line model is likely to be impractical. Consequently, it is important for us to assess how well the linear model fits the data. If the fit is poor, we should discard the linear model and seek another one.

Several methods are used to evaluate the model. In this section, we present two statistics and one test procedure to determine whether a linear model should be employed. They are the **standard error of estimate**, the *t*-test of the slope, and the coefficient of determination. All these methods are based on the sum of squares for error.

15-4a Sum of Squares for Error

The least squares method determines the coefficients that minimise the sum of squared deviations between the points and the line defined by the coefficients. Recall from Section 15-2 that the minimised sum of squared deviations is called the *sum of squares for error,* denoted SSE. In that section, we demonstrated the direct method of calculating SSE. For each value of x, we compute the value of y. In other words, for $i = 1$ to n, we compute

$$\hat{y}_i = b_0 + b_1 x_i$$

For each point, we then compute the difference between the actual value of y and the value calculated at the line, which is the residual. We square each residual and sum the squared values. Table 15.1 shows these calculations for Example 15.1. To calculate SSE manually requires a great deal of arithmetic. Fortunately, there is a shortcut method available that uses the sample variances and the covariance.

Shortcut Calculation of SSE

$$SSE = \sum_{i=1}^{n}(y_i - \hat{y}_i)^2 = (n-1)\left(s_y^2 - \frac{s_{xy}^2}{s_x^2}\right)$$

where s_y^2 is the sample variance of the dependent variable.

15-4b Standard Error of Estimate

In Section 15-3, we pointed out that the error variable ε is normally distributed with mean 0 and standard deviation σ_ε. If σ_ε is large, some of the errors will be large, which implies that the model's fit is poor. If σ_ε is small, the errors tend to be close to the mean (which is 0); as a result, the model fits well. Hence, we could use σ_ε to measure the suitability of using a linear model. Unfortunately, σ_ε is a population parameter and, like most other parameters, is unknown. We can, however, estimate σ_ε from the data. The estimate is based on SSE. The unbiased estimator of the variance of the error variable σ_ε^2 is

$$s_\varepsilon^2 = \frac{SSE}{n-2}$$

The square root of σ_ε^2 is called the *standard error of estimate.*

Standard Error of Estimate

$$s_\varepsilon = \sqrt{\frac{SSE}{n-2}}$$

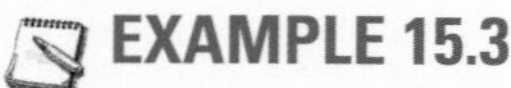

EXAMPLE 15.3

Odometer Reading and Prices of Used Toyota Auris, Part 2

Find the standard error of estimate for Example 15.2 and describe what it tells you about the model's fit.

SOLUTION:

COMPUTE MANUALLY:

To compute the standard error of estimate, we must compute SSE, which is calculated from the sample variances and the covariance. We have already determined the covariance and the variance of x: −2.805 and 111.382, respectively. The sample variance of y (applying the shortcut method) is

$$s_y^2 = \frac{1}{n-1}\left[\sum_{i=1}^{n} y_i^2 - \frac{\left(\sum_{i=1}^{n} y_i\right)^2}{n}\right]$$

$$= \frac{1}{100-1}\left[22\,055.23 - \frac{(1484.1)^2}{100}\right]$$

$$= 0.300$$

$$s_y^2 = \frac{1}{n-1}\left[\sum_{i=1}^{n} y_i^2 - \frac{(\sum_{i=1}^{n} y_i)^2}{n}\right] = \frac{1}{100-1}\left[7\,939.88 - \frac{792\,919.01}{100}\right] = 0.108$$

$$SSE = (n-1)\left(s_y^2 - \frac{s_{xy}^2}{s_x^2}\right) = (100-1)\left(0.108 - \frac{(-2.792)^2}{111.382}\right) = 3.761$$

The standard error of estimate follows:

$$s_\varepsilon = \sqrt{\frac{SSE}{n-2}} = \sqrt{\frac{3.760}{98}} = 0.1959$$

INTERPRET

The smallest value that s_ε can assume is 0, which occurs when SSE = 0; that is, when all the points fall on the regression line. Thus, when s_ε is small, the fit is excellent, and the linear model is likely to be an effective analytical and forecasting tool. If s_ε is large, the model is a poor one, and the statistics practitioner should improve it or discard it.

We judge the value of s_ε by comparing it to the values of the dependent variable y or more specifically to the sample mean $\bar{y}$. In this example, because $s_\varepsilon = 0.1959$ and $\bar{y} = 8.9$, it does appear that the standard error of estimate is small. However, because there is no predefined upper limit on s_ε, it is often difficult to assess the model in this way. In general, the standard error of estimate cannot be used as an absolute measure of the model's utility.

Nonetheless, s_ε is useful in comparing models. If the statistics practitioner has several models from which to choose, the one with the smallest value of s_ε should generally be the one used. As you will see, s_ε is also an important statistic in other procedures associated with regression analysis.

15-4c Testing the Slope

To understand this method of assessing the linear model, consider the consequences of applying the regression technique to two variables that are not at all linearly related. If we could observe the entire population and draw the regression line, we would observe the scatter diagram shown in Figure 15.5. The line is horizontal, which means that no matter what value of x is used, we would estimate the same value for $\hat{y}$; thus, y is not linearly related to x. Recall that a horizontal straight line has a slope of 0, that is, $\beta_1 = 0$.

FIGURE 15.5

Scatter Diagram of Entire Population with $\beta_1 = 0$

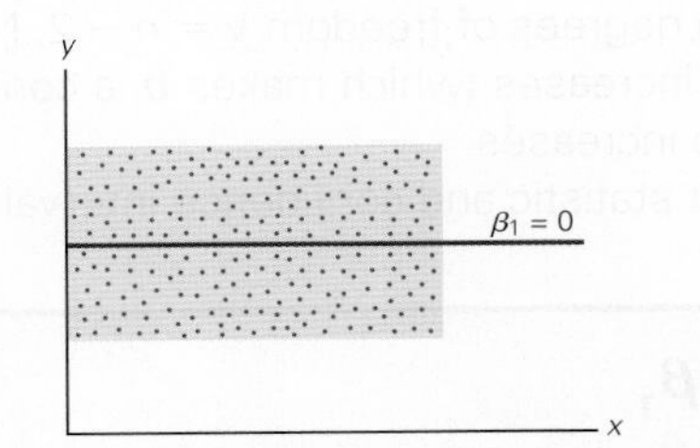

Because we rarely examine complete populations, the parameters are unknown. However, we can draw inferences about the population slope β_1 from the sample slope b_1.

The process of testing hypotheses about β_1 is identical to the process of testing any other parameter. We begin with the hypotheses. The null hypothesis specifies that there is no linear relationship, which means that the slope is 0. Thus, we specify

$$H_0: \ \beta_1 = 0$$

It must be noted that if the null hypothesis is true, it does not necessarily mean that no relationship exists. For example, a quadratic relationship described in Figure 15.6 may exist where $\beta_1 = 0$.

FIGURE 15.6

Quadratic Relationship

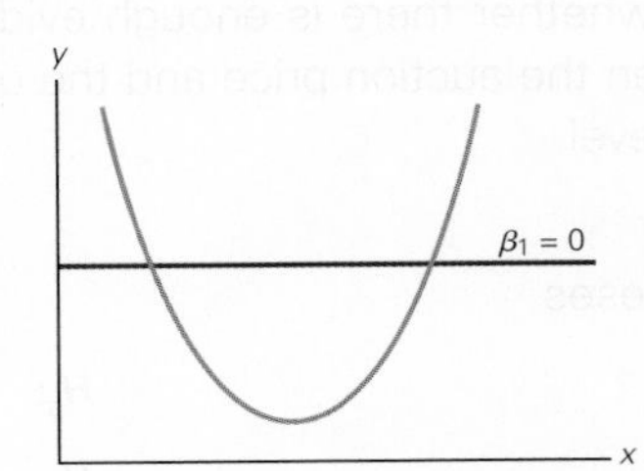

We can conduct one- or two-tail tests of β_1. Most often, we perform a two-tail test to determine whether there is sufficient evidence to infer that a linear relationship exists*. We test the alternative hypothesis.

$$H_1: \ \beta_1 \neq 0$$

15-4d Estimator and Sampling Distribution

In Section 15-2, we pointed out that b_1 is an unbiased estimator of β_1; that is,

$$E(b_1) = \beta_1$$

*If the alternative hypothesis is true, it may be that a linear relationship exists or that a nonlinear relationship exists but that the relationship can be approximated by a straight line.

The estimated standard error of b_1 is

$$s_{b_1} = \frac{s_\varepsilon}{\sqrt{(n-1)s_x^2}}$$

where s_ε is the standard error of estimate and s_x^2 is the sample variance of the independent variable. If the required conditions outlined in Section 15-3 are satisfied, the sampling distribution of the *t*-statistic

$$t = \frac{b_1 - \beta_1}{s_{b_1}}$$

is Student *t* with degrees of freedom $v = n - 2$. Notice that the standard error of b_1 decreases when the sample size increases (which makes b_1 a consistent estimator of β_1) or the variance of the independent variable increases.

Thus, the test statistic and confidence interval estimator are as follows.

Test Statistic for β_1

$$t = \frac{b_1 - \beta_1}{s_{b_1}} \quad v = n - 2$$

Confidence Interval Estimator of β_1

$$b_1 \pm t_{\alpha/2} s_{b_1} \quad v = n - 2$$

EXAMPLE 15.4

Are Odometer Reading and Price of Used Toyota Auris Related?

Test to determine whether there is enough evidence in Example 15.2 to infer that there is a linear relationship between the auction price and the odometer reading for all 3-year-old Toyota Auris. Use a 5% significance level.

SOLUTION:

We test the hypotheses

$$H_0: \ \beta_1 = 0$$

$$H_0: \ \beta_1 \neq 0$$

If the null hypothesis is true, no linear relationship exists. If the alternative hypothesis is true, some linear relationship exists.

COMPUTE

MANUALLY:

To compute the value of the test statistic, we need b_1 and s_{b_1}. In Example 15.2, we found

$$b_1 = -0.0251$$

and

$$s_x^2 = 111.383$$

Thus,

$$s_{b_1} = \frac{s_\varepsilon}{\sqrt{(n-1)s_x^2}} = 0.00185$$

The value of the test statistic is

$$t = \frac{b_1 - \beta_1}{s_{b_1}} = \frac{-0.0251 - 0}{0.00185} = -13.439$$

The rejection region is

$$t < -t_{\alpha/2,v} = -t_{0.025,98} \approx -1.984 \quad \text{or} \quad t > t_{\alpha/2,v} = t_{0.025,98} \approx 1.984$$

EXCEL

	A	B	C	D	E
16		*Coefficients*	*Standard Error*	*t Stat*	*P-value*
17	Intercept	17.25	0.182	94.73	3.57E-98
18	Odometer	−0.0669	0.0050	−13.44	5.75E-24

MINITAB

Predictor	Coef	SE Coef	T	P
Constant	17.2487	0.1821	94.73	0.000
Odometer	−0.066861	0.004975	−13.44	0.000

INTERPRET

The value of the test statistic is $t = -13.44$, with a p-value of 0. (Excel uses scientific notation, which in this case is 5.75×10^{-24}, which is approximately 0.) There is overwhelming evidence to infer that a linear relationship exists. What this means is that the odometer reading may affect the auction selling price of the cars. (See the subsection on cause-and-effect relationship on Section 15-4i.)

As was the case when we interpreted the y-intercept, the conclusion we draw here is valid only over the range of the values of the independent variable. We can infer that there is a relationship between odometer reading and auction price for the 3-year-old Toyota Auris whose odometer readings lie between 30.6 (thousand) and 78.7 (thousand) km (the minimum and maximum values of x in the sample). Because we have no observations outside this range, we do not know how, or even whether, the two variables are related.

Notice that the printout includes a test for β_0. However, as we pointed out before, interpreting the value of the y-intercept can lead to erroneous, if not ridiculous, conclusions. Consequently, we generally ignore the test of β_0.

We can also acquire information about the relationship by estimating the slope coefficient. In this example, the 95% confidence interval estimate (approximating $t_{0.025}$ with 98 degrees of freedom with $t_{0.025}$ with 100 degrees of freedom) is

$$b_1 \pm t_{\alpha/2} s_{b_1} = -0.0251 \pm 1.987(0.00187) = -0.0251 \pm 0.00371$$

We estimate that the slope coefficient lies between −0.0288 and −0.0214.

15-4e One-Tail Tests

If we wish to test for positive or negative linear relationships, we conduct one-tail tests. To illustrate, suppose that in Example 15.2 we wanted to know whether there is evidence of a negative linear relationship between odometer reading and auction selling price. We would specify the hypotheses as

$$H_0: \quad \beta_1 = 0$$

$$H_1: \quad \beta_1 = 0$$

The value of the test statistic would be exactly as computed previously (Example 15.4). However, in this case the *p*-value would be the two-tail *p*-value divided by 2; using Excel's *p*-value, this would be $(5.75 \times 10^{-24})/2 = 2.875 \times 10^{-24}$, which is still approximately 0.

15-4f Coefficient of Determination

The test of β_1 addresses only the question of whether there is enough evidence to infer that a linear relationship exists. In many cases, however, it is also useful to measure the strength of that linear relationship, particularly when we want to compare several different models. The statistic that performs this function is the **coefficient of determination,** which is denoted R^2. Statistics practitioners often refer to this statistic as the '*R*-square'. Recall that we introduced the coefficient of determination in Chapter 4, where we pointed out that this statistic is a measure of the amount of variation in the dependent variable that is explained by the variation in the independent variable. However, we did not describe why we interpret the *R*-square in this way.

Coefficient of Determination

$$R^2 = \frac{s_{xy}^2}{s_x^2 s_y^2}$$

With a little algebra, statisticians can show that

$$R^2 = 1 - \frac{\text{SSE}}{\sum(y_i - \bar{y})^2}$$

We will return to Example 15.1 to learn more about how to interpret the coefficient of determination. In Chapter 13 we partitioned the total sum of squares into two sources of variation. We do so here as well. We begin by adding and subtracting $\hat{y}_i$ from the deviation between y_i from the mean $\bar{y}$; that is,

$$(y_i - y) = (y_i - y) + \hat{y} - \hat{y}_i$$

We observe that by rearranging the terms, the deviation between $\hat{y}_i$ and $\hat{y}_i$ can be decomposed into two parts; that is,

$$(y_i - \bar{y}) = (y_i - \hat{y}_i) + (\hat{y}_i - \bar{y})$$

This equation is represented graphically (for $i = 5$) in Figure 15.7.

FIGURE 15.7

Partitioning the Deviation for $i = 5$

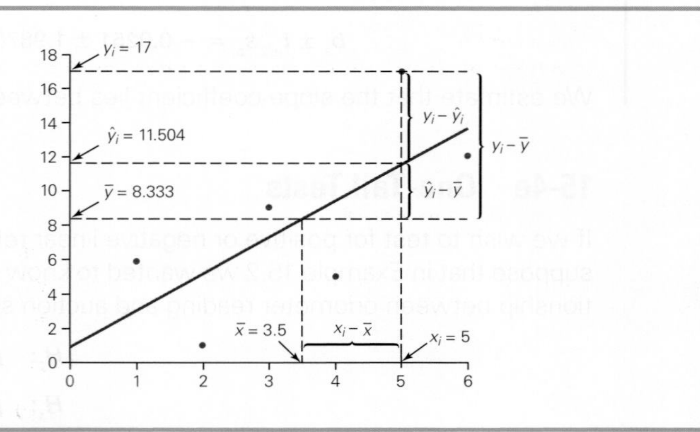

Now we ask why the values of y are different from one another. From Figure 15.7, we see that part of the difference between y_i and $\bar{y}$ is the difference between $\hat{y}_i$ and $\bar{y}$, which is accounted for by the difference between x_i and $\bar{x}$. In other words, some of the variation in y is explained by the changes to x. The other part of the difference between y_i and $\bar{y}$, however, is accounted for by the difference between y_i and $\hat{y}_i$. This difference is the residual, which represents variables not otherwise represented by the model. As a result, we say that this part of the difference is *unexplained* by the variation in x.

If we now square both sides of the equation, sum over all sample points, and perform some algebra, we produce

$$\sum(y_i - \bar{y})^2 = \sum(y_i - \hat{y}_i)^2 + \sum(\hat{y}_i - \bar{y})^2$$

The quantity on the left side of this equation is a measure of the variation in the dependent variable y. The first quantity on the right side of the equation is SSE, and the second term is denoted SSR, for sum of squares for regression. We can rewrite the equation as

$$\text{Variation in } y = \text{SSE} + \text{SSR}$$

As we did in the analysis of variance, we partition the variation of y into two parts: SSE, which measures the amount of variation in y that remains unexplained; and SSR, which measures the amount of variation in y that is explained by the variation in the independent variable x. We can incorporate this analysis into the definition of R^2.

Coefficient of Determination

$$R^2 = 1 - \frac{\text{SSE}}{\sum(y_i - \bar{y})^2} = \frac{\sum(y_i - \bar{y})^2 - \text{SSE}}{\sum(y_i - \bar{y})^2} = \frac{\text{Explained variation}}{\text{Variation in } y}$$

It follows that R^2 measures the proportion of the variation in y that can be explained by the variation in x.

EXAMPLE 15.5

Measuring the Strength of the Linear Relationship between Odometer Reading and Price of Used Toyota Auris

Find the coefficient of determination for Example 15.2 and describe what this statistic tells you about the regression model.

SOLUTION:

COMPUTE

MANUALLY:

We have already calculated all the necessary components of this statistic. In Example 15.2 we found

$$s_{xy} = -2.793$$

$$s_x^2 = 111.383$$

and from Example 15.3

$$s_y^2 = 0.108$$

Thus,

$$R^2 = \frac{s_{xy}^2}{s_x^2 s_y^2} = \frac{(-2.8053)^2}{(111.382)(0.108)} = 0.6483$$

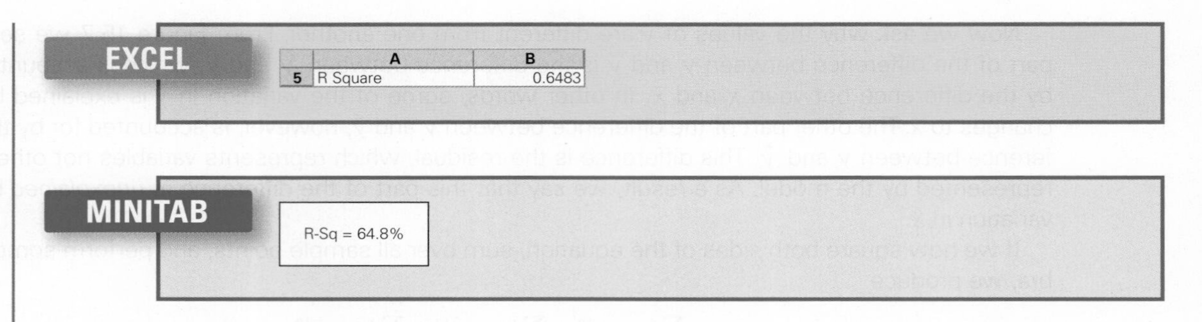

Both Minitab and Excel print a second R^2 statistic called the *coefficient of determination adjusted for degrees of freedom.* We will define and describe this statistic in Chapter 16.

INTERPRET

We found that R^2 is equal to 0.6483. This statistic tells us that 64.83% of the variation in the auction selling prices is explained by the variation in the odometer readings. The remaining 35.17% is unexplained. Unlike the value of a test statistic, the coefficient of determination does not have a critical value that enables us to draw conclusions. In general, the higher the value of R^2, the better the model fits the data. From the *t*-test of β_1 we already know that there is evidence of a linear relationship. The coefficient of determination merely supplies us with a measure of the strength of that relationship. As you will discover in the next chapter, when we improve the model, the value of R^2 increases.

15-4g Other Parts of the Computer Printout

The last part of the printout shown on Example 15-2 relates to our discussion of the interpretation of the value of R^2, when its meaning is derived from the partitioning of the variation in *y*. The values of SSR and SSE are shown in an analysis of variance table similar to the tables introduced in Chapter 13. The general form of the table is shown in Table 15.2. The *F*-test performed in the ANOVA table will be explained in Chapter 16.

TABLE 15.2 General Form of the ANOVA Table in the Simple Linear Regression Model

SOURCE	d.f.	SUMS OF SQUARES	MEAN SQUARES	*F*-STATISTIC
Regression	1	SSR	MSR = SSR/1	F = MSR/MSE
Error	$n - 2$	SSE	MSE = SSE/(n − 2)	
Total	$n - 1$	Variation in *y*		

Note: Excel uses the word 'Residual' to refer to the second source of variation, which we called 'Error'.

15-4h Developing an Understanding of Statistical Concepts

Once again, we encounter the concept of explained variation. We first discussed the concept in Chapter 12 when we introduced the matched pairs experiment, where the experiment was designed to reduce the variation among experimental units. This concept was extended in the analysis of variance, where we partitioned the total variation into two or more sources (depending on the experimental design). And now in regression analysis, we use the concept to measure how the dependent variable is related to the independent variable. We partition the variation of the dependent variable into the sources: the variation explained by the variation in the independent variable and the unexplained variation. The greater the explained variation, the better the model is. We often refer to the coefficient of determination as a measure of the explanatory power of the model.

15-4i Cause-and-Effect Relationship

A common mistake is made by many students when they attempt to interpret the results of a regression analysis when there is evidence of a linear relationship. They imply that changes in the independent variable cause changes in the dependent variable. It must be emphasised that we cannot infer a causal relationship from statistics alone. Any inference about the cause of the changes in the dependent variable must be justified by a reasonable theoretical relationship. For example, statistical tests established that the more one smoked, the greater the probability of developing lung cancer. However, this analysis did not prove that smoking causes lung cancer. It only demonstrated that smoking and lung cancer were somehow related. Only when medical investigations established the connection were scientists able to confidently declare that smoking causes lung cancer.

As another illustration, consider Example 15.2 where we showed that the odometer reading is linearly related to the auction price. Although it seems reasonable to conclude that decreasing the odometer reading would cause the auction price to rise, the conclusion may not be entirely true. It is theoretically possible that the price is determined by the overall condition of the car and that the condition generally worsens when the car is driven longer. Another analysis would be needed to establish the veracity of this conclusion.

Be cautious about the use of the terms *explained variation* and *explanatory power of the model*. Do not interpret the word *explained* to mean *caused*. We say that the coefficient of determination measures the amount of variation in y that is explained (not caused) by the variation in x. Thus, regression analysis can only show that a statistical relationship exists. We cannot infer that one variable causes another.

Recall that we first pointed this out in Chapter 3 using the following sentence:

Correlation is not causation.

15-4j Testing the Coefficient of Correlation

When we introduced the coefficient of correlation (also called the **Pearson coefficient of correlation**) in Chapter 4, we observed that it is used to measure the strength of association between two variables. However, the coefficient of correlation can be useful in another way. We can use it to test for a linear relationship between two variables.

When we are interested in determining *how* the independent variable is related to the dependent variable, we estimate and test the linear regression model. The t-test of the slope presented previously allows us to determine whether a linear relationship actually exists. As we pointed out in Section 15-3, the statistical test requires that for each value of x, there exists a population of values of y that are normally distributed with a constant variance. This condition is required whether the data are experimental or observational.

In many circumstances, we are interested in determining only *whether* a linear relationship exists and not the form of the relationship. When the data are observational and the two variables are bivariate normally distributed (see Section 15-3) we can calculate the coefficient of correlation and use it to test for linear association.

As we noted in Chapter 4, the population coefficient of correlation is denoted ρ (the Greek letter *rho*). Because ρ is a population parameter (which is almost always unknown), we must estimate its value from the sample data. Recall that the sample coefficient of correlation is defined as follows.

Sample Coefficient of Correlation

$$r = \frac{s_{xy}}{s_x s_y}$$

When there is no linear relationship between the two variables, $\rho = 0$. To determine whether we can infer that ρ is 0, we test the hypotheses

$$H_0: \ \rho = 0$$
$$H_1: \ \rho \neq 0$$

The test statistic is defined in the following way.

Test Statistic for Testing $\rho = 0$

$$t = r\sqrt{\frac{n-2}{1-r^2}}$$

is Student t distributed with $\nu = n - 2$ degrees of freedom provided that the variables are bivariate normally distributed.

EXAMPLE 15.6

Are Odometer Reading and Price of Used Toyota Auris Linearly Related? Testing the Coefficient of Correlation

Conduct the t-test of the coefficient of correlation to determine whether odometer reading and auction selling price are linearly related in Example 15.2. Assume that the two variables are bivariate normally distributed.

SOLUTION:

COMPUTE MANUALLY:

The hypotheses to be tested are

$$H_0: \ \rho = 0$$
$$H_1: \ \rho \neq 0$$

In Example 15.2 we found $s_{xy} = -2.793$ and $s_x^2 = 111.383$. In Example 15.5 we determined that $s_y^2 = 0.108$. Thus,

$$s_x = \sqrt{111.38} = 10.55$$

$$s_y = \sqrt{0.108} = 0.328$$

The coefficient of correlation is

$$r = \frac{s_{xy}}{s_x s_y} = \frac{-2.7928}{(10.553)(0.328)} = -0.8052$$

The value of the test statistic is

$$t = r\sqrt{\frac{n-2}{1-r^2}} = -0.8052\sqrt{\frac{100 \quad 2}{1-(-0.8052)^2}} = -13.44$$

Notice that this is the same value we produced in the t-test of the slope in Example 15.4. Because both sampling distributions are Student t with 98 degrees of freedom, the p-value and conclusion are also identical.

EXCEL

	A	B
1	**Correlation**	
2		
3	*Price and Odometer*	
4	Pearson Coefficient of Correlation	–0.8052
5	t Stat	–13.44
6	df	98
7	P(T<=t) one-tail	0
8	t Critical one-tail	1.6606
9	P(T<=t) two-tail	0
10	t Critical two-tail	1.9845

Instructions

1. Type or import the data into two adjacent columns*. (**Open** Xm15-02.)
2. Click **Add-ins, Data Analysis Plus** and **Correlation (Pearson).**
3. Specify the **Variable 1 Input Range** (A1:A101), **Variable 2 Input Range** (B1:B101) and α (0.05).

*If one or both columns contain an empty cell (representing missing data) the row must be removed.

MINITAB

Correlations: Odometer, Price

Pearson correlation of Price and Odometer = –0.805
P-Value = 0.000

Instructions

1. Type or import the data into two adjacent columns. (**Open** Xm15-02.)
2. Click **Stat, Basic Statistics** and **Correlation.**
3. Type the names of the variables in the **Variables** box (Odometer Price).

Notice that the *t*-test of ρ and the βt_1-test of β_1 in Example 15.4 produced identical results. This should not be surprising because both tests are conducted to determine whether there is evidence of a linear relationship. The decision about which test to use is based on the type of experiment and the information we seek from the statistical analysis. If we are interested in discovering the relationship between two variables, or if we have conducted an experiment where we controlled the values of the independent variable (as in Exercise 15.6, the *t*-test of β_1 should be applied. If we are interested only in determining *whether* two random variables that are bivariate normally distributed are linearly related, the *t*-test of ρ should be applied.

As is the case with the *t*-test of the slope, we can also conduct one-tail tests. We can test for a positive or a negative linear relationship.

SURVEY: EDUCATION AND INCOME – HOW ARE THEY RELATED? SOLUTION

IDENTIFY

The problem objective is to analyse the relationship between two interval variables. Because we want to know how education affects income, the independent variable is education (EDUC) and the dependent variable is income (INCOME).

(continued)

COMPUTE

EXCEL

	A	B	C	D	E	F
1	SUMMARY OUTPUT					
2						
3	*Regression Statistics*					
4	Multiple R	0.3790				
5	R Square	0.1436				
6	Adjusted R Square	0.1429				
7	Standard Error	35972				
8	Observations	1189				
9						
10	ANOVA					
11		*df*	*SS*	*MS*	*F*	*Significance F*
12	Regression	1	2.57561E+11	2.58E+11	199.0414	6.70202E-42
13	Residual	1187	1.53599E+12	1.29E+09		
14	Total	1188	1.79355E+12			
15						
16		*Coefficients*	*Standard Error*	*t Stat*	*P-value*	
17	Intercept	–28926	5117	–5.65	1.97E-08	
18	EDUC	5111	362	14.11	6.7E-42	

INTERPRET

The regression equation is $\hat{y} = -28\,296 + 5111x$. The slope coefficient tells us that on average for each additional year of education income increases by \$5111. The intercept is clearly meaningless. We test to determine whether there is evidence of a linear relationship.

$$H_0\text{:}\quad \beta_1 = 0$$

$$H_1\text{:}\quad \beta_1 \neq 0$$

The test statistic is $t = 14.11$ and the p-value is 6.7×10^{-42}, which is virtually 0. The coefficient of determination is $R^2 = 0.1436$, which means that 14.36% of the variation in income is explained by the variation in education, and the remaining 65.64% is not explained.

15-4k Violation of the Required Condition

When the normality requirement is unsatisfied, we can use a non-parametric technique – the Spearman rank correlation coefficient (Chapter 17*) to replace the t-test of ρ.

EXERCISES

Use a 5% significance level for all tests of hypotheses.

15.15 You have been given the following data:

x	1	3	4	6	9	8	10
y	1	8	15	33	75	70	95

a. Draw the scatter diagram. Does it appear that x and y are related? If so, how?

b. Test to determine whether there is evidence of a linear relationship.

15.16 Suppose that you have the following data:

x	3	5	2	6	1	4
y	25	110	9	250	3	71

a. Draw the scatter diagram. Does it appear that x and y are related? If so, how?

b. Test to determine whether there is evidence of a linear relationship.

*Instructors who wish to teach the use of the Spearman rank correlation coefficient here can use the website Appendix Spearman Rank Correlation Coefficient and Test.

15.17 Refer to Exercise 15.2.

a. Determine the standard error of estimate.

b. Is there evidence of a linear relationship between advertising and sales?

c. Estimate β_1 with 95% confidence.

d. Compute the coefficient of determination and interpret this value.

e. Briefly summarise what you have learned in parts (a) through (d).

15.18 Calculate the coefficient of determination and conduct a test to determine whether a linear relationship exists between housing starts and mortgage interest in Exercise 15.3.

15.19 Is there evidence of a linear relationship between the number of hours of television viewing and how overweight the child is in Exercise 15.4?

15.20 Determine whether there is evidence of a negative linear relationship between temperature and the number of beers sold at Camp Nou Stadium in Exercise 15.5.

Exercises 15.21–15.31 require the use of a computer and software.

15.21 Refer to Exercise 15.6.

a. What is the standard error of estimate? Interpret its value.

b. Describe how well the memory test scores and length of television commercial are linearly related.

c. Are the memory test scores and length of commercial linearly related? Test using a 5% significance level.

d. Estimate the slope coefficient with 90% confidence.

15.22 Refer to Exercise 15.7. Use two statistics to measure the strength of the linear association. What do these statistics tell you?

15.23 Is there evidence of a linear relationship between number of cigarettes smoked and number of sick days in Exercise 15.8?

15.24 Refer to Exercise 15.9.

a. Determine the standard error of estimate, and describe what this statistic tells you about the regression line.

b. Can we conclude that the size and price of the apartment building are linearly related?

c. Determine the coefficient of determination and discuss what its value tells you about the two variables.

15.25 Assess fit of the regression line in Exercise 15.10.

15.26 Refer to Exercise 15.11.

a. Determine the coefficient of determination and describe what it tells you.

b. Conduct a test to determine whether there is evidence of a linear relationship between household income and food budget.

15.27 Can we infer that office rents and vacancy rates are linearly related in Exercise 15.12?

15.28 Refer to Exercise 15.13.

a. Compute the coefficient of determination and describe what it tells you.

b. Can we infer that aptitude test scores and percentages of non-defectives are linearly related?

15.29 Repeat Exercise 15.6 using the *t*-test of the coefficient of correlation. Is this result identical to the one you produced in Exercise 15.6?

15.30 Are food budget and household income in Exercise 15.11 linearly related? Employ the *t*-test of the coefficient of correlation to answer the question.

15.31 Refer to Exercise 15.8. Use the *t*-test of the coefficient of correlation to determine whether there is evidence of a positive linear relationship between number of cigarettes smoked and the number of sick days.

15-5 USING THE REGRESSION EQUATION

Using the techniques in Section 15-4, we can assess how well the linear model fits the data. If the model fits satisfactorily, we can use it to forecast and estimate values of the dependent variable. To illustrate, suppose that in Example 15.2, the used-car dealer wanted to predict the selling price of a 3-year-old Toyota Auris with 40 000 miles on the odometer. Using the regression equation, with $x = 64$, we get

$$\hat{y} = 10.349 - 0.0251x = 10.349 - 0.0251(64) = 8.7454$$

We call this value the **point prediction**, and $\hat{y}$ is the point estimate or predicted value for y when $x = 64$. Thus, the dealer would predict that the car would sell for £8745. By itself, however, the point prediction does not provide any information about how closely the value will match the true selling price. To discover that information, we must use an interval. In fact, we can use one of two intervals: the prediction interval of a particular value of *y* or the confidence interval estimator of the expected value of *y*.

15-5a Predicting the Particular Value of *y* for a Given *x*

The first confidence interval we present is used whenever we want to predict a one-time occurrence for a particular value of the dependent variable when the independent variable is a given value x_g. This interval, often called the **prediction interval**, is calculated in the usual way (point estimator ± bound on the error of estimation). Here the point estimate for *y* is $\hat{y}$, and the bound on the error of estimation is shown below.

Prediction Interval

$$\hat{y} \pm t_{\alpha/2,\, n-2}\, s_\varepsilon \sqrt{1 + \frac{1}{n} + \frac{(x_g - \bar{x})^2}{(n-1)s_x^2}}$$

where x_g is the given value of *x* and $\hat{y} = b_0 + b_1 x_g$

15-5b Estimating the Expected Value of *y* for a Given *x*

The conditions described in Section 15-3 imply that, for a given value of *x*, there is a population of values of *y* whose mean is

$$E(y) = \beta_0 + \beta_1 x$$

To estimate the mean of *y* or long-run average value of *y*, we would use the following interval referred to simply as the confidence interval. Again, the point estimator is $\hat{y}$, but the bound on the error of estimation is different from the prediction interval shown below.

Confidence Interval Estimator of the Expected Value of *y*

$$\hat{y} \pm t_{\alpha/2,\, n-2}\, s_\varepsilon \sqrt{\frac{1}{n} + \frac{(x_g - \bar{x})^2}{(n-1)s_x^2}}$$

Unlike the formula for the prediction interval, this formula does not include the 1 under the square-root sign. As a result, the **confidence interval estimate of the expected value of *y*** will be narrower than the prediction interval for the same given value of *x* and confidence level. This is because there is less error in estimating a mean value as opposed to predicting an individual value.

EXAMPLE 15.7

Predicting the Price and Estimating the Mean Price of Used Toyota Auris

a. A used-car dealer is about to bid on a 3-year-old Toyota Auris equipped with all the standard features and with 40 000 ($x_g = 40$) miles on the odometer. To help him decide how much to bid, he needs to predict the selling price.

b. The used-car dealer mentioned in part (a) has an opportunity to bid on a lot of cars offered by a rental company. The rental company has 250 Toyota Auris all equipped with standard features. All the cars in this lot have about 40 000 ($x_g = 40$) miles on their odometers. The dealer would like an estimate of the selling price of all the cars in the lot.

SOLUTION:

IDENTIFY

a. The dealer would like to predict the selling price of a single car. Thus, he must employ the prediction interval

$$\hat{y} \pm t_{\alpha/2,n-2} s_{\varepsilon} \sqrt{1 + \frac{1}{n} + \frac{(x_g - x)^2}{(n-1)s_x^2}}$$

b. The dealer wants to determine the mean price of a large lot of cars, so he needs to calculate the confidence interval estimator of the expected value:

$$\hat{y} \pm t_{\alpha/2,n-2} s_{\varepsilon} \sqrt{\frac{1}{n} + \frac{(x_g - x)^2}{(n-1)s_x^2}}$$

Technically, this formula is used for infinitely large populations. However, we can interpret our problem as attempting to determine the average selling price of all Toyota Auris equipped as described above, all with 40 000 miles on the odometer. The crucial factor in part (b) is the need to estimate the mean price of a number of cars. We arbitrarily select a 95% confidence level.

COMPUTE

MANUALLY:

From previous calculations, we have the following:

$$\hat{y} = 17.250 - 0.0669(40) = 14.574$$

$$s_{\varepsilon} = 0.3265$$

$$s_x^2 = 43.509$$

$$\bar{x} = 36.011$$

From Table 4 in Appendix B, we find

$$t_{\alpha/2} = t_{0.025,98} \approx t_{0.025,100} = 1.984$$

a. The 95% prediction interval is

$$\hat{y} \pm t_{\alpha/2,n-2} s_{\varepsilon} \sqrt{1 + \frac{1}{n} + \frac{(x_g - \bar{x})^2}{(n-1)s_x^2}}$$

$$= 8.745 \pm 1.984$$

$$\times\, 0.196 \sqrt{1 + \frac{1}{100} + \frac{(64 - 57.5176)^2}{(100-1) \times 111.383}}$$

$$= 8.745 \pm 0.391$$

The lower and upper limits of the prediction interval are £8353 and £9136, respectively.

b. The 95% confidence interval estimator of the mean price is

$$\hat{y} \pm t_{\alpha/2,n-2}\, s_{\varepsilon} \sqrt{\frac{1}{n} + \frac{(x_g - \bar{x})^2}{(n-1)s_x^2}}$$

$$= 8.745 \pm 1.984 \times 0.196 \sqrt{\frac{1}{100} + \frac{(64 - 57.5176)^2}{(100-1) \times 111.383}}$$

$$= 8.745 \pm 0.0455$$

The lower and upper limits of the confidence interval estimate of the expected value are £8699 and £8790, respectively.

EXCEL

	A	B	C
1	**Prediction Interval**		
2			
3			
4			Price
5	Predicted Value		8.745
6			
7	Prediction Interval		
8	Lower limit		8.353
9	Upper limit		9.136
10			
11	Interval Estimate of Expected Value		
12	Lower limit		8.699
13	Upper limit		8.790

Instructions

1. Type or import the data into two columns*. (**Open** Xm15-02.)
2. Type the given value of x_g (i.e. 64) into any cell. We suggest the next available row in the column containing the independent variable (i.e. B102).
3. Click **Add-Ins, Data Analysis Plus** and **Prediction Interval.**
4. Specify the **Input *Y* Range** (A1:A101), the **Input X Range** (B1:B101), the **Given X Range** (B102) and the **Confidence Level** (0.95).

*If one or both columns contain an empty cell (representing missing data), the row must be removed.

MINITAB

Predicted Values for New Observations

New Obs	Fit	SE Fit	95% CI	95% PI
1	8.7446	0.0229	(8.6991, 8.7901)	(8.3532, 9.1360)

Values of Predictors for New Observations

New Obs	Odometer
1	64.0

The output includes the predicted value y **(Fit),** the standard deviation of y (SE **Fit),** the 95% confidence interval estimate of the expected value of y (CI), and the 95% prediction interval (PI).

Instructions

1. Type and import the data. (**Open** Xm15-02.)
2. Click **Stat, Regression** and **Regression** ....
3. Type the name of the dependent variable in the **Response** box (Price) and the name of the independent variable in the **Predictor** box (Odometer).
4. Do not click OK. Click **Options** ....
5. Specify the given value of x_g in the **Prediction intervals for new observations** box (64).
6. Specify the confidence level (0.95).

INTERPRET

We predict that one car will sell for between £8353 and £9136. The average selling price of the population of 3-year-old Toyota Auris is estimated to lie between £8699 and £8790. Because predicting the selling price of one car is more difficult than estimating the mean selling price of all similar cars, the prediction interval is wider than the interval estimate of the expected value.

15-5 Effect of the Given Value of x on the Intervals

Calculating the two intervals for various values of x results in the graph in Figure 15.8. Notice that both intervals are represented by curved lines. This is because the farther the given value of x is from $\bar{x}$, the greater the estimated error becomes. This part of the estimated error is measured by

$$\frac{(x_g - \bar{x})^2}{(n-1)s_x^2}$$

which appears in both the prediction interval and the interval estimate of the expected value.

FIGURE 15.8

Interval Estimates and Prediction Intervals

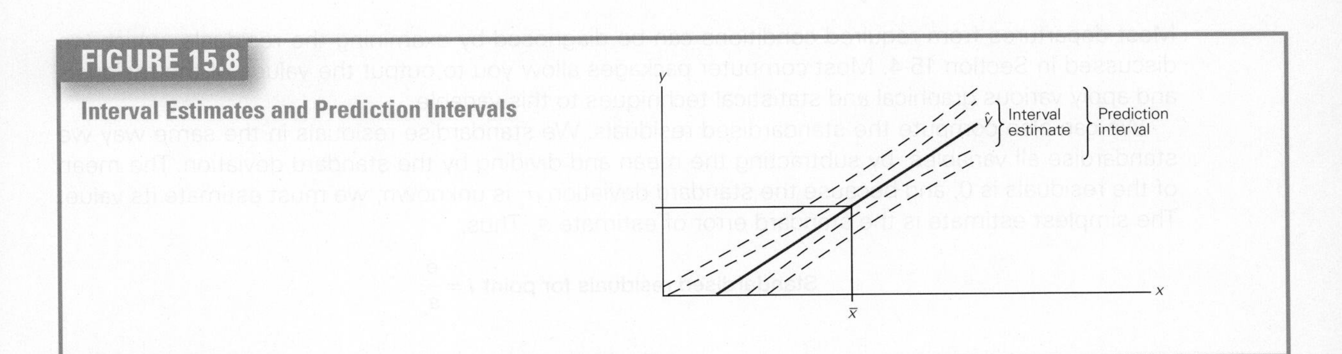

EXERCISES

15.32 Briefly describe the difference between predicting a value of y and estimating the expected value of y.

15.33 Will the prediction interval always be wider than the estimation interval for the same value of the independent variable? Briefly explain.

15.34 Use the regression equation in Exercise 15.2 to predict with 90% confidence the sales when the advertising budget is €80 000.

15.35 Estimate with 90% confidence the mean monthly number of housing starts when the mortgage interest rate is 7% in Exercise 15.3.

15.36 Refer to Exercise 15.4.

- **a.** Predict with 90% confidence the number of pounds overweight for a child who watches 35 hours of television per week.
- **b.** Estimate with 90% confidence the mean number of pounds overweight for children who watch 35 hours of television per week.

15.37 Refer to Exercise 15.5. Predict with 90% confidence the number of beers to be sold when the temperature is 75 degrees.

The next exercises require the use of a computer and software. The answers may be calculated manually.

15.38 Refer to Exercise 15.6.

- **a.** Predict with 95% confidence the memory test score of a viewer who watches a 30-second commercial.
- **b.** Estimate with 95% confidence the mean memory test score of people who watch 30-second commercials.

15.39 Refer to Exercise 15.7. The company has just hired a 22-year-old telemarketer. Predict with 95% confidence how long he will stay with the company.

15.40 Refer to Exercise 15.8. Predict with 95% confidence the number of sick days for individuals who smoke on average 40 cigarettes per day.

15.41 Refer to Exercise 15.9. Estimate with 95% confidence the mean price of 60 000 square metre apartment buildings.

15.42 Refer to Exercise 15.10. Estimate with 90% confidence the mean electricity consumption for households with four occupants.

15.43 Refer to Exercise 15.11. Predict the food budget of a family whose household income is €60 000. Use a 90% confidence level.

15.44 Refer to Exercise 15.11. Predict with 95% confidence the monthly office rent in a city when the vacancy rate is 8%.

15.45 Refer to Exercise 15.13. Estimate with 95% confidence the mean percentage of defectives for workers who score 80 on the dexterity test.

15-6 REGRESSION DIAGNOSTICS-I

In Section 15-3 we described the required conditions for the validity of regression analysis. Simply put, the error variable must be normally distributed with a constant variance, and the errors must be independent of each other. In this section, we show how to diagnose violations. In addition, we discuss how to deal with observations that are unusually large or small. Such observations must be investigated to determine whether an error was made in recording them.

15-6a Residual Analysis

Most departures from required conditions can be diagnosed by examining the residuals, which we discussed in Section 15-4. Most computer packages allow you to output the values of the residuals and apply various graphical and statistical techniques to this variable.

We can also compute the standardised residuals. We standardise residuals in the same way we standardise all variables, by subtracting the mean and dividing by the standard deviation. The mean of the residuals is 0, and because the standard deviation σ_ε is unknown, we must estimate its value. The simplest estimate is the standard error of estimate s_ε. Thus,

$$\text{Standardised residuals for point } i = \frac{e_i}{s_\varepsilon}$$

EXCEL

Excel calculates the standardised residuals by dividing the residuals by the standard deviation of the residuals. (The difference between the standard error of estimate and the standard deviation of the residuals is that in the formula of the former the denominator is $n - 2$, whereas in the formula for the latter, the denominator is $n - 1$.)

Part of the printout (we show only the first five and last five values) for Example 15.2 follows.

	A	B	C	D
23				
24	*Observation*	*Predicted Price*	*Residuals*	*Standard Residuals*
25	1	8.849	−0.089	−0.456
26	2	8.552	−0.092	−0.472
27	3	8.512	−0.112	−0.574
28	4	9.110	0.250	1.285
29	5	9.078	0.282	1.449
119	95	9.090	−0.030	−0.152
120	96	8.897	−0.017	−0.087
121	97	8.977	−0.217	−1.115
122	98	9.017	−0.317	−1.628
123	99	8.777	0.043	0.222
124	100	8.889	−0.309	−1.585

Instructions

1. Type or import data into two columns*, one storing the dependent variable and the other the independent variable. (**Open** Xm15-02.)
2. Click **Data, Data Analysis** and **Regression.**
3. Specify the **Input *Y* Range** (A1:A101) and the **Input *X* Range** (B1:B101).
4. Before clicking OK, select **Residuals** and **Standardised Residuals.** The predicted values, residuals and standardised residuals will be printed.

---

*If one or both columns contain a blank (representing missing data), the row must be deleted.

We can also standardise by computing the standard deviation of each residual. Statisticians have determined that the standard deviation of the residual for observation *i* is defined as follows.

Standard Deviation of the *i*th Residual

$$s_{e_i} = s_\varepsilon \sqrt{1 - b_i}$$

where

$$b_i = \frac{1}{n} + \frac{(x_i - \bar{x})^2}{(n-1)s_x^2}$$

The quantity b_i should look familiar; it was used in the formula for the prediction interval and confidence interval estimate of the expected value of y in Section 15-6. Minitab computes this version of the standardised residuals. Part of the printout (we show only the first five and last five values) for Example 15.2 is shown below.

MINITAB

Obs	Odometer	Price	Fit	SE Fit	Residual	St Resid
1	59.8	8.7600	8.8489	0.0200	−0.0889	−0.46
2	71.7	8.4600	8.5520	0.0327	−0.0920	−0.48
3	73.3	8.4000	8.5119	0.0352	−0.1119	−0.58
4	49.4	9.3600	9.1096	0.0248	0.2504	1.29
5	50.7	9.3600	9.0775	0.0234	0.2825	1.45
95	50.2	9.0600	9.0896	0.0239	−0.0296	−1.15
96	57.9	8.8800	8.8970	0.0196	−0.0170	−0.09
97	54.7	8.7600	8.9773	0.0203	−0.2173	−1.12
98	53.1	8.7000	9.0174	0.0213	−0.3174	−1.63
99	62.7	8.8200	8.7767	0.0218	0.0433	0.22
100	58.2	8.5800	8.8890	0.0196	−0.3090	−1.59

Instructions

1. Type or import the data into two columns. (**Open** Xm15-02.)
2. Click **Stat, Regression** and **Regression….**
3. Type the name of the dependent variable in the **Response** box (Price) and the name of the independent variable in the **Predictors** box (Odometer).
4. After specifying the **Response** and **Predictors**, click **Results…,** and **In addition, the full table of fits and residuals.**

The predicted values, residuals and standardised residuals will be printed.

An analysis of the residuals will allow us to determine whether the error variable is non-normal, whether the error variance is constant and whether the errors are independent. We begin with non-normality.

15-6b Non-normality

As we have done throughout this book, we check for normality by drawing the histogram of the residuals. Figure 15.9 is Excel's version (Minitab's is similar). As you can see, the histogram is bell shaped, leading us to believe that the error is normally distributed.

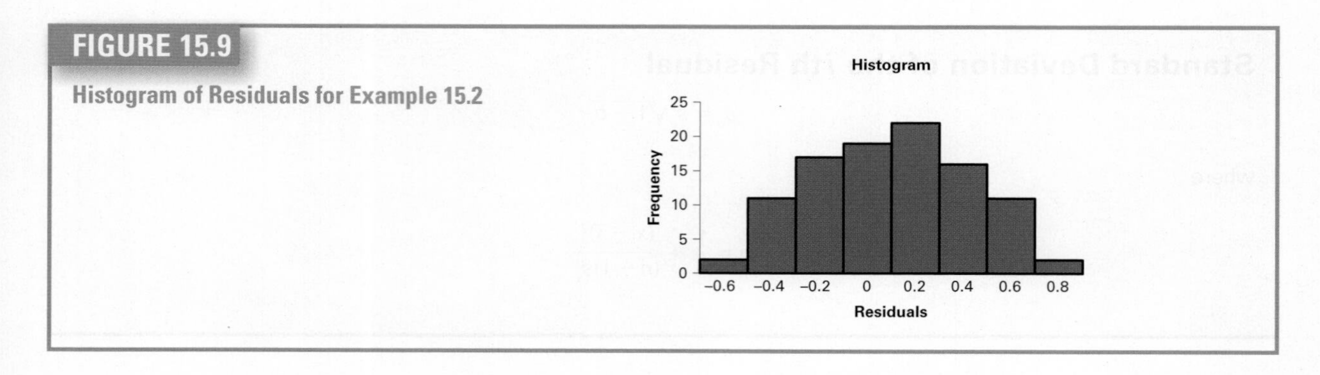

FIGURE 15.9
Histogram of Residuals for Example 15.2

15-6c Heteroscedasticity

The variance of the error variable σ_ε^2 is required to be constant. When this requirement is violated, the condition is called **heteroscedasticity**. (You can impress friends and relatives by using this term. If you cannot pronounce it, try **homoscedasticity**, which refers to the condition where the requirement is satisfied.) One method of diagnosing heteroscedasticity is to plot the residuals against the predicted values of *y*. We then look for a change in the spread of the plotted points*. Figure 15.10 describes such a situation. Notice that in this illustration, σ_ε^2 appears to be small when $\hat{y}$ is small and large when $\hat{y}$ is large. Of course, many other patterns could be used to depict this problem.

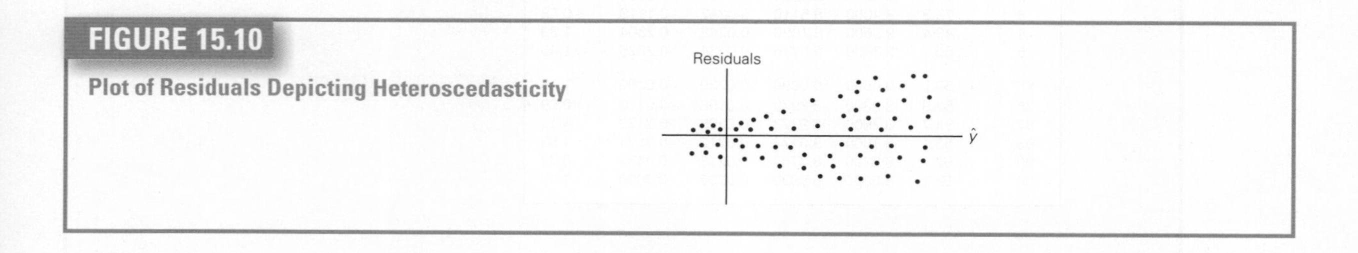

FIGURE 15.10
Plot of Residuals Depicting Heteroscedasticity

Figure 15.11 illustrates a case in which σ_ε^2 is constant. As a result, there is no apparent change in the variation of the residuals.

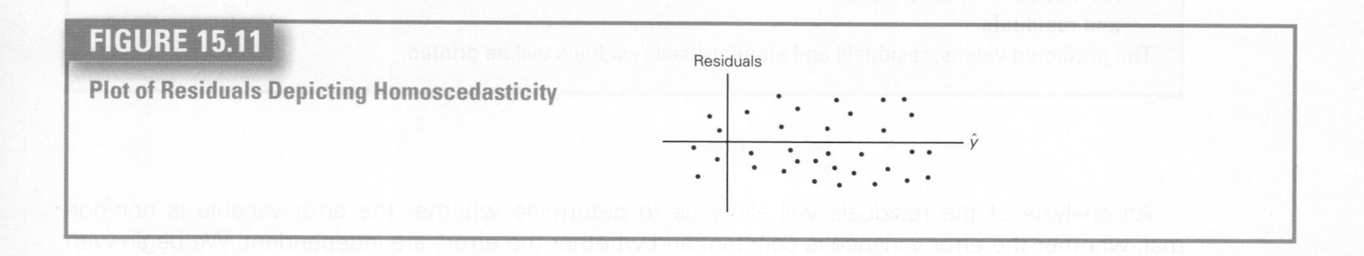

FIGURE 15.11
Plot of Residuals Depicting Homoscedasticity

Excel's plot of the residuals versus the predicted values of *y* for Example 15.2 is shown in Figure 15.12. There is no sign of heteroscedasticity.

*The website Appendix Szroeter's Test describes a test for heteroscedasticity.

FIGURE 15.12

Plot of Predicted Values versus Residuals for Example 15.2

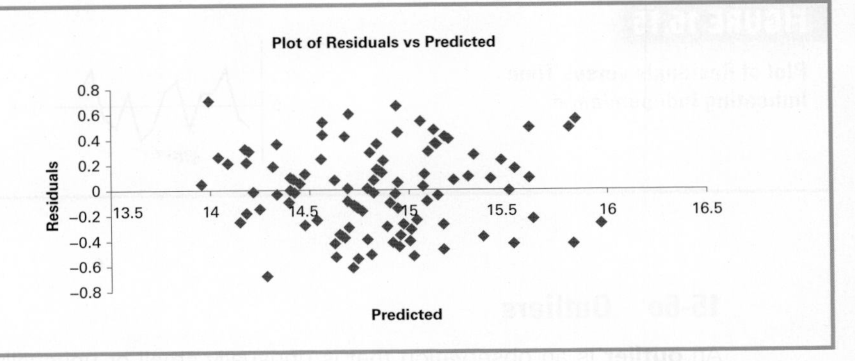

15-6d Non-independence of the Error Variable

In Chapter 3 we briefly described the difference between cross-sectional and time-series data. Cross-sectional data are observations made at approximately the same time, whereas a time series is a set of observations taken at successive points of time. The data in Example 15.2 are cross-sectional because all of the prices and odometer readings were taken at about the same time. If we were to observe the auction price of cars every week for, say, a year, that would constitute a time series.

Condition 4 states that the values of the error variable are independent. When the data are time series, the errors often are correlated. Error terms that are correlated over time are said to be **autocorrelated** or **serially correlated.** For example, suppose that, in an analysis of the relationship between annual gross profits and some independent variable, we observe the gross profits for the years 1991 to 2010.The observed values of y are denoted $y_1, y_2, \ldots y_{20}$; where y_1 is the gross profit for 1991, y_2 is the gross profit for 1992 and so on. If we label the residuals $e_1, e_2, \ldots, e_{20}$, then – if the independence requirement is satisfied – there should be no relationship among the residuals. However, if the residuals are related, it is likely that autocorrelation exists.

We can often detect autocorrelation by graphing the residuals against the time periods. If a pattern emerges, it is likely that the independence requirement is violated. Figures 15.13 (alternating positive and negative residuals) and 15.14 (increasing residuals) exhibit patterns indicating autocorrelation. (Notice that we joined the points to make it easier to see the patterns.) Figure 15.15 shows no pattern (the residuals appear to be randomly distributed over the time periods) and thus likely represent the occurrence of independent errors.

In Chapter 16, we introduce the Durbin–Watson test, which is another statistical test to determine whether one form of autocorrelation is present.

FIGURE 15.13

Plot of Residuals versus Time Indicating Autocorrelation (Alternating)

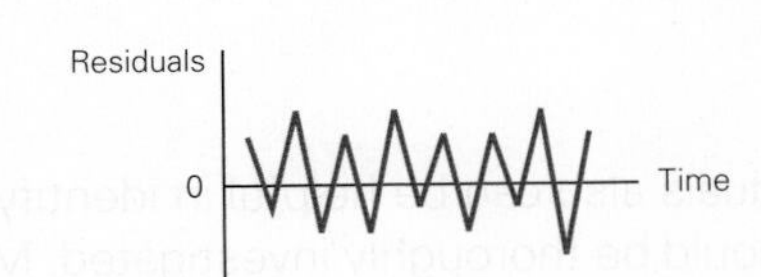

FIGURE 15.14

Plot of Residuals versus Time Indicating Autocorrelation (Increasing)

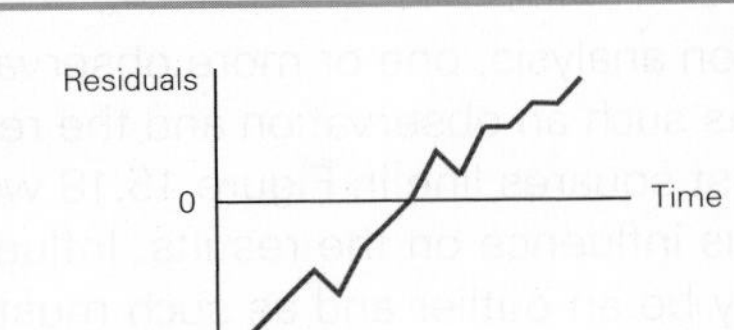

FIGURE 15.15

Plot of Residuals versus Time Indicating Independence

15-6e Outliers

An **outlier** is an observation that is unusually small or unusually large. To illustrate consider Example 15.2, where the range of odometer readings was 30.6 to 78.7 thousand kilometres. If we had observed a value of 5000 km, we would identify that point as an outlier. We need to investigate several possibilities.

1. There was an error in recording the value. To detect an error, we would check the point or points in question. In Example 15.2, we could check the car's odometer to determine whether a mistake was made. If so, we would correct it before proceeding with the regression analysis.
2. The point should not have been included in the sample. Occasionally, measurements are taken from experimental units that do not belong with the sample. We can check to ensure that the car with the 5000 km odometer reading was actually 3 years old. We should also investigate the possibility that the odometer was rolled back. In either case, the outlier should be discarded.
3. The observation was simply an unusually large or small value that belongs to the sample and that was recorded properly. In this case, we would do nothing to the outlier. It would be judged to be valid.

Outliers can be identified from the scatter diagram. Figure 15.16 depicts a scatter diagram with one outlier. The statistics practitioner should check to determine whether the measurement was recorded accurately and whether the experimental unit should be included in the sample.

FIGURE 15.16

Scatter Diagram with One Outlier

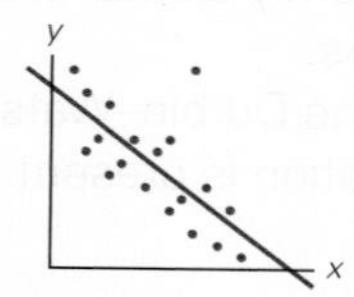

The standardised residuals also can be helpful in identifying outliers. Large absolute values of the standardised residuals should be thoroughly investigated. Minitab automatically reports standardised residuals that are less than –2 and greater than 2.

15-6f Influential Observations

Occasionally, in a regression analysis, one or more observations have a large influence on the statistics. Figure 15.17 describes such an observation and the resulting least squares line. If the point had not been included, the least squares line in Figure 15.18 would have been produced. Obviously, one point has had an enormous influence on the results. Influential points can be identified by the scatter diagram. The point may be an outlier and as such must be investigated thoroughly. Minitab also identities influential observations.

FIGURE 15.17

Scatter Diagram with One Influential Observation

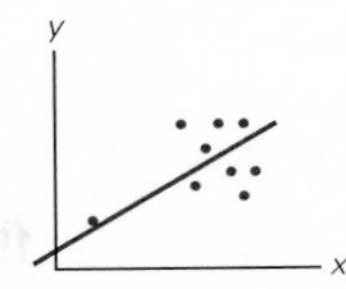

FIGURE 15.18

Scatter Diagram without the Influential Observation

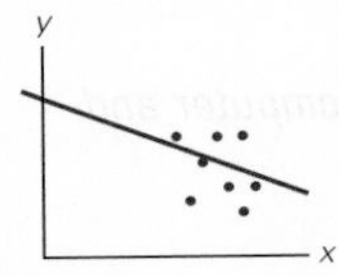

15-6g Procedure for Regression Diagnostics

The order of the material presented in this chapter is dictated by pedagogical requirements. Consequently, we presented the least squares method of assessing the model's fit, predicting and estimating using the regression equation, coefficient of correlation, and finally, the regression diagnostics. In a practical application, the regression diagnostics would be conducted earlier in the process. It is appropriate to investigate violations of the required conditions when the model is assessed and before using the regression equation to predict and estimate. The following steps describe the entire process.

1. Develop a model that has a theoretical basis; that is, for the dependent variable in question, find an independent variable that you believe is linearly related to it.
2. Gather data for the two variables. Ideally, conduct a controlled experiment. If that is not possible, collect observational data.
3. Draw the scatter diagram to determine whether a linear model appears to be appropriate. Identify possible outliers.
4. Determine the regression equation.
5. Calculate the residuals and check the required conditions:

   Is the error variable non-normal?
   Is the variance constant?
   Are the errors independent?
   Check the outliers and influential observations.
6. Assess the model's fit.

   Compute the standard error of estimate.
   Test to determine whether there is a linear relationship. (Test β_1 or ρ.)
   Compute the coefficient of determination.
7. If the model fits the data, use the regression equation to predict a particular value of the dependent variable or estimate its mean (or both).

EXERCISES

15.46 You are given the following six points:

x	−5	−2	0	3	4	7
y	15	9	7	6	4	1

a. Determine the regression equation.

b. Use the regression equation to determine the predicted values of y.

c. Use the predicted and actual values of y to calculate the residuals.

d. Compute the standardised residuals.

e. Identify possible outliers.

15.47 Refer to Exercise 15.2. Calculate the residuals and the predicted values of y.

15.48 Calculate the residuals and predicted values of y in Exercise 15.3.

15.49 Refer to Exercise 15.4.

a. Calculate the residuals.

b. Calculate the predicted values of y.

c. Plot the residuals (on the vertical axis) and the predicted values of y.

15.50 Calculate and plot the residuals and predicted values of y for Exercise 15.5.

The following exercises require the use of a computer and software.

15.51 Refer to Exercise 15.6.

a. Determine the residuals and the standardised residuals.

b. Draw the histgram of the residuals. Does it appear that the errors are normally distributed? Explain.

c. Identify possible outliers.

d. Plot the residuals versus the predicted values of y. Does it appear that heteroscedasticity is a problem? Explain.

15.52 Refer to Exercise 15.7.

a. Determine the residuals and the standardised residuals.

b. Draw the histogram of the residuals. Does it appear that the errors are normally distributed? Explain.

c. Identify possible outliers.

d. Plot the residuals versus the predicted values of y. Does it appear that heteroscedasticity is a problem? Explain.

15.53 Refer to Exercise 15.8. Are the required conditions satisfied?

15.54 Check the required conditions for Exercise 15.9.

15.55 Refer to Exercise 15.10.

a. Determine the residuals and the standardised residuals.

b. Draw the histogram of the residuals. Does it appear that the errors are normally distributed? Explain.

c. Identify possible outliers.

d. Plot the residuals versus the predicted values of y. Does it appear that heteroscedasticity is a problem? Explain.

15.56 Are the required conditions satisfied for Exercise 15.11?

15.57 Check to ensure that the required conditions for Exercise 15.12 are satisfied.

15.58 Perform a complete diagnostic analysis for Exercise 15.13 to determine whether the required conditions are satisfied.

CHAPTER SUMMARY

Simple linear regression and correlation are techniques for analysing the relationship between two interval variables. Regression analysis assumes that the two variables are linearly related. The least squares method produces estimates of the intercept and the slope of the regression line. Considerable effort is expended in assessing how well the linear model fits the data. We calculate the standard error of estimate, which is an estimate of the standard deviation of the error variable. We test the slope to determine whether there is sufficient evidence of a linear relationship. The strength of the linear association is measured by the coefficient of determination. When the model provides a good fit, we can use it to predict the particular value and to estimate the expected value of the dependent variable. We can also use the Pearson correlation coefficient to measure and test the relationship between two bivariate normally distributed variables. We completed this chapter with a discussion of how to diagnose violations of the required conditions.

SYMBOLS:

Symbol	Pronounced	Represents
β_0	Beta sub zero or beta zero	*y*-intercept
β_1	Beta sub one or beta one	Slope coefficient
ε	Epsilon	Error variable
$\hat{y}$	*y* hat	Fitted or calculated value of *y*
b_0	*b* sub zero or *b* zero	Sample *y*-intercept coefficient
b_1	*b* sub one or *b* one	Sample slope coefficient
σ_ε	Sigma sub epsilon or sigma epsilon	Standard deviation of error variable
s_ε	*s* sub epsilon or *s* epsilon	Standard error of estimate
S_{b1}	*s* sub *b* sub one or *s b* one	Standard error of b_1
R^2	*R* squared	Coefficient of determination
x_g	*x* sub *g* or *x g*	Given value of *x*
ρ	Rho	Pearson coefficient of correlation
r		Sample coefficient of correlation
e_i	*e* sub *i* or *e i*	Residual of *i*th point

FORMULAS:

Sample slope

$$b_1 = \frac{s_{xy}}{s_x^2}$$

Sample *y*-intercept

$$b_0 = \bar{y} - b_1\bar{x}$$

Sum of squares for error

$$\text{SSE} = \sum_{i=1}^{n}(y_i - \hat{y}_i)^2$$

Standard error of estimate

$$s_\varepsilon = \sqrt{\frac{\text{SSE}}{n-2}}$$

Test statistic for the slope

$$t = \frac{b_1 - \beta_1}{s_{b_1}}$$

Standard error of

$$s_{b_1} = \frac{s_\varepsilon}{\sqrt{(n-1)s_x^2}}$$

Coefficient of determination

$$R^2 = \frac{s_{xy}^2}{s_x^2 s_y^2} = 1 - \frac{\text{SSE}}{\sum(y_i - \bar{y})^2}$$

Prediction interval

$$\hat{y} \pm t_{\alpha/2,n-2}\, S_\varepsilon \sqrt{1 + \frac{1}{n} + \frac{(x_g - \bar{x})^2}{(n-1)s_x^2}}$$

Confidence interval estimator of the expected value of *y*

$$\hat{y} \pm t_{\alpha/2,n-2}\, s_\varepsilon \sqrt{\frac{1}{n} + \frac{(x_g - x)^2}{(n-1)s_x^2}}$$

Sample coefficient of correlation

$$r = \frac{s_{xy}}{s_x s_y}$$

Test statistic for testing $\rho = 0$

$$t = r\sqrt{\frac{n-2}{1-r^2}}$$

CHAPTER EXERCISES

The following exercises require the use of a computer and software. The answers to some of the questions may be calculated manually. Conduct all tests of hypotheses at the 5% significance level.

15.59 **Xr15-59** The manager of Colonial Furniture has been reviewing weekly advertising expenditures. During the past 6 months, all advertisements for the store have appeared in the local newspaper. The number of ads per week has varied from one to seven. The store's sales staff has been tracking the number of customers who enter the store each week. The number of ads and the number of customers per week for the past 26 weeks were recorded.

a. Determine the sample regression line.

b. Interpret the coefficients.

c. Can the manager infer that the larger the number of ads, the larger the number of customers?

d. Find and interpret the coefficient of determination.

e. In your opinion, is it a worthwhile exercise to use the regression equation to predict the number of customers who will enter the store, given that Colonial intends to advertise five times in the newspaper? If so, find a 95% prediction interval. If not, explain why not.

15.60 **Xr15-60** The managing director of a company that manufactures car seats in Hanover has been concerned about the number and cost of machine breakdowns. The problem is that the machines are old and becoming quite unreliable. However, the cost of replacing them is quite high, and the president is not certain that the cost can be made up in today's slow economy. To help make a decision about replacement, he gathered data about last month's costs for repairs and the ages (in months) of the plant's 20 welding machines.

a. Find the sample regression line.

b. Interpret the coefficients.

c. Determine the coefficient of determination, and discuss what this statistic tells you.

d. Conduct a test to determine whether the age of a machine and its monthly cost of repair are linearly related.

e. Is the fit of the simple linear model good enough to allow the president to predict the monthly repair cost of a welding machine that is 120 months old? If so, find a 95% prediction interval. If not, explain why not.

15.61 **Xr15-61** An agronomist wanted to investigate the factors that determine crop yield. Accordingly, she undertook an experiment wherein a farm was divided into 30 one-acre plots. The amount of fertiliser applied to each plot was varied. Corn was then planted, and the amount of corn harvested at the end of the season was recorded.

a. Find the sample regression line, and interpret the coefficients.

b. Can the agronomist conclude that there is a linear relationship between the amount of fertiliser and the crop yield?

c. Find the coefficient of determination, and interpret its value.

d. Does the simple linear model appear to be a useful tool in predicting crop yield from the amount of fertiliser applied? If so, produce a 95% prediction interval of the crop yield when 135 kg of fertiliser are applied. If not, explain why not.

15.62 **Xr15-62** In South Africa, as in many other countries, levels of tar and nicotine, substances that are hazardous to smokers' health, are printed on cigarette packs. Additionally, the amount of carbon monoxide, which is a by-product of burning tobacco that seriously affects the heart, is also reported. Suppose a random sample of 25 brands was taken.

a. Are the levels of tar and nicotine linearly related?

b. Are the levels of nicotine and carbon monoxide linearly related?

15.63 **Xr15-63** Some critics of television complain that the amount of violence shown on television contributes to violence in our society. Others point out that television also contributes to the high level of obesity among children. We may have to add financial problems to the list. A sociologist theorised that people who watch television frequently are exposed to many commercials, which in turn leads them to buy more, finally resulting in increasing debt. To test this belief, a sample of 430 families was drawn. For each, the total debt and the number of hours the television is turned on per week were recorded. Perform a statistical procedure to help test the theory.

15.64 **Xr15-64** The analysis that the human resources manager performed in Exercise 15.13 indicated that the dexterity test is not a predictor of job performance. However, before discontinuing the test, he decided that the problem is that the statistical analysis was flawed because it examined the relationship between test score and job performance only for those who scored well in the test. (Recall that only those who scored above 70 were hired; applicants who achieved scores below 70 were not hired.) The manager decided to perform another statistical analysis. A sample of 50 job applicants who scored above 50 were hired, and as before the workers' performance was measured. The test scores and percentages of non-defective computers produced were recorded. On the basis of these data, should the manager discontinue the dexterity tests?

15.65 Xr03-27 (Exercise 3.27 revisited.) A very large contribution to profits for a cinema is the sale of popcorn, soft drinks and sweets. A cinema manager speculated that the longer the time between showings of a film, the greater the sales of concessions. To acquire more information the manager conducted an experiment. For a month he varied the amount of time between movie showings and calculated the sales. Can the manager conclude that when the times between movies increase so do sales?

The following exercise employs data files associated with a previous exercise.

15.66 Xr12-66 Exercise 12.66 described a survey that asked people between 18 and 34 years of age, and 35 to 50 years of age, how much time they spent listening to FM radio each day. Also recorded were the amounts spent on music throughout the year. Can we infer that a linear relationship exists between listening times and amounts spent on music?

16 MULTIPLE REGRESSION

16-1 MODEL AND REQUIRED CONDITIONS

16-2 ESTIMATING THE COEFFICIENTS AND ASSESSING THE MODEL

16-3 REGRESSION DIAGNOSTICS-II

16-4 REGRESSION DIAGNOSTICS-III (TIME SERIES)

16-5 POLYNOMIAL MODELS

16-6 NOMINAL INDEPENDENT VARIABLES

SOCIAL SURVEY: VARIABLES THAT AFFECT INCOME

DATA Xm16-00

In Chapter 15's opening example, we showed that income and education are linearly related. This raises the question, what other variables affect one's income? To answer this question, we need to expand the simple linear regression technique used in the previous chapter to allow for more than one independent variable. Your assumption is that more education equals a better job and higher income. Also, other variables, such as age or hours of work per week might be related to income. To find out if these variables are related, you have designed your own online survey using the platform 'Survey Monkey'. Here is the list of the interval variables created for the survey:

Age (AGE)
Year of education of respondent (EDU)
Hours of work per week of respondent and of spouse or partner (HRS and SPHRS)
Income (INCOME)

The goal is to create a regression analysis that includes all variables that you believe affect the level of income that can be attained. **See Section 16-2 for the answer.**

INTRODUCTION

In the previous chapter, we employed the simple linear regression model to analyse how one variable (the dependent variable y) is related to another interval variable (the independent variable x). The restriction of using only one independent variable was motivated by the need to simplify the introduction to regression analysis. Although there are a number of applications where we purposely develop a model with only one independent variable, in general we prefer to include as many independent variables as are believed to affect the dependent variable. Arbitrarily limiting the number of independent variables also limits the usefulness of the model.

In this chapter, we allow for any number of independent variables. In so doing, we expect to develop models that fit the data better than would a simple linear regression model. Describing the multiple regression model and listing the required conditions will be the first step to follow up. Then we let the computer produce the required statistics and use them to assess the model's fit and diagnose violations of the required conditions. We use the model by interpreting the coefficients, predicting the particular value of the dependent variable and estimating its expected value.

In Section 16-5 we introduce models in which the relationship between the dependent variable and the independent variables may not be linear. Section 16-6 introduces indicator variables, which allow us to use nominal independent variables.

16-1 MODEL AND REQUIRED CONDITIONS

We now assume that k independent variables are potentially related to the dependent variable. Thus, the model is represented by the following equation:

$$y = \beta_0 + \beta_1 x_1 + \beta_2 x_2 + \cdots + \beta_k x_k + \varepsilon$$

where y is the dependent variable, $x_1, x_2, \ldots, x_k$, are the independent variables, $\beta_0, \beta_1, \ldots, \beta_k$ are the coefficients and ε is the error variable. The independent variables may actually be functions of other variables. For example, we might define some of the independent variables as follows:

$$x_2 = x_1^2$$

$$x_5 = x_3 x_4$$

$$x_7 = \log(x_6)$$

The error variable is retained because, even though we have included additional independent variables, deviations between predicted values of y and actual values of y will still occur. Incidentally, when there is more than one independent variable in the regression model, we refer to the graphical depiction of the equation as a **response surface** rather than as a straight line. Figure 16.1 depicts a scatter diagram of a response surface with $k = 2$. (When $k = 2$, the regression equation

FIGURE 16.1

Scatter Diagram and Response Surface with $k = 2$

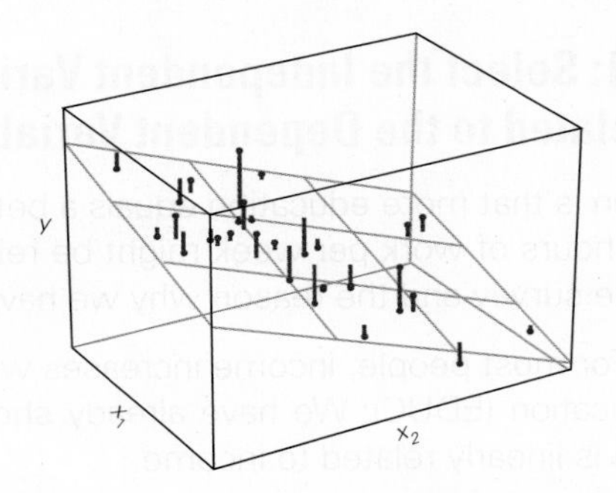

creates a plane.) Of course, whenever k is greater than 2, we can only imagine the response surface; we cannot draw it.

An important part of the regression analysis comprises several statistical techniques that evaluate how well the model fits the data. These techniques require the following conditions, which we introduced in the previous chapter.

Required Conditions for Error Variable

1. The probability distribution of the error variable ε is normal.
2. The mean of the error variable is 0.
3. The standard deviation of ε is σ_ε, which is a constant.
4. The errors are independent.

In Section 15-6, we discussed how to recognise when the requirements are unsatisfied. Those same procedures can be used to detect violations of required conditions in the multiple regression model.

We now proceed as we did in Chapter 15. We discuss how the model's coefficients are estimated and how we assess the model's fit. However, there is one major difference between Chapters 15 and 16. In Chapter 15, we allowed for the possibility that some students will perform the calculations manually. The multiple regression model involves so many computations that it is virtually impossible to conduct the analysis without a computer. All analyses in this chapter will be performed by Excel and Minitab. Your job will be to interpret the output.

16-2 ESTIMATING THE COEFFICIENTS AND ASSESSING THE MODEL

The multiple regression equation is expressed similarly to the simple regression equation. The general form is

$$\hat{y} = b_0 + b_1x_1 + b_2x_2 + \cdots + b_kx_k$$

where k is the number of independent variables.

The procedures introduced in Chapter 15 are extended to the multiple regression model. However, in Chapter 15, we first discussed how to interpret the coefficients and then discussed how to assess the model's fit. In practice, we reverse the process. That is, the first step is to determine how well the model fits. If the model's fit is poor, there is no point in a further analysis of the coefficients of that model. A much higher priority is assigned to the task of improving the model. In this chapter, we show how a regression analysis is performed. We will illustrate the procedure with an example from a sample survey data.

16-2a Step 1: Select the Independent Variables That You Believe May Be Related to the Dependent Variable

Your assumption is that more education equals a better job and higher income. Also, other variables, such as age or hours of work per week might be related to income. Here are the available variables from the sample survey and the reason why we have selected each one:

Age (AGE): For most people, income increases with age

Years of education (EDUC): We have already shown in Chapter 15 in the opening example that education is linearly related to income

Respondent's income (INCOME) (in euros)
Hours of work per week (HRS): Obviously, more hours of work should produce more income
Spouse's hours of work (SPHRS): It is possible that, if one's spouse works more and earns more, the other spouse may choose to work less and thus earn less

You may be wondering why we do not simply include all the interval variables that are available to us. There are three reasons. First, the objective is to determine whether our hypothesised model is valid and whether the independent variables in the model are linearly related to the dependent variable. That is, we should screen the independent variables and include only those that in theory affect the dependent variable.

Second, by including large numbers of independent variables we increase the probability of Type I errors. For example, if we include 100 independent variables, none of which are related to the dependent variable, we are likely to conclude that five of them are linearly related to the dependent variable. This is a problem that we discussed in Chapter 13.

Third, because of a problem called multicollinearity (described in Section 16-3), we may conclude that none of the independent variables are linearly related to the dependent variable when in fact one or more are.

16-2b Step 2: Use a Computer to Compute All the Coefficients and Other Statistics

EXCEL

	A	B	C	D	E	F
1	SUMMARY OUTPUT					
2						
3	*Regression Statistics*					
4	Multiple R	0.30141				
5	R Square	0.09085				
6	Adjusted R Square	0.07840				
7	Standard Error	41444.1				
8	Observations	297				
9						
10	ANOVA					
11		*df*	*SS*	*MS*	*F*	*Significance F*
12	Regression	4	50118240906	1.25E+10	7.294748	1.30334E-05
13	Residual	292	5.01543E+11	1.72E+09		
14	Total	296	5.51661E+11			
15						
16		*Coefficients*	*Standard Error*	*t Stat*	*P-value*	
17	Intercept	−40890	19409.582	−2.11	0.036	
18	AGE	575	227.763	2.52	0.012	
19	EDUC	4101	913.258	4.49	0.000	
20	HRS	77	166.323	0.46	0.646	
21	SPHRS	140	195.490	0.71	0.476	

Instructions

1. Type or import the data so that the independent variables are in adjacent columns.*
2. Click **Data, Data Analysis** and **Regression.**
3. Specify the **Input *Y* Range**, the Input ***X* Range** and a value for α (0.05).

*Rows with empty cells must be entirely removed.

MINITAB

Regression Analysis: INCOME versus AGE, EDUC, HRS, SPHRS

The regression equation is
INCOME = – 40890 + 575 AGE + 4101 EDUC + 77 HRS + 140 SPHRS

Predictor	Coef	SE Coef	T	P
Constant	–40890	19410	–2.11	0.036
AGE	574.9	227.8	2.52	0.012
EDUC	4101.4	913.3	4.49	0.000
HRS	76.5	166.3	0.46	0.646
SPHRS	139.6	195.5	0.71	0.476

S = 41444.1 R-Sq = 9.1% R-Sq(adj) = 7.8%

Analysis of Variance

Source	DF	SS	MS	F	P
Regression	4	50118240906	12529560226	7.29	0.000
Residual Error	292	5.01543E+11	1717613829		
Total	296	5.51661E+11			

Instructions

1. Click **Stat, Regression** and **Regression** ....
2. Specify the dependent variable in the **Response** box and the independent variables in the **Predictors** box.

INTERPRET

The regression model is estimated by:

$$\hat{y}\text{ (INCOME)} = -40\,890 + 575\text{ AGE} + 4101\text{ EDUC} + 77\text{ HRS} + 140\text{ SPHRS}$$

16-2c Step 3: Assess the Model

We assess the model in three ways: the standard error of estimate, the coefficient of determination (both introduced in Chapter 15) and the *F*-test of the analysis of variance (presented subsequently).

16-2d Standard Error of Estimate

Recall that σ_ε is the standard deviation of the error variable ε and that, because σ_ε is a population parameter, it is necessary to estimate its value by using s_ε. In multiple regression, the standard error of estimate is defined as follows:

Standard Error of Estimate

$$s_\varepsilon = \sqrt{\frac{\text{SSE}}{n-k-1}}$$

where n is the sample size and k is the number of independent variables in the model.

As we noted in Chapter 15, each of our software packages reports the standard error of estimate in a different way.

EXCEL **Standard Error = 41 444**

MINITAB S = 41 444

INTERPRET

Recall that we judge the magnitude of the standard error of estimate relative to the values of the dependent variable, and particularly to the mean of *y*. In this example, $\bar{y} = 52\,742$ (not shown in printouts). It appears that the standard error of estimate is quite large.

16-2e Coefficient of Determination

Recall from Chapter 15 that the coefficient of determination is defined as:

$$R^2 = 1 - \frac{SSE}{\sum(y_i - \bar{y})^2}$$

EXCEL R Square = 0.09085

MINITAB R-sq = 9.1%

INTERPRET

This means that 9.1% of the total variation in income is explained by the variation in the four independent variables, whereas 90.9% remains unexplained.

Notice that Excel and Minitab print a second R^2 statistic, called the **coefficient of determination adjusted for degrees of freedom**, which has been adjusted to take into account the sample size and the number of independent variables. The rationale for this statistic is that, if the number of independent variables *k* is large relative to the sample size *n*, the unadjusted R^2 value may be unrealistically high. To understand this point, consider what would happen if the sample size is 2 in a simple linear regression model. The line would fit the data perfectly resulting in $R^2 = 1$ when, in fact, there may be no linear relationship. To avoid creating a false impression, the adjusted R^2 is often calculated. Its formula is as follows:

Coefficient of Determination Adjusted for Degrees of Freedom

$$\text{Adjusted } R^2 = 1 - \frac{SSE/(n-k-1)}{\sum(y_i - \bar{y})^2/(n-1)} = 1 - \frac{MSE}{s_y^2}$$

If *n* is considerably larger than *k*, the unadjusted and adjusted R^2 values will be similar. But if SSE is quite different from 0 and *k* is large compared to *n*, the unadjusted and adjusted values of R^2 will differ substantially. If such differences exist, the analyst should be alerted to a potential problem in interpreting the coefficient of determination. In this example, the adjusted coefficient of determination is 7.8%, indicating that, no matter how we measure the coefficient of determination, the model's fit is not very good.

16-2f Testing the Validity of the Model

In the simple linear regression model, we tested the slope coefficient to determine whether sufficient evidence existed to allow us to conclude that there was a linear relationship between the independent variable and the dependent variable. However, because there is only one independent variable in that model, that same *t*-test was also tested to determine whether that model is valid. When there is more than one independent variable, we need another method to test the overall validity of the model. The technique is a version of the analysis of variance, which we introduced in Chapter 13.

To test the validity of the regression model, we specify the following hypotheses:

$$H_0: \quad \beta_1 = \beta_2 = \cdots = \beta_k = 0$$

$$H_1: \quad \text{At least one } \beta_i \text{ is not equal to 0}$$

If the null hypothesis is true, none of the independent variables $x_1, x_2, \ldots, x_k$ is linearly related to y, and therefore the model is invalid. If at least one β_i is not equal to 0, the model does have some validity.

When we discussed the coefficient of determination in Chapter 15, we noted that the total variation in the dependent variable [measured by $\sum(y_i - \bar{y}^2)$] can be decomposed into two parts: the explained variation (measured by SSR) and the unexplained variation (measured by SSE).That is:

$$\text{Total variation in } y = \text{SSR} + \text{SSE}$$

Furthermore, we established that, if SSR is large relative to SSE, the coefficient of determination will be high – signifying a good model. On the other hand, if SSE is large, most of the variation will be unexplained, which indicates that the model provides a poor fit and consequently has little validity.

The test statistic is the same one we encountered in Section 13-1, where we tested for the equivalence of two or more population means. To judge whether SSR is large enough relative to SSE to allow us to infer that at least one coefficient is not equal to 0, we compute the ratio of the two mean squares. (Recall that the mean square is the sum of squares divided by its degrees of freedom; recall, too, that the ratio of two mean squares is *F*-distributed as long as the underlying population is normal – a required condition for this application.) The calculation of the test statistic is summarised in an analysis of variance (ANOVA) table, whose general form appears in Table 16.1. The Excel and Minitab ANOVA tables are shown next.

TABLE 16.1 Analysis of Variance Table for Regression Analysis

SOURCE OF VARIATION	DEGREES OF FREEDOM	SUMS OF SQUARES	MEAN SQUARES	F-STATISTIC
Regression	k	SSR	MSR = SSR/k	F = MSR/MSE
Residual	$n-k-1$	SSE	MSE = SSE/$(n-k-1)$	
Total	$n-1$	$\sum(Y_i - \bar{Y})^2$		

EXCEL

	A	B	C	D	E	F
1	ANOVA					
2		df	SS	MS	F	Significance F
3	Regression	4	50118240906	1.25E+10	7.294748	1.30334E–05
4	Residual	292	5.01543E+11	1.72E+09		
5	Total	296	5.51661E+11			

MINITAB

Analysis of Variance

Source	DF	SS	MS	F	P
Regression	4	50118240906	12529560226	7.29	0.000
Residual Error	292	5.01543E+11	1717613829		
Total	296	5.51661E+11			

A large value of F indicates that most of the variation in y is explained by the regression equation and that the model is valid. A small value of F indicates that most of the variation in y is unexplained. The rejection region allows us to determine whether F is large enough to justify rejecting the null hypothesis. For this test, the rejection region is:

$$F > F_{a,k,n-k-1}$$

In Example 16.1 the rejection region (assuming $\alpha = 0.05$) is:

$$F > F_{a,k,n-k-1} = F_{0.05,6,335} \approx 2.21$$

As you can see from the printout, $F = 7.29$. The printout also includes the p-value of the test, which is 0. Obviously, there is a great deal of evidence to infer that the model is valid.

Although each assessment measurement offers a different perspective, all agree in their assessment of how well the model fits the data, because all are based on the sum of squares for error, SSE. The standard error of estimate is:

$$s_\varepsilon = \sqrt{\frac{\text{SSE}}{n-k-1}}$$

and the coefficient of determination is:

$$R^2 = 1 - \frac{SSE}{\sum(y_i - \bar{y})^2}$$

When the response surface hits every single point, SSE = 0. Hence, $s_\varepsilon = 0$ and $R^2 = 1$.

If the model provides a poor fit, we know that SSE will be large [its maximum value is $\sum(y_i - \bar{y})^2$] R^2, s_ε will be large, and [since SSE is close to $\sum(y_i - \bar{y})^2$], s_ε will be close to 0.

The F-statistic also depends on SSE. Specifically,

$$F = \frac{\text{MSR}}{\text{MSE}} = \frac{\left(\sum(y_i - \bar{y})^2 - \text{SSE}\right)/k}{\text{SSE}/(n-k-1)}$$

When SSE = 0:

$$F = \frac{\sum(y_i - \bar{y})^2/k}{0/(n-k-1)}$$

which is infinitely large. When SSE is large, SSE is close to $\sum(y_i - \bar{y})^2$ and F is quite small.

The relationship among s_ε, R^2 and F is summarised in Table 16.2.

TABLE 16.2 Relationship Among SSE, s_ε, R^2 and F

SSE	s_ε	R^2	F	ASSESSMENT OF MODEL
0	0	1	∞	Perfect
Small	Small	Close to 1	Large	Good
Large	Large	Close to 0	Small	Poor
$\sum(y_i - \bar{y})^2$	$\sqrt{\frac{\sum(y_i - \bar{y})^2}{n-k-1}}$*	0	0	Useless

*When n is large and k is small, this quantity is approximately equal to the standard deviation of y.

16-2g Interpreting the Coefficients

The coefficients $b_0, b_1, \ldots, b_k$ describe the relationship between each of the independent variables and the dependent variable in the sample. We need to use inferential methods (described below) to draw conclusions about the population. In our example, the sample consists of the 297 observations. The population is composed of all adults.

Intercept The intercept is $b_0 = -40\,890$. This is the average income when all the independent variables are zero. It is often misleading to try to interpret this value, particularly if 0 is outside the range of the values of the independent variables (as is the case here).

Age The relationship between income and age is described by $b_1 = 574$. From this number, we learn that in this model, for each additional year of age, income increases on average by €574, assuming that the other independent variables in this model are held constant.

Education The coefficient $b_2 = 4101$ specifies that in this sample for each additional year of education the income increases on average by €4101, assuming the constancy of the other independent variables.

Hours of Work The relationship between hours of work per week is expressed by $b_3 = 77$. We interpret this number as the average increase in annual income for each additional hour of work per week keeping the other independent variables fixed in this sample.

Spouse's Hours of Work The relationship between annual income and a spouse's hours of work per week is described in this sample $b_4 = 140$, which we interpret to mean that for each additional hour a spouse works per week income increases on average by €140 when the other variables are constant.

16-2h Testing the Coefficients

In Chapter 15 we described how to test to determine whether there is sufficient evidence to infer that in the simple linear regression model x and y are linearly related. The null and alternative hypotheses were:

$$H_0: \beta_1 = 0$$
$$H_1: \beta_1 \neq 0$$

The test statistic was:

$$t = \frac{b_1 - \beta_1}{s_{b_1}}$$

which is Student t distributed with $v = n - 2$ degrees of freedom.

In the multiple regression model, we have more than one independent variable. For each such variable, we can test to determine whether there is enough evidence of a linear relationship between it and the dependent variable for the entire population when the other independent variables are included in the model.

Testing the Coefficients

$$H_0: \beta_i = 0$$
$$H_1: \beta_i \neq 0$$

(for $i = 1, 2 \ldots k$); the test statistic is:

$$t = \frac{b_i - \beta_i}{s_{b_j}}$$

which is Student t distributed with $v = n - k - 1$ degrees of freedom.

To illustrate, we test each of the coefficients in the multiple regression model in the chapter-opening example. The tests that follow are performed just as all other tests in this book have been performed. We set up the null and alternative hypotheses, identify the test statistic, and use the computer to calculate the value of the test statistic and its p-value. For each independent variable, we test ($i = 1, 2, 3, 4$):

$$H_0: \beta_i = 0$$
$$H_1: \beta_i \neq 0$$

The output includes the *t*-tests of β_i. The results of these tests pertain to the entire population of a country. It is also important to add that these test results were determined when the other independent variables were included in the model. We add this statement because a simple linear regression will very likely result in different values of the test statistics and possibly the conclusion.

Test of β_1 (Coefficient of age)

Value of the test statistic: $t = 2.52$; *p*-value = 0.012

Test of β_2 (Coefficient of education)

Value of the test statistic: $t = 4.49$; *p*-value = 0

Test of β_3 (Coefficient of number of hours of work per week)

Value of the test statistic: $t = 0.46$; *p*-value = 0.646

Test of β_4 (Coefficient of spouse's number of hours of work per week)

Value of the test statistic: $t = 0.71$; *p*-value = 0.476

There is sufficient evidence at the 5% significance level to infer that each of the following variables is linearly related to income:

Age
Education

In this model, there is not enough evidence to conclude that each of the following variables is linearly related to income:

Number of hours of work per week
Spouse's number of hours of work per week

Note that this may mean that there is no evidence of a linear relationship between these three independent variables. However, it may also mean that there is a linear relationship between the two variables, but because of a condition called *multicollinearity,* some *t*-tests of β_i revealed no linear relationship. We will discuss multicollinearity in Section 16-3.

16-2i A Cautionary Note about Interpreting the Results

Care should be taken when interpreting the results of this and other regression analyses. We might find that in one model there is enough evidence to conclude that a particular independent variable is linearly related to the dependent variable, but that in another model no such evidence exists. Consequently, whenever a particular *t*-test is *not* significant, we state that there is not enough evidence to infer that the independent and dependent variable are linearly related *in this model.* The implication is that another model may yield different conclusions.

Furthermore, if one or more of the required conditions are violated, the results may be invalid. In Section 16-3, we introduced the procedures that allow the statistics practitioner to examine the model's requirements. We also remind you that it is dangerous to extrapolate far outside the range of the observed values of the independent variables.

16-2j *t*-Tests and the Analysis of Variance

The *t*-tests of the individual coefficients allow us to determine whether $\beta_i \neq 0$ (for $i = 1, 2, \ldots, k$), which tells us whether a linear relationship exists between x_i and y. There is a *t*-test for each independent variable. Consequently, the computer automatically performs k *t*-tests. (It actually conducts $k + 1$ *t*-tests, including the one for the intercept β_0, which we usually ignore.) The *F*-test in the analysis of variance combines these *t*-tests into a single test. That is, we test all the β_i at one time to determine whether at least one of them is not equal to 0. The question naturally arises, why do we need the *F*-test if it is nothing more than the combination of the previously performed *t*-tests? Recall that we addressed this issue earlier. In Chapter 13 we pointed out that we can replace the analysis of variance by a series of t-tests of the difference between two means. However, by doing so, we increase the probability of making a Type I error. Which means that even when there is no linear relationship between each of the independent variables and the dependent variable, multiple t-tests will likely show some are significant. As a result, you will conclude erroneously that, since at least one is not equal to 0, the model is valid. The F-test, on the other hand, is performed only once. Because the probability that a Type I error will occur in a single trial is equal to α, the chance of erroneously concluding that the model is valid is substantially less with the F-test than with multiple t-tests.

There is another reason that the F-test is superior to multiple t-tests. Because of a commonly occurring problem called *multicollinearity*, the t-tests may indicate that some independent variables are not linearly related to the dependent variable, when in fact they are. The problem of multicollinearity does not affect the F-test, nor does it inhibit us from developing a model that fits the data well. Multicollinearity is discussed in Section 16-3.

16-2k The F-Test and the t-Test in the Simple Linear Regression Model

It is useful for you to know that we can use the F-test to test the validity of the simple linear regression model. However, this test is identical to the t-test of β_1. The t-test of β_1 in the simple linear regression model tells us whether that independent variable is linearly related to the dependent variable. However, because there is only one independent variable, the t-test of β_1 also tells us whether the model is valid, which is the purpose of the F-test.

The relationship between the t-test of β_1 and the F-test can be explained mathematically. Statisticians can show that if we square a t-statistic with v degrees of freedom we produce an F-statistic with 1 and v degrees of freedom. (We briefly discussed this relationship in Chapter 13.) To illustrate, consider Example 15.2. We found the t-test of β_1 to be −13.44, with degrees of freedom equal to 98. The p-value was 5.75×10^{-24}. The output included the analysis of variance table where $F = 180.64$ and the p-value was 5.75×10^{-24}. The t-statistic squared is $t^2 = (-13.44)^2 = 180.63$. (The difference is due to rounding errors.) Notice that the degrees of freedom of the F-statistic are 1 and 98. Thus, we can use either test to test the validity of the simple linear regression model.

16-2l Using the Regression Equation

As was the case with simple linear regression, we can use the multiple regression equation in two ways: we can produce the prediction interval for a particular value of y, and we can produce the confidence interval estimate of the expected value of y. Like the other calculations associated with multiple regression, we call on the computer to do the work.

To illustrate, we will predict the income of a 50-year-old, with 12 years of education, who works 40 hours per week, and whose spouse also works 40 hours per week.

As you discovered in the previous chapter, both Excel and Minitab output the prediction interval and interval estimate of the expected (average) operating margin for all sites with the given variables.

EXCEL

	A	B	C	D
1	**Prediction Interval**			
2				
3			INCOME	
4				
5	Predicted value		45720.45	
6				
7	Prediction Interval			
8	Lower limit		−36134	
9	Upper limit		127574.9	
10				
11	Interval Estimate of Expected Value			
12	Lower limit		38866.79	
13	Upper limit		52574.11	

Instructions

See the instructions on Section 15-5b. In cells B299 to E299, we input the values 50 12 40 40 , respectively. We specified 95% confidence.

MINITAB

```
Predicted Values for New Observations

New Obs     Fit  SE Fit         95% CI              95% PI
      1   45720    3482  (38867, 52574)  (-36134, 127575)

Values of Predicted for New Observations

New Obs   AGE  EDUC   HRS  SPHRS
      1  50.0  12.0  40.0   40.0
```

Instructions

See the instructions on Example 15.7. We input the values 50 12 40 40. We specified 95% confidence.

INTERPRET

The prediction interval is −36 134, 127 575. It is so wide as to be completely useless. To be useful in predicting values, the model must be considerably better. The confidence interval estimate of the expected income of a population is 38 867, 52 574.

EXERCISES

The following exercises require the use of a computer and statistical software. Exercises 16.1–16.3 can be solved manually. Use a 5% significance level for all tests.

16.1 Xr16-01 A developer who specialises in summer cottage properties is considering purchasing a large tract of land adjoining a lake. The current owner of the tract has already subdivided the land into separate building lots and has prepared the lots by removing some of the trees. The developer wants to forecast the value of each lot. From previous experience, she knows that the most important factors affecting the price of a lot are size, number of mature trees and distance to the lake. From a nearby area, she gathers the relevant data for 60 recently sold lots.

a. Find the regression equation.

b. What is the standard error of estimate? Interpret its value.

c. What is the coefficient of determination? What does this statistic tell you?

d. What is the coefficient of determination, adjusted for degrees of freedom? Why does this value differ from the coefficient of determination? What does this tell you about the model?

e. Test the validity of the model. What does the *p*-value of the test statistic tell you?

f. Interpret each of the coefficients.

g. Test to determine whether each of the independent variables is linearly related to the price of the lot in this model.

h. Predict with 90% confidence the selling price of a 4000 square metre lot that has 50 mature trees and is 7.5 metres from the lake.

i. Estimate with 90% confidence the average selling price of 5000 square metre lots that have 10 mature trees and are 21 metres from the lake.

16.2 Xr16-02 Pat Statsdud, a student ranking near the bottom of the statistics class, decided that a certain amount of studying could actually improve final grades. However, too much studying would not be warranted because Pat's ambition (if that's what one could call it) was to ultimately graduate with the absolute minimum level of work. Pat was registered in a statistics course that had only 3 weeks to go before the final exam and for which the final grade was determined in the following way:

$$\begin{aligned}\text{Total mark} &= 20\%\ (\text{Assignment})\\ &+ 30\%\ (\text{Midterm test})\\ &+ 50\%\ (\text{Final exam})\end{aligned}$$

To determine how much work to do in the remaining 3 weeks, Pat needed to be able to predict the final exam mark on the basis of the assignment mark (worth 20 points) and the midterm mark (worth 30 points). Pat's marks on these were 12/20 and 14/30, respectively. Accordingly, Pat undertook the following analysis. The final exam mark, assignment mark and midterm test mark for 30 students who took the statistics course last year were collected.

a. Determine the regression equation.

b. What is the standard error of estimate? Briefly describe how you interpret this statistic.

c. What is the coefficient of determination? What does this statistic tell you?

d. Test the validity of the model.

e. Interpret each of the coefficients.

f. Can Pat infer that the assignment mark is linearly related to the final grade in this model?

g. Can Pat infer that the midterm mark is linearly related to the final grade in this model?

h. Predict Pat's final exam mark with 95% confidence.

i. Predict Pat's final grade with 95% confidence.

16.3 Xr16-03 The president of a company that manufactures drywall wants to analyse the variables that affect demand for his product. Drywall is used to construct walls in houses and offices. Consequently, the president decides to develop a regression model in which the dependent variable is monthly sales of drywall (in hundreds of 4 × 8 sheets) and the independent variables are:

Number of building permits issued in the county
Five-year mortgage rates (in percentage points)
Vacancy rate in apartments (in percentage points)
Vacancy rate in office buildings (in percentage points)

To estimate a multiple regression model, he took monthly observations from the past 2 years.

a. Analyse the data using multiple regression.

b. What is the standard error of estimate? Can you use this statistic to assess the model's fit? If so, how?

c. What is the coefficient of determination, and what does it tell you about the regression model?

d. Test the overall validity of the model.

e. Interpret each of the coefficients.

f. Test to determine whether each of the independent variables is linearly related to drywall demand in this model.

g. Predict next month's drywall sales with 95% confidence if the number of building permits is 50, the 5-year mortgage rate is 9.0%, and the vacancy rates are 3.6% in apartments and 14.3% in office buildings.

APPLICATIONS IN HUMAN RESOURCES MANAGEMENT

Severance Pay

In most firms, the entire issue of compensation falls into the domain of the human resources manager. The manager must ensure that the method used to determine compensation contributes to the firm's objectives. Moreover, the firm needs to ensure that discrimination or bias of any kind is not a factor. Another function of the personnel manager is to develop severance packages for employees whose services are no longer needed because of downsizing or merger. The size and nature of severance are rarely part of any working agreement and must be determined by a variety of factors. Regression analysis is often useful in this area.

16.4 Xr16-04 When one company buys another company, it is not unusual that some workers are made redundant. The severance benefits offered to the laid-off workers are often the subject of dispute. Suppose that the Laurier Company recently bought the Western Company and subsequently terminated 20 of Western's employees. As part of the buyout agreement, it was promised that the severance packages offered to the former Western employees would be equivalent to those offered to Laurier employees who had been terminated in the past year. Thirty-six-year-old Pierre Blanc, a Western employee for the past 10 years, earning €32 000 per year, was one of those let go. His severance package included an offer of 5 weeks severance pay. Pierre complained that this offer was less than that offered to Laurier's employees when they were laid off, in contravention of the buyout agreement. A statistician was called in to settle the dispute. The statistician was told that severance is determined by three factors: age, length of service with the company and pay. To determine how generous the severance package had been, a random sample of 50 Laurier ex-employees was taken. For each, the following variables were recorded:

Number of weeks of severance pay
Age of employee
Number of years with the company
Annual pay (in thousands of euros)

a. Determine the regression equation.
b. Comment on how well the model fits the data.
c. Do all the independent variables belong in the equation? Explain.
d. Perform an analysis to determine whether Bill is correct in his assessment of the severance package.

16.5 Xr16-05 The admissions officer of a university is trying to develop a formal system to decide which students to admit to the university. She believes that determinants of success include the standard variables – high school grades and SAT scores. However, she also believes that students who have participated in extracurricular activities are more likely to succeed than those who have not. To investigate the issue, she randomly sampled 100 fourth-year students and recorded the following variables:

GPA for the first 3 years at the university (range: 0 to 12)
GPA from high school (range: 0 to 12)
SAT score (range: 400 to 1600)
Number of hours on average spent per week in organised extracurricular activities in the last year of high school

a. Develop a model that helps the admissions officer decide which students to admit and use the computer to generate the usual statistics.
b. What is the coefficient of determination? Interpret its value.
c. Test the overall validity of the model.
d. Test to determine whether each of the independent variables is linearly related to the dependent variable in this model.
e. Determine the 95% interval of the GPA for the first 3 years of university for a student whose high school GPA is 10, whose SAT score is 1200, and who worked an average of 2 hours per week on organised extracurricular activities in the last year of high school.
f. Find the 90% interval of the mean GPA for the first 3 years of university for all students whose high school GPA is 8, whose SAT score is 1100, and who worked an average of 10 hours per week on organised extracurricular activities in the last year of high school.

16.6 Xr16-06 The marketing manager for a chain of hardware stores needed more information about the effectiveness of the three types of advertising that the chain used. These are localised direct mailing (in which flyers describing sales and featured products are distributed to homes in the area surrounding a store), newspaper advertising and local television advertisements. To determine which type is most effective, the manager collected 1 week's data from 100 randomly selected stores. For each store, the following variables were recorded:

Weekly gross sales
Weekly expenditures on direct mailing
Weekly expenditures on newspaper advertising
Weekly expenditures on television commercials

All variables were recorded in thousands of dollars.

a. Find the regression equation.
b. What are the coefficient of determination and the coefficient of determination adjusted for degrees of freedom? What do these statistics tell you about the regression equation?
c. What does the standard error of estimate tell you about the regression model?
d. Test the validity of the model.
e. Which independent variables are linearly related to weekly gross sales in this model? Explain.
f. Compute the 95% interval of the week's gross sales if a local store spent £800 on direct mailing, £1200 on newspaper advertisements and £2000 on television commercials.
g. Calculate the 95% interval of the mean weekly gross sales for all stores that spend £800 on direct mailing, £1200 on newspaper advertising and £2000 on television commercials.
h. Discuss the difference between the two intervals found in parts (f) and (g).

16.7 Xr16-07 For many cities around the world, rubbish is an increasing problem. In 2012 it was reported that one of Dubai's two landfill sites was full and the other will reach capacity in less than 7 years. A consultant for the waste management department decided to gather data about the problem. She took a random sample of houses and determined the following:

Y = The amount of rubbish per average week (kg)
X_1 = Size of the house (square metre)
X_2 = Number of children
X_3 = Number of adults who are usually home during the day

a. Conduct a regression analysis.
b. Is the model valid?
c. Interpret each of the coefficients.
d. Test to determine whether each of the independent variables is linearly related to the dependent variable.

16.8 Xr16-08 The administrator of a school board in a large county was analysing the average mathematics test scores in the schools under her control. She noticed that there were dramatic differences in scores among the schools. In an attempt to improve the scores of all the schools, she attempted to determine the factors that account for the differences. Accordingly, she took a random sample of 40 schools across the county and, for each, determined the mean test score last year, the percentage of teachers in each school who have at least one university degree in mathematics, the mean age, and the mean annual income (in £000s) of the mathematics teachers.

a. Conduct a regression analysis to develop the equation.
b. Is the model valid?
c. Interpret and test the coefficients.
d. Predict with 95% confidence the test score at a school where 50% of the mathematics teachers have mathematics degrees, the mean age is 43 and the mean annual income is £48 300.

16.9 Xr16-09 Life insurance companies are keenly interested in predicting how long their customers will live because their premiums and profitability depend on such numbers. An actuary for one insurance company gathered data from 100 recently deceased male customers. He recorded the age at death of the customer plus the ages at death of his mother and father, the mean ages at death of his grandmothers, and the mean ages at death of his grandfathers.

a. Perform a multiple regression analysis on these data.
b. Is the model valid?
c. Interpret and test the coefficients.
d. Determine the 95% interval of the longevity of a man whose parents lived to the age of 70, whose grandmothers averaged 80 years, and whose grandfathers averaged 75 years.
e. Find the 95% interval of the mean longevity of men whose mothers lived to 75 years, whose fathers lived to 65 years, whose grandmothers averaged 85 years, and whose grandfathers averaged 75 years.

16.10 Xr16-10 University students often complain that universities reward lecturers for research but not for teaching, and they argue that lecturers react to this situation by devoting more time and energy to the publication of their findings and less time and energy to classroom activities. Lecturers counter that research and teaching go hand in hand: More research makes better teachers. A student organisation at one university decided to investigate the issue. It randomly selected 50 economics lecturers who are employed by a multi-campus university. The students recorded the salaries (in €1000s) of the lecturers, their average teaching evaluations (on a 10-point scale), and the total number of journal articles published in their careers. Perform a complete analysis (produce the regression equation, assess it, and report your findings).

16.11 Xr16-11 Lotteries have become important sources of revenue for governments. Many people have criticised lotteries, however, referring to them as a tax on the poor and uneducated. In an examination of the issue, a random sample of 100 adults was asked how much they spend on lottery tickets and was interviewed about various socioeconomic variables. The purpose of this study is to test the following beliefs:

1. Relatively uneducated people spend more on lotteries than do relatively educated people.
2. Older people buy more lottery tickets than younger people.
3. People with more children spend more on lotteries than people with fewer children.
4. Relatively poor people spend a greater proportion of their income on lotteries than relatively rich people.

The following data were recorded:

Amount spent on lottery tickets as a percentage of total household income
Number of years of education
Age
Number of children
Personal income (in thousands of euros)

a. Develop the multiple regression equation.

b. Is the model valid?

c. Test each of the beliefs. What conclusions can you draw?

16.12 Xr16-12 The MBA programme at a large university is facing a pleasant problem – too many applicants. The current admissions policy requires students to have completed at least 3 years of work experience and an undergraduate degree with a B-average or better. Until 3 years ago, the school admitted any applicant who met these requirements. However, because the programme recently converted from a 2-year programme (four semesters) to a 1-year programme (three semesters), the number of applicants has increased substantially. The dean, who teaches statistics courses, wants to raise the admissions standards by developing a method that more accurately predicts how well an applicant will perform in the MBA programme. She believes that the primary determinants of success are the following:

Undergraduate grade point average (GPA)
Graduate Management Admissions Test (GMAT) score
Number of years of work experience

She randomly sampled students who completed the MBA and recorded their MBA programme GPA, as well as the three variables listed here.

a. Develop a multiple regression model.

b. Test the model's validity.

c. Test to determine which of the independent variables is linearly related to MBA GPA.

16-3 REGRESSION DIAGNOSTICS-II

In Section 15-6 we discussed how to determine whether the required conditions are unsatisfied. The same procedures can be used to diagnose problems in the multiple regression model. Here is a brief summary of the diagnostic procedure we described in Chapter 15.
Calculate the residuals and check the following:

1. *Is the error variable non-normal?* Draw the histogram of the residuals.
2. *Is the error variance constant?* Plot the residuals versus the predicted values of y.
3. *Are the errors independent (time-series data)?* Plot the residuals versus the time periods.
4. *Are there observations that are inaccurate or do not belong to the target population?* Double-check the accuracy of outliers and influential observations.

If the error is non-normal and/or the variance is not a constant, several remedies can be attempted. These are beyond the level of this book.

Outliers and influential observations are checked by examining the data in question to ensure accuracy.

Non-independence of a time series can sometimes be detected by graphing the residuals and the time periods and looking for evidence of autocorrelation. In Section 16-4, we introduce the Durbin–Watson test, which tests for one form of autocorrelation. We will offer a corrective measure for non-independence.

There is another problem that is applicable to multiple regression models only. *Multicollinearity* is a condition wherein the independent variables are highly correlated.

Multicollinearity distorts the t-tests of the coefficients, making it difficult to determine whether any of the independent variables are linearly related to the dependent variable. It also makes interpreting the coefficients problematic. We will discuss this condition and its remedy next.

16-3a Multicollinearity

Multicollinearity (also called *collinearity* and *intercorrelation*) is a condition that exists when the independent variables are correlated with one another. The adverse effect of multicollinearity is that the estimated regression coefficients of the independent variables that are correlated tend to have large sampling errors. There are two consequences of multicollinearity. First, because the variability of the coefficients is large, the sample coefficient may be far from the actual population parameter, including the possibility that the statistic and parameter may have opposite signs. Second, when the coefficients are tested, the t-statistics will be small, which leads to the inference that there is no linear relationship between the affected independent variables and the dependent variable. In some cases, this inference will be wrong. Fortunately, multicollinearity does not affect the F-test of the analysis of variance.

Consider the chapter-opening example. When we tested the coefficient of correlation between income and number of children, we found it to be statistically significant. The Excel printout is shown below. How do we explain the apparent contradiction between the multiple regression t-test of the coefficient of the number of children and the result of the t-test of the correlation coefficient? The answer is multicollinearity.

	A	B
1	**Correlation (Pearson)**	
2		
3	*INCOME and AGE*	
4	Pearson Coefficient of Correlation	0.1883
5	t Stat	3.2083
6	df	280
7	P(T<=t) one tail	0.0007
8	t Critical one tail	1.6503
9	P(T<=t) two tail	0.0015
10	t Critical two tail	1.9685

There is a relatively high degree of correlation between number of family members who earn income and number of children. The result of the t-test of the correlation between number of earners and number of children is shown next. The result should not be surprising, as more earners in a family are very likely children.

	A	B
1	**Correlation (Pearson)**	
2		
3	*INCOME and CUREMPYR*	
4	Pearson Coefficient of Correlation	0.1972
5	t Stat	3.3652
6	df	280
7	P(T<=t) one tail	0.0004
8	t Critical one tail	1.6503
9	P(T<=t) two tail	0.0009
10	t Critical two tail	1.9685

Multicollinearity affected the result of the multiple regression t-test so that it appeared that number of children is not significantly related to income, when in fact it is.

Another problem caused by multicollinearity is the interpretation of the coefficients. We interpret the coefficients as measuring the change in the dependent variable when the corresponding independent variable increases by one unit while all the other independent variables are held constant. This interpretation may be impossible when the independent variables are highly correlated, because when the independent variable increases by one unit, some or all of the other independent variables will change.

This raises two important questions for the statistics practitioner. First, how do we recognise the problem of multicollinearity when it occurs and, second, how do we avoid or correct it?

Multicollinearity exists in virtually all multiple regression models. In fact, finding two completely uncorrelated variables is rare. The problem becomes serious, however, only when two or more independent variables are highly correlated. Unfortunately, we do not have a critical value that indicates when the correlation between two independent variables is large enough to cause problems. To complicate the issue, multicollinearity also occurs when a combination of several independent variables is correlated with another independent variable or with a combination of other independent variables. Consequently, even with access to all the correlation coefficients, determining when the multicollinearity problem has reached the serious stage may be extremely difficult. A good indicator of the problem is a large F-statistic, but small t-statistics.

Minimising the effect of multicollinearity is often easier than correcting it. The statistics practitioner must try to include independent variables that are independent of each other.

EXERCISES

The following exercises require a computer and software.

16.13 Compute the residuals and the predicted values for the regression analysis in Exercise 16.1.

a. Is the normality requirement violated? Explain.

b. Is the variance of the error variable constant? Explain.

16.14 Calculate the coefficients of correlation for each pair of independent variables in Exercise 16.1. What do these statistics tell you about the independent variables and the *t*-tests of the coefficients?

16.15 Refer to Exercise 16.2.

a. Determine the residuals and predicted values.

b. Does it appear that the normality requirement is violated? Explain.

c. Is the variance of the error variable constant? Explain.

d. Determine the coefficient of correlation between the assignment mark and the midterm mark. What does this statistic tell you about the *t*-tests of the coefficients?

16.16 Compute the residuals and predicted values for the regression analysis in Exercise 16.3.

a. Does it appear that the error variable is not normally distributed?

b. Is the variance of the error variable constant?

c. Is multicollinearity a problem?

16.17 Calculate the residuals and predicted values for the regression analysis in Exercise 16.5.

a. Does the error variable appear to be normally distributed?

b. Is the variance of the error variable constant?

c. Is multicollinearity a problem?

16.18 Are the required conditions satisfied in Exercise 16.5?

16.19 Refer to Exercise 16.6.

a. Conduct an analysis of the residuals to determine whether any of the required conditions are violated.

b. Does it appear that multicollinearity is a problem?

c. Identify any observations that should be checked for accuracy.

16.20 Are the required conditions satisfied for the regression analysis in Exercise 16.7?

16.21 Determine whether the required conditions are satisfied in Exercise 16.8.

16.22 Refer to Exercise 16.9. Calculate the residuals and predicted values.

a. Is the normality requirement satisfied?

b. Is the variance of the error variable constant?

c. Is multicollinearity a problem?

16.23 Determine whether there are violations of the required conditions in the regression model used in Exercise 16.10.

16.24 Refer to Exercise 16.11.

a. Are the required conditions satisfied?

b. Is multicollinearity a problem? If so, explain the consequences.

16.25 Refer to Exercise 16.12. Are the required conditions satisfied?

16-4 REGRESSION DIAGNOSTICS-III (TIME SERIES)

In Chapter 15 we pointed out that, in general, we check to see whether the errors are independent when the data constitute a *times series* – data gathered sequentially over a series of time periods. In Section 15-6, we described the graphical procedure for determining whether the required condition that the errors are independent is violated. We plot the residuals versus the time periods and look for patterns. In this section, we augment that procedure with the **Durbin–Watson test**.

16-4a Durbin–Watson Test

The Durbin–Watson test allows the statistics practitioner to determine whether there is evidence of **first-order autocorrelation** – a condition in which a relationship exists between consecutive residuals e_i and e_{i-1} where i is the time period. The Durbin–Watson statistic is defined as

$$d = \frac{\sum_{i=2}^{n}(e_i - e_{i-1})^2}{\sum_{i=1}^{n} e_i^2}$$

The range of the values of d is

$$0 \leq d \leq 4$$

where small values of $d(d < 2)$ indicate a positive first-order autocorrelation and large values of $d(d > 2)$ imply a negative first-order autocorrelation. Positive first-order autocorrelation is a common occurrence in business and economic time series. It occurs when consecutive residuals tend to be similar. In that case $(e_i - e_{i-1})^2$ will be small, producing a small value for d. Negative first-order autocorrelation occurs when consecutive residuals differ widely. For example, if positive and negative residuals generally alternate, $(e_i - e_{i-1})^2$ will be large; as a result, d will be greater than 2. Figures 16.2 and 16.3 depict positive first-order autocorrelation, whereas Figure 16.4 illustrates negative autocorrelation. Notice that in Figure 16.2, the first residual is a small number; the second residual, also a small number, is somewhat larger; and that trend continues. In Figure 16.3, the first residual is large and, in general, succeeding residuals decrease. In both figures, consecutive residuals are similar. In Figure 16.4, the first residual is a positive number and is followed by a negative residual. The remaining residuals follow this pattern (with some exceptions). Consecutive residuals are quite different.

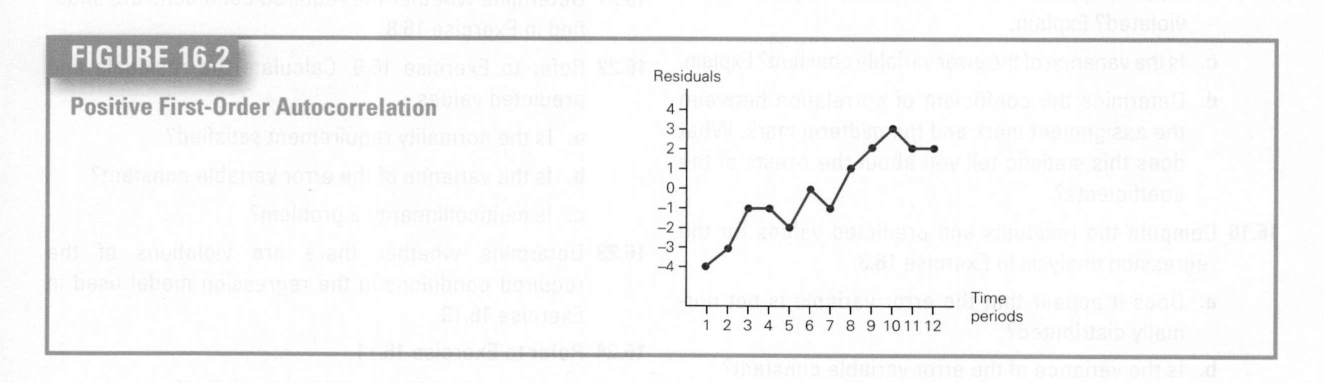

FIGURE 16.2
Positive First-Order Autocorrelation

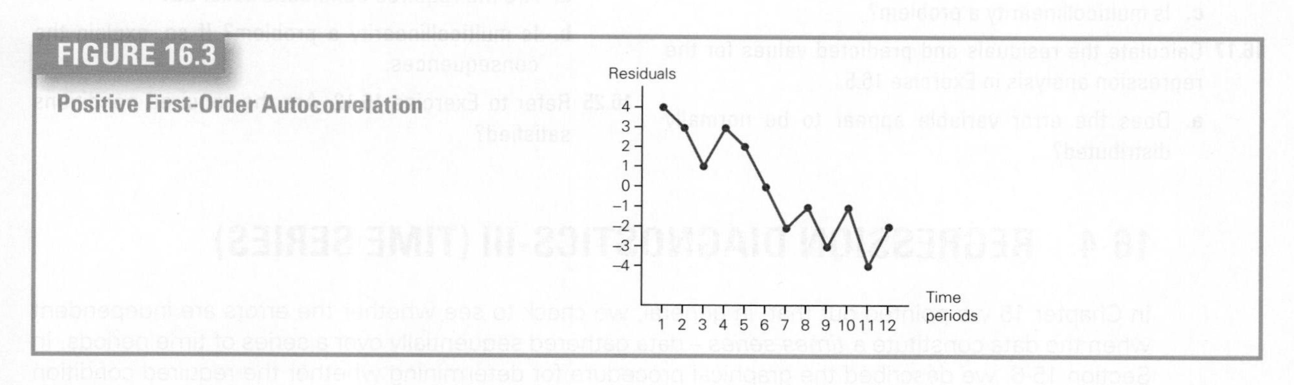

FIGURE 16.3
Positive First-Order Autocorrelation

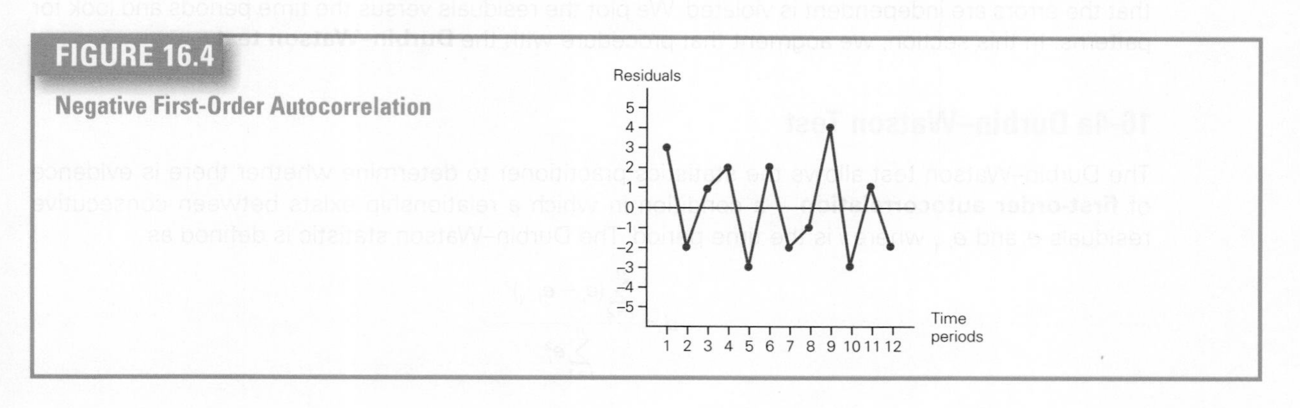

FIGURE 16.4
Negative First-Order Autocorrelation

Table 8 in Appendix A is designed to test for positive first-order autocorrelation by providing values of d_L and d_U for a variety of values of n and k and for $\alpha = 0.01$ and 0.05.

The decision is made in the following way. If $d < d_L$, we conclude that there is enough evidence to show that positive first-order autocorrelation exists. If $d > d_U$, we conclude that there is not enough evidence to show that positive first-order autocorrelation exists. And if $d_L < d < d_U$, the test is inconclusive. The recommended course of action when the test is inconclusive is to continue testing with more data until a conclusive decision can be made.

For example, to test for positive first-order autocorrelation with $n = 20$, $k = 3$, and $\alpha = 0.05$, we test the following hypotheses:

H_0: **There is no first-order autocorrelation**

H_1: **There is positive first-order autocorrelation**

The decision is made as follows:

If $d < d_L = 1.00$, reject the null hypothesis in favour of the alternative hypothesis. If $d > d_U = 1.68$, do not reject the null hypothesis.

If $1.00 \leq d \leq 1.68$, the test is inconclusive.

To test for negative first-order autocorrelation, we change the critical values. If $d > 4 - d_L$, we conclude that negative first-order autocorrelation exists. If $d < 4 - d_U$, we conclude that there is not enough evidence to show that negative first-order autocorrelation exists. If $4 - d_U \leq d \leq 4 - d_L$, the test is inconclusive.

We can also test simply for first-order autocorrelation by combining the two one-tail tests. If $d < d_L$ or $d > 4 - d_L$, we conclude that autocorrelation exists. If $d_L \leq d \leq 4 - d_U$, we conclude that there is no evidence of autocorrelation. If $d_L \leq d \leq d_U$ or $4 - d_U \leq d \leq -d_L$, the test is inconclusive. The significance level will be 2α (where α is the one-tail significance level). Figure 16.5 describes the range of values of d and the conclusion for each interval.

For time-series data, we add the Durbin–Watson test to our list of regression diagnostics. In other words, we determine whether the error variable is normally distributed with constant variance (as we did in Section 16-3), we identify outliers and (if our software allows it) influential observations that should be verified, and we conduct the Durbin–Watson test.

FIGURE 16.5

Durbin–Watson Test

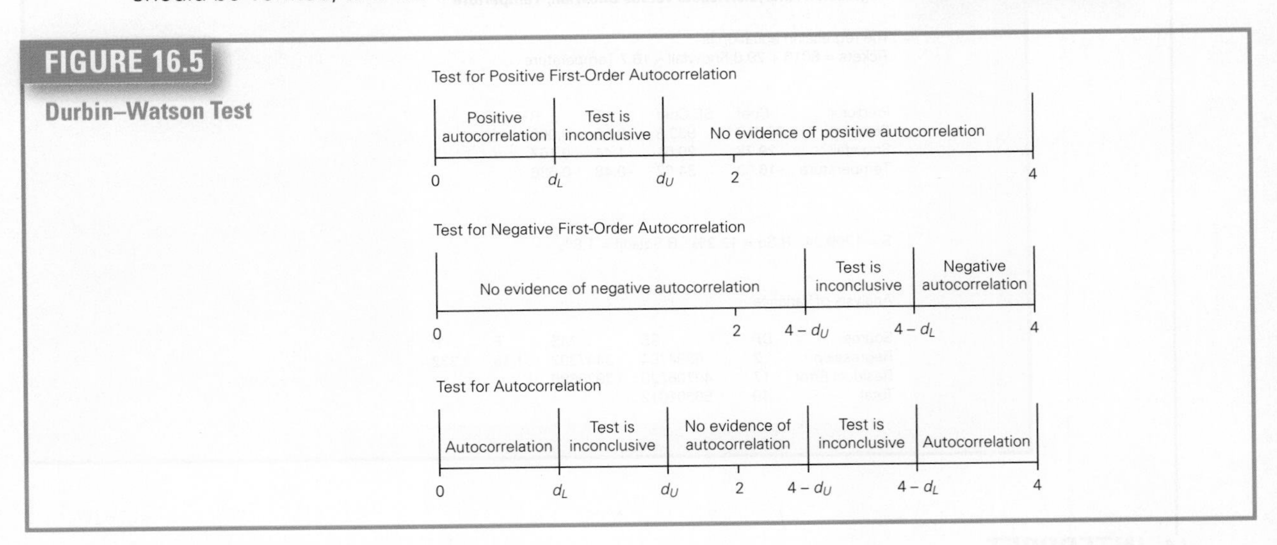

EXAMPLE 16.1

DATA Xm16-01

Christmas Week Ski Lift Sales

Christmas week is a critical period for most ski resorts. Because many students and adults are free from other obligations, they are able to spend several days indulging in their favourite pastime, skiing.

A large proportion of gross revenue is earned during this period. A ski resort in Vermont wanted to determine the effect that weather had on its sales of lift tickets. The manager of the resort collected data on the number of lift tickets sold during Christmas week (y), the total snowfall in centimetres (x_1), and the average temperature in degrees Celsius (x_2) for the past 20 years. Develop the multiple regression model and diagnose any violations of the required conditions.

SOLUTION:

The model is

$$y = \beta_0 + \beta_1 x_1 + \beta_2 x_2 + \varepsilon$$

EXCEL

	A	B	C	D	E	F
1	SUMMARY OUTPUT					
2						
3	*Regression Statistics*					
4	Multiple R	0.3490				
5	R Square	0.1218				
6	Adjusted R Square	0.0185				
7	Standard Error	1710				
8	Observations	20				
9						
10	ANOVA					
11		*df*	*SS*	*MS*	*F*	*Significance F*
12	Regression	2	6894784.447	3447392	1.18	0.3315
13	Residual	17	49706227.75	2923896		
14	Total	19	56601012.2			
15						
16		*Coefficients*	*Standard Error*	*t Stat*	*P-value*	
17	Intercept	8016	934	8.58	1.38E-07	
18	Snowfall	29.78	20.61	1.44	0.1667	
19	Temperature	−16.69	34.62	−0.48	0.6360	

MINITAB

Regression Analysis:Tickets versus Snowfall, Temperture

The regression equation is
Tickets = 8016 + 29.8 Snowfall – 16.7 Temperature

Predictor	Coef	SE Coef	T	P
Constant	8015.8	933.8	8.58	0.000
Snowfall	29.78	20.61	1.44	0.167
Temperature	−16.69	34.62	−0.48	0.636

S = 1709.94 R-Sq = 12.2% R-Sq(adj) = 1.8%

Analysis of Variance

Source	DF	SS	MS	F	P
Regression	2	6894784	3447392	1.18	0.332
Residual Error	17	49706228	2923896		
Total	19	56601012			

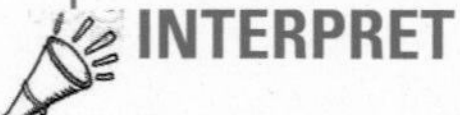

INTERPRET

As you can see, the coefficient of determination is small (R^2 = 12%) and the *p*-value of the *F*-test is 0.3315, both of which indicate that the model is poor. We used Excel to draw the histogram (Figure 16.6) of the residuals and plot the predicted values of *y* versus the residuals in Figure 16.7. Because the observations constitute a time series, we also used Excel to plot the time periods (years) versus the residuals (Figure 16.8).

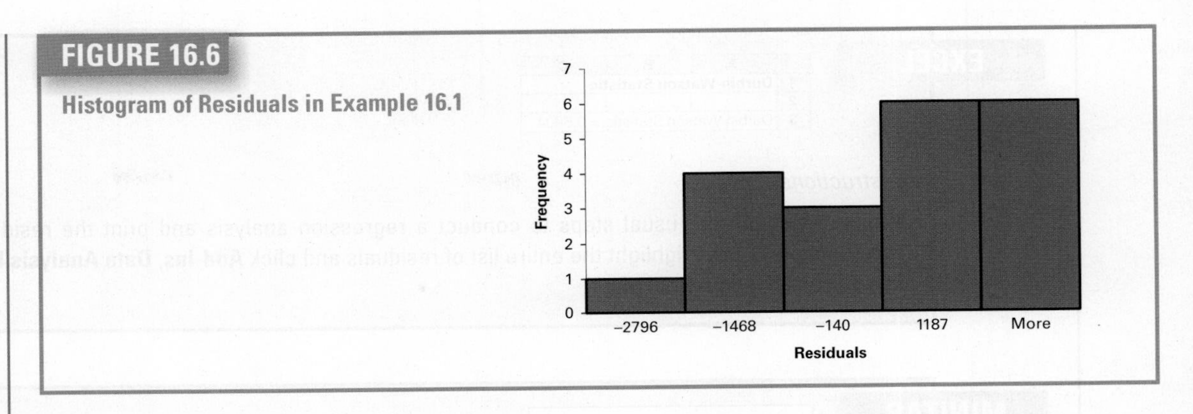

FIGURE 16.6
Histogram of Residuals in Example 16.1

The histogram reveals that the error may be normally distributed.

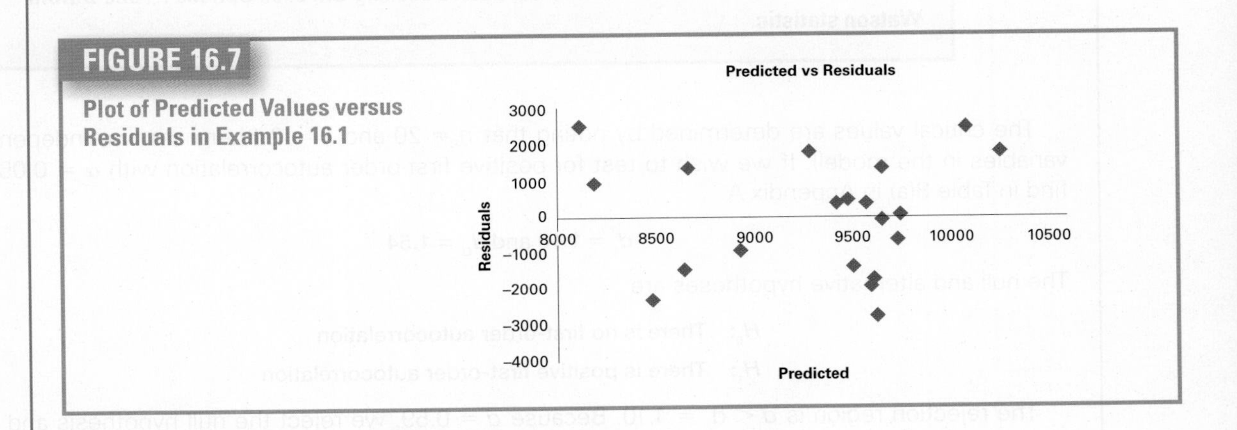

FIGURE 16.7
Plot of Predicted Values versus Residuals in Example 16.1

There does not appear to be any evidence of heteroscedasticity.

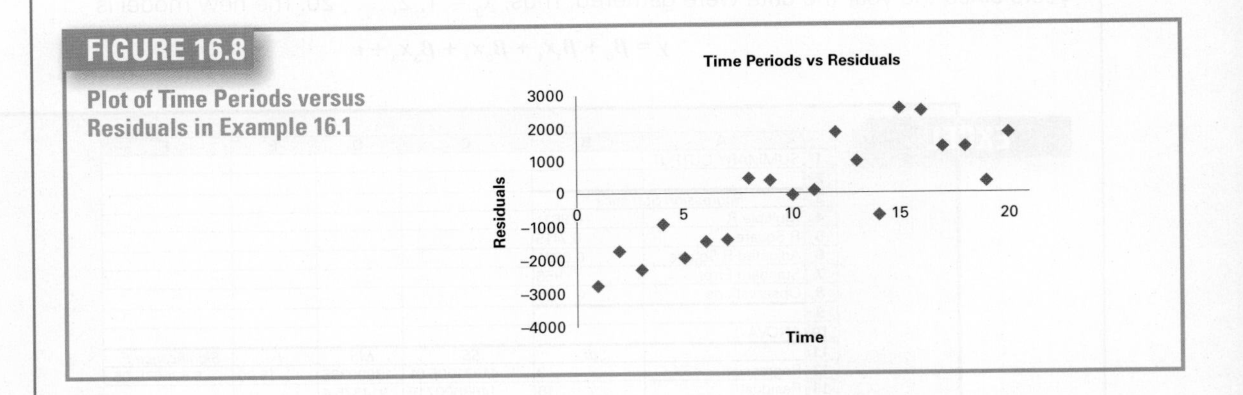

FIGURE 16.8
Plot of Time Periods versus Residuals in Example 16.1

This graph reveals a serious problem. There is a strong relationship between consecutive values of the residuals, which indicates that the requirement that the errors are independent has been violated. To confirm this diagnosis, we instructed Excel and Minitab to calculate the Durbin–Watson statistic.

EXCEL

	A	B	C
1	**Durbin-Watson Statistic**		
2			
3	Durbin-Watson Statistic = 0.5904		

Instructions

Proceed through the usual steps to conduct a regression analysis and print the residuals (see Section 15-6a). Highlight the entire list of residuals and click **Add-Ins, Data Analysis Plus** and **Durbin–Watson Statistic**.

MINITAB

Durbin-Watson statistic = 0.590390

Instructions

Follow the instructions on Section 15-5a. Before clicking **OK**, click **Options** . . . and **Durbin–Watson statistic**.

The critical values are determined by noting that $n = 20$ and $k = 2$ (there are two independent variables in the model). If we wish to test for positive first-order autocorrelation with $\alpha = 0.05$, we find in Table 8(a) in Appendix A

$$d_L = 1.10 \text{ and } d_U = 1.54$$

The null and alternative hypotheses are

H_0: There is no first-order autocorrelation

H_1: There is positive first-order autocorrelation

The rejection region is $d < d_L = 1.10$. Because $d = 0.59$, we reject the null hypothesis and conclude that there is enough evidence to infer that positive first-order autocorrelation exists.

Autocorrelation usually indicates that the model needs to include an independent variable that has a time-ordered effect on the dependent variable. The simplest such independent variable represents the time periods. To illustrate, we included a third independent variable that records the number of years since the year the data were gathered. Thus, $x_3 = 1, 2, \ldots, 20$. The new model is

$$y = \beta_0 + \beta_1 x_1 + \beta_2 x_2 + \beta_3 x_3 + \varepsilon$$

EXCEL

	A	B	C	D	E	F
1	SUMMARY OUTPUT					
2						
3	*Regression Statistics*					
4	Multiple R	0.8611				
5	R Square	0.7415				
6	Adjusted R Square	0.6931				
7	Standard Error	956				
8	Observations	20				
9						
10	ANOVA					
11		*df*	*SS*	*MS*	*F*	*Significance F*
12	Regression	3	41971005.01	13990335	15.30	5.83667E-05
13	Residual	16	14630007.19	914375.4		
14	Total	19	56601012.2			
15						
16		*Coefficients*	*Standard Error*	*t Stat*	*P-value*	
17	Intercept	5670	645.1	8.79	1.6E-07	
18	Snowfall	28.03	11.53	2.43	0.0272	
19	Temperature	−16.62	19.36	−0.86	0.4033	
20	Time	229.7	37.09	6.19	1.29E-05	

MINITAB

Regression Analysis: Tickets versus Snowfall, Temperature, Time

The regression equation is
Tickets = 5670 + 28.0 Snowfall – 16.6 Temperature + 230 Time

Predictor	Coef	SE Coef	T	P
Constant	5669.9	645.1	8.79	0.000
Snowfall	28.03	11.53	2.43	0.027
Temperature	–16.62	19.36	–0.86	0.403
Time	229.73	37.09	6.19	0.000

S = 956.230 R-Sq = 74.2% R-Sq(adj) = 69.3%

Analysis of Variance

Source	DF	SS	MS	F	P
Regression	3	41971005	13990335	15.30	0.000
Residual Error	16	14630007	914375		

As we did before, we calculate the residuals and conduct regression diagnostics using Excel. The results are shown in Figure 16.9–16.11.

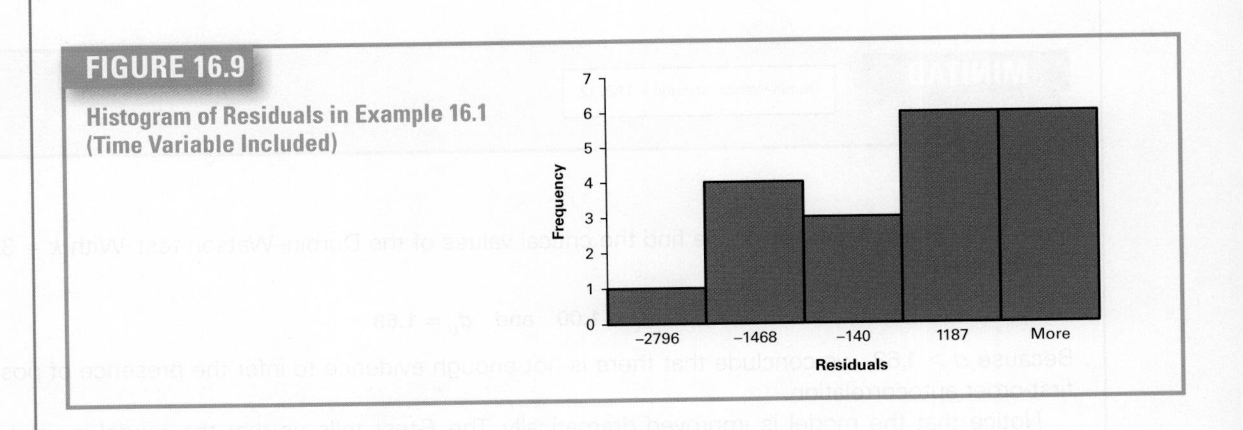

FIGURE 16.9

Histogram of Residuals in Example 16.1 (Time Variable Included)

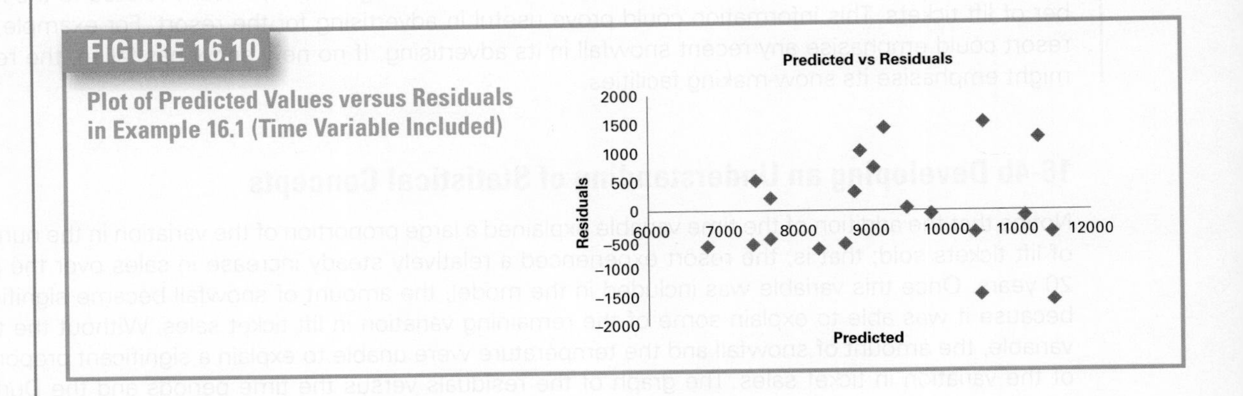

FIGURE 16.10

Plot of Predicted Values versus Residuals in Example 16.1 (Time Variable Included)

The error variable variance appears to be constant.

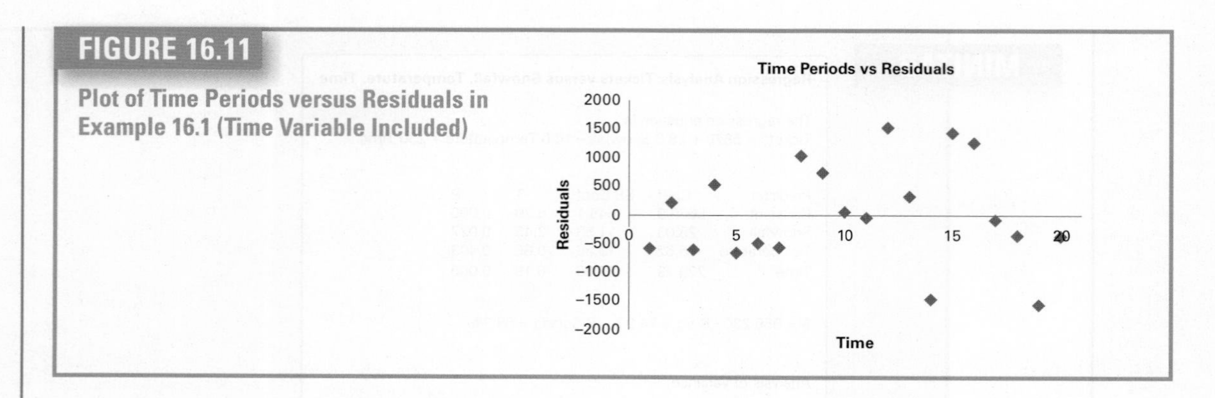

FIGURE 16.11

Plot of Time Periods versus Residuals in Example 16.1 (Time Variable Included)

There is no sign of autocorrelation. To confirm our diagnosis, we conducted the Durbin–Watson test.

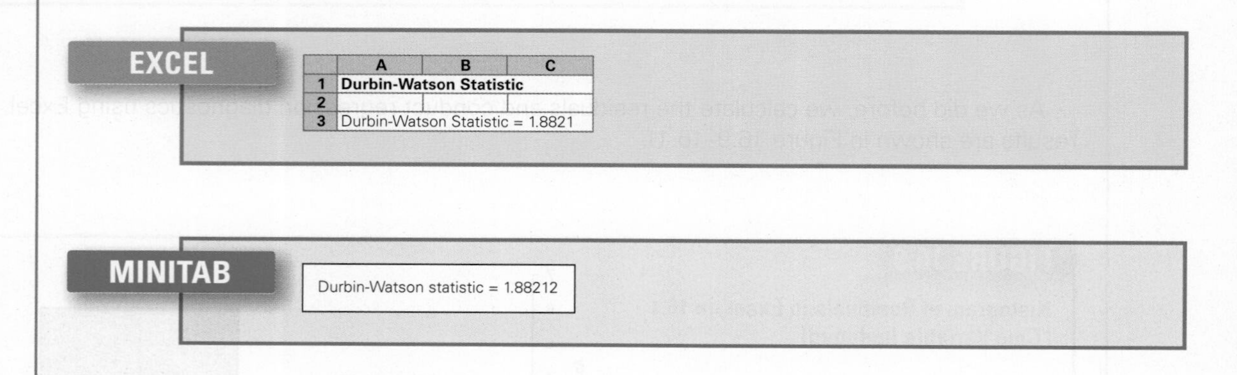

EXCEL

	A	B	C
1	**Durbin-Watson Statistic**		
2			
3	Durbin-Watson Statistic = 1.8821		

MINITAB

Durbin-Watson statistic = 1.88212

From Table 8(a) in Appendix A, we find the critical values of the Durbin–Watson test. With $k = 3$ and $n = 20$, we find

$$d_L = 1.00 \quad \text{and} \quad d_U = 1.68$$

Because $d > 1.68$, we conclude that there is not enough evidence to infer the presence of positive first-order autocorrelation.

Notice that the model is improved dramatically. The *F*-test tells us that the model is valid. The *t*-tests tell us that both the amount of snowfall and time are significantly linearly related to the number of lift tickets. This information could prove useful in advertising for the resort. For example, the resort could emphasise any recent snowfall in its advertising. If no new snow has fallen, the resort might emphasise its snow-making facilities.

16-4b Developing an Understanding of Statistical Concepts

Notice that the addition of the time variable explained a large proportion of the variation in the number of lift tickets sold; that is, the resort experienced a relatively steady increase in sales over the past 20 years. Once this variable was included in the model, the amount of snowfall became significant because it was able to explain some of the remaining variation in lift ticket sales. Without the time variable, the amount of snowfall and the temperature were unable to explain a significant proportion of the variation in ticket sales. The graph of the residuals versus the time periods and the Durbin–Watson test enabled us to identify the problem and correct it. In overcoming the autocorrelation problem, we improved the model so that we identified the amount of snowfall as an important variable in determining ticket sales. This result is quite common. Correcting a violation of a required condition will frequently improve the model.

EXERCISES

16.26 Perform the Durbin–Watson test at the 5% significance level to determine whether positive first-order autocorrelation exists when $d = 1.10$, $n = 25$, and $k = 3$.

16.27 Determine whether negative first-order autocorrelation exists when $d = 2.85$, $n = 50$ and $k = 5$. (Use a 1% significance level.)

16.28 Given the following information, perform the Durbin–Watson test to determine whether first-order autocorrelation exists:

$$n = 25, \quad k = 5, \quad \alpha = 0.10, \quad d = 0.90$$

16.29 Test the following hypotheses with $\alpha = 0.05$.

H_0: There is no first-order autocorrelation

H_1: There is positive first-order autocorrelation

$n = 50, \quad k = 2, \quad d = 1.38$

The following exercises require a computer and software.

16.30 Xr16-30 Observations of variables y, x_1 and x_2 were taken over 100 consecutive time periods.

a. Conduct a regression analysis of these data.

b. Plot the residuals versus the time periods. Describe the graph.

c. Perform the Durbin–Watson test. Is there evidence of autocorrelation? Use $\alpha = 0.10$.

d. If autocorrelation was detected in part (c), propose an alternative regression model to remedy the problem. Use the computer to generate the statistics associated with this model.

e. Redo parts (b) and (c). Compare the two models.

16.31 Xr16-31 Weekly sales of a company's product (y) and those of its main competitor (x) were recorded for 1 year.

a. Conduct a regression analysis of these data.

b. Plot the residuals versus the time periods. Does there appear to be autocorrelation?

c. Perform the Durbin–Watson test. Is there evidence of autocorrelation? Use $\alpha = 0.10$.

d. If autocorrelation was detected in part (c), propose an alternative regression model to remedy the problem. Use the computer to generate the statistics associated with this model.

e. Redo parts (b) and (c). Compare the two models.

16.32 Refer to Exercise 16.3. Is there evidence of positive first-order autocorrelation?

16-5 POLYNOMIAL MODELS

At the beginning of this chapter we introduced the multiple regression model:

$$y = \beta_0 + \beta_1 x_1 + \beta_2 x_2 + \cdots + \beta_k x_k + \varepsilon$$

We included variables $x_1, x_2, \ldots, x_k$ because we believed that these variables were each linearly related to the dependent variable. In this section, we discuss models where the independent variables may be functions of a smaller number of predictor variables. The simplest form of the **polynomial model** is described in the box.

Polynomial Model with One Predictor Variable

$$y = \beta_0 + \beta_1 x + \varepsilon$$

Technically, this is a linear regression model with one independent variable. A linear regression model is a polynomial regression model of order 1 (x has power 1).

However, in many applications, the relationship cannot be represented by a straight line, and therefore must be captured by an appropriate regression model (e.g. of different orders: second-order or quadratic, third-order or cubic, etc.). If all independent variables are based on only one variable, which we label the **predictor variable**; that is x, so that the polynomial model will be

$$y = \beta_0 + \beta_1 x + \beta_2 x^2 + \ldots \beta_p x^p + \varepsilon$$

In this model, p is the **order** of the equation. For reasons that we discuss later, we rarely propose a model whose order is greater than 3. However, it is worthwhile to devote individual attention to situations where $p = 1$, 2 and 3.

16-5a First-Order Model

When $p = 1$, we have the now-familiar simple linear regression model introduced in Chapter 15. It is also called the **first-order** polynomial model.

$$y = \beta_0 + \beta_1 x + \varepsilon$$

Obviously, this model is chosen when the statistics practitioner believes that there is a straight-line relationship between the dependent and independent variables over the range of the values of x.

16-5b Second-Order Model

With $p = 2$, the polynomial model is

$$y = \beta_0 + \beta_1 x + \beta_2 x^2 + \varepsilon$$

When we plot x versus y, the graph is shaped like a parabola, as shown in Figures 16.12 and 16.13. The coefficient β_0 represents the intercept where the response surface strikes the y-axis. The signs of β_1 and β_2 control the position of the parabola relative to the y-axis. If $\beta_1 = 0$, for example, the parabola is symmetric and centred around $x = 0$. If β_1 and β_2 have the same sign, the parabola shifts to the left. If β_1 and β_2 have opposite signs, the parabola shifts to the right. The coefficient β_2 describes the curvature. If $\beta_2 = 0$, there is no curvature. If β_2 is negative, the graph is concave (as in Figure 16.12). If β_2 is positive, the graph is convex (as in Figure 16.13). The greater the absolute value of β_2, the greater the rate of curvature, as can be seen in Figure 16.14.

FIGURE 16.12

Second-Order Model with $\beta_2 < 0$

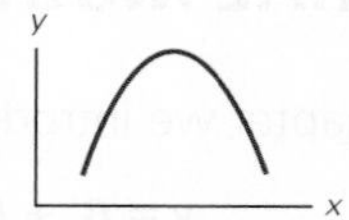

FIGURE 16.13

Second-Order Model with $\beta_2 > 0$

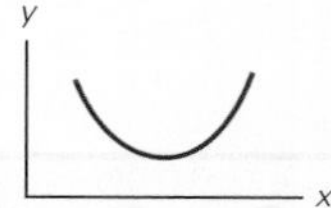

FIGURE 16.14

Second-Order Model with Various Values of β_2

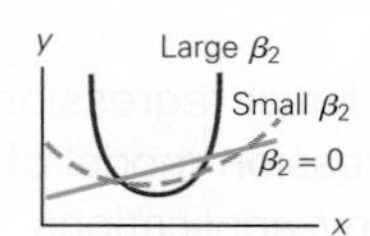

The second-order regression model with one predictor variable is appropriate when the slope (capturing the influence of x on y) changes in the magnitude as well as sign.

16-5c Third-Order Model

By setting $p = 3$, we produce the third-order model

$$y = \beta_0 + \beta_1 x + \beta_2 x^2 + \beta_3 x^3 + \varepsilon$$

Figures 16.15 and 16.16 depict this equation, whose curvature can change twice.

FIGURE 16.15

Third-Order Model with $\beta_3 < 0$

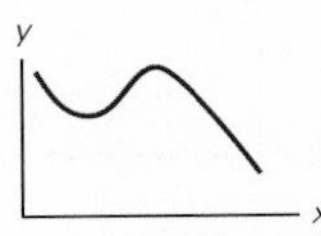

FIGURE 16.16

Third-Order Model with $\beta_3 > 0$

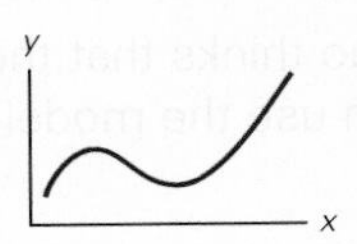

As you can see, when β_3 is negative, y is decreasing over the range of x, and when β_3 is positive, y increases. The other coefficients determine the position of the curvature changes and the point at which the curve intersects the y-axis.

The number of real-life applications of this model is quite small. Statistics practitioners rarely encounter problems involving more than one curvature reversal. Therefore, we will not discuss any higher order models.

16-5d Polynomial Models with Two Predictor Variables

If we believe that two predictor variables influence the dependent variable, we can use one of the following polynomial models. The general form of this model is rather cumbersome, so we will not show it. Instead we discuss several specific examples.

16-5e First-Order Model

The first-order model is represented by

$$y = \beta_0 + \beta_1 x_1 + \beta_2 x_2 + \varepsilon$$

This model is used whenever the statistics practitioner believes that, on average, y is linearly related to each of x_1 and x_2, and the predictor variables do not interact. (Recall that we introduced interaction in Chapter 13.) This means that the effect of one predictor variable on y is independent of the value of the second predictor variable. For example, suppose that the sample regression line of the first-order model is

$$\hat{y} = 5 + 3x_1 + 4x_2$$

If we examine the relationship between y and x_1 for several values of x_2 (say, $x_2 = 1$, 2 and 3), we produce the following equations.

x_2	$\hat{y} = 5 + 3x_1 + 4x_2$
1	$\hat{y} = 9 + 3x_1$
2	$\hat{y} = 13 + 3x_1$
3	$\hat{y} = 17 + 3x_1$

The only difference in the three equations is the intercept. (See Figure 16.17.) The coefficient of x_1 remains the same, which means that the effect of x_1 on y remains the same no matter what the value of x_2. (We could also have shown that the effect of x_2 on y remains the same no matter what the value of x_1.) As you can see from Figure 16.17, the first-order model with no interaction produces parallel straight lines.

FIGURE 16.17

First-Order Model with Two Independent Variables: No Interaction

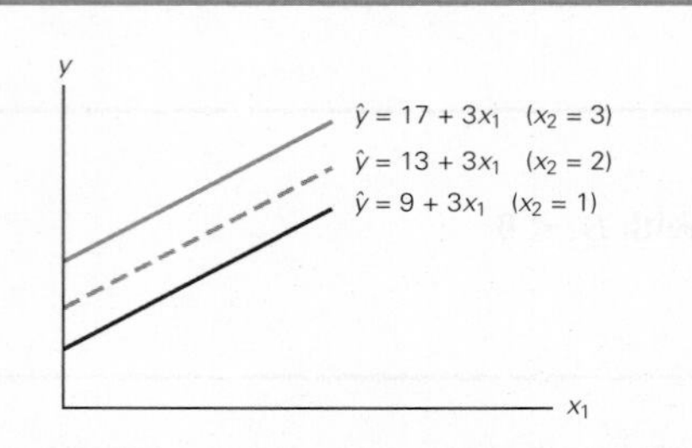

A statistics practitioner who thinks that the effect of one predictor variable on y is influenced by the other predictor variable can use the model described next.

16-5f First-Order Model with Two Predictor Variables and Interaction

Interaction means that the effect of x_1 on y is influenced by the value of x_2. (It also means that the effect of x_2 on y is influenced by x_1.)

First-Order Model with Interaction

$$y = \beta_0 + \beta_1 x_1 + \beta_2 x_2 + \beta_3 x_1 x_2 + \varepsilon$$

Suppose that the sample regression line is

$$\hat{y} = 5 + 3x_1 + 4x_2 - 2x_1 x_2$$

If we examine the relationship between y and x_1 for $x_2 = 1$, 2 and 3, we produce the following table of equations:

x_2	$\hat{y} = 5 + 3x_1 + 4x_2 - 2x_1x_2$
1	$\hat{y} = 9 + x_1$
2	$\hat{y} = 13 - x_1$
3	$\hat{y} = 17 - 3x_1$

As you can see, not only is the intercept different but also the coefficient of x_1 varies. Obviously, the effect of x_1 on y is influenced by the value of x_2. Figure 16.18 depicts these equations. The straight lines are clearly not parallel.

FIGURE 16.18

First-Order Model with Interaction

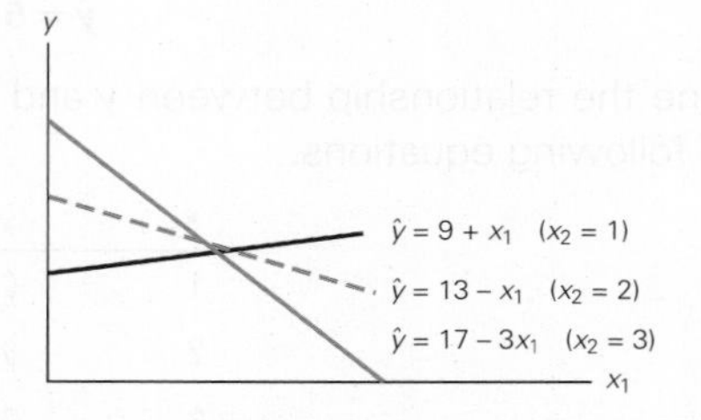

16-5g Second-Order Model with Interaction

A statistics practitioner who believes that a **quadratic relationship** exists between y and each of x_1 and x_2 and that the predictor variables interact in their effect on y can use the following **second-order** model.

Second-Order Model with Interaction

$$y = \beta_0 + \beta_1 x_1 + \beta_2 x_2 + \beta_3 x_1^2 + \beta_4 x_2^2 + \beta_5 x_1 x_2 + \varepsilon$$

Figures 16.19 and 16.20, respectively, depict this model without and with the interaction term.

FIGURE 16.19

Second-Order Model without Interaction

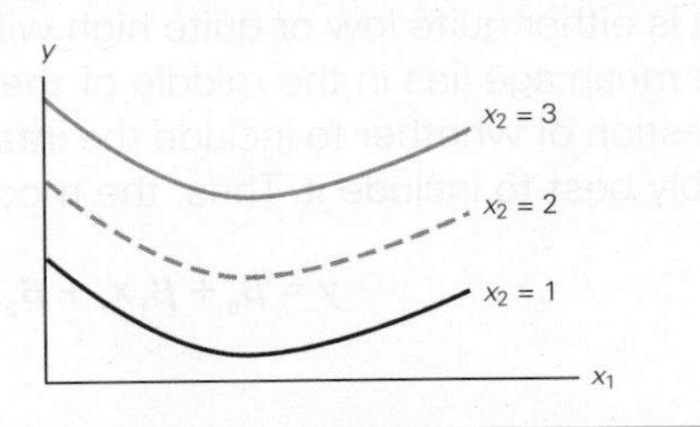

FIGURE 16.20

Second-Order Model with Interaction

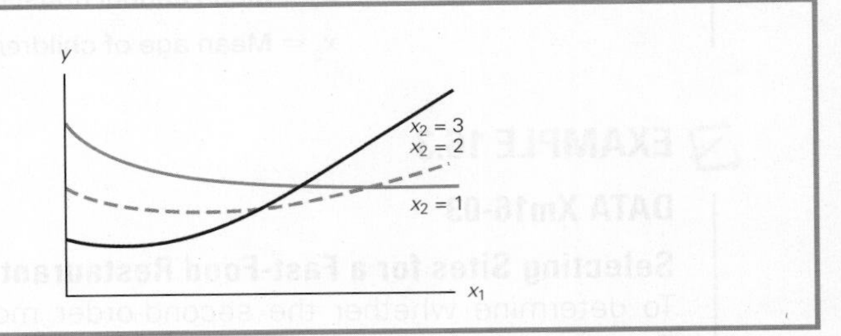

Now that we have introduced several different models, how do we know which model to use? The answer is that we employ a model based on our knowledge of the variables involved and then test that model using the statistical techniques presented in this and the preceding chapters.

EXAMPLE 16.2

Selecting Sites for a Fast-Food Restaurant, Part 1

In trying to find new locations for their restaurants, fast-food restaurant chains like McDonald's and Wendy's usually consider a number of factors. Suppose that an analyst working for a fast-food restaurant chain has been asked to construct a regression model that will help identify new locations that are likely to be profitable. The analyst knows that this type of restaurant has, as its primary market, middle-income adults and their children, particularly those between the ages of 5 and 12. Which model should the analyst propose?

SOLUTION:

The dependent variable is gross revenue or net profit. The predictor variables will be mean annual household income and the mean age of children in the restaurant's neighbourhood. The relationship between the dependent variable and each predictor variable is probably quadratic. In other words, members of relatively poor or relatively affluent households are less likely to eat at this chain's restaurants because the restaurants attract mostly middle-income customers. Figure 16.21 depicts the hypothesised relationship.

FIGURE 16.21

Relationship between Annual Gross Revenue and Mean Household Income

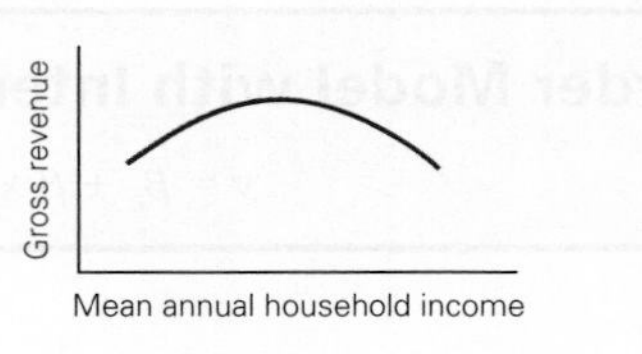

A similar relationship can be proposed for revenue and age. Neighbourhoods where the mean age of children is either quite low or quite high will probably produce lower revenues than in similar areas where the mean age lies in the middle of the 5-to-12 range.

The question of whether to include the interaction term is more difficult to answer. When in doubt, it is probably best to include it. Thus, the model to be tested is

$$y = \beta_0 + \beta_1 x_1 + \beta_2 x_2 + \beta_3 x_1^2 + \beta_4 x_2^2 + \beta_5 x_1 x_2 + \varepsilon$$

where

y = Annual gross sales

x_1 = Mean annual household income in the neighborhood

x_2 = Mean age of children in the neighbourhood

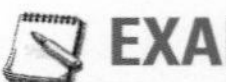

EXAMPLE 16.3

DATA Xm16-03

Selecting Sites for a Fast-Food Restaurant, Part 2

To determine whether the second-order model with interaction is appropriate, the analyst in Example 16.2 selected 25 areas at random. Each area consists of approximately 5000 households, as well as one of her employer's restaurants and three competing fast-food restaurants. The previous year's annual gross sales, the mean annual household income, and the mean age of children (the latter two figures are available from the latest census) were recorded; some of these data are listed here (the file also contains x_1^2, x_2^2 and $x_1 x_2$ What conclusions can be drawn from these data?

Area	Annual Gross Revenue (€1000s) y	Mean Annual Household (€1000s) x_1	Mean Age of Children x_2
1	1 128	23.5	10.5
2	1 005	17.6	7.2
3	1 212	26.3	7.6
⋮	⋮	⋮	⋮
25	950	17.8	6.1

SOLUTION:

Both Excel and Minitab were employed to produce the regression analysis shown here.

EXCEL

	A	B	C	D	E	F
1	SUMMARY OUTPUT					
2						
3	*Regression Statistics*					
4	Multiple R	0.9521				
5	R Square	0.9065				
6	Adjusted R Square	0.8819				
7	Standard Error	44.70				
8	Observations	25				
9						
10	ANOVA					
11		*df*	*SS*	*MS*	*F*	*Significance F*
12	Regression	5	368,140	73,628	36.86	3.86E-09
13	Residual	19	37,956	1,998		
14	Total	24	406,096			
15						
16		*Coefficients*	*Standard Error*	*t Stat*	*P-value*	
17	Intercept	−1134.0	320.0	−3.54	0.0022	
18	Income	173.20	28.20	6.14	6.66E-06	
19	Age	23.55	32.23	0.73	0.4739	
20	Income sq	−3.726	0.542	−6.87	1.48E-06	
21	Age sq	−3.869	1.179	−3.28	0.0039	
22	(Income)(Age)	1.967	0.944	2.08	0.0509	

MINITAB

Regression Analysis: Revenue Versus Income, Age, Inc-sq, Age-sq, Inc-Age

The regression equation is
Revenue = − 1134 + 173 Income + 23.5 Age − 3.73 Inc-sq − 3.87 Age-sq
+ 1.97 Inc-Age

Predictor	Coef	SE Coef	T	P
Constant	−1134.0	320.0	−3.54	0.002
Income	173.20	28.20	6.14	0.000
Age	23.55	32.23	0.73	0.474
Inc-sq	−3.7261	0.5422	−6.87	0.000
Age-sq	−3.869	1.179	−3.28	0.004
Inc-Age	1.9673	0.9441	2.08	0.051

S = 44.70 R-Sq = 90.7% R-Sq(adj) = 88.2%

Analysis of Variance

Source	DF	SS	MS	F	P
Regression	5	368,140	73,628	36.86	0.000
Residual Error	19	37,956	1,998		
Total	24	406,096			

INTERPRET

From the computer output, we determine that the value of the coefficient of determination (R^2) is 90.65%, which tells us that the model fits the data quite well. The value of the F-statistic is 36.86, which has a p-value of approximately 0. This confirms that the model is valid.

Care must be taken when interpreting the t-tests of the coefficients in this type of model. Not surprisingly, each variable will be correlated with its square, and the interaction variable will be correlated with both of its components. As a consequence, multicollinearity distorts the t-tests of the coefficients in some cases, making it appear that some of the components should be eliminated from the model. In fact, in Example 16.3, Minitab warns (not shown) the analyst that multicollinearity is a problem. However, in most such applications, the objective is to forecast the dependent variable and multicollinearity does not affect the model's fit or forecasting capability.

EXERCISES

The following two exercises require the use of a computer and statistical software. Use a 5% significance level.

16.33 Xr15-09 Exercise 15.9 addressed the problem of determining the relationship between the price of apartment buildings and number of square metres. Hoping to improve the predictive capability of the model, the estate agent also recorded the number of apartments, the age, and the number of floors.

a. Calculate the regression equation.

b. Is the model valid?

c. Compare your answer with that of Exercise 15.9.

16.34 Xr15–12 In Exercise 15.12, a statistics practitioner examined the relationship between office rents and the city's office vacancy rate. The model appears to be quite poor. It was decided to add another variable that measures the state of the economy. The city's unemployment rate was chosen for this purpose.

a. Determine the regression equation.

b. Determine the coefficient of determination and describe what this value means.

c. Test the model's validity in explaining office rent.

d. Determine which of the two independent variables is linearly related to rents.

e. Determine whether the error is normally distributed with a constant variance.

f. Determine whether there is evidence of autocorrelation.

g. Predict with 95% confidence the office rent in a city whose vacancy rate is 10% and whose unemployment rate is 7%.

16.35 Graph y versus x_1 for $x_2 = 1$, 2 and 3 for each of the following equations.

a. $y = 1 + 2x_1 + 4x_2$

b. $y = 1 + 2x_1 + 4x_2 - x_1x_2$

16.36 Graph y versus x_1 for $x_2 = 2$, 4 and 5 for each of the following equations.

a. $y = 0.5 + 1x_1 - 0.7x_2 - 1.2x_1^2 + 1.5x_2^2$

b. $y = 0.5 + 1x_1 - 0.7x_2 - 1.2x_1^2 + 1.5x_2^2 + 2x_1x_2$

The following exercises require the use of a computer and software.

16.37 Xr16-37 The general manager of a supermarket chain believes that sales of a product are influenced by the amount of space the product is allotted on shelves. If true, this would have great significance, because the more profitable items could be given more shelf space. The manager realises that sales volume would probably increase with more space only up to a certain point. Beyond that point, sales would likely flatten and perhaps decrease (because customers are often dismayed by very large displays). To test his belief, the manager records the number of boxes of detergent sold during 1 week in 25 stores in the chain. For each store, he records the shelf space (in inches) allotted to the detergent.

a. Write the equation that represents the model.

b. Discuss how well the model fits.

16.38 Xr16-12 Refer to Exercise 16.12. The dean of the school of business wanted to improve the regression model, which was developed to describe the relationship between MBA programme GPA and undergraduate GPA, GMAT score, and years of work experience. The dean now believes that an interaction effect may exist between undergraduate GPA and the GMAT test score.

a. Write the equation that describes the model.

b. Use a computer to generate the regression statistics. Use whatever statistics you deem necessary to assess the model's fit. Is this model valid?

c. Compare your results with those achieved in the original example.

16.39 Xr16-39 The manager of a large hotel on the Riviera in southern France wanted to forecast the monthly vacancy rate (as a percentage) during the peak season. After considering a long list of potential variables, she identified two variables that she believed were most closely related to the vacancy rate: the average daily temperature and the value of the currency in American dollars. She collected data for 25 months.

a. Perform a regression analysis using a first-order model with interaction.

b. Perform a regression analysis using a second-order model with interaction.

c. Which model fits better? Explain.

16.40 Xr16-40 The production manager of a chemical plant wants to determine the roles that temperature and pressure play in the yield of a particular chemical produced at the plant. From past experience, she believes that when pressure is held constant, lower and higher temperatures tend to reduce the yield. When temperature is held constant, higher and lower pressures tend to increase the yield.

She does not have any idea about how the yield is affected by various combinations of pressure and temperature. She observes 80 batches of the chemical in which the pressure and temperature were allowed to vary.

a. Which model should be used? Explain.

b. Conduct a regression analysis using the model you specified in part (a).

c. Assess how well the model fits the data.

16-6 NOMINAL INDEPENDENT VARIABLES

When we introduced regression analysis, we pointed out that all the variables must be interval. But in many real-life cases, one or more independent variables are nominal. For example, suppose that the used-car dealer in Example 15.2 believed that the colour of a car is a factor in determining its auction price. Colour is clearly a nominal variable. If we assign numbers to each possible colour, these numbers will be completely arbitrary, and using them in a regression model will usually be pointless. For example, suppose the dealer believes the colours that are most popular, white and silver, are likely to lead to different prices for other colours. Accordingly, he assigns a code of 1 to white cars, a code of 2 to silver cars, and a code of 3 to all other colours. If we now conduct a multiple regression analysis using odometer reading and colour as independent variables, the following results would be obtained. (File Xm15-02 contains these data. Interested readers can produce the following regression equation.)

$$\hat{y} = 10.4 - 0.0252x_1 - 0.0260x_2$$

Aside from the inclusion of the variable x_2, this equation is very similar to the one we produced in the simple regression model ($\hat{y} = 10.349 - 0.0250x_1$). The *t*-test of colour (*t*-statistic $= -1.11$ and *p*-value $= 0.2694$) indicates that there is not enough evidence to infer that colour is not linearly related to price. There are two possible explanations for this result. First, there is no relationship between colour and price. Second, colour is a factor in determining the car's price, but the way in which the dealer assigned the codes to the colours made detection of that fact impossible – that is, the dealer treated the nominal variable, colour, as an interval variable. To further understand why we cannot use nominal data in regression analysis, try to interpret the coefficient of colour. Such an effort is similar to attempting to interpret the mean of a sample of nominal data. It is futile. Even though this effort failed, it is possible to include nominal variables in the regression model. This is accomplished through the use of *indicator variables.*

An **indicator variable** (also called a **dummy variable**) is a variable that can assume either one of only two values (usually 0 and 1), where 1 represents the existence of a certain condition and 0 indicates that the condition does not hold. In this illustration, we would create two indicator variables to represent the colour of the car:

$$I_1 = \begin{cases} 1 & \text{(if color is white)} \\ 0 & \text{(if color is not white)} \end{cases}$$

and

$$I_2 = \begin{cases} 1 & \text{(if color is silver)} \\ 0 & \text{(if color is not silver)} \end{cases}$$

Notice that we need only two indicator variables to represent the three categories. A white car is represented by $I_1 = 1$ and $I_2 = 0$. A silver car is represented by $I_1 = 0$ and $I_2 = 1$. Because cars that are painted some other colour are neither white nor silver, they are represented by $I_1 = 0$ and $I_2 = 0$. It should be apparent that we cannot have $I_1 = 1$ and $I_2 = 1$, as long as we assume that no Toyota Auris is two-toned.

The effect of using these two indicator variables is to create three equations, one for each of the three colours. As you're about to discover, we can use the equations to determine how the car's colour relates to its auction selling price.

In general, to represent a nominal variable with m categories, we must create $m - 1$ indicator variables. The last category represented by $I_1 = I_2 = \cdots = I_{m-1} = 0$ is called the **omitted category.**

16-6a Interpreting and Testing the Coefficients of Indicator Variables

In file Xm15-02a, we stored the values of I_1 and I_2. We then performed a multiple regression analysis using the variables odometer reading (x), I_1 and I_2.

EXCEL

	A	B	C	D	E	F
1	SUMMARY OUTPUT					
2						
3	*Regression Statistics*					
4	Multiple R	0.8371				
5	R Square	0.7008				
6	Adjusted R Square	0.6914				
7	Standard Error	0.1826				
8	Observations	100				
9						
10	ANOVA					
11		*df*	*SS*	*MS*	*F*	*Significance F*
12	Regression	3	7.49	2.50	74.95	4.65244E-25
13	Residual	96	3.20	0.033		
14	Total	99	10.69			
15						
16		*Coefficients*	*Standard Error*	*t Stat*	*P-value*	
17	Intercept	10.102	0.118	85.42	2.28E-92	
18	Odometer	−0.0222	0.0019	−11.67	4.04E-20	
19	I-1	0.0547	0.0437	1.25	0.214257	
20	I-2	0.1982	0.0490	4.05	0.0001	

MINITAB

Regression Analysis: Price versus Odometer, I-1, I-2

The regression equation is
Price = 10.1 – 0.0222 Odometer + 0.0547 I-1 + 0.198 I-2

Predictor	Coef	SE Coef	T	P
Constant	10.1023	0.1183	85.42	0.000
Odometer	–0.022171	0.001899	–11.67	0.000
I-1	0.05468	0.04373	1.25	0.214
I-2	0.19822	0.04899	4.05	0.000

S = 0.182555 R-Sq = 70.1% R-Sq(adj) = 69.1%

Analysis of Variance

Source	DF	SS	MS	F	P
Regression	3	7.4934	2.4978	74.95	0.000
Residual Error	96	3.1993	0.0333		
Total	99	10.6927			

The regression equation is

$$\hat{y} = 16.837 - 0.0591x + 0.0911I_1 + 0.3304I_2$$

$$\hat{y} = 10.1 - 0.0222x_1 + 0.0547I_1 + 0.198I_2$$

The intercept and the coefficient of odometer reading are interpreted in the usual manner. The intercept ($b_0 = 10.1$) is meaningless in the context of this problem. The coefficient of the odometer reading ($b_1 = -0.0222$) tells us that for each additional mile on the odometer, the auction price decreases an average of 2.22 pence, holding the colour constant. Now examine the remaining two coefficients:

$$b_2 = 0.0547$$

$$b_3 = 0.198$$

Recall that we interpret the coefficients in a multiple regression model by holding the other variables constant. In this example, we interpret the coefficient of I_1 as follows. In this sample, a white Auris sells for 0.0547 thousand or £54.70 on average more than other colours (non-white, non-silver) with the same odometer reading. A silver car sells for £198 on average more than other colours with the same odometer reading. The reason both comparisons are made with other colours is that such cars are represented by $I_1 = I_2 = 0$. Thus, for a non-white and non-silver car, the equation becomes

$$\hat{y} = 10.1 - 0.0222x_1 + 0.0547(0) + 0.198(0)$$

which is

$$\hat{y} = 10.1 - 0.0222x_1$$

For a white car ($I_1 = 1$ and $I_2 = 0$), the regression equation is

$$\hat{y} = 10.1 - 0.0222x_1 + 0.0547(1) + 0.198(0)$$

which is

$$\hat{y} = 10.1547 - 0.0222x_1$$

Finally, for a silver car ($I_1 = 0$ and $I_2 = 1$), the regression equation is

$$\hat{y} = 10.1 - 0.0222x_1 + 0.0547(0) + 0.198(1)$$

which simplifies to

$$\hat{y} = 10.298 - 0.0222\,x_1$$

Figure 16.22 depicts the graph of price versus odometer reading for the three different colour categories. Notice that the three lines are parallel (with slope = −0.0222) while the intercepts differ.

FIGURE 16.22

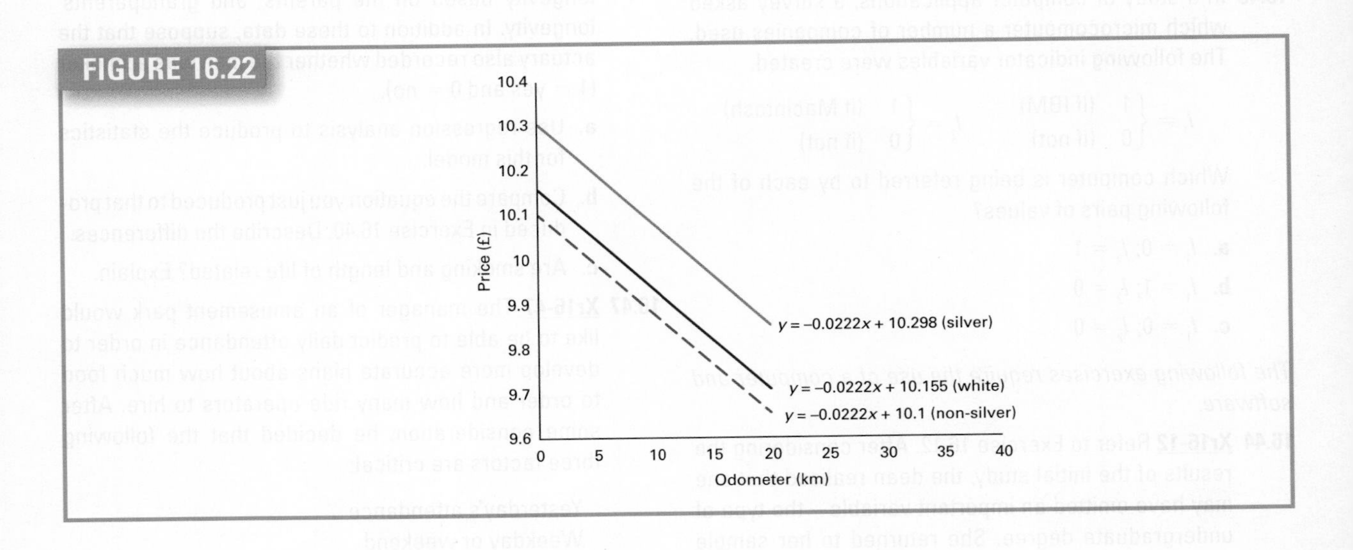

We can also perform *t*-tests on the coefficients of I_1 and I_2. However, because the variables I_1 and I_2 represent different groups (the three colour categories), these *t*-tests allow us to draw inferences about the differences in auction selling prices between the groups for the entire population of similar 3-year-old Toyota Auris.

The test of the coefficient of I_1, which is β_2, is conducted as follows:

$$H_0: \quad \beta_2 = 0$$

$$H_1: \quad \beta_2 \neq 0$$

Test statistic: $t = 1.25$, p-value $= 0.2143$

There is insufficient evidence to infer that white Auris have a different mean selling price than do Auris in the omitted category in the population of 3-year-old Auris with the same odometer reading.

To determine whether silver-coloured Auris sell for a different price than Auris in the other colour category, we test

$$H_0: \beta_3 = 0$$
$$H_1: \beta_3 \neq 0$$

Test statistic: $t = 4.05$, p-value $= 0.0001$.

We can conclude that there are differences in the mean auction selling prices between all 3-year-old, silver-coloured Auris and the omitted colour category with the same odometer readings.

EXERCISES

16.41 How many indicator variables must be created to represent a nominal independent variable that has five categories?

16.42 Create and identify indicator variables to represent the following nominal variables.

a. Religious affiliation (Catholic, Protestant and others)

b. Working shift (8 am to 4 pm, 4 pm to 12 midnight, and 12 midnight to 8 am)

c. Supervisor (Jack Jones, Mary Brown, George Fosse, and Elaine Smith)

16.43 In a study of computer applications, a survey asked which microcomputer a number of companies used. The following indicator variables were created.

$$I_1 = \begin{cases} 1 & \text{(if IBM)} \\ 0 & \text{(if not)} \end{cases} \qquad I_2 = \begin{cases} 1 & \text{(if Macintosh)} \\ 0 & \text{(if not)} \end{cases}$$

Which computer is being referred to by each of the following pairs of values?

a. $I_1 = 0$; $I_2 = 1$

b. $I_1 = 1$; $I_2 = 0$

c. $I_1 = 0$; $I_2 = 0$

The following exercises require the use of a computer and software.

16.44 Xr16-12 Refer to Exercise 16.12. After considering the results of the initial study, the dean realised that she may have omitted an important variable – the type of undergraduate degree. She returned to her sample of students and recorded the type of undergraduate degree using the following codes:

1 = BA
2 = BBA (including similar business or management degrees)
3 = BEng or BSc
4 = Other (including no undergraduate degree)

These data were included with the data from the original example. Can the dean conclude that the undergraduate degree is a factor in determining how well a student performs in the MBA programme?

16.45 Xr16-12 Refer to Exercise 16.12.

a. Predict with 95% confidence the MBA programme GPA of a BEng whose undergraduate GPA was 9.0, whose GMAT score was 700, and who has had 10 years of work experience.

b. Repeat part (a) for a BA student.

16.46 Xr16-09 Refer to Exercise 16.9, where a multiple regression analysis was performed to predict men's longevity based on the parents' and grandparents' longevity. In addition to these data, suppose that the actuary also recorded whether the man was a smoker (1 = yes and 0 = no).

a. Use regression analysis to produce the statistics for this model.

b. Compare the equation you just produced to that produced in Exercise 16.40. Describe the differences.

c. Are smoking and length of life related? Explain.

16.47 Xr16-47 The manager of an amusement park would like to be able to predict daily attendance in order to develop more accurate plans about how much food to order and how many ride operators to hire. After some consideration, he decided that the following three factors are critical:

Yesterday's attendance
Weekday or weekend
Predicted weather

He then took a random sample of 40 days. For each day, he recorded the attendance, the previous day's attendance, day of the week, and weather forecast. The first independent variable is interval, but the other two are nominal. Accordingly, he created the following sets of indicator variables:

$$I_1 = \begin{cases} 1 & \text{(if weekend)} \\ 0 & \text{(if not)} \end{cases}$$

$$I_2 = \begin{cases} 1 & \text{(if mostly sunny is predicted)} \\ 0 & \text{(if not)} \end{cases}$$

$$I_3 = \begin{cases} 1 & \text{(if rain is predicted)} \\ 0 & \text{(if not)} \end{cases}$$

a. Conduct a regression analysis.

b. Is this model valid? Explain.

c. Can we conclude that weather is a factor in determining attendance?

d. Do these results provide sufficient evidence that weekend attendance is, on average, larger than weekday attendance?

16.48 Xr16-48 Profitable banks are ones that make good decisions on loan applications. *Credit scoring* is the statistical technique that helps banks make that decision. However, many branches overturn credit scoring recommendations, whereas other banks do not use the technique. In an attempt to determine the factors that affect loan decisions, a statistics practitioner surveyed 100 banks and recorded the percentage of bad loans (any loan that is not completely repaid), the average size of the loan, and whether a scorecard isused, and, if so, whether scorecard recommendations are overturned more than 10% of the time. These results are stored in columns 1 (percentage good loans), 2 (average loan) and 3 (code 1 = no scorecard, 2 = scorecard overturned more than 10% of the time and 3 = scorecard overturned less than 10% of the time).

a. Create indicator variables to represent the codes.

b. Perform a regression analysis.

c. How well does the model fit the data?

d. Is multicollinearity a problem?

e. Interpret and test the coefficients. What does this tell you?

f. Predict with 95% confidence the percentage of bad loans for a bank whose average loan is £10 000 and that does not use a scorecard.

CHAPTER SUMMARY

The multiple regression model extends the model introduced in Chapter 15. The statistical concepts and techniques are similar to those presented in simple linear regression. We assess the model in three ways: standard error of estimate, the coefficient of determination (and the coefficient of determination adjusted for degrees of freedom), and the *F*-test of the analysis of variance. We can use the *t*-tests of the coefficients to determine whether each of the independent variables is linearly related to the dependent variable. As we did in Chapter 15, we showed how to diagnose violations of the required conditions and to identify other problems. We introduced multicollinearity and demonstrated its effect and its remedy. Finally, we presented the Durbin–Watson test to detect first-order autocorrelation and the polynomial models with one and two independent variables.

SYMBOLS:

Symbol	Pronounced	Represents
β_i	Beta sub *i* or beta *i*	Coefficient of *i*th independent variable
b_i	*b* sub *i* or *b i*	Sample coefficient

FORMULAS:

Standard error of estimate

$$s_\varepsilon = \sqrt{\frac{\text{SSE}}{n-k-1}}$$

Test statistic for β_1

$$t = \frac{b_i - \beta_i}{s_{b_i}}$$

i

Coefficient of determination

$$R^2 = \frac{s_{xy}^2}{s_x^2 s_y^2} = 1 - \frac{SSE}{\sum(y_i - \bar{y})^2}$$

Adjusted coefficient of determination

$$\text{Adjusted } R^2 = 1 - \frac{SSE/(n-k-1)}{\sum(y_i - \bar{y})^2/(n-1)}$$

Mean square for error

$$MSE = SSK/k$$

Mean square for regression

$$MSR = SSR/(n-k-1)$$

***F*-statistic**

$$F = MSR/MSE$$

Durbin–Watson statistic

$$d = \frac{\sum_{i=1}^{m}(e_i - e_{i-1})^2}{\sum_{i=1}^{n} e_i^2}$$

PART FIVE

NONPARAMETRIC STATISTICS FOR ANALYSIS, CONTROL AND FORECASTING

17 Nonparametric Statistics
18 Time-Series Analysis and Forecasting
19 Statistical Process Control
20 Decision Analysis
21 Conclusion

17 NONPARAMETRIC STATISTICS

17-1 WILCOXON RANK SUM TEST

17-2 SIGN TEST AND WILCOXON SIGNED RANK SUM TEST

17-3 KRUSKAL–WALLIS TEST AND FRIEDMAN TEST

17-4 SPEARMAN RANK CORRELATION COEFFICIENT

SOCIAL SURVEY: DOES READING BUSINESS NEWSPAPERS IMPROVE STUDENTS' EXAM SCORES IN STATISTICS?

DATA Xm17-00

A researcher has heard that reading certain business newspapers can have a positive effect of increasing exam scores in those students studying statistics. The researcher has identified a well-known business newspaper and recruited a group of undergraduate students with similar exam results and separated them in three groups (Group A, Group B and Group C). One of the questions asked in the survey was:

How often do you read the newspaper (1 = Every day, 2 = A few times a week, 3 = Once a week, 4 = Less than once a week, 5 = Never)?

After we introduce the appropriate statistical technique, we will provide our answer. **See Section 17-3 for the answer.**

INTRODUCTION

Throughout this book, we have presented statistical techniques that are used when the data are either interval or nominal. In this chapter statistical techniques that deal with ordinal data will be introduced. We will present three methods that compare two populations, two procedures used to compare two or more populations and a technique to analyse the relationship between two variables. As you have seen when we compare two or more populations of interval data, we measure the difference between means. However, as we discussed in Chapter 2, when the data are ordinal, the mean is not an appropriate measure of location. As a result, the methods in this chapter do not enable us to test the difference in population means; instead, we will test characteristics of populations without referring to specific parameters. For this reason, these techniques are called **nonparametric techniques**. Rather than testing to determine whether the population means differ, we will test to determine whether the *population locations* differ. Although nonparametric methods are designed to test ordinal data, they have another area of application. The statistical tests described in Sections 12-4 and 12-6 and in Chapter 13 require that the populations be normally distributed. If the data are extremely non-normal, the *t*-tests and *F*-test are invalid. Fortunately, nonparametric techniques can be used instead. For this reason, nonparametric procedures are often (perhaps more accurately) called **distribution-free statistics**. The techniques presented here can be used when the data are interval and the required condition of normality is unsatisfied. In such circumstances, we will treat the interval data as if they were ordinal. For this reason, even when the data are interval and the mean is the appropriate measure of location, we will choose instead to test population locations.

Figure 17.1 depicts the distributions of two populations when their locations are the same. Notice that because we do not know (or care) anything about the shape of the distributions, we represent them as non-normal. Figure 17.2 describes a circumstance when the location of population 1 is to the right of the location of population 2. The location of population 1 is to the left of the location of population 2 in Figure 17.3.

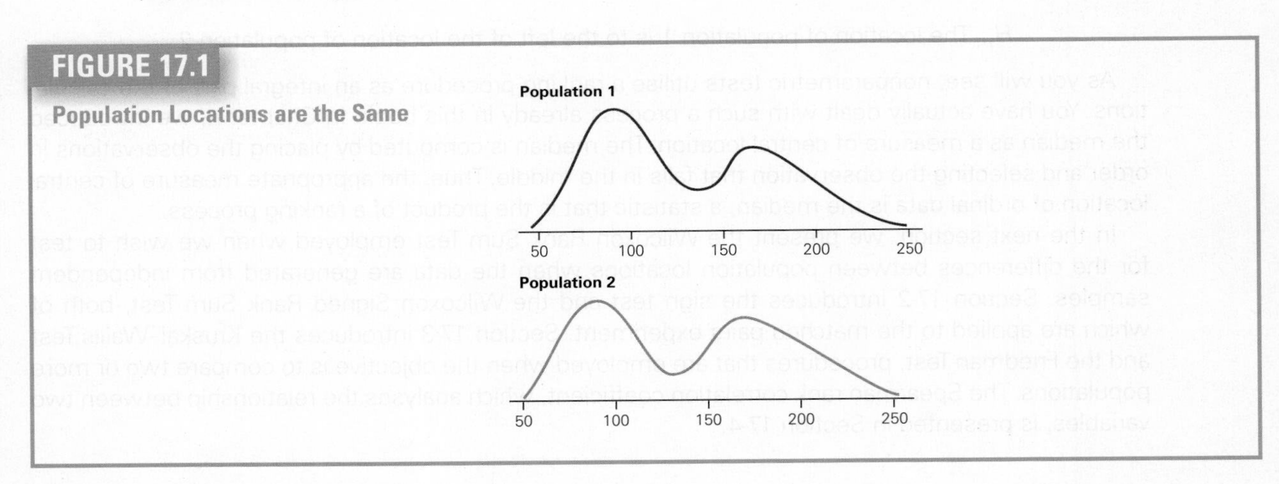

FIGURE 17.1

Population Locations are the Same

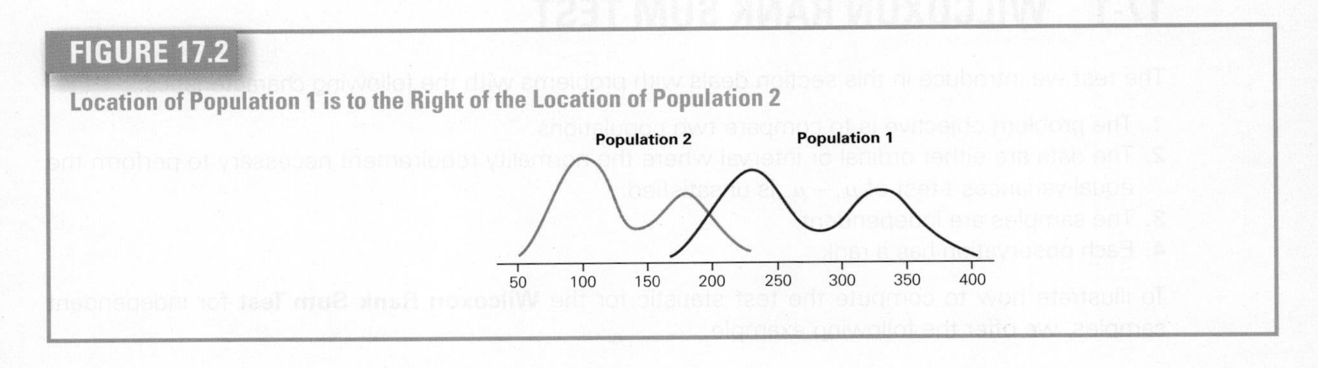

FIGURE 17.2

Location of Population 1 is to the Right of the Location of Population 2

FIGURE 17.3

Location of Population 1 is to the Left of the Location of Population 2

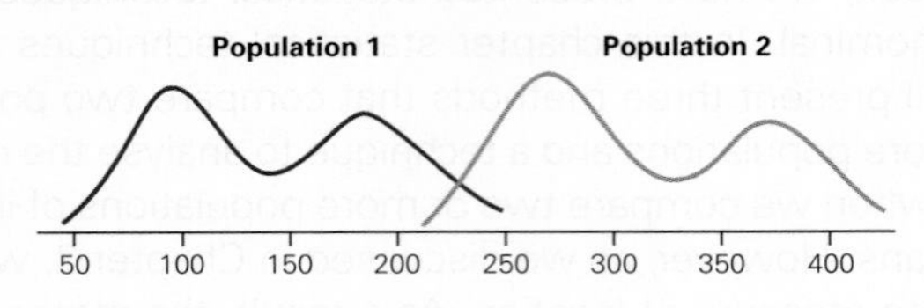

When the problem objective is to compare two populations, the null hypothesis will state

H_0: The two population locations are the same

The alternative hypothesis can take on any one of the following three forms:

1. If we want to know whether there is sufficient evidence to infer that there is a difference between the two populations, the alternative hypothesis is

   H_1: The location of population 1 is different from the location of population 2

2. If we want to know whether we can conclude that the random variable in population 1 is larger in general than the random variable in population 2 (see Figure 17.2), the alternative hypothesis is

   H_1: The location of population 1 is to the right of the location of population 2

3. If we want to know whether we can conclude that the random variable in population 1 is smaller in general than the random variable in population 2 (see Figure 17.3), the alternative hypothesis is

   H_1: The location of population 1 is to the left of the location of population 2

As you will see, nonparametric tests utilise a ranking procedure as an integral part of the calculations. You have actually dealt with such a process already in this book. In Chapter 4, we introduced the median as a measure of central location. The median is computed by placing the observations in order and selecting the observation that falls in the middle. Thus, the appropriate measure of central location of ordinal data is the median, a statistic that is the product of a ranking process.

In the next section, we present the Wilcoxon Rank Sum Test employed when we wish to test for the differences between population locations when the data are generated from independent samples. Section 17-2 introduces the sign test and the Wilcoxon Signed Rank Sum Test, both of which are applied to the matched pairs experiment. Section 17-3 introduces the Kruskal–Wallis Test and the Friedman Test, procedures that are employed when the objective is to compare two or more populations. The Spearman rank correlation coefficient, which analyses the relationship between two variables, is presented in Section 17-4.

17-1 WILCOXON RANK SUM TEST

The test we introduce in this section deals with problems with the following characteristics:

1. The problem objective is to compare two populations.
2. The data are either ordinal or interval where the normality requirement necessary to perform the equal-variances *t*-test of $\mu_1 - \mu_2$ is unsatisfied.
3. The samples are independent.
4. Each observation has a rank.

To illustrate how to compute the test statistic for the **Wilcoxon Rank Sum Test** for independent samples, we offer the following example.

EXAMPLE 17.1

Wilcoxon Rank Sum Test

Suppose that we want to determine whether the following observations drawn from two populations allow us to conclude at the 5% significance level that the location of population 1 is to the left of the location of population 2.

Sample 1:	22	23	20
Sample 2:	18	27	26

We want to test the following hypotheses:

H_0: The two population locations are the same

H_1: The location of population 1 is to the left of the location of population 2

Test Statistic

The first step is to rank all six observations, with rank 1 assigned to the smallest observation and rank 6 to the largest.

Sample 1	Rank	Sample 2	Rank
22	3	18	1
23	4	27	6
20	2	26	5
	$T_1 = 9$		$T_2 = 12$

Observe that 18 is the smallest number, so it receives a rank of 1; 20 is the second-smallest number, and it receives a rank of 2. We continue until rank 6 is assigned to 27, which is the largest of the observations. In case of ties, we average the ranks of the tied observations. The second step is to calculate the sum of the ranks of each sample. The rank sum of sample 1, denoted as T_1 is 9. The rank sum of sample 2, denoted as T_2, is 12. (Note that T_1 plus T_2 must equal the sum of the integers from 1 to 6, which is 21.) We can use either rank sum as the test statistic. We arbitrarily select T_1 as the test statistic and label it T. The value of the test statistic in this example is $T = T_1 = 9$.

Sampling Distribution of the Test Statistic

A small value of T indicates that most of the smaller observations are in sample 1 and that most of the larger observations are in sample 2. This would imply that the location of population 1 is to the left of the location of population 2. Therefore, in order for us to conclude statistically that this is the case, we need to show that T is small. The definition of 'small' comes from the sampling distribution of T. As we did in Section 9-1 when we derived the sampling distribution of the sample mean, we can derive the sampling distribution of T by listing all possible values of T. In Table 17.1, we show all possible rankings of two samples of size 3.

TABLE 17.1 All Possible Ranks and Rank Sums of Two Samples of Size 3

RANKS OF SAMPLE 1	RANK SUM	RANKS OF SAMPLE 2	RANK SUM
1,2,3	6	4,5,6	15
1,2,4	7	3,5,6	14
1,2,5	8	3,4,6	13
1,2,6	9	3,4,5	12
1,3,4	8	2,5,6	13
1,3,5	9	2,4,6	12
1,3,6	10	2,4,5	11

(continued)

RANKS OF SAMPLE 1	RANK SUM	RANKS OF SAMPLE 2	RANK SUM
1,4,5	10	2,3,6	11
1,4,6	11	2,3,5	10
1,5,6	12	2,3,4	9
2,3,4	9	1,5,6	12
2,3,5	10	1,4,6	11
2,3,6	11	1,4,5	10
2,4,5	11	1,3,6	10
2,4,6	12	1,3,5	9
2,5,6	13	1,3,4	8
3,4,5	12	1,2,6	9
3,4,6	13	1,2,5	8
3,5,6	14	1,2,4	7
4,5,6	15	1,2,3	6

If the null hypothesis is true and the two population locations are identical, then it follows that each possible ranking is equally likely. Because there are 20 different possibilities, each value of T has the same probability, namely 1/20. Notice that there is one value of 6, one value of 7, two values of 8, and so on. Table 17.2 summarises the values of T and their probabilities, and Figure 17.4 depicts this sampling distribution.

TABLE 17.2 Sampling Distribution of T with Two Samples of Size 3

T	$P(T)$
6	1/20
7	1/20
8	2/20
9	3/20
10	3/20
11	3/20
12	3/20
13	2/20
14	1/20
15	1/20
Total	1

FIGURE 17.4 Sampling Distribution of T with Two Samples of Size 3

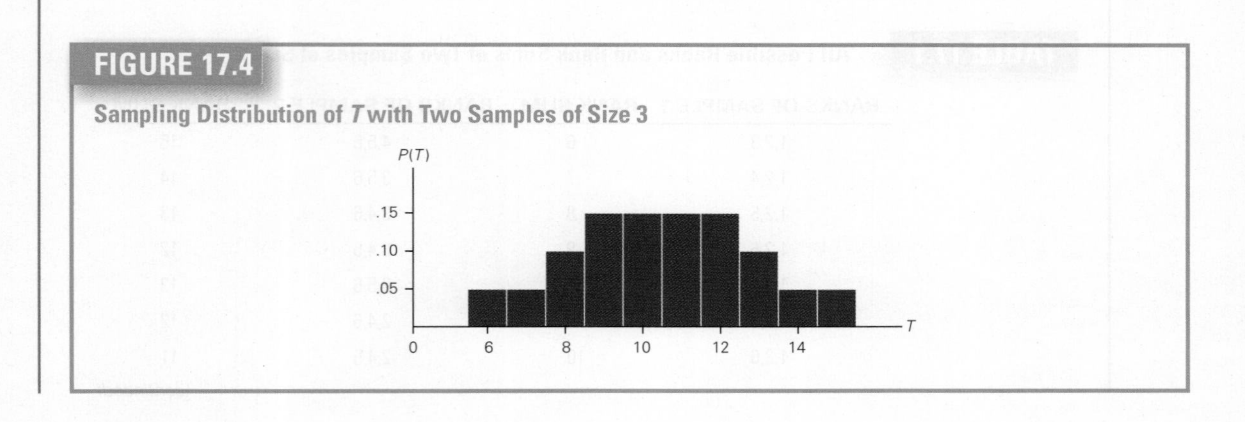

From this sampling distribution we can see that $P(T \leq 6) = P(T = 6) = 1/20 = 0.05$. Because we are trying to determine whether the value of the test statistic is small enough for us to reject the null hypothesis at the 5% significance level, we specify the rejection region as $T \leq 6$. Because $T = 9$, we cannot reject the null hypothesis.

Statisticians have generated the sampling distribution of T for various combinations of sample sizes. The critical values are provided in Table 9 in Appendix A and reproduced here as Table 17.3. Table 17.3 provides values of T_L and T_U for sample sizes between three and ten (n_1 is the size of sample 1, and n_2 is the size of sample 2). The values of T_L and T_U in part (a) of the table are such that

$$P(T \leq T_L) = P(T \geq T_U) = 0.025$$

The values of T_L and T_U in part (b) of the table are such that

$$P(T \leq T_L) = P(T \geq T_U) = 0.05$$

Part (a) is used either in a two-tail test with $\alpha = 0.05$ or in a one-tail test with $\alpha = 0.025$. Part (b) is employed either in a two-tail test with $\alpha = 0.10$ or in a one-tail test with $\alpha = 0.05$. Because no other values are provided, we are restricted to those values of α.

TABLE 17.3 **Critical Values of the Wilcoxon Rank Sum Test**

(a) $\alpha = 0.025$ one-tail; $\alpha = 0.05$ two-tail

n_1/n_2	3		4		5		6		7		8		9		10	
	T_L	T_U	T_L	T_U	T_L	T_U	T_L	T_U	T_L	T_U	T_L	T_U	T_L	T_U	T_L	T_U
4	6	18	11	25	17	33	23	43	31	53	40	64	50	76	61	89
5	6	21	12	28	18	37	25	47	33	58	42	70	52	83	64	96
6	7	23	12	32	19	41	26	52	35	63	44	76	55	89	66	104
7	7	26	13	35	20	45	28	56	37	68	47	81	58	95	70	110
8	8	28	14	38	21	49	29	61	39	73	49	87	60	102	73	117
9	8	31	15	41	22	53	31	65	41	78	51	93	63	108	76	124
10	9	33	16	44	24	56	32	70	43	83	54	98	66	114	79	131

(b) $\alpha = 0.05$ one-tail; $\alpha = 0.10$ two-tail

n_1/n_2	3		4		5		6		7		8		9		10	
	T_L	T_U	T_L	T_U	T_L	T_U	T_L	T_U	T_L	T_U	T_L	T_U	T_L	T_U	T_L	T_U
3	6	15	11	21	16	29	23	37	31	46	39	57	49	68	60	80
4	7	17	12	24	18	32	25	41	33	51	42	62	52	74	63	87
5	7	20	13	27	19	36	26	46	35	56	45	67	55	80	66	94
6	8	22	14	30	20	40	28	50	37	61	47	73	57	87	69	101
7	9	24	15	33	22	43	30	54	39	66	49	79	60	93	73	107
8	9	27	16	36	24	46	32	58	41	71	52	84	63	99	76	114
9	10	29	17	39	25	50	33	63	43	76	54	90	66	105	79	121
10	22	31	18	42	26	54	35	67	46	80	57	95	69	111	83	127

Although it is possible to derive the sampling distribution of the test statistic for any other sample sizes, the process can be quite tedious. Fortunately it is also unnecessary. Statisticians have shown that when the sample sizes are larger than ten, the test statistic is approximately normally distributed with mean $E(T)$ and standard deviation σ_T where

$$E(T) = \frac{n_1(n_1 + n_2 + 1)}{2}$$

And

$$\sigma_T = \sqrt{\frac{n_1 n_2 (n_1 + n_2 + 1)}{12}}$$

Thus, the standardised test statistic is

$$z = \frac{T - E(T)}{\sigma_T}$$

$$z = \frac{T - E(T)}{\sigma_T}$$

EXAMPLE 17.2

DATA Xm17-02

Comparing Pharmaceutical Painkillers

A pharmaceutical company is planning to introduce a new painkiller. In a preliminary experiment to determine its effectiveness, 30 people were randomly selected, of whom 15 were given the new painkiller and 15 were given aspirin. All 30 were told to use the drug when headaches or other minor pains occurred and to indicate which of the following statements most accurately represented the effectiveness of the drug they took:

5 = The drug was extremely effective.
4 = The drug was quite effective.
3 = The drug was somewhat effective.
2 = The drug was slightly effective.
1 = The drug was not at all effective.

The responses are listed here using the codes. Can we conclude at the 5% significance level that the new painkiller is perceived to be more effective?

New painkiller: 3, 5, 4, 3, 2, 5, 1, 4, 5, 3, 3, 5, 5, 5, 4
Aspirin: 4, 1, 3, 2, 4, 1, 3, 4, 2, 2, 2, 4, 3, 4, 5

SOLUTION:

IDENTIFY

The objective is to compare two populations: the perceived effectiveness of the new painkiller and of aspirin. We recognise that the data are ordinal; except for the order of the codes, the numbers used to record the results are arbitrary. Finally, the samples are independent. These factors tell us that the appropriate technique is the Wilcoxon Rank Sum Test. We denote the effectiveness scores of the new painkiller as sample 1 and the effectiveness scores of aspirin as sample 2. Because we want to know whether the new painkiller is better than aspirin, the alternative hypothesis is

H_1: The location of population 1 is to the right of the location of population 2

We specify the null hypothesis as

H_0: The two population locations are the same

COMPUTE

MANUALLY:

If the alternative hypothesis is true, the location of population 1 will be located to the right of the location of population 2. It follows that T and z would be large. Our job is to determine whether z is large enough to reject the null hypothesis in favour of the alternative hypothesis. Thus, the rejection region is

$$z > z_\alpha = z_{0.05} = 1.645$$

We compute the test statistic by ranking all the observations.

New Painkiller	Rank	Aspirin	Rank
3	12	4	19.5
5	27	1	2
4	19.5	3	12
3	12	2	6
2	6	4	19.5
5	27	1	2
1	2	3	12
4	19.5	4	19.5
5	27	2	6
3	12	2	6
3	12	2	6
5	27	4	19.5
5	27	3	12
5	27	4	19.5
4	19.5	5	27
	$T_1 = 276.5$		$T_2 = 188.5$

Notice that three 'ones' occupy ranks 1, 2 and 3. The average is 2. Thus, each 'one' is assigned a rank of 2. There are five 'twos' whose ranks are 4, 5, 6, 7 and 8, the average of which is 6. We continue until all the observations have been similarly ranked. The rank sums are computed with $T_1 = 276.5$ and $T_2 = 188.5$. The unstandardised test statistic is $T = T_1 = 276.5$. To standardise, we determine $E(T)$ and σ_T as follows.

$$E(T) = \frac{n_1(n_1 + n_2 + 1)}{2} = \frac{15(31)}{2} = 232.5$$

$$\sigma_T = \sqrt{\frac{n_1 n_2(n_1 + n_2 + 1)}{12}} = \sqrt{\frac{(15)(15)(31)}{12}} = 24.1$$

The standardised test statistic is calculated next:

$$z = \frac{T - E(T)}{\sigma_T} = \frac{276.5 - 232.5}{24.1} = 1.83$$

The p-value of the test is

$$p\text{-value} = P(Z > 1.83) = 1 - 0.9664 = 0.0336$$

EXCEL

	A	B	C	D	E
1	**Wilcoxon Rank Sum Test**				
2					
3			Rank Sum	Observations	
4	New		276.5	15	
5	Aspirin		188.5	15	
6	z Stat		1.83		
7	P(Z<=z) one-tail		0.0340		
8	z Critical one-tail		1.6449		
9	P(Z<=z) two-tail		0.0680		
10	z Critical two-tail		1.96		

Instructions

1. Type or import the data into two adjacent columns. (**Open** Xm17-02.)
2. Click **Add-Ins, Data Analysis Plus** and **Wilcoxon Rank Sum Test.**
3. Specify the **Variable 1 Range** (A1:A16), the **Variable 2 Range** (B1:B16) and the value of α (0.05).

MINITAB

```
Mann–Whitney Test and CI: New, Aspirin

          N  Median
New      15   4.000
Aspirin  15   3.000


Point estimate for ETA1-ETA2 is 1.000
95.4 Percent CI for ETA1-ETA2 is (0.001,2.000)
W = 276.5
Test of ETA1 = ETA2 vs ETA1 > ETA2 is significant at 0.0356
The test is significant at 0.0321 (adjusted for ties)
```

Minitab performs the Mann–Whitney Test rather than the Wilcoxon Test. However, the tests are equivalent. In the output, **ETA** represents the population median. The test statistic $W = 276.5$ is the value of the Wilcoxon Rank Sum statistic; that is $T = W = 276.5$. The output includes the p-value (0.0356) and another p-value calculated by adjusting for tied observations (0.0321). We will report the first p-value only.

Instructions

1. Type or import the data into two columns. (**Open** Xm17-02.)
2. Click **Stat**, **Nonparametrics** and **Mann–Whitney** ....
3. Type the variable name for the **First Sample** (New), the variable name of the **Second Sample** (Aspirin) and click one of **less than, not equal to**, or **greater than** in the **Alternative** box (greater than).

INTERPRET

The data provide sufficient evidence to infer that the new painkiller is perceived to be more effective than aspirin. We note that the data were generated from a controlled experiment; that is, the subjects were assigned to take either the new painkiller or aspirin. (When subjects decide for themselves which medication to take, the data are observational.) This factor helps support the claim that the new painkiller is indeed more effective than aspirin. Factors that weaken the argument are small sample sizes and the inexactness of the responses. There may be methods to measure the effectiveness less subjectively. In addition, a double-blind experiment should have been conducted.

As we pointed out in the introduction to this chapter, the Wilcoxon Rank Sum Test is used to compare two populations when the data are either ordinal or interval. Example 17.2 illustrated the use of the Wilcoxon Rank Sum Test when the data are ordinal. In the next example we demonstrate its use when the data are interval.

EXAMPLE 17.3

DATA Xm17-03

Retaining Workers

Because of the high cost of hiring and training new employees, employers would like to ensure that they retain highly qualified workers. To help develop a hiring programme, the human resources manager of a large company wanted to compare how long business and non-business university graduates worked for the company before resigning to accept a position elsewhere. The manager selected a random sample of 25 business and 20 non-business graduates who had been hired 5 years ago. The number of months each had worked for the company was recorded. (Those who had not quit were recorded as having worked for 60 months.) The data are listed below. Can the human resources manager conclude at the 5% significance level that a difference in duration of employment exists between business and non-business graduates?

Duration of Employment (Months) Business Graduates													Non-Business Graduates										
60	11	18	19	5	25	60	7	8	17	37	4	8	25	60	22	24	23	36	39	15	35	16	28
28	27	11	60	25	5	13	22	11	17	9	4		9	60	29	16	22	60	17	60	32		

SOLUTION:

IDENTIFY

The problem objective is to compare two populations whose data are interval. The samples are independent. Thus, the appropriate parametric technique is the *t*-test of $\mu_1 - \mu_2$, which requires that the populations be normally distributed. However, when the histograms are drawn (see Figures 17.5 and 17.6), it becomes clear that this requirement is unsatisfied. It follows that the correct statistical procedure is the Wilcoxon Rank Sum Test. The null and alternative hypotheses are

H_0: The two population locations are the same

H_1: The location of population 1 (business graduates) is different from the location of population 2 (non-business graduates)

FIGURE 17.5

Histogram of Length of Employment of Business Graduates in Retaining Workers Example

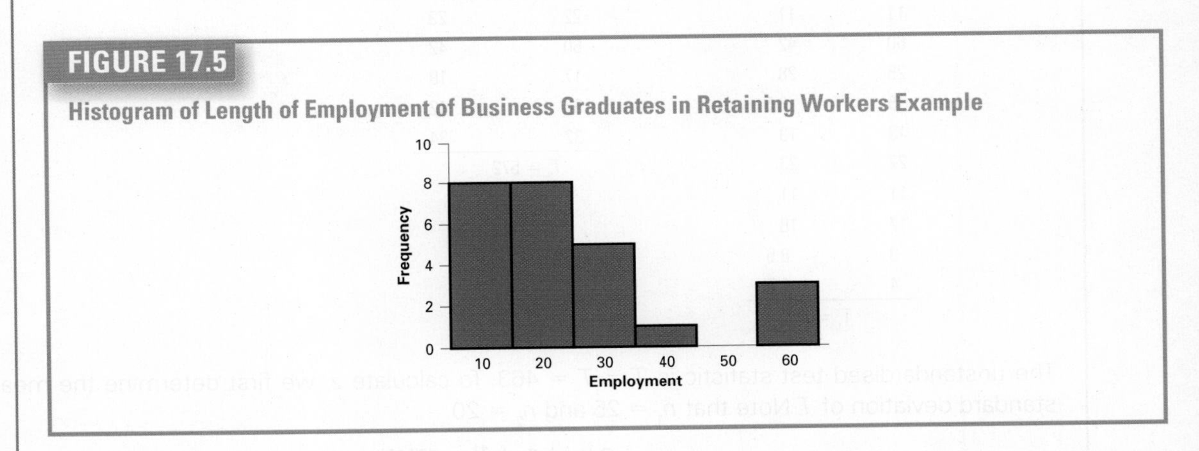

FIGURE 17.6

Histogram of Length of Employment of Non-Business Graduates in Retaining Workers Example

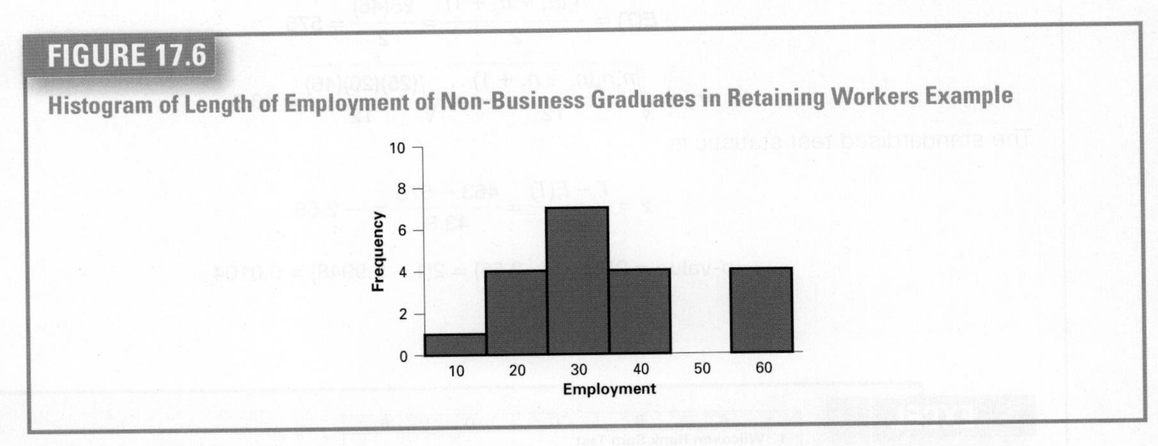

COMPUTE

MANUALLY:

The rejection region is

$$z < -z_{\alpha/2} = -z_{0.025} = -1.96 \quad \text{or} \quad z > z_{\alpha/2} = z_{0.025} = 1.96$$

We calculate the value of the test statistic in the following way.

Business	Rank	Non-Business	Rank
60	42	25	28
11	11	60	42
18	20	22	23
19	21	24	26
5	3.5	23	25
25	28	36	36
60	42	39	38
7	5	15	14
8	6.5	35	35
17	18	16	15.5
37	37	28	31.5
4	1.5	9	8.5
8	6.5	60	42
28	31.5	29	33
27	30	16	15.5
11	11	22	23
60	42	60	42
25	28	17	18
5	3.5	60	42
13	13	32	34
22	23		$T_2 = 572$
11	11		
17	18		
9	8.5		
4	1.5		
	$T_1 = 463$		

The unstandardised test statistic is $T = T_1 = 463$. To calculate z, we first determine the mean and standard deviation of T. Note that $n_1 = 25$ and $n_2 = 20$.

$$E(T) = \frac{n_1(n_1 + n_2 + 1)}{2} = \frac{25(46)}{2} = 575$$

$$\sigma_T = \sqrt{\frac{n_1 n_2(n_1 + n_2 + 1)}{12}} = \sqrt{\frac{(25)(20)(46)}{12}} = 43.8$$

The standardised test statistic is

$$z = \frac{T - E(T)}{\sigma_T} = \frac{463 - 575}{43.8} = -2.56$$

$$p\text{-value} = 2P(Z < -2.56) = 2(1 - 0.9948) = 0.0104$$

EXCEL

	A	B	C	D	E
1	**Wilcoxon Rank Sum Test**				
2					
3			Rank Sum	Observations	
4	Business		463	25	
5	Non-Business		572	20	
6	z Stat		−2.56		
7	P(Z<=z) one-tail		0.0053		
8	z Critical one-tail		1.6449		
9	P(Z<=z) two-tail		0.0106		
10	z Critical two-tail		1.96		

MINITAB

```
Mann-Whitney Test and CI: Business, Non-Business

               N   Median
Business       25  17.00
Non-Business   20  26.50

Point estimate for ETA1-ETA2 is -11.00
95.2 Percent CI for ETA1-ETA2 is (-19.00,-3.00)
W = 463.0
Test of ETA1 = ETA2 vs ETA1 not = ETA2 is significant at 0.0109
The test is significant at 0.0107 (adjusted for ties)
```

INTERPRET

There is strong evidence to infer that the duration of employment is different for business and nonbusiness graduates. The data cannot tell us the cause of this conclusion. For example, we do not know whether business graduates are in greater demand, making it more likely that such employees will leave for better jobs, or whether nonbusiness graduates are more satisfied with their jobs and thus remain longer. Moreover, we do not know what the results would have been had we surveyed employees 10 years after they were employed.

17-1a Required Conditions

The Wilcoxon Rank Sum Test (like the other nonparametric tests presented in this chapter) actually tests to determine whether the population *distributions* are identical. This means that it tests not only for identical locations but also for identical spreads (variances) and shapes (distributions). Unfortunately, this means that the rejection of the null hypothesis may not necessarily signify a difference in population locations. The rejection of the null hypothesis may result instead from a difference in distribution shapes or spreads. To avoid this problem, we will require that the two probability distributions be identical except with respect to location, which then becomes the sole focus of the test. This requirement is made for the tests introduced in the next two sections (sign test, Wilcoxon Signed Rank Sum Test, Kruskal–Wallis Test and Friedman Test).

Both histograms (Figures 17.5 and 17.6) are approximately bimodal. Although there are differences between them it would appear that the required condition for the use of the Wilcoxon Rank Sum Test is roughly satisfied in the example about retaining workers.

17-1b Developing an Understanding of Statistical Concepts

When applying nonparametric techniques, we do not perform any calculations using the original data. Instead, we perform computations only on the ranks. (We determine the rank sums and use them to make our decision.) As a result, we do not care about the actual distribution of the data (hence the name *distribution-free techniques)* and we do not specify parameters in the hypotheses (hence the name *nonparametric techniques).* Although there are other techniques that do not specify parameters in the hypotheses, we use the term *nonparametric* for procedures that feature these concepts. Here is a summary of how to identify the Wilcoxon Rank Sum Test.

Factors that Identify the Wilcoxon Rank Sum

1. **Problem objective:** Compare two populations.
2. **Data type:** Ordinal or interval but non-normal.
3. **Experimental design:** Independent samples.

EXERCISES

Developing an Understanding of Statistical Concepts

Exercises 17.1 and 17.2 are 'what-if' analyses designed to determine what happens to the test statistics and p-values when elements of the statistical inference change. These problems can be solved manually or by creating an Excel spreadsheet.

17.1 **a.** Given the following statistics calculate the value of the test statistic to determine whether the population locations differ.

$T_1 = 250$ $n_1 = 15$
$T_2 = 215$ $n_2 = 15$

b. Repeat part (a) with $T_1 = 275$ and $T_2 = 190$.

c. Describe the effect on the test statistic of increasing T_1 to 275.

17.2 **a.** From the following statistics, test (with $\alpha = 0.05$) to determine whether the location of population 1 is to the right of the location of population 2.

$T_1 = 1205$ $n_1 = 30$
$T_2 = 1280$ $n_2 = 40$

b. Repeat part (a) with $T_1 = 1065$.

c. Discuss the effect on the test statistic and *p*-value of decreasing T_1 to 1065.

17.3 **Xr17-03** Use the Wilcoxon Rank Sum Test on the following data to determine whether the location of population 1 is to the left of the location of population 2. (Use $\alpha = 0.05$.)

Sample 1:	75	60	73	66	81
Sample 2:	90	72	103	82	78

17.4 **Xr17-04** Use the Wilcoxon Rank Sum Test on the following data to determine whether the two population locations differ. (Use a 10% significance level.)

Sample 1:	15	7	22	20	32	18	26	17	23	30
Sample 2:	8	27	17	25	20	16	21	17	10	18

Exercises 17.5 to 17.13 require the use of a computer and software. Conduct tests of hypotheses at the 5% significance level.

17.5 **a.** **Xr17-05a** In a taste test of a new beer, 25 people rated the new beer and another 25 rated the leading brand on the market. The possible ratings were Poor, Fair, Good, Very Good and Excellent. The responses for the new beer and the leading beer were stored using a 1-2-3-4-5 coding system. Can we infer that the new beer is less highly rated than the leading brand?

b. **Xr17-05b** The responses were recoded so that 3 = Poor, 8 = Fair, 22 = Good, 37 = Very Good and 55 = Excellent. Can we infer that the new beer is less highly rated than the leading brand?

c. What does this exercise tell you about ordinal data?

17.6 **a.** **Xr17-06a** To determine whether the satisfaction rating of an airline differs between business class and economy class, a survey was performed. Random samples of both groups were asked to rate their satisfaction with the quality of service using the following responses:

Very satisfied
Quite satisfied
Somewhat satisfied
Neither satisfied nor dissatisfied
Somewhat dissatisfied
Quite dissatisfied
Very dissatisfied

Using a 7-6-5-4-3-2-1 coding system, the results were recorded. Can we infer that business and economy class differ in their degree of satisfaction with the service?

b. **Xr17-06b** The responses were recoded using the values 88-67-39-36-25-21-18. Can we infer that business and economy class differ in their degree of satisfaction with the service?

c. What is the effect of changing the codes? Why was this expected?

17.7 **a.** **Xr17-07** Refer to Example 17.2. Suppose that the responses were coded as follows:

100 = The drug was extremely effective.
60 = The drug was quite effective.
40 = The drug was somewhat effective.
35 = The drug was slightly effective.
10 = The drug was not at all effective.

Determine whether we can infer that the new painkiller is more effective than aspirin.

b. Why are the results of Example 17.2 and part (a) identical?

Applications

17.8 **Xr17-08** A survey of statistics lecturers asked them to rate the importance of teaching nonparametric techniques. The possible responses are

Very important
Quite important
Somewhat important
Not too important
Not important at all

The lecturers were classified as either a member of the Mathematics Department or a member of some other department. The responses were coded (codes 5, 4, 3, 2 and 1, respectively) and recorded. Can we infer that members of the Mathematics Department rate nonparametric techniques as more important than do members of other departments?

17.9 Xr17-09 In recent years, insurance companies offering medical coverage have given discounts to companies that are committed to improving the health of their employees. To help determine whether this policy is reasonable, the general manager of one large insurance company organised a study of a random sample of 30 workers who regularly participate in their company's lunchtime exercise programme and 30 workers who do not. Over a 2-year period he observed the total amount of medical expenses for each individual. Can the manager conclude that companies that provide exercise programmes be given discounts?

17.10 Xr17-10 Feminist organisations in some countries use the issue of who does the housework in two-career families as a gauge of equality. Suppose that a study was undertaken and a random sample of 125 two-career families was taken. The wives were asked to report the number of hours of housework they performed the previous week. The results, together with the responses from a survey performed last year (with a different sample of two-career families), were recorded. Can we conclude that women are doing less housework today than last year?

17.11 Xr17-11 Certain drugs differ in their side-effects depending on the gender of the patient. In a study to determine whether men or women suffer more serious side-effects when taking a powerful penicillin substitute, 50 men and 50 women were given the drug. Each was asked to evaluate the level of stomach upset on a 4-point scale, where 4 = extremely upset, 3 = somewhat upset, 2 = not too upset and 1 = not upset at all. Can we conclude that men and women experience different levels of stomach upset from the drug?

17.12 Xr17-12 The president of Tastee Inc., a baby-food producer, claims that her company's product is superior to that of her leading competitor because babies gain weight faster with her product. As an experiment, 40 healthy newborn infants are randomly selected. For 2 months, 15 of the babies are fed Tastee baby food and the other 25 are fed the competitor's product. Each baby's weight gain (in ounces) was recorded. If we use weight gain as our criterion, can we conclude that Tastee baby food is indeed superior? (This exercise is identical to Exercise 12.64 except for the data.)

17.13 Xr17-13 The Falafel House takeaway restaurant in Riyadh, Saudi Arabia, regularly surveys customers to determine how well it is doing. Suppose that a survey asked customers to rate (among other things) the speed of service. The responses are

1 = Poor
2 = Good
3 = Very good
4 = Excellent

The responses for the day shift and night shift were recorded. Can we infer that night shift customers rate the service differently than the day shift?

17-2 SIGN TEST AND WILCOXON SIGNED RANK SUM TEST

In the preceding section, we discussed the nonparametric technique for comparing two populations of data that are either ordinal or interval (non-normal) and where the data are independently drawn. In this section, the problem objective and data type remain as they were in Section 17-1, but we will be working with data generated from a matched pairs experiment. We have dealt with this type of experiment before. In Section 12-6 we dealt with the mean of the paired differences represented by the parameter μ_D. In this section, we introduce two nonparametric techniques that test hypotheses in problems with the following characteristics:

1. The problem objective is to compare two populations.
2. The data are either ordinal or interval (where the normality requirement necessary to perform the parametric test is unsatisfied).
3. The samples are matched pairs.

To extract all the potential information from a matched pairs experiment, we must create die matched pair differences. Recall that we did so when conducting the *t*-test and estimate of μ_D. We then calculated the mean and standard deviation of these differences and determined die test statistic and confidence interval estimator. The first step in both nonparametric methods presented here is the same: compute the differences for each pair of observations. However, if the data are ordinal, we cannot perform any calculations on those differences because differences between ordinal values have no meaning.

To understand this point, consider comparing two populations of responses of people rating a product or service. The responses are 'excellent', 'good', 'fair', and 'poor'. Recall that we can assign any numbering system as long as the order is maintained. The simplest system is 4-3-2-1. However, any other system such as 66-38-25-11 (or another set of numbers of decreasing order) is equally valid. Now suppose that in one matched pair the sample 1 response was 'excellent' and the sample 2 response was 'good'. Calculating the matched pairs difference under the 4-3-2-1 system gives a difference of 4-3 = 1. Using the 66-38-25-11 system gives a difference of 66 – 38 = 28. If we treat this and other differences as real numbers, we are likely to produce different results depending on which numbering system we used. Thus, we cannot use any method that uses the actual differences. However, we can use the sign of the differences. In fact, when the data are ordinal that is the only method that is valid. In other words, no matter what numbering system is used we know that 'excellent' is better than 'good'. In the 4-3-2-1 system the difference between 'excellent' and 'good' is 4-1. In the 66-38-25-11 system the difference is +28. If we ignore the magnitude of the number and record only the sign, the two numbering systems (and all other systems where the rank order is maintained) will produce exactly the same result.

As you will shortly discover, the sign test uses only the sign of the differences. That's why it's called the *sign test*.

When the data are interval, however, differences have real meaning. Although we can use the sign test when the data are interval, doing so results in a loss of potentially useful information. For example, knowing that the difference in sales between two matched used-car salespeople is 25 cars, is much more informative than simply knowing that the first salesperson sold more cars than the second salesperson. As a result, when the data are interval, but not normal, we will use the *Wilcoxon Signed Rank Sum Test,* which incorporates not only the sign of the difference (hence the name) but also the magnitude.

17-2a Sign Test

The **sign test** is employed in the following situations:

1. The problem objective is to compare two populations.
2. The data are ordinal.
3. The experimental design is matched pairs.

17-2b Test Statistic and Sampling Distribution

The sign test is quite simple. For each matched pair, we calculate the difference between the observation in sample 1 and the related observation in sample 2. We then count the number of positive differences and the number of negative differences. If the null hypothesis is true, we expect the number of positive differences to be approximately equal to the number of negative differences. Expressed another way, we expect the number of positive differences and the number of negative differences each to be approximately equal to half the total sample size. If either number is too large or too small, we reject the null hypothesis. By now you know that the determination of what is too large or too small comes from the sampling distribution of the test statistic. We will arbitrarily choose the test statistic to be the number of positive differences, which we denote x. The test statistic x is a binomial random variable, and under the null hypothesis, the binomial proportion is $p = 0.5$. Thus, the sign test is none other than the z-test of p introduced in Section 12-3.

Recall from Sections 7-4 and 9-2 that x is binomially distributed and that, for sufficiently large n, x is approximately normally distributed with mean $\mu = np$ and standard deviation $\sqrt{np(1-p)}$. Thus, the standardised test statistic is

$$z = \frac{x - np}{\sqrt{np(1-p)}}$$

The null hypothesis

H_0: The two populations are the same is equivalent for testing

H_0: $p = 0.5$

Therefore, the test statistic, assuming that the null hypothesis is true, becomes

$$z = \frac{x - np}{\sqrt{np(1-p)}} = \frac{x - 0.5n}{\sqrt{n(0.5)(0.5)}} = \frac{x - 0.5n}{0.5\sqrt{n}}$$

The normal approximation of the binomial distribution is valid when $np \geq 5$ and $n(1-p) \geq 5$. When $p = 0.5$,

$$np = n(0.5) \geq 5$$

and

$$n(1-p) = n(1-0.5) = n(0.5) \geq 5$$

implies that n must be greater than or equal to 10. Thus, this is one of the required conditions of the sign test. However, the quality of the inference with very small sample size is poor. Larger sample sizes are recommended and will be used in the examples and exercises that follow.

It is common practice in this type of test to eliminate the matched pairs of observations when the differences equal 0. Consequently, n equals the number of non-zero differences in the sample.

EXAMPLE 17.4

DATA Xm17-04

Comparing the Comfort of Two Large Family Cars

In an experiment to determine which of two cars is perceived to have the more comfortable ride, 25 people rode (separately) in the back seat of an expensive European model and also in the back seat of a North American large family car. Each of the 25 people was asked to rate the ride on the following 5-point scale:

1 = Ride is very uncomfortable.

2 = Ride is quite uncomfortable.

3 = Ride is neither uncomfortable nor comfortable.

4 = Ride is quite comfortable.

5 = Ride is very comfortable.

The results are shown here. Do these data allow us to conclude at the 5% significance level that the European car is perceived to be more comfortable than the North American car?

	Comfort Ratings	
Respondent	**European Car**	**North American Car**
1	3	4
2	2	1
3	5	4
4	3	2

(continued)

	Comfort Ratings	
Respondent	European Car	North American Car
5	2	1
6	5	3
7	2	3
8	4	2
9	4	2
10	2	2
11	2	1
12	3	4
13	2	1
14	3	4
15	2	1
16	4	3
17	5	4
18	2	3
19	5	4
20	3	1
21	4	2
22	3	3
23	3	4
24	5	2
25	5	3

SOLUTION:

IDENTIFY

The problem objective is to compare two populations of ordinal data. Because the same 25 people rated both cars, we recognise the experimental design as matched pairs. The sign test is applied, with the following hypotheses:

H_0: The two population locations are the same

H_1: The location of population 1 (European car rating) is to the right of the location of population 2 (North American car rating)

COMPUTE

MANUALLY:

The rejection region is

$$z > z_a = z_{0.05} = 1.645$$

To calculate the value of the test statistic, we calculate the paired differences and count the number of positive, negative and zero differences. The matched pairs differences are

−1	1	1	1	1	2	−1	2	2	0	1	−1	1
−1	1	1	1	−1	1	2	2	0	−1	3	2	

There are seventeen positive, six negative, and two zero differences. Thus, $x = 17$ and $n = 23$. The value of the test statistic is

$$z = \frac{x - 0.5n}{0.5\sqrt{n}} = \frac{17 - 0.5(23)}{0.5\sqrt{23}} = 2.29$$

Because the test statistic is normally distributed, we can calculate the p-value of the test:

$$p\text{-value} = P(Z > 2.29) = 1 - 0.9890 = 0.0110.$$

EXCEL

	A	B	C	D	E
1	**Sign Test**				
2					
3	Difference			*European vs American*	
4					
5	Positive Differences			17	
6	Negative Differences			6	
7	Zero Differences			2	
8	z Stat			2.29	
9	P(Z<=z) one-tail			0.0109	
10	z Critical one-tail			1.6449	
11	P(Z<=z) two-tail			0.0218	
12	z Critical two-tail			1.96	

Instructions

1. Type or import the data into two adjacent columns*. (**Open** Xm17-04.)
2. Click **Add-Ins, Data Analysis Plus** and **Sign Test.**
3. Specify the **Variable 1 Range** (A1:A26), the **Variable 2 Range** (B1:B26) and the value of α (0.05).

MINITAB

Sign Test for Median: Differences

Sign test of median = 0.00000 versus > 0.00000

	N	Below	Equal	Above	P	Median
Differences	25	6	2	17	0.0173	1.000

Minitab prints the number of differences that are negative (Below), the number of zero differences (Equal) and the number of positive differences (Above). The p-value 0.0173 is based on the actual distribution of the number of positive differences, which is binomial.

Instructions

1. Type or import the data into two columns. (**Open** Xm17-04.)
2. Create a new variable, the paired difference.
3. Click **Stat, Nonparametrics** and **1-Sample Sign** ....
4. Type or select the new variable in the **Variables** box, select **Test median** and type 0, and specify the alternative hypothesis in the **Alternative** box (greater than).

There is relatively strong evidence to indicate that people perceive the European car as providing a more comfortable ride than the North American car. There are, however, two aspects of the experiment that may detract from that conclusion. First, did the respondents know in which car they were riding? If so, they may have answered on their preconceived bias that European cars are more expensive and therefore better. If the subjects were blindfolded, we would be more secure in our conclusion. Second, was the order in which each subject rode the two cars varied? If all the subjects rode in the North American car first and the European car second, that may have influenced their ratings. The experiment should have been conducted so that the car each subject rode in first was randomly determined.

*If one or both columns contain an empty cell (representing missing data), the row must be removed.

17-2c Checking the Required Conditions

As we noted in Section 17-1, the sign test requires that the populations be identical in shape and spread. The histogram of the ratings for the European car (Figure 17.7) suggests that the ratings may be uniformly distributed between 2 and 5. The histogram of the ratings for the North American car (Figure 17.8) seems to indicate that the ratings are uniformly distributed between 1 and 4. Thus, both sets of ratings have the same shape and spread but their locations differ. The other condition is that the sample size exceeds ten.

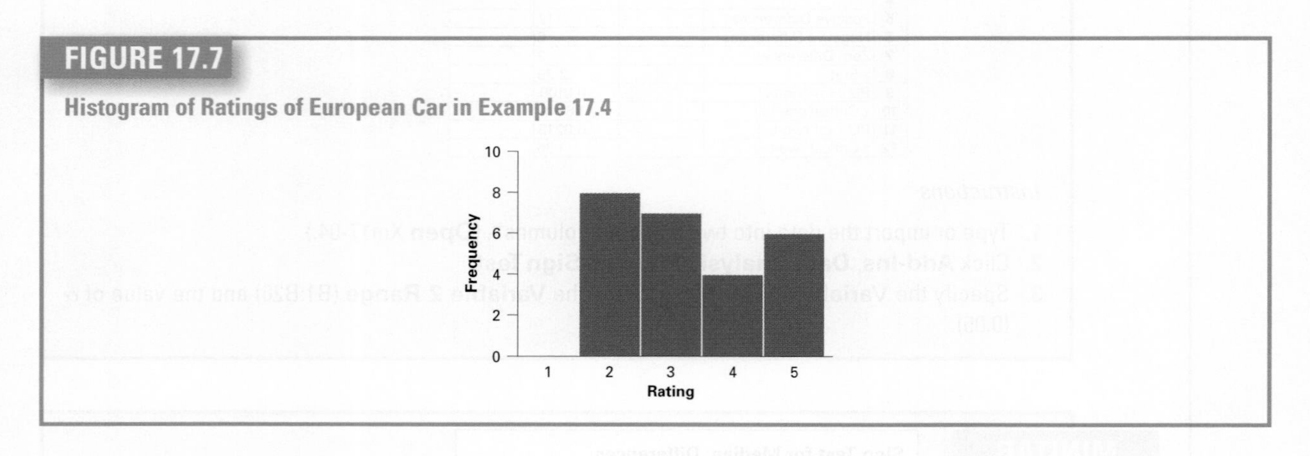

FIGURE 17.7

Histogram of Ratings of European Car in Example 17.4

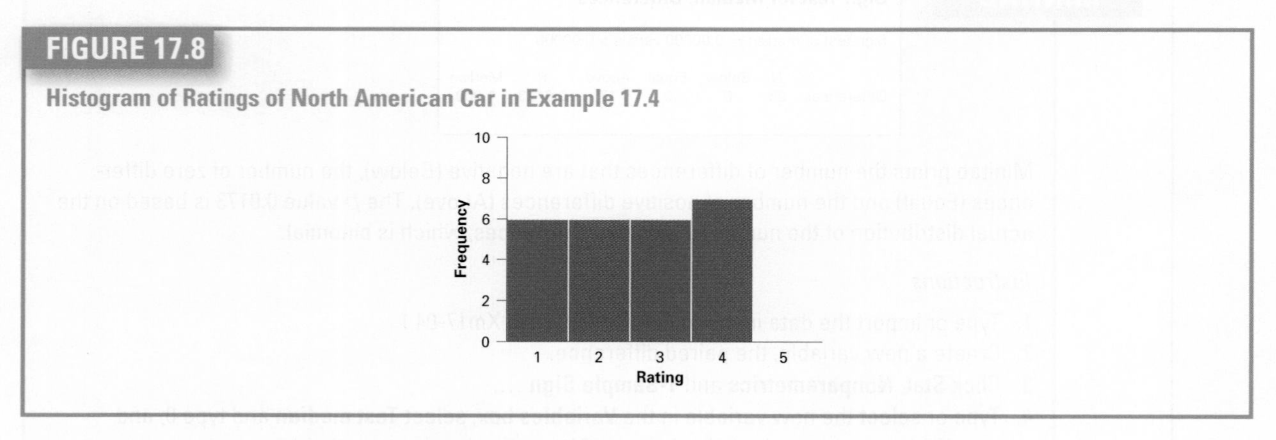

FIGURE 17.8

Histogram of Ratings of North American Car in Example 17.4

17-2d Wilcoxon Signed Rank Sum Test

The **Wilcoxon Signed Rank Sum Test** is used under the following circumstances:

1. The problem objective is to compare two populations.
2. The data (matched pairs differences) are interval, but not normally distributed.
3. The samples are matched pairs.

The Wilcoxon Signed Rank Sum Test is the nonparametric counterpart of the *t*-test of μ_D. Because the data are interval, we can refer to the Wilcoxon Signed Rank Sum Test as a test of μ_D. However, to be consistent with the other nonparametric techniques and to avoid confusion, we will express the hypotheses to be tested in the same way as we did in Section 17-1.

17-2e Test Statistic and Sampling Distribution

We begin by computing the paired differences. As we did in the sign test, we eliminate all differences that are equal to 0. Next, we rank the absolute values of the non-zero differences where 1 = smallest value and n = largest value, with n = number of non-zero differences. (We average the ranks of tied observations.) The sum of the ranks of the positive differences (denoted T^+) and the sum of the ranks of the negative differences (denoted T^-) are then calculated. We arbitrarily select T^+, which we label T, as our test statistic.

For relatively small samples, which we define as $n \leq 30$, the critical values of T can be determined from Table 17.4. This table lists values of T_L and T_U for sample sizes between 6 and 30. The values of T_L and T_v in part (a) of the table are such that

$$P(T \leq T_L) = P(T \geq T_U) = 0.025$$

The values of T_L and T_U in part (b) of the table are such that

$$P(T \leq T_L) = P(T \geq T_U) = 0.05$$

Part (a) is used either in a two-tail test with $\alpha = 0.05$ or in a one-tail test with $\alpha = 0.025$. Part(b) is employed either in a two-tail test with $\alpha = 0.10$ or in a one-tail test with $\alpha = 0.05$.

For relatively large sample sizes (we will define this to mean $n > 30$), T is approximately normally distributed with mean

$$E(T) = \frac{n(n+1)}{4}$$

and standard deviation

$$\sigma_T = \sqrt{\frac{n(n+1)(2n+1)}{24}}$$

Thus, the standardised test statistic is

$$z = \frac{T - E(T)}{\sigma_T}$$

TABLE 17.4 Critical Values of the Wilcoxon Signed Rank Sum Test

	(a) α=0.025 one-tail α=0.05 two-tail		(b) α=0.5 one-tail α=0.10 two-tail	
n	T_L	T_U	T_L	T_U
6	1	20	2	19
7	2	26	4	24
8	4	32	6	30
9	6	39	8	37
10	8	47	11	44
11	11	55	14	52
12	14	64	17	61
13	17	74	21	70
14	21	84	26	79
15	25	95	30	90
6	1	20	2	19
7	2	26	4	24
8	4	32	6	30
9	6	39	8	37
10	8	47	11	44
11	11	55	14	52
12	14	64	17	61

(continued)

	(a) α=0.025 one-tail α=0.05 two-tail		(b) α=0.5 one-tail α=0.10 two-tail	
n	T_L	T_U	T_L	T_U
13	17	74	21	70
14	21	84	26	79
15	25	95	30	90
16	30	106	36	100
17	35	118	41	112
18	40	131	47	124
19	46	144	54	136
20	52	158	60	150
21	59	172	68	163
22	66	187	75	178
23	73	203	83	193
24	81	219	92	208
25	90	235	101	224
26	98	253	110	241
27	107	271	120	258
28	117	289	130	276
29	127	308	141	294
30	137	328	152	313

EXAMPLE 17.5

DATA Xm17-05

Comparing Flextime and Fixed Time Schedules

Traffic congestion on roads and highways costs industry huge sums of money annually as workers struggle to get to and from work. According to TomTom, Johannesburg was rated as the most congested city overall in South Africa for the first quarter of 2012. Several suggestions have been made about how to improve this situation, one of which is called *flextime* – workers are allowed to determine their own schedules (provided they work a full shift). Such workers will likely choose an arrival and departure time to avoid rush-hour traffic. In a preliminary experiment designed to investigate such a programme, the general manager of a large company wanted to compare the times it took workers to travel from their homes to work at 8 am with travel time under the flextime programme. A random sample of 32 workers was selected. The employees recorded the time (in minutes) it took to arrive at work at 8 am on a working day of 1 week. The following week, the same employees arrived at work at times of their own choosing. The travel time on the same day of that week was recorded. These results are listed in the following table. Can we conclude at the 5% significance level that travel times under the flextime programme are different from travel times to arrive at work at 8 am?

	Travel Time	
Worker	Arrival at 8:00 am	Flextime Programme
1	34	31
2	35	31
3	43	44
4	46	44
5	16	15
6	26	28
7	68	63

8	38	39
9	61	63
10	52	54
11	68	65
12	13	12
13	69	71
14	18	13
15	53	55
16	18	19
17	41	38
18	25	23
19	17	14
20	26	21
21	44	40
22	30	33
23	19	18
24	48	51
25	29	33
26	24	21
27	51	50
28	40	38
29	26	22
30	20	19
31	19	21
32	42	38

SOLUTION:

IDENTIFY

The objective is to compare two populations; the data are interval and were produced from a matched pairs experiment. If matched pairs differences are normally distributed, we should apply the *t*-test of μ_D. To judge whether the data are normal, we computed the paired differences and drew the histogram (actually Excel did). Figure 17.9 depicts this histogram. Apparently, the normality requirement is not satisfied, indicating that we should employ the Wilcoxon Signed Rank Sum Test.

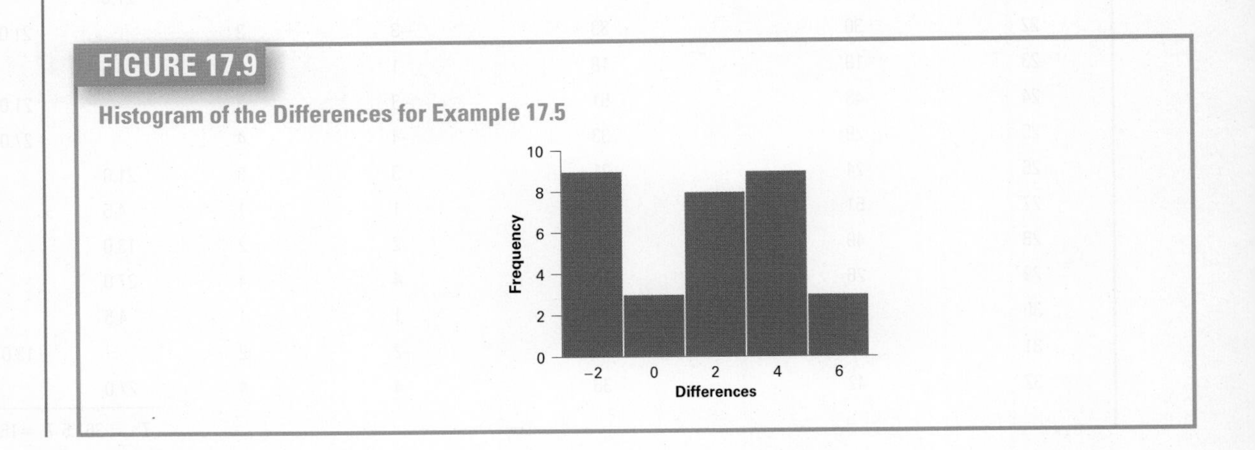

FIGURE 17.9

Histogram of the Differences for Example 17.5

Because we want to know whether the two groups of times differ, we perform a two-tail test whose hypotheses are

H_0: The two population locations are the same

H_1: The location of population 1 (travel times for current work schedule) is different from the location of population 2 (travel times for flextime programme)

COMPUTE MANUALLY:

For each worker, we compute the difference between travel time with arrival at 8 am and travel time under flextime.

	Travel Time					
Worker	**Arrival at 8:00 am**	**Flextime Programme**	**Difference**	**|Difference|**	**Rank**	**|Rank|**
1	34	31	3	3	21.0	
2	35	31	4	4	27.0	
3	43	44	−1	1		4.5
4	46	44	2	2	13.0	
5	16	15	1	1	4.5	
6	26	28	−2	2		13.0
7	68	63	5	5	31.0	
8	38	39	−1	1		4.5
9	61	63	−2	2		13.0
10	52	54	−2	2		13.0
11	68	65	3	3	21.0	
12	13	12	1	1	4.5	
13	69	71	−2	2		13.0
14	18	13	5	5	31.0	
15	53	55	−2	2		13.0
16	18	19	−1	1		4.5
17	41	38	3	3	21.0	
18	25	23	2	2	13.0	
19	17	14	3	3	21.0	
20	26	21	5	5	31.0	
21	44	40	4	4	27.0	
22	30	33	−3	3		21.0
23	19	18	1	1	4.5	
24	48	51	−3	3		21.0
25	29	33	−4	4		27.0
26	24	21	3	3	21.0	
27	51	50	1	1	4.5	
28	40	38	2	2	13.0	
29	26	22	4	4	27.0	
30	20	19	1	1	4.5	
31	19	21	−2	2		13.0
32	42	38	4	4	27.0	
					$T^+ = 367.5$	$T^- = 160.5$

The differences and the absolute values of the differences are calculated. We rank the absolute differences. (If there were any zero differences, we would eliminate them before ranking the absolute differences.) Ties are resolved by calculating the averages. The ranks of the negative differences are offset to facilitate the summing of the ranks. The rank sums of the positive and negative differences are

$$T^+ = 367.5 \quad \text{and} \quad T^- = 160.5$$

The test statistic is

$$z = \frac{T - E(T)}{\sigma_T}$$

where

$$T = T^+ = 367.5$$

$$E(T) = \frac{n(n+1)}{4} = \frac{32(33)}{4} = 264$$

$$\sigma_T = \sqrt{\frac{n(n+1)(2n+1)}{24}} = \sqrt{\frac{32(33)(65)}{24}} = 53.48$$

Thus,

$$z = \frac{T - E(T)}{\sigma_T} = \frac{367.5 - 264}{53.48} = 1.94$$

The rejection region is

$$z < -z_{\alpha/2} = -z_{0.025} = -1.96 \quad \text{or} \quad z > z_{\alpha/2} = z_{0.025} = 1.96$$

The p-value is $2P(Z > 1.94) = 2(1 - 0.9738) = 0.0524$

EXCEL

	A	B	C	D
1	**Wilcoxon Signed Rank Sum Test**			
2				
3	Difference		*8:00 Arrival vs Flextime*	
4				
5	T^+		367.5	
6	T^-		160.5	
7	Observations (for test)		32	
8	z Stat		1.94	
9	P(Z<=z) one-tail		0.0265	
10	z Critical one-tail		1.6449	
11	P(Z<=z) two-tail		0.0530	
12	z Critical two-tail		1.96	

Instructions

1. Type or import the data into two adjacent columns*. (**Open** Xm17-05.)
2. Click **Add-Ins, Data Analysis Plus** and **Wilcoxon Signed Rank Sum Test.**
3. Specify the **Variable 1 Range** (A1:A33), the **Variable 2 Range** (B1:B33) and the value of α (0.05).

MINITAB

Wilcoxon Signed Rank Test: Difference

Test of median = 0.000000 versus median not = 0.000000

	N	N for Test	Wilcoxon Statistic	P	Estimated Median
Difference	32	32	367.5	0.054	1.000

The output includes the original sample size, the number of non-zero differences (*N* for Test), the value of T^+ (Wilcoxon Statistic 367.5) and the p-value (0.054).

(*continued*)

*If one or both columns contain an empty cell (representing missing data), the row must be removed.

Instructions

1. Type or import the data into two columns. (**Open** Xm17-05.)
2. Follow the same steps as in the sign test. At step 2, click **Stat, Nonparametrics** and **1-Sample Wilcoxon. ...**

INTERPRET

There is not enough evidence to infer that flextime commutes are different from the commuting times under the current schedule. This conclusion may be due primarily to the way in which this experiment was performed. All of the drivers recorded their travel time with 8 am arrival on the first day and their flextime travel time on the second day. If the second day's traffic was heavier than usual, that may account for the conclusion reached. As we pointed out in Example 17.4, the order of schedules should have been randomly determined for each employee. In this way, the effect of varying traffic conditions could have been minimised.

Here is how we recognise when to use the two techniques introduced in this section.

Factors that Identify the Sign Test

1. **Problem objective:** Compare two populations.
2. **Data type:** Ordinal.
3. **Experimental design:** Matched pairs.

Factors that Identify the Wilcoxon Signed Rank Sum Test

1. **Problem objective:** Compare two populations.
2. **Data type:** Interval.
3. **Distribution of differences:** Non-normal.
4. **Experimental design:** Matched pairs.

EXERCISES

17.14 In a matched pairs experiment, if we find 30 negative, five zero and 15 positive differences, perform the sign test to determine whether the two population locations differ. (Use a 5% significance level.)

17.15 Suppose that in a matched pairs experiment we find 28 positive differences, seven zero differences and 41 negative differences. Can we infer at the 10% significance level that the location of population 1 is to the left of the location of population 2?

17.16 A matched pairs experiment yielded the following results:

Positive differences: 18

Zero differences: 0

Negative differences: 12

Can we infer at the 5% significance level that the location of population 1 is to the right of the location of population 2?

17.17 Xr17-17 Use the sign test on the following data to determine whether the location of population 1 is to the right of the location of population 2. (Use $\alpha = 0.05$.)

Pair	1	2	3	4	5	6	7	6	9	10	11	12	13	14	15	16
Sample 1	5	3	4	2	3	4	3	5	4	3	4	5	4	5	3	2
Sample 2	3	2	4	3	3	1	3	4	2	5	1	2	2	3	1	2

17.18 Given the following statistics from a matched pairs experiment, perform the Wilcoxon Signed Rank Sum Test to determine whether we can infer at the 5% significance level that the two population locations differ.

$T^+ = 660$ $\quad T^- = 880$ $\quad n = 55$

17.19 A matched pairs experiment produced the following statistics. Conduct a Wilcoxon Signed Rank Sum Test to determine whether the location of population 1 is to the right of the location of population 2. (Use $\alpha = 0.01$.)

$T^+ = 3\,457$ $\quad T^- = 2\,429$ $\quad n = 108$

17.20 Perform the Wilcoxon Signed Rank Sum Test for the following matched pairs to determine whether the two population locations differ. (Use $\alpha = 0.10$.)

Pair	1	2	3	4	5	6
Sample 1	9	12	13	6	7	10
Sample 2	5	10	11	9	3	9

17.21 <u>Xr17-21</u> Perform the Wilcoxon Signed Rank Sum Test to determine whether the location of population 1 differs from the location of population 2 given the data shown here. (Use $\alpha = 0.05$.)

Pair	1	2	3	4	5	6	7	8	9	10	11	12
Sample 1	18.2	14.1	24.5	11.9	9.5	12.1	10.9	16.7	19.6	8.4	21.7	23.4
Sample 2	18.2	14.1	23.6	12.1	9.5	11.3	9.7	17.6	19.4	8.1	21.9	21.6

The following exercises require the use of a computer and software. Use a 5% significance level, unless specified otherwise.

Developing an Understanding of Statistical Concepts

17.22 a. <u>Xr17-22a</u> In a taste test of a new beer, 100 people rated the new beer and the leading brand on the market. The possible ratings were Poor, Fair, Good, Very good and Excellent. The responses for the new beer and the leading beer were recorded using a 1-2-3-4-5 coding system. Can we infer that the new beer is more highly rated than the leading brand?

b. <u>Xr17-22b</u> The responses were recoded so that 3= Poor, 8 = Fair, 22 = Good, 37 = Very good and 55 = Excellent. Can we infer that the new beer is more highly rated than the leading brand?

c. Why are the answers to parts (a) and (b) identical?

17.23 a. <u>Xr17-23a</u> A random sample of 50 people was asked to rate two brands of ice cream, A and B, using the following responses:

Delicious
OK
Not bad
Terrible

The responses were converted to codes 4, 3, 2 and 1, respectively. Can we infer that Brand A is preferred?

b. <u>Xr17-23b</u> The responses were recoded using the values 28-25-16-3. Can we infer that Brand A is preferred?

c. Compare your answers for parts (a) and (b). Are they identical? Explain why?

17.24 <u>Xr17-24</u> Refer to Example 17.4. Suppose that the responses have been recorded in the following way:

6 = Ride is very uncomfortable.
24 = Ride is quite uncomfortable.
28 = Ride is neither uncomfortable nor comfortable.
53 = Ride is quite comfortable.
95 = Ride is very comfortable.

a. Do these data allow us to conclude that the European car is perceived to be more comfortable than the North American car?

b. Compare your answer with that obtained in Example 17.4. Explain why the results are identical.

17.25 a. <u>Xr17-25</u> Data from a matched pairs experiment were recorded. Use the sign test to determine whether the population locations differ.

b. Repeat part (a) using the Wilcoxon Signed Rank Sum Test.

c. Why do the answers to parts (a) and (b) differ?

Applications

17.26 <u>Xr17-26</u> Research scientists at a pharmaceutical company have recently developed a new non-prescription sleeping pill. They decide to test its effectiveness by measuring the time it takes for people to fall asleep after taking the pill. Preliminary analysis indicates that the time to fall asleep varies considerably from one person to another. Consequently they organise the experiment in the following way. A random sample of 100 volunteers who regularly suffer from insomnia is chosen. Each person is given one pill containing the newly developed drug and one placebo. (A placebo is a pill that contains absolutely no medication.) Participants are told to take one pill one night and the second pill one night a week later. (They do not know whether the pill they are taking is the placebo or the new drug, and the order of use is random.) Each participant is fitted with a device that measures the time until sleep occurs. Can we conclude that the new drug is effective? (This exercise is identical to Exercise 12.71, except for the data.)

17.27 <u>Xr17-27</u> Suppose that the housework study referred to in Exercise 17.10 was repeated with some changes. In the revised experiment, 60 women were asked last year and again this year how many hours of housework they perform weekly. Can we conclude at the 1% significance level that women as a group are doing less housework now than last year?

17.28 Xr17-28 A locksmith is in the process of selecting a new key-cutting machine. If there is a difference in key-cutting speed between the two machines under consideration, he will purchase the faster one. If there is no difference, he will purchase the cheaper machine. The times (in seconds) required to cut each of the 35 most common types of keys were recorded. What should he do?

17.29 Xr-17-29 A large sporting-goods store located in Dammam is planning a renovation that will result in an increase in the floor space for one department. The manager of the store has narrowed her choice about which department's floor space to increase to two possibilities: the tennis-equipment department or the swimming-accessories department. The manager would like to enlarge the tennis-equipment department because she believes that this department improves the overall image of the store. She decides, however, that if the swimming-accessories department can be shown to have higher gross sales, she will choose that department. She has collected each of the two departments' weekly gross sales data for the past 32 weeks. Which department should be enlarged?

17.30 Xr-17-30 Does the brand name of an ice cream affect consumers' perceptions of it? The marketing manager of a major dairy pondered this question. She decided to ask 60 randomly selected people to taste the same flavour of ice cream in two different dishes. The dishes contained exactly the same ice cream but were labelled differently. One was given a name that suggested that its maker was European and sophisticated; the other was given a name that implied that the product was domestic and inexpensive. The tasters were asked to rate each ice cream on a 5-point scale, where 1 = Poor, 2 = Fair, 3 = Good, 4 = Very good and 5 = Excellent. Do the results allow the manager to conclude at the 10% significance level that the European brand is preferred?

17.31 Xr17-31 Admissions officers at universities and colleges face the problem of comparing grades achieved at different secondary schools. As a step towards developing a more informed interpretation of such grades, an admissions officer at a large university conducts the following experiment. The records of 100 students from the same local secondary school (high school 1) who just completed their first year at the university were selected. Each of these students was paired (according to average grade in the last year of high school) with a student from another local secondary school (high school 2) who also just completed the first year at the university. For each matched pair, the average letter grades (4 = A, 3 = B, 2 = C, 1 = D or 0 = F) in the first year of university study were recorded. Do these results allow us to conclude that, in comparing two students with the same secondary-school average (one from high school 1 and the other from high school 2), preference in admissions should be given to the student from high school 1?

17-3 KRUSKAL–WALLIS TEST AND FRIEDMAN TEST

In this section we introduce two statistical procedures designed to compare two or more populations. The first test is the **Kruskal–Wallis Test**, which is applied to problems with the following characteristics:

1. The problem objective is to compare two or more populations.
2. The data are either ordinal or interval, but non-normal.
3. The samples are independent.

When the data are interval and normal, we use the one-way analysis of variance *F-test* presented in Section 13-1 to determine whether differences exist. When the data are not normal, we will treat the data as if they were ordinal and employ the Kruskal–Wallis Test. The second procedure is the **Friedman Test**, which is applied to problems with the following characteristics:

1. The problem objective is to compare two or more populations.
2. The data are either ordinal or interval, but not normal.
3. The data are generated from a randomised block experiment.

The parametric counterpart is the two-way analysis of variance, which we use when the data are interval and normal.

17-3a Hypotheses

The null and alternative hypotheses for both tests are similar to those we specified in the analysis of variance. Because the data are ordinal or are treated as ordinal, however, we test population locations instead of population means. In all applications of the Kruskal–Wallis Test and the Friedman Test, the null and alternative hypotheses are

H_0: The locations of all k populations are the same

H_1: At least two population locations differ

Here, k represents the number of populations to be compared.

17-3b Kruskal–Wallis Test

Test Statistic The test statistic is calculated in a way that closely resembles the way in which the Wilcoxon Rank Sum Test was calculated. The first step is to rank all the observations. As before, 1 = smallest observation and n = largest observation, where $n = n_1 + n_2 + \ldots + n_k$. In case of ties, average the ranks.

If the null hypothesis is true, the ranks should be evenly distributed among the k samples. The degree to which this is true is judged by calculating the rank sums (labeled $T_1, T_2, \ldots, T_k$). The last step is to calculate the test statistic, which is denoted H.

Test Statistic for Kruskal–Wallis Test

$$H = \left[\frac{12}{n(n+1)} \sum_{j=1}^{k} \frac{T_j^2}{n_j}\right] - 3(n+1)$$

Although it is impossible to see from this formula, if the rank sums are similar, the test statistic will be small. As a result, a small value of H supports the null hypothesis. Conversely, if considerable differences exist between the rank sums, the test statistic will be large. To judge the value of H, we need to know its sampling distribution.

Sampling Distribution The distribution of the test statistic can be derived in the same way we derived the sampling distribution of the test statistic in the Wilcoxon Rank Sum Test. In other words, we can list all possible combinations of ranks and their probabilities to yield the sampling distribution. A table of critical values can then be determined. However, this is necessary only for small sample sizes. For sample sizes greater than or equal to 5, the test statistic H is approximately chi-squared distributed with $k - 1$ degrees of freedom. Recall that we introduced the chi-squared distribution in Section 8-4.

Rejection Region and p-Value As we noted previously, large values of H are associated with different population locations. Consequently, we want to reject the null hypothesis if H is sufficiently large. Thus, the rejection region is

$$H > \chi^2_{\alpha,k-1}$$

and the p-value is

$$P(\chi^2 > H)$$

Figure 17.10 describes this sampling distribution and the p-value.

FIGURE 17.10

Sampling Distribution of *H*

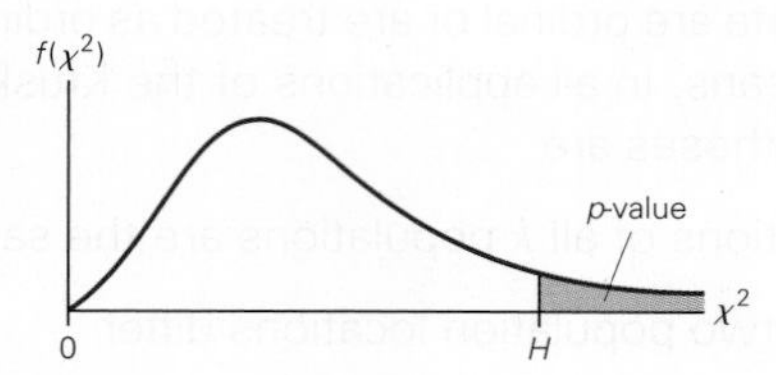

SOCIAL SURVEY: DOES READING BUSINESS NEWSPAPERS IMPROVE STUDENTS' EXAM SCORES IN STATISTICS?

SOLUTION:

IDENTIFY

The problem objective is to compare three populations (Group A, Group B and Group C). The data are ordinal and the samples are independent. These factors are sufficient to justify the use of the Kruskal–Wallis test. The null and alternative hypotheses are:

H_0: The locations of all three populations are the same

H_1: At least two population locations differ

COMPUTE

EXCEL

	A	B	C	D	E
1	**Kruskal–Wallis Test**				
2					
3	Group		Rank Sum	Observations	
4	Group 1		2181.5	41	
5	Group 2		2953	41	
6	Group 3		2491.5	41	
7					
8					
9	H Statistic		5.784412		
10	Degrees of Freedom		2		
11	p-Value		0.055454		
12	Chi-Squared Critical		5.991465		

Instructions

1. Type or import the data into adjacent columns. (**Open** Xm17-00.)
2. Click **Add-Ins**, **Data Analysis Plus** and **Kruskal–Wallis Test**.
3. Specify the **Input Range** and the value of α (0.05).

MINITAB

Kruskal–Wallis Test: READ versus Group

Kruskal–Wallis Test on READ

Group	N	Median	Ave Rank	Z
A	41	1.000	53.2	−1.93
B	41	2.000	72.0	2.21
C	41	2.000	60.8	−0.27
Overall	123		62.0	

H = 5.78 DF = 2 P = 0.055
H = 6.45 DF = 2 P = 0.040 (adjusted for ties)

Instructions

1. The data must be stacked so that the responses are in one column and the codes identifying the shift are in a second column.
2. Click **Stat, Nonparametrics** and **Kruskal–Wallis** ....
3. Select the name of the **Response** variable and the name of **Factor** variable.

INTERPRET

There is not enough evidence at the 5% significance level to infer that a difference in frequency of newspaper reading differs between the three political affiliations.

17-3c Kruskal–Wallis Test and the Wilcoxon Rank Sum Test

When the Kruskal–Wallis Test is used to test for a difference between two populations, it will produce the same outcome as the two-tail Wilcoxon Rank Sum Test. However, the Kruskal–Wallis Test can determine only whether a difference exists. To determine, for example, if one population is located to the right of another, we must apply the Wilcoxon Rank Sum Test.

17-3d Friedman Test

Test Statistic To calculate the test statistic, we first rank each observation within each block, where 1 = smallest observation and k = largest observation, averaging the ranks of ties. Then we compute the rank sums, which we label $T_1, T_2, \ldots, T_k$. The test statistic is defined as follows. (Recall that b = number of blocks.)

Test Statistic for the Friedman Test

$$F_r = \left[\frac{12}{b(k)(k+1)} \sum_{j=1}^{k} T_j^2\right] - 3b(k+1)$$

Sampling Distribution of the Test Statistic The test statistic is approximately chi-squared distributed with $k - 1$ degrees of freedom, provided that either k or b is greater than or equal to 5. As was the case with the Kruskal–Wallis Test, we reject the null hypothesis when the test statistic is large. Hence, the rejection region is

$$F_r > \chi^2_{\alpha,k-1}$$

and the p-value is

$$P(\chi^2 > K)$$

Figure 17.11 depicts the sampling distribution and p-value.

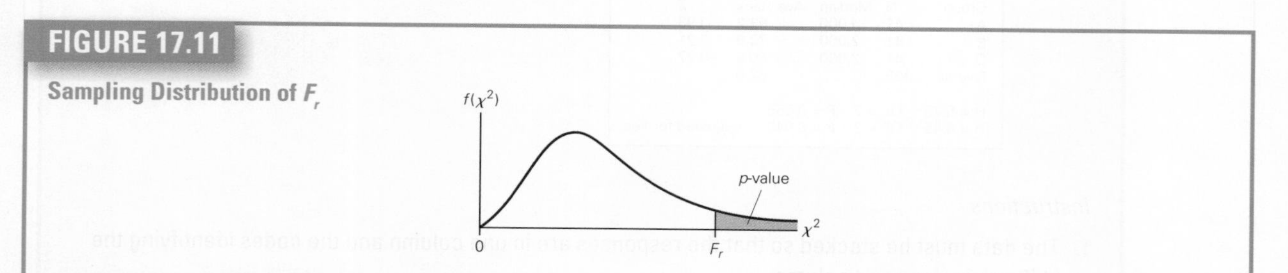

FIGURE 17.11
Sampling Distribution of F_r

This test, like all the other nonparametric tests, requires that the populations being compared be identical in shape and spread.

EXAMPLE 17.6

DATA Xm17-06

Comparing Managers' Evaluations of Job Applicants

The personnel manager of a national accounting firm has been receiving complaints from senior managers about the quality of recent hirings. All new accountants are hired through a process in which four managers interview the candidate and rate her or him on several dimensions, including academic credentials, previous work experience and personal suitability. Each manager then summarises the results and produces an evaluation of the candidate. There are five possibilities:

1. The candidate is in the top 5% of applicants.
2. The candidate is in the top 10% of applicants, but not in the top 5%.
3. The candidate is in the top 25% of applicants, but not in the top 10%.
4. The candidate is in the top 50% of applicants, but not in the top 25%.
5. The candidate is in the bottom 50% of applicants.

The evaluations are then combined in making the final decision. The personnel manager believes that the quality problem is caused by the evaluation system. However, she needs to know whether there is general agreement or disagreement between the interviewing managers in their evaluations. To test for differences between the managers, she takes a random sample of the evaluations of eight applicants. The results are shown below. What conclusions can the personnel manager draw from these data? Employ a 5% significance level.

	Manager			
Applicant	1	2	3	4
1	2	1	2	2
2	4	2	3	2
3	2	2	2	3
4	3	1	3	2
5	3	2	3	5
6	2	2	3	4
7	4	1	5	5
8	3	2	5	3

SOLUTION:

IDENTIFY

The problem objective is to compare the four populations of managers' evaluations, which we can see are ordinal data. This experiment is identified as a randomised block design because the eight applicants were evaluated by all four managers. (The treatments are the managers, and the blocks are the applicants.) The appropriate statistical technique is the Friedman Test. The null and alternative hypotheses are as follows:

H_0: The locations of all four populations are the same

H_1: At least two population locations differ

COMPUTE MANUALLY:

The rejection region is

$$F_r > \chi^2_{\alpha,k-1} = \chi^2_{0.05,3} = 7.81$$

The following table demonstrates how the ranks are assigned and the rank sums calculated. Notice how the ranks are assigned by moving across the rows (blocks) and the rank sums computed by adding down the columns (treatments).

	Manager			
Applicant	1 (Rank)	2 (Rank)	3 (Rank)	4 (Rank)
1	2(3)	KD	2(3)	2(3)
2	4(4)	2(1.5)	3(3)	2(1.5)
3	2(2)	2(2)	2(2)	3(4)
4	3(3.5)	KD	3(3.5)	2(2)
5	3(2.5)	2(1)	3(2.5)	5(4)
6	2(1.5)	2(1.5)	3(3)	4(4)
7	4(2)	KD	5(3.5)	5(3.5)
8	3(2.5)	2(1)	5(4)	3(2.5)
	$T_1 = 21$	$T_2 = 10$	$T_3 = 24.5$	$T_4 = 24.5$

The value of the test statistic is

$$F_r = \left[\frac{12}{b(k)(k+1)}\sum_{j=1}^{k} T_j^2\right] - 3b(k+1)$$

$$= \left[\frac{12}{(8)(4)(5)}(21^2 + 10^2 + 24.5^2 + 24.5^2)\right] - 3(8)(5)$$

$$= 10.61$$

EXCEL

	A	B	C
1	Friedman Test		
2			
3	Group		Rank Sum
4	*Manager 1*		21
5	*Manager 2*		10
6	*Manager 3*		24.5
7	*Manager 4*		24.5
8			
9	Fr Stat		10.61
10	df		3
11	p-value		0.0140
12	chi-squared Critical		7.8147

(continued)

Instructions

1. Type or import the data into adjacent columns*. (**Open** Xm17-06.)
2. Click **Add-Ins**, **Data Analysis Plus** and **Friedman Test**.
3. Specify the **Input Range** (A1:D9) and the value of α (0.05).

MINITAB

Friedman Test: Ratings Versus Manager Blocked by Applicant

S = 10.61 DF = 3 P = 0.014
S = 12.86 DF = 3 P = 0.005 (adjusted for ties)

Manager	N	Est Median	Sum of Ranks
1	8	2.8750	21.0
2	8	2.0000	10.0
3	8	3.0000	24.5
4	8	3.1250	24.5

Grand median = 2.7500

Instructions

1. Type or import the data in stacked format. The responses are stored in one column, the treatment codes are stored in another column and the block codes are stored in a third column.
2. Click **Stat**, **Nonparametrics** and **Friedman** ....
3. Select the **Response** variable, the **Treatment** variable and the **Blocks** variable.

INTERPRET

There appears to be sufficient evidence to indicate that the managers' evaluations differ. The personnel manager should attempt to determine why the evaluations differ. Is there a problem with the way in which the assessments are conducted, or are some managers using different criteria? If it is the latter, those managers may need additional training.

17-3e The Friedman Test and the Sign Test

The relationship between the Friedman and sign tests is the same as the relationship between the Kruskal–Wallis and Wilcoxon Rank Sum Tests; that is, we can use the Friedman Test to determine whether two populations differ. The conclusion will be the same as that produced from the sign test. However, we can use the Friedman Test to determine only whether a difference exists. If we want to determine whether one population is, for example, to the left of another population, we must use the sign test. Here is a list of the factors that tell us when to use the Kruskal–Wallis Test and the Friedman Test.

Factors that Identify the Kruskal–Wallis Test

1. **Problem objective:** Compare two or more populations.
2. **Data type:** Ordinal or interval but not normal.
3. **Experimental design:** Independent samples.

*If one or more columns contain an empty cell (representing missing data), the entire row must be removed.

Factors that Identify the Friedman Test

1. **Problem objective:** Compare two or more populations.
2. **Data type:** Ordinal or interval but not normal.
3. **Experimental design:** Randomised blocks.

EXERCISES

17.32 Conduct the Kruskal–Wallis test on the following statistics. Use a 5% significance level.

$T_1 = 984$ $n_1 = 23$

$T_2 = 1502$ $n_2 = 36$

$T_3 = 1430$ $n_3 = 29$

17.33 From the following statistics, use the Kruskal–Wallis test (with $\alpha = 0.01$) to determine whether the population locations differ.

$T_1 = 1207$ $n_1 = 25$

$T_2 = 1088$ $n_2 = 25$

$T_3 = 1310$ $n_3 = 25$

$T_4 = 1445$ $n_4 = 25$

17.34 Apply the Kruskal–Wallis test and the following statistics to determine whether there is enough statistical evidence at the 10% significance level to infer that the population locations differ.

$T_1 = 3741$ $n_1 = 47$

$T_2 = 1610$ $n_2 = 29$

$T_3 = 4945$ $n_3 = 67$

17.35 Xr17-35 Use the Kruskal–Wallis test on the following data to determine whether the population locations differ. (Use $\alpha = 0.05$.)

Sample 1	27	33	18	29	41	52	75
Sample 2	37	12	17	22	30		
Sample 3	19	12	33	41	28	18	

17.36 Xr17-36 Using the Kruskal–Wallis test, determine whether there is enough evidence provided by the accompanying data to enable us to infer that at least two population locations differ. (Use $\alpha = 0.05$.)

Sample 1	25	15	20	22	23
Sample 2	19	21	23	22	28
Sample 3	27	25	22	29	28

Developing an Understanding of Statistical Concepts

Exercise 17.37 requires the use of a computer and software.

17.37 a. Xr17-37a Four random samples of 50 people each were asked to rate four different computer printers in terms of their ease of use. The responses are:

Very easy to use

Easy to use

Difficult to use

Very difficult to use

The responses were coded using a 4-3-2-1 system. Do these data yield enough evidence at the 5% significance level to infer that differences in ratings exist among the four printers?

b. Xr17-37b The responses were recoded using a 25-22-5-2 system. Do these data yield enough evidence to infer that differences in ratings exist between the four printers?

c. Why are the results of parts (a) and (b) identical?

17.38 Xr17-38 Apply the Friedman Test to the accompanying table of data to determine whether we can conclude that at least two population locations differ. (Use $\alpha = 0.10$.)

	Treatment			
Block	**1**	**2**	**3**	**4**
1	10	12	15	9
2	6	10	11	6
3	13	14	16	11
4	9	9	12	13
5	7	6	14	10

17.39 Xr17-39 The following data were generated from a blocked experiment. Conduct a Friedman Test to determine whether at least two population locations differ. (Use $\alpha = 0.05$.)

	Treatment		
Block	1	2	3
1	7.3	6.9	8.4
2	8.2	7.0	7.3
3	5.7	6.0	8.1
4	6.1	6.5	9.1
5	5.9	6.1	8.0

Developing an Understanding of Statistical Concepts

The following exercises require the use of a computer and software. Use a 5% significance level.

17.40 b. **Xr17-40** Refer to Example 17.6. Suppose that the responses were recoded so that the numbers equalled the midpoint of the range of percentiles. That is:

97.5 = The candidate is in the top 5% of applicants

92.5 = The candidate is in the top 10% of applicants, but not in the top 5%

82.5 = The candidate is in the top 25% of applicants, but not in the top 10%

62.5 = The candidate is in the top 50% of applicants, but not in the top 25%

25 = The candidate is in the bottom 50% of applicants

Can we conclude that differences exist between the ratings assigned by the four professors?

b. Compare your answer in part (a) with the one obtained in Example 17.6. Are they the same? Explain why.

Applications

Unless specified otherwise, use a 5% significance level.

17.41 **Xr17-41** In an effort to determine whether differences exist between three methods of teaching statistics, a professor of business taught his course differently in each of three large sections. In the first section, he taught by lecturing; in the second, he taught by the case method; and in the third, he used a computer software package extensively. At the end of the semester, each student was asked to evaluate the course on a 7-point scale, where 1 = Atrocious, 2 = Poor, 3 = Fair, 4 = Average, 5 = Good, 6 = Very good and 7 = Excellent. From each section, the professor chose 25 evaluations at random. Is there evidence that differences in student satisfaction exist with respect to at least two of the three teaching methods?

17.42 **Xr17-42** In many countries, applicants to MBA programmes must take the Graduate Management Admission Test (GMAT). There are several companies that offer assistance in preparing for the test. To determine whether they work, and if so, which one is best, an experiment was conducted. Several hundred MBA applicants were surveyed and asked to report their GMAT score and which, if any, GMAT preparation course they took. The responses are course A, course B, course C, or no preparatory course. Do these data allow us to infer that there are differences between the four groups of GMAT scores?

17.43 **Xr17-43** Ten judges were asked to test the quality of four different brands of orange juice. The judges assigned scores using a 5-point scale where 1 = Bad, 2 = Poor, 3 = Average, 4 = Good and 5 = Excellent. The results are shown here. Can we conclude at the 5% significance level that there are differences in sensory quality between the four brands of orange juice?

	Orange Juice Brand			
Judge	1	2	3	4
1	3	5	4	3
2	2	3	5	4
3	4	4	3	4
4	3	4	5	2
5	2	4	4	3
6	4	5	5	3
7	3	3	4	4
6	2	3	3	3
9	4	3	5	4
10	2	4	5	3

17.44 **Xr17-44** The manager of a chain of electronic-products retailers is trying to decide on a location for its newest store in Kuwait. After a thorough analysis, the choice has been narrowed to three possibilities. An important factor in the decision is the number of people passing each location. The number of people passing each location per day was counted during 30 days.

a. Which techniques should be considered to determine whether the locations differ? What are the required conditions? How do you select a technique?

b. Can management conclude that there are differences in the numbers of people passing the three locations if the number of people passing each location is not normally distributed?

17.45 **Xr17-45** The manager of a personnel company is in the process of examining her company's advertising

	Newspaper		
Job Advertised	1	2	3
Receptionist	14	17	12
Systems analyst	6	9	6
Junior secretary	25	20	23
Computer programmer	12	15	10
Legal secretary	7	10	5
Office manager	5	9	4

a. What techniques should be considered to apply in reaching a decision? What are the required conditions? How do we determine whether the conditions are satisfied?

b. Assuming that the data are not normally distributed, can we conclude at the 5% significance level that differences exist between the newspapers' abilities to attract **potential employees**?

17.46 Xr17-46 Many adults suffer from high levels of cholesterol, which can lead to heart attacks. For those with very high levels (over 280), doctors prescribe drugs to reduce cholesterol levels. A pharmaceutical company has recently developed three such drugs. To determine whether any differences exist in their benefits, an experiment was organised. The company selected 25 groups of four men, each of whom had cholesterol levels in excess of 280. In each group, the men were matched according to age and weight. The drugs were administered over a 2-month period, and the reduction in cholesterol was recorded. Do these results allow the company to conclude differences exist between the four new drugs? (This exercise is identical to Example 13.3, except for the data.)

17.47 Xr17-47 A well-known soft-drink manufacturer has used the same secret recipe for its product since its introduction over 100 years ago. In response to a decreasing market share, however, the president of the company is contemplating changing the recipe. He has developed two alternative recipes. In a preliminary study, he asked 20 people to taste the original recipe and the two new recipes. He asked each to evaluate the taste of the product on a 5-point scale, where 1 = Awful, 2 = Poor, 3 = Fair, 4 = Good and 5 = Wonderful. The president decides that unless significant differences exist between evaluations of the products, he will not make any changes. Can we conclude that there are differences in the ratings of the three recipes?

17.48 Xr17-48 The management of fast-food restaurants is extremely interested in knowing how their customers rate the quality of food and service and the cleanliness of the restaurants. Customers are given the opportunity to fill out customer comment cards. Suppose that one franchise wanted to compare how customers rate the three shifts (4:00 pm to midnight, midnight to 8:00 am and 8:00 am to 4:00 pm). In a preliminary study, 100 customer cards were randomly selected from each shift. The responses to the question concerning speed of service were recorded, where 4 = Excellent, 3 = Good, 2 = Fair and 1 = Poor, and are listed here. Do these data provide sufficient evidence at the 5% significance level to indicate whether customers perceive the speed of service to be different between the three shifts?

17.49 Xr17-49 A consumer testing service compared the effectiveness of four different brands of drain cleaner. The experiment consisted of using each product on 50 different clogged sinks and measuring the amount of time that elapsed until each drain became unclogged. The recorded times were measured in minutes.

a. Which techniques should be considered as possible procedures to apply to determine whether differences exist? What are the required conditions? How do you decide?

b. If a statistical analysis has shown that the times are not normally distributed, can the service conclude that differences exist between the speeds at which the four brands perform?

17.50 Xr17-50 In anticipation of buying a new scanner, a student turned to a website that reported the results of surveys of users of the different scanners. A sample of 133 responses was listed showing the ease of use of five different brands. The survey responses were:

Very easy
Easy
Not easy
Difficult
Very difficult

The responses were assigned numbers from 1 to 5. Can we infer that there are differences in perceived ease of use between the five brands of scanners?

17-4 SPEARMAN RANK CORRELATION COEFFICIENT

Earlier we introduced the test of the coefficient of correlation, which allows us to determine whether there is evidence of a linear relationship between two interval variables. Recall that the required condition for the *t*-test of ρ is that the variables are bivariate normally distributed. In many situations, however, one or both variables may be ordinal; or if both variables are interval, the normality requirement may not be satisfied. In such cases, we measure and test to determine whether a relationship exists by employing a nonparametric technique, the **Spearman rank correlation coefficient**.

The Spearman rank correlation coefficient is calculated like all of the other previously introduced nonparametric methods by first ranking the data. We then calculate the *Pearson correlation coefficient* of the ranks.

The population Spearman correlation coefficient is labelled ρ_s, and the sample statistic used to estimate its value is labeled r_s.

Sample Spearman Rank Correlation Coefficient

$$r_s = \frac{s_{ab}}{s_a s_b}$$

where *a* and *b* are the ranks of *x* and *y*, respectively, s_{ab} is the covariance of the values *of a* and *b*, s_a is the standard deviation of the values *of a*, and s_b is the standard deviation of the values of *b*.

We can test to determine whether a relationship exists between the two variables. The hypotheses to be tested are

$$H_0: \rho_s = 0$$
$$H_1: \rho_s \neq 0$$

(We also can conduct one-tail tests.) The test statistic is the absolute value of r_s. To determine whether the value of r_s is large enough to reject the null hypothesis, we refer to Table 17.5, which lists the critical values of the test statistic for one-tail tests. To conduct a two-tail test, the value of α must be doubled. The table lists critical values for $\alpha = 0.01$, 0.025 and 0.05 and for $n = 5$ to 30. When *n* is greater than 30, r_s is approximately normally distributed with mean 0 and standard deviation $1/\sqrt{n-1}$. Thus, for $n > 30$, the test statistic is as shown in the box.

Test Statistic for Testing $\rho_s = 0$ when $n > 30$

$$z = \frac{r_s - 0}{1/\sqrt{n-1}} = r_s\sqrt{n-1}$$

which is standard normally distributed.

TABLE 17.5 Critical Values for the Spearman Rank Correlation Coefficient

The α values correspond to a one-tail test of H_0: $\rho_s = 0$. The value should be doubled for two-tail tests.

n	$\alpha = 0.05$	$\alpha = 0.025$	$\alpha = 0.01$
5	0.900	–	–
6	0.829	0.886	0.943
7	0.714	0.786	0.893
8	0.643	0.738	0.833
9	0.600	0.683	0.783
10	0.564	0.648	0.745

EXAMPLE 17.7

DATA Xm17-07

Testing the Relationship between Aptitude Tests and Performance

The production manager of a firm wants to examine the relationship between aptitude test scores given before hiring production-line workers and performance ratings received by the employees 3 months after starting work. The results of the study would allow the firm to decide how much weight to give to these aptitude tests relative to other work-history information obtained, including references. The aptitude test results range from 0 to 100.The performance ratings are as follows:

1 = Employee has performed well below average.

2 = Employee has performed somewhat below average.

3 = Employee has performed at the average level.

4 = Employee has performed somewhat above average.

5 = Employee has performed well above average.

A random sample of 40 production workers yielded the results listed here. Can the firm's manager infer at the 5% significance level that aptitude test scores are correlated with performance rating?

Employee	Aptitude	Performance
1	59	3
2	47	2
3	58	4
4	66	3
5	77	2
6	57	4
7	62	3
8	68	3
9	69	5
10	36	1
11	48	3
12	65	3
13	51	2
14	61	3
15	40	3
16	67	4
17	60	2
18	56	3
19	76	3
20	71	2
21	52	3
22	62	5
23	54	2
24	50	3
25	57	1
26	59	5
27	66	4
28	84	5

(continued)

Employee	Aptitude	Performance
29	56	2
30	61	1
31	53	4
32	76	3
33	42	4
34	59	4
35	58	2
36	66	4
37	58	2
38	53	1
39	63	5
40	85	3

SOLUTION:

IDENTIFY

The problem objective is to analyse the relationship between two variables. The aptitude test score is interval, but the performance rating is ordinal. We will treat the aptitude test score as if it were ordinal and calculate the Spearman rank correlation coefficient. To answer the question, we specify the hypotheses as

$$H_0: \rho_s = 0$$
$$H_1: \rho_s \neq 0$$

COMPUTE

MANUALLY:

We rank each of the variables separately, averaging any ties that we encounter. The original data and ranks are as follows.

Employee	Aptitude	Rank *a*	Performance	Rank *b*
1	59	20	3	20.5
2	47	4	2	9
3	58	17	4	31.5
4	66	30	3	20.5
5	77	38	2	9
6	57	14.5	4	31.5
7	62	25.5	3	20.5
8	68	33	3	20.5
9	69	34	5	38
10	36	1	1	2.5
11	48	5	3	20.5
12	65	28	3	20.5
13	51	7	2	9
14	61	23.5	3	20.5
15	40	2	3	20.5

16	67	32	4	31.5
17	60	22	2	9
18	56	12.5	3	20.5
19	76	36	3	20.5
20	71	35	2	9
21	52	8	3	20.5
22	62	25.5	5	38
23	54	11	2	9
24	50	6	3	20.5
25	57	14.5	1	2.5
26	59	20	5	38
27	66	30	4	31.5
28	84	39	5	38
29	56	12.5	2	9
30	61	23.5	1	2.5
31	53	9.5	4	31.5
32	76	37	3	20.5
33	42	3	4	31.5
34	59	20	4	31.5
35	58	17	2	9
36	66	30	4	31.5
37	58	17	2	9
38	53	9.5	1	2.5
39	63	27	5	38
40	85	40	3	20.5

The next step is to calculate the following sums

$$\sum a_i b_i = 18\,319$$

$$\sum a_i = 820 \quad \sum b_i = 820$$

$$\sum a_i^2 = 22\,131.5$$

$$\sum b_i^2 = 21\,795.5$$

Using the shortcut calculation on Section 4-4a, we determine that the covariance of the ranks is

$$s_{ab} = \frac{1}{n-1}\left(\sum a_i b_i - \frac{\sum a_i \sum b_i}{n}\right) = \frac{1}{40-1}\left[18\,319 - \frac{(820)(820)}{40}\right] = 38.69$$

The sample variances of the ranks (using the short-cut formula in Section 4-2) are

$$s_a^2 = \frac{1}{n-1}\left[\sum a_i^2 - \frac{\left(\sum a_i\right)^2}{n}\right] = \frac{1}{40-1}\left[22\,131.5 - \frac{(820)^2}{40}\right] = 136.45$$

$$s_b^2 = \frac{1}{n-1}\left[\sum b_i^2 - \frac{\left(\sum ab_i\right)^2}{n}\right] = \frac{1}{40-1}\left[22\,795.5 - \frac{(820)^2}{40}\right] = 127.83$$

The standard deviations are

$$s_a = \sqrt{s_a^2} = \sqrt{136.45} = 11.68$$

$$s_b = \sqrt{s_b^2} = \sqrt{127.83} = 11.31$$

Thus,

$$r_s = \frac{s_{ab}}{s_a s_b} = \frac{38.69}{(11.68)(11.31)} = 0.2929$$

The value of the test statistic is

$$z = r_s\sqrt{n-1} = 0.2929\sqrt{40-1} = 1.83$$

$$p\text{-value} = 2P(Z > 1.83) = 2(1 - 0.9664) = 0.0672$$

EXCEL

	A	B	C	D
1	**Spearman Rank Correlation**			
2				
3	*Aptitude and Performance*			
4	Spearman Rank Correlation			0.2930
5	z Stat			1.83
6	P(Z<=z) one-tail			0.0337
7	z Critical one-tail			1.6449
8	P(Z<=z) two-tail			0.0674
9	z Critical two-tail			1.96

Instructions

1. Type or import the data into two adjacent columns*. (**Open** Xm17-07.)
2. Click **Add-Ins, Data Analysis Plus** and **Correlation (Spearman)**.
3. Specify the **Input Range** (A1:B41) and the value of α (0.05).

MINITAB

Correlations: Rank Aptitude, Rank Performance

Pearson correlation of Rank Aptitude and Rank Performance = 0.293
P-Value = 0.067

Instructions

1. Click **Data** and **Rank** ... to rank each variable.
2. Click **Stat, Basic Statistics** and **Correlation.** Select the variables representing the ranks.

INTERPRET

There is not enough evidence to believe that the aptitude test scores and performance ratings are related. This conclusion suggests that the aptitude test should be improved to better measure the knowledge and skill required by a production-line worker. If this proves impossible, the aptitude test should be discarded.

*If one or both columns contain an empty cell (representing missing data), the row must be removed.

EXERCISES

17.51 Test the following hypotheses:

H_0: $\rho_s = 0$

H_1: $\rho_s \neq 0$

$n = 50 \quad r_s = 0.23 \quad \alpha = 0.05$

17.52 Is there sufficient evidence at the 5% significance level to infer that there is a positive relationship between two ordinal variables given that $r_s = 0.15$ and $n = 12$?

17.53 Xr17-53 A statistics student asked seven first-year economics students to report their grades in the required mathematics and economics courses. The results (where 1 = F, 2 = D, 3 = C, 4 = B, 5 = A) are as follows:

Mathematics	4	2	5	4	2	2	1
Economics	5	2	3	5	3	3	2

Calculate the Spearman rank correlation coefficient, and test to determine whether we can infer that a relationship exists between the grades in the two courses. (Use $\alpha = 0.05$.)

17.54 Xr17-54 Does the number of commercials shown during a half-hour television programme affect how viewers rate the show? In a preliminary study eight people were asked to watch a pilot for a situation comedy and rate the show (1 = Terrible, 2 = Bad, 3 = OK, 4 = Good, 5 = Very good). Each person was shown a different number of 30-second commercials. The data are shown here. Calculate the Spearman rank correlation coefficient and test with a 10% significance level to determine whether there is a relationship between the two variables.

Number of commercials	1	2	3	4	5	6	7	8
Rating	4	5	3	3	3	2	3	1

17.55 Xr17-55 The weekly returns of two stocks for a 13-week period were recorded and are listed here. Assuming that the returns are not normally distributed, can we infer at the 5% significance level that the stock returns are correlated?

Stock 1	−7	−4	−7	−3	2	−10	−10
Stock 2	6	6	−4	9	3	−3	7
Stock 1	5	1	−4	2	6	−13	
Stock 2	−3	4	7	9	5	−7	

17.56 Xr17-56 The general manager of an engineering firm wants to know whether a draftsman's experience influences the quality of his work. She selects 24 draftsmen at random and records their years of work experience and their quality rating (as assessed by their supervisors, where 5 = Excellent, 4 = Very good, 3 = Average, 2 = Fair and 1 = Poor). The data are listed here. Can we infer from these data that years of work experience is a factor in determining the quality of work performed? (Use $\alpha = 0.05$.)

Draftsman	Experience	Rating	Draftsman	Experience	Rating
1	1	1	13	8	2
2	17	4	14	20	5
3	20	4	15	21	3
4	9	5	16	19	2
5	2	2	17	1	1
6	13	4	18	22	3
7	9	3	19	20	4
8	23	5	20	11	3
9	7	2	21	18	5
10	10	5	22	14	4
11	12	5	23	21	3
12	24	2	24	21	1

The following exercises require the use of a computer and software. Use a 5% significance level.

17.57 Xm15-02 Refer to Example 15.2. If the required condition is not satisfied conduct another more appropriate test to determine whether odometer reading and price are related.

17.58 Xr17-58 At the completion of most courses in universities and colleges a course evaluation is undertaken. Some professors believe that the way in which students fill out the evaluations is based on how well the student is doing in the course. To test this theory, a random sample of course evaluations was selected. Two answers were recorded. The questions and answers are:

a. How would you rate the course?

1. Poor 2. Fair 3. Good 4. Very good 5. Excellent

b. What grade do you expect in this course?

1. F 2. D 3. C 4. B 5. A

Is there enough evidence to conclude that the theory is correct?

17.59 Xr17-59 Many people suffer from heartburn. It appears, however, that the problem may increase with age. A researcher for a pharmaceutical company wanted to determine whether age and the incidence and extent of heartburn are related. A random sample of

325 adults was drawn. Each person was asked to give his or her age and to rate the severity of heartburn (1 = Low, 2 = Moderate, 3 = High, 4 = Very high). Do these data provide sufficient evidence to indicate that older people suffer more severe heartburn?

17.60 Xr15-06 Assume that the conditions for the test conducted in Exercise 15.6 are not met. Do the data allow us to conclude that the longer the commercial, the higher the memory test score will be?

CHAPTER SUMMARY

Nonparametric statistical tests are applied to problems where the data are either ordinal or interval but not normal. The Wilcoxon Rank Sum Test is used to compare two populations of ordinal or interval data when the data are generated from independent samples. The sign test is used to compare two populations of ordinal data drawn from a matched pairs experiment. The Wilcoxon Signed Rank Sum Test is employed to compare two populations of non-normal interval data taken from a matched pairs experiment. When the objective is to compare two or more populations of independently sampled ordinal or interval non-normal data, the Kruskal–Wallis Test is employed. The Friedman Test is used instead of the Kruskal–Wallis Test when the samples are blocked. To determine whether two variables are related we employ the test of the Spearman rank correlation coefficient.

SYMBOLS:

Symbol	Pronounced	Represents
T_i	T sub i or Ti	Rank sum of sample i ($i = 1, 2, \dots, K$)
T^+	T plus	Rank sum of positive differences
T^-	T minus	Rank sum of negative differences
σT	Sigma sub T or sigma T	Standard deviation of the sampling distribution of T
ρ_s	Rho sub s or rho s	Spearman rank correlation coefficient

FORMULAS:

Wilcoxon Rank Sum Test

$$T = T_1$$

$$E(T) = \frac{n_1(n_1 + n_2 + 1)}{2}$$

$$\sigma_T = \sqrt{\frac{n_1 n_2(n_1 + n_2 + 1)}{12}}$$

$$z = \frac{T - E(T)}{\sigma_T}$$

Sign test

x = number of positive differences

$$z = \frac{x - 0.5n}{0.5\sqrt{n}}$$

Wilcoxon Signed Rank Sum Test

$$T = T^+$$

$$E(T) = \frac{n(n+1)}{4}$$

$$\sigma_T = \sqrt{\frac{n(n+1)(2n+1)}{24}}$$

$$z = \frac{T - E(T)}{\sigma_T}$$

Kruskal–Wallis Test

$$H = \left[\frac{12}{n(n+1)} \sum_{j=1}^{k} \frac{T_j^2}{n_j}\right] - 3(n+1)$$

Friedman Test

$$F_r = \left[\frac{12}{b(k)(k+1)}\sum_{j=1}^{k} T_j^2\right] - 3b(k+1)$$

Spearman rank correlation coefficient

$$r_s = \frac{s_{ab}}{s_a s_b}$$

Spearman test statistic for $n > 30$

$$z = r_s\sqrt{n-1}$$

CHAPTER EXERCISES

The following exercises require the use of a computer and software. Use a 5% significance level.

17.61 Xr17-61 Are education and income related? To answer this question, a random sample of people was selected and each was asked to indicate into which of the following categories of education they belonged:

1. Less than high school
2. High school graduate
3. Some college or university but no degree
4. University degree
5. Postgraduate degree

Additionally, respondents were asked for their annual income group from the following choices:

1. Under £25 000
2. £25 000 up to but not including £40 000
3. £40 000 up to but not including £60 000
4. £60 000 up to £100 000
5. Greater than £100 000

Conduct a test to determine whether more education and higher incomes are linked.

17.62 Xr17-62 In a study to determine which of two teaching methods is perceived to be better, two sections of an introductory marketing course were taught in different ways by the same professor. At the course's completion, each student rated the course on a boring/stimulating spectrum, with 1 = Very boring, 2 = Somewhat boring, 3 = A little boring, 4 = Neither boring nor stimulating, 5 = A little stimulating, 6 = Somewhat stimulating and 7 = Very stimulating. Can we conclude that the ratings of the two teaching methods differ?

17.63 Xr17-63 The researchers at a large carpet manufacturer have been experimenting with a new dyeing process in hopes of reducing the streakiness that frequently occurs with the current process. As an experiment, 15 carpets are dyed using the new process and another 15 are dyed using the existing method. Each carpet is rated on a 5-point scale of streakiness, where 5 is Extremely streaky, 4 is Quite streaky, 3 is Somewhat streaky, 2 is A little streaky and 1 is Not streaky at all. Is there enough evidence to infer that the new method is better?

17.64 Xr17-64 The editor of the student newspaper was in the process of making some major changes in the newspaper's layout. He was also contemplating changing the typeface of the print used. To help make a decision, he set up an experiment in which 20 individuals were asked to read four newspaper pages, with each page printed in a different typeface. If the reading speed differed, the typeface that was read fastest would be used. However, if there was not enough evidence to allow the editor to conclude that such differences exist, the current typeface would be continued. The times (in seconds) to completely read one page were recorded. We have determined that the times are not normally distributed. Determine the course of action the editor should follow.

17.65 Xr17-65 Large potential profits for pharmaceutical companies exist in the area of hair growth drugs. The head chemist for a large pharmaceutical company is conducting experiments to determine which of two new drugs is more effective in growing hair among balding men. One experiment was conducted as follows. A total of 30 pairs of men – each pair of which was matched according to their degree of baldness – was selected. One man used drug A, and the other used drug B. After 10 weeks, the men's new hair growth was examined, and the new growth was judged using the following ratings:

0 = No growth

1 = Some growth

2 = Moderate growth

Do these data provide sufficient evidence that drug B is more effective?

17.66 Xr17-66 The printing department of a publishing company wants to determine whether there are differences in durability between three types of book bindings. Twenty-five books with each type of binding were selected and placed in machines that continually opened and closed them. The numbers of openings and closings until the pages separated from the binding were recorded.

a. What techniques should be considered to determine whether differences exist between the types of bindings? What are the required conditions? How do you decide which technique to use?

b. If we know that the number of openings and closings is not normally distributed, test to determine whether differences exist between the types of bindings.

17.67 Xr17-67 In recent years, consumers have become more safety conscious, particularly about children's products. A manufacturer of children's pyjamas is looking for material that is as non-flammable as possible. In an experiment to compare a new fabric with the kind now being used, 50 pieces of each kind were exposed to an open flame, and the number of seconds until the fabric burst into flames was recorded. Because the new material is much more expensive than the current material, the manufacturer will switch only if the new material can be shown to be better. On the basis of these data, what should the manufacturer do?

17.68 Xr17-68 Ahmed's is a chain of family restaurants. Like many other service companies, Ahmed's surveys its customers on a regular basis to monitor their opinions. Two questions (among others) asked in the survey are as follows:

a. While you were at Ahmed's, did you find the service Slow (1), Moderate (2) or Fast (3)?

b. What day was your visit to Ahmed's?

The responses of a random sample of 269 customers were recorded. Can the manager infer that there are differences in customer perceptions of the speed of service between the days of the week?

17.69 Xr17-69 It is common practice in many MBA programmes to require applicants to arrange for a letter of reference. Some universities have their own forms in which referees assess the applicant using the following categories:

5: The candidate is in the top 5% of applicants

4: The candidate is in the top 10% of applicants, but not in the top 5%

3: The candidate is in the top 25% of applicants, but not in the top 10%

2: The candidate is in the top 50% of applicants, but not in the top 25%

1: The candidate is in the bottom 50% of applicants

However, the question arises, Are the referees' ratings related to how well the applicant performs in the MBA programme? To answer the question, a random sample of recently graduated MBAs was drawn. For each, the rating of the referee and the MBA grade-point average (GPA) were recorded. Do these data present sufficient evidence to infer that the letter of reference and the MBA GPA are related?

17.70 Xr17-70 The increasing number of travelling businesswomen represents a large potential clientele for the hotel industry. Many hotel chains have made changes designed to attract more women. To help direct these changes, a hotel chain commissioned a study to determine whether major differences exist between male and female business travellers. A total of 100 male and 100 female executives were questioned on a variety of topics, one of which was the number of trips they had taken in the previous 12 months. We would like to know whether these data provide enough evidence to allow us to conclude that businesswomen and businessmen differ in the number of business trips taken per year.

17.71 Xr17-71 To examine the effect that a tough midterm test has on student evaluations of professors, a statistics professor had her class evaluate her teaching effectiveness before the midterm test. The questionnaire asked for opinions on a number of dimensions, but the last question is considered the most important. It is 'How would you rate the overall performance of the instructor?' The possible responses are 1 = poor, 2 = fair, 3 = good and 4 = excellent. After a difficult test, the evaluation was redone. The evaluation scores before and after the test for each of the 40 students in the class were recorded. Do the data allow the professor to conclude that the results of the midterm negatively influence student opinion?

17.72 Xr17-72 The town of Stratford-upon-Avon has many famous events, including Shakespeare's birthday celebrations. Thousands of people visit Stratford-upon-Avon to attend one or more Shakespearean plays and spend money in hotels, restaurants and gift shops. As a consequence, any sign that the number of visitors will decrease in the future is cause for concern. Two years ago, a survey of 100 visitors asked how likely it was that they would return within the next 2 years. This year the survey was repeated with another 100 visitors. The likelihood of returning within 2 years was measured as: 4 = Very likely, 3 = Somewhat likely, 2 = Somewhat unlikely and 1 = Very unlikely. Conduct whichever statistical procedures you deem necessary to determine whether the citizens of Stratford should be concerned about the results of the two surveys.

17.73 Xr17-73 How does gender affect teaching evaluations? Several researchers addressed this question during the past decade. In one study several female and male professors in the same department with similar backgrounds were selected. A random sample of 100 female students was drawn. Each student evaluated a female professor and a male professor. A sample of 100 male students was drawn and each also evaluated a female professor and a male professor. The ratings were based on a 4-point scale (where 1 = Poor, 2 = Fair, 3 = Good and 4 = Excellent). The evaluations were recorded in the following way:

Column 1 = Female student
Column 2 = Female professor rating
Column 3 = Male professor rating
Column 4 = Male student
Column 5 = Female professor rating
Column 6 = Male professor rating

a. Can we infer that female students rate female professors higher than they rate male professors?

b. Can we infer that male students rate male professors higher than they rate female professors?

18 TIME-SERIES ANALYSIS AND FORECASTING

18-1 TIME-SERIES COMPONENTS

18-2 SMOOTHING TECHNIQUES

18-3 TREND AND SEASONAL EFFECTS

18-4 INTRODUCTION TO FORECASTING

18-5 FORECASTING MODELS

NEW HOME DEVELOPMENT: HOUSING 'STARTS'

DATA Xm18-00

At the end of 2005 a major housing developer in the UK wanted to predict the number of housing units that would be started in 2006. This information would be extremely useful in determining a variety of variables, including housing demand, availability of labour and the price of building materials. To help develop an accurate forecasting model the government collected data on the number of housing starts (in thousands) for the previous 20 quarters (2001–2005). Forecast the number of housing starts for the 4 quarters of 2006. We have chosen the new housing construction data from 2001 because starts were strongly affected by the economic downturn between 2008 and 2009. As a result, statistical forecasting tools became wildly inaccurate. **See Section 18-5c for answer.**

INTRODUCTION

Any variable that is measured over time in sequential order is called a **time series**. We introduced time series in Chapter 3 and demonstrated how we use a line chart to graphically display the data. Our objective in this chapter is to analyse time series in order to detect patterns that will enable us to forecast future values of the time series. There is an almost unlimited number of such applications in management and economics. Some examples follow.

1. Governments want to know future values of interest rates, unemployment rates and percentage increases in the cost of living.
2. Housing industry economists must forecast mortgage interest rates, demand for housing and the cost of building materials.
3. Many companies attempt to predict the demand for their products and their share of the market.
4. Universities and colleges often try to forecast the number of students who will be applying for acceptance at post-secondary school institutions.

Forecasting is a common practice among managers and government decision-makers. This chapter focuses on time-series forecasting, which uses historical time-series data to predict future values of variables such as sales or unemployment rates. This entire chapter is an application tool for both economists and managers in all functional areas of business because forecasting is such a vital factor in decision-making in these areas.

For example, the starting point for aggregate production planning by operations managers is to forecast the demand for the company's products. These forecasts will make use of economists' forecasts of macroeconomic variables (such as gross domestic product, disposable income and housing starts) as well as the marketing managers' internal forecasts of their customers' future needs. Not only are these sales forecasts critical to production planning but also they are the key to accurate pro forma (i.e. forecasted) financial statements, which are produced by the accounting and financial managers to assist in their planning for future financial needs such as borrowing. Likewise, the human resources department will find such forecasts of a company's growth prospects to be invaluable in their planning for future worker requirements.

There are many different forecasting techniques. Some are based on developing a model that attempts to analyse the relationship between a dependent variable and one or more independent variables. We presented some of these methods in the chapters on regression analysis (Chapters 15 and 16). The forecasting methods to be discussed in this chapter are all based on time series, which we discuss in the next section. In Sections 18-2 and 18-3 we deal with methods for detecting and measuring which time-series components exist. After we uncover this information, we can develop forecasting tools. We will only scratch the surface of this topic. Our objective is to expose you to the concepts of forecasting and to introduce some of the simpler techniques. The level of this text precludes the investigation of more complicated methods.

18-1 TIME-SERIES COMPONENTS

A time series can consist of four different components as described in the box.

Time-Series Components

1. Long-term trend.
2. Cyclical variation.
3. Seasonal variation.
4. Random variation.

A **trend** (also known as a **secular trend**) is a long-term, relatively smooth pattern or direction exhibited by a series. Its duration is more than 1 year. For example, the world population exhibited a trend of relatively steady growth from 2.5 billion in 1950 to 7.2 billion in 2014. Figure 18.1 exhibits the line chart.

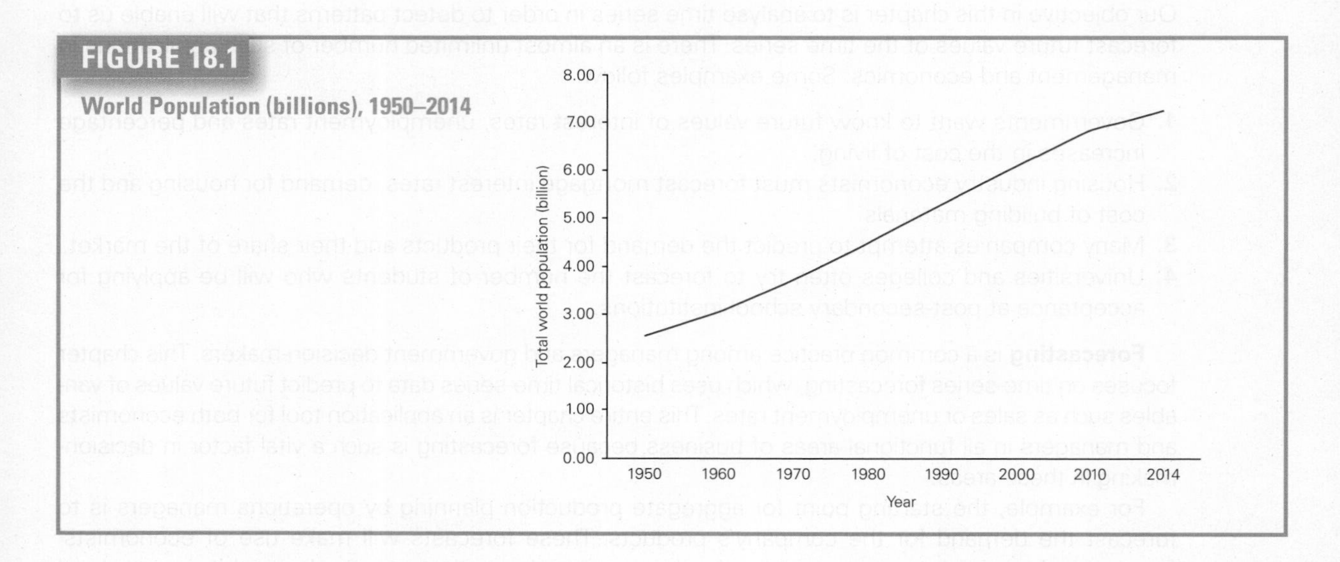

FIGURE 18.1
World Population (billions), 1950–2014

The trend of a time series is not always linear. For example, Figure 18.2 describes US annual retail book sales in $millions. As you can see, sales increased from 1992 to 2008 and has decreased since then.

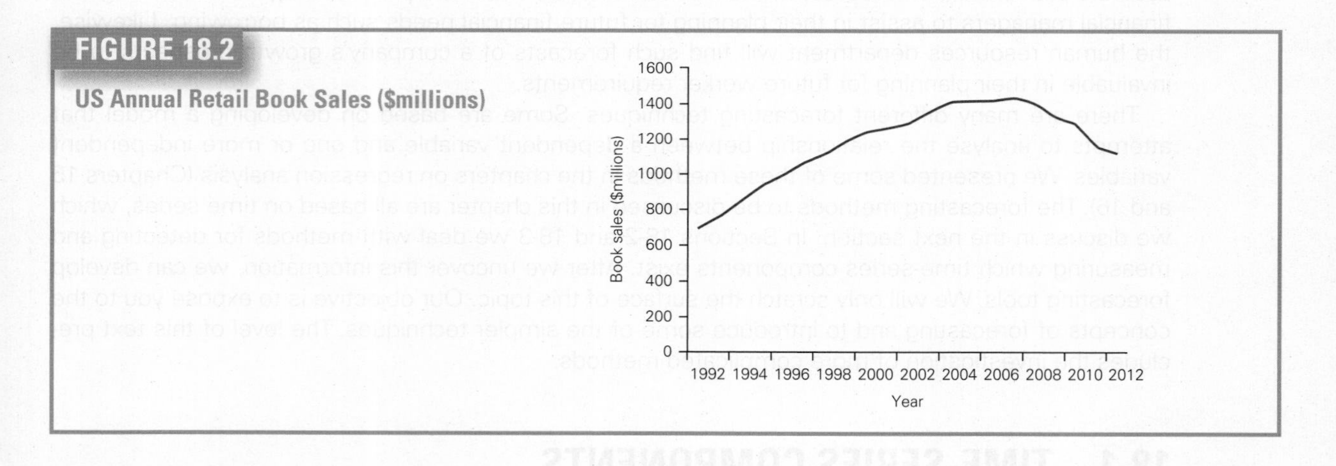

FIGURE 18.2
US Annual Retail Book Sales ($millions)

Cyclical variation is a wavelike pattern describing a long-term trend that is generally apparent over a number of years, resulting in a cyclical effect. By definition, it has duration of more than 1 year. Examples include business cycles that record periods of economic recession and inflation, long-term product-demand cycles, and cycles in monetary and financial sectors. However, cyclical patterns that are consistent and predictable are quite rare. For practical purposes, we will ignore this type of variation.

Seasonal variation refers to cycles that occur over short repetitive calendar periods and, by definition, have a duration of less than 1 year. The term *seasonal variation* may refer to the four traditional seasons or to systematic patterns that occur during a month, a week or even 1 day. Demand for restaurants feature 'seasonal' variation throughout the day.

An illustration of seasonal variation is provided in Figure 18.3, which graphs monthly practical motorcycle test pass rates in Great Britain from 1 January 2011 to 31 December 2013.

FIGURE 18.3

Practical motorcycle test pass rates in Great Britain

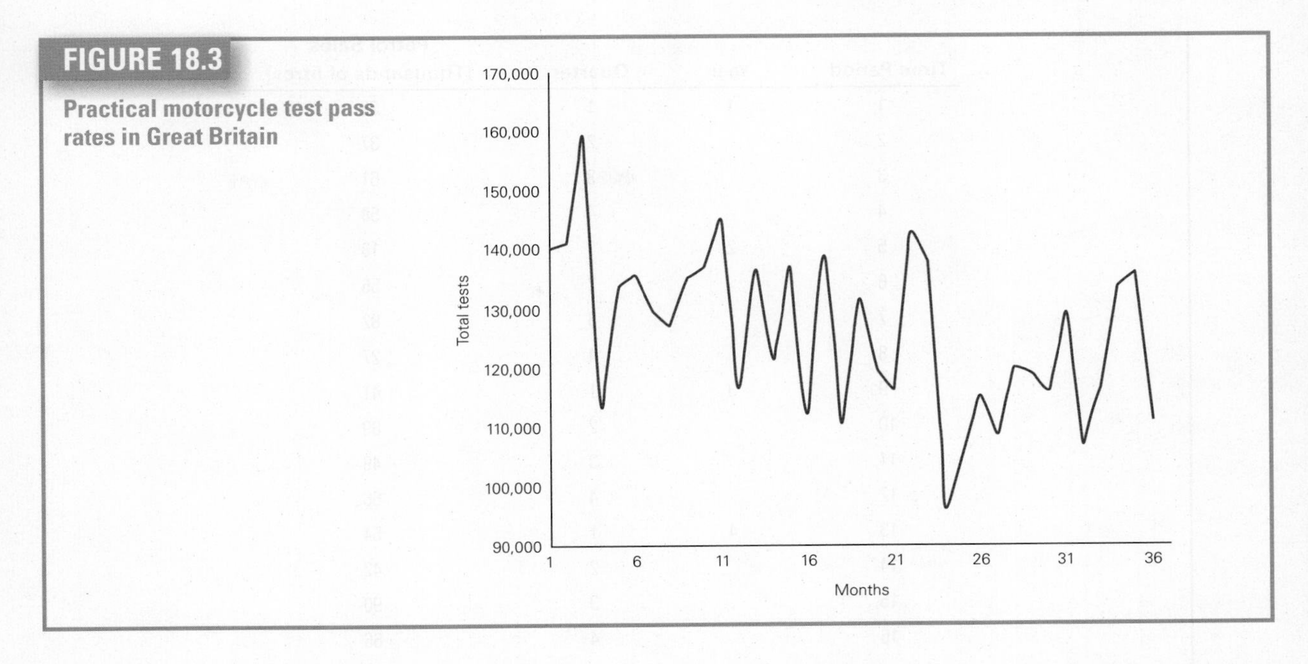

Random variation is caused by irregular and unpredictable changes in a time series that are not caused by any other components. It tends to mask the existence of the other more predictable components. Because random variation exists in almost all time series, one of the objectives of this chapter is to introduce ways to reduce the random variation, which will enable statistics practitioners to describe and measure the other components. By doing so, we hope to be able to make accurate predictions of the time series.

18-2 SMOOTHING TECHNIQUES

If we can determine which components actually exist in a time series, we can develop better forecasts. Unfortunately, the existence of random variation often makes the task of identifying components difficult. One of the simplest ways to reduce random variation is to smooth the time series. In this section, we introduce two methods: *moving averages* and *exponential smoothing.*

18-2a Moving Averages

A moving average for a time period is the arithmetic mean of the values in that time period and those close to it. For example, to compute the three-period moving average for any time period, we would average the time-series values in that time period, the previous period and the following period. We compute the three-period moving averages for all time periods except the first and the last. To calculate the five-period moving average, we average the value in that time period, the values in the two preceding periods and the values in the two following time periods. We can choose any number of periods with which to calculate the moving averages.

 EXAMPLE 18.1

Data Xm18-01

Petrol Sales, Part 1

As part of an effort to forecast future sales, an operator of five independent petrol stations recorded the quarterly petrol sales (in thousands of litres) for the past 4 years. These data are shown below. Calculate the three-quarter and five-quarter moving averages. Draw graphs of the time series and the moving averages.

Time Period	Year	Quarter	Petrol Sales (Thousands of litres)
1	1	1	39
2		2	37
3		3	61
4		4	58
5	2	1	18
6		2	56
7		3	82
8		4	27
9	3	1	41
10		2	69
11		3	49
12		4	66
13	4	1	54
14		2	42
15		3	90
16		4	66

SOLUTION:

COMPUTE MANUALLY:

To compute the first three-quarter moving average, we group the petrol sales in periods 1,2 and 3, and then average them. Thus, the first moving average is

$$\frac{39 + 37 + 61}{3} = \frac{137}{3} = 45.7$$

The second moving average is calculated by dropping the first period's sales (39), adding the fourth period's sales (58) and then computing the new average. Thus, the second moving average is

$$\frac{37 + 61 + 58}{3} = \frac{156}{3} = 52.0$$

The process continues as shown in the following table. Similar calculations are made to produce the five-quarter moving averages (also shown in the table).

Time Period	Petrol Sales	Three-Quarter Moving Average	Five-Quarter Moving Average
1	39	–	
2	37	45.7	–
3	61	52.0	42.6
4	58	45.7	46.0
5	18	44.0	55.0
6	56	52.0	48.2
7	82	55.0	44.8
8	27	50.0	55.0
9	41	45.7	53.6
10	69	53.0	50.4
11	49	61.3	55.8

Time Period	Petrol Sales	Three-Quarter Moving Average	Five-Quarter Moving Average
12	66	56.3	56.0
13	54	54.0	60.2
14	42	62.0	63.6
15	90	66.0	–
16	66	–	–

Notice that we place the moving averages in the centre of the group of values being averaged. It is for this reason that we prefer to use an odd number of periods in the moving averages. Later in this section, we discuss how to deal with an even number of periods.

Figure 18.4 displays the line chart for petrol sales, and Figure 18.5 shows the three-period and five-period moving averages.

FIGURE 18.4

Quarterly Petrol Sales

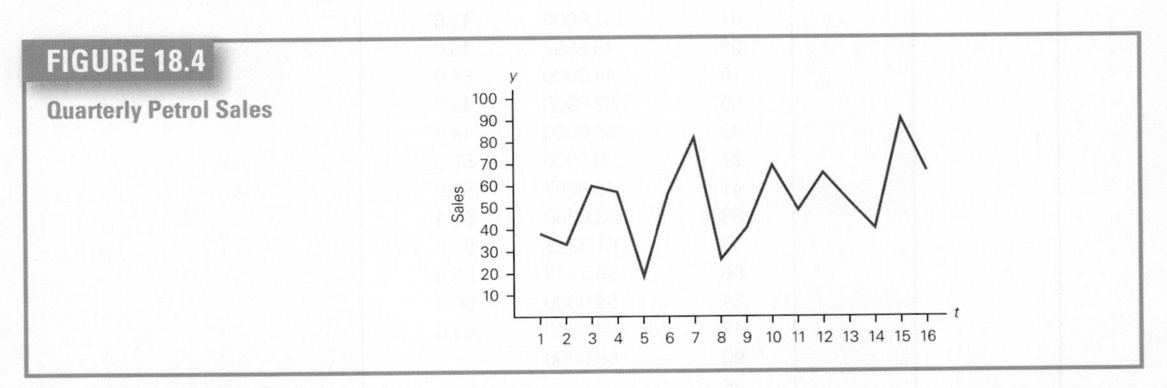

FIGURE 18.5

Quarterly Petrol Sales and Three-Quarter and Five-Quarter Moving Averages

EXCEL

	A	B
1	Petrol Sales	Moving Average
2	39	
3	37	45.7
4	61	52.0
5	58	45.7
6	18	44.0
7	56	52.0
8	82	55.0
9	27	50.0
10	41	45.7
11	69	53.0
12	49	61.3
13	66	56.3
14	54	54.0
15	42	62.0
16	90	66.0
17	66	

(continued)

Instructions

1. Type or import the data into one column. (**Open** Xm18-01.)
2. Click **Data, Data Analysis** and **Moving Average.**
3. Specify the **Input Range** (A1:A17). Specify the number of periods (3) and the **Output Range** (B1).
4. Delete the cells containing N/A.
5. To draw the line charts, follow the instructions on Section 3-2.

MINITAB

Petrol Sales	AVER1	AVER2
39	*	*
37	45.6667	*
61	52.0000	42.6
58	45.6667	46.0
18	44.0000	55.0
56	52.0000	48.2
82	55.0000	44.8
27	50.0000	55.0
41	45.6667	53.6
69	53.0000	50.4
49	61.3333	55.8
66	56.3333	56.0
54	54.0000	60.2
42	62.0000	63.6
90	66.0000	
66		

Moving Average Plot for Petrol Sales

Variable
Actual
Smoothed

Moving Average
Length 3

Accuracy Measures
MAPE 34.708
MAD 14.452
MSD 318.659

Petrol sales

Index

Moving Average Plot for Petrol Sales

Variable
Actual
Smoothed

Moving Average
Length 5

Accuracy Measures
MAPE 33.497
MAD 14.067
MSD 287.633

Petrol sales

Index

Instructions

1. Type or import the data into one column. (**Open** Xm18-01.)
2. Click **Stat, Time series** and **Moving Average** . . . .
3. Type or select the variable in the **Variable** box (Petrole sales) and type the number of periods in the **MA length** box **(3)**. Click **Centre the moving averages, Storage** . . . and **Moving averages.**
4. To draw the graph click **Graphs . . .** and **Plot smoothed vs actual.**

INTERPRET

To see how the moving averages remove some of the random variation, examine Figures 18.4 and 18.5. Figure 18.4 depicts the quarterly petrol sales. Discerning any of the time-series components is difficult because of the large amount of random variation. Now consider the three-quarter moving average in Figure 18.5. You should be able to detect the seasonal pattern that exhibits peaks in the third quarter of each year (periods 3, 7, 11 and 15) and valleys in the first quarter of the year (periods 5, 9 and 13). There is also a small but discernible long-term trend of increasing sales.

Notice also in Figure 18.5 that the five-quarter moving average produces more smoothing than the three-quarter moving average. In general, the longer the time period over which we average, the smoother the series becomes. Unfortunately, in this case we have smoothed too much – the seasonal pattern is no longer apparent in the five-quarter moving average. All we can see is the long-term trend. It is important to realise that our objective is to smooth the time series sufficiently to remove the random variation and to reveal the other components (trend, cycle or season) present. With too little smoothing, the random variation disguises the real pattern. With too much smoothing, however, some or all of the other effects may be eliminated along with the random variation.

18-2b Centred Moving Averages

Using an even number of periods to calculate the moving averages presents a problem about where to place the moving averages in a graph or table. For example, suppose that we calculate the four-period moving average of the following time series:

Period	Time Series
1	15
2	27
3	20
4	14
5	25
6	11

The first moving average is

$$\frac{15 + 27 + 20 + 14}{4} = 19.0$$

However, because this value represents time periods 1, 2, 3 and 4, we must place it between periods 2 and 3 . The next moving average is

$$\frac{27 + 20 + 14 + 25}{4} = 21.5$$

and it must be placed between periods 3 and 4. The moving average that falls between periods 4 and 5 is

$$\frac{20 + 14 + 25 + 11}{4} = 17.5$$

There are several problems that result from placing the moving averages between time periods, including graphing difficulties. Centring the moving average corrects the problem. We do this by computing the two-period moving average of the four-period moving average. Thus, the centred moving average for period 3 is

$$\frac{19.0 + 21.5}{2} = 20.25$$

The centred moving average for period 4 is

$$\frac{21.5 + 17.5}{2} = 19.50$$

The following table summarises these results.

Period	Time Series	Four-Period Moving Average	Four-Period Centred Moving Average
1	15	—	—
2	27	19.0	—
3	20	21.5	20.25
4	14	17.5	19.50
5	25	—	—
6	11	—	—

Minitab centres the moving averages on command. Excel does not centre the moving averages.

18-2c Exponential Smoothing

Two drawbacks are associated with the moving average method of smoothing time series. First, we do not have moving averages for the first and last sets of time periods. If the time series has few observations, the missing values can represent an important loss of information. Second, the moving average 'forgets' most of the previous time-series values. For example, in the five-quarter moving average described in Example 18.1, the average for quarter 4 reflects quarters 2, 3, 4, 5 and 6 but is not affected by quarter 1. Similarly, the moving average for quarter 5 forgets quarters 1 and 2 . Both of these problems are addressed by **exponential smoothing.**

Exponentially Smoothed Time Series

$$S_t = wy_t + (1 - w)S_{t-1} \text{ for } t \geq 2$$

where

S_t = Exponentially smoothed time series at time period t

y_t = Time series at time period t

S_{t-1} = Exponentially smoothed time series at time period $t - 1$

w = Smoothing constant, where $0 \leq w \leq 1$

We begin by setting

$$S_1 = y_1$$

Then

$$S_2 = wy_2 + (1 - w)S_1$$
$$= wy_2 + (1 - w)y_1$$
$$S_3 = wy_3 + (1 - w)S_2$$
$$= wy_3 + (1 - w)[wy_2 + (1 - w)y_1]$$
$$= wy_3 + w(1 - w)y_2 + (1 - w)^2y_1$$

and so on. In general, we have

$$S_t = wy_t + w(1 - w)y_{t-1} + w(1 - w)^2y_{t-2} + \cdots + (1 - w)^{t-1}y_1$$

This formula states that the smoothed time series in period t depends on all the previous observations of the time series.

The smoothing constant w is chosen on the basis of how much smoothing is required. A small value of w produces a great deal of smoothing. A large value of w results in very little smoothing. Figure 18.6 depicts a time series and two exponentially smoothed series with $w = 0.1$ and $w = 0.5$.

FIGURE 18.6

Time Series and Two Exponentially Smoothed Series

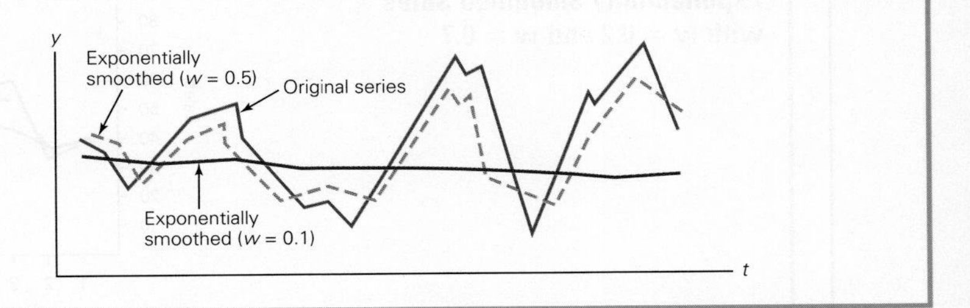

EXAMPLE 18.2

Petrol Sales, Part 2

Apply the exponential smoothing technique with $w = 0.2$ and $w = 0.7$ to the data in Example 18.1, and graph the results.

SOLUTION:

COMPUTE

MANUALLY:

The exponentially smoothed values are calculated from the formula

$$S_t = wy_t + (1 - w)S_{t-1}$$

The results with $w = 0.2$ and $w = 0.7$ are shown in the following table.

Time Period	Petrol Sales	Exponentially Smoothed with $w = 0.2$	Exponentially Smoothed with $w = 0.7$
1	39	39.0	39.0
2	37	38.6	37.6
3	61	43.1	54.0
4	58	46.1	56.8
5	18	40.5	29.6
6	56	43.6	48.1
7	82	51.2	71.8
8	27	46.4	40.4

(continued)

Time Period	Petrol Sales	Exponentially Smoothed with $w = 0.2$	Exponentially Smoothed with $w = 0.7$
9	41	45.3	40.8
10	69	50.1	60.6
11	49	49.8	52.5
12	66	53.1	61.9
13	54	53.3	56.4
14	42	51.0	46.3
15	90	58.8	76.9
16	66	60.2	69.3

Figure 18.7 shows the exponentially smoothed time series.

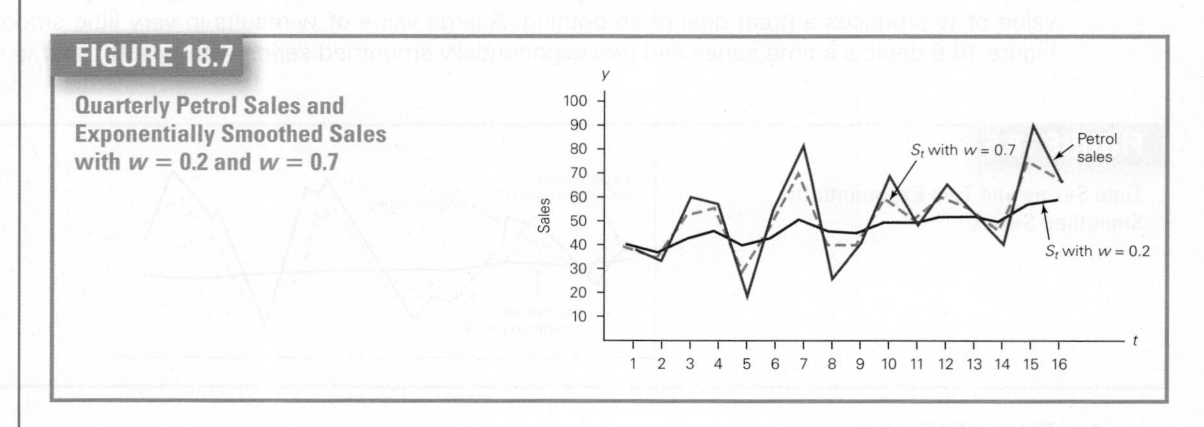

FIGURE 18.7 Quarterly Petrol Sales and Exponentially Smoothed Sales with $w = 0.2$ and $w = 0.7$

EXCEL

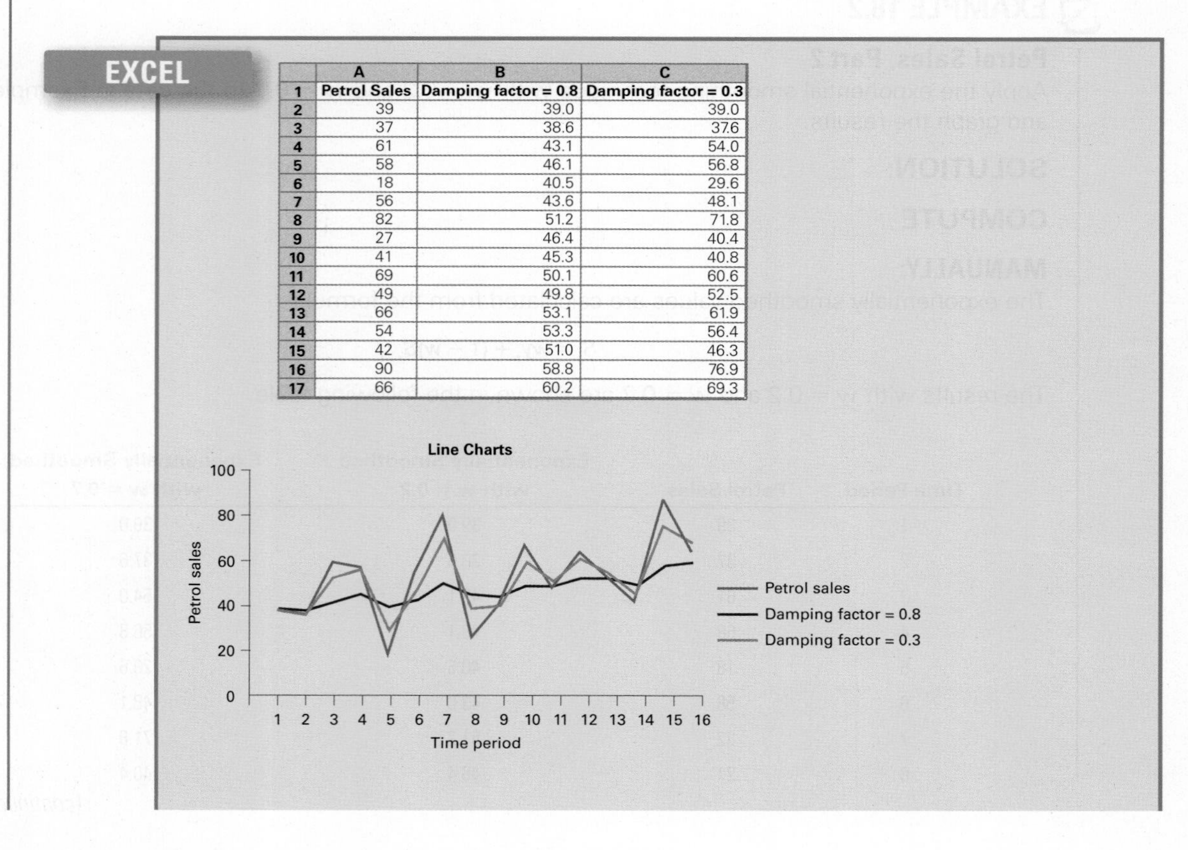

	A	B	C
1	Petrol Sales	Damping factor = 0.8	Damping factor = 0.3
2	39	39.0	39.0
3	37	38.6	37.6
4	61	43.1	54.0
5	58	46.1	56.8
6	18	40.5	29.6
7	56	43.6	48.1
8	82	51.2	71.8
9	27	46.4	40.4
10	41	45.3	40.8
11	69	50.1	60.6
12	49	49.8	52.5
13	66	53.1	61.9
14	54	53.3	56.4
15	42	51.0	46.3
16	90	58.8	76.9
17	66	60.2	69.3

Instructions

1. Type or import the data into one column. (**Open** Xm18-01.)
2. Click **Data, Data Analysis** and **Exponential Smoothing.**
3. Specify the **Input Range** (A1:A17). Type the **Damping factor,** which is 1 − *w* (0.8). Specify the **Output Range** (B1). To calculate the second exponentially smoothed time series specify 1 − *w* (0.3) **Output Range** (C1).

To modify the table so that the smoothed values appear the way we calculated, manually click the cell containing the last smoothed value displayed here (58.8) and drag it to the cell below to reveal the final smoothed value (60.2 and 69.3).

MINITAB

Petrol Sales	SMOO1	SMOO2
39	39.0000	39.0000
37	38.6000	37.6000
61	43.0800	53.9800
58	46.0640	56.7940
18	40.4512	29.6382
56	43.5610	48.0915
82	51.2488	71.8274
27	46.3990	40.4482
41	45.3192	40.8345
69	50.0554	60.5503
49	49.8443	52.4651
66	53.0754	61.9395
54	53.2603	56.3819
42	51.0083	46.3146
90	58.8066	76.8944
66	60.2453	69.2683

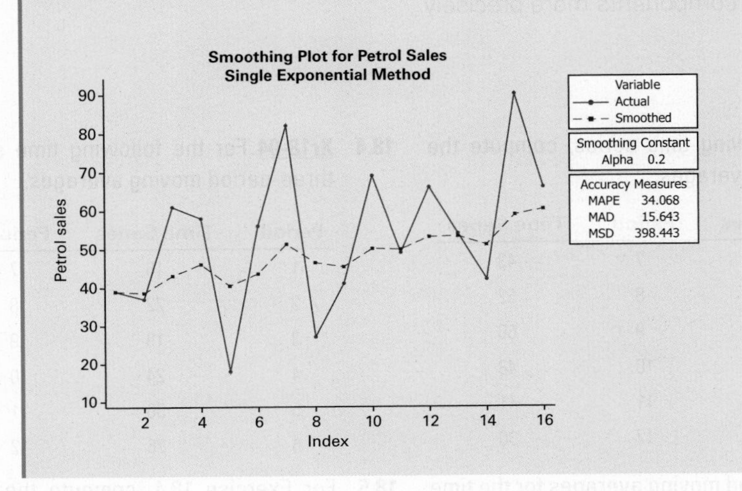

(*continued*)

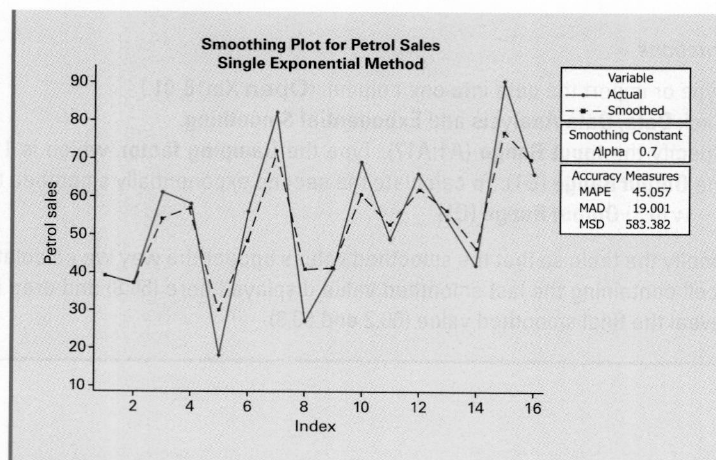

Instructions

1. Type or import the data into one column. (**Open** Xm18-01.)
2. Click **Stat, Time-series** and **Single Exp Smoothing** . . . .
3. Type or select the time variable in the **Variable box** (Petrol Sales). Click **Use** under **Weight to use in Smoothing** and type the value of *w* (0.2) (0.7). Click **Options** . . . .
4. Type 1 in the **Use average of first *k* observations K** = box.
5. Click **Storage** . . . and **Smoothed data.**

To draw the graph, click **Graphs** . . . and **Plot smoothed vs actual.**

INTERPRET

Figure 18.7 depicts the graph of the original time series and the exponentially smoothed series. As you can see, $w = 0.7$ results in very little smoothing, whereas $w = 0.2$ results in perhaps too much smoothing. In both smoothed time series, it is difficult to discern the seasonal pattern that we detected by using moving averages. A different value of *w* (perhaps $w = 0.5$) would be likely to produce more satisfactory results.

Moving averages and exponential smoothing are relatively crude methods of removing the random variation to discover the existence of other components. In the next section, we attempt to measure these components more precisely.

EXERCISES

18.1 Xr18-01 For the following time series, compute the three-period moving averages.

Period	Time Series	Period	Time Series
1	48	7	43
2	41	8	52
3	37	9	60
4	32	10	48
5	36	11	41
6	31	12	30

18.2 Compute the five-period moving averages for the time series in Exercise 18.1.

18.3 For Exercises 18.1 and 18.2, graph the time series and the two moving averages.

18.4 Xr18-04 For the following time series, compute the three-period moving averages.

Period	Time Series	Period	Time Series
1	16	7	24
2	22	8	29
3	19	9	21
4	24	10	23
5	30	11	19
6	26	12	15

18.5 For Exercise 18.4, compute the five-period moving averages.

18.6 For Exercises 18.4 and 18.5, graph the time series and the two moving averages.

18.7 **Xr18-07** Apply exponential smoothing with $w = 0.1$ to help detect the components of the following time series.

Period	1	2	3	4	5
Time Series	12	18	16	24	17
Period	6	7	8	9	10
Time Series	16	25	21	23	14

18.8 Repeat Exercise 18.7 with $w = 0.8$.

18.9 For Exercises 18.7 and 18.8, draw the time series and the two sets of exponentially smoothed values. Does there appear to be a trend component in the time series?

18.10 **Xr18-10** The following daily sales figures have been recorded in a medium-size merchandising firm.

	Week			
Day	**1**	**2**	**3**	**4**
Monday	43	51	40	64
Tuesday	45	41	57	58
Wednesday	22	37	30	33
Thursday	25	22	33	38
Friday	31	25	37	25

a. Compute the three-day moving averages.

b. Plot the time series and the moving averages on a graph.

c. Does there appear to be a seasonal (weekly) pattern?

18.11 For Exercise 18.10, compute the 5-day moving averages, and superimpose these on the same graph. Does this help you answer part (c) of Exercise 18.10?

18.12 **Xr18-12** The following quarterly sales of a department store chain were recorded for the years 2010–2013.

	Year			
Quarter	**2010**	**2011**	**2012**	**2013**
1	18	33	25	41
2	22	20	36	33
3	27	38	44	52
4	31	26	29	45

a. Calculate the four-quarter centred moving averages.

b. Graph the time series and the moving averages.

c. What can you conclude from your time-series smoothing?

18.13 Repeat Exercise 18.12, using exponential smoothing with $w = 0.4$.

18.14 Repeat Exercise 18.12, using exponential smoothing with $w = 0.8$.

18-3 TREND AND SEASONAL EFFECTS

In the previous section, we described how smoothing a time series can give us a clearer picture of which components are present. In order to forecast, however, we often need more precise measurements of the time-series components.

18-3a Trend Analysis

A trend can be linear or non-linear and, indeed, can take on a whole host of functional forms. The easiest way of measuring the long-term trend is by regression analysis, where the independent variable is time. If we believe that the long-term trend is approximately linear, we will use the linear model introduced in Chapter 15:

$$y = \beta_0 + \beta_1 t + \varepsilon$$

If we believe that the trend is non-linear, we can use one of the polynomial models. For example, the quadratic model is

$$y = \beta_0 + \beta_1 t + \beta_2 t^2 + \varepsilon$$

In most realistic applications, the linear model is used. We will demonstrate how the long-term trend is measured and applied later in this section.

18-3b Seasonal Analysis

Seasonal variation may occur within a year or within shorter intervals, such as a month, week or day. To measure the seasonal effect, we compute seasonal indexes, which gauge the degree to which the seasons differ from one another. One requirement necessary to calculate seasonal indexes is a time series sufficiently long enough to allow us to observe the variable over several seasons. For example, if the seasons are defined as the quarters of a year, we need to observe the time series for at least 4 years. The seasonal indexes are computed in the following way.

18-3c Procedure for Computing Seasonal Indexes

1. Remove the effect of seasonal and random variation by regression analysis; that is, compute the sample regression line

$$\hat{y}_t = b_0 + b_1 t$$

2. For each time period compute the ratio

$$\frac{y_t}{\hat{y}_t}$$

   This ratio removes most of the trend variation.
3. For each type of season, compute the average of the ratios in step 2. This procedure removes most (but seldom all) of the random variation, leaving a measure of seasonality.
4. Adjust the averages in step 3 so that the average of all the seasons is 1 (if necessary).

EXAMPLE 18.3

Data Xm18-03

Hotel Quarterly Occupancy Rates

The tourist industry is subject to seasonal variation. In most resorts, the spring and summer seasons are considered the 'high' seasons. Autumn and winter (except for Christmas and New Year) are 'low' seasons. A hotel in Bermuda has recorded the occupancy rate for each quarter for the past 5 years. These data are shown here. Measure the seasonal variation by computing the seasonal indexes.

Year	Quarter	Occupancy Rate
2009	1	0.561
	2	0.702
	3	0.800
	4	0.568
2010	1	0.575
	2	0.738
	3	0.868
	4	0.605
2011	1	0.594
	2	0.738
	3	0.729
	4	0.600
2012	1	0.622
	2	0.708
	3	0.806
	4	0.632
2013	1	0.665
	2	0.835
	3	0.873
	4	0.670

SOLUTION:

COMPUTE MANUALLY:

We performed a regression analysis with y = occupancy rate and t = time period 1, 2, . . . , 20. The regression equation is

$$\hat{y} = 0.639368 + 0.005246t$$

For each time period, we computed the ratio

$$\frac{y_t}{\hat{y}_t}$$

In the next step, we collected the ratios associated with each quarter and computed the average. We then computed the seasonal indexes by adjusting the average ratios so that they summed to 4.0, if necessary. In this example, it was not necessary.

Year	Quarter	t	y_t	$\hat{y} = 0.639368 + 0.005246t$	Ratio $\frac{y_t}{\hat{y}_t}$
2009	1	1	0.561	0.645	0.870
	2	2	0.702	0.650	1.080
	3	3	0.800	0.655	1.221
	4	4	0.568	0.660	0.860
2010	1	5	0.575	0.666	0.864
	2	6	0.738	0.671	1.100
	3	7	0.868	0.676	1.284
	4	8	0.605	0.681	0.888
2011	1	9	0.594	0.687	0.865
	2	10	0.738	0.692	1.067
	3	11	0.729	0.697	1.046
	4	12	0.600	0.702	0.854
2012	1	13	0.622	0.708	0.879
	2	14	0.708	0.713	0.993
	3	15	0.806	0.718	1.122
	4	16	0.632	0.723	0.874
2013	1	17	0.665	0.729	0.913
	2	18	0.835	0.734	1.138
	3	19	0.873	0.739	1.181
	4	20	0.670	0.744	0.900

	Quarter			
Year	1.	2	3	4
2009	0.870	1.080	1.221	0.860
2010	0.864	1.100	1.284	0.888
2011	0.865	1.067	1.046	0.854
2012	0.879	0.993	1.122	0.874
2013	0.913	1.138	1.181	0.900
Average	0.878	1.076	1.171	0.875
Index	0.878	1.076	1.171	0.875

EXCEL

	A	B
1	Seasonal Indexes	
2		
3	Season	Index
4	1	0.8782
5	2	1.0756
6	3	1.1709
7	4	0.8753

Instructions

1. Type or import the time series in chronological order into one column and the codes representing the seasons into the next column. (**Open** Xm18-03.)
2. Click **ADD-INS Data Analysis Plus** and **Seasonal Indexes**.
3. Specify the **Input Range** (A1:B21).

MINITAB

Time Series Decomposition for Rate

Fitted Trend Equation

Yt = 0.638747 + 0.00535513*t

Seasonal Indices

Period	Index
1	0.87684
2	1.07939
3	1.17764
4	0.86613

Note: Some output has been omitted.

Instructions

1. Type or import the data into one column. (**Open** file Xm18-03.)
2. Click **Stat, Time-series** and **Decomposition** . . . .
3. Select the **variable** (Rate), specify the **Seasonal length** (4) and click **Multiplicative** under **Model Type** and **Trend plus seasonal** under **Model Components.** Click **Options** . . . .
4. Type 1 in the **First obs. is in period box.**
5. Click **Storage** . . . and **Seasonals.**

INTERPRET

Note that Minitab's results differ from those computed manually and by Excel. (Minitab uses a technique other than ordinary least squares to produce the trend line.) In this example, the differences are small. However, in other cases there can be substantial differences. The discussion that follows uses Excel's figures.

The seasonal indexes tell us that, on average, the occupancy rates in the first and fourth quarters are below the annual average, and the occupancy rates in the second and third quarters are above the annual average. Using Excel's figures (the manually calculated seasonal indexes are the same, but Minitab's indexes differ slightly), we expect the occupancy rate in the first quarter to be 12.2% (100% – 87.8%) below the annual rate. The second and third quarters' rates are expected to be 7.6% and 17.1%, respectively, above the annual rate. The fourth quarter's rate is 12.5% below the annual rate. Figure 18.8 depicts the time series and the regression trend line.

FIGURE 18.8

Time Series and Trend for Example 18.3

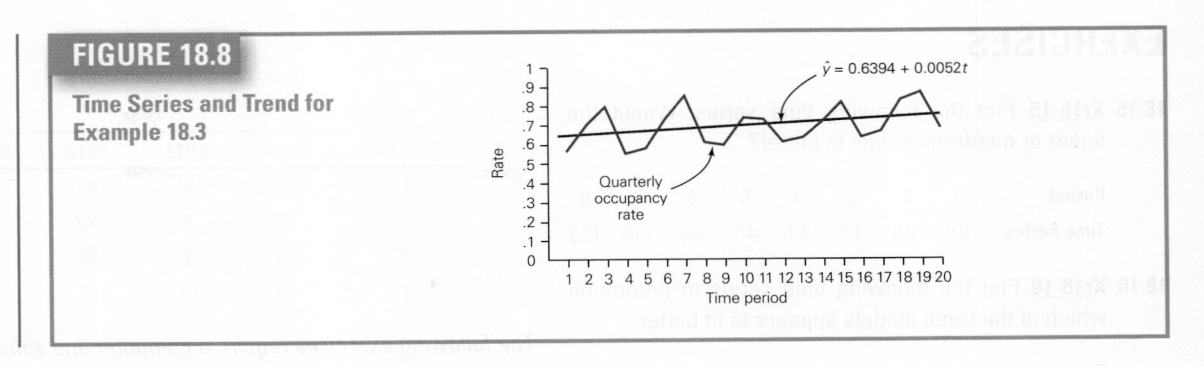

18-3d Deseasonalising a Time Series

One application of seasonal indexes is to remove the seasonal variation in a time series. The process is called **deseasonalising**, and the result is called a **seasonally adjusted time series**. Often this allows the statistics practitioner to more easily compare the time series across seasons. For example, the unemployment rate varies according to the season. During the winter months, unemployment usually rises: it falls in the spring and summer. The seasonally adjusted unemployment rate allows economists to determine whether unemployment has increased or decreased over the previous months. The process is easy: Simply divide the time series by the seasonal indexes. To illustrate, we have deseasonalised the occupancy rates in Example 18.3 (using the seasonal indexes produced by Excel). The results are shown next.

Year	Quarter	Occupancy Rate y_t	Seasonal Index	Seasonally Adjusted Occupancy Rate
2009	1	0.561	0.878	0.639
	2	0.702	1.076	0.652
	3	0.800	1.171	0.683
	4	0.568	0.875	0.649
2010	1	0.575	0.878	0.655
	2	0.738	1.076	0.686
	3	0.868	1.171	0.741
	4	0.605	0.875	0.691
2011	1	0.594	0.878	0.677
	2	0.738	1.076	0.686
	3	0.729	1.171	0.623
	4	0.600	0.875	0.686
2012	1	0.622	0.878	0.708
	2	0.708	1.076	0.658
	3	0.806	1.171	0.688
	4	0.632	0.875	0.722
2013	1	0.665	0.878	0.757
	2	0.835	1.076	0.776
	3	0.873	1.171	0.746
	4	0.670	0.875	0.766

By removing the seasonality, we can see when there has been a 'real' increase or decrease in the occupancy rate. This enables the statistics practitioner to examine the factors that produced the rate change. We can more easily see that there has been an increase in the occupancy rate over the 5-year period.

In the next section, we show how to forecast with seasonal indexes.

EXERCISES

18.15 Xr18-15 Plot the following time series. Would the linear or quadratic model fit better?

Period	1	2	3	4	5	6	7	8
Time Series	0.5	0.6	1.3	2.7	4.1	6.9	10.8	19.2

18.16 Xr18-16 Plot the following time series to determine which of the trend models appears to fit better.

Period	1	2	3	4	5
Time Series	55	57	53	49	47
Period	6	7	8	9	10
Time Series	39	41	33	28	20

18.17 Refer to Exercise 18.15. Use regression analysis to calculate the linear and quadratic trends. Which line fits better?

18.18 Refer to Exercise 18.16. Use regression analysis to calculate the linear and quadratic trends. Which line fits better?

18.19 Xr18-19 For the following time series, compute the seasonal (daily) indexes.

The regression line is

$$\hat{y} = 16.8 + 0.366t \,(t = 1, 2, \ldots, 20)$$

Day	Week 1	2	3	4
Monday	12	11	14	17
Tuesday	18	17	16	21
Wednesday	16	19	16	20
Thursday	25	24	28	24
Friday	31	27	25	32

Applications

18.20 Xr18-20 The quarterly earnings (in €millions) of a large soft-drink manufacturer have been recorded for the years 2010 to 2013. These data are listed here. Compute the seasonal indexes given the regression line

$$\hat{y} = 61.75 + 1.18t \,(t = 1, 2, \ldots, 16)$$

Quarter	Year 2010	2011	2012	2013
1	52	57	60	66
2	67	75	77	82
3	85	90	94	98
4	54	61	63	67

The following exercises require a computer and software.

18.21 Xr18-21 Student enrolments on Higher Education (HE) courses in the UK have increased in recent years. To help forecast future enrolment, we can look at the available data showing the total number of undergraduate HE enrolments at UK HE institutions from 2007–08 to 2011–12. The data (in thousands) are listed here.

Year	2007/08	2008/09	2009/10	2010/11	2011/12
Enrolment	1 804 970	1 859 240	1 914 715	1 912 580	1 928 140

a. Plot the time series.

b. Use regression analysis to determine the trend.

18.22 Xr18-22 The number of cable television subscribers has increased over the past 5 years. The marketing manager for a cable company has recorded the numbers of subscribers for the past 24 quarters.

a. Plot the numbers.

b. Compute the seasonal (quarterly) indexes.

18.23 Xr18-23 The owner of a pizzeria wants to forecast the number of pizzas she will sell each day. She recorded the numbers sold daily for the past 4 weeks. Calculate the seasonal (daily) indexes.

18.24 Xr18-24 A manufacturer of ski equipment is in the process of reviewing his accounts receivable. He noticed that there appears to be a seasonal pattern with the accounts receivable increasing in the winter months and decreasing during the summer. The quarterly accounts receivable (in €millions) were recorded. Compute the seasonal (quarterly) indexes.

18-4 INTRODUCTION TO FORECASTING

Many different forecasting methods are available for the statistics practitioner. One factor to be considered in choosing among them is the type of component that makes up the time series. Even then, however, we have several different methods from which to choose. One way of deciding which method to apply is to select the technique that achieves the greatest forecast accuracy.

The most commonly used measures of forecast accuracy are **mean absolute deviation (MAD)** and the **sum of squares for forecast errors (SSE)**.

Mean Absolute Deviation

$$\text{MAD} = \frac{\sum_{i=1}^{n} |y_t - F_t|}{n}$$

where

y_t = Actual value of the time series at time period t

F_t = Forecasted value of die time series at time period t

n = Number of time periods

Sum of Squares for Forecast Error

$$\text{SSE} = \sum_{i=1}^{n} (y_t - F_t)^2$$

MAD averages the absolute differences between the actual and forecast values; SSE is the sum of the squared differences. Which measure to use in judging forecast accuracy depends on the circumstances. If avoiding large errors is important, SSE should be used because it penalises large deviations more heavily than does MAD. Otherwise, use MAD.

It is probably best to use some of the observations of the time series to develop several competing forecasting models and then forecast for the remaining time periods. Afterwards, compute MAD or SSE for the forecasts. For example, if we have 5 years of monthly observations, use the first 4 years to develop the forecasting models and then use them to forecast the fifth year. Because we know the actual values in the fifth year, we can choose the technique that results in the most accurate forecast using either MAD or SSE.

EXAMPLE 18.4

Comparing Forecasting Models

Annual data from 1976 to 2009 were used to develop three different forecasting models. Each model was used to forecast the time series for 2010, 2011, 2012 and 2013. The forecasted and actual values for these years are shown here. Use MAD and SSE to determine which model performed best.

Year	Actual Time Series	1	2	3
2010	129	136	118	130
2011	142	148	141	146
2012	156	150	158	170
2013	183	175	163	180

SOLUTION:

For model 1, we have

$$\text{MAD} = \frac{|129 - 136| + |142 - 148| + |156 - 150| + |183 - 175|}{4}$$

$$= \frac{7 + 6 + 6 + 8}{4} = 6.75$$

$$\text{SSE} = (129 - 136)^2 + (142 - 148)^2 + (156 - 150)^2 + (183 - 175)^2$$
$$= 49 + 36 + 36 + 64 = 185$$

For model 2, we compute

$$\text{MAD} = \frac{|129 - 118| + |142 - 141| + |156 - 158| + |183 - 163|}{4}$$
$$= \frac{11 + 1 + 2 + 20}{4} = 8.5$$
$$\text{SSE} = (129 - 118)^2 + (142 - 141)^2 + (156 - 158)^2 + (183 - 163)^2$$
$$= 121 + 1 + 4 + 400 = 526$$

The measures of forecast accuracy for model 3 are

$$\text{MAD} = \frac{|129 - 130| + |142 - 146| + |156 - 170| + |183 - 180|}{4}$$
$$= \frac{1 + 4 + 14 + 3}{4} = 5.5$$
$$\text{SSE} = (129 - 130)^2 + (142 - 146)^2 + (156 - 170)^2 + (183 - 180)^2$$
$$= 1 + 16 + 196 + 9 = 222$$

Model 2 is inferior to both models 1 and 3, no matter how we measure forecast accuracy. Using MAD, model 3 is best – but using SSE, model 1 is most accurate. The choice between model 1 and model 3 should be made on the basis of whether we prefer a model that consistently produces moderately accurate forecasts (model 1) or one whose forecasts come quite close to most actual values but is less sensitive to outliers because it is based on the absolute differences between the actual and forecast values (model 3).

EXERCISES

18.25 For the actual and forecast values of a time series shown here, calculate MAD and SSE.

Period	1	2	3	4	5
Forecast	173	186	192	211	223
Actual Value	166	179	195	214	220

18.26 Two forecasting models were used to predict the future values of a time series. These are shown here together with the actual values. Compute MAD and SSE for each model to determine which was more accurate.

Period	1	2	3	4
Forecast (Model 1)	7.5	6.3	5.4	8.2
Forecast (Model 2)	6.3	6.7	7.1	7.5
Actual	6.0	6.6	7.3	9.4

18.27 Three forecasting techniques were used to predict the values of a time series. These values are given in the following table. Compute MAD and SSE for each technique to determine which was most accurate.

Period	1	2	3	4	5
Forecast (Model 1)	21	27	29	31	35
Forecast (Model 2)	22	24	26	28	30
Forecast (Model 3)	17	20	25	31	39
Actual	19	24	28	32	38

18-5 FORECASTING MODELS

There are a large number of different forecasting techniques available to statistics practitioners. However, many are beyond the level of this book. In this section, we present three models. Similar to the method of choosing the correct statistical inference technique in Chapters 12 to 17, the choice of model depends on the time-series components.

18-5a Forecasting with Exponential Smoothing

If the time series displays a gradual trend or no trend and no evidence of seasonal variation, exponential smoothing can be effective as a forecasting method. Suppose that t represents the most recent time period and we have computed the exponentially smoothed value S_t. This value is then the forecasted value at time $t + 1$; that is,

$$F_{t+1} = S_t$$

If we wish, we can forecast two or three or any number of periods into the future:

$$F_{t+2} = S_t \quad \text{or} \quad F_{t+3} = S_t$$

It must be understood that the accuracy of the forecast decreases rapidly for predictions more than one time period into the future. However, as long as we are dealing with time series with no cyclical or seasonal variation, we can produce reasonably accurate predictions for the next time period.

18-5b Forecasting with Seasonal Indexes

If the time series is composed of seasonal variation and long-term trend, we can use seasonal indexes and the regression equation to forecast.

Forecast of Trend and Seasonality

The forecast for time period t is

$$F_t = [b_0 + b_1 t] \times SI_t$$

where

F_t = Forecast for period t

$b_0 + b_1 t$ = Regression equation

SI_t = Seasonal index for period t

EXAMPLE 18.5

Forecasting Hotel Occupancy Rates

Forecast hotel occupancy rates for next year in Example 18.3.

SOLUTION:

In the process of computing the seasonal indexes, we computed the trend line. It is

$$\hat{y} = 0.639 + 0.00525t$$

For t = 21, 22, 23 and 24, we calculate the forecasted trend values.

Quarter	t	$\hat{y} = 0.639 + 0.00525t$
1	21	$0.639 + 0.00525(21) = 0.749$
2	22	$0.639 + 0.00525(22) = 0.755$
3	23	$0.639 + 0.00525(23) = 0.760$
4	24	$0.639 + 0.00525(24) = 0.765$

We now multiply the forecasted trend values by the seasonal indexes calculated in Example 18.3. The seasonalised forecasts are as follows:

Quarter	t	Trend Value $\hat{y}_t$	Seasonal Index	Forecast $F_t = \hat{y}_t \times SI_t$
1	21	0.749	0.878	$0.749 \times 0.878 = 0.658$
2	22	0.755	1.076	$0.755 \times 1.076 = 0.812$
3	23	0.760	1.171	$0.760 \times 1.171 = 0.890$
4	24	0.765	0.875	$0.765 \times 0.875 = 0.670$

INTERPRET

We forecast that the quarterly occupancy rates during the next year will be 0.658, 0.812, 0.890 and 0.670.

18-5c Autoregressive Model

In Chapter 16 we discussed autocorrelation wherein the errors are not independent of one another. The existence of strong autocorrelation indicates that the model has been misspecified, which usually means that until we improve the regression model, it will not provide an adequate fit. However, autocorrelation also provides us with an opportunity to develop another forecasting technique. If there is no obvious trend or seasonality and we believe that there is a correlation between consecutive residuals, the **autoregressive model** may be most effective.

Autoregressive Model

$$y_t = \beta_0 + \beta_1 y_{t-1} + \varepsilon$$

The model specifies that consecutive values of the time series are correlated. We estimate the coefficient in the usual way. The estimated regression line is defined as

$$\hat{y}_t = b_0 + b_1 y_{t-1}$$

EXAMPLE 18.6

Data Xm18-06

Forecasting Changes to the Consumer Price Index

The consumer price index (CPI) is used as a general measure of inflation. It is an important measure because a high rate of inflation often influences governments to take corrective measures. The table below lists the consumer price index from 1980 to 2013 and the annual percentage increases in the CPI for South Africa. Forecast next year's change in the CPI.

Year	CPI	%Change	Year	CPI	%Change
1980	5.4		1997	41.9	8.8
1981	6.2	14.8	1998	44.8	6.9
1982	7.1	14.5	1999	47.1	5.1
1983	8.0	12.7	2000	49.6	5.3
1984	8.9	11.3	2001	52.4	5.6
1985	10.3	15.7	2002	57.2	9.2

Year	CPI	%Change	Year	CPI	%Change
1986	12.3	19.4	2003	60.5	5.8
1987	14.2	15.4	2004	61.4	1.5
1988	16.1	13.4	2005	63.4	3.3
1989	18.4	14.3	2006	66.4	4.7
1990	21.0	14.1	2007	71.1	7.1
1991	24.3	15.7	2008	79.3	11.5
1992	27.7	14.0	2009	84.6	6.7
1993	30.3	9.4	2010	88.2	4.3
1994	33.0	8.9	2011	92.6	5.0
1995	35.9	8.8	2012	97.8	5.6
1996	38.5	7.2	2013	103.4	5.7

SOLUTION:

Notice that we included the CPI for 1980 because we wanted to determine the percentage change ($\%\,Change = \dfrac{Current\ year\ CPI - Prior\ year\ CPI}{Prior\ year\ CPI} \times 100$) for 1981. We will use the percentage changes for 1981 to 2008 as the independent variable and the percentage change from 1982 to 2013 as the dependent variable. File Xm18-06 stores the data in the format necessary to determine the autoregressive model.

EXCEL

	A	B	C	D	E
16		*Coefficients*	*Standard Error*	*t Stat*	*P-value*
17	Intercept	1.3156	1.0250	1.28	0.2092
18	% Change X	0.8328	0.0970	8.58	0.0000

MINITAB

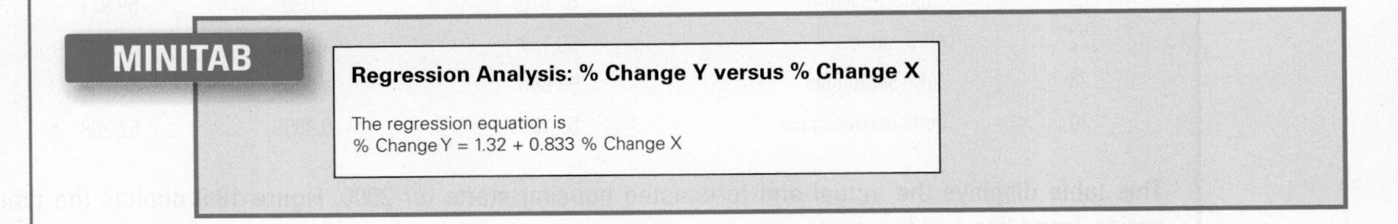

Regression Analysis: % Change Y versus % Change X

The regression equation is
% Change Y = 1.32 + 0.833 % Change X

INTERPRET

The regression line is

$$\hat{y}_t = 1.3156 + 0.08328y_{t-1}$$

Our forecast for 2014 is

$$\hat{y}_{2014} = 1.3156 + 0.08328(5.73\%) = 6.09\%$$

The autoregressive model forecasts a 6.09% increase in the CPI for the year 2014.

NEW HOME DEVELOPMENT: HOUSING 'STARTS': SOLUTION

A preliminary examination of the data reveals that there is a very small upward trend over the 5-year period. Moreover, the number of housing starts varies by month. The presence of these components suggests that we determine the linear trend and seasonal indexes.

EXCEL

With housing starts as the dependent variable and the season as the independent variable, Excel yielded the following regression line:

$$\hat{y} = 47014 + 507.43t, \quad t = 1, 2, \ldots, 20$$

The seasonal indexes were computed as follows. (See the instructions for Example 18.3.)

	A	B	C
1	Seasonal Indexes		
2			
3	Season		Index
4	1		1.0382
5	2		1.0543
6	3		1.0240
7	4		0.8835

The regression equation was used again to predict the number of housing starts based on the linear trend

$$\hat{y} = 47\,014 + 507.43\,t \quad (t = 21, 22, 23, 24)$$

These figures were multiplied by the seasonal indexes, which resulted in the following forecasts.

Period	Months	$\hat{y} = 47\,014 + 507.43t$	Index	Forecasts
21	January–March	57 670	1.0382	59 871
22	April–June	58 117	1.0543	61 335
23	July–September	58 685	1.0240	60 095
24	October–December	59 192	0.8835	52 298

This table displays the actual and forecasted housing starts for 2006. Figure 18.9 depicts the time series, trend line and forecasts.

Period	Month	Forecasts	Actual
21	January–March	59 871	65 500
22	April–June	61 335	58 980
23	July–September	60 095	52 810
24	October–December	52 298	46 680

The size of the error was measured by MAD and SSE. They are:

$$\text{MAD} = 5221.65$$

$$\text{SSE} = 121\,863\,264$$

FIGURE 18.9

Time Series, Trend and Forecasts of Housing Starts

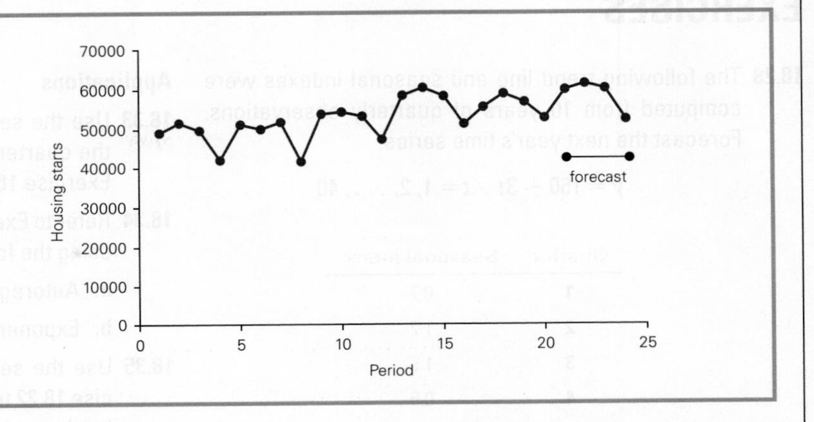

MINITAB

Time Series Decomposition for Starts

Multiplicative Model

Data	Starts
Length	20
NMissing	0

Fitted Trend Equation

Yt = 45774 + 633*t

Seasonal Indices

Period	Index
1	1.05203
2	1.05223
3	1.02516
4	0.87059

Accuracy Measures

MAPE	3
MAD	1449
MSD	3190719

Forecasts

Period	Forecast
21	62129.3
22	62806.5
23	61839.2
24	53065.8

Instructions

Follow the instructions presented in Example 18.3. For **Seasonal Length** type 4, check **Generate forecasts**, type 4 for the **Number of forecasts** and type 20 for **Starting from origin**.

EXERCISES

18.28 The following trend line and seasonal indexes were computed from 10 years of quarterly observations. Forecast the next year's time series.

$$\hat{y} = 150 + 3t \quad t = 1, 2, \ldots, 40$$

Quarter	Seasonal Index
1	0.7
2	1.2
3	1.5
4	0.6

18.29 The following trend line and seasonal indexes were computed from 4 weeks of daily observations. Forecast the 7 values for next week.

$$\hat{y} = 120 + 2.3t \quad t = 1, 2, \ldots, 28$$

Day	Seasonal Index
Sunday	1.5
Monday	0.4
Tuesday	0.5
Wednesday	0.6
Thursday	0.7
Friday	1.4
Saturday	1.9

18.30 Use the following autoregressive equation to forecast the next value of the time series if the last observed value is 65.

$$\hat{y} = 625 + 1.3y_{t-1}$$

18.31 Apply exponential smoothing with $w = 0.4$ to forecast the next four quarters in Exercise 18.12.

18.32 Use the seasonal indexes and trend line to forecast the time series for the next 5 days in Exercise 18.19.

Applications

18.33 Use the seasonal indexes and trend line to forecast the quarterly earnings for the years 2014 and 2015 in Exercise 18.20.

18.34 Refer to Exercise 18.21. Forecast next year's enrolment using the following methods.

a. Autoregressive forecasting model

b. Exponential smoothing method with $w = 0.5$

18.35 Use the seasonal indexes and trend line from Exercise 18.22 to forecast the number of cable subscribers for the next four quarters.

18.36 Refer to Exercise 18.23. Use the seasonal indexes and trend line to forecast the number of pizzas to be sold for each of the next 7 days.

18.37 Apply the trend line and seasonal indexes from Exercise 18.24 to forecast accounts receivable for the next four quarters.

18.38 Xr18-38 The revenues (in €millions) of a chain of ice-cream stores are listed for each quarter during the previous 5 years.

	Year				
Quarter	2009	2010	2011	2012	2013
1	16	14	17	18	21
2	25	27	31	29	30
3	31	32	40	45	52
4	24	23	27	24	32

a. Plot the time series.

b. Discuss why exponential smoothing is not recommended as a forecasting tool in this problem.

c. Use regression analysis to determine the trend line.

d. Determine the seasonal indexes.

e. Using the seasonal indexes and trend line, forecast revenues for the next four quarters.

CHAPTER SUMMARY

In this chapter we discussed the classical time series and its decomposition into trend, seasonal and random variation. Moving averages and exponential smoothing were used to remove some of the random variation, making it easier to detect trend and seasonality. The long-term trend was measured by regression analysis. Seasonal variation was measured by computing the seasonal indexes. Three forecasting techniques were described in this chapter: exponential smoothing, forecasting with seasonal indexes and the autoregressive model.

SYMBOLS:

Symbol	Represents
y_y	Time series
s_t	Exponentially smoothed time series
w	Smoothing constant
F_t	Forecasted time series

FORMULAS:

Exponential smoothing

$$S_t = wy_t + (1 - w)S_{t-1}$$

Mean absolute deviation

$$\text{MAD} = \frac{\sum_{i=1}^{n} |y_t - F_t|}{n}$$

Sum of squares for error

$$\text{SSE} = \sum_{i=1}^{n} (y_t - F_t)^2$$

Forecast of trend and seasonality

$$F_t = [b_0 + b_1 t] \times SI_t$$

Autoregressive model

$$y_t = \beta_0 + \beta_1 y_{t-1} + \varepsilon$$

19 STATISTICAL PROCESS CONTROL

19-1 PROCESS VARIATION

19-2 CONTROL CHARTS

19-3 CONTROL CHARTS FOR VARIABLES: $\bar{x}$ AND S CHARTS

19-4 CONTROL CHARTS FOR ATTRIBUTES: P CHART

DETECTING THE SOURCE OF DEFECTIVE DISCS

Data Xm19-00

A company that produces compact discs (CDs) has been receiving complaints from its customers about the large number of discs that will not store data properly. Company management has decided to institute statistical process control to remedy the problem. Every hour, a random sample of 200 discs is taken, and each disc is tested to determine whether it is defective. The number of defective discs in the samples of size 200 for the first 40 hours is shown here (in chronological order). Using these data, draw a *p* chart to monitor the production process. Was the process out of control when the sample results were generated?

19 5 16 20 6 12 18 6 13 15 10 6 7 10 18 20 13 6 8 3
8 7 4 19 3 19 9 10 10 18 15 16 5 14 3 10 19 13 19 9

See Section 19-4b for answer.

INTRODUCTION

Operations managers are responsible for developing and maintaining the production processes that deliver quality products and services. The goal is to select the methods, materials, machines and personnel (workers) that combine to yield the production process that features the smallest amount of variation at a reasonable cost. Once the production process is operating, it is necessary to constantly monitor the process to ensure that it functions the way it was designed. The statistical methods we are about to introduce are the most common application of statistics. At any point in time, there are literally thousands of firms applying these methods. This chapter deals with the subject of **statistical process control or SPC** (formerly called **quality control**). There are two general approaches to the management of quality. The first approach is to produce the product and, at the completion of the production process, inspect the unit to determine whether it conforms to specifications; if it doesn't, the unit is either discarded or repaired. This approach has several drawbacks. Foremost among them is producing substandard products that are later discarded or fixed, which is costly. In recent years, this approach has been employed by a decreasing number of companies. Instead, many firms have adopted the **prevention approach**. Using the concepts of hypothesis testing, statistics practitioners concentrate on the production process. Rather than inspect the product, they inspect the process to determine when the process starts producing units that do not conform to specifications. This allows them to correct the production process before it creates a large number of defective products.

In Section 19-1, we discuss the problem of process variation and why it is often the key to the management of quality. In Section 19-2, we also introduce the concept and logic of control charts and show why they work. In the rest of the chapter, we introduce three specific control charts.

19-1 PROCESS VARIATION

All production processes result in variation; that is, no product is exactly the same as another. You can see for yourself that this is true by weighing, for example, two boxes of breakfast cereal that are supposed to weigh 500 grams each. They not only will not weigh exactly 500 grams but also will not even have equal weights. All products exhibit some degree of variation. There are two sources of variation: *chance* and *assignable variation.* **Chance or common variation** is caused by a number of randomly occurring events that are part of the production process and, in general, cannot be eliminated without changing the process. In effect, chance variation was built into the product when the production process was first set up, perhaps as a result of a statistical analysis that attempted to minimise but not necessarily eliminate such variation.

Assignable or special variation is caused by specific events or factors that are frequently temporary and that can usually be identified and eliminated. To illustrate, consider a paint company that produces and sells paint in 1-litre cans. The cans are filled by an automatic valve that regulates the amount of paint in each can. The designers of the valve acknowledge that there will be some variation in the amount of paint even when the valve is working as it was designed to work. This is chance variation. Occasionally the valve will malfunction, causing the variation in the amount delivered to each can to increase. This increase is the assignable variation.

Perhaps the best way to understand what is happening is to consider the volume of paint in each can as a random variable. If the only sources of variation are caused by chance, then each can's volume is drawn from identical distributions; that is, each distribution has the same shape, mean and standard deviation. Under such circumstances, the production process is said to be **under control**. In recognition of the fact that variation in output will occur even when the process is under control and operating properly, most processes are designed so that their products will fall within designated **specification limits** or 'specs'. For example, the process that fills the paint cans may be designed so that the cans contain between 0.99 and 1.01 litres. Inevitably, some event or combination of factors in a production process will cause the process distribution to change. When it does, the process is said to be **out of control**. There are several possible ways for the process to go out of control. Here is a list of the most commonly occurring possibilities and their likely assignable causes.

1. **Level shift.** This is a change in the mean of the process distribution. Assignable causes include machine breakdown, new machine or operator, or a change in the environment. In the paint-can illustration, a temperature or humidity change may affect the density of the paint, resulting in less paint in each can.
2. **Instability.** This is the name we apply to the process when the standard deviation increases. (As we discuss later, a decrease in the standard deviation is desirable.) This may be caused by a machine in need of repair, defective materials, worn tools or a poorly trained operator. Suppose, for example that a part in the valve that controls the amount of paint wears down, causing greater variation than normal.
3. **Trend.** When there is a slow steady shift (either up or down) in the process distribution mean, the result is a trend. This is frequently the result of less-than-regular maintenance, operator fatigue, residue or dirt build-up, or gradual loss of lubricant. If the paint-control valve becomes increasingly clogged, we would expect to see a steady decrease in the amount of paint delivered.
4. **Cycle.** This is a repeated series of small observations followed by large observations. Likely assignable causes include environmental changes, worn parts or operator fatigue. If there are changes in the voltage in the electricity that runs the machines in the paint can example, we might see series of overfilled cans and series of underfilled cans.

The key to quality is to detect when the process goes out of control so that we can correct the malfunction and restore control. The control chart is the statistical method that we use to detect problems.

EXERCISES

19.1 What is meant by *chance variation*?

19.2 Provide two examples of production processes and their associated chance variation.

19.3 What is meant by *special variation*?

19.4 Your education as a statistics practitioner can be considered a production process overseen by the course instructor. The variable we measure is the grade achieved by each student.

a. Discuss chance variation – that is, describe the sources of variation that the instructor has no control over.

b. Discuss special variation.

19-2 CONTROL CHARTS

A **control chart** is a plot of statistics over time. For example, an $\overline{x}$ **chart** plots a series of sample means taken over a period of time. Each control chart contains a **centreline** and *control limits*. The control limit above the centreline is called the **upper control limit** and that below the centreline is called the **lower control limit**. If, when the sample statistics are plotted, all points are randomly distributed between the control limits, we conclude that the process is under control. If the points are not randomly distributed between the control limits, we conclude that the process is out of control.

To illustrate the logic of control charts, let us suppose that in the paint can example described previously we want to determine whether the central location of the distribution has changed from one period to another. We will draw our conclusion from an x chart. For the moment, let us assume that we know the mean μ and standard deviation σ of the process when it is under control. We can construct the $\overline{x}$ chart as shown in Figure 19.1. The chart is drawn so that the vertical axis plots the values of $\overline{x}$ that will be calculated and the horizontal axis tracks the samples in the order in which they are drawn. The centreline is the value of μ. The control limits are set at three standard errors from the centreline. Recall that the standard error of $\bar{x}$ is $\sigma/\sqrt{n}$. Hence, we define the control limits as follows:

$$\text{Lower control limit} = \mu - 3\frac{\sigma}{\sqrt{n}}$$

$$\text{Upper control limit} = \mu + 3\frac{\sigma}{\sqrt{n}}$$

FIGURE 19.1

$\bar{x}$ Chart: μ and σ Known

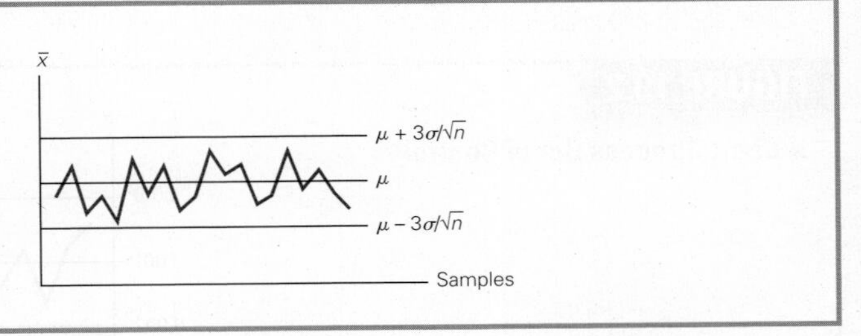

After we have constructed the chart by drawing the centreline and control limits, we use it to plot the sample means, which are joined to make it easier to interpret. The principles underlying control charts are identical to the principles of hypothesis testing. The null and alternative hypotheses are

H_0: The process is under control

H_1: The process is out of control

For an $\bar{x}$ chart, the test statistic is the sample mean $\bar{x}$. However, because we are dealing with a dynamic process rather than a fixed population, we test a series of sample means: we compute the mean for each of a continuing series of samples taken over time. For each series of samples, we want to determine whether there is sufficient evidence to infer that the process mean has changed. We reject the null hypothesis if at any time the sample mean falls outside the control limits. It is logical to ask why we use 3 standard errors and not 2 or 1.96 or 1.645 as we did when we tested hypotheses about a population mean in Chapter 11. The answer lies in the way in which all tests are conducted. Because test conclusions are based on sample data, there are two possible errors. In statistical process control, a Type I error occurs if we conclude that the process is out of control when, in fact, it is not. This error can be quite expensive because the production process must be stopped and the causes of the variation found and repaired. Consequently, we want the probability of a Type I error to be small. With control limits set at 3 standard errors from the mean, the probability of a Type I error for each sample is

$$\alpha = P(|z| > 3) = 0.0026$$

Recall that a small value of α results in a relatively large value of the probability of a Type II error. A Type II error occurs when at any sample we do not reject a false null hypothesis. This means that, for each sample, we are less likely to recognise when the process goes out of control. However, because we will be performing a series of tests (one for each sample), we will eventually discover that the process is out of control and take steps to rectify the problem.

Suppose that in order to test the production process that fills 1-litre paint cans, we choose to take a sample of size 4 every hour. Let us also assume that we know the mean and standard deviation of the process distribution of the amount of paint when the process is under control, say, $\mu = 1.001$ and $\sigma = 0.006$. (This means that when the valve is working the way it was designed, the amount of paint put into each can is a random variable whose mean is 1.001 litres and whose standard deviation is 0.006 litres.) Thus,

$$\text{Centreline} = \mu = 1.001$$

$$\text{Lower control limit} = \mu - 3\frac{\sigma}{\sqrt{n}} = 1.001 - 3\frac{0.006}{\sqrt{4}} = 1.001 - 0.009 = 0.992$$

$$\text{Upper control limit} = \mu + 3\frac{\sigma}{\sqrt{n}} = 1.001 + 3\frac{0.006}{\sqrt{4}} = 1.001 + 0.009 = 1.010$$

Figure 19.2 depicts a situation in which the first 15 samples were taken when the process was under control. However, after the 15th sample was drawn, the process went out of control and produced sample means outside the control limits. We conclude that the process distribution has changed because the data display variability beyond that predicted for a process with the specified mean and standard deviation. This means that the variation is assignable and that the cause must be identified and corrected.

FIGURE 19.2

$\bar{x}$ Chart: Process Out of Control

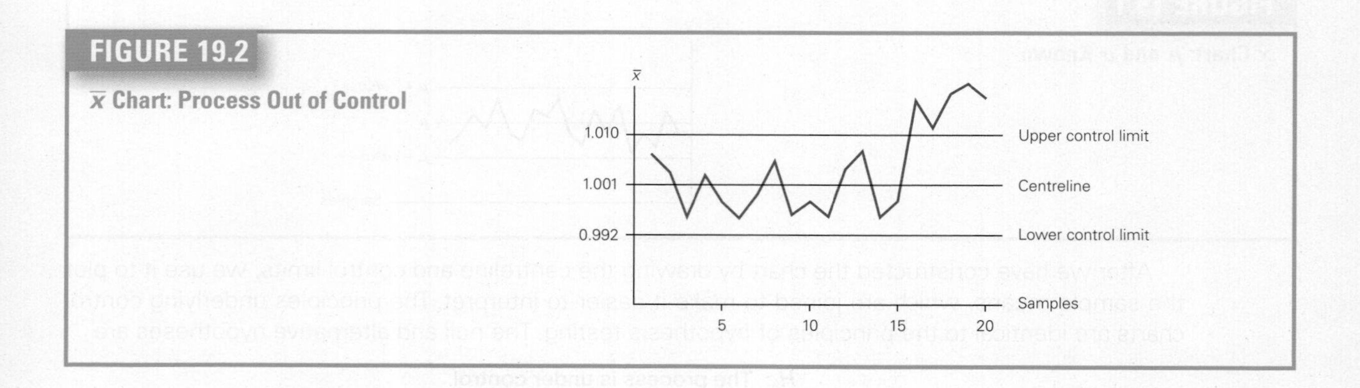

As we stated previously, Statistical Process Control (SPC) is a slightly different form of hypothesis testing. The concept is the same, but there are differences that you should be aware of. The most important difference is that when we tested means and proportions in Chapters 11 and 12, we were dealing with fixed but unknown parameters of populations. For instance, in Example 11.1 the population we dealt with was the account balances of the department store customers. The population mean balance was a constant value that we simply did not know. The purpose of the test was to determine whether there was enough statistical evidence to allow us to infer that the mean balance was greater than €170. So we took one sample and based the decision on the sample mean. When dealing with a production process, it's important to realise that the process distribution itself is variable; that is, at any time, the process distribution of the amount of paint fill may change if the valve malfunctions. Consequently, we do not simply take one sample and make the decision. Instead, we plot a number of statistics over time in the control chart. Simply put, in Chapters 11 to 16 we assumed static population distributions with fixed but unknown parameters, whereas in this chapter we assume a dynamic process distribution with parameters subject to possible shifts.

19-2a Sample Size and Sampling Frequency

In designing a control chart, the statistics practitioner must select a sample size and a sampling frequency. These decisions are based on several factors, including the costs of making Type I and Type II errors, the length of the production run, and the typical change in the process distribution when the process goes out of control. A useful aid in making the decision is the operating characteristic (OC) curve.

Operating Characteristic Curve Recall that in Chapter 11 we drew the operating characteristic (OC) curve that plotted the probabilities of Type II errors and population means. Here is how the OC curve for the $\bar{x}$ chart is drawn.

Suppose that when the production process is under control the mean and standard deviation of the process variable are μ_0 and σ, respectively. For specific values of α and n, we can compute the probability of a Type II error when the process mean changes to $\mu_1 = \mu_0 + k\sigma$. A Type II error occurs when a sample mean falls between the control limits when the process is out of control. In other words, the probability of a Type II error is the probability that the $\overline{x}$ chart will be unable to detect a shift of $k\sigma$ in the process mean on the first sample after the shift has occurred. Figure 19.3 depicts the OC curve for $n = 2, 3, 4$ and 5. Figure 19.4 is the OC curve for $n = 10, 15, 20$ and 25. (We drew two sets of curves because one alone would not provide the precision we need.) We can use the OC curves to help determine the sample size we should use.

Figure 19.3 tells us that for small shifts in the mean of 1 standard deviation or less, samples of size 2 to 5 produce probabilities of not detecting shifts that range between 0.8 and 0.95 (approximately). To appreciate the effect of large probabilities of Type II errors, consider the paint can illustration. Suppose that when the process goes out of control it shifts the mean by about 1 standard deviation. The probability that the first sample after the shift will not detect this shift is approximately 0.85. The probability that it will not detect the shift for the first m samples after the shift is 0.85^m. Thus, for $m = 5$ the probability of not detecting the shift for the first five samples after the shift is 0.44. If the process fills 1000 cans per hour, a large proportion of the 5000 cans filled will be overfilled or underfilled (depending on the direction of the shift). Figure 19.4 suggests that when the shift moves the process mean by 1 standard deviation, samples of size 15 or 20 are recommended. For $n = 15$, the probability that a shift of 1 standard deviation will not be detected by the first sample is approximately 0.2.

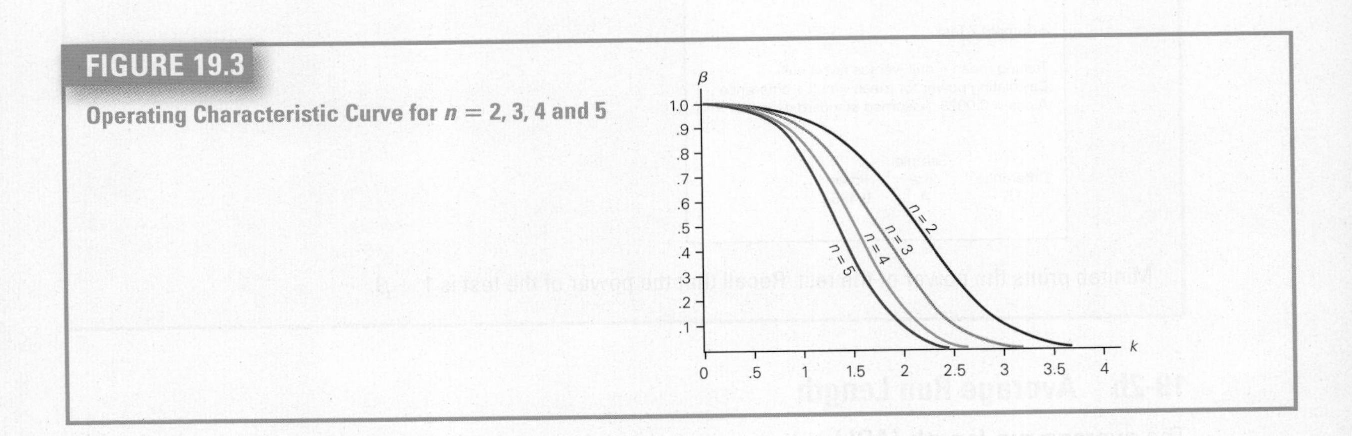

FIGURE 19.3

Operating Characteristic Curve for $n = 2, 3, 4$ and 5

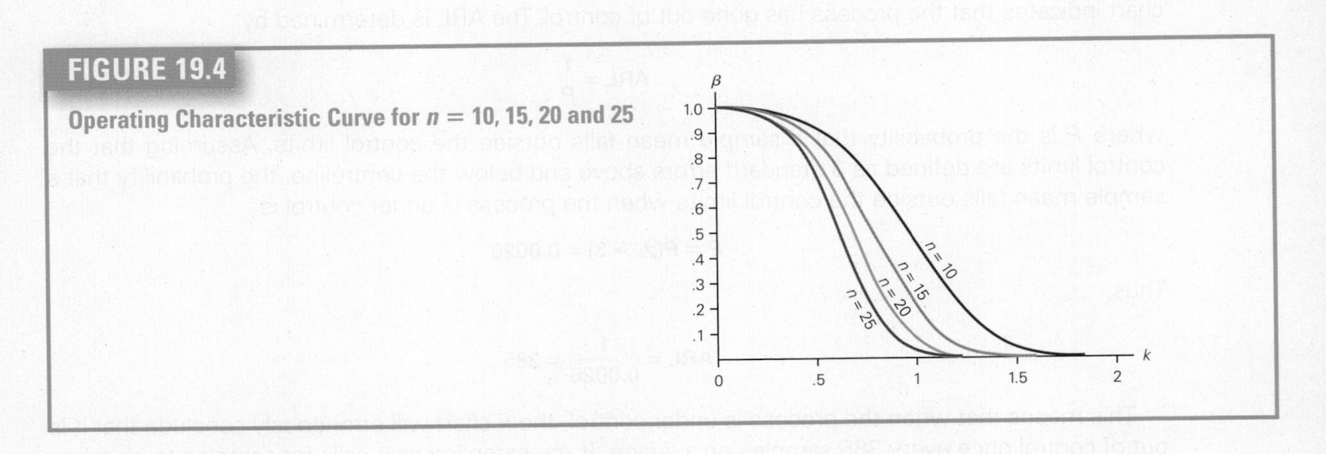

FIGURE 19.4

Operating Characteristic Curve for $n = 10, 15, 20$ and 25

If the typical shift is 2 or more standard deviations, samples of size 4 or 5 will likely suffice. For $n = 5$, the probability of a Type II error is about 0.07.

Using the Computer

EXCEL

As we did in Chapter 11, we can use the computer to determine the probability of a Type II error. Here is the printout for $k = 1.5$ and $n = 4$. (See Section 11-3d for instructions.)

	A	B	C	D
1	Type II Error			
2				
3	H0: MU	0	Critical values	−1.51
4	SIGMA	1		1.51
5	Sample size	4	Prob(Type II error)	0.5046
6	ALPHA	0.0026	Power of the test	0.4954
7	H1: MU	1.5		

MINITAB

As we did in Chapter 11, we can use the computer to determine the probability of a Type II error. Here is the printout for $k = 1.5$ and $n = 4$. (See Section 11-3d for instructions.)

Power and Sample Size

1-Sample Z Test

Testing mean = null (versus not = null)
Calculating power for mean = null + difference
Alpha = 0.0026 Assumed standard deviation = 1

Difference	Sample Size	Power
1.5	4	0.4954

Minitab prints the power of the test. Recall that the power of the test is $1 - \beta$.

19-2b Average Run Length

The **average run length (ARL)** is the expected number of samples that must be taken before the chart indicates that the process has gone out of control. The ARL is determined by

$$\text{ARL} = \frac{1}{P}$$

where P is the probability that a sample mean falls outside the control limits. Assuming that the control limits are defined as 3 standard errors above and below the centreline, the probability that a sample mean falls outside the control limits when the process is under control is

$$P = P(|z| > 3) = 0.0026$$

Thus,

$$\text{ARL} = \frac{1}{0.0026} = 385$$

This means that when the process is under control, the $\bar{x}$ chart will erroneously conclude that it is out of control once every 385 samples on average. If the sampling plan calls for samples to be taken every hour, on average there will be a *false alarm* once every 385 hours.

We can use the OC curve to determine the average run length until the $\bar{x}$ chart detects a process that is out of control. Suppose that when the process goes out of control, it typically shifts the process mean 1.5 standard deviations to the right or left. From Figure 19.3, we can see that for $n = 4$ and $k = 1.5$ the probability of a Type II error is approximately 0.5; that is, the probability that a mean falls between the control limits, which indicates that the process is under control when there has been a shift of 1.5 standard deviations, is 0.5.The probability that the sample mean falls outside the control limits is $P = 1 - 0.5 = 0.5$. Thus, the average run length is

$$\text{ARL} = \frac{1}{0.5} = 2$$

This means that the control chart will require two samples on average to detect a shift of 1.5 standard deviations. Suppose that a shift of this magnitude results in an unacceptably high number of nonconforming cans. We can reduce that number in two ways: by sampling more frequently or increasing the sample size. For example, if we take samples of size 4 every half hour, then on average it will take 1 hour to detect the shift and make repairs. If we take samples of size 10 every hour, Figure 19.4 indicates that the probability of a Type II error when the shift is 1.5 standard deviations is about 0.05. Thus, $P = 1 - 0.05 = 0.95$ and

$$\text{ARL} = \frac{1}{0.95} = 1.05$$

This tells us that a sample size of 10 will allow the statistics practitioner to detect a shift of 1.5 standard deviations about twice as quickly as a sample of size 4.

19-2c Changing the Control Limits

Another way to decrease the probability of a Type II error is to increase the probability of making a Type I error. Thus, we may define the control limits so that they are two standard errors above and below the centreline. To judge whether this is advisable, it is necessary to draw the OC curve for this plan.

In our demonstration of the logic of control charts, we resorted to traditional methods of presenting inferential methods; we assumed that the process parameters were known. When the parameters are unknown, we estimate their values from the sample data. In the next two sections, we discuss how to construct and use control charts in more realistic situations. In Section 19-3, we present control charts when the data are interval. In the context of statistical process control, we call these **control charts for variables**. Section 19-4 demonstrates the use of control charts that record whether a unit is defective or non-defective. These are called **control charts for attributes**.

EXERCISES

19.5 If the control limits of an $\bar{x}$ chart are set at 2.5 standard errors from the centreline, what is the probability that on any sample the control chart will indicate that the process is out of control when it is under control?

19.6 Refer to Exercise 19.5. What is the average run length until the $\bar{x}$ chart signals that the process is out of control when it is under control?

19.7 The control limits of an $\bar{x}$ chart are set at 2 standard errors from the centreline. Calculate the probability that on any sample the control chart will indicate that the process is out of control when it is under control.

19.8 Refer to Exercise 19.7. Determine the ARL until the $\bar{x}$ chart signals that the process is out of control when it is under control.

Exercises 19.9 to 19.15 are based on the following scenario.

A production facility produces 100 units per hour and uses an $\bar{x}$ chart to monitor its quality. The control limits are set at 3 standard errors from the mean. When the process goes out of control, it usually shifts the mean by 1.5 standard deviations. Sampling is conducted once per hour with a sample size of 3.

19.9 On average, how many units will be produced until the control chart signals that the process is out of control when it is under control?

19.10 Refer to Exercise 19.9.

a. Find the probability that the $\bar{x}$ chart does not detect a shift of 1.5 standard deviations on the first sample after the shift occurs.

b. Compute the probability that the $\bar{x}$ chart will not detect the shift for the first eight samples after the shift.

19.11 Refer to Exercise 19.10. Find the average run length to detect the shift.

19.12 The operations manager is unsatisfied with the current sampling plan. He changes it to samples of size 2 every half hour. What is the average number of units produced until the chart indicates that the process is out of control when it is not?

19.13 Refer to Exercise 19.12.

a. Find the probability that the $\bar{x}$ chart does not detect a shift of 1.5 standard deviations on the first sample after the shift occurs.

b. Compute the probability that the $\bar{x}$ chart will not detect the shift for the first eight samples after the shift.

19.14 Refer to Exercise 19.13. What is the average run length to detect the shift?

19.15 Write a brief report comparing the sampling plans described in Exercises 19.9 and 19.12. Discuss the relative costs of the two plans and the frequency of Type I and Type II errors.

Exercises 19.16 to 19.22 are based on the following scenario.

A firm that manufactures notebook computers uses statistical process control to monitor all its production processes. For one component, the company draws samples of size 10 every 30 minutes. The company makes 4000 of these components per hour. The control limits of the $\bar{x}$ chart are set at 3 standard errors from the mean. When the process goes out of control, it usually shifts the mean by 0.75 standard deviation.

19.16 On average, how many units will be produced until the control chart signals that the process is out of control when it is under control?

19.17 Refer to Exercise 19.16.

a. Find the probability that the $\bar{x}$ chart does not detect a shift of 0.75 standard deviation on the first sample after the shift occurs.

b. Compute the probability that the $\bar{x}$ chart will not detect the shift for the first four samples after the shift.

19.18 Refer to Exercise 19.17. Find the average run length to detect the shift.

19.19 The company is considering changing the sampling plan so that 20 components are sampled every hour. What is the average number of units produced until the chart indicates that the process is out of control when it is not?

19.20 Refer to Exercise 19.19.

a. Find the probability that the $\bar{x}$ chart does not detect a shift of 0.75 standard deviation on the first sample after the shift occurs.

b. Compute the probability that the $\bar{x}$ chart will not detect the shift for the first four samples after the shift.

19.21 Refer to Exercise 19.20. What is the average run length to detect the shift?

19.22 Write a brief report comparing the sampling plans described in Exercises 19.16 and 19.19. Discuss the relative costs of the two plans and the frequency of Type I and Type II errors.

19-3 CONTROL CHARTS FOR VARIABLES: $\bar{x}$ AND *S* CHARTS

There are several ways to judge whether a change in the process distribution has occurred when the data are interval. To determine whether the distribution means have changed, we employ the $\bar{x}$ chart. To determine whether the process distribution standard deviation has changed, we can use the *S* (which stands for *standard deviation*) chart or the *R* (which stands for *range*) chart.

Throughout this textbook, we have used the sample standard deviation to estimate the population standard deviation. However, for a variety of reasons, SPC frequently employs the range instead of the standard deviation. This is primarily because computing the range is simpler than computing the standard deviation. Because many practitioners conducting SPC perform calculations by hand (with the assistance of a calculator), they select the computationally simple range as the method to estimate the process standard deviation. In this section, we will introduce control charts that feature the sample standard deviation. In the website Appendix Control Charts for Variables *X* and *R* Charts, we employ the sample range to construct our charts.

19-3a $\bar{x}$ Chart

In Section 19-2 we determined the centreline and control limits of an $\bar{x}$ chart using the mean and standard deviation of the process distribution. However, it is unrealistic to believe that the mean and standard deviation of the process distribution are known. Thus, to construct the $\bar{x}$ chart, we need to estimate the relevant parameters from the data.

We begin by drawing samples when we have determined that the process is under control. The sample size must lie between 2 and 25. We discuss later how to determine that the process is under control. For each sample, we compute the mean and the standard deviation. The estimator of the mean of the distribution is the mean of the sample means (denoted $\bar{\bar{x}}$):

$$\bar{\bar{x}} = \frac{\sum_{j=1}^{k} \bar{x}_j}{k}$$

where $\bar{x}_j$ is the mean of the *jth* sample and there are *k* samples. (Note that $\bar{\bar{x}}$ is simply the average of all *k* observations.)

To estimate the standard deviation of the process distribution, we calculate the sample variance s_j^2 for each sample. We then compute the pooled standard deviation,* which we denote *S* and define as

$$S = \sqrt{\frac{\sum_{j=1}^{k} s_j^2}{k}}$$

In the previous section, where we assumed that the process distribution mean and variance were known, the centreline and control limits were defined as

$$\text{Centreline} = \mu$$

$$\text{Lower control limit} = \mu - 3\frac{\sigma}{\sqrt{n}}$$

$$\text{Upper control limit} = \mu + 3\frac{\sigma}{\sqrt{n}}$$

Because the values of μ and σ are unknown, we must use the sample data to estimate them. The estimator of μ is $\bar{\bar{x}}$, and the estimator of μ is *S*. Therefore, the centreline and control limits are as shown in the box.

Centreline and Control Limits for $\bar{x}$ Chart

$$\text{Centreline} = \bar{\bar{x}}$$

$$\text{Lower control limit} = \bar{\bar{x}} - 3\frac{S}{\sqrt{n}}$$

$$\text{Upper control limit} = \bar{\bar{x}} + 3\frac{S}{\sqrt{n}}$$

EXAMPLE 19.1

Data Xm19-01

Statistical Process Control at Lear Seating, Part 1

Lear Seating of Kitchener, Ontario, manufactures seats for Chrysler, Ford and General Motors cars. Several years ago, Lear instituted statistical process control, which has resulted in improved quality and lower costs. One of the components of a front-seat cushion is a wire spring produced from 4-mm steel wire. A machine is used to bend the wire so that the spring's length is 500 mm. If the springs are longer than 500 mm, they will loosen and eventually fall out. If they are too short, they won't easily fit into position. (In fact, in the past, when there were a relatively large number of short

*This formula requires that the sample size be the same for all samples, a condition that is imposed throughout this chapter.

springs, workers incurred arm and hand injuries when attempting to install the springs.) To determine whether the process is under control, random samples of four springs are taken every hour. The last 25 samples are shown here. Construct an $\bar{x}$ chart from these data.

Sample				
1	501.02	501.65	504.34	501.10
2	499.80	498.89	499.47	497.90
3	497.12	498.35	500.34	499.33
4	500.68	501.39	499.74	500.41
5	495.87	500.92	498.00	499.44
6	497.89	499.22	502.10	500.03
7	497.24	501.04	498.74	503.51
8	501.22	504.53	499.06	505.37
9	499.15	501.11	497.96	502.39
10	498.90	505.99	500.05	499.33
11	497.38	497.80	497.57	500.72
12	499.70	500.99	501.35	496.48
13	501.44	500.46	502.07	500.50
14	498.26	495.54	495.21	501.27
15	497.57	497.00	500.32	501.22
16	500.95	502.07	500.60	500.44
17	499.70	500.56	501.18	502.36
18	501.57	502.09	501.18	504.98
19	504.20	500.92	500.02	501.71
20	498.61	499.63	498.68	501.84
21	499.05	501.82	500.67	497.36
22	497.85	494.08	501.79	501.95
23	501.08	503.12	503.06	503.56
24	500.75	501.18	501.09	502.88
25	502.03	501.44	498.76	499.39

SOLUTION:

COMPUTE MANUALLY:

The means and standard deviations for each sample were computed and are listed in Table 19.1. We then calculated the mean of the means (which is also the mean of all 100 numbers) and the pooled standard deviation:

$$\bar{\bar{x}} = 500.296$$

$$S = 1.971$$

Thus, the centreline and control limits are

$$\text{Centreline} = \bar{\bar{x}} = 500.296$$

$$\text{Lower control limit} = \bar{\bar{x}} - 3\frac{S}{\sqrt{n}} = 500.296 - 3\frac{1.971}{\sqrt{4}} = 497.340$$

$$\text{Upper control limit} = \bar{\bar{x}} + 3\frac{S}{\sqrt{n}} = 500.296 + 3\frac{1.971}{\sqrt{4}} = 503.253$$

The centreline and control limits are drawn and the sample means plotted in the order in which they occurred. The manually drawn chart is identical to the Excel and Minitab versions shown here.

TABLE 19.1 Means and Standard Deviations of Samples in Example 19.1

Sample					$\bar{x}_i$	s_i
1	501.02	501.65	504.34	501.10	502.03	1.567
2	499.80	498.89	499.47	497.90	499.02	0.833
3	497.12	498.35	500.34	499.33	498.79	1.376
4	500.68	501.39	499.74	500.41	500.56	0.683
5	495.87	500.92	498.00	499.44	498.56	2.152
6	497.89	499.22	502.10	500.03	499.81	1.763
7	497.24	501.04	498.74	503.51	500.13	2.741
8	501.22	504.53	499.06	505.37	502.55	2.934
9	499.15	501.11	497.96	502.39	500.15	1.978
10	498.90	505.99	500.05	499.33	501.07	3.316
11	497.38	497.80	497.57	500.72	498.37	1.578
12	499.70	500.99	501.35	496.48	499.63	2.216
13	501.44	500.46	502.07	500.50	501.12	0.780
14	498.26	495.54	495.21	501.27	497.57	2.820
15	497.57	497.00	500.32	501.22	499.03	2.059
16	500.95	502.07	500.60	500.44	501.02	0.735
17	499.70	500.56	501.18	502.36	500.95	1.119
18	501.57	502.09	501.18	504.98	502.46	1.724
19	504.20	500.92	500.02	501.71	501.71	1.796
20	498.61	499.63	498.68	501.84	499.69	1.507
21	499.05	501.82	500.67	497.36	499.73	1.943
22	497.85	494.08	501.79	501.95	498.92	3.741
23	501.08	503.12	503.06	503.56	502.71	1.106
24	500.75	501.18	501.09	502.88	501.48	0.955
25	502.03	501.44	498.76	499.39	500.41	1.576

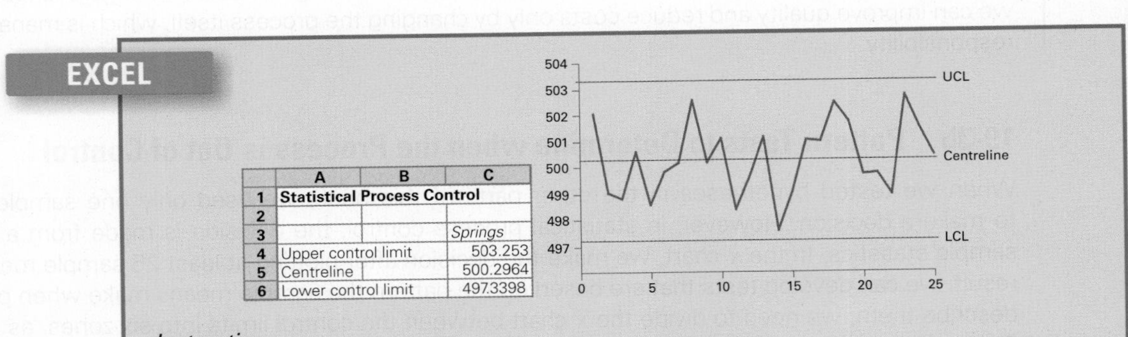

Instructions

1. Type or import the data into one column. (**Open** Xm19-01.)
2. Click **Add-Ins, Data Analysis Plus** and **Statistical Process Control.**
3. Specify the **Input Range** (A1:101) and the **Sample Size** (4). Click **XBAR (Using S).**

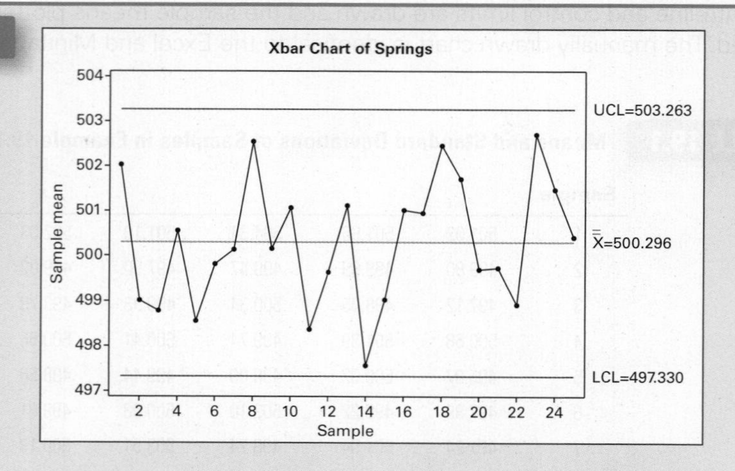

Instructions

1. Type or import the data into one column. (**Open** Xm19-01.)
2. Click **Stat, Control Charts, Variable charts for subgroups** and **Xbar** ....
3. Specify **All observations for a chart are in one column.** Type or select the variable (Springs). Type the **Subgroup sizes (4). Click Xbar Options** ....
4. Click **Estimate** .... Under **Method for estimating standard deviation,** click **Pooled standard deviation.**

INTERPRET

As you can see, no point lies outside the control limits. We conclude that the variation in the lengths of the springs is caused by chance – that is, there is not enough evidence to infer that the process is out of control. No remedial action by the operator is called for.

We stress that statistical process control allows us to detect assignable variation only. In Example 19.1, we determined that the process is under control, which means that there are no detectable sources of assignable variation. However, this does not mean that the process is a good one. It may well be that the production process yields a large proportion of defective units because the amount of chance variation is large. Recall that the chance variation decreases product quality and increases costs. If the costs of producing defective units are high because of large chance variation, we can improve quality and reduce costs only by changing the process itself, which is management's responsibility.

19-3b Pattern Tests to Determine when the Process is Out of Control

When we tested hypotheses in the other parts of this book, we used only one sample statistic to make a decision. However, in statistical process control, the decision is made from a series of sample statistics. In the $\bar{x}$ chart, we make the decision after plotting at least 25 sample means. As a result, we can develop tests that are based on the pattern the sample means make when plotted. To describe them, we need to divide the $\bar{x}$ chart between the control limits into six zones, as shown in Figure 19.5. The C zones represent the area within one standard error of the centreline. The B zones are the regions between one and two standard errors from the centreline. The spaces between two and three standard errors from the centreline are defined as A zones.

FIGURE 19.5

Zones of $\bar{x}$ Chart

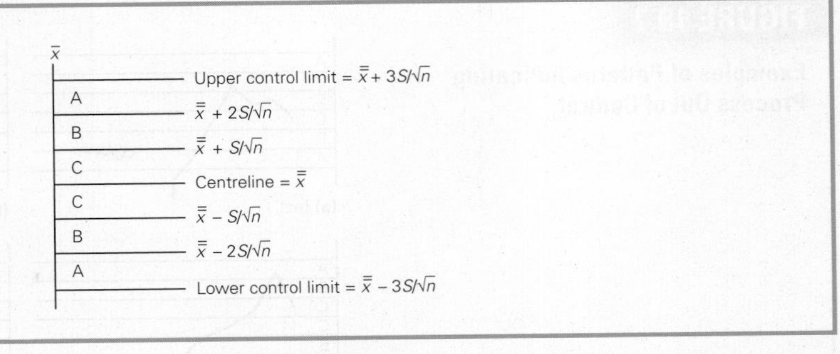

The width of the zones is one standard error of $\bar{x}(S/\sqrt{n})$. If the calculations were performed manually, the value of S will be known. However, if a computer was used, the centreline and control limits are the only statistics printed. We can calculate $S/\sqrt{n}$ by finding the difference between the upper and lower control limits and dividing the difference by 6; that is,

$$S/\sqrt{n} = \frac{(\bar{\bar{x}} + 3S/\sqrt{n}) - (\bar{\bar{x}} - 3S/\sqrt{n})}{6} = \frac{503.253 - 497.340}{6} = 0.9855$$

Figure 19.6 describes the centreline, control limits, and zones for Example 19.1.

FIGURE 19.6

Zones of $\bar{x}$ Chart: Example 19.1

$\bar{x}$
A
B
C
C
B
A
503.253
502.268
501.282
500.296
499.311
498.326
497.340

Several pattern tests can be applied. We list eight tests that are conducted by Minitab and by Data Analysis Plus.

Test 1: One point beyond zone A. This is the method discussed previously, where we conclude that the process is out of control if any point is outside the control limits.

Test 2: Nine points in a row in zone C or beyond (on the same side of the centreline).

Test 3: Six increasing or six decreasing points in a row.

Test 4: Fourteen points in a row alternating up and down.

Test 5: Two out of three points in a row in zone A or beyond (on the same side of the centreline).

Test 6: Four out of five points in a row in zone B or beyond (on the same side of the centreline).

Test 7: Fifteen points in a row in zone C (on both sides of the centreline).

Test 8: Eight points in a row beyond zone C (on both sides of the centreline).

In the examples shown in Figure 19.7, each of the eight tests indicates a process out of control.

FIGURE 19.7

Examples of Patterns Indicating Process Out of Control

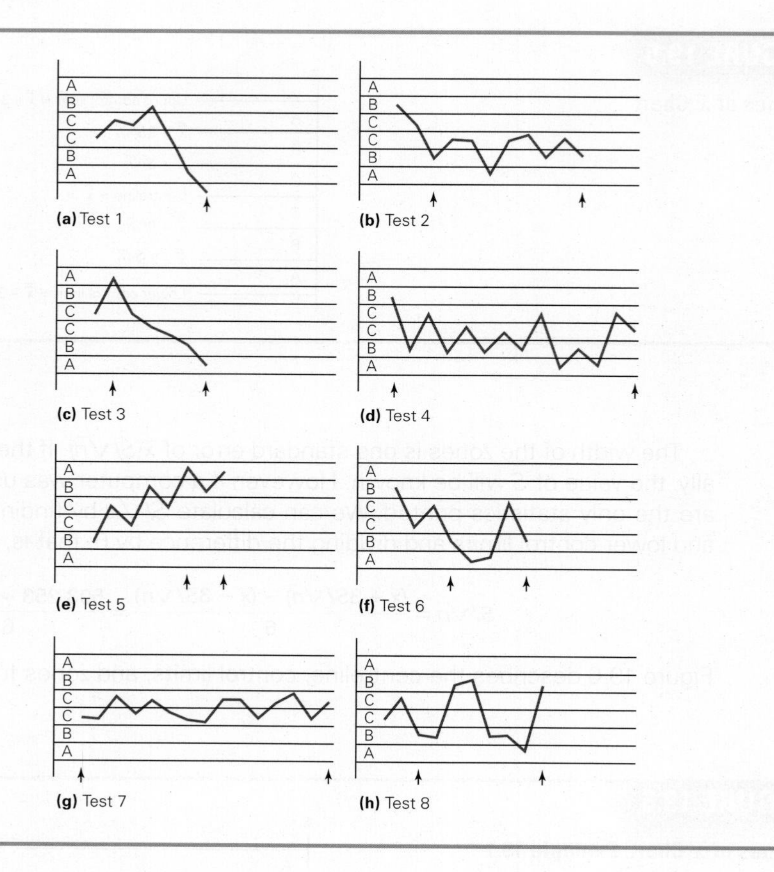

All eight tests are based on the same concepts used to test hypotheses throughout this book. In other words, each pattern is a rare event that is unlikely to occur when a process is under control. Thus, when any one of these patterns is recognised, the statistics practitioner has reason to believe that the process is out of control. In fact, it is often possible to identify the cause of the problem from the pattern in the control chart.

Figure 19.8 depicts the zones and the means for Example 19.1. After checking each of the eight pattern tests, we conclude that the process is under control.

FIGURE 19.8

$\bar{x}$ Chart with Zones: Example 19.1

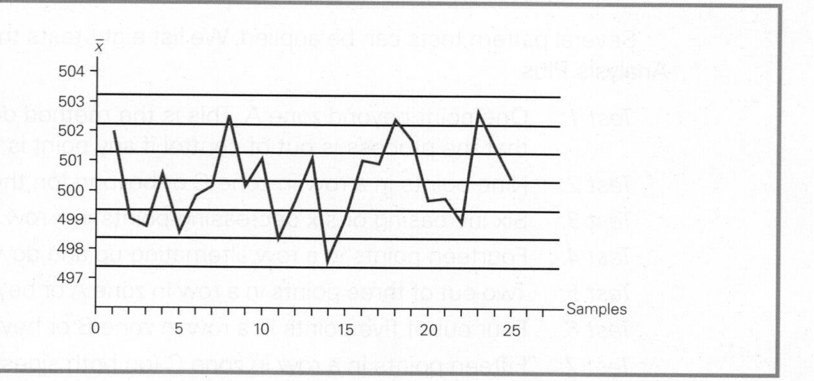

EXCEL Excel automatically performs all eight tests. Any that fail will be reported on the spreadsheet.

MINITAB Minitab will perform any or all of the pattern tests and report any that fail.

Instructions

After clicking **Xbar Options** ... click **Tests** and **click Perform all tests for special causes.** To draw the zones, click **S Limits**... and type **1 2 3** in the **Display control limits at** box.

19-3c Pattern Tests in Practice

There appears to be a great deal of disagreement among statisticians with regard to pattern tests. Some authors and statistical software packages apply eight tests, whereas others employ a different number. In addition, some statisticians apply pattern tests to $\overline{x}$ charts, but not to other charts. Rather than joining the debate with our own opinions, we will follow Minitab's rules. There are eight pattern tests for $\overline{x}$ charts, no pattern tests for *S* and *R* charts, and four pattern tests for the chart presented in Section 19-4 (*p* charts). The same rules apply to Data Analysis Plus.

19-3d *S* Charts

The ***S* chart** graphs sample standard deviations to determine whether the process distribution standard deviation has changed. The format is similar to that of the $\overline{x}$ chart: The *S* chart will display a centreline and control limits. However, the formulas for the centreline and control limits are more complicated than those for the $\overline{x}$ chart. Consequently, we will not display the formulas; instead we will let the computer do all the work.

EXAMPLE 19.2

Statistical Process Control at Lear Seating, Part 2

Using the data provided in Example 19.1, determine whether there is evidence to indicate that the process distribution standard deviation has changed over the period when the samples were taken.

SOLUTION:

COMPUTE

EXCEL

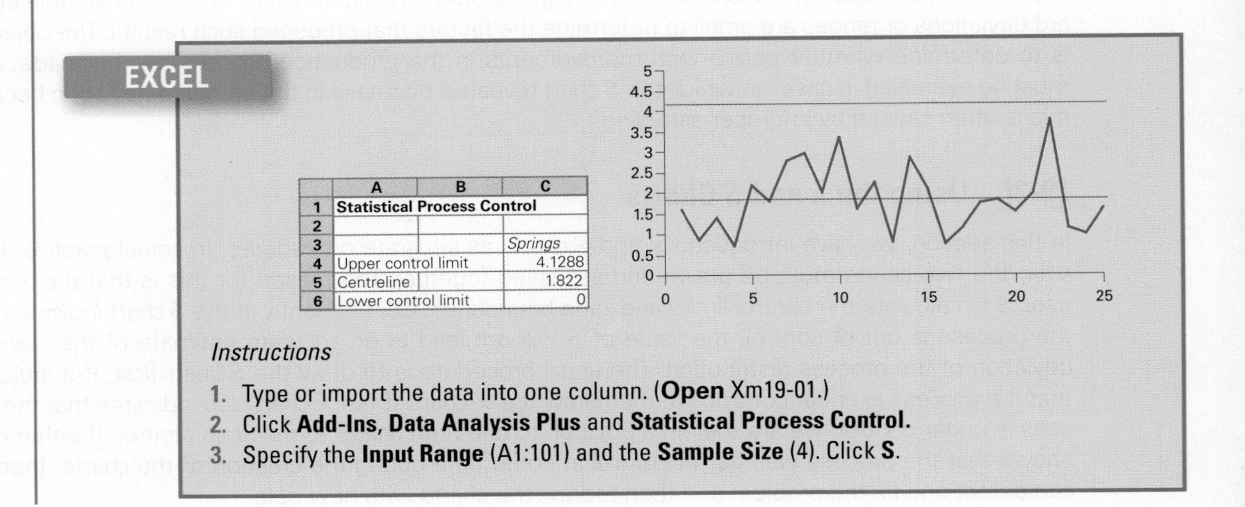

	A	B	C
1	Statistical Process Control		
2			
3			Springs
4	Upper control limit		4.1288
5	Centreline		1.822
6	Lower control limit		0

Instructions

1. Type or import the data into one column. (**Open** Xm19-01.)
2. Click **Add-Ins, Data Analysis Plus** and **Statistical Process Control.**
3. Specify the **Input Range** (A1:101) and the **Sample Size** (4). Click **S.**

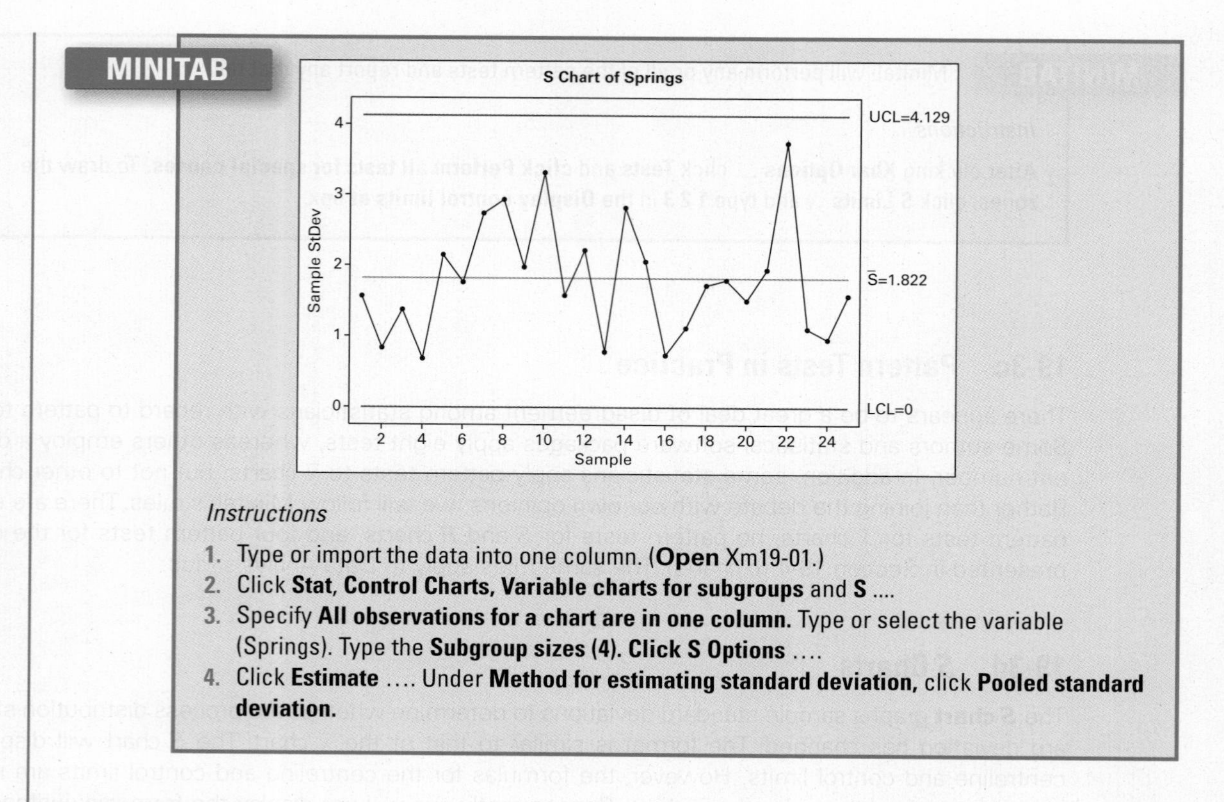

Instructions

1. Type or import the data into one column. (**Open** Xm19-01.)
2. Click **Stat, Control Charts, Variable charts for subgroups** and **S** ....
3. Specify **All observations for a chart are in one column.** Type or select the variable (Springs). Type the **Subgroup sizes (4). Click S Options** ....
4. Click **Estimate** .... Under **Method for estimating standard deviation,** click **Pooled standard deviation.**

INTERPRET

There are no points outside the control limits. Because we do not apply any of the pattern tests, we conclude that there is no evidence to believe that the standard deviation has changed over this period.

19-3e Good News and Bad News About *S* Charts

In analysing *S* charts, we would conclude that the process distribution has changed if we observe points outside the control limits. Obviously, points above the upper control limit indicate that the process standard deviation has increased – an undesirable situation. Points below the lower control limit also indicate that the process standard deviation has changed. However, cases in which the standard deviation has decreased are welcome occurrences because reducing the variation generally leads to improvements in quality. The operations manager should investigate cases where the sample standard deviations or ranges are small to determine the factors that produced such results. The objective is to determine whether permanent improvements in the production process can be made. Care must be exercised in cases in which the *S* chart reveals a decrease in the standard deviation because this is often caused by improper sampling.

19-3f Using the $\overline{x}$ and *S* Charts

In this section, we have introduced $\overline{x}$ and *S* charts as separate procedures. In actual practice, however, the two charts must be drawn and assessed together. The reason for this is that the $\overline{x}$ chart uses *S* to calculate the control limits and zone boundaries. Consequently, if the *S* chart indicates that the process is out of control, the value of *S* will not lead to an accurate estimate of the standard deviation of the process distribution. The usual procedure is to draw the *S* chart first. If it indicates that the process is under control, we then draw the $\overline{x}$ chart. If the $\overline{x}$ chart also indicates that the process is under control, we are then in a position to use both charts to maintain control. If either chart shows that the process was out of control at some time during the creation of the charts, then we can detect and fix the problem and then redraw the charts with new data.

19-3g Monitoring the Production Process

When the process is under control, we can use the control chart limits and centreline to monitor the process in the future. We do so by plotting all future statistics on the control chart.

EXAMPLE 19.3

Statistical Process Control at Lear Seating, Part 3

After determining that the process is under control, the company in Example 19.1 began using the statistics generated in the creation of the $\bar{x}$ and S charts to monitor the production process. The sampling plan calls for samples of size 4 every hour. The following table lists the lengths of the springs taken during the first 6 hours.

Sample				
1	502.653	498.354	502.209	500.080
2	501.212	494.454	500.918	501.855
3	500.086	500.826	496.426	503.591
4	502.994	500.481	502.996	503.113
5	500.549	498.780	502.480	499.836
6	500.441	502.666	502.569	503.248

SOLUTION:

After each sample is taken, the mean and standard deviation are computed. The standard deviations are plotted on the S chart using the previously determined control limits when the process variation was deemed to be in control. The sample means are plotted on the $\bar{x}$ chart, again using the zone limits determined when the process was deemed to be in control, and the pattern tests are checked after each point is plotted. The first six samples are shown in Figure 19.9. After the standard

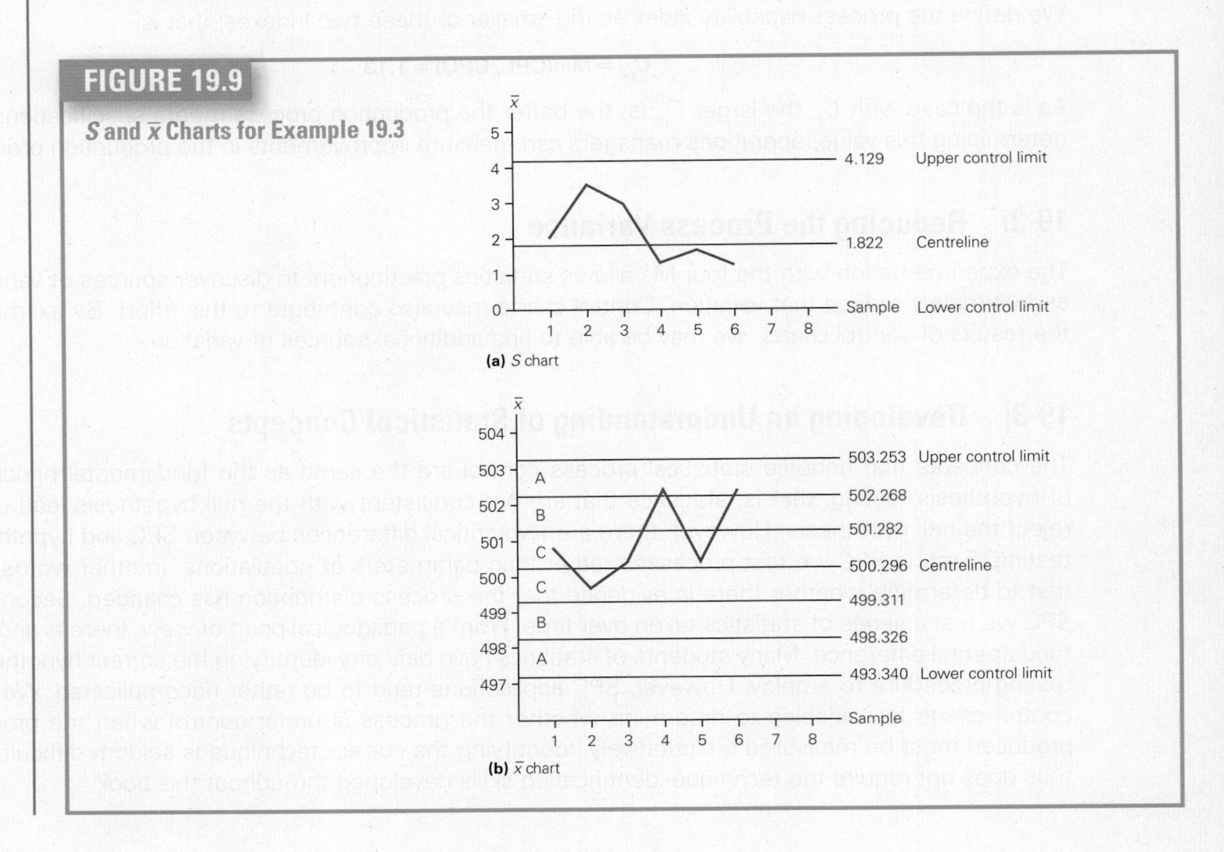

FIGURE 19.9

***S* and $\bar{x}$ Charts for Example 19.3**

deviation and mean of the sixth sample are plotted, the technician would stop the production process. Although the process variation still appears to be in control, the fourth and sixth means on the $\bar{x}$ chart combine to indicate that test 5 has failed; there are two out of three points in a row that are in zone A or beyond. Thus, it appears that the process mean has shifted upwards. Technicians need to find the source of the problem and make repairs. After repairs are completed, production resumes and new control charts and their centrelines and control limits are recalculated.

19-3h Process Capability Index

The **process capability index** measures the capability of the process to produce units whose dimensions fall within the specifications. We defined the index as

$$C_p = \frac{\text{USL} - \text{LSL}}{6\sigma}$$

where USL and LSL are the upper and lower specification limits, respectively. To compute the process capability index, we need to know these limits and the process standard deviation. The standard deviation is a population parameter that is generally unknown. Thus, C_p measures the theoretical or potential process capability. To produce a measure of the process's actual capability, we must use statistics computed in the construction of the control chart. Suppose that in Example 19.1 the operations manager determined that the springs will fit, provided that their lengths fall between the lower specification limit LSL = 493 and the upper specification limit USL = 507. In Example 19.1 we found $\bar{\bar{x}} = 500.296$ and $S = 1.971$. We define the following:

$$\text{CPL} = \frac{\bar{\bar{x}} - \text{LSL}}{3S} = \frac{500.296 - 493}{3(1.971)} = 1.23$$

$$\text{CPU} = \frac{\text{USL} - \bar{\bar{x}}}{3S} = \frac{507 - 500.296}{3(1.971)} = 1.13$$

We define the process capability index as the smaller of these two indexes; that is,

$$C_{pk} = \text{Min(CPL, CPU)} = 1.13$$

As is the case with C_p, the larger C_{pk} is, the better the production process meets specifications. By determining this value, operations managers can measure improvements in the production process.

19-3i Reducing the Process Variance

The experimentation with the four Ms allows statistics practitioners to discover sources of variation and ultimately reduce that variation. Control charts may also contribute to this effort. By examining the results of control charts, we may be able to find additional sources of variation.

19-3j Developing an Understanding of Statistical Concepts

The concepts that underlie statistical process control are the same as the fundamental principles of hypothesis testing; that is, statistics that are not consistent with the null hypothesis lead us to reject the null hypothesis. However, there are two critical differences between SPC and hypothesis testing. First, in SPC we test processes rather than parameters of populations. In other words, we test to determine whether there is evidence that the process distribution has changed. Second, in SPC we test a series of statistics taken over time. From a pedagogical point of view, there is another fundamental difference. Many students of statistics have difficulty identifying the correct hypothesis-testing procedure to employ. However, SPC applications tend to be rather uncomplicated. We use control charts for variables to determine whether the process is under control when the product produced must be measured quantitatively. Identifying the correct technique is seldom difficult and thus does not require the technique-identification skills developed throughout this book.

EXERCISES

19.23 Given the following statistics drawn from 30 samples of size 4, calculate the centreline and control limits for the $\bar{x}$ chart.

$$\bar{\bar{x}} = 453.6 \quad S = 12.5$$

19.24 The mean of the sample means and the pooled standard deviation of 40 samples of size 9 taken from a production process under control are shown here. Compute the centreline, control limits and zone boundaries for the $\bar{x}$ chart.

$$\bar{\bar{x}} = 181.1 \quad S = 11.0$$

19.25 Twenty-five samples of size 4 were taken from a production process. The sample means are listed in chronological order below. The mean of the sample means and the pooled standard deviation are $\bar{\bar{x}} = 13.3$ and $S = 3.8$, respectively.

14.5	10.3	17.0	9.4	13.2	9.3	17.1
5.5	5.3	16.3	10.5	11.5	8.8	12.6
10.5	16.3	8.7	9.4	11.4	17.6	20.5
21.1	16.3	18.5	20.9			

a. Find the centreline and control limits for the $\bar{x}$ chart.

b. Plot the sample means on the $\bar{x}$ chart.

c. Is the process under control? Explain.

The following exercises require a computer and statistical software.

19.26 Xr19-26 Thirty samples of size 4 were drawn from a production process.

a. Construct an S chart.

b. Construct an $\bar{x}$ chart.

c. Do the charts allow you to conclude that the process is under control?

d. If the process went out of control, which of the following is the likely cause: level shift, instability, trend or cycle?

19.27 Xr19-27 The fence of a saw is set so that it automatically cuts 2-by-4 boards into 2.44 m lengths needed to produce prefabricated homes. To ensure that the lumber is cut properly, three pieces of wood are measured after each 100 cuts are made. The measurements in inches for the last 40 samples were recorded.

a. Do these data indicate that the process is out of control?

b. If so, when did it go out of control? What is the likely cause: level shift, instability, trend or cycle?

c. Speculate on how the problem could be corrected.

19.28 Xr19-28 An arc extinguishing unit (AEU) is used in the high-voltage electrical industry to eliminate the occurrence of electrical flash from one live 25 000-volt switch contact to another. A small but important component of an AEU is a non-conductive sliding bearing called a (ST-90811) pin guide. The dimensional accuracy of this pin guide is critical to the overall operation of the AEU. If any one of its dimensions is 'out of spec' (specification), the part will bind within the AEU, causing failure. This would cause the complete destruction of both the AEU and the 25 000 volt-switch contacts, resulting in a power blackout. A pin guide has a square shape with a circular hole in the centre, as shown below with its specified dimensions. The specification limits are LSL = 0.4335 and USL = 0.4435.

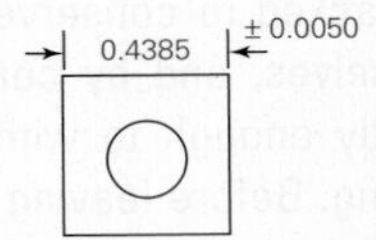

Because of the critical nature of the dimensions of the pin guide, statistical process control is used during long production runs to check that the production process is under control. Suppose that samples of five pin guides are drawn every hour. The results of the last 25 samples were recorded. Do these data allow the technician to conclude that the process is out of control?

19.29 Refer to Exercise 19.28. Find the process capability index C_{pk}.

19.30 Xr19-30 KW Paints is a company that manufactures various kinds of paints and sells them in 1- and 4-litre cans. The cans are filled on an assembly line with an automatic valve regulating the amount of paint. If the cans are overfilled, paint and money will be wasted. If the cans are underfilled, customers will complain. To ensure that the proper amount of paint goes into each can, statistical process control is used. Every hour, five cans are opened, and the volume of paint is measured. The results from the last 30 hours from the 1-litre production line were recorded. To avoid rounding errors, we recorded the volumes in cubic centimetres (cc) after subtracting 1000. Thus, the file contains the amounts of overfill and underfill. Draw the $\bar{x}$ and S charts to determine whether the process is under control.

19.31 Refer to Exercise 19.30. If the lower and upper specification limits are 995 cc and 1005 cc, respectively, what is C_{pk}?

19.32 Xr19-32 The degree to which nuts and bolts are tightened in numerous places on a car is often important. For example, in Toyota cars, a nut holds the rear signal light. If the nut is not tightened sufficiently, it will loosen and fall off; if it is too tight, the light may break. The nut is tightened with a torque wrench with a set clutch. The target torque is 8 kgf/cm (kilogram-force per centimetre) with specification limits LSL = 7kgf/cm and USL = 9 kgf/cm. Statistical process control is employed to constantly check the process. Random samples of size 4 are drawn after every 200 nuts are tightened. The data from the last 25 samples were recorded.

a. Determine whether the process is under control.

b. If it is out of control, identify when this occurred and the likely cause.

19.33 Xr19-33 Motor oil is packaged and sold in plastic bottles. The bottles are often handled quite roughly in delivery to the stores (bottles are packed in boxes, which are stacked to conserve truck space), in the stores themselves, and by consumers. The bottles must be hardy enough to withstand this treatment without leaking. Before leaving the plant, the bottles undergo statistical process control procedures. Five out of every 10 000 bottles are sampled. The burst strength (the pressure required to burst the bottle) is measured in pounds per square inch (psi). The process is designed to produce bottles that can withstand as much as 800 psi. The burst strengths of the last 30 samples were recorded.

a. Draw the appropriate control chart(s).

b. Does it appear that the process went out of control? If so, when did this happen, and what are the likely causes and remedies?

19.34 Xr19-34 Almost all computer hardware and software producers offer a toll-free telephone number to solve problems associated with their products. The ability to work quickly to resolve difficulties is critical. One software maker's policy is that all calls must be answered by a software consultant within 120 seconds. (All calls are initially answered by computer and the caller is put on hold until a consultant attends to the caller.) To help maintain the quality of the service, four calls per day are monitored. The amount of time before the consultant responds to the calls was recorded for the last 30 days.

a. Draw the appropriate control chart(s).

b. Does it appear that the process went out of control? If so, when did this happen, and what are the likely causes and remedies?

19-4 CONTROL CHARTS FOR ATTRIBUTES: *P* CHART

In this section, we introduce a control chart that is used to monitor a process whose results are categorised as either defective or non-defective. We construct a ***p* chart** to track the proportion of defective units in a series of samples.

19-4a *p* Chart

We draw the *p* chart in a way similar to the construction of the $\overline{x}$ chart. We draw samples of size *n* from the process at a minimum of 25 time periods. For each sample, we calculate the sample proportion of defective units, which we label $\hat{p}_j$. We then compute the mean of the sample proportions, which is labeled $\overline{p}$; that is,

$$\overline{p} = \frac{\sum_{j=1}^{k} \hat{p}_j}{k}$$

The centreline and control limits are as follows.

Centreline and Control Limits for the *p* Chart

$$\text{Centreline} = \overline{p}$$

$$\text{Lower control limit} = \overline{p} - 3\sqrt{\frac{\overline{p}(1-\overline{p})}{n}}$$

$$\text{Upper control limit} = \overline{p} + 3\sqrt{\frac{\overline{p}(1-\overline{p})}{n}}$$

If the lower limit is negative, set it equal to 0.

19-4b Pattern Tests

As we did in the previous section, we use Minitab's pattern tests. Minitab performs only tests 1 through 4, which are as follows:

Test 1: One point beyond zone A.

Test 2: Nine points in a row in zone C or beyond (on the same side of the centreline).

Test 3: Six increasing or six decreasing points in a row.

Test 4: Fourteen points in a row alternating up and down.

We'll demonstrate this technique using the chapter-opening example.

DETECTING THE SOURCE OF DEFECTIVE DISCS: SOLUTION

For each sample, we compute the proportion of defective discs and calculate the mean sample proportion, which is $\bar{p} = 0.05762$. Thus,

$$\text{Centreline} = \bar{p} = 0.05762$$

$$\text{Lower control limit} = \bar{p} - 3\sqrt{\frac{\bar{p}(1-\bar{p})}{n}}$$

$$= 0.05762 - 3\sqrt{\frac{(0.05762)(1-0.05762)}{200}}$$

$$= 0.008188$$

$$\text{Upper control limit} = \bar{p} + 3\sqrt{\frac{\bar{p}(1-\bar{p})}{n}}$$

$$= 0.05762 + 3\sqrt{\frac{(0.05762)(1-0.05762)}{200}}$$

$$= 0.1071$$

Because

$$\sqrt{\frac{\bar{p}(1-\bar{p})}{n}} = \sqrt{\frac{(0.05762)(1-0.05762)}{200}} = 0.01648$$

The boundaries of the zones are as follows:

$$\text{Zone C: } 0.05762 \pm 0.01\,648 = (0.04114, 0.0741)$$

$$\text{Zone B: } 0.05762 \pm 2(0.01\,648) = (0.02467, 0.09057)$$

$$\text{Zone A: } 0.05762 \pm 3(0.01648) = (0.0081\,88, 0.1\,071)$$

The following output exhibits this p chart.

(continued)

EXCEL

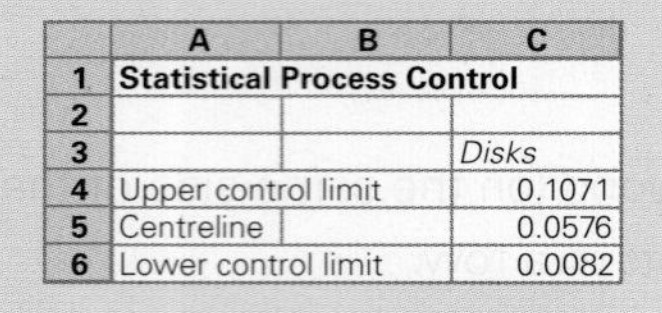

	A	B	C
1	Statistical Process Control		
2			
3			Disks
4	Upper control limit		0.1071
5	Centreline		0.0576
6	Lower control limit		0.0082

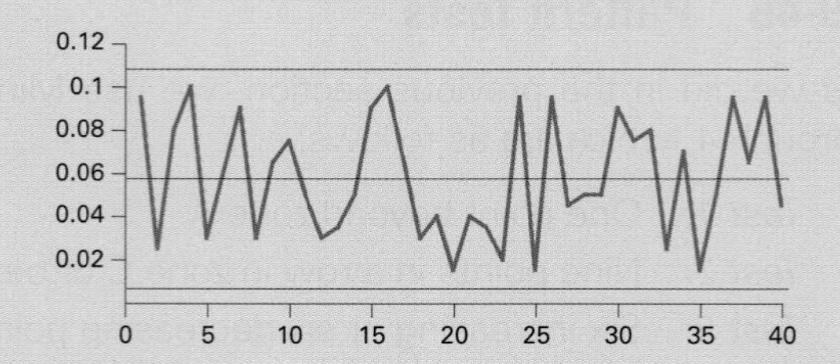

Instructions

1. Type or import the data into one column. (**Open** Xm19-00.)
2. Click **Add-Ins, Data Analysis Plus** and **Statistical Process Control**.
3. Specify the **Input Range** (A1:41) and the **Sample Size** (200). Click P.

MINITAB

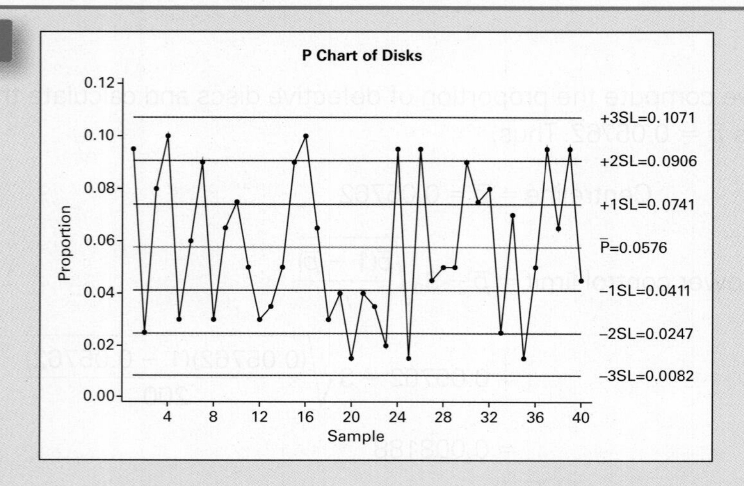

Instructions

1. Type or import the data into one column. (**Open** Xm19-00.)
2. Click **Stat, Control Charts, Attribute Charts** and **P** ....
3. Type or select the variable (Discs). Type the sample size in the **Subgroup sizes** box (200). Click **P Chart Options** ....
4. Click **S Limits** and type 12 3.
5. Click **Tests** and **Perform all four tests.**

INTERPRET

None of the points lie outside the control limits (test 1), and the other test results are negative. There is no evidence to infer that the process is out of control. However, this does not mean that 5.76% is an acceptable proportion of defects. Management should continually improve the process to reduce the defective rate and improve the process.

The comment we made about *S* charts is also valid for *p* charts: Sample proportions that are less than the lower control limit indicate a change in the process that we would like to make permanent. We need to investigate the reasons for such a change just as vigorously as we investigate the causes of large proportions of defects.

EXERCISES

19.35 To ensure that a manufacturing process is under control, 40 samples of size 1000 were drawn, and the number of defectives in each sample was counted. The mean sample proportion was 0.035. Compute the centreline and control limits for the *p* chart.

19.36 Xr19-36 Random samples of 200 copier machines were taken on an assembly line every hour for the past 25 hours. The number of defective machines is shown here. Are there any points beyond the control limits? If so, what do they tell you about the production process?

3 5 3 2 2 11 12 6 7 5 0 7 8
2 10 6 4 2 10 5 4 11 10 13 14

19.37 Xr19-37 A plant in Bangkok produces 1000 cordless telephones daily. A random sample of 100 telephones is inspected each day. After 30 days, the following number of defectives were found. Construct a *p* chart to determine whether the process is out of control.

5 0 4 3 0 3 1 1 5 0 2 1 6 0 3
0 5 5 8 5 0 1 9 6 11 6 6 4 5 10

19.38 Xr19-38 The Publishers' Association of South Africa (PASA) is the largest publishing industry body in South Africa and produces millions of books containing hundreds of millions of pages each year. To ensure the quality of the printed page, PASA uses statistical process control. In each production run, 1000 pages are randomly inspected. The examiners look for print clarity and whether the material is centred on the page properly. The numbers of defective pages in the last 40 production runs are listed here. Draw the *p* chart. Using the pattern tests, can we conclude that the production process is under control?

11 9 17 19 15 15 18 21 18 6 27 14 7 18
18 19 17 15 7 16 17 22 12 12 12 16 12
9 21 17 20 17 17 18 23 29 24 27 23 21

The following exercises require the use of a computer and statistical software.

19.39 Xr19-39 A company that manufactures batteries employs statistical process control to ensure that its product functions properly. The sampling plan for the D-cell batteries calls for samples of 500 batteries to be taken and tested. The numbers of defective batteries in the last 30 samples were recorded. Determine whether the process is under control.

19.40 Xr19-40 Optical scanners are used in all supermarkets to speed the checkout process. Whenever the scanner fails to read the bar code on the product, the cashier is required to manually punch the code into the register. Obviously, unreadable bar codes slow the checkout process. Statistical process control is used to determine whether the scanner is working properly. Once a day at each checkout counter, a sample of 500 scans is taken, and the number of times the scanner is unable to read the bar code is determined. (The sampling process is performed automatically by the cash register.) The results for one checkout counter for the past 25 days were recorded.

a. Draw the appropriate control chart(s).

b. Does it appear that the process went out of control? If so, identify when this happened and suggest several possible explanations for the cause.

CHAPTER SUMMARY

In this chapter, we introduced statistical process control and explained how it contributes to the maintenance of quality. We discussed how control charts detect changes in the process distribution and introduced the $\overline{x}$ chart, *S* chart and *p* chart.

SYMBOLS:

Symbol	Pronounced	Represents
S		Pooled standard deviation
s_j	*s-sub-j*	Standard deviation of the j'th sample
$\hat{p}_j$	*p-hat-sub-j*	Proportion of defectives in j'th sample
$\bar{p}$	*p-bar*	Mean proportion of defectives

FORMULAS:

Centreline and control limits for $\bar{x}$ chart using S

$$\text{Centreline} = \bar{x}$$

$$\text{Lower control limit} = \bar{x} - 3\frac{S}{\sqrt{n}}$$

$$\text{Upper control limit} = \bar{x} + 3\frac{S}{\sqrt{n}}$$

Centreline and control limits for the p chart

$$\text{Centreline} = \bar{p}$$

$$\text{Lower control limit} = \bar{p} - 3\sqrt{\frac{\bar{p}(1-\bar{p})}{n}}$$

$$\text{Upper control limit} = \bar{p} + 3\sqrt{\frac{\bar{p}(1-\bar{p})}{n}}$$

20 DECISION ANALYSIS

20-1 DECISION PROBLEM

20-2 ACQUIRING, USING AND EVALUATING ADDITIONAL INFORMATION

ACCEPTANCE SAMPLING

A factory produces a small but important component used in computers. The factory manufactures the component in 1000-unit lots. Because of the relatively advanced technology, the manufacturing process results in a large proportion of defective units. In fact, the operations manager has observed that the percentage of defective units per lot has been either 15% or 35%. In the past year, 60% of the lots have had 15% defectives, and 40% have had 35% defectives. The current policy of the company is to send the lot to the customer, replace all defectives and pay any additional costs. The total cost of replacing a defective unit that has been sent to the customer is €10/unit. Because of the high costs, the company management is considering inspecting all units and replacing the defective units before shipment. The sampling cost is €2/unit and the replacement cost is €0.50/unit. Each unit sells for €5.

a. Based on the history of the past year, should the company adopt the 100% inspection plan?
b. Is it worthwhile to take a sample of size 2 from the lot before deciding whether to inspect 100%?

See Section 20-2b for the answer.

INTRODUCTION

In previous chapters, we dealt with techniques for summarising data in order to make decisions about population parameters and population characteristics. Our focus in this chapter is also on decision-making, but the types of problems we deal with here differ in several ways. First, the technique for hypothesis testing concludes with either rejecting or not rejecting some hypothesis concerning a dimension of a population. In decision analysis, we deal with the problem of selecting one alternative from a list of several possible decisions. Second, in hypothesis testing, the decision is based on the statistical evidence available. In decision analysis, there may be no statistical data, or if there are data, the decision may depend only partly on them. Third, costs (and profits) are only indirectly considered (in the selection of a significance level or in interpreting the *p-value*) in the formulation of a hypothesis test. Decision analysis directly involves profits and losses. Because of these major differences, the only topics covered previously in the text that are required for an understanding of decision analysis are probability (including Bayes's Law) and expected value.

20-1 DECISION PROBLEM

You would think that, by this point in the text, we would already have introduced all the necessary concepts and terminology. Unfortunately, because decision analysis is so radically different from statistical inference, several more terms must be defined. They will be introduced in the following example.

EXAMPLE 20.1

An Investment Decision

A man wants to invest €1 million for 1 year. After analysing and eliminating numerous possibilities, he has narrowed his choice to one of three alternatives. These alternatives are referred to as **acts** and are denoted a_i.

a_1: Invest in a guaranteed income certificate paying 3%.
a_2: Invest in a bond with a coupon value of 2%.
a_3: Invest in a well-diversified portfolio of stocks.

He believes that the payoffs associated with the last two acts depend on a number of factors, foremost among which is interest rates. He concludes that there are three possible **states of nature**, denoted s_j.

s_1: Interest rates increase.
s_2: Interest rates stay the same.
s_3: Interest rates decrease.

After further analysis, he determines the amount of profit he will make for each possible combination of an act and a state of nature. Of course, the payoff for the guaranteed income certificate will be €30 000 no matter which state of nature occurs. The profits from each alternative investment are summarised in Table 20.1, in what is called a **payoff table**. Notice that, for example, when the decision is a_2 and the state of nature is s_1, the investor would suffer a €15 000 loss, which is represented by a €15 000 payoff.

TABLE 20.1 Payoff Table for Example 20.1

STATES OF NATURE	a_1 (GIC)	a_2 (BOND)	a_3 (STOCKS)
s_1 (interest rates increase)	€30 000	€15 000	€40 000
s_2 (interest rates stay the same)	€30 000	€20 000	€27 500
s_3 (interest rates decrease)	€30 000	€60 000	€15 000

Another way of expressing the consequence of an act involves measuring the opportunity loss associated with each combination of an act and a state of nature. An **opportunity loss** is the difference between what the decision-maker's profit for an act is and what the profit could have been had the best decision been made. For example, consider the first row of Table 20.1. If s_1 is the state of nature that occurs and the investor chooses act a_1, he makes a profit of €30 000. However, had he chosen act a_3, he would have made a profit of €40 000. The difference between what he could have made (€40 000) and what he actually made (€30 000) is the opportunity loss. Thus, given that s_1 is the state of nature, the opportunity loss of act a_1 is €10 000. The opportunity loss of act a_2 is €25 000, which is the difference between €40 000 and €15 000. The opportunity loss of act a_3 is 0, because there is no opportunity loss when the best alternative is chosen. In a similar manner, we can compute the remaining opportunity losses for this example (see Table 20.2). Notice that we can never experience a negative opportunity loss.

TABLE 20.2 Opportunity Loss Table for Example 20.1

STATES OF NATURE	a_1 (GIC)	a_2 (BOND)	a_3 (STOCKS)
s_1 (interest rates increase)	€10 000	€55 000	0
s_2 (interest rates stay the same)	0	€10 000	€2 500
s_3 (interest rates decrease)	€30 000	0	€45 000

Decision Trees

Most problems involving a simple choice of alternatives can readily be resolved by using the payoff table (or the opportunity loss table). In other situations, however, the decision-maker must choose between sequences of acts. In Section 20-2, we introduce one form of such situations. In these cases, a payoff table will not suffice to determine the best alternative; instead, we require a **decision tree**.

In Chapter 6 we suggested the probability tree as a useful device for computing probabilities. In this type of tree, all the branches represent stages of events. In a decision tree, however, the branches represent both acts and events (states of nature). We distinguish between them in the following way: A square node represents a point where a decision is to be made; a point where a state of nature occurs is represented by a round node. Figure 20.1 depicts the decision tree for Example 20.1.

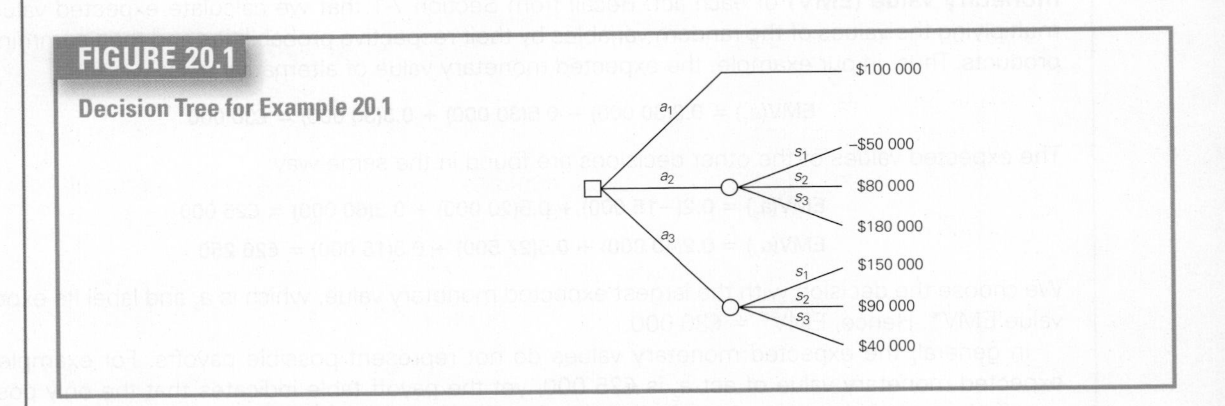

FIGURE 20.1 Decision Tree for Example 20.1

The tree begins with a square node; that is, we begin by making a choice among a_1, a_2 and a_3. The branches emanating from the square node represent these alternatives. At the ends of branches a_2 and a_3, we reach round nodes representing the occurrence of some state of nature. These are depicted as branches representing s_1, s_2 and s_3. At the end of branch a_1, we don't really have a state of nature, because the payoff is fixed at €30 000 no matter what happens to interest rates.

At the ends of the branches, the payoffs are shown (alternatively, we could have worked with opportunity losses instead of with payoffs). These are, of course, the same values that appear in Table 20.1.

Up to this point, all we have done is set up the problem; we have not made any attempt to determine the decision. It should be noted that in many real-life problems, determining the payoff table or decision tree can be a formidable task in itself. Many managers, however, have observed that this task is often extremely helpful in decision-making.

Expected Monetary Value Decision

In many decision problems, it is possible to assign probabilities to the states of nature. For example, if the decision involves trying to decide whether to draw to an inside straight in the game of poker, the probability of succeeding can easily be determined by the use of simple rules of probability. If we must decide whether to replace a machine that has broken down frequently in the past, we can assign probabilities on the basis of the relative frequency of the breakdowns. In many other instances, however, formal rules and techniques of probability cannot be applied. In Example 20.1, the historical relative frequencies of the ups and downs of interest rates will supply scant useful information to help the investor assign probabilities to the behaviour of interest rates during the coming year. In such cases, probabilities must be assigned subjectively. In other words, the determination of the probabilities must be based on the experience, knowledge and (perhaps) guesswork of the decision-maker.

If, in Example 20.1, the investor has some knowledge about a number of economic variables, he might have a reasonable guess about what will happen to interest rates in the next year. Suppose, for example, that our investor believes that future interest rates are most likely to remain essentially the same as they are today and that (of the remaining two states of nature) rates are more likely to decrease than to increase. He might then guess the following probabilities:

$$P(s_1) = 0.2, P(s_2) = 0.5, P(s_3) = 0.3$$

Because the probabilities are subjective, we would expect another decision-maker to produce a completely different set of probabilities. In fact, if this were not true, we would rarely have buyers and sellers of stocks (or any other investment), because everyone would be a buyer (and there would be no sellers) or everyone would be a seller (with no buyers).

After determining the probabilities of the states of nature, we can address the *expected monetary value decision*. We now calculate what we expect will happen for each decision. Because we generally measure the consequences of each decision in monetary terms, we compute the **expected monetary value (EMV)** of each act. Recall from Section 7-1 that we calculate expected values by multiplying the values of the random variables by their respective probabilities and then summing the products. Thus, in our example, the expected monetary value of alternative a_1 is:

$$\text{EMV}(a_1) = 0.2(30\,000) - 0.5(30\,000) + 0.3(30\,000) = €30\,000$$

The expected values of the other decisions are found in the same way:

$$\text{EMV}(a_2) = 0.2(-15\,000) + 0.5(20\,000) + 0.3(60\,000) = €25\,000$$

$$\text{EMV}(a_3) = 0.2(40\,000) + 0.5(27\,500) + 0.3(15\,000) = €26\,250$$

We choose the decision with the largest expected monetary value, which is a_1 and label its expected value EMV*. Hence, EMV* = €30 000.

In general, the expected monetary values do not represent possible payoffs. For example, the expected monetary value of act a_2 is €25 000, yet the payoff table indicates that the only possible payoffs from choosing a_2 are €15 000, €20 000 and €60 000. Of course, the expected monetary value of act a_1 (€30 000) is possible, because that is the only payoff of the act.

What, then, does the expected monetary value represent? If the investment is made a large number of times, with exactly the same payoffs and probabilities, the expected monetary value is the average payoff per investment. That is, if the investment is repeated an infinite number of times with act a_2, 20% of the investments will result in a €15 000 loss, 50% will result in a €20 000 profit and

30% will result in a €60 000 profit. The average of all these investments is the expected monetary value, €25 000. If act a_3 is chosen, the average payoff in the long run will be €26 250.

An important point is raised by the question of how many investments are going to be made. The answer is one. Even if the investor intends to make the same type of investment annually, the payoffs and the probabilities of the states of nature will undoubtedly change from year to year. Hence, we are faced with having determined the expected monetary value decision on the basis of an infinite number of investments, when there will be only one investment. We can rationalise this apparent contradiction in two ways. First, the expected value decision is the only method that allows us to combine the two most important factors in the decision process – the payoffs and their probabilities. It seems inconceivable that, where both factors are known, the investor would want to ignore either one. (There are processes that make decisions on the basis of the payoffs alone; however, these processes assume no knowledge of the probabilities, which is not the case with our example.) Second, typical decision-makers make a large number of decisions over their lifetimes. By using the expected value decision, the decision-maker should perform at least as well as anyone else. Thus, despite the problem of interpretation, we advocate the expected monetary value decision.

Expected Opportunity Loss Decision

We can also calculate the **expected opportunity loss (EOL)** of each act. From the opportunity loss table (Table 20.2), we get the following values:

$$\text{EOL}(a_1) = 0.2(10\ 000) + 0.5(0) + 0.3(30\ 000) = €11\ 000$$
$$\text{EOL}(a_2) = 0.2(55\ 000) + 0.5(10\ 000) + 0.3(0) = €16\ 000$$
$$\text{EOL}(a_3) = 0.2(0) + 0.5(2\ 500) + 0.3(45\ 000) = €14\ 750$$

Because we want to minimise losses, we choose the act that produces the smallest expected opportunity loss, which is a_1. We label its expected value EOL. Observe that the EMV decision is the same as the EOL decision. This is not a coincidence – the opportunity loss table was produced directly from the payoff table.

Rollback Technique for Decision Trees

Figure 20.2 presents the decision tree for Example 20.1, with the probabilities of the states of nature included. The process of determining the EMV decision is called the **rollback technique**; it operates as follows. Beginning at the end of the tree (right-hand side), we calculate the expected monetary value at each round node. The numbers above the round nodes in Figure 20.2 specify these expected monetary values.

At each square node, we make a decision by choosing the branch with the largest EMV. In our example, there is only one square node. Our optimal decision is, of course, a_1.

FIGURE 20.2

Rollback Technique for Example 20.1

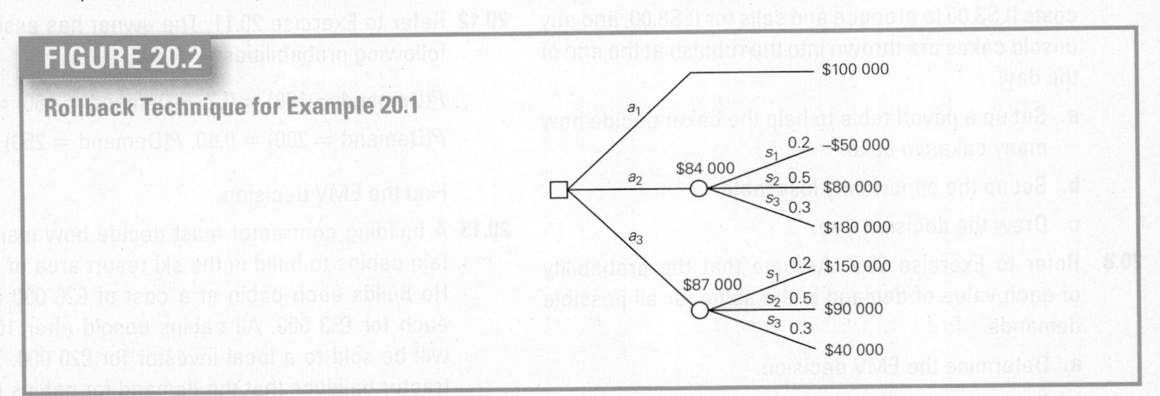

Using the Computer

There is yet another Excel add-in called Tree plan that enables Excel to draw the decision tree and produce the optimal solution. The Appendix Tree plan on the accompanying website provides a brief introduction to get you started.

EXERCISES

20.1 Set up the opportunity loss table from the following payoff table:

	a_1	a_2
s_1	55	26
s_2	43	38
s_3	29	43
s_4	15	51

20.2 Draw the decision tree for Exercise 20.1.

20.3 If we assign the following probabilities to the states of nature in Exercise 20.1, determine the EMV decision:

$$P(s_1) = 0.4, P(s_2) = 0.1, P(s_3) = 0.3, P(s_4) = 0.2$$

20.4 Given the following payoff table, draw the decision tree:

	a_1	a_2	a_3
s_1	20	5	−1
s_2	8	5	4
s_3	−10	5	10

20.5 Refer to Exercise 20.4. Set up the opportunity loss table.

20.6 If we assign the following probabilities to the states of nature in Exercise 20.5, determine the EOL decision:

$$P(s_1) = 0.2, P(s_2) = 0.6, P(s_3) = 0.2$$

Applications

20.7 A baker is going to open a new bakery in Cairo and he must decide how many specialty cakes to bake each morning. From past experience, he knows that the daily demand for cakes ranges from 0 to 3. Each cake costs ILS3.00 to produce and sells for ILS8.00, and any unsold cakes are thrown into the rubbish at the end of the day.

a. Set up a payoff table to help the baker decide how many cakes to bake.

b. Set up the opportunity loss table.

c. Draw the decision tree.

20.8 Refer to Exercise 20.7. Assume that the probability of each value of demand is the same for all possible demands.

a. Determine the EMV decision.

b. Determine the EOL decision.

20.9 The manager of a large shopping centre in Bergen is in the process of deciding on the type of snow-clearing service to hire for his parking lot. Two services are available. The White Christmas Company will clear all snowfalls for a flat fee of NOK4000 for the entire winter season. The Snow White Company charges NOK1800 for each snowfall it clears. Set up the payoff table to help the manager decide, assuming that the number of snowfalls per winter season ranges from 0 to 4.

20.10 Refer to Exercise 20.9. Using subjective assessments the manager has assigned the following probabilities to the number of snowfalls. Determine the optimal decision.

$$P(0) = 0.05, P(1) = 0.15, P(2) = 0.30, P(3) = 0.40, P(4) = 0.10$$

20.11 The owner of a clothing store in Birmingham must decide how many men's shirts to order for the new season. For a particular type of shirt, she must order in quantities of 100 shirts. If she orders 100 shirts, her cost is £10 per shirt; if she orders 200 shirts, her cost is £9 per shirt; and if she orders 300 or more shirts, her cost is £8.50 per shirt. Her selling price for the shirt is £12, but any shirts that remain unsold at the end of the season are sold at her famous 'half-price, end-of-season sale'. For the sake of simplicity, she is willing to assume that the demand for this type of shirt will be 100, 150, 200 or 250 shirts. Of course, she cannot sell more shirts than she stocks. She is also willing to assume that she will suffer no loss of goodwill among her customers if she understocks and the customers cannot buy all the shirts they want. Furthermore, she must place her order today for the entire season; she cannot wait to see how the demand is running for this type of shirt.

a. Construct the payoff table to help the owner decide how many shirts to order.

b. Set up the opportunity loss table.

c. Draw the decision tree.

20.12 Refer to Exercise 20.11. The owner has assigned the following probabilities:

$$P(\text{Demand} = 100) = 0.2, P(\text{Demand} = 150) = 0.25,$$
$$P(\text{Demand} = 200) = 0.40, P(\text{Demand} = 250) = 0.15$$

Find the EMV decision.

20.13 A building contractor must decide how many mountain cabins to build in the ski resort area of Alpbach. He builds each cabin at a cost of €26 000 and sells each for €33 000. All cabins unsold after 10 months will be sold to a local investor for €20 000. The contractor believes that the demand for cabins follows a Poisson distribution, with a mean of 0.5. He assumes that any probability less than 0.01 can be treated as 0. Construct the payoff table and the opportunity loss table for this decision problem.

20.14 An electric company is in the process of building a new power plant. There is some uncertainty regarding the size of the plant to be built. If the community that the plant will service attracts a large number of industries, the demand for electricity will be high. If commercial establishments (offices and retail stores) are attracted, demand will be moderate. If neither industries nor commercial stores locate in the community, the electricity demand will be low. The company can build a small, medium or large plant, but if the plant is too small, the company will incur extra costs. The total costs (in £ millions) of all options are shown in the accompanying table.

	Size of Plant		
Electricity Demand	**Small**	**Medium**	**Large**
Low	220	300	350
Moderate	330	320	350
High	440	390	350

The following probabilities are assigned to the electricity demand:

Demand	*P*(Demand)
Low	0.15
Moderate	0.55
High	0.30

a. Determine the act with the largest expected monetary value. (*Caution:* All the values in the table are costs.)

b. Draw up an opportunity loss table.

c. Calculate the expected opportunity loss for each decision, and determine the optimal decision.

20-2 ACQUIRING, USING AND EVALUATING ADDITIONAL INFORMATION

In this section, we discuss methods of introducing and incorporating additional information into the decision process. Such information generally has value, but it also has attendant costs; that is, we can acquire useful information from consultants, surveys, or other experiments, but we usually must pay for this information. We can calculate the maximum price that a decision-maker should be willing to pay for any information by determining the value of perfect information. We begin by calculating the **expected payoff with perfect information** (EPPI).

If we knew in advance which state of nature would occur, we would certainly make our decisions accordingly. For instance, if the investor in Example 20.1 knew before investing his money what interest rates would do, he would choose the best act to suit that case. Referring to Table 20.1, if he knew that s_1 was going to occur, he would choose act a_3; if s_2 were certain to occur, he would choose a_1, and if a_3 were certain, he would choose a_2. Thus, in the long run, his expected payoff from perfect information would be:

$$\text{EPPI} = 0.2(40\ 000) + 0.5(30\ 000) + 0.3(60\ 000) = €41\ 000$$

Notice that we compute EPPI by multiplying the probability of each state of nature by the largest payoff associated with that state of nature and then summing the products.

This figure, however, does not represent the maximum amount he would be willing to pay for perfect information. Because the investor could make an expected profit of EMV* = €30 000 without perfect information, we subtract EMV* from EPPI to determine the **expected value of perfect information** (EVPI).That is:

$$\text{EVPI} = \text{EPPI} - \text{EMV}^* = €41\ 000 - €30\ 000 = €11\ 000$$

This means that, if perfect information were available, the investor should be willing to pay up to €311 000 to acquire it.

You may have noticed that the expected value of perfect information (EVPI) equals the smallest expected opportunity loss (EOL*). Again, this is not a coincidence – it will always be the case. In future questions, if the opportunity loss table has been determined, you need only calculate EOL* in order to know EVPI.

20-2a Decision-Making with Additional Information

Suppose the investor in our continuing example wants to improve his decision-making capabilities. He learns about Investment Management Consultants (IMC), who, for a fee of €5000, will analyse the economic conditions and forecast the behaviour of interest rates over the next 12 months. The investor, who is quite shrewd (after all, he does have €1 million to invest), asks for some measure of IMC's past successes. IMC has been forecasting interest rates for many years and so provides him with various conditional probabilities (referred to as **likelihood probabilities**), as shown in Table 20.3 which contains the following notation:

I_1: IMC predicts that interest rates will increase.

I_2: IMC predicts that interest rates will stay the same.

I_3: IMC predicts that interest rates will decrease.

TABLE 20.3 **Likelihood Probabilities $P(I_i \mid s_j)$**

	I_1 (PREDICT s_1)	I_2 (PREDICT s_2)	I_3 (PREDICT s_3)
s_1	$P(I_1 \mid s_1) = 0.60$	$P(I_2 \mid s_1) = 0.30$	$P(I_3 \mid s_1) = 0.10$
s_2	$P(I_1 \mid s_2) = 0.10$	$P(I_2 \mid s_2) = 0.80$	$P(I_3 \mid s_2) = 0.10$
s_3	$P(I_1 \mid s_3) = 0.10$	$P(I_2 \mid s_3) = 0.20$	$P(I_3 \mid s_3) = 0.70$

The I_i terms are referred to as **experimental outcomes**, and the process by which we gather additional information is called the **experiment**.

Examine the first line of Table 20.3. When s_1 actually did occur in the past, IMC correctly predicted s_1 60% of the time; 30% of the time, it predicted s_2; and 10% of the time, it predicted s_3. The second row gives the conditional probabilities of I_1, I_2 and I_3 when s_2 actually occurred. The third row shows the conditional probabilities of I_1, I_2 and I_3 when s_3 actually occurred.

The following question now arises: How is the investor going to use the forecast that IMC produces? One approach is simply to assume that whatever IMC forecasts will actually take place and to choose the act accordingly. There are several drawbacks to this approach. Foremost among them is that it puts the investor in the position of ignoring whatever knowledge (in the form of subjective probabilities) he had concerning the issue. Instead the decision-maker should use this information to modify his initial assessment of the probabilities of the states of nature. To incorporate the investor's subjective probabilities with the consultant's forecast requires the use of Bayes's Law, which we introduced in Section 6-4. We will review Bayes's Law in the context of our example.

Suppose that the investor pays IMC the €5000 fee and IMC forecasts that s_1 will occur. We want to revise our estimates for the probabilities of the states of nature, given that I_1 is the outcome of the experiment. That is, we want $P(s_1 \mid I_1)$, $P(s_2 \mid I_1)$ and $P(s_3 \mid I_1)$. Before proceeding, let's develop some terminology.

Recall from Section 6-4 that the original probabilities, $P(s_1)$, $P(s_2)$ and $P(s_3)$, are called **prior probabilities**, because they were determined prior to the acquisition of any additional information. In this example, they were based on the investor's experience. The set of probabilities we want to compute $P(s_1 \mid I_1)$, $P(s_2 \mid I_1)$ and $P(s_3 \mid I_1)$ – are called **posterior or revised probabilities**.

Now we will calculate the posterior probabilities, first by using a probability tree and then by applying a less time-consuming method. Figure 20.3 depicts the probability tree. We begin with the branches of the prior probabilities, which are followed by the likelihood probabilities.

Notice that we label only $P(I_1 \mid s_1)$, $P(I_1 \mid s_2)$ and $P(I_1 \mid s_3)$ because (at this point) we are assuming that I_1 is the experimental outcome. Now recall that conditional probability is defined as:

$$P(A \mid B) = \frac{P(A \text{ and } B)}{P(B)}$$

FIGURE 20.3
Probability Tree to Compute Posterior Probabilities

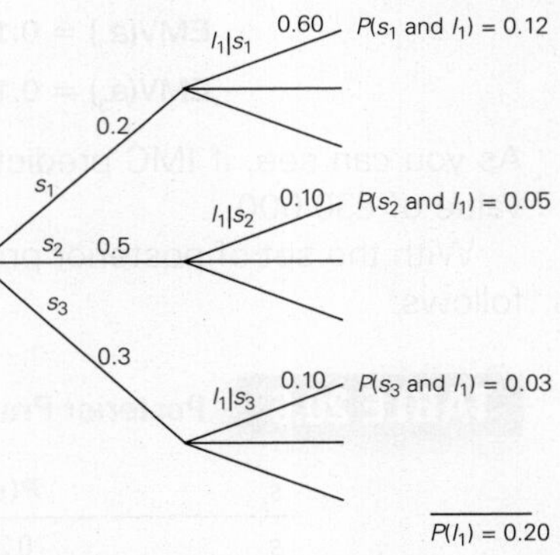

At the end of each branch, we have the joint probability $P(s_j \text{ and } I_1)$. By summing the joint probabilities $P(s_j \text{ and } I_1)$ for $j = 1$, 2 and 3, we calculate $P(I_1)$. Finally:

$$P(s_j|I_1) = \frac{P(S_j \text{ and } I_1)}{P(I_1)}$$

Table 20.4 performs exactly the same calculations as the probability tree except without the tree. So, for example, our revised probability for s_3, which was initially 0.3, is now 0.15.

TABLE 20.4 Posterior Probabilities for I_1

S_j	$P(S_j)$	$P(I_1, S_j)$	$P(s_j \text{ and } I_1)$	$P(s_j \mid I_1)$
s_1	0.2	0.60	(0.2)(0.60) = 0.12	0.12/0.20 = 0.60
s_2	0.5	0.10	(0.5)(0.10) = 0.05	0.05/0.20 = 0.25
s_3	0.3	0.10	(3)(0.10) = 0.03	0.03/0.20 = 0.15
			$P(I_1) = 0.20$	

After the probabilities have been revised, we can use them in exactly the same way we used the prior probabilities. That is, we can calculate the expected monetary value of each act:

$$\text{EMV}(a_1) = 0.60(30\ 000) + 0.25(30\ 000) + 0.15(30\ 000) = €30\ 000$$
$$\text{EMV}(a_2) = 0.60(-15\ 000) + 0.25(20\ 000) + 0.15(60\ 000) = €5000$$
$$\text{EMV}(a_3) = 0.60(40\ 000) + 0.25(27\ 500) + 0.15(15\ 000) = €33\ 125$$

Thus, if IMC forecasts s_1, the optimal act is a_3, and the expected monetary value of the decision is €33 125.

As a further illustration, we now repeat the process for I_2 and I_3 in Tables 20.5 and 20.6, respectively. Applying the posterior probabilities for I_2 from Table 20.5 to the payoff table, we find the following:

TABLE 20.5 Posterior Probabilities for I_2

s_j	$P(s_j)$	$P(I_2 \mid s_j)$	$P(s_j \text{ and } I_2)$	$P(s_j \mid I_2)$
s_1	0.2	0.30	(0.2) (0.30) = 0.06	0.06/0.52 = 0.115
s_2	0.5	0.80	(0.5) (0.80) = 0.40	0.40/0.52 = 0.770
s_3	0.3	0.20	(0.3) (0.20) = 0.06	0.06/0.52 = 0.115
			$P(I_2) = 0.52$	

$$\text{EMV}(a_1) = 0.115(30\ 000) + 0.770(30\ 000) + 0.115(30\ 000) = €30\ 000$$
$$\text{EMV}(a_2) = 0.115(-15\ 000) + 0.770(20\ 000) + 0.115(60\ 000) = €20\ 575$$
$$\text{EMV}(a_3) = 0.115(40\ 000) + 0.770(27\ 500) + 0.115(15\ 000) = €27\ 500$$

As you can see, if IMC predicts that s_2 will occur, the optimal act is a_1 with an expected monetary value of €30 000.

With the set of posterior probabilities for I_3 from Table 20.6, the expected monetary values are as follows:

TABLE 20.6 **Posterior Probabilities for I_3**

s_j	$P(s_j)$	$P(I_3\|s_j)$	$P(s_j \text{ and } I_3)$	$P(s_j\|I_3)$
s_1	0.2	0.10	(0.2)(0.10) = 0.02	0.02/0.28 = 0.071
s_2	0.5	0.10	(0.5)(0.10) = 0.05	0.05/0.28 = 0.179
s_3	0.3	0.70	(0.3)(0.70) = 0.21	0.21/0.28 = 0.750
			$P(I_3) = 0.28$	

$$\text{EMV}(a_1) = 0.071(30\ 000) + 0.179(30\ 000) + 0.750(30\ 000) = €30\ 000$$
$$\text{EMV}(a_2) = 0.071(-15\ 000) + 0.179(20\ 000) + 0.750(60\ 000) = €47\ 515$$
$$\text{EMV}(a_3) = 0.071(40\ 000) + 0.179(27\ 500) + 0.750(15\ 000) = €19\ 013$$

If IMC predicts that s_3 will occur, the optimal act is a_2, with an expected monetary value of €47 515.

At this point, we know the following:

If IMC predicts s_1, then the optimal act is a_3.
If IMC predicts s_2, then the optimal act is a_1.
If IMC predicts s_3, then the optimal act is a_2.

Thus, even before IMC makes its forecast, the investor knows which act is optimal for each of the three possible IMC forecasts. Of course, all these calculations can be performed before paying IMC its €5000 fee. This leads to an extremely important calculation. By performing the computations just described, the investor can determine *whether* he should hire IMC, that is, he can determine whether the value of IMC's forecast exceeds the cost of its information. Such a determination is called a **preposterior analysis**.

20-2b Preposterior Analysis

The objective of a preposterior analysis is to determine whether the value of the prediction is greater or less than the cost of the information. *Posterior* refers to the revision of the probabilities, and the *pre* indicates that this calculation is performed before paying the fee.

We begin by finding the expected monetary value of using the additional information. This value is denoted EMV, which for our example is determined on the basis of the following analysis:

If IMC predicts s_1, then the optimal act is a_3, and the expected payoff is €33 125.
If IMC predicts s_2, then the optimal act is a_1 and the expected payoff is €30 000.
If IMC predicts s_3, then the optimal act is a_2, and the expected payoff is €47 515.

A useful by-product of calculating the posterior probabilities is the set of probabilities of I_1, I_2 and I_3:

$$P(I_1) = 0.20,\ P(I_2) = 0.52,\ P(I_2) = 0.28$$

(Notice that these probabilities sum to 1.) Now imagine that the investor seeks the advice of IMC an infinite number of times. (This is the basis for the expected value decision.) The set of probabilities of I_1, I_2 and I_3 indicates the following outcome distribution: 20% of the time, IMC will predict s_1 and the expected monetary value will be €33 125; 52% of the time, IMC will predict s_2 and the expected monetary value will be €30 000; and 28% of the time, IMC will predict s_3 and the expected monetary value will be €47 515.

The expected monetary value with additional information is the weighted average of the expected monetary values, where the weights are $P(I_1)$, $P(I_2)$ and $P(I_3)$. Hence:

$$\text{EMV}' = 0.20(33\,125) + 0.52(30\,000) + 0.28(47\,515) = €35\,529$$

The value of IMC's forecast is the difference between the expected monetary value with additional information (EMV′) and the expected monetary value without additional information (EMV*). This difference is called the **expected value of sample information** and is denoted EVSI. Thus:

$$\text{EVSI} = \text{EMV}' - \text{EMV}^* = €35\,529 - €30\,000 = €5529$$

By using IMC's forecast, the investor can make an average additional profit of €5529 in the long run. Because the cost of the forecast is only €5000, the investor is advised to hire IMC.

If you review this problem, you'll see that the investor had to make two decisions. The first (chronologically) was whether to hire IMC, and the second was which type of investment to make. A decision tree is quite helpful in describing the acts and states of nature in this question. Figure 20.4 provides the complete tree diagram.

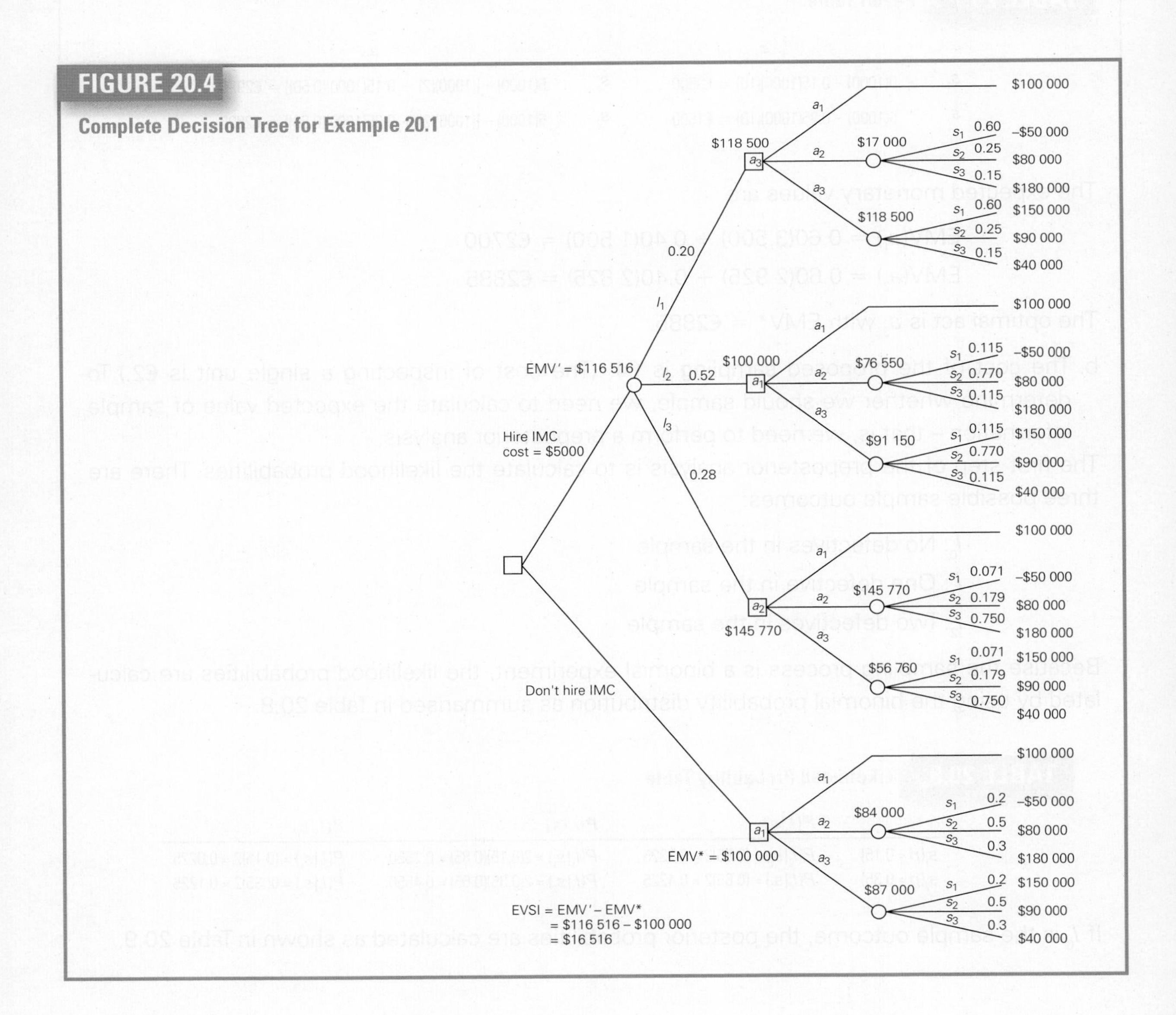

FIGURE 20.4
Complete Decision Tree for Example 20.1

ACCEPTANCE SAMPLING: SOLUTION

a. The two alternatives are

a_1: No inspection (the current policy)

a_2: 100% inspection

The two states of nature are

s_1: The lot contains 15% defectives

s_2: The lot contains 35% defectives

Based on the past year's historical record,

$$P(s_1) = 0.60 \text{ and } P(s_2) = 0.40$$

The payoff table is constructed as shown in Table 20.7.

TABLE 20.7 Payoff Table

a_1		a_2	
s_1	5(1000) – 0.15(1000)(10) = €3500	s_1	5(1000) – [(1000)(2) + 0.15(1000)(0.50)] = €2925
s_2	5(1000) – 0.35(1000)(10) = €1500	s_2	5(1000) – [(1000)(2) + 0.35(1000)(0.50)] = €2825

The expected monetary values are

$$\text{EMV}(a_1) = 0.60(3\ 500) + 0.40(1\ 500) = €2700$$
$$\text{EMV}(a_2) = 0.60(2\ 925) + 0.40(2\ 825) = €2885$$

The optimal act is a_2 with EMV* = €2885.

b. The cost of the proposed sampling is €4. (The cost of inspecting a single unit is €2.) To determine whether we should sample, we need to calculate the expected value of sample information – that is, we need to perform a preposterior analysis.

The first step of the preposterior analysis is to calculate the likelihood probabilities. There are three possible sample outcomes:

I_0: No defectives in the sample

I_1: One defective in the sample

I_2: Two defectives in the sample

Because the sampling process is a binomial experiment, the likelihood probabilities are calculated by using the binomial probability distribution as summarised in Table 20.8.

TABLE 20.8 Likelihood Probability Table

	$P(I_0\|s_j)$	$P(I_1\|s_j)$	$P(I_2\|s_j)$
$s_1(p = 0.15)$	$P(I_0\|s_1)$ = (0.85)2 = 0.7225	$P(I_1\|s_1)$ = 2(0.15)(0.85) = 0.2550	$P(I_2\|s_1)$ = (0.15)2 = 0.0225
$s_2(p = 0.35)$	$P(I_0\|s_2)$ = (0.65)2 = 0.4225	$P(I_1\|s_2)$ = 2(0.35)(0.65) = 0.4550	$P(I_2\|s_2)$ = 0(.35)2 = 0.1225

If I_0 is the sample outcome, the posterior probabilities are calculated as shown in Table 20.9.

TABLE 20.9 Posterior Probabilities for I_0

s_j	$P(s_j)$	$P(I_0 \mid s_j)$	$P(s_j \text{ and } I_0)$	$P(s_j \mid I_0)$
s_1	0.60	0.7225	(0.60)(0.7225) = 0.4335	0.4335/0.6025 = 0.720
s_2	0.40	0.4225	(0.40)(0.4225) = 0.1690 $P(I_0)$ = 0.6025	0.1690/0.6025 = 0.280

The expected monetary values if the sample outcome is I_0 are

$$\text{EMV}(a_1) = 0.457(3500) + 0.543(1500) = €2414$$

$$\text{EMV}(a_2) = 0.457(2925) + 0.543(2825) = €2871$$

Therefore, the optimal act is a_1.
If I_1 is the sample outcome, the posterior probabilities are calculated as shown in Table 20.10.

TABLE 20.10 Posterior Probabilities for I_1

s_j	$P(s_j)$	$P(I_1 \mid s_j)$	$P(s_j \text{ and } I_1)$	$P(s_j \mid I_1)$
s_1	0.60	0.2550	(0.60)(0.2550) = 0.1530	0.1530/0.3350 = 0.457
s_2	0.40	0.4550	(0.40)(0.4550) = 0.1820	0.1820/0.3350 = 0.543
			$P(I_1)$ = 0.3350	

The expected monetary values if the sample outcome is I_2 are

$$\text{EMV}(a_1) = 0.720(3500) + 0.280(1500) = €2940$$

$$\text{EMV}(a_2) = 0.720(2925) + 0.280(2825) = €2897$$

Therefore, the optimal act is a_2.
If I_2 is the sample outcome, the posterior probabilities are calculated as shown in Table 20.11.

TABLE 20.11 Posterior Probabilities for I_2

s_j	$P(s_j)$	$P(I_1 \mid s_j)$	$P(s_j \text{ and } I_1)$	$P(s_j \mid I_1)$
s_1	0.60	0.0225	(0.60)(0.0225) = 0.0135	0.0135/0.0625 = 0.216
s_2	0.40	0.1225	(0.40)(0.1225) = 0.0490	0.0490/0.0625 = 0.784
			$P(I_2)$ = 0.0625	

The expected monetary values if the sample outcome is I_2 are

$$\text{EMV}(a_1) = 0.216(3500) + 0.784(1500) = €1932$$

$$\text{EMV}(a_2) = 0.216(2925) + 0.784(2825) = €2847$$

Therefore, the optimal act is a_2.
We can now summarise these results, as shown in Table 20.12.

TABLE 20.12 Summary of Optimal Acts

Sample Outcome	Probability	Optimal Act	Expected Monetary Value
I_0	0.6025	a_1	€2940
I_1	0.3350	a_2	€2871
I_2	0.0625	a_3	€2847

(continued)

The expected monetary value with additional information is

$$EMV' = 0.6025(2940) + 0.3350(2871) + 0.0625(2847) = €2911$$

The expected value of sample information is

$$EVSI = EMV' - EMV^* = 2911 - 2885 = €26$$

Because the expected value of sample information is €26 and the sampling cost is €4, the company should take a sample of 2 units before deciding whether to inspect 100%. The optimal sequence is as follows:

1. Take a sample of 2 units.
2. If there are no defective units in the sample, continue the current policy of no inspection. If either one or two of the sample units are defective, perform a complete inspection of the lot.

20-2c Bayesian Statistics

In Chapters 10 to 17 we dealt with inference about unknown parameters. In Chapter 10 we pointed out that when interpreting the confidence interval estimate, we cannot make probability statements about parameters because they are not variables. However, Bayesian statistics specifies that parameters are variables, and we can assume various probability distributions. The acceptance sampling example illustrates this concept. The parameter was the proportion p of defective units in the 1000-unit batch. The example was unrealistic because we allowed the parameter to assume one of only two values, 15% and 35%. We assigned prior probabilities using the relative frequency approach; that is, based on historic records we had

$$P(p = 15\%) = 0.60 \text{ and } P(p = 35\%) = 0.40$$

To make the problem more realistic, we let p be a continuous random variable rather than a discrete one. In other words, p can take on any value between 0 and 100%. We assign a density function also based on historical records. We can express the payoffs as a linear function of p. Then, using calculus, we can determine the optimum decision. We can also revise the prior probabilities based on the outcome of the sampling of two units. The technique requires some calculus, but the concept is the same as the one developed in this chapter. It should be noted that there is a parallel universe of Bayesian statistics that more or less matches the material in the inference part of this book. Interested readers can learn more about Bayesian statistics from additional courses dedicated to the subject.

EXERCISES

20.15 Find EPPI, EMV* and EVPI for the accompanying payoff table and probabilities.

	a_1	a_2	a_3
s_1	60	110	75
s_2	40	110	150
s_3	220	120	85
s_4	250	120	130

$P(s_1) = 0.10$, $P(s_2) = 0.25$, $P(s_3) = 0.50$, $P(s_4) = 0.15$

20.16 For Exercise 20.15, determine the opportunity loss table and compute EOL*. Confirm that EOL* = EVPI.

20.17 Given the following payoff table and probabilities, determine EVPI.

	a_1	a_2	a_3	a_4
s_1	65	20	45	30
s_2	70	110	80	95

$P(s_1) = 0.5$, $P(s_2) = 0.5$

20.18 Redo Exercise 20.17, changing the probabilities to the following values.

a. $P(s_1) = 0.75$, $P(s_2) = 0.25$

b. $P(s_1) = 0.95$, $P(s_2) = 0.05$

20.19 What conclusion can you draw about the effect of the probabilities on EVPI from Exercises 20.17 and 20.18?

20.20 Determine the posterior probabilities, given the following prior and likelihood probabilities.

Prior probabilities

$P(s_1) = 0.25$, $P(s_2) = 0.40$, $P(s_3) = 0.35$

Likelihood probabilities

	I_1	I_2	I_3	I_4
s_1	0.40	0.30	0.20	0.10
s_2	0.25	0.25	0.25	0.25
s_3	0	0.30	0.40	0.30

20.21 Calculate the posterior probabilities from the prior and likelihood probabilities that follow.

Prior probabilities

$P(s_1) = 0.5$, $P(s_2) = 0.5$

Likelihood probabilities

	I_1	I_2
s_1	0.98	0.02
s_2	0.05	0.95

20.22 With the accompanying payoff table and the prior and posterior probabilities computed in Exercise 20.21 calculate the following.

a. The optimal act for each experimental outcome.

b. The expected value of sample information.

Payoff table

	a_1	a_2	a_3
s_1	10	18	23
s_2	22	19	15

Applications

20.23 A sporting-goods storeowner has the opportunity to purchase a lot of 50 000 footballs for £100 000. He believes that he can sell some or all by taking out mail-order advertisements in a magazine. Each football will be sold for £6. The advertising cost is £25 000, and the mailing cost per football is £1. He believes that the demand distribution is as follows.

Demand	P(Demand)
10 000	0.2
30 000	0.5
50 000	0.3

What is the maximum price the owner should pay for additional information about demand?

20.24 A radio station that currently directs its programming toward middle-aged listeners is contemplating switching to rock-and-roll music. After analysing advertising revenues and operating costs, the owner concludes that, for each percentage point of market share, revenues increase by €100 000 per year. Fixed annual operating costs are €700 000. The owner believes that, with the change, the station will get a 5%, 10% or 20% market share, with probabilities 0.4, 0.4 and 0.2, respectively. The current annual profit is €285 000.

a. Set up the payoff table.

b. Determine the optimal act.

c. What is the most the owner should be willing to pay to acquire additional information about the market share?

20.25 Exactly what to do with household waste is a growing concern globally, particularly with the rise in population and greater understanding of the wider environmental issues associated with waste disposal. Ultimately, there is too much rubbish to be disposed of and volumes are expected to increase to 2.2 billion tonnes by 2025 according to World Bank figures. As a consequence, the idea of recycling is becoming ever more popular. A waste-management company in Naples, where the issue is especially prevalent, is willing to begin recycling newspapers, aluminium cans and plastic containers. However, it is only profitable to do so if a sufficiently large proportion of households is willing to participate. Suppose that in Naples, 1 million households are potential recyclers. After some analysis, it is determined that, for every 1000 households that participate in the programme, the contribution to profit is €500. It is also discovered that fixed costs are €55 000 per year. It is believed that 50 000, 100 000, 200 000, or 300 000 households will participate, with probabilities 0.5, 0.3, 0.1 and 0.1, respectively. A preliminary survey was performed wherein 25 households were asked whether they would be willing to be part of this recycling programme. Suppose only three of the 25 respond affirmatively, incorporate this information into the decision-making process to decide whether the waste-management company should proceed with the recycling venture.

20.26 Repeat Exercise 20.25, given that 12 out of 100 households respond affirmatively.

20.27 Suppose that in Exercise 20.14 a consultant offers to analyse the problem and predict the amount of electricity required by the new community. To induce the electric company to hire her, the consultant provides the set of likelihood probabilities given here. Perform a preposterior analysis to determine the expected value of the consultant's sample information.

	I_1(Predict low demand)	I_2(Predict moderate demand)	I_3(Predict high demand)
s_1	0.5	0.3	0.2
s_2	0.3	0.6	0.1
s_3	0.2	0.2	0.6

20.28 In Exercise 20.24, suppose that it is possible to survey radio listeners to determine whether they would tune in to the station if the format changed to rock and roll. What would a survey of size 2 be worth?

20.29 Suppose that in Exercise 20.24 a random sample of 25 radio listeners revealed that two people would be regular listeners of the station. What is the optimal decision now?

20.30 The president of an automobile battery company must decide which one of three new types of batteries to produce. The fixed and variable costs of each battery are shown in the accompanying table.

Battery	Fixed Cost (£)	Variable Cost (per Unit) (£)
1	900 000	40
2	1 150 000	34
3	1 400 000	30

The president believes that demand will be 50 000, 100 000 or 150 000 batteries, with probabilities 0.3, 0.3 and 0.4, respectively. The selling price of the battery will be £40.

a. Determine the payoff table.

b. Determine the opportunity loss table.

c. Find the expected monetary value for each act, and select the optimal one.

d. What is the most the president should be willing to pay for additional information about demand?

20.31 Credibility is often the most effective feature of an advertising campaign. Suppose that, for a particular advertisement, 32% of people surveyed currently believe what the ad claims. A marketing manager believes that for each 1-point increase in that percentage, annual sales will increase by €1 million. For each 1-point decrease, annual sales will decrease by €1 million. The manager believes that a change in the advertising approach can influence the ad's credibility. The probability distribution of the potential percentage changes is listed here.

Percentage Change	Probability
−2	0.1
−1	0.1
0	0.2
+1	0.3
+2	0.3

If for each euro of sales the profit contribution is 10 euro cents and the overall cost of changing the ad is €58 000, should the ad be changed?

20.32 Suppose that in Exercise 20.31 it is possible to perform a survey to determine the percentage of people who believe the ad. What would a sample of size 1 be worth?

20.33 Suppose that in Exercise 20.31 a sample of size 5 showed that only one person believes the new ad. In light of this additional information, what should the manager do?

20.34 Max the Bookie is trying to decide how many telephones to install in his new bookmaking operation. Because of heavy police activity, he cannot increase or decrease the number of telephones once he sets up his operation. He has narrowed the possible choices to three. He can install 25, 50 or 100 telephones. His profit for 1 year (the usual length of time he can remain in business before the police close him down) depends on the average number of calls he receives. The number of calls is Poisson distributed. After some deliberation, he concludes that the average number of calls per minute can be 0.5, 1.0 or 1.5, with probabilities of 0.50, 0.25 and 0.25, respectively. Max then produces the payoffs given in the accompanying table.

Payoff Table

	25 Telephones (£)	50 Telephones (£)	100 Telephones (£)
$s_1(\mu = 0.5)$	50 000	30 000	20 000
$s_2(\mu = 1.0)$	50 000	60 000	40 000
$s_3(\mu = 1.5)$	50 000	60 000	80 000

Max's assistant, Lefty (who attended a business school for 2 years), points out that Max may be able to get more information by observing a competitor's similar operation. However, he will be able to watch for only 10 minutes, and doing so will cost him £4000. Max determines that if he counts fewer than eight calls, that would be a low number; at least eight but fewer than 17 would be a medium number; and at least 17 would be a large number of calls. Max also decides that, if the experiment is run, he will record only whether there is a small, medium or large number of calls. Help Max by performing a preposterior analysis to determine whether the sample should be taken. Conclude by specifying clearly what the optimal strategy is.

20.35 A major movie studio has just completed its latest epic, a musical comedy about the life of Attila the Hun. Because the movie is different (no sex or violence), the studio is uncertain about how to distribute it. The studio executives must decide whether to release the movie to North American audiences or to sell it to a European distributor and realise a profit of $12 million. If the movie is shown in North America, the studio profit depends on its level of success, which can be classified as excellent, good or fair. The payoffs and the prior subjective probabilities of the success levels are shown in the accompanying table.

Success Level	Payoff ($Million)	Probability
Excellent	33	0.5
Good	12	0.3
Fair	–15	0.2

Another possibility is to have the movie shown to a random sample of North Americans and use their collective judgement to help the studio make a decision. These judgements are categorised as 'rave review', 'lukewarm response', and 'poor response'. The cost of the sample is $100 000. The sampling process has been used several times in the past. The likelihood probabilities describing the audience judgements and the movie's success level are shown next. Perform a preposterior analysis to determine what the studio executives should do.

	Judgement		
Success Level	Rave Review	Lukewarm Response	Poor Response
Excellent	0.8	0.1	0.1
Good	0.5	0.3	0.2
Fair	0.4	0.3	0.3

CHAPTER SUMMARY

The objective of decision analysis is to select the optimal act from a list of alternative acts. We define as optimal the act with the largest expected monetary value or smallest expected opportunity loss. The expected values are calculated after assigning prior probabilities to the states of nature. The acts, states of nature and their consequences may be presented in a payoff table, an opportunity loss table or a decision tree. We also discussed a method by which additional information in the form of an experiment can be incorporated in the analysis. This method involves combining prior and likelihood probabilities to produce posterior probabilities. The preposterior analysis allows us to decide whether to pay for and acquire the experimental outcome. That decision is based on the expected value of sample information and on the sampling cost.

SYMBOLS:

Symbol	Represents
a_i	Acts
s_j	States of nature
I_i	Experimental outcomes
$P(s_j)$	Prior probability
$P(I_i \mid s_j)$	Likelihood probability
$P(s_j \text{ and } I_i)$	Joint probability
$P(s_j \mid I_i)$	Posterior probability

21 CONCLUSION

21-1 TWELVE KEY STATISTICAL CONCEPTS

We have come to the end of the journey that began with the words 'Statistics is a way to get information from data'. At some point you will probably write the final examination in your statistics course. (We assume that readers of this book are taking a statistics course and not just reading it for fun.) If you believe that this event will be the point where you and statistics part company, you could not be more wrong. In the world into which you are about to graduate, the potential applications of statistical techniques are virtually limitless.

However, if you are unable or unwilling to employ statistics, you cannot consider yourself to be competent. Can you imagine a marketing manager who does not fully understand marketing concepts and techniques? Can an accountant who knows little about accounting principles do his or her job? Similarly, you cannot be a competent decision-maker without a comprehension of statistical concepts and techniques.

In our experience, we have come across far too many people who display an astonishing ignorance of probability and statistics. In some cases, this is displayed in the way they gamble. We have seen managers who regularly make decisions involving millions of dollars who do not understand the fundamental principles that should govern the way decisions are made. The worst may be the managers who have access to vast amounts of information no further away than the nearest computer but don't know how to get it or even know it is there.

This raises the question, what statistical concepts and techniques will you need for your life after the final exam? We do not expect students to remember the formulas (or computer commands) that calculate the confidence interval estimates or test statistics. (Statistics reference books are available for that purpose.) However, you must know what you can and cannot do with statistical techniques. You must remember a number of important principles that were covered in this book. To assist you, we have selected the 12 most important concepts and list them here. We hope that they prove useful to you.

21-1 TWELVE KEY STATISTICAL CONCEPTS

1. Statistical techniques are processes that convert data into information. Descriptive techniques describe and summarise; inferential techniques allow us to make estimates and draw conclusions about populations from samples.
2. We need a large number of techniques because there are numerous objectives and types of data. There are four types of data: interval (values that lie along an evenly dispersed range of real

numbers), ratio (special kind of interval data where there is an absolute zero), nominal (categories that cannot be ordered in any particular way) and ordinal (categories that can be ordered). Each combination of data type and objective requires specific techniques.

3. We gather data by various sampling plans. However, the validity of any statistical outcome is dependent on the validity of the sampling. 'Garbage in, garbage out' very much applies in statistics.
4. The sampling distribution is the source of statistical inference. The confidence interval estimator and the test statistic are derived directly from the sampling distribution. All inferences are actually probability statements based on the sampling distribution.
5. All tests of hypotheses are conducted similarly. We assume that the null hypothesis is true. We then compute the value of the test statistic. If the difference between what we have observed (and calculated) and what we expect to observe is too large, we reject the null hypothesis. The standard that decides what is 'too large' is determined by the probability of a Type I error.
6. In any test of hypothesis (and in most decisions) there are two possible errors: Type I and Type II. The relationship between the probabilities of these errors helps us decide where to set the standard. If we set the standard so high that the probability of a Type I error is very small, we increase the probability of a Type II error. A procedure designed to decrease the probability of a Type II error must have a relatively large probability of a Type I error.
7. We can improve the exactitude of a confidence interval estimator or decrease the probability of a Type II error by increasing the sample size. More data mean more information, which results in narrower intervals or lower probabilities of making mistakes, which in turn leads to better decisions.
8. The sampling distributions that are used for interval data are the Student t and the F. These distributions are related so that the various techniques for interval data are themselves related. We can use the analysis of variance in place of the t-test of two means. We can use regression analysis with indicator variables in place of the analysis of variance. We often build a model to represent relationships among interval variables, including indicator variables.
9. In analysing interval data, we attempt to explain as much of the variation as possible. By doing so, we can learn a great deal about whether populations differ and what variables affect the response (dependent) variable.
10. The techniques used on nominal data require that we count the number of times each category occurs. The counts are then used to compute statistics. The sampling distributions we use for nominal data are the standard normal and the chi-squared. These distributions are related, as are the techniques.
11. The techniques used on ordinal data are based on a ranking procedure. We call these techniques *nonparametric*. Because the requirements for the use of nonparametric techniques are less stringent than those for a parametric procedure, we often use nonparametric techniques in place of parametric ones when the required conditions for the parametric test are not satisfied. To ensure the validity of a statistical technique, we must check the required conditions.
12. We can obtain data through experimentation or by observation. Observational data lend themselves to several conflicting interpretations. Data gathered by an experiment are more likely to lead to a definitive interpretation. In addition to designing experiments, statistics practitioners can also select particular sample sizes to produce the accuracy and confidence they desire.

APPENDIX A

TABLES

TABLE 1 Binomial Probabilities

Tabulated values are $P(X \leq k) = \sum_{x=0}^{k} p(x)$ (Values are rounded to four decimal places.)

$n = 5$

	P														
k	**0.01**	**0.05**	**0.10**	**0.20**	**0.25**	**0.30**	**0.40**	**0.50**	**0.60**	**0.70**	**0.75**	**0.80**	**0.90**	**0.95**	**0.99**
0	0.9510	0.7738	0.5905	0.3277	0.2373	0.1681	0.0778	0.0313	0.0102	0.0024	0.0010	0.0003	0.0000	0.0000	0.0000
1	0.9990	0.9774	0.9185	0.7373	0.6328	0.5282	0.3370	0.1875	0.0870	0.0308	0.0156	0.0067	0.0005	0.0000	0.0000
2	1.0000	0.9988	0.9914	0.9421	0.8965	0.8369	0.6826	0.5000	0.3174	0.1631	0.1035	0.0579	0.0086	0.0012	0.0000
3	1.0000	1.0000	0.9995	0.9933	0.9844	0.9692	0.9130	0.8125	0.6630	0.4718	0.3672	0.2627	0.0815	0.0226	0.0010
4	1.0000	1.0000	1.0000	0.9997	0.9990	0.9976	0.9898	0.9688	0.9222	0.8319	0.7627	0.6723	0.4095	0.2262	0.0490

$n = 6$

	P														
k	**0.01**	**0.05**	**0.10**	**0.20**	**0.25**	**0.30**	**0.40**	**0.50**	**0.60**	**0.70**	**0.75**	**0.80**	**0.90**	**0.95**	**0.99**
0	0.9415	0.7351	0.5314	0.2621	0.1780	0.1176	0.0467	0.0156	0.0041	0.0007	0.0002	0.0001	0.0000	0.0000	0.0000
1	0.9985	0.9672	0.8857	0.6554	0.5339	0.4202	0.2333	0.1094	0.0410	0.0109	0.0046	0.0016	0.0001	0.0000	0.0000
2	1.0000	0.9978	0.9842	0.9011	0.8306	0.7443	0.5443	0.3438	0.1792	0.0705	0.0376	0.0170	0.0013	0.0001	0.0000
3	1.0000	0.9999	0.9987	0.9830	0.9624	0.9295	0.8208	0.6563	0.4557	0.2557	0.1694	0.0989	0.0159	0.0022	0.0000
4	1.0000	1.0000	0.9999	0.9984	0.9954	0.9891	0.9590	0.8906	0.7667	0.5798	0.4661	0.3446	0.1143	0.0328	0.0015
5	1.0000	1.0000	1.0000	0.9999	0.9998	0.9993	0.9959	0.9844	0.9533	0.8824	0.8220	0.7379	0.4686	0.2649	0.0585

$n = 7$

	P														
k	**0.01**	**0.05**	**0.10**	**0.20**	**0.25**	**0.30**	**0.40**	**0.50**	**0.60**	**0.70**	**0.75**	**0.80**	**0.90**	**0.95**	**0.99**
0	0.9321	0.6983	0.4783	0.2097	0.1335	0.0824	0.0280	0.0078	0.0016	0.0002	0.0001	0.0000	0.0000	0.0000	0.0000
1	0.9980	0.9556	0.8503	0.5767	0.4449	0.3294	0.1586	0.0625	0.0188	0.0038	0.0013	0.0004	0.0000	0.0000	0.0000
2	1.0000	0.9962	0.9743	0.8520	0.7564	0.6471	0.4199	0.2266	0.0963	0.0288	0.0129	0.0047	0.0002	0.0000	0.0000
3	1.0000	0.9998	0.9973	0.9667	0.9294	0.8740	0.7102	0.5000	0.2898	0.1260	0.0706	0.0333	0.0027	0.0002	0.0000
4	1.0000	1.0000	0.9998	0.9953	0.9871	0.9712	0.9037	0.7734	0.5801	0.3529	0.2436	0.1480	0.0257	0.0038	0.0000
5	1.0000	1.0000	1.0000	0.9996	0.9987	0.9962	0.9812	0.9375	0.8414	0.6706	0.5551	0.4233	0.1497	0.0444	0.0020
6	1.0000	1.0000	1.0000	1.0000	0.9999	0.9998	0.9984	0.9922	0.9720	0.9176	0.8665	0.7903	0.5217	0.3017	0.0679

TABLE 1 (*Continued*)

$n = 8$

	P														
k	**0.01**	**0.05**	**0.10**	**0.20**	**0.25**	**0.30**	**0.40**	**0.50**	**0.60**	**0.70**	**0.75**	**0.80**	**0.90**	**0.95**	**0.99**
0	0.9227	0.6634	0.4305	0.1678	0.1001	0.0576	0.0168	0.0039	0.0007	0.0001	0.0000	0.0000	0.0000	0.0000	0.0000
1	0.9973	0.9428	0.8131	0.5033	0.3671	0.2553	0.1064	0.0352	0.0085	0.0013	0.0004	0.0001	0.0000	0.0000	0.0000
2	0.9999	0.9942	0.9619	0.7969	0.6785	0.5518	0.3154	0.1445	0.0498	0.0113	0.0042	0.0012	0.0000	0.0000	0.0000
3	1.0000	0.9996	0.9950	0.9437	0.8862	0.8059	0.5941	0.3633	0.1737	0.0580	0.0273	0.0104	0.0004	0.0000	0.0000
4	1.0000	1.0000	0.9996	0.9896	0.9727	0.9420	0.8263	0.6367	0.4059	0.1941	0.1138	0.0563	0.0050	0.0004	0.0000
5	1.0000	1.0000	1.0000	0.9988	0.9958	0.9887	0.9502	0.8555	0.6846	0.4482	0.3215	0.2031	0.0381	0.0058	0.0001
6	1.0000	1.0000	1.0000	0.9999	0.9996	0.9987	0.9915	0.9648	0.8936	0.7447	0.6329	0.4967	0.1869	0.0572	0.0027
7	1.0000	1.0000	1.0000	1.0000	1.0000	0.9999	0.9993	0.9961	0.9832	0.9424	0.8999	0.8322	0.5695	0.3366	0.0773

$n = 9$

	P														
k	**0.01**	**0.05**	**0.10**	**0.20**	**0.25**	**0.30**	**0.40**	**0.50**	**0.60**	**0.70**	**0.75**	**0.80**	**0.90**	**0.95**	**0.99**
0	0.9135	0.6302	0.3874	0.1342	0.0751	0.0404	0.0101	0.0020	0.0003	0.0000	0.0000	0.0000	0.0000	0.0000	0.0000
1	0.9966	0.9288	0.7748	0.4362	0.3003	0.1960	0.0705	0.0195	0.0038	0.0004	0.0001	0.0000	0.0000	0.0000	0.0000
2	0.9999	0.9916	0.9470	0.7382	0.6007	0.4628	0.2318	0.0898	0.0250	0.0043	0.0013	0.0003	0.0000	0.0000	0.0000
3	1.0000	0.9994	0.9917	0.9144	0.8343	0.7297	0.4826	0.2539	0.0994	0.0253	0.0100	0.0031	0.0001	0.0000	0.0000
4	1.0000	1.0000	0.9991	0.9804	0.9511	0.9012	0.7334	0.5000	0.2666	0.0988	0.0489	0.0196	0.0009	0.0000	0.0000
5	1.0000	1.0000	0.9999	0.9969	0.9900	0.9747	0.9006	0.7461	0.5174	0.2703	0.1657	0.0856	0.0083	0.0006	0.0000
6	1.0000	1.0000	1.0000	0.9997	0.9987	0.9957	0.9750	0.9102	0.7682	0.5372	0.3993	0.2618	0.0530	0.0084	0.0001
7	1.0000	1.0000	1.0000	1.0000	0.9999	0.9996	0.9962	0.9805	0.9295	0.8040	0.6997	0.5638	0.2252	0.0712	0.0034
8	1.0000	1.0000	1.0000	1.0000	1.0000	1.0000	0.9997	0.9980	0.9899	0.9596	0.9249	0.8658	0.6126	0.3698	0.0865

$n = 10$

	P														
k	**0.01**	**0.05**	**0.10**	**0.20**	**0.25**	**0.30**	**0.40**	**0.50**	**0.60**	**0.70**	**0.75**	**0.80**	**0.90**	**0.95**	**0.99**
0	0.9044	0.5987	0.3487	0.1074	0.0563	0.0282	0.0060	0.0010	0.0001	0.0000	0.0000	0.0000	0.0000	0.0000	0.0000
1	0.9957	0.9139	0.7361	0.3758	0.2440	0.1493	0.0464	0.0107	0.0017	0.0001	0.0000	0.0000	0.0000	0.0000	0.0000
2	0.9999	0.9885	0.9298	0.6778	0.5256	0.3828	0.1673	0.0547	0.0123	0.0016	0.0004	0.0001	0.0000	0.0000	0.0000
3	1.0000	0.9990	0.9872	0.8791	0.7759	0.6496	0.3823	0.1719	0.0548	0.0106	0.0035	0.0009	0.0000	0.0000	0.0000
4	1.0000	0.9999	0.9984	0.9672	0.9219	0.8497	0.6331	0.3770	0.1662	0.0473	0.0197	0.0064	0.0001	0.0000	0.0000
5	1.0000	1.0000	0.9999	0.9936	0.9803	0.9527	0.8338	0.6230	0.3669	0.1503	0.0781	0.0328	0.0016	0.0001	0.0000
6	1.0000	1.0000	1.0000	0.9991	0.9965	0.9894	0.9452	0.8281	0.6177	0.3504	0.2241	0.1209	0.0128	0.0010	0.0000
7	1.0000	1.0000	1.0000	0.9999	0.9996	0.9984	0.9877	0.9453	0.8327	0.6172	0.4744	0.3222	0.0702	0.0115	0.0001
8	1.0000	1.0000	1.0000	1.0000	1.0000	0.9999	0.9983	0.9893	0.9536	0.8507	0.7560	0.6242	0.2639	0.0861	0.0043
9	1.0000	1.0000	1.0000	1.0000	1.0000	1.0000	0.9999	0.9990	0.9940	0.9718	0.9437	0.8926	0.6513	0.4013	0.0956

$n = 15$

	P														
k	**0.01**	**0.05**	**0.10**	**0.20**	**0.25**	**0.30**	**0.40**	**0.50**	**0.60**	**0.70**	**0.75**	**0.80**	**0.90**	**0.95**	**0.99**
0	0.8601	0.4633	0.2059	0.0352	0.0134	0.0047	0.0005	0.0000	0.0000	0.0000	0.0000	0.0000	0.0000	0.0000	0.0000
1	0.9904	0.8290	0.5490	0.1671	0.0802	0.0353	0.0052	0.0005	0.0000	0.0000	0.0000	0.0000	0.0000	0.0000	0.0000
2	0.9996	0.9638	0.8159	0.3980	0.2361	0.1268	0.0271	0.0037	0.0003	0.0000	0.0000	0.0000	0.0000	0.0000	0.0000
3	1.0000	0.9945	0.9444	0.6482	0.4613	0.2969	0.0905	0.0176	0.0019	0.0001	0.0000	0.0000	0.0000	0.0000	0.0000
4	1.0000	0.9994	0.9873	0.8358	0.6865	0.5155	0.2173	0.0592	0.0093	0.0007	0.0001	0.0000	0.0000	0.0000	0.0000
5	1.0000	0.9999	0.9978	0.9389	0.8516	0.7216	0.4032	0.1509	0.0338	0.0037	0.0008	0.0001	0.0000	0.0000	0.0000
6	1.0000	1.0000	0.9997	0.9819	0.9434	0.8689	0.6098	0.3036	0.0950	0.0152	0.0042	0.0008	0.0000	0.0000	0.0000
7	1.0000	1.0000	1.0000	0.9958	0.9827	0.9500	0.7869	0.5000	0.2131	0.0500	0.0173	0.0042	0.0000	0.0000	0.0000
8	1.0000	1.0000	1.0000	0.9992	0.9958	0.9848	0.9050	0.6964	0.3902	0.1311	0.0566	0.0181	0.0003	0.0000	0.0000
9	1.0000	1.0000	1.0000	0.9999	0.9992	0.9963	0.9662	0.8491	0.5968	0.2784	0.1484	0.0611	0.0022	0.0001	0.0000
10	1.0000	1.0000	1.0000	1.0000	0.9999	0.9993	0.9907	0.9408	0.7827	0.4845	0.3135	0.1642	0.0127	0.0006	0.0000
11	1.0000	1.0000	1.0000	1.0000	1.0000	0.9999	0.9981	0.9824	0.9095	0.7031	0.5387	0.3518	0.0556	0.0055	0.0000
12	1.0000	1.0000	1.0000	1.0000	1.0000	1.0000	0.9997	0.9963	0.9729	0.8732	0.7639	0.6020	0.1841	0.0362	0.0004
13	1.0000	1.0000	1.0000	1.0000	1.0000	1.0000	1.0000	0.9995	0.9948	0.9647	0.9198	0.8329	0.4510	0.1710	0.0096
14	1.0000	1.0000	1.0000	1.0000	1.0000	1.0000	1.0000	1.0000	0.9995	0.9953	0.9866	0.9648	0.7941	0.5367	0.1399

$n = 20$

	P														
k	**0.01**	**0.05**	**0.10**	**0.20**	**0.25**	**0.30**	**0.40**	**0.50**	**0.60**	**0.70**	**0.75**	**0.80**	**0.90**	**0.95**	**0.99**
0	0.8179	0.3585	0.1216	0.0115	0.0032	0.0008	0.0000	0.0000	0.0000	0.0000	0.0000	0.0000	0.0000	0.0000	0.0000
1	0.9831	0.7358	0.3917	0.0692	0.0243	0.0076	0.0005	0.0000	0.0000	0.0000	0.0000	0.0000	0.0000	0.0000	0.0000
2	0.9990	0.9245	0.6769	0.2061	0.0913	0.0355	0.0036	0.0002	0.0000	0.0000	0.0000	0.0000	0.0000	0.0000	0.0000
3	1.0000	0.9841	0.8670	0.4114	0.2252	0.1071	0.0160	0.0013	0.0000	0.0000	0.0000	0.0000	0.0000	0.0000	0.0000
4	1.0000	0.9974	0.9568	0.6296	0.4148	0.2375	0.0510	0.0059	0.0003	0.0000	0.0000	0.0000	0.0000	0.0000	0.0000
5	1.0000	0.9997	0.9887	0.8042	0.6172	0.4164	0.1256	0.0207	0.0016	0.0000	0.0000	0.0000	0.0000	0.0000	0.0000
6	1.0000	1.0000	0.9976	0.9133	0.7858	0.6080	0.2500	0.0577	0.0065	0.0003	0.0000	0.0000	0.0000	0.0000	0.0000
7	1.0000	1.0000	0.9996	0.9679	0.8982	0.7723	0.4159	0.1316	0.0210	0.0013	0.0002	0.0000	0.0000	0.0000	0.0000
8	1.0000	1.0000	0.9999	0.9900	0.9591	0.8867	0.5956	0.2517	0.0565	0.0051	0.0009	0.0001	0.0000	0.0000	0.0000
9	1.0000	1.0000	1.0000	0.9974	0.9861	0.9520	0.7553	0.4119	0.1275	0.0171	0.0039	0.0006	0.0000	0.0000	0.0000
10	1.0000	1.0000	1.0000	0.9994	0.9961	0.9829	0.8725	0.5881	0.2447	0.0480	0.0139	0.0026	0.0000	0.0000	0.0000
11	1.0000	1.0000	1.0000	0.9999	0.9991	0.9949	0.9435	0.7483	0.4044	0.1133	0.0409	0.0100	0.0001	0.0000	0.0000
12	1.0000	1.0000	1.0000	1.0000	0.9998	0.9987	0.9790	0.8684	0.5841	0.2277	0.1018	0.0321	0.0004	0.0000	0.0000
13	1.0000	1.0000	1.0000	1.0000	1.0000	0.9997	0.9935	0.9423	0.7500	0.3920	0.2142	0.0867	0.0024	0.0000	0.0000
14	1.0000	1.0000	1.0000	1.0000	1.0000	1.0000	0.9984	0.9793	0.8744	0.5836	0.3828	0.1958	0.0113	0.0003	0.0000
15	1.0000	1.0000	1.0000	1.0000	1.0000	1.0000	0.9997	0.9941	0.9490	0.7625	0.5852	0.3704	0.0432	0.0026	0.0000
16	1.0000	1.0000	1.0000	1.0000	1.0000	1.0000	1.0000	0.9987	0.9840	0.8929	0.7748	0.5886	0.1330	0.0159	0.0000
17	1.0000	1.0000	1.0000	1.0000	1.0000	1.0000	1.0000	0.9998	0.9964	0.9645	0.9087	0.7939	0.3231	0.0755	0.0010
18	1.0000	1.0000	1.0000	1.0000	1.0000	1.0000	1.0000	1.0000	0.9995	0.9924	0.9757	0.9308	0.6083	0.2642	0.0169
19	1.0000	1.0000	1.0000	1.0000	1.0000	1.0000	1.0000	1.0000	1.0000	0.9992	0.9968	0.9885	0.8784	0.6415	0.1821

(Continued)

TABLE 1 ***(Continued)***

$n = 25$

k	P 0.01	0.05	0.10	0.20	0.25	0.30	0.40	0.50	0.60	0.70	0.75	0.80	0.90	0.95	0.99
0	0.7778	0.2774	0.0718	0.0038	0.0008	0.0001	0.0000	0.0000	0.0000	0.0000	0.0000	0.0000	0.0000	0.0000	0.0000
1	0.9742	0.6424	0.2712	0.0274	0.0070	0.0016	0.0001	0.0000	0.0000	0.0000	0.0000	0.0000	0.0000	0.0000	0.0000
2	0.9980	0.8729	0.5371	0.0982	0.0321	0.0090	0.0004	0.0000	0.0000	0.0000	0.0000	0.0000	0.0000	0.0000	0.0000
3	0.9999	0.9659	0.7636	0.2340	0.0962	0.0332	0.0024	0.0001	0.0000	0.0000	0.0000	0.0000	0.0000	0.0000	0.0000
4	1.0000	0.9928	0.9020	0.4207	0.2137	0.0905	0.0095	0.0005	0.0000	0.0000	0.0000	0.0000	0.0000	0.0000	0.0000
5	1.0000	0.9988	0.9666	0.6167	0.3783	0.1935	0.0294	0.0020	0.0001	0.0000	0.0000	0.0000	0.0000	0.0000	0.0000
6	1.0000	0.9998	0.9905	0.7800	0.5611	0.3407	0.0736	0.0073	0.0003	0.0000	0.0000	0.0000	0.0000	0.0000	0.0000
7	1.0000	1.0000	0.9977	0.8909	0.7265	0.5118	0.1536	0.0216	0.0012	0.0000	0.0000	0.0000	0.0000	0.0000	0.0000
8	1.0000	1.0000	0.9995	0.9532	0.8506	0.6769	0.2735	0.0539	0.0043	0.0001	0.0000	0.0000	0.0000	0.0000	0.0000
9	1.0000	1.0000	0.9999	0.9827	0.9287	0.8106	0.4246	0.1148	0.0132	0.0005	0.0000	0.0000	0.0000	0.0000	0.0000
10	1.0000	1.0000	1.0000	0.9944	0.9703	0.9022	0.5858	0.2122	0.0344	0.0018	0.0002	0.0000	0.0000	0.0000	0.0000
11	1.0000	1.0000	1.0000	0.9985	0.9893	0.9558	0.7323	0.3450	0.0778	0.0060	0.0009	0.0001	0.0000	0.0000	0.0000
12	1.0000	1.0000	1.0000	0.9996	0.9966	0.9825	0.8462	0.5000	0.1538	0.0175	0.0034	0.0004	0.0000	0.0000	0.0000
13	1.0000	1.0000	1.0000	0.9999	0.9991	0.9940	0.9222	0.6550	0.2677	0.0442	0.0107	0.0015	0.0000	0.0000	0.0000
14	1.0000	1.0000	1.0000	1.0000	0.9998	0.9982	0.9656	0.7878	0.4142	0.0978	0.0297	0.0056	0.0000	0.0000	0.0000
15	1.0000	1.0000	1.0000	1.0000	1.0000	0.9995	0.9868	0.8852	0.5754	0.1894	0.0713	0.0173	0.0001	0.0000	0.0000
16	1.0000	1.0000	1.0000	1.0000	1.0000	0.9999	0.9957	0.9461	0.7265	0.3231	0.1494	0.0468	0.0005	0.0000	0.0000
17	1.0000	1.0000	1.0000	1.0000	1.0000	1.0000	0.9988	0.9784	0.8464	0.4882	0.2735	0.1091	0.0023	0.0000	0.0000
18	1.0000	1.0000	1.0000	1.0000	1.0000	1.0000	0.9997	0.9927	0.9264	0.6593	0.4389	0.2200	0.0095	0.0002	0.0000
19	1.0000	1.0000	1.0000	1.0000	1.0000	1.0000	0.9999	0.9980	0.9706	0.8065	0.6217	0.3833	0.0334	0.0012	0.0000
20	1.0000	1.0000	1.0000	1.0000	1.0000	1.0000	1.0000	0.9995	0.9905	0.9095	0.7863	0.5793	0.0980	0.0072	0.0000
21	1.0000	1.0000	1.0000	1.0000	1.0000	1.0000	1.0000	0.9999	0.9976	0.9668	0.9038	0.7660	0.2364	0.0341	0.0001
22	1.0000	1.0000	1.0000	1.0000	1.0000	1.0000	1.0000	1.0000	0.9996	0.9910	0.9679	0.9018	0.4629	0.1271	0.0020
23	1.0000	1.0000	1.0000	1.0000	1.0000	1.0000	1.0000	1.0000	0.9999	0.9984	0.9930	0.9726	0.7288	0.3576	0.0258
24	1.0000	1.0000	1.0000	1.0000	1.0000	1.0000	1.0000	1.0000	1.0000	0.9999	0.9992	0.9962	0.9282	0.7226	0.2222

TABLE 2 Poisson Probabilities

Tabulated values are $P(X \leq k) = \sum_{x=0}^{k} p(xi)$ (Values are rounded to four decimal places.)

	μ															
k	0.10	0.20	0.30	0.40	0.50	1.0	1.5	2.0	2.5	3.0	3.5	4.0	4.5	5.0	5.5	6.0
0	0.9048	0.8187	0.7408	0.6703	0.6065	0.3679	0.2231	0.1353	0.0821	0.0498	0.0302	0.0183	0.0111	0.0067	0.0041	0.0025
1	0.9953	0.9825	0.9631	0.9384	0.9098	0.7358	0.5578	0.4060	0.2873	0.1991	0.1359	0.0916	0.0611	0.0404	0.0266	0.0174
2	0.9998	0.9989	0.9964	0.9921	0.9856	0.9197	0.8088	0.6767	0.5438	0.4232	0.3208	0.2381	0.1736	0.1247	0.0884	0.0620
3	1.0000	0.9999	0.9997	0.9992	0.9982	0.9810	0.9344	0.8571	0.7576	0.6472	0.5366	0.4335	0.3423	0.2650	0.2017	0.1512
4		1.0000	1.0000	0.9999	0.9998	0.9963	0.9814	0.9473	0.8912	0.8153	0.7254	0.6288	0.5321	0.4405	0.3575	0.2851
5				1.0000	1.0000	0.9994	0.9955	0.9834	0.9580	0.9161	0.8576	0.7851	0.7029	0.6160	0.5289	0.4457
6						0.9999	0.9991	0.9955	0.9858	0.9665	0.9347	0.8893	0.8311	0.7622	0.6860	0.6063
7						1.0000	0.9998	0.9989	0.9958	0.9881	0.9733	0.9489	0.9134	0.8666	0.8095	0.7440
8							1.0000	0.9998	0.9989	0.9962	0.9901	0.9786	0.9597	0.9319	0.8944	0.8472
9								1.0000	0.9997	0.9989	0.9967	0.9919	0.9829	0.9682	0.9462	0.9161
10									0.9999	0.9997	0.9990	0.9972	0.9933	0.9863	0.9747	0.9574
11									1.0000	0.9999	0.9997	0.9991	0.9976	0.9945	0.9890	0.9799
12										1.0000	0.9999	0.9997	0.9992	0.9980	0.9955	0.9912
13											1.0000	0.9999	0.9997	0.9993	0.9983	0.9964
14												1.0000	0.9999	0.9998	0.9994	0.9986
15													1.0000	0.9999	0.9998	0.9995
16														1.0000	0.9999	0.9998
17															1.0000	0.9999
18																1.0000
19																
20																

(Continued)

TABLE 2 *(Continued)*

	μ												
k	6.50	7.00	7.50	8.00	8.50	9.00	9.50	10	11	12	13	14	15
0	0.0015	0.0009	0.0006	0.0003	0.0002	0.0001	0.0001	0.0000	0.0000	0.0000	0.0000	0.0000	0.0000
1	0.0113	0.0073	0.0047	0.0030	0.0019	0.0012	0.0008	0.0005	0.0002	0.0001	0.0000	0.0000	0.0000
2	0.0430	0.0296	0.0203	0.0138	0.0093	0.0062	0.0042	0.0028	0.0012	0.0005	0.0002	0.0001	0.0000
3	0.1118	0.0818	0.0591	0.0424	0.0301	0.0212	0.0149	0.0103	0.0049	0.0023	0.0011	0.0005	0.0002
4	0.2237	0.1730	0.1321	0.0996	0.0744	0.0550	0.0403	0.0293	0.0151	0.0076	0.0037	0.0018	0.0009
5	0.3690	0.3007	0.2414	0.1912	0.1496	0.1157	0.0885	0.0671	0.0375	0.0203	0.0107	0.0055	0.0028
6	0.5265	0.4497	0.3782	0.3134	0.2562	0.2068	0.1649	0.1301	0.0786	0.0458	0.0259	0.0142	0.0076
7	0.6728	0.5987	0.5246	0.4530	0.3856	0.3239	0.2687	0.2202	0.1432	0.0895	0.0540	0.0316	0.0180
8	0.7916	0.7291	0.6620	0.5925	0.5231	0.4557	0.3918	0.3328	0.2320	0.1550	0.0998	0.0621	0.0374
9	0.8774	0.8305	0.7764	0.7166	0.6530	0.5874	0.5218	0.4579	0.3405	0.2424	0.1658	0.1094	0.0699
10	0.9332	0.9015	0.8622	0.8159	0.7634	0.7060	0.6453	0.5830	0.4599	0.3472	0.2517	0.1757	0.1185
11	0.9661	0.9467	0.9208	0.8881	0.8487	0.8030	0.7520	0.6968	0.5793	0.4616	0.3532	0.2600	0.1848
12	0.9840	0.9730	0.9573	0.9362	0.9091	0.8758	0.8364	0.7916	0.6887	0.5760	0.4631	0.3585	0.2676
13	0.9929	0.9872	0.9784	0.9658	0.9486	0.9261	0.8981	0.8645	0.7813	0.6815	0.5730	0.4644	0.3632
14	0.9970	0.9943	0.9897	0.9827	0.9726	0.9585	0.9400	0.9165	0.8540	0.7720	0.6751	0.5704	0.4657
15	0.9988	0.9976	0.9954	0.9918	0.9862	0.9780	0.9665	0.9513	0.9074	0.8444	0.7636	0.6694	0.5681
16	0.9996	0.9990	0.9980	0.9963	0.9934	0.9889	0.9823	0.9730	0.9441	0.8987	0.8355	0.7559	0.6641
17	0.9998	0.9996	0.9992	0.9984	0.9970	0.9947	0.9911	0.9857	0.9678	0.9370	0.8905	0.8272	0.7489
18	0.9999	0.9999	0.9997	0.9993	0.9987	0.9976	0.9957	0.9928	0.9823	0.9626	0.9302	0.8826	0.8195
19	1.0000	1.0000	0.9999	0.9997	0.9995	0.9989	0.9980	0.9965	0.9907	0.9787	0.9573	0.9235	0.8752
20			1.0000	0.9999	0.9998	0.9996	0.9991	0.9984	0.9953	0.9884	0.9750	0.9521	0.9170
21				1.0000	0.9999	0.9998	0.9996	0.9993	0.9977	0.9939	0.9859	0.9712	0.9469
22					1.0000	0.9999	0.9999	0.9997	0.9990	0.9970	0.9924	0.9833	0.9673
23						1.0000	0.9999	0.9999	0.9995	0.9985	0.9960	0.9907	0.9805
24							1.0000	1.0000	0.9998	0.9993	0.9980	0.9950	0.9888
25									0.9999	0.9997	0.9990	0.9974	0.9938
26									1.0000	0.9999	0.9995	0.9987	0.9967
27										0.9999	0.9998	0.9994	0.9983
28										1.0000	0.9999	0.9997	0.9991
29											1.0000	0.9999	0.9996
30												0.9999	0.9998
31												1.0000	0.9999
32													1.0000

TABLE 3 Cumulative Standardized Normal Probabilities

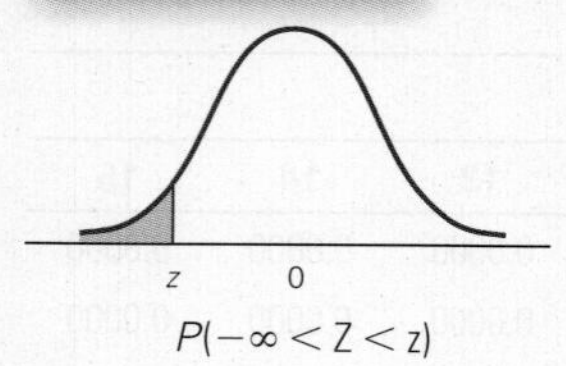

Z	0.00	0.01	0.02	0.03	0.04	0.05	0.06	0.07	0.08	0.09
−3.0	0.0013	0.0013	0.0013	0.0012	0.0012	0.0011	0.0011	0.0011	0.0010	0.0010
−2.9	0.0019	0.0018	0.0018	0.0017	0.0016	0.0016	0.0015	0.0015	0.0014	0.0014
−2.8	0.0026	0.0025	0.0024	0.0023	0.0023	0.0022	0.0021	0.0021	0.0020	0.0019
−2.7	0.0035	0.0034	0.0033	0.0032	0.0031	0.0030	0.0029	0.0028	0.0027	0.0026
−2.6	0.0047	0.0045	0.0044	0.0043	0.0041	0.0040	0.0039	0.0038	0.0037	0.0036
−2.5	0.0062	0.0060	0.0059	0.0057	0.0055	0.0054	0.0052	0.0051	0.0049	0.0048
−2.4	0.0082	0.0080	0.0078	0.0075	0.0073	0.0071	0.0069	0.0068	0.0066	0.0064
−2.3	0.0107	0.0104	0.0102	0.0099	0.0096	0.0094	0.0091	0.0089	0.0087	0.0084
−2.2	0.0139	0.0136	0.0132	0.0129	0.0125	0.0122	0.0119	0.0116	0.0113	0.0110
−2.1	0.0179	0.0174	0.0170	0.0166	0.0162	0.0158	0.0154	0.0150	0.0146	0.0143
−2.0	0.0228	0.0222	0.0217	0.0212	0.0207	0.0202	0.0197	0.0192	0.0188	0.0183
−1.9	0.0287	0.0281	0.0274	0.0268	0.0262	0.0256	0.0250	0.0244	0.0239	0.0233
−1.8	0.0359	0.0351	0.0344	0.0336	0.0329	0.0322	0.0314	0.0307	0.0301	0.0294
−1.7	0.0446	0.0436	0.0427	0.0418	0.0409	0.0401	0.0392	0.0384	0.0375	0.0367
−1.6	0.0548	0.0537	0.0526	0.0516	0.0505	0.0495	0.0485	0.0475	0.0465	0.0455
−1.5	0.0668	0.0655	0.0643	0.0630	0.0618	0.0606	0.0594	0.0582	0.0571	0.0559
−1.4	0.0808	0.0793	0.0778	0.0764	0.0749	0.0735	0.0721	0.0708	0.0694	0.0681
−1.3	0.0968	0.0951	0.0934	0.0918	0.0901	0.0885	0.0869	0.0853	0.0838	0.0823
−1.2	0.1151	0.1131	0.1112	0.1093	0.1075	0.1056	0.1038	0.1020	0.1003	0.0985
−1.1	0.1357	0.1335	0.1314	0.1292	0.1271	0.1251	0.1230	0.1210	0.1190	0.1170
−1.0	0.1587	0.1562	0.1539	0.1515	0.1492	0.1469	0.1446	0.1423	0.1401	0.1379
−0.9	0.1841	0.1814	0.1788	0.1762	0.1736	0.1711	0.1685	0.1660	0.1635	0.1611
−0.8	0.2119	0.2090	0.2061	0.2033	0.2005	0.1977	0.1949	0.1922	0.1894	0.1867
−0.7	0.2420	0.2389	0.2358	0.2327	0.2296	0.2266	0.2236	0.2206	0.2177	0.2148
−0.6	0.2743	0.2709	0.2676	0.2643	0.2611	0.2578	0.2546	0.2514	0.2483	0.2451
−0.5	0.3085	0.3050	0.3015	0.2981	0.2946	0.2912	0.2877	0.2843	0.2810	0.2776
−0.4	0.3446	0.3409	0.3372	0.3336	0.3300	0.3264	0.3228	0.3192	0.3156	0.3121
−0.3	0.3821	0.3783	0.3745	0.3707	0.3669	0.3632	0.3594	0.3557	0.3520	0.3483
−0.2	0.4207	0.4168	0.4129	0.4090	0.4052	0.4013	0.3974	0.3936	0.3897	0.3859
−0.1	0.4602	0.4562	0.4522	0.4483	0.4443	0.4404	0.4364	0.4325	0.4286	0.4247
−0.0	0.5000	0.4960	0.4920	0.4880	0.4840	0.4801	0.4761	0.4721	0.4681	0.4641

(Continued)

TABLE 3 *(Continued)*

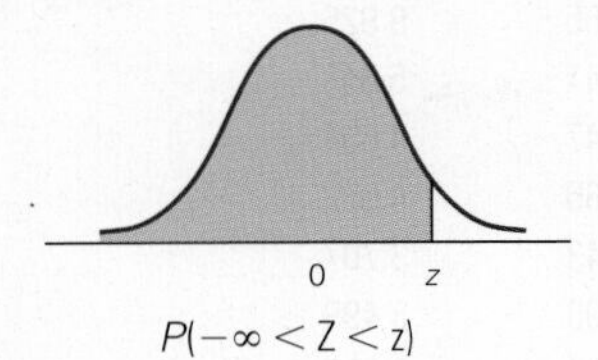

Z	0.00	0.01	0.02	0.03	0.04	0.05	0.06	0.07	0.08	0.09
0.0	0.5000	0.5040	0.5080	0.5120	0.5160	0.5199	0.5239	0.5279	0.5319	0.5359
0.1	0.5398	0.5438	0.5478	0.5517	0.5557	0.5596	0.5636	0.5675	0.5714	0.5753
0.2	0.5793	0.5832	0.5871	0.5910	0.5948	0.5987	0.6026	0.6064	0.6103	0.6141
0.3	0.6179	0.6217	0.6255	0.6293	0.6331	0.6368	0.6406	0.6443	0.6480	0.6517
0.4	0.6554	0.6591	0.6628	0.6664	0.6700	0.6736	0.6772	0.6808	0.6844	0.6879
0.5	0.6915	0.6950	0.6985	0.7019	0.7054	0.7088	0.7123	0.7157	0.7190	0.7224
0.6	0.7257	0.7291	0.7324	0.7357	0.7389	0.7422	0.7454	0.7486	0.7517	0.7549
0.7	0.7580	0.7611	0.7642	0.7673	0.7704	0.7734	0.7764	0.7794	0.7823	0.7852
0.8	0.7881	0.7910	0.7939	0.7967	0.7995	0.8023	0.8051	0.8078	0.8106	0.8133
0.9	0.8159	0.8186	0.8212	0.8238	0.8264	0.8289	0.8315	0.8340	0.8365	0.8389
1.0	0.8413	0.8438	0.8461	0.8485	0.8508	0.8531	0.8554	0.8577	0.8599	0.8621
1.1	0.8643	0.8665	0.8686	0.8708	0.8729	0.8749	0.8770	0.8790	0.8810	0.8830
1.2	0.8849	0.8869	0.8888	0.8907	0.8925	0.8944	0.8962	0.8980	0.8997	0.9015
1.3	0.9032	0.9049	0.9066	0.9082	0.9099	0.9115	0.9131	0.9147	0.9162	0.9177
1.4	0.9192	0.9207	0.9222	0.9236	0.9251	0.9265	0.9279	0.9292	0.9306	0.9319
1.5	0.9332	0.9345	0.9357	0.9370	0.9382	0.9394	0.9406	0.9418	0.9429	0.9441
1.6	0.9452	0.9463	0.9474	0.9484	0.9495	0.9505	0.9515	0.9525	0.9535	0.9545
1.7	0.9554	0.9564	0.9573	0.9582	0.9591	0.9599	0.9608	0.9616	0.9625	0.9633
1.8	0.9641	0.9649	0.9656	0.9664	0.9671	0.9678	0.9686	0.9693	0.9699	0.9706
1.9	0.9713	0.9719	0.9726	0.9732	0.9738	0.9744	0.9750	0.9756	0.9761	0.9767
2.0	0.9772	0.9778	0.9783	0.9788	0.9793	0.9798	0.9803	0.9808	0.9812	0.9817
2.1	0.9821	0.9826	0.9830	0.9834	0.9838	0.9842	0.9846	0.9850	0.9854	0.9857
2.2	0.9861	0.9864	0.9868	0.9871	0.9875	0.9878	0.9881	0.9884	0.9887	0.9890
2.3	0.9893	0.9896	0.9898	0.9901	0.9904	0.9906	0.9909	0.9911	0.9913	0.9916
2.4	0.9918	0.9920	0.9922	0.9925	0.9927	0.9929	0.9931	0.9932	0.9934	0.9936
2.5	0.9938	0.9940	0.9941	0.9943	0.9945	0.9946	0.9948	0.9949	0.9951	0.9952
2.6	0.9953	0.9955	0.9956	0.9957	0.9959	0.9960	0.9961	0.9962	0.9963	0.9964
2.7	0.9965	0.9966	0.9967	0.9968	0.9969	0.9970	0.9971	0.9972	0.9973	0.9974
2.8	0.9974	0.9975	0.9976	0.9977	0.9977	0.9978	0.9979	0.9979	0.9980	0.9981
2.9	0.9981	0.9982	0.9982	0.9983	0.9984	0.9984	0.9985	0.9985	0.9986	0.9986
3.0	0.9987	0.9987	0.9987	0.9988	0.9988	0.9989	0.9989	0.9989	0.9990	0.9990

TABLE 4

Critical Values of the Student *t* Distribution

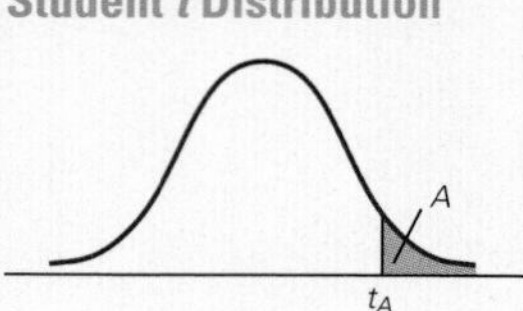

Degrees of Freedom	$t_{.100}$	$t_{.050}$	$t_{.025}$	$t_{.010}$	$t_{.005}$
1	3.078	6.314	12.706	31.821	63.657
2	1.886	2.920	4.303	6.965	9.925
3	1.638	2.353	3.182	4.541	5.841
4	1.533	2.132	2.776	3.747	4.604
5	1.476	2.015	2.571	3.365	4.032
6	1.440	1.943	2.447	3.143	3.707
7	1.415	1.895	2.365	2.998	3.499
8	1.397	1.860	2.306	2.896	3.355
9	1.383	1.833	2.262	2.821	3.250
10	1.372	1.812	2.228	2.764	3.169
11	1.363	1.796	2.201	2.718	3.106
12	1.356	1.782	2.179	2.681	3.055
13	1.350	1.771	2.160	2.650	3.012
14	1.345	1.761	2.145	2.624	2.977
15	1.341	1.753	2.131	2.602	2.947
16	1.337	1.746	2.120	2.583	2.921
17	1.333	1.740	2.110	2.567	2.898
18	1.330	1.734	2.101	2.552	2.878
19	1.328	1.729	2.093	2.539	2.861
20	1.325	1.725	2.086	2.528	2.845
21	1.323	1.721	2.080	2.518	2.831
22	1.321	1.717	2.074	2.508	2.819
23	1.319	1.714	2.069	2.500	2.807
24	1.318	1.711	2.064	2.492	2.797
25	1.316	1.708	2.060	2.485	2.787
26	1.315	1.706	2.056	2.479	2.779
27	1.314	1.703	2.052	2.473	2.771
28	1.313	1.701	2.048	2.467	2.763
29	1.311	1.699	2.045	2.462	2.756
30	1.310	1.697	2.042	2.457	2.750
35	1.306	1.690	2.030	2.438	2.724
40	1.303	1.684	2.021	2.423	2.704
45	1.301	1.679	2.014	2.412	2.690
50	1.299	1.676	2.009	2.403	2.678
55	1.297	1.673	2.004	2.396	2.668
60	1.296	1.671	2.000	2.390	2.660
65	1.295	1.669	1.997	2.385	2.654
70	1.294	1.667	1.994	2.381	2.648
75	1.293	1.665	1.992	2.377	2.643
80	1.292	1.664	1.990	2.374	2.639
85	1.292	1.663	1.988	2.371	2.635
90	1.291	1.662	1.987	2.368	2.632
95	1.291	1.661	1.985	2.366	2.629
100	1.290	1.660	1.984	2.364	2.626
110	1.289	1.659	1.982	2.361	2.621
120	1.289	1.658	1.980	2.358	2.617
130	1.288	1.657	1.978	2.355	2.614
140	1.288	1.656	1.977	2.353	2.611
150	1.287	1.655	1.976	2.351	2.609
160	1.287	1.654	1.975	2.350	2.607
170	1.287	1.654	1.974	2.348	2.605
180	1.286	1.653	1.973	2.347	2.603
190	1.286	1.653	1.973	2.346	2.602
200	1.286	1.653	1.972	2.345	2.601
∞	1.282	1.645	1.960	2.326	2.576

TABLE 5 Critical Values of the χ^2 Distribution

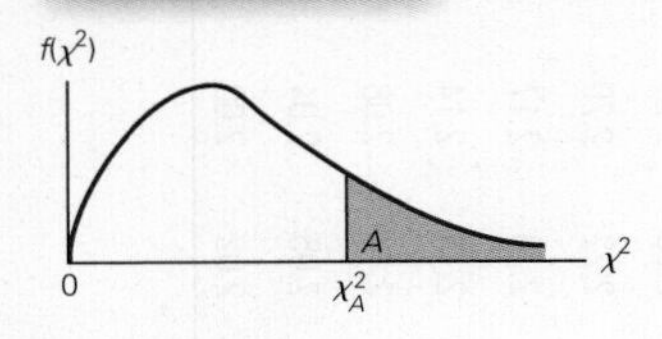

Degrees of Freedom	$\chi^2_{.995}$	$\chi^2_{.990}$	$\chi^2_{.975}$	$\chi^2_{.950}$	$\chi^2_{.900}$	$\chi^2_{.100}$	$\chi^2_{.050}$	$\chi^2_{.025}$	$\chi^2_{.010}$	$\chi^2_{.005}$
1	0.000039	0.000157	0.000982	0.00393	0.0158	2.71	3.84	5.02	6.63	7.88
2	0.0100	0.0201	0.0506	0.103	0.211	4.61	5.99	7.38	9.21	10.6
3	0.072	0.115	0.216	0.352	0.584	6.25	7.81	9.35	11.3	12.8
4	0.207	0.297	0.484	0.711	1.06	7.78	9.49	11.1	13.3	14.9
5	0.412	0.554	0.831	1.15	1.61	9.24	11.1	12.8	15.1	16.7
6	0.676	0.872	1.24	1.64	2.20	10.6	12.6	14.4	16.8	18.5
7	0.989	1.24	1.69	2.17	2.83	12.0	14.1	16.0	18.5	20.3
8	1.34	1.65	2.18	2.73	3.49	13.4	15.5	17.5	20.1	22.0
9	1.73	2.09	2.70	3.33	4.17	14.7	16.9	19.0	21.7	23.6
10	2.16	2.56	3.25	3.94	4.87	16.0	18.3	20.5	23.2	25.2
11	2.60	3.05	3.82	4.57	5.58	17.3	19.7	21.9	24.7	26.8
12	3.07	3.57	4.40	5.23	6.30	18.5	21.0	23.3	26.2	28.3
13	3.57	4.11	5.01	5.89	7.04	19.8	22.4	24.7	27.7	29.8
14	4.07	4.66	5.63	6.57	7.79	21.1	23.7	26.1	29.1	31.3
15	4.60	5.23	6.26	7.26	8.55	22.3	25.0	27.5	30.6	32.8
16	5.14	5.81	6.91	7.96	9.31	23.5	26.3	28.8	32.0	34.3
17	5.70	6.41	7.56	8.67	10.1	24.8	27.6	30.2	33.4	35.7
18	6.26	7.01	8.23	9.39	10.9	26.0	28.9	31.5	34.8	37.2
19	6.84	7.63	8.91	10.1	11.7	27.2	30.1	32.9	36.2	38.6
20	7.43	8.26	9.59	10.9	12.4	28.4	31.4	34.2	37.6	40.0
21	8.03	8.90	10.3	11.6	13.2	29.6	32.7	35.5	38.9	41.4
22	8.64	9.54	11.0	12.3	14.0	30.8	33.9	36.8	40.3	42.8
23	9.26	10.2	11.7	13.1	14.8	32.0	35.2	38.1	41.6	44.2
24	9.89	10.9	12.4	13.8	15.7	33.2	36.4	39.4	43.0	45.6
25	10.5	11.5	13.1	14.6	16.5	34.4	37.7	40.6	44.3	46.9
26	11.2	12.2	13.8	15.4	17.3	35.6	38.9	41.9	45.6	48.3
27	11.8	12.9	14.6	16.2	18.1	36.7	40.1	43.2	47.0	49.6
28	12.5	13.6	15.3	16.9	18.9	37.9	41.3	44.5	48.3	51.0
29	13.1	14.3	16.0	17.7	19.8	39.1	42.6	45.7	49.6	52.3
30	13.8	15.0	16.8	18.5	20.6	40.3	43.8	47.0	50.9	53.7
40	20.7	22.2	24.4	26.5	29.1	51.8	55.8	59.3	63.7	66.8
50	28.0	29.7	32.4	34.8	37.7	63.2	67.5	71.4	76.2	79.5
60	35.5	37.5	40.5	43.2	46.5	74.4	79.1	83.3	88.4	92.0
70	43.3	45.4	48.8	51.7	55.3	85.5	90.5	95.0	100	104
80	51.2	53.5	57.2	60.4	64.3	96.6	102	107	112	116
90	59.2	61.8	65.6	69.1	73.3	108	113	118	124	128
100	67.3	70.1	74.2	77.9	82.4	118	124	130	136	140

TABLE 6(a) Critical Values of the F-Distribution: $A = 0.05$

f(F)

A

F

0

F_A

v_1	NUMERATOR DEGREES OF FREEDOM																			
v_2	1	2	3	4	5	6	7	8	9	10	11	12	13	14	15	16	17	18	19	20
DENOMINATOR DEGREES OF FREEDOM																				
1	161	199	216	225	230	234	237	239	241	242	243	244	245	245	246	246	247	247	248	2 48
2	18.5	19.0	19.2	19.2	19.3	19.3	19.4	19.4	19.4	19.4	19.4	19.4	19.4	19.4	19.4	19.4	19.4	19.4	19.4	19.4
3	10.1	9.55	9.28	9.12	9.01	8.94	8.89	8.85	8.81	8.79	8.76	8.74	8.73	8.71	8.70	8.69	8.68	8.67	8.67	8.66
4	7.71	6.94	6.59	6.39	6.26	6.16	6.09	6.04	6.00	5.96	5.94	5.91	5.89	5.87	5.86	5.84	5.83	5.82	5.81	5.80
5	6.61	5.79	5.41	5.19	5.05	4.95	4.88	4.82	4.77	4.74	4.70	4.68	4.66	4.64	4.62	4.60	4.59	4.58	4.57	4.56
6	5.99	5.14	4.76	4.53	4.39	4.28	4.21	4.15	4.10	4.06	4.03	4.00	3.98	3.96	3.94	3.92	3.91	3.90	3.88	3.87
7	5.59	4.74	4.35	4.12	3.97	3.87	3.79	3.73	3.68	3.64	3.60	3.57	3.55	3.53	3.51	3.49	3.48	3.47	3.46	3.44
8	5.32	4.46	4.07	3.84	3.69	3.58	3.50	3.44	3.39	3.35	3.31	3.28	3.26	3.24	3.22	3.20	3.19	3.17	3.16	3.15
9	5.12	4.26	3.86	3.63	3.48	3.37	3.29	3.23	3.18	3.14	3.10	3.07	3.05	3.03	3.01	2.99	2.97	2.96	2.95	2.94
10	4.96	4.10	3.71	3.48	3.33	3.22	3.14	3.07	3.02	2.98	2.94	2.91	2.89	2.86	2.85	2.83	2.81	2.80	2.79	2.77
11	4.84	3.98	3.59	3.36	3.20	3.09	3.01	2.95	2.90	2.85	2.82	2.79	2.76	2.74	2.72	2.70	2.69	2.67	2.66	2.65
12	4.75	3.89	3.49	3.26	3.11	3.00	2.91	2.85	2.80	2.75	2.72	2.69	2.66	2.64	2.62	2.60	2.58	2.57	2.56	2.54
13	4.67	3.81	3.41	3.18	3.03	2.92	2.83	2.77	2.71	2.67	2.63	2.60	2.58	2.55	2.53	2.51	2.50	2.48	2.47	2.46
14	4.60	3.74	3.34	3.11	2.96	2.85	2.76	2.70	2.65	2.60	2.57	2.53	2.51	2.48	2.46	2.44	2.43	2.41	2.40	2.39
15	4.54	3.68	3.29	3.06	2.90	2.79	2.71	2.64	2.59	2.54	2.51	2.48	2.45	2.42	2.40	2.38	2.37	2.35	2.34	2.33
16	4.49	3.63	3.24	3.01	2.85	2.74	2.66	2.59	2.54	2.49	2.46	2.42	2.40	2.37	2.35	2.33	2.32	2.30	2.29	2.28
17	4.45	3.59	3.20	2.96	2.81	2.70	2.61	2.55	2.49	2.45	2.41	2.38	2.35	2.33	2.31	2.29	2.27	2.26	2.24	2.23
18	4.41	3.55	3.16	2.93	2.77	2.66	2.58	2.51	2.46	2.41	2.37	2.34	2.31	2.29	2.27	2.25	2.23	2.22	2.20	2.19
19	4.38	3.52	3.13	2.90	2.74	2.63	2.54	2.48	2.42	2.38	2.34	2.31	2.28	2.26	2.23	2.21	2.20	2.18	2.17	2.16
20	4.35	3.49	3.10	2.87	2.71	2.60	2.51	2.45	2.39	2.35	2.31	2.28	2.25	2.22	2.20	2.18	2.17	2.15	2.14	2.12
22	4.30	3.44	3.05	2.82	2.66	2.55	2.46	2.40	2.34	2.30	2.26	2.23	2.20	2.17	2.15	2.13	2.11	2.10	2.08	2.07
24	4.26	3.40	3.01	2.78	2.62	2.51	2.42	2.36	2.30	2.25	2.22	2.18	2.15	2.13	2.11	2.09	2.07	2.05	2.04	2.03
26	4.23	3.37	2.98	2.74	2.59	2.47	2.39	2.32	2.27	2.22	2.18	2.15	2.12	2.09	2.07	2.05	2.03	2.02	2.00	1.99

(Continued)

v_1 / v_2 (DENOMINATOR DEGREES OF FREEDOM)	NUMERATOR DEGREES OF FREEDOM 1	2	3	4	5	6	7	8	9	10	11	12	13	14	15	16	17	18	19	20
28	4.20	3.34	2.95	2.71	2.56	2.45	2.36	2.29	2.24	2.19	2.15	2.12	2.09	2.06	2.04	2.02	2.00	1.99	1.97	1.96
30	4.17	3.32	2.92	2.69	2.53	2.42	2.33	2.27	2.21	2.16	2.13	2.09	2.06	2.04	2.01	1.99	1.98	1.96	1.95	1.93
35	4.12	3.27	2.87	2.64	2.49	2.37	2.29	2.22	2.16	2.11	2.07	2.04	2.01	1.99	1.96	1.94	1.92	1.91	1.89	1.88
40	4.08	3.23	2.84	2.61	2.45	2.34	2.25	2.18	2.12	2.08	2.04	2.00	1.97	1.95	1.92	1.90	1.89	1.87	1.85	1.84
45	4.06	3.20	2.81	2.58	2.42	2.31	2.22	2.15	2.10	2.05	2.01	1.97	1.94	1.92	1.89	1.87	1.86	1.84	1.82	1.81
50	4.03	3.18	2.79	2.56	2.40	2.29	2.20	2.13	2.07	2.03	1.99	1.95	1.92	1.89	1.87	1.85	1.83	1.81	1.80	1.78
60	4.00	3.15	2.76	2.53	2.37	2.25	2.17	2.10	2.04	1.99	1.95	1.92	1.89	1.86	1.84	1.82	1.80	1.78	1.76	1.75
70	3.98	3.13	2.74	2.50	2.35	2.23	2.14	2.07	2.02	1.97	1.93	1.89	1.86	1.84	1.81	1.79	1.77	1.75	1.74	1.72
80	3.96	3.11	2.72	2.49	2.33	2.21	2.13	2.06	2.00	1.95	1.91	1.88	1.84	1.82	1.79	1.77	1.75	1.73	1.72	1.70
90	3.95	3.10	2.71	2.47	2.32	2.20	2.11	2.04	1.99	1.94	1.90	1.86	1.83	1.80	1.78	1.76	1.74	1.72	1.70	1.69
100	3.94	3.09	2.70	2.46	2.31	2.19	2.10	2.03	1.97	1.93	1.89	1.85	1.82	1.79	1.77	1.75	1.73	1.71	1.69	1.68
120	3.92	3.07	2.68	2.45	2.29	2.18	2.09	2.02	1.96	1.91	1.87	1.83	1.80	1.78	1.75	1.73	1.71	1.69	1.67	1.66
140	3.91	3.06	2.67	2.44	2.28	2.16	2.08	2.01	1.95	1.90	1.86	1.82	1.79	1.76	1.74	1.72	1.70	1.68	1.66	1.65
160	3.90	3.05	2.66	2.43	2.27	2.16	2.07	2.00	1.94	1.89	1.85	1.81	1.78	1.75	1.73	1.71	1.69	1.67	1.65	1.64
180	3.89	3.05	2.65	2.42	2.26	2.15	2.06	1.99	1.93	1.88	1.84	1.81	1.77	1.75	1.72	1.70	1.68	1.66	1.64	1.63
200	3.89	3.04	2.65	2.42	2.26	2.14	2.06	1.98	1.93	1.88	1.84	1.80	1.77	1.74	1.72	1.69	1.67	1.66	1.64	1.62
∞	3.84	3.00	2.61	2.37	2.21	2.10	2.01	1.94	1.88	1.83	1.79	1.75	1.72	1.69	1.67	1.64	1.62	1.60	1.59	1.57

v_1	NUMERATOR DEGREES OF FREEDOM																			
v_2 (DENOMINATOR DEGREES OF FREEDOM)	22	24	26	28	30	35	40	45	50	60	70	80	90	100	120	140	160	180	200	∞
1	249	249	249	250	250	251	251	251	252	252	252	253	253	253	253	253	254	254	254	254
2	19.5	19.5	19.5	19.5	19.5	19.5	19.5	19.5	19.5	19.5	19.5	19.5	19.5	19.5	19.5	19.5	19.5	19.5	19.5	19.5
3	8.65	8.64	8.63	8.62	8.62	8.60	8.59	8.59	8.58	8.57	8.57	8.56	8.56	8.55	8.55	8.55	8.54	8.54	8.54	8.53
4	5.79	5.77	5.76	5.75	5.75	5.73	5.72	5.71	5.70	5.69	5.68	5.67	5.67	5.66	5.66	5.65	5.65	5.65	5.65	5.63
5	4.54	4.53	4.52	4.50	4.50	4.48	4.46	4.45	4.44	4.43	4.42	4.41	4.41	4.41	4.40	4.39	4.39	4.39	4.39	4.37
6	3.86	3.84	3.83	3.82	3.81	3.79	3.77	3.76	3.75	3.74	3.73	3.72	3.72	3.71	3.70	3.70	3.70	3.69	3.69	3.67
7	3.43	3.41	3.40	3.39	3.38	3.36	3.34	3.33	3.32	3.30	3.29	3.29	3.28	3.27	3.27	3.26	3.26	3.25	3.25	3.23
8	3.13	3.12	3.10	3.09	3.08	3.06	3.04	3.03	3.02	3.01	2.99	2.99	2.98	2.97	2.97	2.96	2.96	2.95	2.95	2.93
9	2.92	2.90	2.89	2.87	2.86	2.84	2.83	2.81	2.80	2.79	2.78	2.77	2.76	2.76	2.75	2.74	2.74	2.73	2.73	2.71
10	2.75	2.74	2.72	2.71	2.70	2.68	2.66	2.65	2.64	2.62	2.61	2.60	2.59	2.59	2.58	2.57	2.57	2.57	2.56	2.54
11	2.63	2.61	2.59	2.58	2.57	2.55	2.53	2.52	2.51	2.49	2.48	2.47	2.46	2.46	2.45	2.44	2.44	2.43	2.43	2.41
12	2.52	2.51	2.49	2.48	2.47	2.44	2.43	2.41	2.40	2.38	2.37	2.36	2.36	2.35	2.34	2.33	2.33	2.33	2.32	2.30
13	2.44	2.42	2.41	2.39	2.38	2.36	2.34	2.33	2.31	2.30	2.28	2.27	2.27	2.26	2.25	2.25	2.24	2.24	2.23	2.21
14	2.37	2.35	2.33	2.32	2.31	2.28	2.27	2.25	2.24	2.22	2.21	2.20	2.19	2.19	2.18	2.17	2.17	2.16	2.16	2.13
15	2.31	2.29	2.27	2.26	2.25	2.22	2.20	2.19	2.18	2.16	2.15	2.14	2.13	2.12	2.11	2.11	2.10	2.10	2.10	2.07
16	2.25	2.24	2.22	2.21	2.19	2.17	2.15	2.14	2.12	2.11	2.09	2.08	2.07	2.07	2.06	2.05	2.05	2.04	2.04	2.01
17	2.21	2.19	2.17	2.16	2.15	2.12	2.10	2.09	2.08	2.06	2.05	2.03	2.03	2.02	2.01	2.00	2.00	1.99	1.99	1.96
18	2.17	2.15	2.13	2.12	2.11	2.08	2.06	2.05	2.04	2.02	2.00	1.99	1.98	1.98	1.97	1.96	1.96	1.95	1.95	1.92
19	2.13	2.11	2.10	2.08	2.07	2.05	2.03	2.01	2.00	1.98	1.97	1.96	1.95	1.94	1.93	1.92	1.92	1.91	1.91	1.88
20	2.10	2.08	2.07	2.05	2.04	2.01	1.99	1.98	1.97	1.95	1.93	1.92	1.91	1.91	1.90	1.89	1.88	1.88	1.88	1.84
22	2.05	2.03	2.01	2.00	1.98	1.96	1.94	1.92	1.91	1.89	1.88	1.86	1.86	1.85	1.84	1.83	1.82	1.82	1.82	1.78
24	2.00	1.98	1.97	1.95	1.94	1.91	1.89	1.88	1.86	1.84	1.83	1.82	1.81	1.80	1.79	1.78	1.78	1.77	1.77	1.73
26	1.97	1.95	1.93	1.91	1.90	1.87	1.85	1.84	1.82	1.80	1.79	1.78	1.77	1.76	1.75	1.74	1.73	1.73	1.73	1.69
28	1.93	1.91	1.90	1.88	1.87	1.84	1.82	1.80	1.79	1.77	1.75	1.74	1.73	1.73	1.71	1.71	1.70	1.69	1.69	1.65
30	1.91	1.89	1.87	1.85	1.84	1.81	1.79	1.77	1.76	1.74	1.72	1.71	1.70	1.70	1.68	1.68	1.67	1.66	1.66	1.62
35	1.85	1.83	1.82	1.80	1.79	1.76	1.74	1.72	1.70	1.68	1.66	1.65	1.64	1.63	1.62	1.61	1.61	1.60	1.60	1.56
40	1.81	1.79	1.77	1.76	1.74	1.72	1.69	1.67	1.66	1.64	1.62	1.61	1.60	1.59	1.58	1.57	1.56	1.55	1.55	1.51
45	1.78	1.76	1.74	1.73	1.71	1.68	1.66	1.64	1.63	1.60	1.59	1.57	1.56	1.55	1.54	1.53	1.52	1.52	1.51	1.47
50	1.76	1.74	1.72	1.70	1.69	1.66	1.63	1.61	1.60	1.58	1.56	1.54	1.53	1.52	1.51	1.50	1.49	1.49	1.48	1.44
60	1.72	1.70	1.68	1.66	1.65	1.62	1.59	1.57	1.56	1.53	1.52	1.50	1.49	1.48	1.47	1.46	1.45	1.44	1.44	1.39
70	1.70	1.67	1.65	1.64	1.62	1.59	1.57	1.55	1.53	1.50	1.49	1.47	1.46	1.45	1.44	1.42	1.42	1.41	1.40	1.35
80	1.68	1.65	1.63	1.62	1.60	1.57	1.54	1.52	1.51	1.48	1.46	1.45	1.44	1.43	1.41	1.40	1.39	1.38	1.38	1.33
90	1.66	1.64	1.62	1.60	1.59	1.55	1.53	1.51	1.49	1.46	1.44	1.43	1.42	1.41	1.39	1.38	1.37	1.36	1.36	1.30
100	1.65	1.63	1.61	1.59	1.57	1.54	1.52	1.49	1.48	1.45	1.43	1.41	1.40	1.39	1.38	1.36	1.35	1.35	1.34	1.28
120	1.63	1.61	1.59	1.57	1.55	1.52	1.50	1.47	1.46	1.43	1.41	1.39	1.38	1.37	1.35	1.34	1.33	1.32	1.32	1.26
140	1.62	1.60	1.57	1.56	1.54	1.51	1.48	1.46	1.44	1.41	1.39	1.38	1.36	1.35	1.33	1.32	1.31	1.30	1.30	1.23
160	1.61	1.59	1.57	1.55	1.53	1.50	1.47	1.45	1.43	1.40	1.38	1.36	1.35	1.34	1.32	1.31	1.30	1.29	1.28	1.22
180	1.60	1.58	1.56	1.54	1.52	1.49	1.46	1.44	1.42	1.39	1.37	1.35	1.34	1.33	1.31	1.30	1.29	1.28	1.27	1.20
200	1.60	1.57	1.55	1.53	1.52	1.48	1.46	1.43	1.41	1.39	1.36	1.35	1.33	1.32	1.30	1.29	1.28	1.27	1.26	1.19
∞	1.54	1.52	1.50	1.48	1.46	1.42	1.40	1.37	1.35	1.32	1.29	1.28	1.26	1.25	1.22	1.21	1.19	1.18	1.17	1.00

TABLE 6(b) Values of the F-Distribution: $A = 0.025$

ν_2 \ ν_1	1	2	3	4	5	6	7	8	9	10	11	12	13	14	15	16	17	18	19	20
	NUMERATOR DEGREES OF FREEDOM																			
1	648	799	864	900	922	937	948	957	963	969	973	977	980	983	985	987	989	990	992	993
2	38.5	39.0	39.2	39.2	39.3	39.3	39.4	39.4	39.4	39.4	39.4	39.4	39.4	39.4	39.4	39.4	39.4	39.4	39.4	39.4
3	17.4	16.0	15.4	15.1	14.9	14.7	14.6	14.5	14.5	14.4	14.4	14.3	14.3	14.3	14.3	14.2	14.2	14.2	14.2	14.2
4	12.2	10.6	10.0	9.60	9.36	9.20	9.07	8.98	8.90	8.84	8.79	8.75	8.71	8.68	8.66	8.63	8.61	8.59	8.58	8.56
5	10.0	8.43	7.76	7.39	7.15	6.98	6.85	6.76	6.68	6.62	6.57	6.52	6.49	6.46	6.43	6.40	6.38	6.36	6.34	6.33
6	8.81	7.26	6.60	6.23	5.99	5.82	5.70	5.60	5.52	5.46	5.41	5.37	5.33	5.30	5.27	5.24	5.22	5.20	5.18	5.17
7	8.07	6.54	5.89	5.52	5.29	5.12	4.99	4.90	4.82	4.76	4.71	4.67	4.63	4.60	4.57	4.54	4.52	4.50	4.48	4.47
8	7.57	6.06	5.42	5.05	4.82	4.65	4.53	4.43	4.36	4.30	4.24	4.20	4.16	4.13	4.10	4.08	4.05	4.03	4.02	4.00
9	7.21	5.71	5.08	4.72	4.48	4.32	4.20	4.10	4.03	3.96	3.91	3.87	3.83	3.80	3.77	3.74	3.72	3.70	3.68	3.67
10	6.94	5.46	4.83	4.47	4.24	4.07	3.95	3.85	3.78	3.72	3.66	3.62	3.58	3.55	3.52	3.50	3.47	3.45	3.44	3.42
11	6.72	5.26	4.63	4.28	4.04	3.88	3.76	3.66	3.59	3.53	3.47	3.43	3.39	3.36	3.33	3.30	3.28	3.26	3.24	3.23
12	6.55	5.10	4.47	4.12	3.89	3.73	3.61	3.51	3.44	3.37	3.32	3.28	3.24	3.21	3.18	3.15	3.13	3.11	3.09	3.07
13	6.41	4.97	4.35	4.00	3.77	3.60	3.48	3.39	3.31	3.25	3.20	3.15	3.12	3.08	3.05	3.03	3.00	2.98	2.96	2.95
14	6.30	4.86	4.24	3.89	3.66	3.50	3.38	3.29	3.21	3.15	3.09	3.05	3.01	2.98	2.95	2.92	2.90	2.88	2.86	2.84
15	6.20	4.77	4.15	3.80	3.58	3.41	3.29	3.20	3.12	3.06	3.01	2.96	2.92	2.89	2.86	2.84	2.81	2.79	2.77	2.76
16	6.12	4.69	4.08	3.73	3.50	3.34	3.22	3.12	3.05	2.99	2.93	2.89	2.85	2.82	2.79	2.76	2.74	2.72	2.70	2.68
17	6.04	4.62	4.01	3.66	3.44	3.28	3.16	3.06	2.98	2.92	2.87	2.82	2.79	2.75	2.72	2.70	2.67	2.65	2.63	2.62
18	5.98	4.56	3.95	3.61	3.38	3.22	3.10	3.01	2.93	2.87	2.81	2.77	2.73	2.70	2.67	2.64	2.62	2.60	2.58	2.56
19	5.92	4.51	3.90	3.56	3.33	3.17	3.05	2.96	2.88	2.82	2.76	2.72	2.68	2.65	2.62	2.59	2.57	2.55	2.53	2.51
20	5.87	4.46	3.86	3.51	3.29	3.13	3.01	2.91	2.84	2.77	2.72	2.68	2.64	2.60	2.57	2.55	2.52	2.50	2.48	2.46
22	5.79	4.38	3.78	3.44	3.22	3.05	2.93	2.84	2.76	2.70	2.65	2.60	2.56	2.53	2.50	2.47	2.45	2.43	2.41	2.39
24	5.72	4.32	3.72	3.38	3.15	2.99	2.87	2.78	2.70	2.64	2.59	2.54	2.50	2.47	2.44	2.41	2.39	2.36	2.35	2.33
26	5.66	4.27	3.67	3.33	3.10	2.94	2.82	2.73	2.65	2.59	2.54	2.49	2.45	2.42	2.39	2.36	2.34	2.31	2.29	2.28
28	5.61	4.22	3.63	3.29	3.06	2.90	2.78	2.69	2.61	2.55	2.49	2.45	2.41	2.37	2.34	2.32	2.29	2.27	2.25	2.23

DENOMINATOR DEGREES OF FREEDOM

v_1	NUMERATOR DEGREES OF FREEDOM																			
v_2	1	2	3	4	5	6	7	8	9	10	11	12	13	14	15	16	17	18	19	20
30	5.57	4.18	3.59	3.25	3.03	2.87	2.75	2.65	2.57	2.51	2.46	2.41	2.37	2.34	2.31	2.28	2.26	2.23	2.21	2.20
35	5.48	4.11	3.52	3.18	2.96	2.80	2.68	2.58	2.50	2.44	2.39	2.34	2.30	2.27	2.23	2.21	2.18	2.16	2.14	2.12
40	5.42	4.05	3.46	3.13	2.90	2.74	2.62	2.53	2.45	2.39	2.33	2.29	2.25	2.21	2.18	2.15	2.13	2.11	2.09	2.07
45	5.38	4.01	3.42	3.09	2.86	2.70	2.58	2.49	2.41	2.35	2.29	2.25	2.21	2.17	2.14	2.11	2.09	2.07	2.04	2.03
50	5.34	3.97	3.39	3.05	2.83	2.67	2.55	2.46	2.38	2.32	2.26	2.22	2.18	2.14	2.11	2.08	2.06	2.03	2.01	1.99
60	5.29	3.93	3.34	3.01	2.79	2.63	2.51	2.41	2.33	2.27	2.22	2.17	2.13	2.09	2.06	2.03	2.01	1.98	1.96	1.94
70	5.25	3.89	3.31	2.97	2.75	2.59	2.47	2.38	2.30	2.24	2.18	2.14	2.10	2.06	2.03	2.00	1.97	1.95	1.93	1.91
80	5.22	3.86	3.28	2.95	2.73	2.57	2.45	2.35	2.28	2.21	2.16	2.11	2.07	2.03	2.00	1.97	1.95	1.92	1.90	1.88
90	5.20	3.84	3.26	2.93	2.71	2.55	2.43	2.34	2.26	2.19	2.14	2.09	2.05	2.02	1.98	1.95	1.93	1.91	1.88	1.86
100	5.18	3.83	3.25	2.92	2.70	2.54	2.42	2.32	2.24	2.18	2.12	2.08	2.04	2.00	1.97	1.94	1.91	1.89	1.87	1.85
120	5.15	3.80	3.23	2.89	2.67	2.52	2.39	2.30	2.22	2.16	2.10	2.05	2.01	1.98	1.94	1.92	1.89	1.87	1.84	1.82
140	5.13	3.79	3.21	2.88	2.66	2.50	2.38	2.28	2.21	2.14	2.09	2.04	2.00	1.96	1.93	1.90	1.87	1.85	1.83	1.81
160	5.12	3.78	3.20	2.87	2.65	2.49	2.37	2.27	2.19	2.13	2.07	2.03	1.99	1.95	1.92	1.89	1.86	1.84	1.82	1.80
180	5.11	3.77	3.19	2.86	2.64	2.48	2.36	2.26	2.19	2.12	2.07	2.02	1.98	1.94	1.91	1.88	1.85	1.83	1.81	1.79
200	5.10	3.76	3.18	2.85	2.63	2.47	2.35	2.26	2.18	2.11	2.06	2.01	1.97	1.93	1.90	1.87	1.84	1.82	1.80	1.78
∞	5.03	3.69	3.12	2.79	2.57	2.41	2.29	2.19	2.11	2.05	1.99	1.95	1.90	1.87	1.83	1.80	1.78	1.75	1.73	1.71

v_1	NUMERATOR DEGREES OF FREEDOM																			
v_2 (DENOMINATOR DEGREES OF FREEDOM)	22	24	26	28	30	35	40	45	50	60	70	80	90	100	120	140	160	180	200	∞
1	995	997	999	1000	1001	1004	1006	1007	1008	1010	1011	1012	1013	1013	014	1015	1015	1015	1016	1018
2	39.5	39.5	39.5	39.5	39.5	39.5	39.5	39.5	39.5	39.5	39.5	39.5	39.5	39.5	39.5	39.5	39.5	39.5	39.5	39.5
3	14.1	14.1	14.1	14.1	14.1	14.1	14.0	14.0	14.0	14.0	14.0	14.0	14.0	14.0	13.9	13.9	13.9	13.9	13.9	13.9
4	8.53	8.51	8.49	8.48	8.46	8.43	8.41	8.39	8.38	8.36	8.35	8.33	8.33	8.32	8.31	8.30	8.30	8.29	8.29	8.26
5	6.30	6.28	6.26	6.24	6.23	6.20	6.18	6.16	6.14	6.12	6.11	6.10	6.09	6.08	6.07	6.06	6.06	6.05	6.05	6.02
6	5.14	5.12	5.10	5.08	5.07	5.04	5.01	4.99	4.98	4.96	4.94	4.93	4.92	4.92	4.90	4.90	4.89	4.89	4.88	4.85
7	4.44	4.41	4.39	4.38	4.36	4.33	4.31	4.29	4.28	4.25	4.24	4.23	4.22	4.21	4.20	4.19	4.18	4.18	4.18	4.14
8	3.97	3.95	3.93	3.91	3.89	3.86	3.84	3.82	3.81	3.78	3.77	3.76	3.75	3.74	3.73	3.72	3.71	3.71	3.70	3.67
9	3.64	3.61	3.59	3.58	3.56	3.53	3.51	3.49	3.47	3.45	3.43	3.42	3.41	3.40	3.39	3.38	3.38	3.37	3.37	3.33
10	3.39	3.37	3.34	3.33	3.31	3.28	3.26	3.24	3.22	3.20	3.18	3.17	3.16	3.15	3.14	3.13	3.13	3.12	3.12	3.08
11	3.20	3.17	3.15	3.13	3.12	3.09	3.06	3.04	3.03	3.00	2.99	2.97	2.96	2.96	2.94	2.94	2.93	2.92	2.92	2.88
12	3.04	3.02	3.00	2.98	2.96	2.93	2.91	2.89	2.87	2.85	2.83	2.82	2.81	2.80	2.79	2.78	2.77	2.77	2.76	2.73
13	2.92	2.89	2.87	2.85	2.84	2.80	2.78	2.76	2.74	2.72	2.70	2.69	2.68	2.67	2.66	2.65	2.64	2.64	2.63	2.60
14	2.81	2.79	2.77	2.75	2.73	2.70	2.67	2.65	2.64	2.61	2.60	2.58	2.57	2.56	2.55	2.54	2.54	2.53	2.53	2.49
15	2.73	2.70	2.68	2.66	2.64	2.61	2.59	2.56	2.55	2.52	2.51	2.49	2.48	2.47	2.46	2.45	2.44	2.44	2.44	2.40
16	2.65	2.63	2.60	2.58	2.57	2.53	2.51	2.49	2.47	2.45	2.43	2.42	2.40	2.40	2.38	2.37	2.37	2.36	2.36	2.32
17	2.59	2.56	2.54	2.52	2.50	2.47	2.44	2.42	2.41	2.38	2.36	2.35	2.34	2.33	2.32	2.31	2.30	2.29	2.29	2.25
18	2.53	2.50	2.48	2.46	2.44	2.41	2.38	2.36	2.35	2.32	2.30	2.29	2.28	2.27	2.26	2.25	2.24	2.23	2.23	2.19
19	2.48	2.45	2.43	2.41	2.39	2.36	2.33	2.31	2.30	2.27	2.25	2.24	2.23	2.22	2.20	2.19	2.19	2.18	2.18	2.13
20	2.43	2.41	2.39	2.37	2.35	2.31	2.29	2.27	2.25	2.22	2.20	2.19	2.18	2.17	2.16	2.15	2.14	2.13	2.13	2.09
22	2.36	2.33	2.31	2.29	2.27	2.24	2.21	2.19	2.17	2.14	2.13	2.11	2.10	2.09	2.08	2.07	2.06	2.05	2.05	2.00
24	2.30	2.27	2.25	2.23	2.21	2.17	2.15	2.12	2.11	2.08	2.06	2.05	2.03	2.02	2.01	2.00	1.99	1.99	1.98	1.94
26	2.24	2.22	2.19	2.17	2.16	2.12	2.09	2.07	2.05	2.03	2.01	1.99	1.98	1.97	1.95	1.94	1.94	1.93	1.92	1.88
28	2.20	2.17	2.15	2.13	2.11	2.08	2.05	2.03	2.01	1.98	1.96	1.94	1.93	1.92	1.91	1.90	1.89	1.88	1.88	1.83
30	2.16	2.14	2.11	2.09	2.07	2.04	2.01	1.99	1.97	1.94	1.92	1.90	1.89	1.88	1.87	1.86	1.85	1.84	1.84	1.79
35	2.09	2.06	2.04	2.02	2.00	1.96	1.93	1.91	1.89	1.86	1.84	1.82	1.81	1.80	1.79	1.77	1.77	1.76	1.75	1.70
40	2.03	2.01	1.98	1.96	1.94	1.90	1.88	1.85	1.83	1.80	1.78	1.76	1.75	1.74	1.72	1.71	1.70	1.70	1.69	1.64
45	1.99	1.96	1.94	1.92	1.90	1.86	1.83	1.81	1.79	1.76	1.74	1.72	1.70	1.69	1.68	1.66	1.66	1.65	1.64	1.59
50	1.96	1.93	1.91	1.89	1.87	1.83	1.80	1.77	1.75	1.72	1.70	1.68	1.67	1.66	1.64	1.63	1.62	1.61	1.60	1.55
60	1.91	1.88	1.86	1.83	1.82	1.78	1.74	1.72	1.70	1.67	1.64	1.63	1.61	1.60	1.58	1.57	1.56	1.55	1.54	1.48
70	1.88	1.85	1.82	1.80	1.78	1.74	1.71	1.68	1.66	1.63	1.60	1.59	1.57	1.56	1.54	1.53	1.52	1.51	1.50	1.44
80	1.85	1.82	1.79	1.77	1.75	1.71	1.68	1.65	1.63	1.60	1.57	1.55	1.54	1.53	1.51	1.49	1.48	1.47	1.47	1.40
90	1.83	1.80	1.77	1.75	1.73	1.69	1.66	1.63	1.61	1.58	1.55	1.53	1.52	1.50	1.48	1.47	1.46	1.45	1.44	1.37
100	1.81	1.78	1.76	1.74	1.71	1.67	1.64	1.61	1.59	1.56	1.53	1.51	1.50	1.48	1.46	1.45	1.44	1.43	1.42	1.35
120	1.79	1.76	1.73	1.71	1.69	1.65	1.61	1.59	1.56	1.53	1.50	1.48	1.47	1.45	1.43	1.42	1.41	1.40	1.39	1.31
140	1.77	1.74	1.72	1.69	1.67	1.63	1.60	1.57	1.55	1.51	1.48	1.46	1.45	1.43	1.41	1.39	1.38	1.37	1.36	1.28
160	1.76	1.73	1.70	1.68	1.66	1.62	1.58	1.55	1.53	1.50	1.47	1.45	1.43	1.42	1.39	1.38	1.36	1.35	1.35	1.26
180	1.75	1.72	1.69	1.67	1.65	1.61	1.57	1.54	1.52	1.48	1.46	1.43	1.42	1.40	1.38	1.36	1.35	1.34	1.33	1.25
200	1.74	1.71	1.68	1.66	1.64	1.60	1.56	1.53	1.51	1.47	1.45	1.42	1.41	1.39	1.37	1.35	1.34	1.33	1.32	1.23
∞	1.67	1.64	1.61	1.59	1.57	1.52	1.49	1.46	1.43	1.39	1.36	1.33	1.31	1.30	1.27	1.25	1.23	1.22	1.21	1.00

TABLE 6(C) Values of the F-Distribution: $A = 0.01$

v_2 \ v_1	1	2	3	4	5	6	7	8	9	10	11	12	13	14	15	16	17	18	19	20
	NUMERATOR DEGREES OF FREEDOM																			
1	4052	4999	5403	5625	5764	5859	5928	5981	6022	6056	6083	6106	6126	6143	6157	6170	6181	6192	6201	6209
2	98.5	99.0	99.2	99.2	99.3	99.3	99.4	99.4	99.4	99.4	99.4	99.4	99.4	99.4	99.4	99.4	99.4	99.4	99.4	99.4
3	34.1	30.8	29.5	28.7	28.2	27.9	27.7	27.5	27.3	27.2	27.1	27.1	27.0	26.9	26.9	26.8	26.8	26.8	26.7	26.7
4	21.2	18.0	16.7	16.0	15.5	15.2	15.0	14.8	14.7	14.5	14.5	14.4	14.3	14.2	14.2	14.2	14.1	14.1	14.0	14.0
5	16.3	13.3	12.1	11.4	11.0	10.7	10.5	10.3	10.2	10.1	9.96	9.89	9.82	9.77	9.72	9.68	9.64	9.61	9.58	9.55
6	13.7	10.9	9.78	9.15	8.75	8.47	8.26	8.10	7.98	7.87	7.79	7.72	7.66	7.60	7.56	7.52	7.48	7.45	7.42	7.40
7	12.2	9.55	8.45	7.85	7.46	7.19	6.99	6.84	6.72	6.62	6.54	6.47	6.41	6.36	6.31	6.28	6.24	6.21	6.18	6.16
8	11.3	8.65	7.59	7.01	6.63	6.37	6.18	6.03	5.91	5.81	5.73	5.67	5.61	5.56	5.52	5.48	5.44	5.41	5.38	5.36
9	10.6	8.02	6.99	6.42	6.06	5.80	5.61	5.47	5.35	5.26	5.18	5.11	5.05	5.01	4.96	4.92	4.89	4.86	4.83	4.81
10	10.0	7.56	6.55	5.99	5.64	5.39	5.20	5.06	4.94	4.85	4.77	4.71	4.65	4.60	4.56	4.52	4.49	4.46	4.43	4.41
11	9.65	7.21	6.22	5.67	5.32	5.07	4.89	4.74	4.63	4.54	4.46	4.40	4.34	4.29	4.25	4.21	4.18	4.15	4.12	4.10
12	9.33	6.93	5.95	5.41	5.06	4.82	4.64	4.50	4.39	4.30	4.22	4.16	4.10	4.05	4.01	3.97	3.94	3.91	3.88	3.86
13	9.07	6.70	5.74	5.21	4.86	4.62	4.44	4.30	4.19	4.10	4.02	3.96	3.91	3.86	3.82	3.78	3.75	3.72	3.69	3.66
14	8.86	6.51	5.56	5.04	4.69	4.46	4.28	4.14	4.03	3.94	3.86	3.80	3.75	3.70	3.66	3.62	3.59	3.56	3.53	3.51
15	8.68	6.36	5.42	4.89	4.56	4.32	4.14	4.00	3.89	3.80	3.73	3.67	3.61	3.56	3.52	3.49	3.45	3.42	3.40	3.37
16	8.53	6.23	5.29	4.77	4.44	4.20	4.03	3.89	3.78	3.69	3.62	3.55	3.50	3.45	3.41	3.37	3.34	3.31	3.28	3.26
17	8.40	6.11	5.18	4.67	4.34	4.10	3.93	3.79	3.68	3.59	3.52	3.46	3.40	3.35	3.31	3.27	3.24	3.21	3.19	3.16
18	8.29	6.01	5.09	4.58	4.25	4.01	3.84	3.71	3.60	3.51	3.43	3.37	3.32	3.27	3.23	3.19	3.16	3.13	3.10	3.08
19	8.18	5.93	5.01	4.50	4.17	3.94	3.77	3.63	3.52	3.43	3.36	3.30	3.24	3.19	3.15	3.12	3.08	3.05	3.03	3.00
20	8.10	5.85	4.94	4.43	4.10	3.87	3.70	3.56	3.46	3.37	3.29	3.23	3.18	3.13	3.09	3.05	3.02	2.99	2.96	2.94
22	7.95	5.72	4.82	4.31	3.99	3.76	3.59	3.45	3.35	3.26	3.18	3.12	3.07	3.02	2.98	2.94	2.91	2.88	2.85	2.83
24	7.82	5.61	4.72	4.22	3.90	3.67	3.50	3.36	3.26	3.17	3.09	3.03	2.98	2.93	2.89	2.85	2.82	2.79	2.76	2.74
26	7.72	5.53	4.64	4.14	3.82	3.59	3.42	3.29	3.18	3.09	3.02	2.96	2.90	2.86	2.81	2.78	2.75	2.72	2.69	2.66
28	7.64	5.45	4.57	4.07	3.75	3.53	3.36	3.23	3.12	3.03	2.96	2.90	2.84	2.79	2.75	2.72	2.68	2.65	2.63	2.60

(Continued)

v_1 / v_2	NUMERATOR DEGREES OF FREEDOM																			
DENOMINATOR DEGREES OF FREEDOM	1	2	3	4	5	6	7	8	9	10	11	12	13	14	15	16	17	18	19	20
30	7.56	5.39	4.51	4.02	3.70	3.47	3.30	3.17	3.07	2.98	2.91	2.84	2.79	2.74	2.70	2.66	2.63	2.60	2.57	2.55
35	7.42	5.27	4.40	3.91	3.59	3.37	3.20	3.07	2.96	2.88	2.80	2.74	2.69	2.64	2.60	2.56	2.53	2.50	2.47	2.44
40	7.31	5.18	4.31	3.83	3.51	3.29	3.12	2.99	2.89	2.80	2.73	2.66	2.61	2.56	2.52	2.48	2.45	2.42	2.39	2.37
45	7.23	5.11	4.25	3.77	3.45	3.23	3.07	2.94	2.83	2.74	2.67	2.61	2.55	2.51	2.46	2.43	2.39	2.36	2.34	2.31
50	7.17	5.06	4.20	3.72	3.41	3.19	3.02	2.89	2.78	2.70	2.63	2.56	2.51	2.46	2.42	2.38	2.35	2.32	2.29	2.27
60	7.08	4.98	4.13	3.65	3.34	3.12	2.95	2.82	2.72	2.63	2.56	2.50	2.44	2.39	2.35	2.31	2.28	2.25	2.22	2.20
70	7.01	4.92	4.07	3.60	3.29	3.07	2.91	2.78	2.67	2.59	2.51	2.45	2.40	2.35	2.31	2.27	2.23	2.20	2.18	2.15
80	6.96	4.88	4.04	3.56	3.26	3.04	2.87	2.74	2.64	2.55	2.48	2.42	2.36	2.31	2.27	2.23	2.20	2.17	2.14	2.12
90	6.93	4.85	4.01	3.53	3.23	3.01	2.84	2.72	2.61	2.52	2.45	2.39	2.33	2.29	2.24	2.21	2.17	2.14	2.11	2.09
100	6.90	4.82	3.98	3.51	3.21	2.99	2.82	2.69	2.59	2.50	2.43	2.37	2.31	2.27	2.22	2.19	2.15	2.12	2.09	2.07
120	6.85	4.79	3.95	3.48	3.17	2.96	2.79	2.66	2.56	2.47	2.40	2.34	2.28	2.23	2.19	2.15	2.12	2.09	2.06	2.03
140	6.82	4.76	3.92	3.46	3.15	2.93	2.77	2.64	2.54	2.45	2.38	2.31	2.26	2.21	2.17	2.13	2.10	2.07	2.04	2.01
160	6.80	4.74	3.91	3.44	3.13	2.92	2.75	2.62	2.52	2.43	2.36	2.30	2.24	2.20	2.15	2.11	2.08	2.05	2.02	1.99
180	6.78	4.73	3.89	3.43	3.12	2.90	2.74	2.61	2.51	2.42	2.35	2.28	2.23	2.18	2.14	2.10	2.07	2.04	2.01	1.98
200	6.76	4.71	3.88	3.41	3.11	2.89	2.73	2.60	2.50	2.41	2.34	2.27	2.22	2.17	2.13	2.09	2.06	2.03	2.00	1.97
∞	6.64	4.61	3.78	3.32	3.02	2.80	2.64	2.51	2.41	2.32	2.25	2.19	2.13	2.08	2.04	2.00	1.97	1.94	1.91	1.88

v_1	NUMERATOR DEGREES OF FREEDOM																			
v_2	22	24	26	28	30	35	40	45	50	60	70	80	90	100	120	140	160	180	200	∞
DENOMINATOR DEGREES OF FREEDOM																				
1	6223	6235	6245	6253	6261	6276	6287	6296	6303	6313	6321	6326	6331	6334	6339	6343	6346	6348	6350	6366
2	99.5	99.5	99.5	99.5	99.5	99.5	99.5	99.5	99.5	99.5	99.5	99.5	99.5	99.5	99.5	99.5	99.5	99.5	99.5	99.5
3	26.6	26.6	26.6	26.5	26.5	26.5	26.4	26.4	26.4	26.3	26.3	26.3	26.3	26.2	26.2	26.2	26.2	26.2	26.2	26.1
4	14.0	13.9	13.9	13.9	13.8	13.8	13.7	13.7	13.7	13.7	13.6	13.6	13.6	13.6	13.6	13.5	13.5	13.5	13.5	13.5
5	9.51	9.47	9.43	9.40	9.38	9.33	9.29	9.26	9.24	9.20	9.18	9.16	9.14	9.13	9.11	9.10	9.09	9.08	9.08	9.02
6	7.35	7.31	7.28	7.25	7.23	7.18	7.14	7.11	7.09	7.06	7.03	7.01	7.00	6.99	6.97	6.96	6.95	6.94	6.93	6.88
7	6.11	6.07	6.04	6.02	5.99	5.94	5.91	5.88	5.86	5.82	5.80	5.78	5.77	5.75	5.74	5.72	5.72	5.71	5.70	5.65
8	5.32	5.28	5.25	5.22	5.20	5.15	5.12	5.09	5.07	5.03	5.01	4.99	4.97	4.96	4.95	4.93	4.92	4.92	4.91	4.86
9	4.77	4.73	4.70	4.67	4.65	4.60	4.57	4.54	4.52	4.48	4.46	4.44	4.43	4.41	4.40	4.39	4.38	4.37	4.36	4.31
10	4.36	4.33	4.30	4.27	4.25	4.20	4.17	4.14	4.12	4.08	4.06	4.04	4.03	4.01	4.00	3.98	3.97	3.97	3.96	3.91
11	4.06	4.02	3.99	3.96	3.94	3.89	3.86	3.83	3.81	3.78	3.75	3.73	3.72	3.71	3.69	3.68	3.67	3.66	3.66	3.60
12	3.82	3.78	3.75	3.72	3.70	3.65	3.62	3.59	3.57	3.54	3.51	3.49	3.48	3.47	3.45	3.44	3.43	3.42	3.41	3.36
13	3.62	3.59	3.56	3.53	3.51	3.46	3.43	3.40	3.38	3.34	3.32	3.30	3.28	3.27	3.25	3.24	3.23	3.23	3.22	3.17
14	3.46	3.43	3.40	3.37	3.35	3.30	3.27	3.24	3.22	3.18	3.16	3.14	3.12	3.11	3.09	3.08	3.07	3.06	3.06	3.01
15	3.33	3.29	3.26	3.24	3.21	3.17	3.13	3.10	3.08	3.05	3.02	3.00	2.99	2.98	2.96	2.95	2.94	2.93	2.92	2.87
16	3.22	3.18	3.15	3.12	3.10	3.05	3.02	2.99	2.97	2.93	2.91	2.89	2.87	2.86	2.84	2.83	2.82	2.81	2.81	2.75
17	3.12	3.08	3.05	3.03	3.00	2.96	2.92	2.89	2.87	2.83	2.81	2.79	2.78	2.76	2.75	2.73	2.72	2.72	2.71	2.65
18	3.03	3.00	2.97	2.94	2.92	2.87	2.84	2.81	2.78	2.75	2.72	2.70	2.69	2.68	2.66	2.65	2.64	2.63	2.62	2.57
19	2.96	2.92	2.89	2.87	2.84	2.80	2.76	2.73	2.71	2.67	2.65	2.63	2.61	2.60	2.58	2.57	2.56	2.55	2.55	2.49
20	2.90	2.86	2.83	2.80	2.78	2.73	2.69	2.67	2.64	2.61	2.58	2.56	2.55	2.54	2.52	2.50	2.49	2.49	2.48	2.42
22	2.78	2.75	2.72	2.69	2.67	2.62	2.58	2.55	2.53	2.50	2.47	2.45	2.43	2.42	2.40	2.39	2.38	2.37	2.36	2.31
24	2.70	2.66	2.63	2.60	2.58	2.53	2.49	2.46	2.44	2.40	2.38	2.36	2.34	2.33	2.31	2.30	2.29	2.28	2.27	2.21
26	2.62	2.58	2.55	2.53	2.50	2.45	2.42	2.39	2.36	2.33	2.30	2.28	2.26	2.25	2.23	2.22	2.21	2.20	2.19	2.13
28	2.56	2.52	2.49	2.46	2.44	2.39	2.35	2.32	2.30	2.26	2.24	2.22	2.20	2.19	2.17	2.15	2.14	2.13	2.13	2.07
30	2.51	2.47	2.44	2.41	2.39	2.34	2.30	2.27	2.25	2.21	2.18	2.16	2.14	2.13	2.11	2.10	2.09	2.08	2.07	2.01
35	2.40	2.36	2.33	2.30	2.28	2.23	2.19	2.16	2.14	2.10	2.07	2.05	2.03	2.02	2.00	1.98	1.97	1.96	1.96	1.89
40	2.33	2.29	2.26	2.23	2.20	2.15	2.11	2.08	2.06	2.02	1.99	1.97	1.95	1.94	1.92	1.90	1.89	1.88	1.87	1.81
45	2.27	2.23	2.20	2.17	2.14	2.09	2.05	2.02	2.00	1.96	1.93	1.91	1.89	1.88	1.85	1.84	1.83	1.82	1.81	1.74
50	2.22	2.18	2.15	2.12	2.10	2.05	2.01	1.97	1.95	1.91	1.88	1.86	1.84	1.82	1.80	1.79	1.77	1.76	1.76	1.68
60	2.15	2.12	2.08	2.05	2.03	1.98	1.94	1.90	1.88	1.84	1.81	1.78	1.76	1.75	1.73	1.71	1.70	1.69	1.68	1.60
70	2.11	2.07	2.03	2.01	1.98	1.93	1.89	1.85	1.83	1.78	1.75	1.73	1.71	1.70	1.67	1.65	1.64	1.63	1.62	1.54
80	2.07	2.03	2.00	1.97	1.94	1.89	1.85	1.82	1.79	1.75	1.71	1.69	1.67	1.65	1.63	1.61	1.60	1.59	1.58	1.50
90	2.04	2.00	1.97	1.94	1.92	1.86	1.82	1.79	1.76	1.72	1.68	1.66	1.64	1.62	1.60	1.58	1.57	1.55	1.55	1.46
100	2.02	1.98	1.95	1.92	1.89	1.84	1.80	1.76	1.74	1.69	1.66	1.63	1.61	1.60	1.57	1.55	1.54	1.53	1.52	1.43
120	1.99	1.95	1.92	1.89	1.86	1.81	1.76	1.73	1.70	1.66	1.62	1.60	1.58	1.56	1.53	1.51	1.50	1.49	1.48	1.38
140	1.97	1.93	1.89	1.86	1.84	1.78	1.74	1.70	1.67	1.63	1.60	1.57	1.55	1.53	1.50	1.48	1.47	1.46	1.45	1.35
160	1.95	1.91	1.88	1.85	1.82	1.76	1.72	1.68	1.66	1.61	1.58	1.55	1.53	1.51	1.48	1.46	1.45	1.43	1.42	1.32
180	1.94	1.90	1.86	1.83	1.81	1.75	1.71	1.67	1.64	1.60	1.56	1.53	1.51	1.49	1.47	1.45	1.43	1.42	1.41	1.30
200	1.93	1.89	1.85	1.82	1.79	1.74	1.69	1.66	1.63	1.58	1.55	1.52	1.50	1.48	1.45	1.43	1.42	1.40	1.39	1.28
∞	1.83	1.79	1.76	1.73	1.70	1.64	1.59	1.56	1.53	1.48	1.44	1.41	1.38	1.36	1.33	1.30	1.28	1.26	1.25	1.00

TABLE 6(d) Values of the *F*-Distribution: $A = 0.005$

ν_2 \ ν_1	1	2	3	4	5	6	7	8	9	10	11	12	13	14	15	16	17	18	19	20
	NUMERATOR DEGREES OF FREEDOM																			
DENOMINATOR DEGREES OF FREEDOM																				
1	16211	19999	21615	22500	23056	23437	23715	23925	24091	24224	24334	24426	24505	24572	24630	24681	24727	24767	24803	24836
2	199	199	199	199	199	199	199	199	199	199	199	199	199	199	199	199	199	199	199	199
3	55.6	49.8	47.5	46.2	45.4	44.8	44.4	44.1	43.9	43.7	43.5	43.4	43.3	43.2	43.1	43.0	42.9	42.9	42.8	42.8
4	31.3	26.3	24.3	23.2	22.5	22.0	21.6	21.4	21.1	21.0	20.8	20.7	20.6	20.5	20.4	20.4	20.3	20.3	20.2	20.2
5	22.8	18.3	16.5	15.6	14.9	14.5	14.2	14.0	13.8	13.6	13.5	13.4	13.3	13.2	13.1	13.1	13.0	13.0	12.9	12.9
6	18.6	14.5	12.9	12.0	11.5	11.1	10.8	10.6	10.4	10.3	10.1	10.0	9.95	9.88	9.81	9.76	9.71	9.66	9.62	9.59
7	16.2	12.4	10.9	10.1	9.52	9.16	8.89	8.68	8.51	8.38	8.27	8.18	8.10	8.03	7.97	7.91	7.87	7.83	7.79	7.75
8	14.7	11.0	9.60	8.81	8.30	7.95	7.69	7.50	7.34	7.21	7.10	7.01	6.94	6.87	6.81	6.76	6.72	6.68	6.64	6.61
9	13.6	10.1	8.72	7.96	7.47	7.13	6.88	6.69	6.54	6.42	6.31	6.23	6.15	6.09	6.03	5.98	5.94	5.90	5.86	5.83
10	12.8	9.43	8.08	7.34	6.87	6.54	6.30	6.12	5.97	5.85	5.75	5.66	5.59	5.53	5.47	5.42	5.38	5.34	5.31	5.27
11	12.2	8.91	7.60	6.88	6.42	6.10	5.86	5.68	5.54	5.42	5.32	5.24	5.16	5.10	5.05	5.00	4.96	4.92	4.89	4.86
12	11.8	8.51	7.23	6.52	6.07	5.76	5.52	5.35	5.20	5.09	4.99	4.91	4.84	4.77	4.72	4.67	4.63	4.59	4.56	4.53
13	11.4	8.19	6.93	6.23	5.79	5.48	5.25	5.08	4.94	4.82	4.72	4.64	4.57	4.51	4.46	4.41	4.37	4.33	4.30	4.27
14	11.1	7.92	6.68	6.00	5.56	5.26	5.03	4.86	4.72	4.60	4.51	4.43	4.36	4.30	4.25	4.20	4.16	4.12	4.09	4.06
15	10.8	7.70	6.48	5.80	5.37	5.07	4.85	4.67	4.54	4.42	4.33	4.25	4.18	4.12	4.07	4.02	3.98	3.95	3.91	3.88
16	10.6	7.51	6.30	5.64	5.21	4.91	4.69	4.52	4.38	4.27	4.18	4.10	4.03	3.97	3.92	3.87	3.83	3.80	3.76	3.73
17	10.4	7.35	6.16	5.50	5.07	4.78	4.56	4.39	4.25	4.14	4.05	3.97	3.90	3.84	3.79	3.75	3.71	3.67	3.64	3.61
18	10.2	7.21	6.03	5.37	4.96	4.66	4.44	4.28	4.14	4.03	3.94	3.86	3.79	3.73	3.68	3.64	3.60	3.56	3.53	3.50
19	10.1	7.09	5.92	5.27	4.85	4.56	4.34	4.18	4.04	3.93	3.84	3.76	3.70	3.64	3.59	3.54	3.50	3.46	3.43	3.40
20	9.94	6.99	5.82	5.17	4.76	4.47	4.26	4.09	3.96	3.85	3.76	3.68	3.61	3.55	3.50	3.46	3.42	3.38	3.35	3.32
22	9.73	6.81	5.65	5.02	4.61	4.32	4.11	3.94	3.81	3.70	3.61	3.54	3.47	3.41	3.36	3.31	3.27	3.24	3.21	3.18
24	9.55	6.66	5.52	4.89	4.49	4.20	3.99	3.83	3.69	3.59	3.50	3.42	3.35	3.30	3.25	3.20	3.16	3.12	3.09	3.06
26	9.41	6.54	5.41	4.79	4.38	4.10	3.89	3.73	3.60	3.49	3.40	3.33	3.26	3.20	3.15	3.11	3.07	3.03	3.00	2.97
28	9.28	6.44	5.32	4.70	4.30	4.02	3.81	3.65	3.52	3.41	3.32	3.25	3.18	3.12	3.07	3.03	2.99	2.95	2.92	2.89
30	9.18	6.35	5.24	4.62	4.23	3.95	3.74	3.58	3.45	3.34	3.25	3.18	3.11	3.06	3.01	2.96	2.92	2.89	2.85	2.82

ν_2 \ ν_1	NUMERATOR DEGREES OF FREEDOM																			
	1	2	3	4	5	6	7	8	9	10	11	12	13	14	15	16	17	18	19	20
35	8.98	6.19	5.09	4.48	4.09	3.81	3.61	3.45	3.32	3.21	3.12	3.05	2.98	2.93	2.88	2.83	2.79	2.76	2.72	2.69
40	8.83	6.07	4.98	4.37	3.99	3.71	3.51	3.35	3.22	3.12	3.03	2.95	2.89	2.83	2.78	2.74	2.70	2.66	2.63	2.60
45	8.71	5.97	4.89	4.29	3.91	3.64	3.43	3.28	3.15	3.04	2.96	2.88	2.82	2.76	2.71	2.66	2.62	2.59	2.56	2.53
50	8.63	5.90	4.83	4.23	3.85	3.58	3.38	3.22	3.09	2.99	2.90	2.82	2.76	2.70	2.65	2.61	2.57	2.53	2.50	2.47
60	8.49	5.79	4.73	4.14	3.76	3.49	3.29	3.13	3.01	2.90	2.82	2.74	2.68	2.62	2.57	2.53	2.49	2.45	2.42	2.39
70	8.40	5.72	4.66	4.08	3.70	3.43	3.23	3.08	2.95	2.85	2.76	2.68	2.62	2.56	2.51	2.47	2.43	2.39	2.36	2.33
80	8.33	5.67	4.61	4.03	3.65	3.39	3.19	3.03	2.91	2.80	2.72	2.64	2.58	2.52	2.47	2.43	2.39	2.35	2.32	2.29
90	8.28	5.62	4.57	3.99	3.62	3.35	3.15	3.00	2.87	2.77	2.68	2.61	2.54	2.49	2.44	2.39	2.35	2.32	2.28	2.25
100	8.24	5.59	4.54	3.96	3.59	3.33	3.13	2.97	2.85	2.74	2.66	2.58	2.52	2.46	2.41	2.37	2.33	2.29	2.26	2.23
120	8.18	5.54	4.50	3.92	3.55	3.28	3.09	2.93	2.81	2.71	2.62	2.54	2.48	2.42	2.37	2.33	2.29	2.25	2.22	2.19
140	8.14	5.50	4.47	3.89	3.52	3.26	3.06	2.91	2.78	2.68	2.59	2.52	2.45	2.40	2.35	2.30	2.26	2.22	2.19	2.16
160	8.10	5.48	4.44	3.87	3.50	3.24	3.04	2.88	2.76	2.66	2.57	2.50	2.43	2.38	2.33	2.28	2.24	2.20	2.17	2.14
180	8.08	5.46	4.42	3.85	3.48	3.22	3.02	2.87	2.74	2.64	2.56	2.48	2.42	2.36	2.31	2.26	2.22	2.19	2.15	2.12
200	8.06	5.44	4.41	3.84	3.47	3.21	3.01	2.86	2.73	2.63	2.54	2.47	2.40	2.35	2.30	2.25	2.21	2.18	2.14	2.11
∞	7.88	5.30	4.28	3.72	3.35	3.09	2.90	2.75	2.62	2.52	2.43	2.36	2.30	2.24	2.19	2.14	2.10	2.07	2.03	2.00

(Continued)

v_1	NUMERATOR DEGREES OF FREEDOM																			
v_2 (DENOMINATOR DEGREES OF FREEDOM)	22	24	26	28	30	35	40	45	50	60	70	80	90	100	120	140	160	180	200	∞
1	24892	24940	24980	25014	25044	25103	25148	25183	25211	25253	25283	25306	25323	25337	25359	25374	25385	25394	25401	25464
2	199	199	199	199	199	199	199	199	199	199	199	199	199	199	199	199	199	199	199	199
3	42.7	42.6	42.6	42.5	42.5	42.4	42.3	42.3	42.2	42.1	42.1	42.1	42.0	42.0	42.0	42.0	41.9	41.9	41.9	41.8
4	20.1	20.0	20.0	19.9	19.9	19.8	19.8	19.7	19.7	19.6	19.6	19.5	19.5	19.5	19.5	19.4	19.4	19.4	19.4	19.3
5	12.8	12.8	12.7	12.7	12.7	12.6	12.5	12.5	12.5	12.4	12.4	12.3	12.3	12.3	12.3	12.3	12.2	12.2	12.2	12.1
6	9.53	9.47	9.43	9.39	9.36	9.29	9.24	9.20	9.17	9.12	9.09	9.06	9.04	9.03	9.00	8.98	8.97	8.96	8.95	8.88
7	7.69	7.64	7.60	7.57	7.53	7.47	7.42	7.38	7.35	7.31	7.28	7.25	7.23	7.22	7.19	7.18	7.16	7.15	7.15	7.08
8	6.55	6.50	6.46	6.43	6.40	6.33	6.29	6.25	6.22	6.18	6.15	6.12	6.10	6.09	6.06	6.05	6.04	6.03	6.02	5.95
9	5.78	5.73	5.69	5.65	5.62	5.56	5.52	5.48	5.45	5.41	5.38	5.36	5.34	5.32	5.30	5.28	5.27	5.26	5.26	5.19
10	5.22	5.17	5.13	5.10	5.07	5.01	4.97	4.93	4.90	4.86	4.83	4.80	4.79	4.77	4.75	4.73	4.72	4.71	4.71	4.64
11	4.80	4.76	4.72	4.68	4.65	4.60	4.55	4.52	4.49	4.45	4.41	4.39	4.37	4.36	4.34	4.32	4.31	4.30	4.29	4.23
12	4.48	4.43	4.39	4.36	4.33	4.27	4.23	4.19	4.17	4.12	4.09	4.07	4.05	4.04	4.01	4.00	3.99	3.98	3.97	3.91
13	4.22	4.17	4.13	4.10	4.07	4.01	3.97	3.94	3.91	3.87	3.84	3.81	3.79	3.78	3.76	3.74	3.73	3.72	3.71	3.65
14	4.01	3.96	3.92	3.89	3.86	3.80	3.76	3.73	3.70	3.66	3.62	3.60	3.58	3.57	3.55	3.53	3.52	3.51	3.50	3.44
15	3.83	3.79	3.75	3.72	3.69	3.63	3.58	3.55	3.52	3.48	3.45	3.43	3.41	3.39	3.37	3.36	3.34	3.34	3.33	3.26
16	3.68	3.64	3.60	3.57	3.54	3.48	3.44	3.40	3.37	3.33	3.30	3.28	3.26	3.25	3.22	3.21	3.20	3.19	3.18	3.11
17	3.56	3.51	3.47	3.44	3.41	3.35	3.31	3.28	3.25	3.21	3.18	3.15	3.13	3.12	3.10	3.08	3.07	3.06	3.05	2.99
18	3.45	3.40	3.36	3.33	3.30	3.25	3.20	3.17	3.14	3.10	3.07	3.04	3.02	3.01	2.99	2.97	2.96	2.95	2.94	2.87
19	3.35	3.31	3.27	3.24	3.21	3.15	3.11	3.07	3.04	3.00	2.97	2.95	2.93	2.91	2.89	2.87	2.86	2.85	2.85	2.78
20	3.27	3.22	3.18	3.15	3.12	3.07	3.02	2.99	2.96	2.92	2.88	2.86	2.84	2.83	2.81	2.79	2.78	2.77	2.76	2.69
22	3.12	3.08	3.04	3.01	2.98	2.92	2.88	2.84	2.82	2.77	2.74	2.72	2.70	2.69	2.66	2.65	2.63	2.62	2.62	2.55
24	3.01	2.97	2.93	2.90	2.87	2.81	2.77	2.73	2.70	2.66	2.63	2.60	2.58	2.57	2.55	2.53	2.52	2.51	2.50	2.43
26	2.92	2.87	2.84	2.80	2.77	2.72	2.67	2.64	2.61	2.56	2.53	2.51	2.49	2.47	2.45	2.43	2.42	2.41	2.40	2.33
28	2.84	2.79	2.76	2.72	2.69	2.64	2.59	2.56	2.53	2.48	2.45	2.43	2.41	2.39	2.37	2.35	2.34	2.33	2.32	2.25
30	2.77	2.73	2.69	2.66	2.63	2.57	2.52	2.49	2.46	2.42	2.38	2.36	2.34	2.32	2.30	2.28	2.27	2.26	2.25	2.18
35	2.64	2.60	2.56	2.53	2.50	2.44	2.39	2.36	2.33	2.28	2.25	2.22	2.20	2.19	2.16	2.15	2.13	2.12	2.11	2.04
40	2.55	2.50	2.46	2.43	2.40	2.34	2.30	2.26	2.23	2.18	2.15	2.12	2.10	2.09	2.06	2.05	2.03	2.02	2.01	1.93
45	2.47	2.43	2.39	2.36	2.33	2.27	2.22	2.19	2.16	2.11	2.08	2.05	2.03	2.01	1.99	1.97	1.95	1.94	1.93	1.85
50	2.42	2.37	2.33	2.30	2.27	2.21	2.16	2.13	2.10	2.05	2.02	1.99	1.97	1.95	1.93	1.91	1.89	1.88	1.87	1.79
60	2.33	2.29	2.25	2.22	2.19	2.13	2.08	2.04	2.01	1.96	1.93	1.90	1.88	1.86	1.83	1.81	1.80	1.79	1.78	1.69
70	2.28	2.23	2.19	2.16	2.13	2.07	2.02	1.98	1.95	1.90	1.86	1.84	1.81	1.80	1.77	1.75	1.73	1.72	1.71	1.62
80	2.23	2.19	2.15	2.11	2.08	2.02	1.97	1.94	1.90	1.85	1.82	1.79	1.77	1.75	1.72	1.70	1.68	1.67	1.66	1.57
90	2.20	2.15	2.12	2.08	2.05	1.99	1.94	1.90	1.87	1.82	1.78	1.75	1.73	1.71	1.68	1.66	1.64	1.63	1.62	1.52
100	2.17	2.13	2.09	2.05	2.02	1.96	1.91	1.87	1.84	1.79	1.75	1.72	1.70	1.68	1.65	1.63	1.61	1.60	1.59	1.49
120	2.13	2.09	2.05	2.01	1.98	1.92	1.87	1.83	1.80	1.75	1.71	1.68	1.66	1.64	1.61	1.58	1.57	1.55	1.54	1.43
140	2.11	2.06	2.02	1.99	1.96	1.89	1.84	1.80	1.77	1.72	1.68	1.65	1.62	1.60	1.57	1.55	1.53	1.52	1.51	1.39
160	2.09	2.04	2.00	1.97	1.93	1.87	1.82	1.78	1.75	1.69	1.65	1.62	1.60	1.58	1.55	1.52	1.51	1.49	1.48	1.36
180	2.07	2.02	1.98	1.95	1.92	1.85	1.80	1.76	1.73	1.68	1.64	1.61	1.58	1.56	1.53	1.50	1.49	1.47	1.46	1.34
200	2.06	2.01	1.97	1.94	1.91	1.84	1.79	1.75	1.71	1.66	1.62	1.59	1.56	1.54	1.51	1.49	1.47	1.45	1.44	1.32
∞	1.95	1.90	1.86	1.82	1.79	1.72	1.67	1.63	1.59	1.54	1.49	1.46	1.43	1.40	1.37	1.34	1.31	1.30	1.28	1.00

TABLE 7(a) Critical Values of the Studentized Range, $\alpha = 0.05$

ν	k																		
	2	3	4	5	6	7	8	9	10	11	12	13	14	15	16	17	18	19	20
1	18.0	27.0	32.8	37.1	40.4	43.1	45.4	47.4	49.1	50.6	52.0	53.2	54.3	55.4	56.3	57.2	58.0	58.8	59.6
2	6.08	8.33	9.80	10.9	11.7	12.4	13.0	13.5	14.0	14.4	14.7	15.1	15.4	15.7	15.9	16.1	16.4	16.6	16.8
3	4.50	5.91	6.82	7.50	8.04	8.48	8.85	9.18	9.46	9.72	9.95	10.2	10.3	10.5	10.7	10.8	11.0	11.1	11.2
4	3.93	5.04	5.76	6.29	6.71	7.05	7.35	7.60	7.83	8.03	8.21	8.37	8.52	8.66	8.79	8.91	9.03	9.13	9.23
5	3.64	4.60	5.22	5.67	6.03	6.33	6.58	6.80	6.99	7.17	7.32	7.47	7.60	7.72	7.83	7.93	8.03	8.12	8.21
6	3.46	4.34	4.90	5.30	5.63	5.90	6.12	6.32	6.49	6.65	6.79	6.92	7.03	7.14	7.24	7.34	7.43	7.51	7.59
7	3.34	4.16	4.68	5.06	5.36	5.61	5.82	6.00	6.16	6.30	6.43	6.55	6.66	6.76	6.85	6.94	7.02	7.10	7.17
8	3.26	4.04	4.53	4.89	5.17	5.40	5.60	5.77	5.92	6.05	6.18	6.29	6.39	6.48	6.57	6.65	6.73	6.80	6.87
9	3.20	3.95	4.41	4.76	5.02	5.24	5.43	5.59	5.74	5.87	5.98	6.09	6.19	6.28	6.36	6.44	6.51	6.58	6.64
10	3.15	3.88	4.33	4.65	4.91	5.12	5.30	5.46	5.60	5.72	5.83	5.93	6.03	6.11	6.19	6.27	6.34	6.40	6.47
11	3.11	3.82	4.26	4.57	4.82	5.03	5.20	5.35	5.49	5.61	5.71	5.81	5.90	5.98	6.06	6.13	6.20	6.27	6.33
12	3.08	3.77	4.20	4.51	4.75	4.95	5.12	5.27	5.39	5.51	5.61	5.71	5.80	5.88	5.95	6.02	6.09	6.15	6.21
13	3.06	3.73	4.15	4.45	4.69	4.88	5.05	5.19	5.32	5.43	5.53	5.63	5.71	5.79	5.86	5.93	5.99	6.05	6.11
14	3.03	3.70	4.11	4.41	4.64	4.83	4.99	5.13	5.25	5.36	5.46	5.55	5.64	5.71	5.79	5.85	5.91	5.97	6.03
15	3.01	3.67	4.08	4.37	4.59	4.78	4.94	5.08	5.20	5.31	5.40	5.49	5.57	5.65	5.72	5.78	5.85	5.90	5.96
16	3.00	3.65	4.05	4.33	4.56	4.74	4.90	5.03	5.15	5.26	5.35	5.44	5.52	5.59	5.66	5.73	5.79	5.84	5.90
17	2.98	3.63	4.02	4.30	4.52	4.70	4.86	4.99	5.11	5.21	5.31	5.39	5.47	5.54	5.61	5.67	5.73	5.79	5.84
18	2.97	3.61	4.00	4.28	4.49	4.67	4.82	4.96	5.07	5.17	5.27	5.35	5.43	5.50	5.57	5.63	5.69	5.74	5.79
19	2.96	3.59	3.98	4.25	4.47	4.65	4.79	4.92	5.04	5.14	5.23	5.31	5.39	5.46	5.53	5.59	5.65	5.70	5.75
20	2.95	3.58	3.96	4.23	4.45	4.62	4.77	4.90	5.01	5.11	5.20	5.28	5.36	5.43	5.49	5.55	5.61	5.66	5.71
24	2.92	3.53	3.90	4.17	4.37	4.54	4.68	4.81	4.92	5.01	5.10	5.18	5.25	5.32	5.38	5.44	5.49	5.55	5.59
30	2.89	3.49	3.85	4.10	4.30	4.46	4.60	4.72	4.82	4.92	5.00	5.08	5.15	5.21	5.27	5.33	5.38	5.43	5.47
40	2.86	3.44	3.79	4.04	4.23	4.39	4.52	4.63	4.73	4.82	4.90	4.98	5.04	5.11	5.16	5.22	5.27	5.31	5.36
60	2.83	3.40	3.74	3.98	4.16	4.31	4.44	4.55	4.65	4.73	4.81	4.88	4.94	5.00	5.06	5.11	5.15	5.20	5.24
120	2.80	3.36	3.68	3.92	4.10	4.24	4.36	4.47	4.56	4.64	4.71	4.78	4.84	4.90	4.95	5.00	5.04	5.09	5.13
∞	2.77	3.31	3.63	3.86	4.03	4.17	4.29	4.39	4.47	4.55	4.62	4.68	4.74	4.80	4.85	4.89	4.93	4.97	5.01

TABLE 7(b) Critical Values of the Studentized Range, $\alpha = 0.01$

ν	k = 2	3	4	5	6	7	8	9	10	11	12	13	14	15	16	17	18	19	20
1	90.0	135	164	186	202	216	227	237	246	253	260	266	272	277	282	286	290	294	298
2	14.0	19.0	22.3	24.7	26.6	28.2	29.5	30.7	31.7	32.6	33.4	34.1	34.8	35.4	36.0	36.5	37.0	37.5	37.9
3	8.26	10.6	12.2	13.3	14.2	15.0	15.6	16.2	16.7	17.1	17.5	17.9	18.2	18.5	18.8	19.1	19.3	19.5	19.8
4	6.51	8.12	9.17	9.96	10.6	11.1	11.5	11.9	12.3	12.6	12.8	13.1	13.3	13.5	13.7	13.9	14.1	14.2	14.4
5	5.70	6.97	7.80	8.42	8.91	9.32	9.67	9.97	10.2	10.5	10.7	10.9	11.1	11.2	11.4	11.6	11.7	11.8	11.9
6	5.24	6.33	7.03	7.56	7.97	8.32	8.61	8.87	9.10	9.30	9.49	9.65	9.81	9.95	10.1	10.2	10.3	10.4	10.5
7	4.95	5.92	6.54	7.01	7.37	7.68	7.94	8.17	8.37	8.55	8.71	8.86	9.00	9.12	9.24	9.35	9.46	9.55	9.65
8	4.74	5.63	6.20	6.63	6.96	7.24	7.47	7.68	7.87	8.03	8.18	8.31	8.44	8.55	8.66	8.76	8.85	8.94	9.03
9	4.60	5.43	5.96	6.35	6.66	6.91	7.13	7.32	7.49	7.65	7.78	7.91	8.03	8.13	8.23	8.32	8.41	8.49	8.57
10	4.48	5.27	5.77	6.14	6.43	6.67	6.87	7.05	7.21	7.36	7.48	7.60	7.71	7.81	7.91	7.99	8.07	8.15	8.22
11	4.39	5.14	5.62	5.97	6.25	6.48	6.67	6.84	6.99	7.13	7.25	7.36	7.46	7.56	7.65	7.73	7.81	7.88	7.95
12	4.32	5.04	5.50	5.84	6.10	6.32	6.51	6.67	6.81	6.94	7.06	7.17	7.26	7.36	7.44	7.52	7.59	7.66	7.73
13	4.26	4.96	5.40	5.73	5.98	6.19	6.37	6.53	6.67	6.79	6.90	7.01	7.10	7.19	7.27	7.34	7.42	7.48	7.55
14	4.21	4.89	5.32	5.63	5.88	6.08	6.26	6.41	6.54	6.66	6.77	6.87	6.96	7.05	7.12	7.20	7.27	7.33	7.39
15	4.17	4.83	5.25	5.56	5.80	5.99	6.16	6.31	6.44	6.55	6.66	6.76	6.84	6.93	7.00	7.07	7.14	7.20	7.26
16	4.13	4.78	5.19	5.49	5.72	5.92	6.08	6.22	6.35	6.46	6.56	6.66	6.74	6.82	6.90	6.97	7.03	7.09	7.15
17	4.10	4.74	5.14	5.43	5.66	5.85	6.01	6.15	6.27	6.38	6.48	6.57	6.66	6.73	6.80	6.87	6.94	7.00	7.05
18	4.07	4.70	5.09	5.38	5.60	5.79	5.94	6.08	6.20	6.31	6.41	6.50	6.58	6.65	6.72	6.79	6.85	6.91	6.96
19	4.05	4.67	5.05	5.33	5.55	5.73	5.89	6.02	6.14	6.25	6.34	6.43	6.51	6.58	6.65	6.72	6.78	6.84	6.89
20	4.02	4.64	5.02	5.29	5.51	5.69	5.84	5.97	6.09	6.19	6.29	6.37	6.45	6.52	6.59	6.65	6.71	6.76	6.82
24	3.96	4.54	4.91	5.17	5.37	5.54	5.69	5.81	5.92	6.02	6.11	6.19	6.26	6.33	6.39	6.45	6.51	6.56	6.61
30	3.89	4.45	4.80	5.05	5.24	5.40	5.54	5.65	5.76	5.85	5.93	6.01	6.08	6.14	6.20	6.26	6.31	6.36	6.41
40	3.82	4.37	4.70	4.93	5.11	5.27	5.39	5.50	5.60	5.69	5.77	5.84	5.90	5.96	6.02	6.07	6.12	6.17	6.21
60	3.76	4.28	4.60	4.82	4.99	5.13	5.25	5.36	5.45	5.53	5.60	5.67	5.73	5.79	5.84	5.89	5.93	5.98	6.02
120	3.70	4.20	4.50	4.71	4.87	5.01	5.12	5.21	5.30	5.38	5.44	5.51	5.56	5.61	5.66	5.71	5.75	5.79	5.83
∞	3.64	4.12	4.40	4.60	4.76	4.88	4.99	5.08	5.16	5.23	5.29	5.35	5.40	5.45	5.49	5.54	5.57	5.61	5.65

TABLE 8(a) Critical Values for the Durbin–Watson Statistic, $\alpha = 0.05$

n	$k=1$		$k=2$		$k=3$		$k=4$		$k=5$	
	d_L	d_U	d_L	d_U	d_L	d_U	d_L	d_U	d_L	d_U
15	1.08	1.36	.95	1.54	.82	1.75	.69	1.97	.56	2.21
16	1.10	1.37	.98	1.54	.86	1.73	.74	1.93	.62	2.15
17	1.13	1.38	1.02	1.54	.90	1.71	.78	1.90	.67	2.10
18	1.16	1.39	1.05	1.53	.93	1.69	.82	1.87	.71	2.06
19	1.18	1.40	1.08	1.53	.97	1.68	.86	1.85	.75	2.02
20	1.20	1.41	1.10	1.54	1.00	1.68	.90	1.83	.79	1.99
21	1.22	1.42	1.13	1.54	1.03	1.67	.93	1.81	.83	1.96
22	1.24	1.43	1.15	1.54	1.05	1.66	.96	1.80	.86	1.94
23	1.26	1.44	1.17	1.54	1.08	1.66	.99	1.79	.90	1.92
24	1.27	1.45	1.19	1.55	1.10	1.66	1.01	1.78	.93	1.90
25	1.29	1.45	1.21	1.55	1.12	1.66	1.04	1.77	.95	1.89
26	1.30	1.46	1.22	1.55	1.14	1.65	1.06	1.76	.98	1.88
27	1.32	1.47	1.24	1.56	1.16	1.65	1.08	1.76	1.01	1.86
28	1.33	1.48	1.26	1.56	1.18	1.65	1.10	1.75	1.03	1.85
29	1.34	1.48	1.27	1.56	1.20	1.65	1.12	1.74	1.05	1.84
30	1.35	1.49	1.28	1.57	1.21	1.65	1.14	1.74	1.07	1.83
31	1.36	1.50	1.30	1.57	1.23	1.65	1.16	1.74	1.09	1.83
32	1.37	1.50	1.31	1.57	1.24	1.65	1.18	1.73	1.11	1.82
33	1.38	1.51	1.32	1.58	1.26	1.65	1.19	1.73	1.13	1.81
34	1.39	1.51	1.33	1.58	1.27	1.65	1.21	1.73	1.15	1.81
35	1.40	1.52	1.34	1.58	1.28	1.65	1.22	1.73	1.16	1.80
36	1.41	1.52	1.35	1.59	1.29	1.65	1.24	1.73	1.18	1.80
37	1.42	1.53	1.36	1.59	1.31	1.66	1.25	1.72	1.19	1.80
38	1.43	1.54	1.37	1.59	1.32	1.66	1.26	1.72	1.21	1.79
39	1.43	1.54	1.38	1.60	1.33	1.66	1.27	1.72	1.22	1.79
40	1.44	1.54	1.39	1.60	1.34	1.66	1.29	1.72	1.23	1.79
45	1.48	1.57	1.43	1.62	1.38	1.67	1.34	1.72	1.29	1.78
50	1.50	1.59	1.46	1.63	1.42	1.67	1.38	1.72	1.34	1.77
55	1.53	1.60	1.49	1.64	1.45	1.68	1.41	1.72	1.38	1.77
60	1.55	1.62	1.51	1.65	1.48	1.69	1.44	1.73	1.41	1.77
65	1.57	1.63	1.54	1.66	1.50	1.70	1.47	1.73	1.44	1.77
70	1.58	1.64	1.55	1.67	1.52	1.70	1.49	1.74	1.46	1.77
75	1.60	1.65	1.57	1.68	1.54	1.71	1.51	1.74	1.49	1.77
80	1.61	1.66	1.59	1.69	1.56	1.72	1.53	1.74	1.51	1.77
85	1.62	1.67	1.60	1.70	1.57	1.72	1.55	1.75	1.52	1.77
90	1.63	1.68	1.61	1.70	1.59	1.73	1.57	1.75	1.54	1.78
95	1.64	1.69	1.62	1.71	1.60	1.73	1.58	1.75	1.56	1.78
100	1.65	1.69	1.63	1.72	1.61	1.74	1.59	1.76	1.57	1.78

TABLE 8(b) **Critical Values for the Durbin–Watson Statistic, $\alpha = 0.01$**

	k=1		k=2		k=3		k=4		k=5	
n	d_L	d_U	d_L	d_U	d_L	d_U	d_L	d_U	d_L	d_U
15	.81	1.07	.70	1.25	.59	1.46	.49	1.70	.39	1.96
16	.84	1.09	.74	1.25	.63	1.44	.53	1.66	.44	1.90
17	.87	1.10	.77	1.25	.67	1.43	.57	1.63	.48	1.85
18	.90	1.12	.80	1.26	.71	1.42	.61	1.60	.52	1.80
19	.93	1.13	.83	1.26	.74	1.41	.65	1.58	.56	1.77
20	.95	1.15	.86	1.27	.77	1.41	.68	1.57	.60	1.74
21	.97	1.16	.89	1.27	.80	1.41	.72	1.55	.63	1.71
22	1.00	1.17	.91	1.28	.83	1.40	.75	1.54	.66	1.69
23	1.02	1.19	.94	1.29	.86	1.40	.77	1.53	.70	1.67
24	1.04	1.20	.96	1.30	.88	1.41	.80	1.53	.72	1.66
25	1.05	1.21	.98	1.30	.90	1.41	.83	1.52	.75	1.65
26	1.07	1.22	1.00	1.31	.93	1.41	.85	1.52	.78	1.64
27	1.09	1.23	1.02	1.32	.95	1.41	.88	1.51	.81	1.63
28	1.10	1.24	1.04	1.32	.97	1.41	.90	1.51	.83	1.62
29	1.12	1.25	1.05	1.33	.99	1.42	.92	1.51	.85	1.61
30	1.13	1.26	1.07	1.34	1.01	1.42	.94	1.51	.88	1.61
31	1.15	1.27	1.08	1.34	1.02	1.42	.96	1.51	.90	1.60
32	1.16	1.28	1.10	1.35	1.04	1.43	.98	1.51	.92	1.60
33	1.17	1.29	1.11	1.36	1.05	1.43	1.00	1.51	.94	1.59
34	1.18	1.30	1.13	1.36	1.07	1.43	1.01	1.51	.95	1.59
35	1.19	1.31	1.14	1.37	1.08	1.44	1.03	1.51	.97	1.59
36	1.21	1.32	1.15	1.38	1.10	1.44	1.04	1.51	.99	1.59
37	1.22	1.32	1.16	1.38	1.11	1.45	1.06	1.51	1.00	1.59
38	1.23	1.33	1.18	1.39	1.12	1.45	1.07	1.52	1.02	1.58
39	1.24	1.34	1.19	1.39	1.14	1.45	1.09	1.52	1.03	1.58
40	1.25	1.34	1.20	1.40	1.15	1.46	1.10	1.52	1.05	1.58
45	1.29	1.38	1.24	1.42	1.20	1.48	1.16	1.53	1.11	1.58
50	1.32	1.40	1.28	1.45	1.24	1.49	1.20	1.54	1.16	1.59
55	1.36	1.43	1.32	1.47	1.28	1.51	1.25	1.55	1.21	1.59
60	1.38	1.45	1.35	1.48	1.32	1.52	1.28	1.56	1.25	1.60
65	1.41	1.47	1.38	1.50	1.35	1.53	1.31	1.57	1.28	1.61
70	1.43	1.49	1.40	1.52	1.37	1.55	1.34	1.58	1.31	1.61
75	1.45	1.50	1.42	1.53	1.39	1.56	1.37	1.59	1.34	1.62
80	1.47	1.52	1.44	1.54	1.42	1.57	1.39	1.60	1.36	1.62
85	1.48	1.53	1.46	1.55	1.43	1.58	1.41	1.60	1.39	1.63
90	1.50	1.54	1.47	1.56	1.45	1.59	1.43	1.61	1.41	1.64
95	1.51	1.55	1.49	1.57	1.47	1.60	1.45	1.62	1.42	1.64
100	1.52	1.56	1.50	1.58	1.48	1.60	1.46	1.63	1.44	1.65

TABLE 9 Critical Values for the Wilcoxon Rank Sum Test

(a) $\alpha = 0.025$ one-tail; $\alpha = 0.05$ two-tail

n_1	3		4		5		6		7		8		9		10	
n_2	T_L	T_U	T_L	T_U	T_L	T_U	T_L	T_U	T_L	T_U	T_L	T_U	T_L	T_U	T_L	T_U
4	6	18	11	25	17	33	23	43	31	53	40	64	50	76	61	89
5	6	11	12	28	18	37	25	47	33	58	42	70	52	83	64	96
6	7	23	12	32	19	41	26	52	35	63	44	76	55	89	66	104
7	7	26	13	35	20	45	28	56	37	68	47	81	58	95	70	110
8	8	28	14	38	21	49	29	61	39	63	49	87	60	102	73	117
9	8	31	15	41	22	53	31	65	41	78	51	93	63	108	76	124
10	9	33	16	44	24	56	32	70	43	83	54	98	66	114	79	131

(b) $\alpha = 0.05$ one-tail; $\alpha = 10$ two-tail

n_1	3		4		5		6		7		8		9		10	
n_2	T_L	T_U	T_L	T_U	T_L	T_U	T_L	T_U	T_L	T_U	T_L	T_U	T_L	T_U	T_L	T_U
3	6	15	11	21	16	29	23	37	31	46	39	57	49	68	60	80
4	7	17	12	24	18	32	25	41	33	51	42	62	52	74	63	87
5	7	20	13	27	19	37	26	46	35	56	45	67	55	80	66	94
6	8	22	14	30	20	40	28	50	37	61	47	73	57	87	69	101
7	9	24	15	33	22	43	30	54	39	66	49	79	60	93	73	107
8	9	27	16	36	24	46	32	58	41	71	52	84	63	99	76	114
9	10	29	17	39	25	50	33	63	43	76	54	90	66	105	79	121
10	11	31	18	42	26	54	35	67	46	80	57	95	69	111	83	127

TABLE 10 Critical Values for the Wilcoxon Signed Rank Sum Test

	(a) $\alpha = 0.025$ one-tail; $\alpha = 0.05$ two-tail		(b) $\alpha = 0.05$ one-tail; $\alpha = 0.10$ two-tail	
n	T_L	T_U	T_L	T_U
6	1	20	2	19
7	2	26	4	24
8	4	32	6	30
9	6	39	8	37
10	8	47	11	44
11	11	55	14	52
12	14	64	17	61
13	17	74	21	70
14	21	84	26	79
15	25	95	30	90
16	30	106	36	100
17	35	118	41	112
18	40	131	47	124
19	46	144	54	136
20	52	158	60	150
21	59	172	68	163
22	66	187	75	178
23	73	203	83	193
24	81	219	92	208
25	90	235	101	224
26	98	253	110	241
27	107	271	120	258
28	117	289	130	276
29	127	308	141	294
30	137	328	152	313

TABLE 11 Critical Values for the Spearman Rank Correlation Coefficient

The α values correspond to a one-tail test of H_0: $\rho_s = 0$.
The value should be doubled for two-tail tests.

n	$\alpha = 0.05$	$\alpha = 0.025$	$\alpha = 0.01$
5	.900	—	—
6	.829	.886	.943
7	.714	.786	.893
8	.643	.738	.833
9	.600	.683	.783
10	.564	.648	.745
11	.523	.623	.736
12	.497	.591	.703
13	.475	.566	.673
14	.457	.545	.646
15	.441	.525	.623
16	.425	.507	.601
17	.412	.490	.582
18	.399	.476	.564
19	.388	.462	.549
20	.377	.450	.534
21	.368	.438	.521
22	.359	.428	.508
23	.351	.418	.496
24	.343	.409	.485
25	.336	.400	.475
26	.329	.392	.465
27	.323	.385	.456
28	.317	.377	.448
29	.311	.370	.440
30	.305	.364	.432

TABLE 12 Control Chart Constants

SAMPLE SIZE n	A_2	d_2	d_3	D_3	D_4
2	1.880	1.128	.853	.000	3.267
3	1.023	1.693	.888	.000	2.575
4	.729	2.059	.880	.000	2.282
5	.577	2.326	.864	.000	2.115
6	.483	2.534	.848	.000	2.004
7	.419	2.704	.833	.076	1.924
8	.373	2.847	.820	.136	1.864
9	.337	2.970	.808	.184	1.816
10	.308	3.078	.797	.223	1.777
11	.285	3.173	.787	.256	1.744
12	.266	3.258	.778	.284	1.716
13	.249	3.336	.770	.308	1.692
14	.235	3.407	.762	.329	1.671
15	.223	3.472	.755	.348	1.652
16	.212	3.532	.749	.364	1.636
17	.203	3.588	.743	.379	1.621
18	.194	3.640	.738	.392	1.608
19	.187	3.689	.733	.404	1.596
20	.180	3.735	.729	.414	1.586
21	.173	3.778	.724	.425	1.575
22	.167	3.819	.720	.434	1.566
23	.162	3.858	.716	.443	1.557
24	.157	3.895	.712	.452	1.548
25	.153	3.931	.709	.459	1.541

CREDITS

Unless indicated otherwise below, all figures, tables, images, data and other items are © Cengage Learning.

The publisher would like to thank all the various copyright holders for granting permission to reproduce material throughout the text. Every effort has been made to trace all copyright holders. However, if anything remains outstanding, the publisher will be pleased to rectify this at the first opportunity. Please contact the publisher directly.

Thank you to the following companies and individuals for granting permission to reproduce their copyrighted material:

- Table 17.3: 'Critical Values of the Wilcoxon Rank Sum Test', from F. Wilcoxon and R.A. Wilcox, *Some Rapid Approximate Statistical Procedures* (1964), p. 28. Reproduced with the permission of American Cyanamid Company.
- Appendix A, Table 7(b): 'Critical Values of the Studentised Range, $\alpha = 0.01$', from E.S. Pearson and H.O. Hartley, *Biometrika Tables for Statisticians* 1:176–77; Table 8(a): 'Critical Values for the Durbin–Watson Statistic, $\alpha = 0.05$', from J. Durbin and G.S. Watson, 'Testing for Serial Correlation in Least Squares Regression, II', *Biometrika* 30 (1951):159–78; Table 8(b): 'Critical Values for the Durbin–Watson Statistic, $\alpha = 0.01$', from J. Durbin and G.S. Watson, 'Testing for Serial Correlation in Least Squares Regression, II', *Biometrika* 30 (1951): 159–78; Table 9: 'Critical Values for the Wilcoxon Rank Sum Test', from F. Wilcoxon and R.A. Wilcox, *Some Rapid Approximate Statistical Procedures* (1964), p. 28; Table 10: 'Critical Values for the Wilcoxon Signed Rank Sum Test', from F. Wilcoxon and R.A. Wilcox, *Some Rapid Approximate Statistical Procedures* (1964), p. 28. Reproduced with the permission of American Cyanamid Company; Table 11: 'Critical Values for the Spearman Rank Correlation Coefficient', from E.G. Olds, 'Distribution of Sums of Squares of Rank Differences for Small Samples', *Annals of Mathematical Statistics* 9 (1938). Reproduced with the permission of the Institute of Mathematical Statistics; Table 12: 'Control Chart Constants', from E.S. Pearson', The Percentage Limits for the Distribution of Range in Samples from a Normal Population', *Biometrika* 24 (1932): 416. Reproduced by permission of the Biometrika Trustees.

General Citations

Chapter	Item	Source
1	Example 1.1: Global Barometer	PSR: Poll Number 50 (Centre for Palestine Research and Studies)
2	Opening Example box: Do Male and Female British Voters Differ in their Party Affiliation?	ESS Round 6: European Social Survey Round 6 Data (2012). Data file edition 1.2. United Kingdom Social Science Data Services, United Kingdom – Data Archive and distributor of ESS data.
2	Example 2.1: Work Status in the Census 2011 Survey	Census2011. Adapted from data from the Office of National Statistics licensed under the Open Government Licence v.1.0.
2	Exercise 2.11	CIA World Factbook 2012.
2	Exercise 2.13	CIA World Factbook 2012.
2	Exercise 2.14	California Energy Commission, based on 2004 data.
2	Exercise 2.15	*Statistical Abstract of the United States,* 2012, Table 1389.
2	Exercise 2.16	World Steel Association, 2011 (worldsteel)
2	Exercise 2.17	Eurostat:ten00110
2	Exercise 2.18	*Statistical Abstract of the United States,* 2012, Table 937.
2	Exercise 2.26	*Statistical Abstract of the United States,* 2012, Table 593.
2	Exercise 2.29	Labour Force Survey – Office for National Statistics – Adapted from data from the Office of National Statistics licensed under the Open Government Licence v.1.0.

Chapter	Item	Source
2	Table 2.3: Energy Consumption in Europe and Eurasia by Source, 2012	BP Statistical Review of World Energy June 2013
2	Table 2.4: Per Capita Beer Consumption, 2012	www.beerinfo.com.
3	Opening Example box: Were Oil Companies Gouging Customers 2000–2012?	US Department of Energy; CIA World Factbook
3	Exercise 3.15	UNESCO Institute of Statistics (UIS), 2013
3	Exercise 3.16	Publication: PGDP Preliminary Estimate Of GDP – Adapted from data from the Office of National Statistics licensed under the Open Government Licence v.1.0.
3	Exercise 3.17	*Output in the Construction Industry: December and Q4 2013*, Table 10 – Adapted from data from the Office of National Statistics licensed under the Open Government Licence v.1.0.
3	Exercise 3.18	*Board of Governors of the Federal Reserve System* – US/UK Foreign Exchange Rate
3	Exercise 3.26	Adapted from *Statistical Abstract of the United States*, 2012, Table 612.
3	Exercise 3.28	Bank of England and Office of National Statistics, Statistical Bulletin Office for National Statistics, Labour Market Statistics, January 2014 – Adapted from data from the Office of National Statistics licensed under the Open Government Licence v.1.0.
3	Exercise 3.32	'Drivers in Reported Accidents by Gender, Number Injured, Road User Type and Age, Great Britain, 2012', DfT STATS19-RAS20002 – Adapted from data from the Office of National Statistics licensed under the Open Government Licence v.1.0.
3	Exercise 3.33	*Statistical Abstract of the United States*, 2012, Table 267.
3	Exercise 3.34	*Labour Market Statistics, January 2014* – Adapted from data from the Office of National Statistics licensed under the Open Government Licence v.1.0.
3	Exercise 3.35	Fxtop.com
3	Exercise 3.41	*Statistical Abstract of the United States*, 2009, Table 1036.
3	Case 3.1: The Question of Global Warming	National Climatic Data Center (NCDC) – monthly temperature anomalies from 1880 to 2012.
3	Case 3.2	CIA Factbook
3	Figure 3.13: Chart Depicting Napoleon's Invasion and Retreat from Russia in 1812	Edward Tufte, *The Visual Display of Quantitative Information* (Cheshire, CT: Graphics Press, 1983), p. 41.
4	Exercise 4.46	National Post Business.
4	Exercise 4.47	US Department of Energy.
6	Example 6.1: Determinants of Success among Mutual Fund Managers—Part 1	This example is adapted from 'Are Some Mutual Fund Managers Better than Others? Cross-Sectional Patterns in Behavior and Performance', by Judith Chevalier and Glenn Ellison, Working paper 5852, National Bureau of Economic Research.
6	Exercise 6.44	*BES 2005–2010* (University of Essex)
7	Example 7.1: Probability Distribution of Persons per Household	*2011 Census: Population Estimates by Five-Year Age Bands, and Household Estimates, for Local Authorities in the United Kingdom. Office of National Statistics, 2011*, Table H01UK – Adapted from data from the Office of National Statistics licensed under the Open Government Licence v.1.0.
7	Exercise 7.80	*Labour Force Survey* – Adapted from data from the Office of National Statistics licensed under the Open Government Licence v.1.0.
9	Exercise 9.15	*Height Chart of Men and Women in Different Countries* (http://www.disabled-world.com/artman/publish/height-chart.shtml)
12	Opening Example box: TV Ratings	http://www.tvsa.co.za
12	Example 12.2: Income Tax	HMRC's Annual Survey of Personal Incomes (SPI) – Adapted from data from the Office of National Statistics licensed under the Open Government Licence v.1.0.
12	Example 12.6: Direct and Broker-Purchased Mutual Funds	D. Bergstresser, J. Chalmers and P. Tufano, 'Assessing the Costs and Benefits of Brokers in the Mutual Fund Industry'.
12	Example 12.7: Effect of New CEO in Family-Run Businesses	M. Bennedsen and K. Nielsen, Copenhagen Business School and D. Wolfenzon, New York University.

Chapter	Item	Source
13	Example 13.1: Proportion of Total Assets Invested in Stocks	Adapted from US Census Bureau, 'Asset Ownership of Households, May 2003', *Statistical Abstract of the United States,* 2006, Table 700.
13	Exercise 13.15	*Adapted from Statistical Abstract of the United States* 2012, Table 1371.
13	Case 13.1: Comparing Three Methods of Treating Childhood Ear Infections	This case is adapted from the *British Medical Journal,* February 2004.
14	Example box: Continuous Household Surveys Has Level of Concern for the Environment Changed Since 2006?	Continuous Household Survey, NISRA. Licensed under the Open Government Licence v.1.0.
14	Exercise 14.31	*New England Journal of Medicine* ('Conflict of Interest in the Debate over Calcium Channel Antagonists', January 8, 1998, p. 101).
14	Exercise 14.33	Adapted from *Statistical Abstract of the United States* 2012, Table 1338.
14	Exercise 14.34	Adapted from *Statistical Abstract of the United States* 2012, Table 1338.
14	Exercise 14.35	Adapted from *Statistical Abstract of the United States* 2012, Table 1342.
14	Exercise 14.36	Adapted from *Statistical Abstract of the United States* 2012, Table 1343.
14	Exercise 14.37	Adapted from *Statistical Abstract of the United States* 2012, Table 1343.
18	Opening Example box: New House Development: Housing 'Starts'	'Live Tables on House Building' (updated 15 May 2014) https://www.gov.uk/government/statistical-data-sets/live-tables-on-house-building – Table 211 – Adapted from data from the Office of National Statistics licensed under the Open Government Licence v.1.0.
18	Figure 18.3	'Monthly Practical Motorcycles Test Pass Rates in Great Britain from 1 January 2011 to 31 December 2013', DSA/DfT, 13 March 2014 – Adapted from data from the Office of National Statistics licensed under the Open Government Licence v.1.0.
18	Exercise 18.21	'Students by level of study 2000/01–2012/2013', Higher Education Statistics Agency (HESA).
18	Example 18.6: Forecasting Changes to the Consumer Price Index	Table B1 – CPI headline index numbers, Table B2 – CPI headline year-on-year rates3 Statistics South Africa (http://beta2.statssa.gov.za/publications/P0141/CPIHistory.pdf).

INDEX

acceptance sampling 653, 664–6
add-ins 274
addition rule of probability 154–6
advertising 284
alkaline batteries 232–3
alternative hypothesis 290
analysis of variance (ANOVA) 399–443
  complete factorial experiments 440–1
  experimental designs 422–3
  fixed-effects 423
  multi-factor experiments 422
  multiple comparisons 413–21
  one-way 399–410
  random-effects 423
  randomised block 422–9
  repeated measures designs 422
  two-factor 431–41
  two-way 422–9
Analysis ToolPak 274
ANOVA *see* analysis of variance
aptitude tests 591–4
arithmetic mean *see* mean
ARL *see* average run length
assignable variation 629
AT&T 304–6
auditing 140, 162
auto-correlation 507
autocorrelation 531–3
autoregressive model for forecasting 622
average *see* mean
average run length (ARL) 634–5

balanced design 433
bar charts 18–22
bar exams 157
batteries 232–3
Bayesian statistics 666
Bayes's Law 159–67
beer consumption 24–5
Bernoulli processes 195
between-treatments variations 401
billing systems 293–4
binomial distribution 195–201
binomial experiments 196
binomial random variables 195, 196–7
binomial tables 198–200
birthdays 171
bivariate analysis 28
bivariate distributions 182–5
bond valuation 44–5
Bonferroni adjustment 416–17
bonuses 478–9
bottling 254, 380–2
box plots 97–100
breakeven analysis 106
business newspapers 554, 582–3
business school graduates 248, 255
business statistics marks 47–8

Cape Computer Company 275–7
car acceleration 398, 409–10
car bumpers 414–15
car seats 637–40, 643–4, 645–6
cars 479–82, 488, 490–1, 493–4, 496–7, 500–2, 569–71
categorical data 12
causal relationships 495
censuses 6, 16–21, 28, 55, 126, 135–6
central limit theorem 253
central location, measures of 79–85
centred moving averages 607–8
chance variation 629
Chebysheff's Theorem 91–2
chi-squared density function 239
chi-squared distribution 239–42
chi-squared statistics 334
chi-squared tests 449–69
  contingency tables 455–62
  goodness-of-fit 449–53
  for normality 466–9
cholesterol-lowering drugs 425–7
CHS (Continuous Household Survey) 448, 461
class intervals 42–3
classical approach to probability assignment 142
cluster sampling 135
Coca-Cola 3–4
coefficient of correlation 103–4, 114, 184
coefficient of determination 111–12, 114, 492–3
coefficient of variation 92, 93
colleges 27
collinearity 529–30
common variation 629
compact discs 628, 649–50
complement rule of probability 153
complete factorial experiments 433, 440–1
completely randomised design 406
computers 6–7, 275–7
conditional probability 148, 660–2
confidence interval estimates 277–80
confidence interval estimator of μ 273–4, 352, 373
confidence interval estimators 307
confidence levels 5
consistency 271–2
Consumer Price Index (CPI) 58, 622–3
container-filling machines 335–9
contingency tables 455–62
continuity correction factor 261
continuous distributions 214, 235–45
Continuous Household Survey (CHS) 448, 461
continuous random variables 174
control charts 630–5
  for attributes 648–9
  for variables 636–46
control limits 630, 635
correlation 113
covariance 102–3, 104, 114, 184
CPI *see* Consumer Price Index
Critical Path Method (CPM) 187–8, 231
cross-classification tables 28, 455
cross-sectional data 55
cumulative line graphs 52–3
cumulative probability 198
cumulative relative frequency distributions 52
cyclical variations 602

data
  collection 127–9
  definition 12–13
  exploration 122
  hierarchy of 14
  sets 6
decision analysis 654–66
decision problems 654–7
decision trees 655–6, 657
degrees 455–6
degrees of freedom 331, 379
denominator degrees of freedom 379
department stores 293–4
dependent variables 474
derivations 320
descriptive statistics 3–4
deseasonalisation 617
deterministic models 475
deviation 87
direct observation 127
discrete probability distributions 174–7
discrete random variables 174
distribution-free statistics 555–96
Down syndrome 171
drivers 398, 409–10
dummy variables 547
Durbin–Watson test 531–3

ear infections 447
economic freedom 77
education and income 473, 497–8
educational attainment 431–4
empirical rule 91
EMV *see* expected monetary value
energy economics 22–3
environmental issues 448, 461
EOL *see* expected opportunity loss
EPPI *see* expected payoff with perfect information
equal-variances test statistics 352, 365
error of estimation 285
error variables 476, 485–6, 507
errors, types of 290–1
estimation 270–87
events 143
EVPI *see* expected value of perfect information
Excel 6–7, 274
exhaustive outcomes 141–2
exit polls 342–4
expected frequency 450
expected monetary value (EMV) 656–7
expected opportunity loss (EOL) 657
expected payoff with perfect information (EPPI) 659
expected value of perfect information (EVPI) 659
expected values 177, 178
experimental data 128, 486
experimental outcomes 660
experimental units 400
experiments 660
exponential distribution 231–4
exponential probability density function 231
exponential random variables 232
exponential smoothing 608–12, 621

F density function 242
F distribution 242–5
face-to-face interviews 128
factorial experiments 431, 433
factors 400–1
false-negative results 163, 166–7
false-positive results 163, 166–7
family cars 569–71
family-run businesses 358–62
fast-food restaurants 98–100, 543–5
Federal Express (FedEx) 289
finite population correction factor 253
first-order autocorrelation 531–3
first-order linear model 476
Fisher's least significant difference method 415–16
fixed costs 107–9
fixed-effects analysis of variance 423
flextime 574–8
football 78, 112–13
forecasting 601, 618–23
formulas 123
frequency distributions 16
Friedman Test 580, 583–7

geometric mean 84–5
Global Barometer 4–5
global warming 76–7
grand mean 401–2
graphical deception 70–4
graphical excellence 68–70
graphical techniques and numerical techniques 119–21

heteroscedasticity 506
hierarchy of data 14
histograms 40, 43–4
homoscedasticity 506
hotel occupancy 614–17, 621–2
house prices 61–4
house sales 183, 186
housing starts 600, 624–5
hypergeometric distribution 154–6
hypothesis testing 290–320
  one-sided confidence interval estimators 307
  operating characteristic (OC) curve 315–16
  power of 314–15
  and sample statistics 308

income 514
  and education 473, 498–9
income tax 327–9
income tax returns 132–3
independent events 149
independent samples 360–5, 374–5
independent variables 474
indicator variables 547
inference *see* statistical inference
inferential statistics 4
inflation 58
information 15
insurance, medical 162–5
interactions 433
intercorrelation 529–30
internet time spent 81, 82, 95
interquartile range 96–7, 100
intersection of events 145–7
interval data 12–13, 38–53, 319
interval estimation 270–1, 280–1
interval variables 15, 61–7
interviews 128
inventory management 227, 275
investment 172, 222, 399–406, 654–7

job applicant evaluation 584–6
job applicant testing 484
job tenure 431–4
joint probability 145–7, 168

key statistical concepts 670–1
Kruskal–Wallis Test 580–3, 586

Lear Seating 637–40, 643–4, 645–6
least significant difference (LSD) method 415–16
least squares line 106, 114
least squares method 105–6, 476
Let's Make a Deal 170
likelihood probabilities 160, 165–7, 660
line charts 56–7
linear 476
linear relationships 64–5, 101–14, 109–11
long-distance telephone bills 38–42
LSD *see* least significant difference method

macroeconomics 22
MAD *see* mean absolute deviation
mail surveys 128–9
marginal probability 147, 183
market models 115–18
market shares 450
marketing 38
mathematical derivations 320
mathematical statistics marks 48–9
MBA graduates 265, 367–74
MBA programmes 159–60, 209, 225, 455–6
mean absolute deviation (MAD) 88, 619
mean squares 403
mean(s) 79–81, 83–4; *see also* population mean
  approximation from grouped data 92
  binomial distribution 200–1

difference between two independent samples 350–65
matched pairs experiment 367–77
formulas 80
sampling distribution of 249–57
standard error of 252–3
measures of central location 79–85
measures of relative standing 94–100
measures of variability 86–92
median 81–2, 83–4
medical experiments 393
medical insurance 162–5
medicine 162–5
Microsoft Excel 6–7, 274
Minitab 6–7
missing data 344
mode 82–4
monthly profits 179
moving averages 603–8
multicollinearity 529–30
multinomial experiments 449
multiple comparisons 413–21
multiple regression 515–50
coefficients 516–19, 521–4
diagnostics 529–30, 531–9
equation 524–5
model 515–16
assessment of 518–21
nominal independent variables 547–50
polynomial models 539–45
required conditions 516
result interpretation 523
time-series data 531–9
multiple-choice questions 197–8, 200–1
multiplication rule of probability 153
mutual funds 146–50, 353–8
mutually exclusive outcomes 141–2

negative linear relationships 65
newspaper recycling 324–7
newspapers 28–31
95% confidence interval estimator of μ 274
nominal data 12, 16–25
calculations 13–14
comparisons 32–4
measures of central location 84
summary of tests 464–6
variability measures 92
nominal independent variables 547–50
nominal variables 15, 28–34
nonparametric techniques 363, 555–96
non-response errors 137
non-sampling errors 137
normal density function 216
normal distribution 216–28
normal probabilities 217–22
null hypothesis 290
numerator degrees of freedom 379
numerical data 12
numerical techniques and graphical techniques 119–21

observational data 127, 486
observed frequencies 450
OC *see* operating characteristic curve
ogives 52–3
oil prices 66–7
oil reserves 25–7, 37
one-sided confidence interval estimators 307
one-tail tests 304–7
one-way analysis of variance (ANOVA) 399–410
operating characteristic (OC) curve 315–16, 632–3
opportunity losses 655
ordinal data 12–13, 25
calculations 14
measures of central location 84
relative standing 100
variability measures 92, 100
ordinal variables 15
out of control production processes 629–30, 640–2
outliers 97, 508

p charts 648–9
package designs 387–91
painkillers 560–2
parameters 5
pattern tests 640–3, 649
payment periods 302–4
payoff tables 654
percentiles 94–6, 100, 226–8
performance 591–4
personal interviews 128
persons per household 178
PERT *see* Project Evaluation and Review Technique
petrol prices 56–60
petrol sales 213, 217–18, 603–7, 609–12
pharmaceutical experiments 393
pharmaceuticals 560–2
pie charts 18–22
point estimators 270
point prediction 499–500
Poisson distribution 201–5
Poisson experiments 202
Poisson random variables 201, 202
Poisson tables 203–5
political surveys 262–3
polynomial models 539–45
pooled proportion estimate 385
pooled variance estimator 351
population mean 177; *see also* mean(s)
and population standard deviation 273–5
sample median 281
statistical inference about, standard deviation unknown 324–31
population proportions 340–8, 383–91
population standard deviation 178, 273–5
population variance 87, 177, 334–9, 353
populations 5, 176–8
portfolio diversification 189–93
positive linear relationships 65
posterior probabilities 160, 165–7, 660–2
prediction intervals 500
preposterior analysis 662–3
prevention approach to quality 629
pricing 38
prior probabilities 160, 165–7, 660
probabilistic models 475
probability 140–71
assignment to events 141–4
conditional 148
cumulative 198
of events 143
interpretation 143–4
joint 145–7, 168
marginal 147, 183
requirements of 142
rules 153–6
probability density functions 210–14
probability distributions 174–8
probability trees 156–7
problem objectives 15, 320
process capability index 646
process variation 629, 646
Project Evaluation and Review Technique (PERT) 187–8, 231
prosperity 77
prostate cancer 163–5
p-values 296–9

qualitative data 12
quality 335
quality control 629–50
quantitative data 12
quartiles 94, 96–7, 100
questionnaires 128–9
queues 206, 233

random experiments 141
random sampling 131–5
random variables 173–4
random variations 603
random-effects analysis of variance 423
randomised block analysis of variance 422–9
randomised design 406
range 87, 93
ratio data 12
rectangular probability distribution 212–13
recycling 324–7
regression analysis 474–5
cause-and-effect relationship 495
coefficient of determination 492–3
coefficients 476–82
diagnostics 504–9
procedure 509
equation 499–503
error variables 485–6
first-order linear model 476
influential observations 508
model assessment 487–98
multiple regression 515–50
one-tail tests 491–2
outliers 508
rejection regions 294–5
relative efficiency 272
relative frequency approach to probability assignment 142
relative frequency distributions 16, 51–2
relative standing, measures of 94–100
reorder points 227–8
repeated measures designs 422
replicates 433
research hypothesis 290
residuals 504–5
response rates 128
response surfaces 515–16
response variables 400
retention of workers 483, 562–5
return on investment 45–7, 119–21
revised probabilities 160, 660–2
risk 222
rollback technique 657
rule of five 460

S charts 643–4
salaries of MBA graduates 265
sample mean 264–5
sample median 281
sample size 135, 285–7, 311–17, 346–8
sample spaces 142
sample variance 87
sampled populations 130
samples 5
sampling 130–7
sampling distribution
of the difference between two sample means 264–5
empirical creation 254–5
formula 320
inference from 256–7
of the mean 249–57
of two dice 249–53
of a proportion 259–63
of the sample mean 250, 254
of a sample proportion 262–3
and statistical inference 266–7
sampling errors 136–7
scatter diagrams 61, 64, 104
seasonal analysis 613–14
seasonal indexes 621
seasonal variations 602
seasonally adjusted time series 617
secular trends 602
selection bias 137
self-administered questionnaires 128–9
self-selected samples 130–1
serial correlation 507
severance pay 526–7
sign test 567–78, 586
significance levels 5, 290
simple linear regression model 476, 524
simple random sampling 131–3
ski lifts 533–8
slope coefficient 117
software packages 6–7
SPC *see* statistical process control
Spearman rank correlation coefficient 589–94
special variation 629
specification limits 629
spreadsheets 6–7
SSB *see* sum of squares for blocks
SSE *see* sum of squares for error
SSE *see* sum of squares for forecast errors
SST *see* sum of squares for treatments
SS(Total) 405
standard deviation 87–8, 90–2, 93, 200–1
standard error of a proportion 262
standard error of estimate 487
standard error of the difference between two means 264
standard error of the mean 252–3
standard normal random variables 217, 222–8
standardised test statistics 295–6
statistical applications 5–6
statistical concepts 670–1
statistical inference 5, 141
about a population mean 324–31
purpose of 319
and sampling distribution 266–7
statistical process control (SPC) 629–50
statistical significance 296, 299
statisticians 2
statistics 5
statistics practitioners 2
stem-and-leaf displays 49–51
stock investment 399–406
stock market indexes 116
stock valuation 44–5
stratified random sampling 133–5
student selection 154
Student t density function 235
Student t distribution 235–9, 324
subjective approach to probability assignment 143
sum of squares for blocks (SSB) 423–4
sum of squares for error (SSE) 402, 479, 487
sum of squares for forecast errors (SSE) 619
sum of squares for treatments (SST) 401
sum of two variables 185–6
summation 79
supermarket checkout counters 233–4
surveys 128–9
symbols 122–3

target populations 130
tax returns 140, 162
telephone bills 38–42, 95–6, 97–8
telephone surveys 128
test drivers 398, 409–10
test marketing 386–91
time-series analysis 601–17
time-series components 601–3
time-series data 55–60, 531–9
Toyota Auris 479–82, 488, 490–1, 493–4, 496–7, 500–2
tree diameters 269, 286
trend analysis 613
trends 602
Tukey's multiple comparison method 417–20
TV ratings 322–3, 345–6
twelve key statistical concepts 670–1
two-factor analysis of variance 431–41
two-tail tests 304–7
two-way analysis of variance 422–9
type I errors 290–1, 316–17
type II errors 290–1, 311–17
typographical errors 202–3

unbiased estimators 271
under control production processes 629
undergraduate degrees 455–6
unequal-variances test statistics 352, 365
uniform distribution 212–13
union of events 149
univariate analysis 28
universities 27
used cars 479–82, 488, 490–1, 493–4, 496–7, 500–2

values, definition 11
variability, measures of 86–92
variable costs 107–9
variables, definition 11
variance(s) 87–9, 93; *see also* covariance; population variance; sample variance
approximation from grouped data 92
binomial distribution 200–1
law of 179
ratio of two 378–82
shortcut method 89

Wilcoxon Rank Sum Test 556–65, 583
Wilcoxon Signed Rank Sum Test 567–78
Wilson estimators 348
within-treatments variation 402
worker turnover 483

x-bar charts 636–7, 644

y-intercept 123

Z (standard normal random variables) 217, 222–8